# Organic Reactions

# Organic Reactions

## VOLUME 14



JOHN WILEY & SONS, INC.

*NEW YORK · LONDON · SYDNEY*

Library of Congress Catalog Card Number: 42-20265

PRINTED IN THE UNITED STATES OF AMERICA

## PREFACE TO THE SERIES

In the course of nearly every program of research in organic chemistry the investigator finds it necessary to use several of the better-known synthetic reactions. To discover the optimum conditions for the application of even the most familiar one to a compound not previously subjected to the reaction often requires an extensive search of the literature; even then a series of experiments may be necessary. When the results of the investigation are published, the synthesis, which may have required months of work, is usually described without comment. The background of knowledge and experience gained in the literature search and experimentation is thus lost to those who subsequently have occasion to apply the general method. The student of preparative organic chemistry faces similar difficulties. The textbooks and laboratory manuals furnish numerous examples of the application of various syntheses, but only rarely do they convey an accurate conception of the scope and usefulness of the processes.

For many years American organic chemists have discussed these problems. The plan of compiling critical discussions of the more important reactions thus was evolved. The volumes of *Organic Reactions* are collections of chapters each devoted to a single reaction, or a definite phase of a reaction, of wide applicability. The authors have had experience with the processes surveyed. The subjects are presented from the preparative viewpoint, and particular attention is given to limitations, interfering influences, effects of structure, and the selection of experimental techniques. Each chapter includes several detailed procedures illustrating the significant modifications of the method. Most of these procedures have been found satisfactory by the author or one of the editors, but unlike those in *Organic Syntheses* they have not been subjected to careful testing in two or more laboratories. When all known examples of the reaction are not mentioned in the text, tables are given to list compounds which have been prepared by or subjected to the reaction. Every effort has been made to include in the tables all such compounds and references; however, because of the very nature of the reactions discussed and their frequent use as one of the several steps of syntheses in which not all of the intermediates have been isolated, some instances may well have been missed. Nevertheless, the investigator will be able to use the tables and their accompanying bibliographies in place of most or all of the literature search so often required.

Because of the systematic arrangement of the material in the chapters and the entries in the tables, users of the books will be able to find information desired by reference to the table of contents of the appropriate chapter. In the interest of economy the entries in the indices have been kept to a minimum, and, in particular, the compounds listed in the tables are not repeated in the indices.

The success of this publication, which will appear periodically, depends upon the cooperation of organic chemists and their willingness to devote time and effort to the preparation of the chapters. They have manifested their interest already by the almost unanimous acceptance of invitations to contribute to the work. The editors will welcome their continued interest and their suggestions for improvements in *Organic Reactions*.

# CONTENTS

# CHAPTER 1

# THE CHAPMAN REARRANGEMENT

J. W. Schulenberg and S. Archer
*Sterling-Winthrop Research Institute, Rensselaer, New York*

## CONTENTS

## INTRODUCTION

The thermal conversion of aryl N-arylbenzimidates to N-aroyldiphenyl-

$$\mathrm{Ar''{-}C(OAr){=}N{-}Ar'} \rightarrow \mathrm{Ar''{-}C({=}O){-}N(Ar)(Ar')}$$

amines is known as the Chapman rearrangement. The rearrangement was discovered by Mumm, Hesse, and Volquartz who, in 1915, described the rearrangement of the parent compound, phenyl N-phenylbenzimidate, to N-benzoyldiphenylamine.[1] No further examples of the reaction were reported until 1925, when Chapman's first study was published.[2] Since he not only studied the mechanism of the reaction[3] but also first used it as a general synthesis of diphenylamines,[4] his name is associated with this reaction.*

The starting materials for the Chapman rearrangement have been referred to as imino ethers, imido esters, imidates, and by a miscellany of other names. We agree with the recommendations of Roger and Neilson[6] and have named all compounds as imidates, i.e., as esters of the parent imidic acid. Thus compound **1**, when named as a derivative of

[1] Mumm, Hesse, and Volquartz, *Ber.*, **48,** 379 (1915).

[2] Chapman, *J. Chem. Soc.*, **127,** 1992 (1925).

[3] Chapman, *J. Chem. Soc.*, **1927,** 1743.

[4] Chapman, *J. Chem. Soc.*, **1929,** 569.

* A broader definition which includes the rearrangement of alkyl imidates has been used.[5,6] We prefer the more limited one for the following reasons. (1) It is in better accord with general literature usage; Müller[7] and Gowan and Wheeler[8] employ the more restricted version. (2) Mechanistically, the rearrangement of aryl benzimidates is distinctly different from that of the alkyl esters. (3) Chapman himself confined his investigations, with one exception, to aryl imidates.

[5] *The Merck Index*, Merck & Co., Inc., Rahway, N.J., 7th ed. 1960, p. 1413.

[6] Roger and Neilson, *Chem. Rev.*, **61,** 179 (1961).

[7] Müller, *Methoden der Organischen Chemie*, XI/1, Georg Thieme Verlag, Stuttgart, 1957, p. 910; (*a*) p. 250; (*b*) p. 117.

[8] Gowan and Wheeler, *Name Index of Organic Reactions*, 2nd ed., Interscience Division, John Wiley and Sons, New York, 1960, p. 50; (*a*) p. 136; (*b*) p. 244; (*c*) p.104.

benzimidic acid **(2)**, is called *p*-chlorophenyl N-(*p*-bromophenyl)benzim-

$$\underset{\mathbf{1}}{C_6H_5\overset{OC_6H_4Cl\text{-}p}{\overset{|}{C}}{=}NC_6H_4Br\text{-}p} \qquad \underset{\mathbf{2}}{C_6H_5\overset{OH}{\overset{|}{C}}{=}NH}$$

idate, in accordance with *Chemical Abstracts* usage. When named as an imino ether, the compound is called N-(*p*-bromophenyl)benzimino *p*-chlorophenyl ether.

We have not named the rearrangement products according to *Chemical Abstracts*, which lists the compounds as derivatives of benzanilide or benzamide. Since the Chapman rearrangement is used as a preparative method for diarylamines, we have stressed this fact by calling the initial products aroyldiarylamines, providing a uniform nomenclature. Similarly, we have named all the compounds obtained by amide hydrolysis as diarylamines, even though *Chemical Abstracts* often uses other names. For example, **3** is named 2′-chlorodiphenylamine-2-carboxylic acid rather

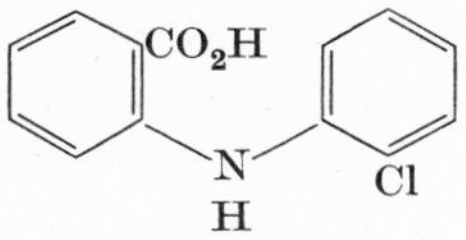

$CO_2H$ N H Cl

**3**

than N-(*o*-chlorophenyl)anthranilic acid. Acridones are numbered according to *Chemical Abstracts*.

## MECHANISM

The Chapman rearrangement has been shown to be an intramolecular reaction in which a 1,3 shift of an aryl group from oxygen to nitrogen occurs. The reaction, which proceeds via a 4-membered transition state, may be considered as a nucleophilic attack by nitrogen on the migrating aryl group.

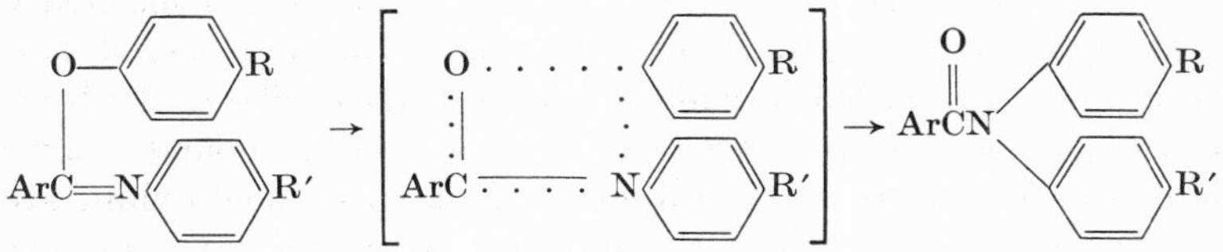

O R ArC=N R′ **4** → [O ····· R ArC ····· N R′] → ArC(=O)N R R′ **5**

Wiberg and Rowland showed that the reaction is intramolecular by heating a mixture of phenyl N-phenylbenzimidate (**4**, R = R′ = H) and *p*-chlorophenyl N-(*p*-chlorophenyl)benzimidate (**4,** R = R′ = Cl).[9] The infrared spectra and x-ray powder patterns of the reaction mixtures showed that the composition was the same as those obtained when the two imidates were heated separately and the products mixed. Earlier, Chapman had performed a similar experiment with a mixture of **4** (R = R′ = H) and **4** (R = R′ = $CH_3$). The absence of the mixed product **5** (R = H, R′ = $CH_3$) in the pyrolysis product, which was determined only from the freezing point of the mixture, led Chapman to conclude correctly that the reaction is intramolecular.[2]

Chapman[2] also used freezing point data to determine that the rearrangement of phenyl N-phenylbenzimidate followed first-order kinetics. In a later paper, however, in which a large number of imidates were pyrolyzed, he determined the percentages of the unreacted basic imidates in the mixtures by titration with acid.[3] Since his rate data were obtained using molten imidates (no solvent), the results were probably not very accurate. Wiberg and Rowland conducted kinetic experiments on a variety of imidates in diphenyl ether, under carefully controlled conditions, and confirmed Chapman's conclusion that the reaction is unimolecular.[9]

Chapman studied the rates of rearrangement of several imidates **6,** which differed only in the substituents on the aryloxy ring.[3] By far the most reactive was **6** (R = *o*-$NO_2$) since over 40% was rearranged in 90 minutes at 162–163°. The next most reactive compound, **6** (R = 2,4,6-$Cl_3$), was unchanged at this temperature, but 87% conversion to the amide occurred when it was heated for 90 minutes at 200–201°. The

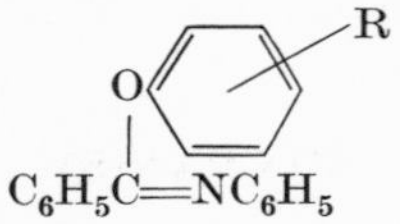

R
O
$C_6H_5C{=}NC_6H_5$
**6**

parent compound **6** (R = H) required a temperature of 266° for 90 minutes to effect 74% conversion to amide. The least reactive compound studied, **6** (R = *p*-$CH_3O$) required 90 minutes at 266° for 44% rearrangement.

Chapman concluded that the rate of the reaction was associated with the acidity of the phenol from which the imidate was derived and that the rearrangement proceeded via a four-membered intermediate. He believed that the aryl group migrated with the pair of electrons bound to oxygen. The first suggestion that the reaction is initiated by nucleophilic attack of the unshared electrons of the nitrogen atom was made in *Annual Reports*.[10] Later Wiberg and Rowland[9] concluded from their kinetic

[9] Wiberg and Rowland, *J. Am. Chem. Soc.*, **77,** 2205 (1955).

[10] Bennett, *Ann. Rept. Progr. Chem.*, (*Chem. Soc. London*), **26,** 123 (1929).

study that the reaction is indeed essentially a nucleophilic displacement on an aromatic ring. Electron-attracting groups on the aryloxy ring accelerate the reaction by helping to accommodate the partial negative charge induced on the ring by the nitrogen electrons. Alkyl imidates do not undergo the normal Chapman rearrangement (see p. 26) presumably because the saturated alkyl group, in contrast to an aryl group, cannot accommodate the additional negative charge.

Wiberg and Rowland studied a larger number of imidates, **6,** where R is an *ortho* or a *para* substituent. In all but one case, the *ortho*-substituted compound reacted more rapidly than the *para*, confirming Chapman's earlier observations. Wiberg attributed this to an entropy effect; the *ortho* group lessened the entropy decrease in going from the reactant to the four-membered transition state. The one exception involved the *o*-*t*-butyl group, where steric compressibility is important. For a fuller explanation of the entropy effects, as well as a discussion of the geometric configurations of the imidates, the original paper should be consulted.[9] In a later paper Wiberg discussed the various types of reactions in which intramolecular 1,3 shifts take place.[11]

Chapman also studied several imidates which differed only in the substituents on the arylimino ring.[3] He found that compound **7** (R = *p*-$CH_3O$) rearranged faster than compound **7** (R = H). The imidate **7** (R = 2,4,6-$Cl_3$) was considerably less reactive than the parent compound. Thus electron-withdrawing groups make the electrons on nitrogen less available for nucleophilic attack and slow down the reaction. Derivatives of the imidate **8** exhibited the same behavior, but the effect

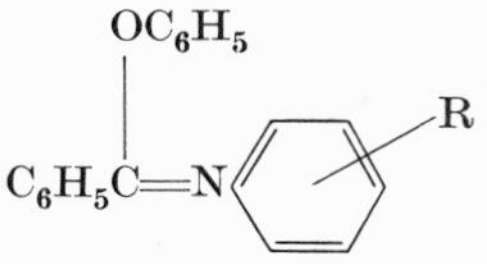

$OC_6H_5$ | $C_6H_5C{=}N$–$C_6H_4$–R

**7**

$OC_6H_5$ | R–$C_6H_4$–$C{=}NC_6H_5$

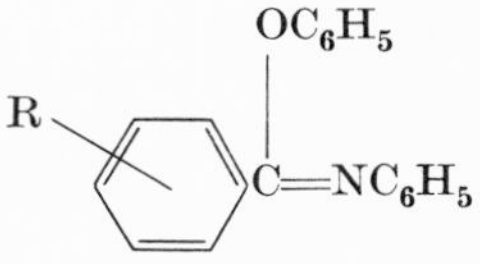

**8**

was less marked.[3] Compound **7** (R = *o*-$NO_2$) could not be rearranged successfully even at 220°; above this temperature decomposition occurred.[3] Owing to resonance, the electrons of arylamines are less available than those of alkylamines, and it would seem reasonable that N-alkyl imidates would rearrange readily. However, in the only case reported, the N-methyl analog of **7** decomposed without isomerizing.[3]

The imidate **6** (R = *p*-Br) rearranged more rapidly than the corresponding *p*-chloro compound, which in turn rearranged more rapidly than

[11] Wiberg and Shryne, *J. Am. Chem. Soc.*, **77**, 2774 (1955).

the parent imidate **9** (R = H).[9] The parent imidate **9** (R = H) was

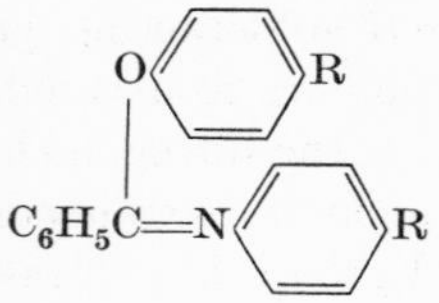

O R

$C_6H_5C{=}N$ R

**9**

converted to N-benzoyldiphenylamine in 80% yield by heating for 1 hour at 280°.[12] The dibromo analog **9** (R = Br) required heating for 2 hours to effect a comparable conversion. It therefore appears that in this case the effect of the substituent in the nitrogen-bonded ring is greater than that in the oxygen-bonded ring. A more dramatic effect is reported for the difluoro compound **9** (R = F), heating for 18 hours at 280° being required for 80% conversion.[12] This is attributed to the great electron-withdrawing ability of fluorine, the halide atom in the nitrogen-bonded ring again being the dominant factor. Kinetic data on a variety of compounds substituted in the nitrogen-bonded ring alone and in both rings would be of great interest for determining the relative effects of substituents in the two rings. Both monofluoro imidates **6** (R = *p*-F) and **7** (R = *p*-F) have been converted in over 80% yield to the amides upon heating at 305–310° for 1 hour.[13] However, no kinetic data are available. Rate data on the imidate **6** (R = *p*-F) would be of special interest since a *p*-fluorine atom may either accelerate or decelerate (relative to *p*-H) nucleophilic displacement.[14]

## SCOPE AND LIMITATIONS

The objective of the early work on the pyrolysis of aryl imidates was to obtain information about the mechanism of the reaction, rather than to develop it for preparative purposes.[1–3] In 1929 it was pointed out that the rearrangement could serve as a general method for preparing diphenylamines by hydrolysis of the primary reaction products,[4] and a variety of substituted diarylamines has since been prepared in this way. The Chapman rearrangement has been carried out successfully with most of the imidates tried, and yields have generally been high. Unfortunately, yields have not been reported for many reactions, especially those reported in the earlier literature. The conversion of phenyl N-phenylbenzimidate to N-benzoyldiphenylamine was said to be complete after 2 hours at 270–300° and no by-products were formed, but the yield of crystallized product was not reported.[2] Similarly it has been implied that various chloro-substituted imidates rearranged quantitatively, but yields were not given.[3]

[12] Benington, Shoop, and Poirier, *J. Org. Chem.*, **18**, 1506 (1953).

[13] Schulenberg and Archer, unpublished results.

[14] Sauer and Huisgen, *Angew. Chem.*, **72**, 309 (1960).

The reaction can be used also to prepare aroyldiphenylamines substituted in the aroyl ring, e.g., the compounds **10** where R is *o*-chloro, *p*-chloro, *p*-nitro, *p*-methoxy, and 2,4,6-trichloro,[3] and the even more highly substituted aroylamine **11**.[15] An attempt to rearrange **12** was unsuccessful.[3]

$$R\text{-}C_6H_4\text{-}C(=O)\text{—}N(C_6H_5)_2$$
**10**

$$p\text{-}CH_3C_6H_4C(=O)\text{—}N(C_6H_4Cl\text{-}p)(C_6H_3Cl_2\text{-}2,4)$$
**11**

$$o\text{-}O_2NC_6H_4C(OC_6H_5){=}NC_6H_5$$
**12**

The usefulness of the Chapman rearrangement obviously depends on the availability of the aryl imidates, which are generally prepared by the following route.

$$H_2N\text{-}C_6H_4\text{-}R' \rightarrow C_6H_5C(=O)\text{—}NH\text{-}C_6H_4\text{-}R' \rightarrow C_6H_5C(Cl){=}N\text{-}C_6H_4\text{-}R' \xrightarrow[\text{Base}]{HO\text{-}C_6H_4\text{-}R} C_6H_5C(O\text{-}C_6H_4\text{-}R){=}N\text{-}C_6H_4\text{-}R'$$

For the preparation of unsymmetrically substituted diarylamines **13**, either of two imidates, **14** or **15**, may be utilized. The choice between

$$\underset{\mathbf{14}}{C_6H_5C(O\text{-}C_6H_4\text{-}R){=}N\text{-}C_6H_4\text{-}R'} \rightarrow \underset{\mathbf{13}}{C_6H_5C(=O)\text{—}N(C_6H_4\text{-}R)(C_6H_4\text{-}R')} \leftarrow \underset{\mathbf{15}}{C_6H_5C(O\text{-}C_6H_4\text{-}R'){=}N\text{-}C_6H_4\text{-}R}$$

**14** and **15** is often dictated by the relative availabilities of the amine and the phenolic components. When both sets of reagents are available, the imidate selected is usually the one which will rearrange at the lower temperature. Therefore the more acidic phenol (with electron-withdrawing substituents) should be chosen.[4] In several cases, both imidates have

[15] Jamison and Turner, *J. Chem. Soc.*, **1937**, 1954.

been used successfully. For example, all six possible monochloroimidates have been prepared and pyrolyzed to the three isomeric N-benzoylchlorodiphenylamines.[2,3] On the other hand, while the imidate **14** (R = $o$-$NO_2$, R′ = H), was converted in 40% yield to N-benzoyl-$o$-nitrodiphenylamine in 1 hour at 165°, the isomeric compound **14** (R = H, R′ = $o$-$NO_2$) did not rearrange.[3]

A great variety of halogenated diarylamines has been prepared by the Chapman rearrangement: the monochlorodiphenylamines previously mentioned and a large number of dichloro-, trichloro- and tetrachlorodiphenylamines.[3,4,15–20] The pentachloro imidate **14** (R = 2,4,6-$Cl_3$, R′ = 2,4-$Cl_2$) was converted to the N-benzoyldiarylamine in 81% yield, and this product was then hydrolyzed to 2,2′,4,4′,6-pentachlorodiphenylamine in 92% yield.[15] The hexachloro compound **14** (R = R′ = 2,4,6-$Cl_3$) has also been rearranged successfully.[3]

N-Benzoyl-2,2′-dibromodiphenylamine[21] and its 4,4′-dibromo analog[22] have been prepared and hydrolyzed to the diphenylamines. In the former case the rearranged product was obtained in 86% yield, and its hydrolysis product in 87% yield. The iodine-containing imidates **16*a*** and **16*b*** were rearranged to the N-benzoyldiphenylamines in 56% and 95% yields, respectively.[23]

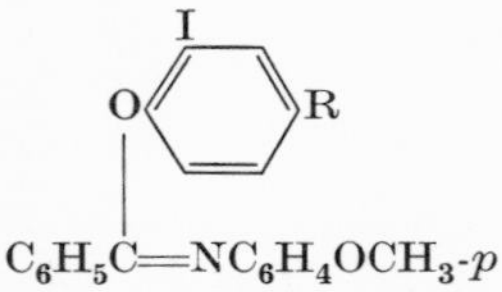

I

O R

$C_6H_5C{=}NC_6H_4OCH_3$-$p$

**16**

(*a*, R = $CH_3$; *b*, R = $CO_2CH_3$)

Pyrolysis of the fluorine-containing imidate, **14** (R = R′ = $p$-F), has been reported to furnish N-benzoyl-4,4′-difluorodiphenylamine (80% yield) which was hydrolyzed almost quantitatively to 4,4′-difluorodiphenylamine.[12]

The first alkyl-substituted aryl imidate to be pyrolyzed was the dimethyl derivative **14** (R = R′ = $p$-$CH_3$) which was converted to N-benzoyl-4,4′-dimethyldiphenylamine.[2] A little later the synthesis of several monomethyl- and dimethyl-diphenylamines via the Chapman route was

[16] Chapman, *J. Chem. Soc.*, **1930**, 2458.

[17] Chapman and Perrott, *J. Chem. Soc.*, **1932**, 1770.

[18] Chapman and Perrott, *J. Chem. Soc.*, **1932**, 1775.

[19] Elson and Gibson, *J. Chem. Soc.*, **1931**, 294.

[20] Smith and Kalenda, *J. Org. Chem.*, **23**, 1599 (1958).

[21] Jones and Mann, *J. Chem. Soc.*, **1956**, 786.

[22] Crounse and Raiford, *J. Am. Chem. Soc.*, **67**, 875 (1945).

[23] Cookson, *J. Chem. Soc.*, **1953**, 643.

reported.[24,25] Two of these, 3-methyldiphenylamine[24] and 2,3′-dimethyldiphenylamine,[25] were each prepared via both isomeric imidates. 4-Methyldiphenylamine was synthesized in an overall yield of 84 % based on the imidate **14** (R = H, R′ = $p$-$CH_3$).[26]

Wiberg and Rowland prepared and rearranged two ethyl-substituted imidates, **14** (R = $o$-$C_2H_5$, R′ = H) and **14** (R = $p$-$C_2H_5$, R′ = H), and the corresponding isopropyl compounds but did not isolate any of the four alkylbenzoyldiphenylamines.[9] The same workers pyrolyzed a large number of imidates for their kinetic study, but the only product isolated was the $t$-butyl compound **17** (40% yield).[9] N-Benzoyl-4,4′-di-$t$-butyldiphenylamine was reported much earlier as resulting in "almost quantitative" yield by heating the corresponding imidate at 300°.[27] Benzyl groups are stable at the temperature required for the Chapman rearrangement. Both benzylphenyl imidates **18** were converted to the desired aroyldiphenylamines in over 80% yield by heating at 275°.[28]

$C_6H_5C(=O)$—$N(C_6H_4C(CH_3)_3$-$o)(C_6H_5)$

**17**

$C_6H_5C(OC_6H_4CH_2C_6H_5)$=$NC_6H_5$

**18**

($a$ = $o$-$C_6H_5CH_2$; $b$ = $p$-$C_6H_5CH_2$)

In addition to alkyl- and halogen-substituted imidates, compounds bearing methoxy, nitro, acetyl, benzoyl, cyano, and carbalkoxy groups have been rearranged. The three isomers of **14** (R′ = H, R = $OCH_3$), as well as **14** (R = H, R′ = $p$-$OCH_3$), were converted to N-benzoyldiphenylamines.[3] Since then, the iodo-methoxy imidates[23] **16** and the dicyanomethoxy imidate[29] **19** have been rearranged.

$C_6H_5C(O$—$C_6H_3(OCH_3)(CN))$=$NC_6H_4CN$-$p$

**19**

$C_6H_5C(O$—$C_6H_4R)$=$N$—$C_6H_4R'$

**14**

[24] Gibson and Johnson, *J. Chem. Soc.*, **1929,** 1473.

[25] Gibson and Johnson, *J. Chem. Soc.*, **1929,** 2743.

[26] Hey and Moynehan, *J. Chem. Soc.*, **1959,** 1563.

[27] Craig, *J. Am. Chem. Soc.*, **57,** 195 (1935).

[28] Waters and Watson, *J. Chem. Soc.*, **1957,** 253.

[29] Easson, *J. Chem. Soc.*, **1961,** 1029.

Compounds **14** (R = *o*-$NO_2$, R′ = H[3] and R = *o*-$NO_2$, R′ = 2-Br, 4-$CH_3$[30]) were converted to the desired benzoyldiarylamines by heating for 1 hour at less than 200°, but under similar treatment **14** (R = *p*-$NO_2$, R′ = H and R = H, R′ = *o*-$NO_2$) decomposed to tarry materials.[3] No *m*-nitro compounds have been studied.

Only three acyl-substituted imidates have been subjected to pyrolysis. The *para*-substituted compounds **14** (R = *p*-$CH_3CO$, R′ = H[3] and R = *p*-$C_6H_5CO$, R′ = H[31]) reacted normally. The latter gave N,4-dibenzoyldiphenylamine (90% yield), which on hydrolysis lost the N-benzoyl group and furnished the diarylamine in 89% yield. The *o*-acetyl compound **20** when pyrolyzed at 267° gave the quinoline derivative **22**.[3] The normal Chapman rearrangement product, **21**, was an intermediate in this reaction.

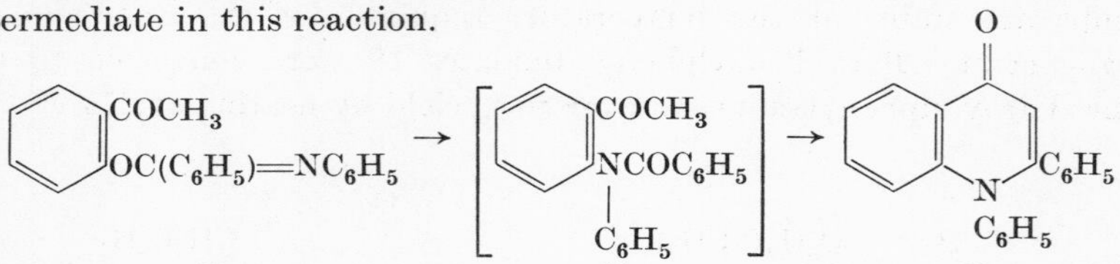

$COCH_3$ / $OC(C_6H_5){=}NC_6H_5$ → [ $COCH_3$ / $NCOC_6H_5$ / $C_6H_5$ ] → O / $C_6H_5$ / N / $C_6H_5$

**20** **21** **22**

Pyrolysis of the appropriate imidate at 280–300° furnished N-benzoyl-4,4′-dicyanodiphenylamine (**23**).[32] Hydrolysis with sodium hydroxide in ethylene glycol afforded 4,4′-dicyanodiphenylamine (**25**). Ammonolysis of the nitrile **23** gave the diamidino compound **24** which, when heated, gave benzamide and the diarylamine **25** in 76% yield.

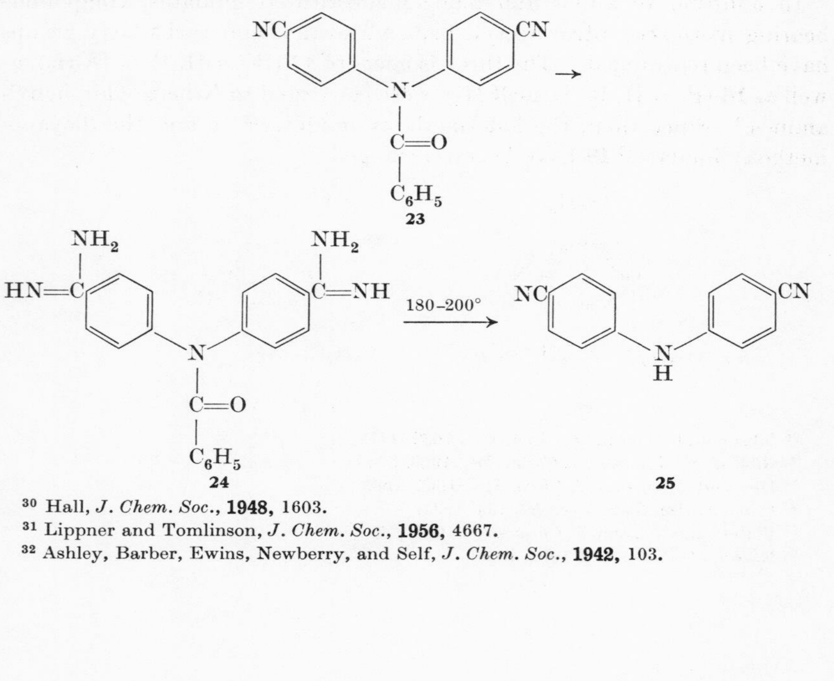

NC / CN / N / C=O / $C_6H_5$ →

**23**

$NH_2$ / HN=C / $NH_2$ / C=NH / N / C=O / $C_6H_5$ —180–200°→ NC / CN / N / H

**24** **25**

[30] Hall, *J. Chem. Soc.*, **1948**, 1603.

[31] Lippner and Tomlinson, *J. Chem. Soc.*, **1956**, 4667.

[32] Ashley, Barber, Ewins, Newberry, and Self, *J. Chem. Soc.*, **1942**, 103.

Several derivatives of the dinitrile **23** substituted in the *ortho* positions have been prepared by the Chapman rearrangement in high yield.[29] For example, the imidate **14** (R = 2-Cl, 4-CN, R′ = *p*-CN) heated in boiling Dowtherm furnished the *o*-chloro analog of **23** in 84% yield. Hydrolysis in ethylene glycol gave 2-chloro-4,4′-dicyanodiphenylamine in 60% yield. The 2-methoxy compounds, analogs of **23** and **25,** were similarly prepared in yields of 88% and 86%, respectively. The imidate **14** (R = 2-$NO_2$, 4-CN, R′ = *p*-CN) was quantitatively rearranged in either boiling anisole or pyridine, but the product could not be hydrolyzed cleanly to 4,4′-dicyano-2-nitrodiphenylamine. The last-named compound was synthesized by nitration of the dicyanodiphenylamine **25**.[29] Only two cyano-containing imidates, **14** (R = *p*-CN, R′ = 2-$NO_2$, 4-CN and R = R′ = 2-$NO_2$, 4-CN), that could not be rearranged to the appropriate diphenylamine have been reported.[29]

The conversion of **14** (R = *p*-$CO_2CH_3$, R′ = H) to N-benzoyl-4-carbomethoxydiphenylamine in high yield has been reported by two groups of investigators.[26,33] Hydrolysis to diphenylamine-4-carboxylic acid proceeded smoothly in 73% yield.[26] Attempts to isomerize the *meta*-substituted ester analog (**14,** R = *m*-$CO_2CH_3$, R′ = H),[33] the corresponding carboxylic acid,[33] or the ester **26** failed.[34] The rearrangement of derivatives of salicylic acid is discussed on pp. 13–15.

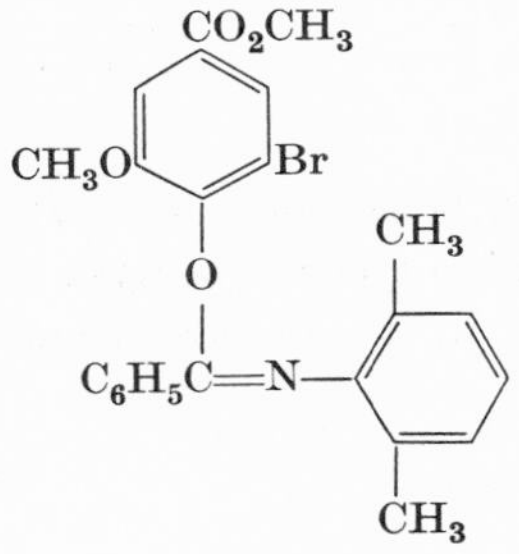

$CO_2CH_3$ / $CH_3O$ / Br / $CH_3$ / O / $C_6H_5C{=}N$— / $CH_3$

**26**

The imidates **27** containing aldehyde groups could not be rearranged.[23,33] However, the anil **28** pyrolyzed at 270° to furnish presumably the expected

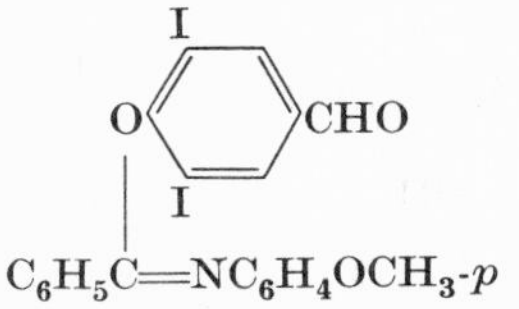

I / O / CHO / I / $C_6H_5C{=}NC_6H_4OCH_3$-*p*

**27a**

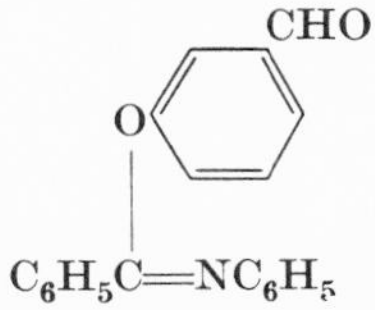

CHO / O / $C_6H_5C{=}NC_6H_5$

**27b**

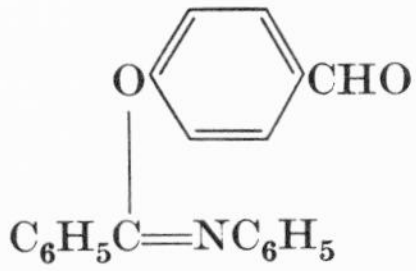

O / CHO / $C_6H_5C{=}NC_6H_5$

**27c**

[33] Brown, Carter, and Tomlinson, *J. Chem. Soc.*, **1958,** 1843.

[34] Harris, Potter, and Turner, *J. Chem. Soc.*, **1955,** 145.

rearrangement product **29**, which, however, could not be hydrolyzed to N-benzoyldiphenylamine-4-carboxaldehyde.[33]

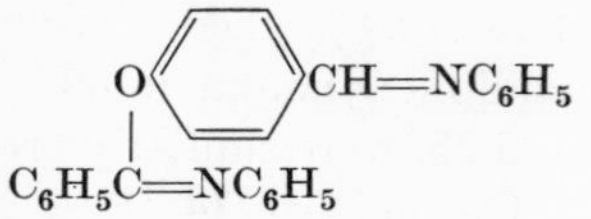

**28** 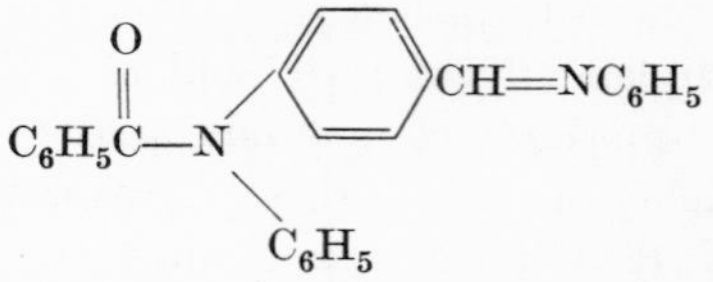 **29**

Only one abnormal Chapman rearrangement has been reported, excluding those examples where the normal product results and then reacts further.[35] The carbomethoxymethyl derivative **30** on pyrolysis gave a mixture from which two products, each obtained in 29% yield, were identified. One was shown to be the normal rearrangement product **31**; hydrolysis with base followed by acidification gave the expected N-phenyloxindole (**32**).

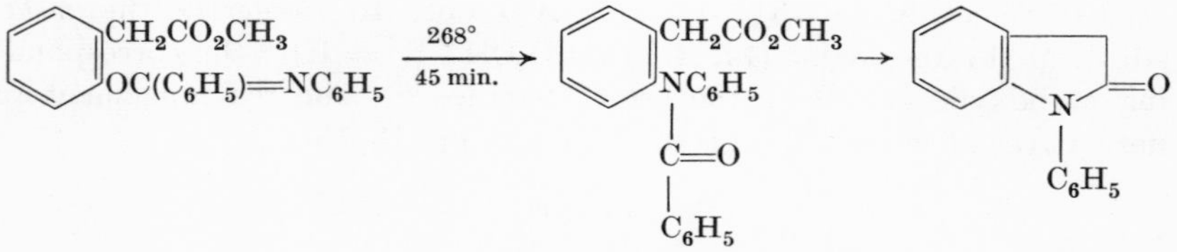

**30** $\xrightarrow[\text{45 min.}]{268°}$ **31** → **32**

The other product was 3-benzanilimino-2(3*H*)benzofuranone (**34**). The following mechanism involving basic catalysis (imidates are weakly basic) and rearrangement of the anion **33** was proposed.[35] The imidate **30** on

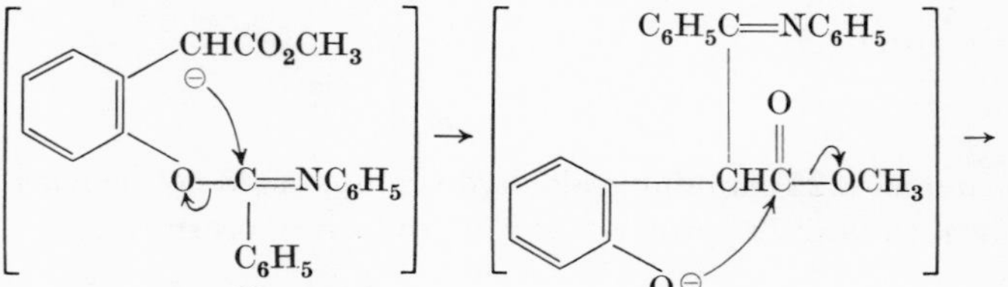

**33** 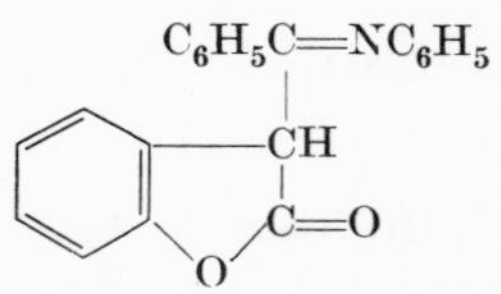 → **34**

[35] Schulenberg and Archer, *J. Am. Chem. Soc.*, **82**, 2035 (1960).

treatment with sodium methoxide in boiling benzene furnished the furanone **34** in 56% yield, showing the feasibility of the proposed base-catalyzed reaction.[35] Other imidates with active hydrogen atoms on the *ortho* substituent could also be expected to give abnormal pyrolysis products. In these cases the isomeric imidates should be used to prepare the desired aroyldiphenylamines.

Relatively few imidates containing aryl groups other than phenyl or substituted phenyl groups have been rearranged. Chapman pyrolyzed the three naphthyl derivatives **35**. The first two gave N-benzoyl-N-phenyl-α-naphthylamine; the third gave the corresponding β-naphthylamine.[3]

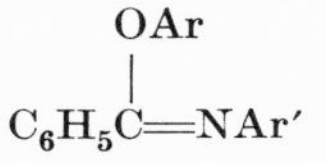

$C_6H_5C(OAr){=}NAr'$

**35**

(*a*, Ar = $C_6H_5$, Ar′ = α-$C_{10}H_7$)
(*b*, Ar = α-$C_{10}H_7$, Ar′ = $C_6H_5$)
(*c*, Ar = β-$C_{10}H_7$, Ar′ = $C_6H_5$)

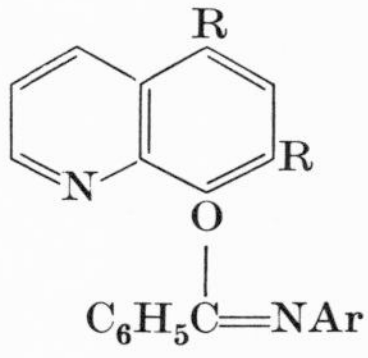

**36**

(*a*, R = H, Ar = $C_6H_5$)
(*b*, R = H, Ar = *o*-$CH_3C_6H_4$)
(*c*, R = Cl, Ar = $C_6H_5$)
(*d*, R = H, Ar = *o*-$ClC_6H_4$)
(*e*, R = H, Ar = 2-Br-4-$CH_3C_6H_3$)
(*f*, R = H, Ar = 2-Br-4, 6-$(CH_3)_2C_6H_2$)

Hall pyrolyzed the six 8-quinolinyl imidates **36**; the first in 72% yield, the second and third in unspecified yield, and the remaining three unsuccessfully.[30]

Hey and Moynehan rearranged three imidates **37** derived from phenanthridine;[26] the phenoxy derivative (R = H) to the phenanthridone **38** in 60% yield, the analogs (R = Cl) and (R = $CH_3$) in 75% and 70% yields, respectively.

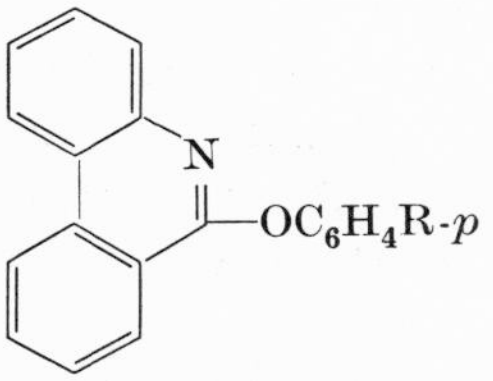

**37** (N, $OC_6H_4R$-*p*)

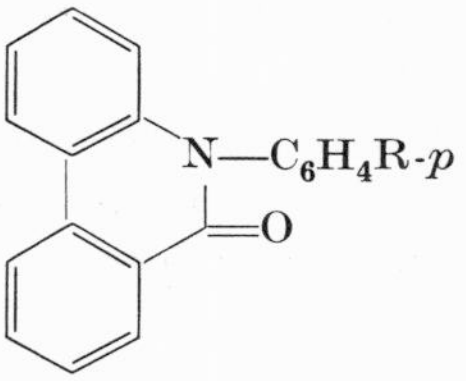

**38** (N—$C_6H_4R$-*p*, =O)

## Diphenylamine-2-carboxylic Acids

The Chapman rearrangement of imidates in which the phenolic component is a salicylic acid derivative serves as a synthesis of derivatives of diphenylamine-2-carboxylic acid (N-phenylanthranilic acid). The development of this synthesis is due to Jamison and Turner who showed that the parent imidate **39** (R = R′ = H), prepared from methyl salicylate and N-phenylbenzimidoyl chloride, rearranged readily to the aroylamine **40** in 73% yield on heating for 10 minutes at 270–275°. Hydrolysis of the aroylamine with 1 mole of base gave the N-benzoyl acid **41** in 76% yield, and excess base gave diphenylamine-2-carboxylic acid (**42**) in 96% yield.[15] (In a later paper the yield was reported as 78%.[36])

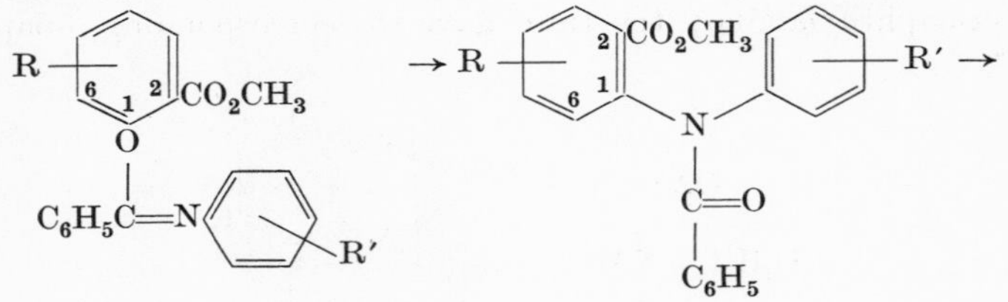

The rearrangement of a variety of substituted imidates also proceeded well, usually in better than 80% yield. Among these were chloro, bromo, methoxy, and methyl analogs of **39**.[15] Later a large number of imidates derived from 3-methylsalicylic acid were rearranged, for example, the imidates **39** in which R = 6-$CH_3$ and R′ = H, *o*-F, *o*-Cl, *o*-Br, and *o*-$CH_3$.[34,36,37] The rearrangement products were all partially hydrolyzed to the corresponding N-benzoyl acids **41**,[15,34,36–38] many of which were partially or completely resolved and their optical stabilities studied.[34,37–39] The only imidate that could not be isomerized to the desired aroyldiphenylamine was **39** in which R = 6-$CH_3$ and R′ = 2,4,6-$(CH_3)_3$.[34]

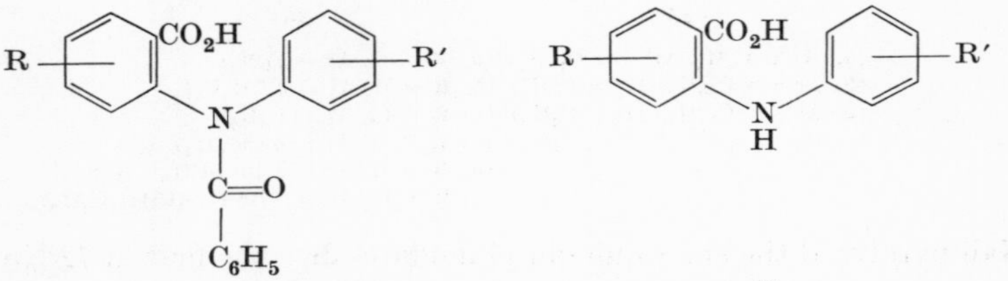

Singh et al, rearranged the four imidates **39** in which R = 5-$CH_3O$ and R′ = *o*-Cl,[40] *m*-Cl,[41] *p*-Cl,[42] and *p*-I;[43] the *p*-chloro compound in 80% yield, the other three in unspecified yields.

[36] Hall and Turner, *J. Chem. Soc.*, **1945,** 694.
[37] Jamison and Turner, *J. Chem. Soc.*, **1940,** 264.
[38] Brooks, Harris, and Howlett, *J. Chem., Soc.*, **1957,** 2380.
[39] Jamison and Turner, *J. Chem. Soc.*, **1938,** 1646.

The imidate **30** derived from *o*-hydroxyphenylacetic acid gave a mixture of a benzofuranone and the normal rearrangement product on pyrolysis. (See p. 12.) However, the compounds **39** (R = H, R′ = *o*-$CH_2CO_2CH_3$ or *o*-$CH_2CO_2C_2H_5$), containing a carbalkoxymethyl substituent in the phenyl group bonded to nitrogen, rearranged normally in 83% and 91% yield, respectively.[44]

## Acridones

Jamison and Turner were also the first to use the products of the Chapman rearrangement as intermediates for the synthesis of acridones.[15] 2-Chloroacridone (**46**) was prepared by them in three different ways. The aroylamine **43,** (which had been prepared by heating the appropriate imidate at 270°) was pyrolyzed at 320°, yielding methyl benzoate and the acridone **46**. The yield was "not quantitative." The second route involved partial hydrolysis to the acid **44** which was heated for 5 minutes at 250°. Benzoic acid was eliminated and the acridone **46** obtained in 90% yield. The third path utilized the completely hydrolyzed acid **45** which was converted to **46** by treatment with phosphorus oxychloride and hydrolysis of the resulting 2,9-dichloroacridine;[45] the over-all yield was 87%.[15]

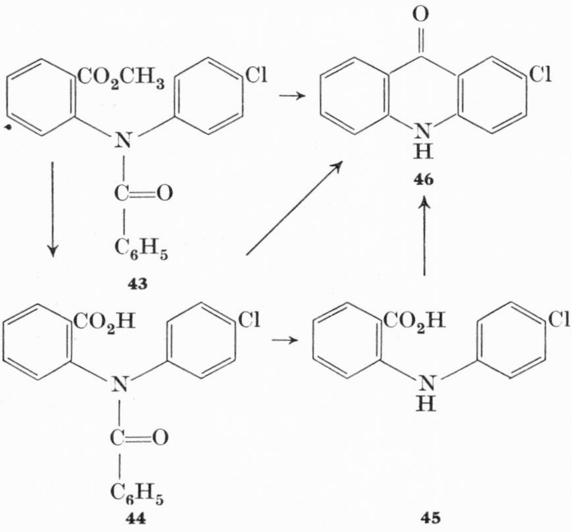

[40] G. Singh, S. Singh, A. Singh, and M. Singh, *J. Indian Chem. Soc.*, **29,** 783 (1952).

[41] G. Singh, A. Singh, S. Singh, and M. Singh, *J. Indian Chem. Soc.*, **28,** 698 (1951).

[42] G. Singh, S. Singh, A. Singh, and M. Singh, *J. Indian Chem. Soc.*, **28,** 459 (1951).

[43] S. Singh, *J. Sci. Ind. Res. (India)*, **10B,** 82–6 (1951) [*C.A.*, **46,** 2547 (1952)].

[44] Schulenberg and Archer, *J. Am. Chem. Soc.*, **83,** 3091 (1961).

[45] N. S. Drozdov and S. S. Drozdov, *J. Gen. Chem. USSR (Engl. Transl.)*, **4,** 1 (1934) [*C.A.*, **28,** 5456 (1934)]; N. S. Drozdov, *ibid.*, **4,** 117 (1934) [*C.A.*, **28,** 5456 (1934)].

Surprisingly, the parent compound **47** (R = R′ = H) did not give acridone at temperatures as high as 350°.[15] On the other hand, the acid **47** (R = H, R′ = 2,4($CH_3$)$_2$) formed 2,4-dimethylacridone in 71% yield when it was pyrolyzed at 300° for a few minutes, but the corresponding ester did not give the acridone when heated at 350°.[15]

Pyrolysis of compounds **47** (R = 4-Cl, R′ = *p*-$OCH_3$) and **47** (R = 6-$CH_3$, R′ = *o*-Cl) gave, respectively, the acridones **48,** in almost quantitative yield,[15] and **49** in 86% yield.[36] The phosphorus oxychloride method was used for the preparation of the acridone **50** in 91% yield.[15]

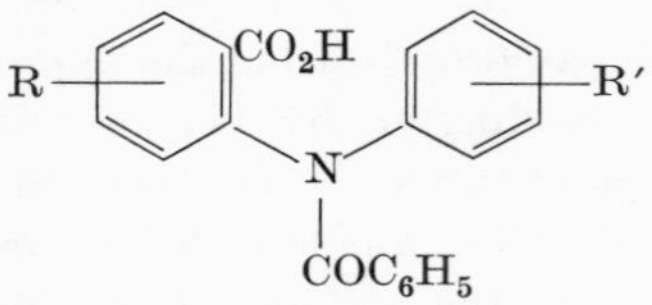

**47** 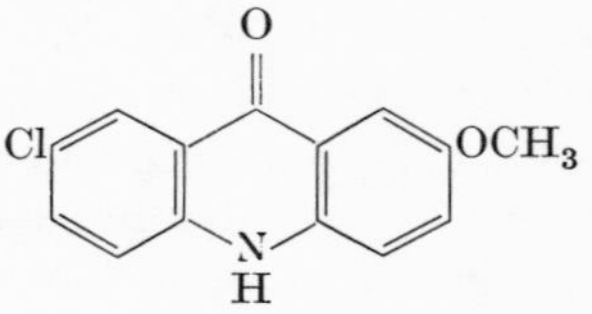 **48**

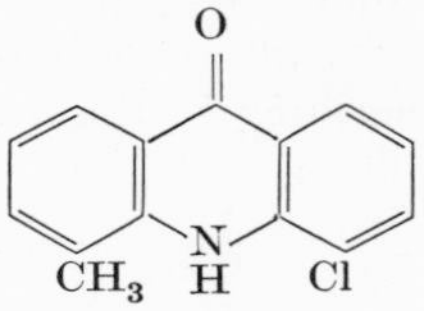

**49** 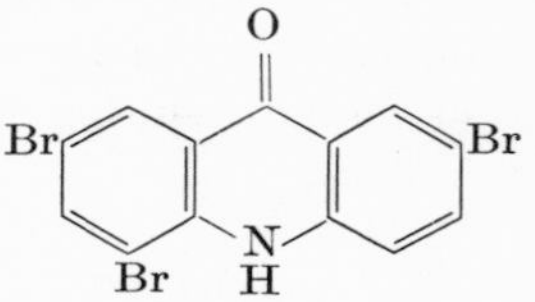 **50**

2-Chloro-6-methoxyacridone was prepared in unspecified yields by all three routes.[42] The isomeric 5-chloro-3-methoxyacridone was prepared by the two pyrolytic methods,[40] while an acridone, believed to be the 1-chloro-6-methoxy isomer, was synthesized by the phosphorus oxychloride route.[41] 2-Iodo-6-methoxyacridone was also synthesized by pyrolysis of the ester and by the phosphorus oxychloride method.[43] In all cases the starting imidate was that derived from 4-methoxysalicyclic acid.[40–43]

1,9-Dichloro-7-methoxyacridine **(51)** was prepared as shown in the

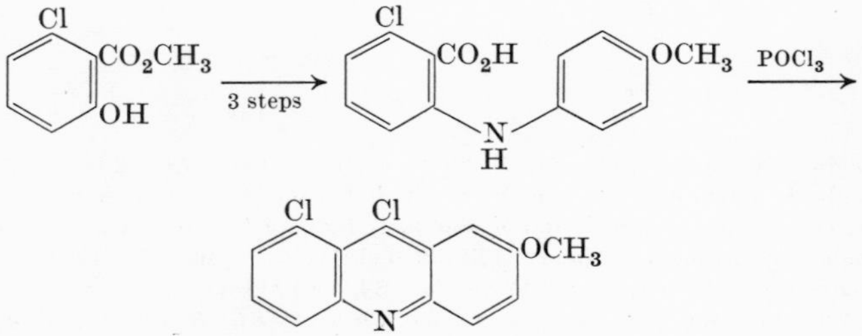

3 steps $POCl_3$

**51**

accompanying formulation.[46] The over-all yield was 28% based on methyl 6-chlorosalicylate. Ionescu synthesized 2,4,5,7-tetrachloroacridone (**53**) in 10% yield by heating the imidate **52** for 10 minutes at 260° and, without isolating the product, heating for 1 minute at 200° with concentrated sulfuric acid.[47]

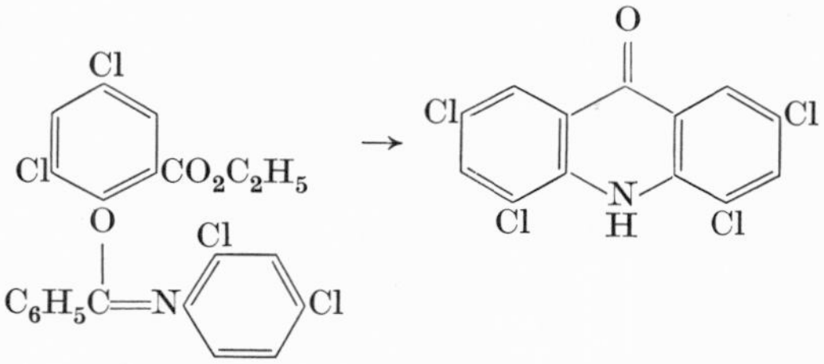

**52** → **53**

The synthesis of acridones with additional fused rings, using the Chapman rearrangement, was first accomplished by Cymerman-Craig and Loder.[48] They pyrolyzed the imidate **54** which rearranged and lost methyl benzoate to form the benzacridone **55** in 72% yield. The intermediate benzoyldiarylamine was not obtained.

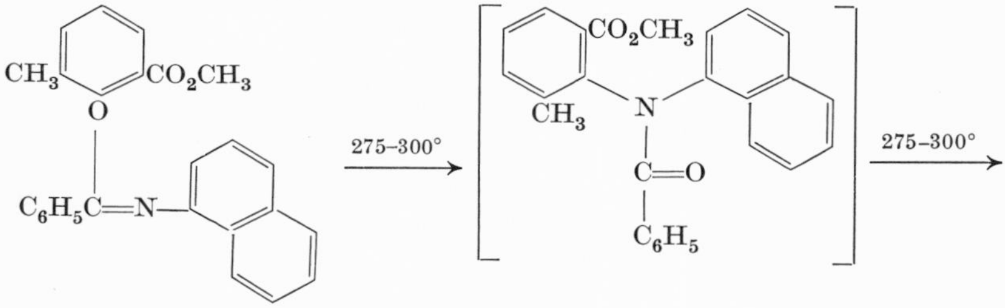

**54** —275–300°→ [intermediate] —275–300°→ **55**

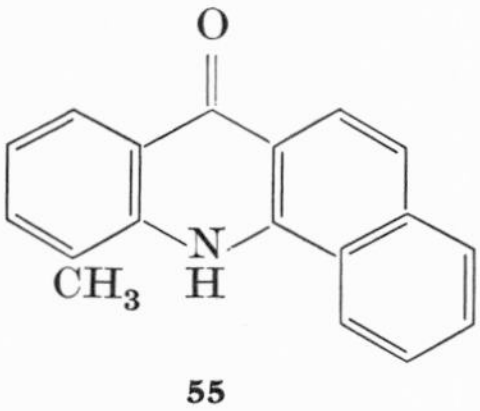

Similarly the β-naphthyl analog of **54** gave the acridone **56** in 38% yield on pyrolysis for 30 minutes at 360°, and the imidate **57** gave the pentacyclic acridone **58** in 47% yield on pyrolysis.[48]

[46] Dauben and Hodgson, *J. Am. Chem. Soc.*, **72**, 3479 (1950).

[47] Ionescu, Goia, and Felmeri, *Acad. Rep. Populare Romine, Filiala Cluj, Studii Cercetari Chem.*, **8**, 351 (1957) [*C.A.*, **54**, 4587 (1960)]; Ionescu and Goia, *Rev. Chim. Acad. Rep. Populaire Roumaine*, **5**, No. 1, 85 (1960) [*C.A.*, **55**, 9402 (1961)].

[48] Cymerman-Craig and Loder, *J. Chem. Soc.*, **1955**, 4309.

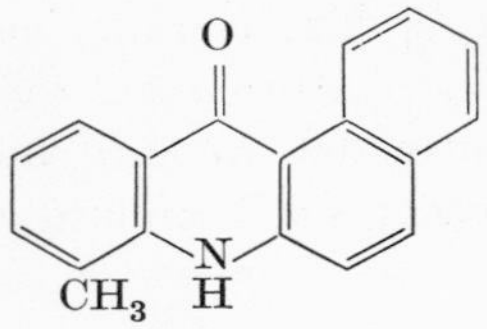

$CH_3$ H

**56**

$CO_2CH_3$ O $C_6H_5C{=}N$— 280–300° 30 min. O N H

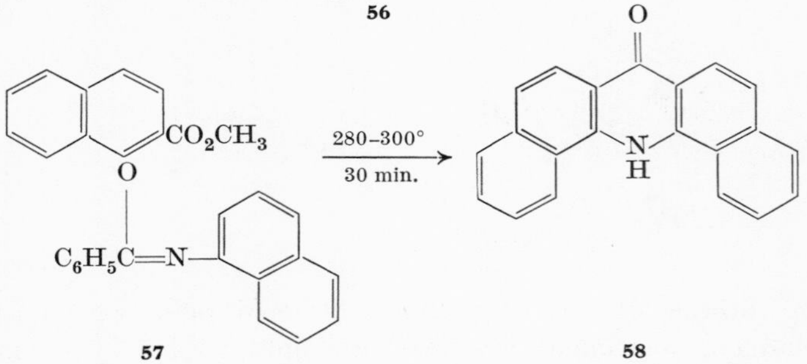

**57** **58**

Two other pentacyclic acridones, **59**[48] and **60**,[49] were synthesized in 54% and 80% yield, respectively, as indicated in the equations, but an attempt to prepare the hexacyclic acridone **61** failed.[48]

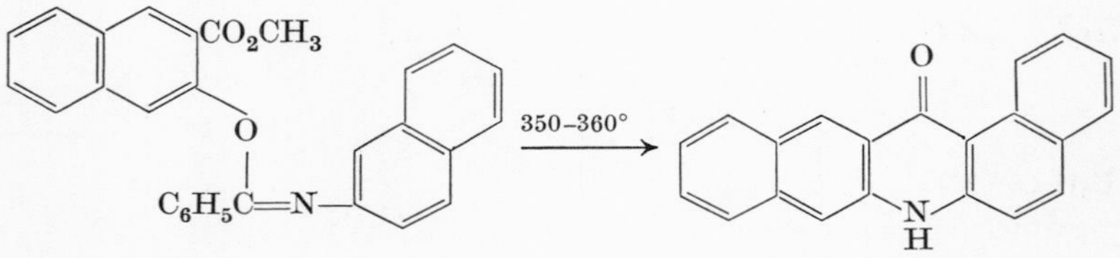

$CO_2CH_3$ O $C_6H_5C{=}N$— 350–360° O N H

**59**

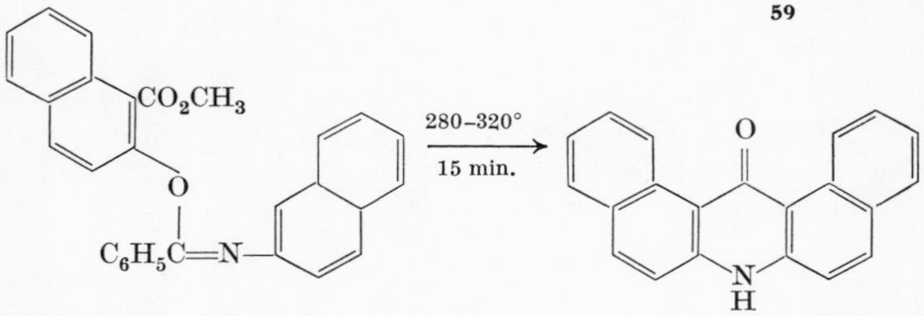

$CO_2CH_3$ O $C_6H_5C{=}N$— 280–320° 15 min. O N H

**60**

O N H

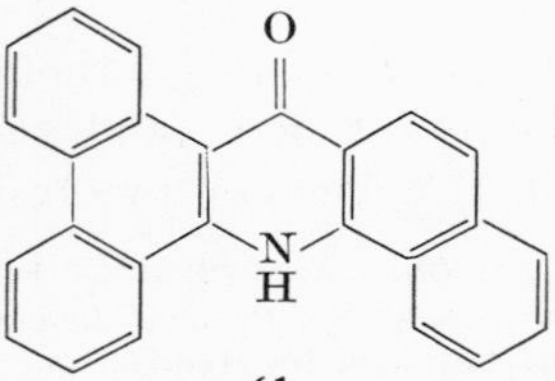

**61**

[49] Chatterjea, *J. Indian Chem. Soc.*, **35**, 41 (1958).

## COMPARISON WITH OTHER METHODS

The majority of substituted diphenylamines reported in the literature has been prepared by direct nucleophilic displacement of halide by aromatic amines. The uncatalyzed reaction, discovered by Jourdan,

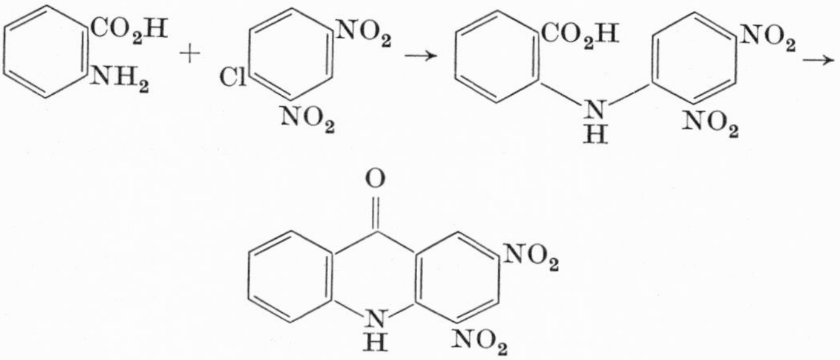

proceeds only with highly activated halides such as 2,4-dinitrochlorobenzene.[8a,50] The copper-catalyzed reaction, known as the Ullmann condensation,[8b,51,52]

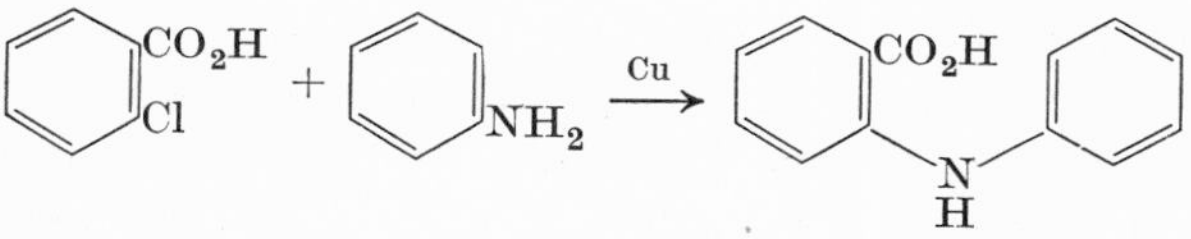

is of much greater general utility.[50,53,54] This reaction has been applied chiefly to the synthesis of diphenylamines substituted with nitro or carboxyl groups. If diphenylamines that do not contain nitro or carboxyl groups are desired, a two-step procedure may be used: an Ullmann reaction between aniline or a substituted aniline and an *o*-chlorobenzoic acid, followed by thermal decarboxylation of the product. The Goldberg diphenylamine synthesis,

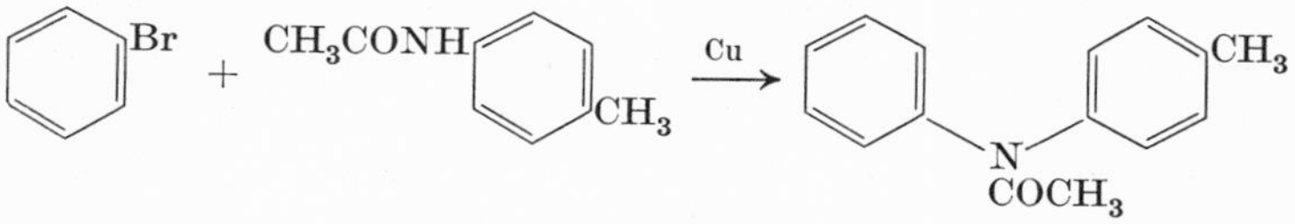

[50] Albert, *The Acridines*, E. Arnold and Co., London, 1951, p. 42.

[51] Surrey, *Name Reactions in Organic Chemistry*, Academic Press, New York, 1961, p. 236.

[52] Ullmann, *Ber.*, **36**, 2382 (1903).

[53] Bunnett and Zahler, *Chem. Rev.*, **49**, 392 (1951).

[54] Acheson, *Acridines*, Interscience Division, John Wiley and Sons, New York, 1956, p. 148; (*a*) p. 157; (*b*) p. 105.

the reaction of N-acylanilines with aryl halides followed by hydrolysis, is a copper-catalyzed reaction also.[8c,55–57]

The Chapman synthesis involves five steps from the starting aniline: benzanilide formation, conversion to imidoyl chloride, imidate synthesis, rearrangement, and hydrolysis. All five steps are easily carried out and two of the intermediates, the imidoyl chloride and the rearrangement product, are often not isolated. However, the Ullmann synthesis will generally require less time and, if other factors are equal, is to be preferred.

It is difficult to compare the yields by these two routes since, for most of the examples of the Chapman method, yields are not given in all steps. This is especially true in the early literature where the preparation of most of the simpler diphenylamines is reported. Furthermore, many of the N-benzoyldiphenylamines were not hydrolyzed. Although a huge number of derivatives of diphenylamine-2-carboxylic acid has been prepared by the Ullmann condensation, only a small percentage has been decarboxylated; most have been converted to acridine derivatives.[50,54] Some of the yields reported in the early literature are questionable. Ullmann reported that *o*-toluidine and *o*-chlorobenzoic acid could be converted to 2-methyldiphenylamine in an over-all yield of 87% (the decarboxylation was claimed to be quantitative).[58] In a recent paper an over-all yield of 33% was reported.[59] The corresponding *o*-chloro and *o*-methoxy compounds were obtained in 69% and 52% over-all yields, respectively.[59] The corresponding yields via the Chapman synthesis are not available.

Yields are reported for imidate formation,[9,24] rearrangement,[24] and hydrolysis[24] in the preparation of 3-methyldiphenylamine. When the higher yield reported for the imidate synthesis is used,[9] an over-all yield of 35% can be estimated. Ullmann reported a 62% yield in the reaction between *o*-chlorobenzoic acid and *m*-toluidine, but no yield was given for the decarboxylation.[58] Ullmann claimed that the corresponding 4-methyl acid, obtained in 89% yield from *p*-toluidine, was decarboxylated quantitatively.[58] 4-Methyldiphenylamine was synthesized from phenol and benz-*p*-toluidide by the Chapman route in 74% over-all yield.[26]

A large number of chloro-substituted diphenylamines was prepared by Elson and Gibson by Ullmann condensation of *o*-chlorobenzoic acid and various aniline derivatives.[19] The products from 2,5-dichloroaniline and 3,4-dichloroaniline were decarboxylated in 85% and 75% yields, respectively, but no yields were reported for the condensation reaction.[19]

[55] Goldberg, *Ber.*, **40,** 4541 (1907).
[56] Davis and Ashdown, *J. Am. Chem. Soc.*, **46,** 1051 (1924).
[57] Weston and Adkins, *J. Am. Chem. Soc.*, **50,** 859 (1928).
[58] Ullmann, *Ann.*, **355,** 312 (1907).
[59] Massie and Kadaba, *J. Org. Chem.*, **21,** 347 (1956).

Over-all yields (three steps) for several dihalodiphenylamines prepared by the Chapman route are very similar. 4,4′-Difluorodiphenylamine was synthesized in 48% yield,[12] the corresponding 4,4′-dibromo compound was obtained in 43% yield,[22] and the dichloro analog was prepared in 47% yield.[15] Good over-all yields were obtained for the following more highly substituted compounds: 2,4,4′-trichlorodiphenylamine, 62%;[15] 2-methyl-4,4′-dicyanodiphenylamine, 61%;[29] and 2,2′,4,4′,6-pentachlorodiphenylamine, 66%.[15] Although direct comparison of the two methods is not possible, it seems evident that at least comparable yields are attainable by the longer route.

Frequently the choice between the two methods will depend on availability of starting materials. Often the needed *o*-chlorobenzoic acid will be difficult to obtain, whereas the corresponding phenol will be available. In this case the Chapman route will obviously be preferred. It must be kept in mind that there are always *at least* two sets of reagents leading to a diphenylamine via the Ullmann condensation and decarboxylation. To illustrate, the diarylamine **62** can be prepared in the following ways.

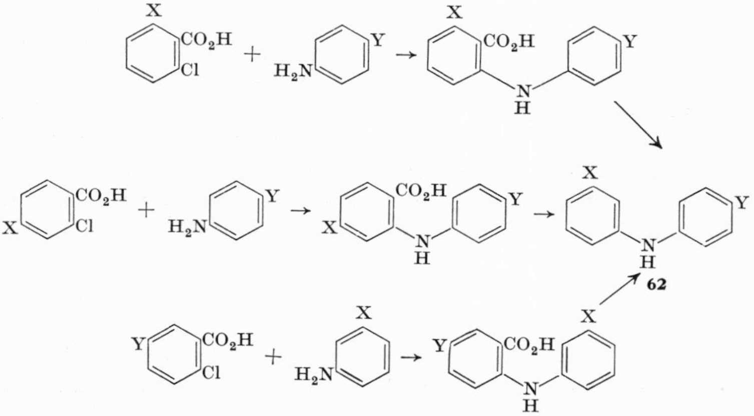

If Y were a *meta* or *ortho* substituent, a fourth possibility would exist. Furthermore, anthranilic acids will react with derivatives of bromobenzene to give diphenylamine-2-carboxylic acids.[60] Although chloro compounds are generally unreactive and bromo compounds must be used, the method has the advantage that some substituted anthranilic acids are commercially available while the corresponding *o*-chloro acids are not.[50] Therefore the following three sets of reagents must also be considered for the synthesis of **62**.

[60] Goldberg, *Ber.*, **39**, 1691 (1906).

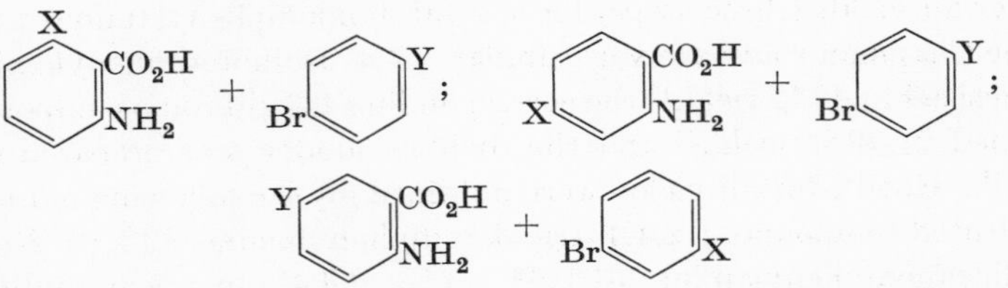

For the Chapman synthesis only two sets of compounds must be considered.

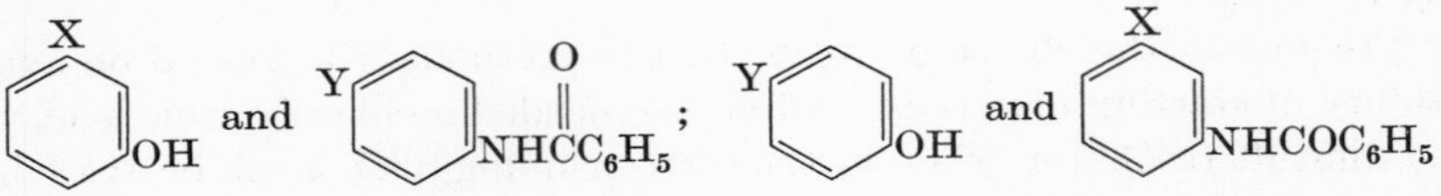

The Goldberg synthesis appears to be the preferred method for some of the simpler compounds. When *p*-chloroacetanilide and bromobenzene were treated with copper, potassium iodide, and iodine and the resulting product was hydrolyzed, 4-chlorodiphenylamine resulted in 72% yield.[61] Earlier, Goldberg[55] reported the conversion of *m*-nitroacetanilide and bromobenzene to 3-nitrodiphenylamine in 80% yield. Weston and Adkins studied thoroughly the reaction between acet-*p*-toluidide and bromobenzene, determining the effects of changes in catalyst and solvent.[57] Davis and Ashdown synthesized a large number of nitrodiphenylamines by the Goldberg procedure but gave no yields.[56] They also reported, without yields, the synthesis of nitrodiphenylamines from the unacetylated nitroanilines.[56] Earlier Goldberg had reported the direct conversion of *o*-nitroaniline and bromobenzene to 2-nitrodiphenylamine in 75–80% yield.[55]

The Ullmann condensation suffers some drawbacks, a major one being that reductive dehalogenation may sometimes become the major reaction.[53] Limitations of the reaction are discussed by Acheson.[54a]

When the appropriate *o*-chlorobenzoic acid is readily available, the Ullmann route will generally be preferred for the preparation of diphenylamine-2-carboxylic acids and the acridones derived from them. The parent acid itself has been prepared in 82–93% yield by the Ullmann procedure (one step),[62] while the three-step Chapman method from methyl salicylate and N-phenylbenzimidoyl chloride gave an over-all yield of 57%.[15] In a later paper lower yields were reported for the same Chapman route.[36] However, 4′-chlorodiphenylamine-2-carboxylic acid was prepared in about 75% over-all yield using the Chapman rearrangement, so the two methods may give comparable yields in some cases.[15] If the

[61] Wieland and Wecker, *Ber.*, **55**, 1804 (1922).

[62] Allen and McKee, *Org. Syntheses*, Col. Vol. **2**, 15 (1943).

necessary *o*-chloro acid is not available, the Chapman rearrangement may well be preferable since it proceeds particularly well with derivatives of methyl salicylate.[15,34,36,37]

Some Chapman rearrangement products can be converted directly to acridones by heating, but this does not appear to be a general reaction.[15] When the thermal reaction is possible, acridones may be synthesized in two steps from derivatives of methyl salicylate, so this route may in some cases be the preferred one, particularly in the synthesis of acridones with additional fused rings.[48] For example, 1,2,7,8-dibenzoacridone (**60**) (p. 18) was prepared by this method in 80% yield from the imidate after the Ullmann condensation between $\beta$-naphthylamine and 2-chloro-1-naphthoic acid was found to give only 1–2% of the desired diarylamine.[49]

In addition to the Ullmann and Chapman routes, a few other methods for synthesizing diphenylamines may be mentioned. Some compounds may be prepared by substitution of diphenylamine, e.g., 4,4′-dibromodiphenylamine by direct bromination.[63] This compound has also been prepared via the Chapman rearrangement.[22] Certain symmetrical compounds may be prepared by the reaction of equimolar amounts of an amine and its hydrochloride. Diphenylamine itself is best prepared in this way from aniline and aniline hydrochloride.[7a] Occasionally, unsymmetrical amines are also prepared by this method, but the mixed product must be separated from the three-component mixture that is obtained.[7a]

Diphenylamines may also be prepared by the reaction between an amine and a phenol.[7b] The parent compound has been synthesized by heating aniline and phenol at 250° with antimony trichloride as the catalyst.[64] The reaction between nitrobenzene and phenylmagnesium bromide is reported to give diphenylamine in 58% yield.[65] Similarly, methyl *p*-nitrobenzoate and phenylmagnesium bromide gave 4-carbomethoxydiphenylamine in 45% yield.[66] In contrast, the Ullmann condensation between iodobenzene and methyl *p*-aminobenzoate gave the same product in only 27% yield.[67] The corresponding acid has also been synthesized in 46% yield via the Chapman route starting with methyl *p*-hydroxybenzoate and N-phenylbenzimidoyl chloride.[26]

Finally, one example should be cited in which the Chapman method gave a good yield of the desired amine after several other routes failed completely. Jones and Mann[21] were unable to synthesize 2,2′-dibromodiphenylamine from *o*-chloronitrobenzene and *o*-bromoaniline, from

[63] Galatis and Megaloikonomos, *Prakt. Akad. Athenon*, **9,** 20 (1934) [*Chem. Zentr.*, **105,** II, 2974 (1934)].

[64] Buch, *Ber.*, **17,** 2634 (1884).

[65] Gilman and McCracken, *J. Am. Chem. Soc.*, **51,** 821 (1929).

[66] Curtin and Kauer, *J. Am. Chem. Soc.*, **75,** 6041 (1953).

[67] Gilman and Brown, *J. Am. Chem. Soc.*, **62,** 3208 (1940).

*o*-bromoaniline and its hydrochloride, from *o*-bromophenol and *o*-bromoaniline (with a zinc chloride catalyst), or from *o*-chloronitrobenzene and the sodium salt of *o*-bromoacetanilide. The Chapman route gave the desired product in 36% over-all yield.[21]

## RELATED REACTIONS

There are several imidate-amide rearrangements which present many features of interest but which do not have the usefulness of the Chapman rearrangement. In this section these related reactions are discussed, but completeness of coverage is not attempted.

### Catalyzed Rearrangements of Alkyl Imidates

The conversion of alkyl imidates to amides has been reported by many investigators, but the first careful study was that of Lander who found that the reaction proceeds well when catalyzed by small amounts of an alkyl halide. The rearrangement shown in the accompanying equation is typical of many others described by Lander.[68]

$$C_6H_5C(OCH_3){=}NCH_3 \xrightarrow[\text{Trace } CH_3I]{100^\circ;\ 8\text{ hr.}} C_6H_5C({=}O){-}N(CH_3)_2$$

Dialkyl sulfates have been used as catalysts for the conversion of O-alkyl lactims to N-alkyl lactams.

$$\underset{\mathbf{63}}{\text{(CH}_2)_5\text{N}{=}\text{COCH}_3 \text{ (seven-membered ring)}} \xrightarrow[\text{Trace } (CH_3)_2SO_4]{\text{Boiling } C_6H_6} \text{(CH}_2)_5\text{NCH}_3\text{—C}{=}\text{O (seven-membered ring)} \qquad \text{(Ref. 69)}$$

$$\text{(CH}_2)_4\text{N}{=}\text{COCH}_3 \text{ (six-membered ring)} \xrightarrow[5\%\ (CH_3)_2SO_4]{172^\circ;\ 1/4\text{ hr.}} \text{(CH}_2)_4\text{NCH}_3\text{—C}{=}\text{O (six-membered ring)} \qquad \text{(Ref. 70)}$$

When O-ethyl caprolactim (the ethyl homolog of **63**) was heated with dimethyl sulfate, a mixture of the N-methyl and N-ethyl lactams resulted.[70] Acylic analogs gave similar results.[68] In contrast to the Chapman reaction, the Lander rearrangement is intermolecular. Arbuzov has recently made use of the intermolecular nature of the reaction by treating alkyl imidates with various halides.[71] The following reaction has been carried out in 90% yield.[71]

[68] Lander, *J. Chem. Soc.*, **83,** 406 (1903).

[69] Benson and Cairns, *J. Am. Chem. Soc.*, **70,** 2115 (1948).

[70] Ralls and Elliger, *Chem. & Ind.* (*London*), **1961,** 20.

[71] Arbuzov and Shishkin, *Dokl. Akad. Nauk. SSSR*, **141,** 349, 611 (1961) [*C.A.*, **56,** 11491, 15424 (1962)].

$$C_6H_5C(OC_2H_5){=}NC_6H_5 + C_6H_5CH_2I \rightarrow C_6H_5C({=}O){-}N(C_6H_5)CH_2C_6H_5$$

Roberts and Vogt studied the rearrangement of alkyl imidates in the presence of sulfuric acid. The accompanying reaction was carried out in 41–53% yield.[72] The reaction proceeded in better yield (74–85%)

$$\underset{\mathbf{64}}{HC(OC_2H_5){=}NC_6H_5} \xrightarrow[H_2SO_4]{170\text{–}180^0;\ 1/2\ \text{hr.}} \underset{\mathbf{65}}{HC({=}O){-}N(C_2H_5)(C_6H_5)}$$

when a half mole of triethyl orthoformate was present in the reaction mixture. The *p*-chloro analog of **64** was rearranged in 63% yield without the use of additional orthoformate while the O-methyl analog gave N-methylformanilide in 61% yield.[72] Treatment of **64** with sulfuric acid and triisoamyl orthoformate furnished a mixture of the N-ethyl **(65)** and N-isoamyl amides.[72] Thus it is clear that the reaction is intermolecular and analogous to the Lander rearrangement. Presumably, the sulfuric acid generates an alkylating species from the ortho ester (and from the imidate) which is equivalent to the alkyl halide catalyst used by Lander.

Since ethyl N-phenylformimidate **(64)** is prepared from aniline and triethyl orthoformate by an acid-catalyzed reaction and rearranged to N-ethylformanilide **(65)** in the presence of acids, it is possible to prepare the anilide **65** from aniline and triethyl orthoformate in a single step without isolating the imidate. The anilide **65** is readily hydrolyzed to N-ethylaniline in high yield. Roberts has suggested these reactions as a practical synthesis of the monoalkylanilines.[72]

Cramer and Hennrich studied the rearrangement of trichloroacetimidates and found that boron trifluoride is an excellent catalyst for the reaction.[73] The methyl ester **66** gave the amide **67** in 91% yield, while the corresponding ethyl ester rearranged in 96% yield. The benzyl ester

$$\underset{\mathbf{66}}{CCl_3C(OCH_3){=}NH} \rightarrow \underset{\mathbf{67}}{CCl_3C({=}O){-}NHCH_3}$$

gave only 15% of N-benzyltrichloroacetamide, the remainder being trichloroacetamide.[73]

[72] Roberts and Vogt, *J. Am. Chem. Soc.*, **78,** 4778 (1956).

[73] Cramer and Hennrich, *Chem. Ber.*, **94,** 976 (1961).

Although boron trifluoride and sulfuric acid can be used to catalyze the rearrangements of imidates, acids which furnish good nucleophiles cannot. When imidate hydrochlorides are heated, cleavage occurs instead of rearrangement. This reaction, known as the Pinner fission,

$$\mathrm{RC(OR')\!=\!NH{\cdot}HCl \rightarrow RC(=O)NH_2 + R'Cl}$$

has been studied by McElvain[74] and Cramer.[75] It involves nucleophilic attack by chloride ion on the protonated imidate.

Aryl imidates do not undergo the Pinner fission. Phenyl N-phenylbenzimidate on heating with hydrogen chloride yielded phenol and N-phenylbenzimidoyl chloride.[76]

$$\mathrm{C_6H_5C(OC_6H_5)\!=\!NC_6H_5 \xrightarrow{HCl} C_6H_5OH + C_6H_5C(Cl)\!=\!NC_6H_5}$$

## Thermal Rearrangement of Alkyl Imidates

There have been several reports of the thermal isomerization of alkyl imidates, but, except for the methyl, benzyl, and allyl derivatives which will be considered later, when the reactions were re-examined it was found that the pure imidates did not rearrange. Two examples will suffice.

The imidate **68** was reported to rearrange to the amide **69** in 35% yield on heating for 6 hours at 270–280°.[77] Lander, in the first careful study of the thermal behavior of alkyl imidates, showed that the carefully purified imidate **68** underwent negligible isomerization after 8 hours at 250–270° and 3 hours at 300°.[68] Similarly the imidate **70** was reported to

$$\underset{\mathbf{68}}{\mathrm{C_6H_5C(OCH_3)\!=\!NC_6H_5}} \qquad \underset{\mathbf{69}}{\mathrm{C_6H_5C(=O)N(CH_3)C_6H_5}}$$

rearrange to the lactam **71** in 85% yield after 2 hours at 180–250°.[69] Later it was shown that the rearrangement of **70** was due to the presence of an impurity, probably diethyl sulfate.[70]

[74] McElvain and Tate, *J. Am. Chem. Soc.*, **73,** 2233 (1951).

[75] Cramer and Baldauf, *Chem. Ber.*, **92,** 370 (1959).

[76] Chapman, *J. Chem. Soc.*, **123,** 1150 (1923).

[77] Wislicenus and Goldschmidt, *Ber.*, **33,** 1467 (1900).

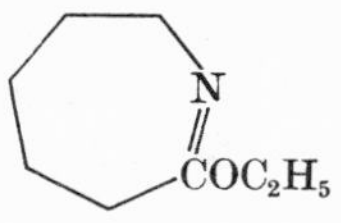

$$\text{(cycloheptane ring)}\;N{=}C\text{-}OC_2H_5 \qquad \text{(cycloheptane ring)}\;NC_2H_5\text{-}CO$$

**70** 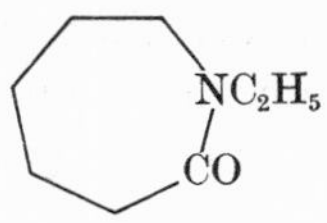 **71**

When the alkyl group in an alkyl imidate can undergo elimination, this, rather than isomerization, appears to be the normal pyrolysis pattern. Thus the ethyl and *sec*-butyl analogs of **68** furnished benzanilide and ethylene[68] and 2-butene,[9] respectively, and the imidate **70** yielded caprolactam and ethylene.[70] Pyrolysis of alkyl imidates has been suggested as a way of dehydrating secondary alcohols.[9]

Since elimination cannot occur, pyrolysis of methyl imidates can apparently give rearrangement products; e.g., the methyl imidate **68** gave the amide **69** in 25% yield when heated at 300–330°,[3] and conversion of **72** to the amides **73** occurred in 20–40% yields at about 300°.[78]

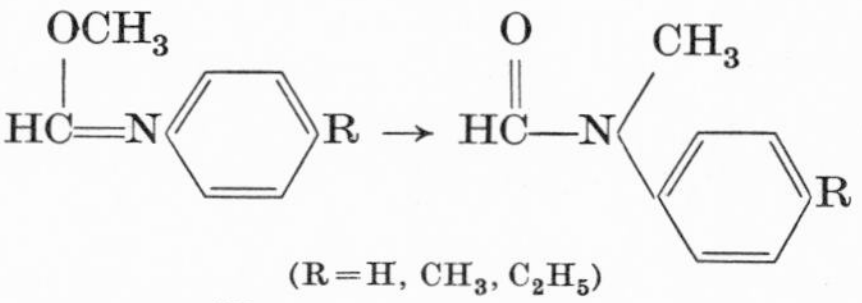

$$HC(OCH_3){=}N\text{-}C_6H_4R \rightarrow HC(=O)\text{-}N(CH_3)\text{-}C_6H_4R$$

$(R = H, CH_3, C_2H_5)$

**72** **73**

The temperatures needed are higher than those generally required for the Chapman rearrangement of aryl imidates, and the yields are lower. The reaction has been shown to be intermolecular, in contrast to the Chapman rearrangement which is intramolecular.[78] Wiberg[9] believes that the thermal rearrangement of alkyl imidates is a free radical process.

Benzyl imidates, like the methyl compounds, cannot undergo the elimination reaction and would be expected to rearrange. The thermal conversion shown in the accompanying equation has been reported without reference to reaction conditions.[79]

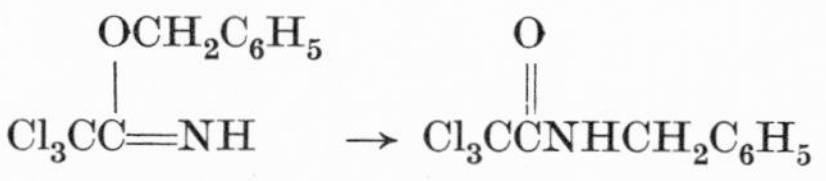

$$Cl_3CC(OCH_2C_6H_5){=}NH \rightarrow Cl_3CC(=O)NHCH_2C_6H_5$$

The cyclic benzyl imidate **74** furnished the isoindole derivative **75** in 70% yield when heated for 2 hours at 300°.[80] On steric grounds the four-membered transition state of the Chapman rearrangement cannot exist,

[78] Wiberg, Shryne, and Kintner, *J. Am. Chem. Soc.*, **79**, 3160 (1957).

[79] Cramer, Pawelzik, and Kupper, *Angew. Chem.*, **68**, 649 (1956).

[80] Stirling, *J. Chem. Soc.*, **1960**, 255.

and it seems likely that the reaction proceeds via the intermediate diradical shown in the formulation. Attempts to rearrange the analogous N-benzyl and N-cyclohexyl compounds at 300° failed.[80]

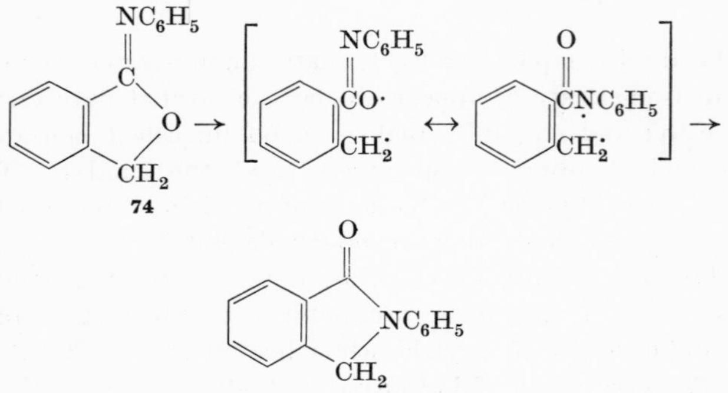

$NC_6H_5$ C O $CH_2$ **74** → [ $NC_6H_5$ CO· $CH_2$· ↔ O $CNC_6H_5$ $CH_2$· ] →

O $NC_6H_5$ $CH_2$ **75**

Allyl imidates undergo thermal rearrangement more readily than the methyl or benzyl derivatives. The reaction was first observed by Mumm and Möller[81] with the allyl imidate **76.**

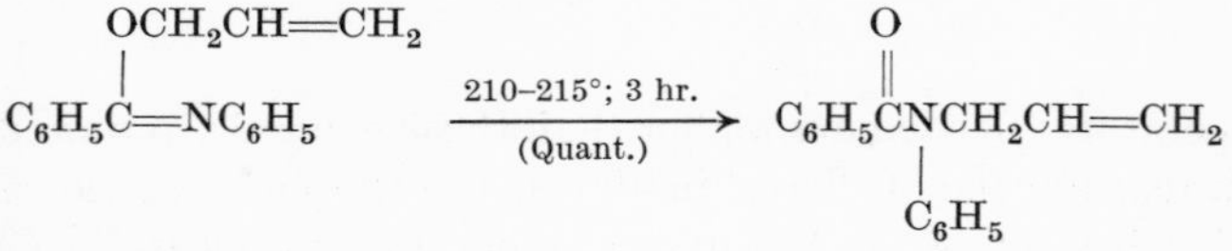

$$\underset{\mathbf{76}}{C_6H_5\overset{OCH_2CH{=}CH_2}{\overset{|}{C}}{=}NC_6H_5} \xrightarrow[\text{(Quant.)}]{210\text{–}215°;\ 3\ hr.} C_6H_5\overset{O}{\overset{\|}{C}}\underset{C_6H_5}{\underset{|}{N}}CH_2CH{=}CH_2$$

This rearrangement could conceivably occur either via a four-membered or a six-membered transition state. Mumm showed that the latter was the route indicating that the reaction resembles the Claisen rather than the Chapman rearrangement. The imidate **77** gave the amide **79** stereospecifically, presumably via **78.**[81]

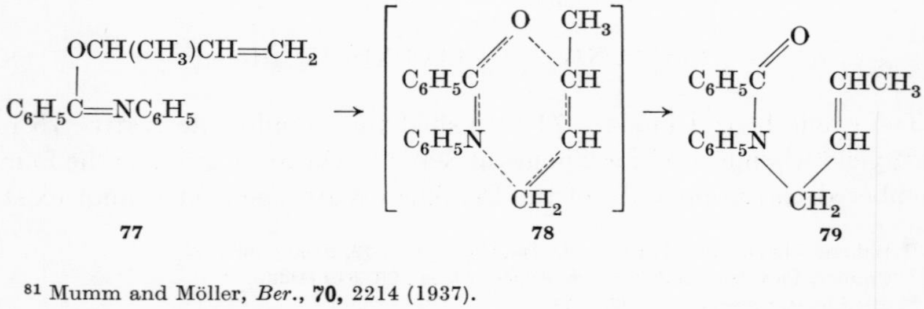

$OCH(CH_3)CH{=}CH_2$ $C_6H_5C{=}NC_6H_5$ **77** → [ O $CH_3$ $C_6H_5C$ CH $C_6H_5N$ CH $CH_2$ **78** ] → $C_6H_5C{=}O$ $CHCH_3$ $C_6H_5N$ CH $CH_2$ **79**

[81] Mumm and Möller, *Ber.*, **70,** 2214 (1937).

Similarly, the imidate **80** yielded the amide **81**.[81] If the Chapman mechanism had been operative, the expected product would be **79.**

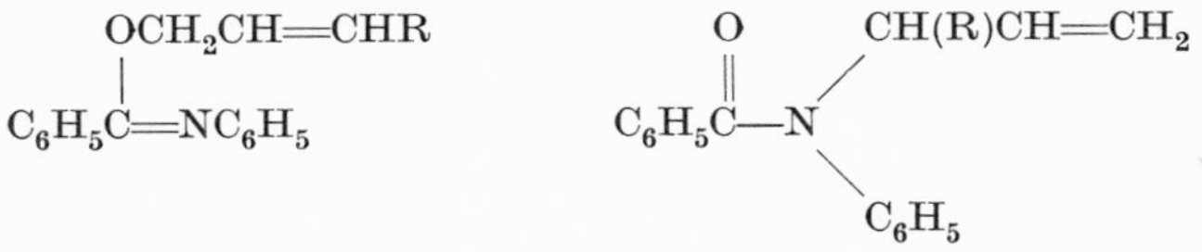

$OCH_2CH{=}CHR$

$C_6H_5C{=}NC_6H_5$

**80** ($R{=}CH_3$)
**82** ($R{=}C_2H_5$)

$C_6H_5C(=O){-}N(C_6H_5)CH(R)CH{=}CH_2$

**81** ($R{=}CH_3$)
**83** ($R{=}C_2H_5$)

The higher homolog **82** rearranged at 235° to furnish **83**.[82] When either **82** or **83** was heated for 3 hours at 290°, a different product, **85,** was obtained.[82] Evidently this resulted via a six-membered transition state **84** strictly analogous to that postulated for the Claisen rearrangement.*

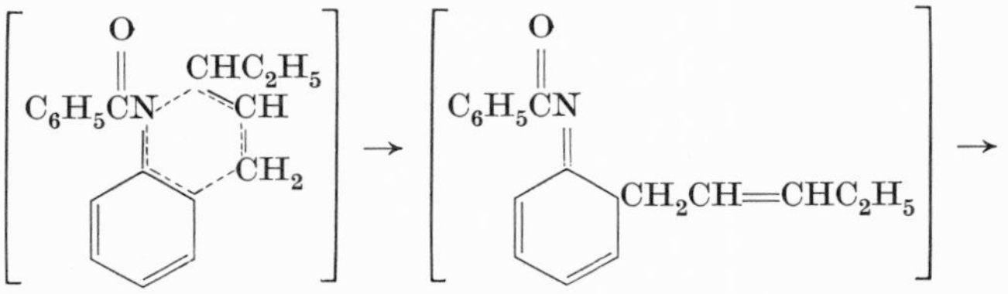

**84** → [$C_6H_5C(=O)N{=}$cyclohexadiene$-CH_2CH{=}CHC_2H_5$] →

$C_6H_5C(=O)NH-C_6H_4-CH_2CH{=}CHC_2H_5$

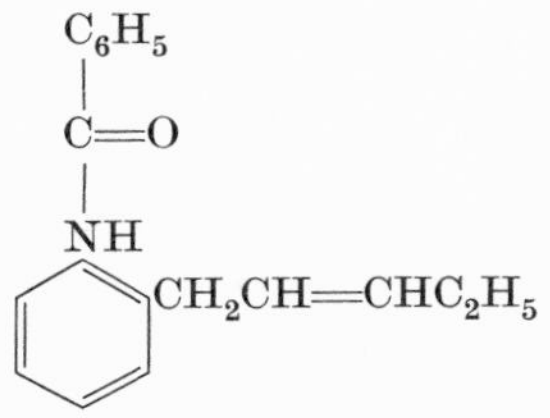

**85**

Earlier the rearrangement of the terminally substituted imidate **86** at 225–260° was shown to result in a 75% conversion to **87.** The possible intermediate **89** (corresponding to **83**) was not obtained.[84]

[82] Lauer and Benton, *J. Org. Chem.*, **24,** 804 (1959).

* Recently the conversion of N-allyl-1-naphthylamine to 2-allyl-1-naphthylamine was claimed to be the first example of a Claisen rearrangement from a nitrogen atom.[83] Apparently the earlier work was overlooked.

[83] Marcinkiewicz, Green, and Mamalis, *Tetrahedron*, **14,** 208 (1961).

[84] Lauer and Lockwood, *J. Am. Chem. Soc.*, **76,** 3974 (1954).

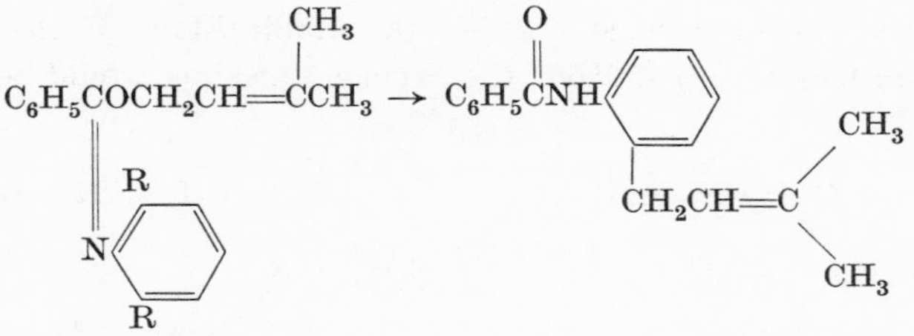

$$C_6H_5COCH_2CH{=}C(CH_3)CH_3 \;(=N\text{-}C_6H_3R_2) \rightarrow C_6H_5C(=O)NH\text{-}C_6H_4\text{-}CH_2CH{=}C(CH_3)_2$$

**86** (R=H)
**88** (R=$CH_3$)

**87**

The analogous *o,o-'*disubstituted imidate **88,** which could not lead to aryl substitution, gave only benz-2,6-dimethylanilide on pyrolysis; **90** was not obtained.[82] Lauer therefore concluded that the rearrangement of **86** to **87** is probably not a two-step reaction and does not proceed via **89**. Instead, a one-step mechanism proceeding via the transition state **91** was postulated.[82,84] However, the steric effect exerted by the two *ortho* methyl groups in **88** could conceivably prevent formation of **90** even if **89** were formed normally. Furthermore, it seems unlikely that **86**

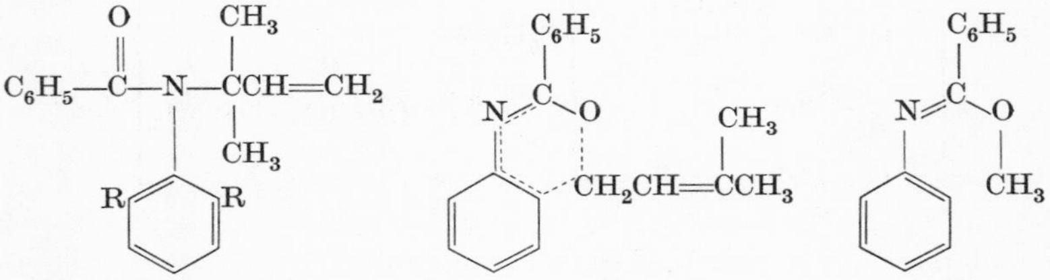

$$C_6H_5\text{—}C(=O)\text{—}N(C_6H_3R_2)\text{—}C(CH_3)_2CH{=}CH_2$$

**89** (R=H)
**90** (R=$CH_3$)

**91**

**92**

would rearrange readily via **91** at 260° since the imidate **92** is stable at 300°.[68] The mechanism of this change is doubtful at present.

Although the preceding thermal reactions proceed stereospecifically and intramolecularly, the acid-catalyzed rearrangment of allyl imidates gives mixtures, undoubtedly by way of an ionic mechanism. Roberts and Hussein found that the thermal rearrangement of the imidate **93,** for example, gave **94**.[85] In contrast, when the same imidate, **93,** was heated

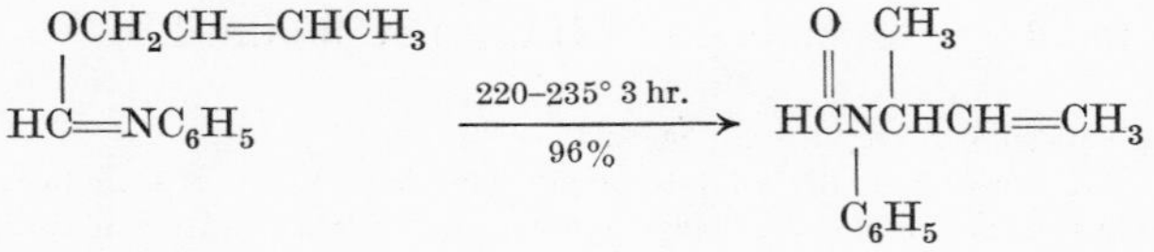

$$HC(OCH_2CH{=}CHCH_3){=}NC_6H_5 \xrightarrow[96\%]{220\text{–}235^\circ\ 3\ \text{hr.}} HC(=O)N(C_6H_5)CH(CH_3)CH{=}CH_3$$

**93** **94**

[85] Roberts and Hussein, *J. Am. Chem. Soc.*, **78,** 4778 (1956).

with concentrated sulfuric acid, reaction began at 100° and a mixture of many components, including **94** and its allylic isomer, resulted.[85]

## Rearrangements of Acyl Imidates

The Mumm rearrangement, which apparently involves the rearrangement of acyl imidates (imino anhydrides), was discovered before the Chapman rearrangement. When N-phenylbenzimidoyl chloride **(95)** was treated with sodium *m*-nitrobenzoate, the expected product **96** was not isolated; instead, the diacylaniline **97** was obtained.[1] N-Phenyl-*m*-

$$C_6H_5CCl{=}NC_6H_5 \ (\mathbf{95}) + m\text{-}NO_2C_6H_4CO_2Na \rightarrow [m\text{-}NO_2C_6H_4CO_2C(C_6H_5){=}NC_6H_5] \ (\mathbf{96}) \rightarrow$$

$$m\text{-}NO_2C_6H_4CON(C_6H_5)C(=O)C_6H_5 \ (\mathbf{97}) \leftarrow [m\text{-}NO_2C_6H_4C(OCOC_6H_5){=}NC_6H_5] \ (\mathbf{99}) \leftarrow m\text{-}NO_2C_6H_4CCl{=}NC_6H_5 \ (\mathbf{98})$$

nitrobenzimidoyl chloride **98** and sodium benzoate also gave **97**, presumably via the acyl imidate **99**.[1] In no case could an O-acyl compound be isolated. As a result of this failure Mumm was led to study the behavior of a stable aryl imidate and, as a result, discovered the reaction now known as the Chapman rearrangement.

Recently Cramer and Baer attempted to prepare acyl imidates from the reaction of imidoyl chlorides, carboxylic acids, and triethylamine.[86] As before, no intermediates could be isolated; the imides were obtained in almost quantitative yield. When the triethylamine was omitted, a reaction analogous to the Pinner fission resulted, and acyl halides and amides were isolated. For example, benzoyl chloride was obtained in 90% yield according to the accompanying formulation.[85]

$$C_6H_5CCl{=}NC_6H_5 + C_6H_5CO_2H \rightarrow \left[C_6H_5C(=O){-}O{-}C(C_6H_5){=}\overset{\oplus}{N}HC_6H_5 \quad Cl^{\ominus}\right] \rightarrow$$

$$C_6H_5COCl + C_6H_5CONHC_6H_5$$

[86] Cramer and Baer, *Chem. Ber.*, **93**, 1231 (1960).

Stevens and Munk[87] prepared imides **(102)** from diphenylketene *p*-tolylimine **(100)** and carboxylic acids, but the anticipated acyl imidates **(101)** were not isolated. The authors suggested that the reaction proceeded via a four-membered transition state; if so, it is mechanistically analogous to the Chapman rearrangement. The same authors also briefly mention other related reactions which involve acyl migration from oxygen to nitrogen, probably via four-membered transition states.[87] Recently a compound believed to be a stable acyl imidate **(103)** has been prepared.[88]

$$(C_6H_5)_2C{=}C{=}N{-}C_6H_4{-}CH_3 \;(\mathbf{100}) + RCO_2H \rightarrow \left[ RC(=O)OC(CH(C_6H_5)_2){=}N{-}C_6H_4{-}CH_3 \right] (\mathbf{101}) \rightarrow$$

$$RC(=O){-}N(C_6H_4CH_3){-}C(=O){-}CH(C_6H_5)_2 \;(\mathbf{102})$$

$$\text{1-isoquinolyl}{-}C(OCOC_6H_5){=}NC_6H_5 \;(\mathbf{103})$$

## Miscellaneous Rearrangements

The conversion of the chloroethyl imidate **104** to the amide **105** takes place at 130°.[89] The low temperature indicates that neither the normal Chapman rearrangment nor a dissociation into radicals occurs. Most likely the reaction proceeds via internal displacement of halogen to give the five-membered intermediate shown. Nucleophilic attack of chloride ion would then give the amide.

$$C_6H_5CH_2C(OCH_2CH_2Cl){=}NH \;(\mathbf{104}) \rightarrow$$

$$\left[ C_6H_5CH_2C{=}\overset{\oplus}{N}H \text{ (ring O–CH}_2\text{–CH}_2\text{)} \leftrightarrow C_6H_5CH_2C({=}\overset{\oplus}{O}){-}NH \text{ (ring)} \right] Cl^{\ominus} \rightarrow$$

$$C_6H_5CH_2C(=O){-}NHCH_2CH_2Cl \;(\mathbf{105})$$

[87] Stevens and Munk, *J. Am. Chem. Soc.*, **80,** 4065 (1958).

[88] Walters, Podrebarac, and McEwen, *J. Org. Chem.*, **26,** 1161 (1961).

[89] Partridge and Turner, *J. Chem. Soc.*, **1949,** 1308.

The quinazoline **106** is converted to the tricyclic compound **108** upon distillation at 211°.[90] It has been suggested that the reaction proceeds by way of **107**. The mechanism for the formation of **107** is uncertain. A route analogous to that suggested for the rearrangement of **104** seems unlikely, since this would involve formation of an ion pair containing the diethylamide anion.

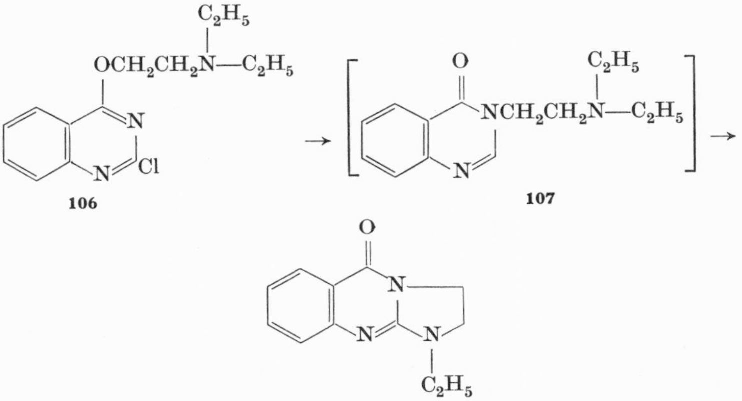

$C_2H_5$

$OCH_2CH_2N—C_2H_5$

N

N Cl

**106**

→

O $C_2H_5$

$NCH_2CH_2N—C_2H_5$

N

**107**

→

O

N

N N

$C_2H_5$

**108**

The thermal rearrangment of 4-aryloxy-2-phenylquinazolines **109** to furnish 3-aryl-2-phenyl-4(3H)-quinazolinones **110** proceeds smoothly. The reaction follows first order kinetics and is favored by electronegative groups and *ortho* substitution in the phenol. It is therefore very similar to and a potentially very useful variant of the Chapman rearrangement.[91]

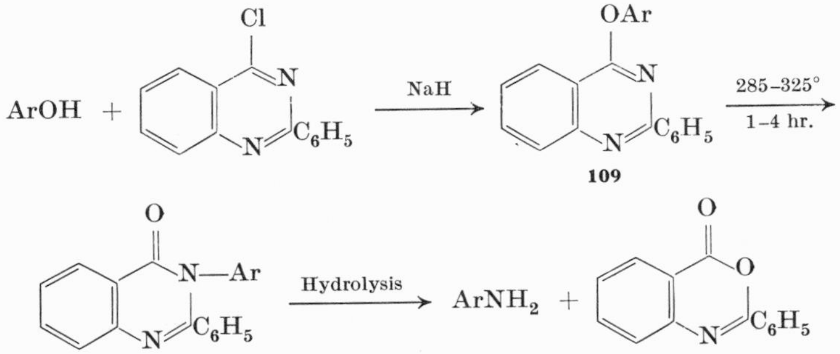

Cl

ArOH + N

N $C_6H_5$

NaH →

OAr

N

N $C_6H_5$

**109**

285–325° / 1–4 hr. →

O

N—Ar

N $C_6H_5$

**110**

Hydrolysis → $ArNH_2$ +

O

O

N $C_6H_5$

[90] Grout and Partridge, *J. Chem. Soc.*, **1960,** 3551.

[91] Scherrer, Abstracts American Chemical Society Meeting, New York, Sept. 9, 1963, p. 33Q.

The cyclic imine **111** has been converted to the urethan **112** by heating with lithium chloride.[92] The reaction probably proceeds by halide ion attack to give the intermediate shown, which then closes in the alternative sense to the final product.

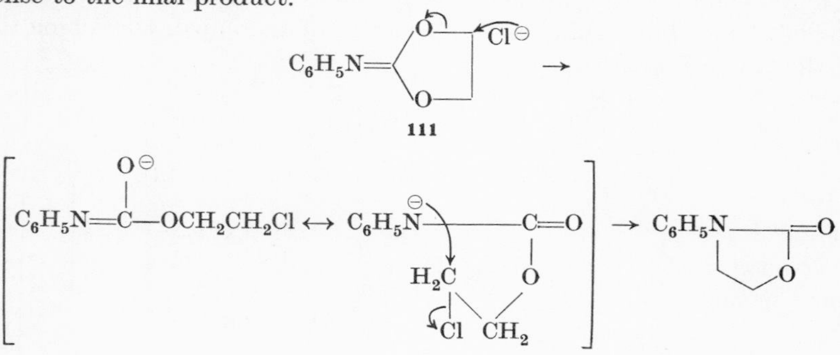

$C_6H_5N$= (cyclic O–CH₂–CH₂–O) + $Cl^{\ominus}$ →

**111**

$$\left[ C_6H_5N{=}C(O^{\ominus}){-}OCH_2CH_2Cl \leftrightarrow C_6H_5\overset{\ominus}{N}{-}C({=}O){-}O{-}CH_2{-}CH_2Cl \right] \rightarrow C_6H_5N\langle CH_2CH_2O \rangle C{=}O$$

**112**

The tetrazole **113** is converted to the hydrazide **115** in refluxing phenol.[93] The reaction presumably proceeded through the intermediacy of **114** which, Huisgen suggested, rearranges via a five-membered cyclic transition state. This route, which would involve a hydride shift, supposedly took precedence over the Chapman rearrangement because of the larger ring involved in the transition state.[93]

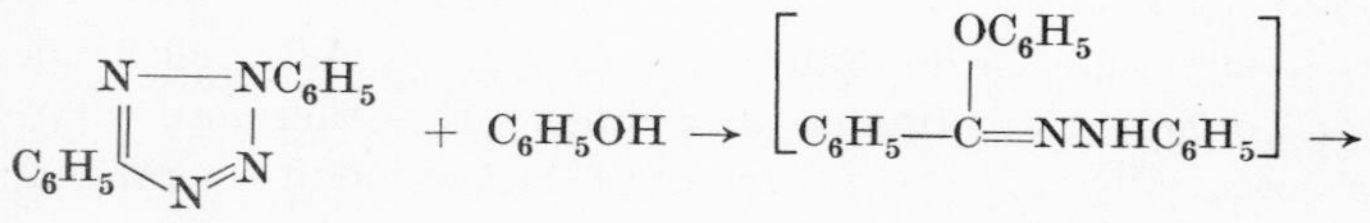

$$\text{113} + C_6H_5OH \rightarrow \left[ C_6H_5{-}C(OC_6H_5){=}NNHC_6H_5 \right] \rightarrow$$

**113** **114**

$$\left[ C_6H_5{-}C(O{-}C_6H_5){=}N{-}N(H)C_6H_5 \right] \rightarrow C_6H_5CONHN(C_6H_5)_2$$

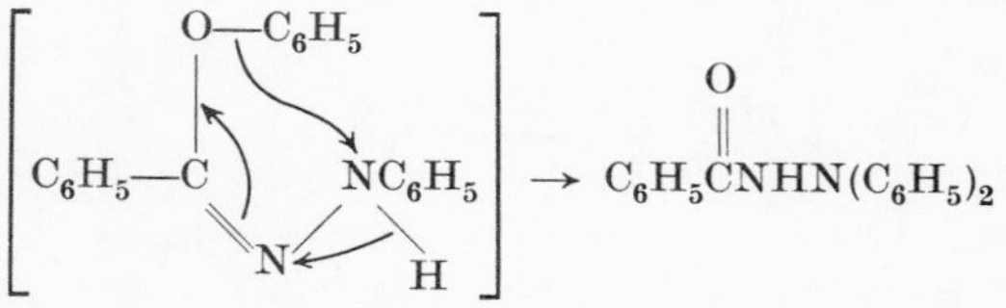

**115**

As a natural extension of his study of aryl imidates, Chapman examined sulfur analogs such as **116**. At 280–290° for 2 hours, isomerization to **117** took place only to a small extent. At 320°, both **116** and **117** gave a mixture containing diphenyl sulfide, benzonitrile, thiophenol, and the

[92] Gulbins and Hamann, *Angew Chem.*, **73,** 434 (1961).

[93] Huisgen, Sauer, and Seidel, *Chem. Ber.*, **94,** 2503 (1961).

benzthiazole **118**. Since both reactants gave the same mixture, Chapman suggested that the rearrangement is reversible.[94]

$$\underset{\mathbf{116}}{C_6H_5C(SC_6H_5){=}NC_6H_5} \rightarrow \underset{\mathbf{117}}{C_6H_5C({=}S){-}N(C_6H_5)_2} \qquad \underset{\mathbf{118}}{\text{2-}C_6H_5\text{-benzthiazole}}$$

In a series of papers Chapman reported on the analogous amidines.[16–18,95,96] Here, as expected, the rearrangement was shown to be reversible. Either **119** or **120** on heating gave the equilibrium mixture shown in the equation.[95]

$$\underset{\substack{\mathbf{119}\\(60\text{–}65\%)}}{C_6H_5NC(C_6H_5){=}NC_6H_5 \ (N\text{-}p\text{-}CH_3C_6H_4)} \underset{}{\overset{322\text{–}323°}{\rightleftharpoons}} \underset{\substack{\mathbf{120}\\(35\text{–}40\%)}}{(C_6H_5)_2NC(C_6H_5){=}NC_6H_4CH_3\text{-}p}$$

Chapman also heated a mixture of triphenylbenzamidine and tri-*p*-tolylbenzamidine. Since the mixture after heating had the same melting point as before, it was concluded that no mixed amidines had been formed and that the reaction is intramolecular.[16] The mechanism is probably completely analogous to that of the Chapman imidate rearrangement. In the amidine rearrangment, if all three N-aryl groups are different, an equilibrium mixture of three components should result, but, if the three aryl groups are identical, aryl migration leads only to starting material.

Chapman[3] believed that the thermal rearrangement of aryl imidates was reversible, but, since heating aroyldiphenylamines failed to produce detectable amounts of imidates, he concluded that the equilibrium lies so far to the amide side that reversibility is not appreciable. No examples of the reverse reaction have been reported thus far.

The thermal rearrangement of aroylaziridines such as **121** to cyclic imidates **122** has been reported. Since this reaction is readily catalyzed

[94] Chapman, *J. Chem. Soc.*, **1926**, 2296.
[95] Chapman, *J. Chem. Soc.*, **1929**, 2133.
[96] Chapman and Perrott, *J. Chem. Soc.*, **1930**, 2462.

by acids or nucleophiles (iodide ion), it is uncertain whether the uncatalyzed reaction really occurs.[97]

$$\mathrm{ArC(=O)N(CH_2)_2} \rightarrow \mathrm{Ar\text{-}C{=}N\text{-}CH_2CH_2\text{-}O\ (ring)}$$

**121** **122**

## EXPERIMENTAL CONDITIONS

### N-Arylbenzimidoyl Chlorides

The conversion of amides to imidoyl chlorides has generally been carried out by heating with an equimolecular amount of phosphorus pentachloride, usually without solvent.[98,99] The reaction often begins spontaneously at room temperature, and, after removal of the resulting phosphorus oxychloride, the product may be isolated either by crystallization or distillation, the latter being preferred for low-melting solids. Frequently, after removal of the phosphorus oxychloride, the crude product is used directly without purification.[12,20,26,44] Occasionally, carbon tetrachloride[29] or toluene[100] has been used as a solvent for the reaction.

Thionyl chloride has been used less frequently but has given equally good results,[82,101,102] as exemplified by the preparation of N-phenylbenzimidoyl chloride in 95% or better yield.[82,101] Phosphorus oxychloride has been used for the preparation of N-(9-phenanthryl)benzimidoyl chloride, but no yield was given.[48] The authors of this chapter have found this reagent distinctly inferior to phosphorus pentachloride or thionyl chloride in the preparation of N-(4-fluorophenyl)benzimidoyl chloride.[13]

### Aryl N-Arylbenzimidates

Phenyl N-phenylbenzimidate was first prepared by Hantzsch from a suspension of sodium phenoxide and N-phenylbenzimidoyl chloride in ether.[103] Since then, almost all imidates have been synthesized by treating the phenol with sodium ethoxide in ethanol and then adding

[97] Heine and Proctor, *J. Org. Chem.*, **23**, 1554 (1958); Heine, *Angew. Chem., Intern. Ed.*, **1**, 528 (1962).

[98] Wallach, *Ann.*, **184**, 79 (1877).

[99] Just, *Ber.*, **19**, 979 (1886).

[100] Jaunin and Sechaud, *Helv. Chim. Acta*, **39**, 1257 (1956).

[101] vonBraun and Pinkernelle, *Ber.*, **67**, 1218 (1934).

[102] Levy and Stephen, *J. Chem. Soc.*, **1956**, 985.

[103] Hantzsch, *Ber.*, **26**, 926 (1893).

the imidoyl chloride in ether.[9,15,104] The reaction, which is often run under nitrogen, takes place readily at room temperature. The imidates are usually crystalline and are easily isolated, generally in high yield. Commercial sodium methoxide in methanol gives equally good results.[35,44] A slight excess of the phenol is often used.

Dioxane[23] and ethyl acetate[13] may be used instead of ether for dissolving the chlorides. Cookson has also prepared imidates from phenols and imidoyl chlorides in pyridine, the last reagent serving both as base and solvent.[23] Easson has treated sodium salts of cyanophenols in pyridine with either the molten imidoyl chloride or the chloride in pyridine.[29] Good results were also obtained from phenols, chlorides, and triethylamine in ether or dioxane.[29]

## N-Aroyldiarylamines

The Chapman rearrangement has usually been carried out by heating the imidate without a solvent. In most cases, temperatures of 250–300° are used. However, imidates derived from acidic phenols will rearrange at lower temperatures, *o*-nitrophenyl N-phenylbenzimidate being converted to N-benzoyl-*o*-nitrodiphenylamine in 1 hour at 165°.[3] Usually heating periods of less than 3 hours are sufficient. Imidates derived from methyl salicylate are usually rearranged by heating at about 280° for about 10 minutes.[15] The reaction is exothermic and the internal temperature may go above 300° if a large quantity of imidate is pyrolyzed. In experiments where temperature control is not critical, the vessel used for the rearrangement may be heated by a Wood's metal bath. If more careful control is desired, a constant external temperature may be held by heating the vessel with the vapors of a high-boiling material such as bibenzyl.[35]

In a few reactions, solvents have been used. Cookson was unable to obtain crystalline rearrangement products by heating certain iodine-containing imidates without solvent. However, when nitrobenzene (b.p. 209°), diphenyl ether (b.p. 259°), biphenyl (b.p. 254°), or *o*-dichlorobenzene (b.p. 180°) was used as solvent, the desired product could be obtained in reasonable yield in some cases.[23] Dowtherm was used successfully as the solvent for rearranging many cyanoimidates.[29] Imidates derived from *o*-nitrophenol have been rearranged in boiling anisole (b.p. 155°) and even pyridine (115°).[29] In Wiberg's[9] kinetic study of the Chapman rearrangement, diphenyl ether was used as the solvent.

[104] Chapman, *J. Chem. Soc.*, **121**, 1676 (1922).

## Hydrolysis of Rearrangement Products

Hydrolysis of the rearrangement products was first carried out by Chapman who heated under reflux a mixture of 10 g. of N-benzoyldiarylamine, 50 ml. of 50% aqueous potassium hydroxide, and 125 ml. of ethanol for 2 hours.[4] Essentially the same method of basic hydrolysis has been used in almost all cases reported, the only changes being in the relative proportions of reagents. Sodium hydroxide in aqueous ethylene glycol has been employed to hydrolyze a series of cyano-substituted N-benzoylamines.[29]

Hydrolysis of the rearrangement products derived from methyl salicylate to derivatives of diphenylamine-2-carboxylic acids is also carried out with excess base in aqueous ethanol[15] or aqueous dioxane.[44] Essentially the same procedure, but with equimolar amounts of alkali and N-benzoyl ester, is used for the partial hydrolysis of these compounds to the N-benzoyl acids.[15]

## Acridones

Although acridones have been prepared directly from benzimidates and from derivatives of N-benzoyldiphenylamine-2-carboxylic acid, very few examples have been reported. Because detailed experimental directions have rarely been given, little can be said about experimental conditions. For the preparation of acridones via other routes, the reader should consult Acheson.[54b]

## EXPERIMENTAL PROCEDURES

**N-Phenylbenzimidoyl Chloride.**[35,98] A mixture of 296 g. (1.5 moles) of benzanilide and 312 g. (1.5 moles) of phosphorus pentachloride is gently heated. After the initial vigorous reaction has subsided, the resulting liquid is heated under reflux for 2 hours. The phosphorus oxychloride is then removed by distilling in vacuum using the water pump. The product is then distilled with the aid of an oil pump and 305 g. (94%) of light yellow liquid, b.p. 115–120°/0.3 mm., is collected. The material crystallizes in the receiving flask to an almost white solid, m.p. 38–40°.

**N-(*p*-Fluorophenyl)benzimidoyl Chloride.**[12,13] *Thionyl Chloride.* A mixture of 38.7 g. (0.18 mole) of 4′-fluorobenzanilide[12] and 75 ml. of thionyl chloride is heated to reflux while being stirred with a magnetic stirrer. Hydrogen chloride and sulfur dioxide are evolved and the solid soon dissolves. The yellow solution is heated under reflux for 4 hours, the excess reagent is removed in vacuum, and the residue is distilled rapidly at 110–116°/0.15 mm., through a short Vigreux column. Rapid distillation is necessary to prevent the yellow distillate from crystallizing in the apparatus. The yield is 41.6 g. (98%) of almost white solid, m.p. 58–63°.

*Phosphorus Pentachloride.* (*a*) A mixture of 10.8 g. (0.05 mole) of 4′-fluorobenzanilide and 10.4 g. (0.05 mole) of phosphorus pentachloride is heated under reflux for 1 hour. After removal of the phosphorus oxychloride at water pump vacuum, the product is distilled using an oil pump to give 10.6 g. (91%) of product.

(*b*) The phosphorus oxychloride is removed as described above, and the residue is crystallized from hexane to give 9.3 g. (79%) of white solid. The product is less pure than the distilled material and gives a cloudy melt. Recystallization from hexane gives the product, m.p. 62–65°, in 54% yield.

***o*-Carbomethoxyphenyl N-Phenylbenzimidate.**[15,36] A 250-ml. three-necked flask, equipped with a Hershberg stirrer and a delivery tube arranged so that a slow stream of nitrogen passes through the liquid, is charged with 100 ml. of absolute ethanol to which is added 1.9 g. (0.083 g. atom) of sodium. After the conversion to sodium ethoxide is complete, the solution is cooled to room temperature and a solution of 15.2 g. (0.10 mole) of methyl salicylate in 15 ml. of absolute ethanol is added quickly. A solution of 17.3 g. (0.08 mole) of N-phenylbenzimidoyl chloride in 30 ml. of dry diethyl ether is then added within a few minutes. The reaction mixture is stirred overnight at room temperature under nitrogen, most of the solvent is removed in vacuum, and the residue is mixed with water. The insoluble solid is recrystallized from absolute ethanol to give 23.5 g. (89%) of white prisms, m.p. 114–117°. Ultraviolet (ethanol): $\lambda_{max}$ 227 m$\mu$ ($\epsilon$ 27,345) and 275 m$\mu$ ($\epsilon$ 11,720). Infrared (KBr): 5.82 $\mu$, 6.00 $\mu$.

***p*-Fluorophenyl N-(*p*-Fluorophenyl)benzimidate.**[12] In a 250-ml. flask equipped with a magnetic stirrer, 6.5 g. (0.12 mole) of sodium methoxide (Matheson, Coleman and Bell) is dissolved in 125 ml. of methanol and the solution is cooled to about 20°. Then 13.4 g. (0.12 mole) of *p*-fluorophenol (Aldrich) is added all at once. In the course of the next few minutes a solution of 23.4 g. (0.10 mole) of N-(*p*-fluorophenyl)benzimidoyl chloride in 50 ml. of dry ethyl acetate is added. The mixture turns cloudy at once, and the temperature rises from 25° to 37°.

The mixture is stirred for 3 hours, after which time the solvent is removed in vacuum. Water is added and the white insoluble solid filtered and washed with water. After air-drying, 27 g. (m.p. 99–106°) of the product is obtained which upon recrystallization from absolute ethanol gives 25.5 g. (83%) of material which melts at 105–109°. Ultraviolet (ethanol): $\lambda_{max}$ 220 m$\mu$ (shoulder, $\epsilon$ 16,300), 229 m$\mu$ (shoulder, $\epsilon$ 15,500), 266 m$\mu$ ($\epsilon$ 6560), and 272 m$\mu$ ($\epsilon$ 6500). Infrared (KBr): 6.08 $\mu$.

***o*-Carbomethoxyphenyl N-[(*o*-Carbomethoxymethyl)phenyl]-benzimidate.**[44] N[(*o*-Carbomethoxymethyl)phenyl]benzimidoyl chloride is prepared by mixing 56.5 g. (0.21 mole) of methyl *o*-benzamidophenyl-acetate[44] and 43.7 g. (0.21 mole) of phosphorus pentachloride in a 500-ml. flask. The reaction begins spontaneously at room temperature, and hydrogen chloride evolution occurs with extensive foaming. The mixture is then heated gently on the steam bath until the gas evolution virtually ceases. The phosphorus oxychloride is removed in vacuum below 50°, toluene is added, and the solvent once again removed in vacuum to leave the crude imidoyl chloride as a dark red oil, which is used without purification.

Meanwhile a solution of 12.4 g. (0.23 mole) of sodium methoxide in 200 ml. of methanol is flushed with nitrogen and cooled with an ice bath. To this is added with stirring a solution of 35.0 g. (0.23 mole) of methyl salicylate in 50 ml. of methanol. The crude imidoyl chloride in 65 ml. of anhydrous diethyl ether is then added during 5 minutes. The ice bath is removed and the cloudy tan mixture stirred for 3 hours at room temperature, after which water is added and the product taken up in ether. The red extracts are dried and the solvent is removed to give a dark oil which solidifies and is crystallized from methanol-hexane to furnish 59.2 g. (70%) of yellow prisms, m.p. 60–64°. This material is pure enough to be used directly in the Chapman rearrangement. Two recrystallizations from methanol give a white solid, m.p. 62–65°. Ultraviolet (ethanol): $\lambda_{max}$ 228 m$\mu$ ($\epsilon$ 24,000), 278 m$\mu$ ($\epsilon$ 5400). Infrared (KBr): 5.80 $\mu$ and 5.97 $\mu$.

**N-Benzoyldiphenylamine.**[2,12] Five grams of phenyl N-phenylbenzimidate is placed in a small pear-shaped flask carrying a thermometer that dips into the solid. The flask is immersed for an hour in a Wood's metal bath held at 312–315°. The melt, whose temperature during the heating is 305–310°, crystallizes on cooling. The solid is taken up in warm absolute ethanol. After cooling, filtering, washing the tan solid with ethanol and air-drying, 4.6 g. (92%) of product, m.p. 177–181° is obtained. One recrystallization from absolute ethanol (charcoal) gives 4.0 g. (80%) of white needles, m.p. 178–182°. Ultraviolet (ethanol): $\lambda_{max}$ 235 m$\mu$ (shoulder $\epsilon$ 13,100), 271 m$\mu$ ($\epsilon$ 7220). Infrared (KBr): 6.09 $\mu$.

**Methyl N-Benzoyldiphenylamine-2-carboxylate.**[15,36] A 125-ml. Erlenmeyer flask equipped with a thermometer and containing 30 g. of *o*-carbomethoxyphenyl N-phenylbenzimidate is heated in a Wood's metal bath. When the bath temperature reaches 240°, an exothermic reaction begins and the internal temperature quickly rises to 260°. After about a minute, the internal temperature begins to drop. The bath is then heated for 10 minutes at 280°. The dark melt is cooled somewhat, then poured into 160 ml. of hot absolute ethanol. On cooling, the product crystallizes to give 24.5 g. (82%) of slightly yellow solid, m.p. 133–135.5°. Ultraviolet (ethanol): inflection at 273 m$\mu$, $\epsilon$ 6830. Infrared (KBr): 5.80 $\mu$, 6.04 $\mu$.

**Methyl 2,6-Diiodo-4′-methoxydiphenylamine-4-carboxylate.**[23] A solution of 4.0 g. of 4-carbomethoxy-2,6-diiodophenyl-N-*p*-methoxyphenylbenzimidate[23] in 12 ml. of *o*-dichlorobenzene is heated under reflux for 80 minutes. Petroleum ether (b.p. 100–120°) is added until crystals begin to separate from the boiling solution. On cooling, 3.8 g. (95%) of the product is obtained. After recrystallization from toluene-petroleum ether (b.p. 100–120°), the pure material melts at 205°.

**4-Fluorodiphenylamine.**[13] A mixture of 10.9 g. (0.037 mole) of N-benzoyl-4-fluorodiphenylamine, 30 g. of potassium hydroxide in 30 ml. of water, and 125 ml. of ethanol is heated under reflux for 2 hours. Water is added and the mixture is concentrated in vacuum in order to remove most of the the ethanol. The cloudy mixture is then extracted three times with diethyl ether and the extracts dried. Removal of the solvent gives a dark red oil which is vacuum-distilled with a small Vigreux column. The product is obtained as a yellow liquid, b.p. 78–80°/0.01 mm., which crystallizes on cooling, m.p. 33.5–35.5°; yield 5.9 g. (85%). Ultraviolet (ethanol): $\lambda_{max}$ 241 m$\mu$ (shoulder $\epsilon$ 3860), 281 m$\mu$ ($\epsilon$ 18,600). Infrared (chloroform): 2.99 $\mu$.

**N-Benzoyldiphenylamine-2-carboxylic Acid.**[15] A mixture of 24.5 g. (0.074 mole) of methyl N-benzoyldiphenylamine-2-carboxylate, 3.8 g. (0.070 mole) of sodium methoxide, 180 ml. of ethanol, and 75 ml. of water is heated under reflux for 2 hours. The resulting pale orange alkaline solution is diluted with water, washed once with ether, and acidified with excess hydrochloric acid. The precipitate is filtered and washed with water. After drying for 1 hour at 95°, 21 g. of crude product is obtained which, after recrystallization from acetone-pentane, affords white crystals, m.p. 190–193°. The yield is 18.9 g. (80% based on N-benzoyl ester).

**Diphenylamine-2-carboxylic Acid.**[15,36] Two grams (0.006 mole) of methyl N-benzoyldiphenylamine-2-carboxylate is treated with a solution of 8 g. of potassium hydroxide in 20 ml. of water and 60 ml. of methanol. The mixture is heated under reflux for 2.5 hours, diluted with water, and a trace of insoluble material is removed by filtration. Excess hydrochloric acid is then added and the mixture is heated to boiling to dissolve benzoic acid. The insoluble product is filtered hot and washed with boiling water to give a white powder, m.p. 169–179°. One recrystallization from ethanol-water gives 1.0 g. (77%) of product, m.p. 188.5–191°. Ultraviolet (ethanol): $\lambda_{max}$ 220 m$\mu$ ($\epsilon$ 21,800), 287 m$\mu$ ($\epsilon$ 13,900), 348 m$\mu$ ($\epsilon$ 6500). Infrared (KBr): 3.00 $\mu$, 6.02 $\mu$.

**2,4,7-Tribromoacridone.**[105] (*a*) N-Benzoyl-4,4′,6-tribromodiphenylamine-2-carboxylic acid (5.0 g., 0.009 mole)[15,105] is heated at 300–350° for a few minutes. The material is boiled with a little aqueous sodium hydroxide, then with water. Recrystallization from *m*-cresol furnishes 2.17 g. (56%) of pale yellow needles, m.p. >340°.

(*b*) A mixture of 1.0 g. (0.0018 mole) of methyl N-benzoyl-4,4′,6-tribromodiphenylamine-2-carboxylate[15,105] and 2.10 ml. of concentrated sulfuric acid is heated at 160–200° for about a minute. On cooling, the product crystallizes to give 0.66 g. (87%) of the acridone.

**5-Methyl-1,2-benzacridone.**[48] 2-Carbomethoxy-6-methylphenyl N-$\beta$-naphthylbenzimidate[48] (22 g., 0.056 mole) is heated at 360° under nitrogen for 0.5 hour. Methyl benzoate is formed and boils away. The residual solid is cooled, washed with benzene, and recrystallized from nitrobenzene to give 5.5 g. (38%) of yellow needles, m.p. 264–265°.

## TABULAR SURVEY

In listing benzimidates

```
       OR
       |
R″—C=NR′
```

compounds with R = R′ = phenyl are named first. Next, derivatives of phenol (R = phenyl, R′ varies) are listed, and, last, compounds where R varies are listed. Within each group, monohalo derivatives are named first, followed by nitro-, hydroxy-, methoxy-, and alkyl-substituted compounds. Imidates with one carbonyl-containing substituent are named next. Finally, compounds with more than one substituent on the key radical are listed, followed by imidates derived from polycyclic and heterocyclic phenols.

In the tables a dash (—) indicates that a compound has been prepared but no yield was given. Omission of both the yield and the dash shows that the reaction leading to the compound in question was not attempted.

The literature coverage includes July, 1963.

All references to a particular listing are located in the last column. When multiple references occur, individual citations appear in parentheses under the appropriate heading.

[105] Acheson and Robinson, *J. Chem. Soc.*, **1953**, 232.

# TABLE I

## Aryl N-Arylbenzimidates, N-Aroyldiarylamines, and Diarylamines

| Aryl N-Arylbenzimidate: Substituents in $R''-C(OR)=NR'$ R | R′ | R″ | Yield, % (Ref.) | Conditions for Rearrangement (Ref.) | N-Aroyldiarylamine $R''-C(=O)-NRR'$ Yield, % (Ref.) | Diarylamine RNHR′ Yield, % (Ref.) | References |
|---|---|---|---|---|---|---|---|
| $C_6H_5$ | $C_6H_5$ | $C_6H_5$ | 64 (9) | 240°; 1 hr. (1) | —* | | 1–3, 9 |
| | | | | 270–300°; 2 hr. (2) | Complete transformation (2) | | |
| | | | | 266°; 1.5 hr. (3) | 71† | | |
| $C_6H_5$ | $C_6H_5$ | $o$-$ClC_6H_4$ | — | 270°; 1.5 hr. | 63† | | 3 |
| $C_6H_5$ | $C_6H_5$ | $p$-$ClC_6H_4$ | — | 270°; 1.5 hr. | 62† | | 3 |
| $C_6H_5$ | $C_6H_5$ | $o$-$O_2NC_6H_4$ | — | | ‡ | | 3 |
| $C_6H_5$ | $C_6H_5$ | $p$-$O_2NC_6H_4$ | — | 270°; 1.5 hr. | 49† | | 3 |
| $C_6H_5$ | $C_6H_5$ | $p$-$CH_3OC_6H_4$ | — | 267°; 1.5 hr. | 85† | | 3 |
| $C_6H_5$ | $C_6H_5$ | 2,4,6-$Cl_3C_6H_2$ | — | 270°; 1.5 hr. | <55† | | 3 |
| $C_6H_5$ | $p$-$FC_6H_4$ | $C_6H_5$ | 54 | 305–318°; 1 hr. | 91 | 85 | 13 |
| $C_6H_5$ | $o$-$ClC_6H_4$ | $C_6H_5$ | — | 270°; 1.5 hr. | 39† | | 3 |
| $C_6H_5$ | $m$-$ClC_6H_4$ | $C_6H_5$ | — | 270°; 1.5 hr. | 49† | | 3 |
| $C_6H_5$ | $p$-$ClC_6H_4$ | $C_6H_5$ | — | 270°; 1.5 hr. (3) | 61† | — (26) | 3, 26 |
| $C_6H_5$ | $o$-$O_2NC_6H_4$ | $C_6H_5$ | — | | ‡ | | 3 |
| $C_6H_5$ | $p$-$CH_3OC_6H_4$ | $C_6H_5$ | — | 270°; 1.5 hr. | 85† | | 3 |
| $C_6H_5$ | $o$-$CH_3C_6H_4$ | $C_6H_5$ | 68 | 280–300°; 2 hr. | — | Satisfactory | 25 |
| $C_6H_5$ | $m$-$CH_3C_6H_4$ | $C_6H_5$ | 28 | 280–300°; 2 hr. | — | 68 | 24 |
| $C_6H_5$ | $p$-$CH_3C_6H_4$ | $C_6H_5$ | 88 | 290°; 2 hr. | High | 84§ | 27 |
| $C_6H_5$ | 2,4-$Cl_2C_6H_3$ | $C_6H_5$ | — | 270°; 1.5 hr. | 36† | | 3 |
| $C_6H_5$ | 3,5-$Cl_2C_6H_3$ | $C_6H_5$ | — | 300–310°; 2 hr. | — | — | 17 |
| $C_6H_5$ | 2,4,6-$Cl_3C_6H_2$ | $C_6H_5$ | — | 267°; 1.5 hr. | <17† | | 3 |
| $C_6H_5$ | $\alpha$-$C_{10}H_7$ | $C_6H_5$ | — | 270°; 1.5 hr. | 63† | | 3 |

*Note:* References 106 and 107 are on p. 51.

* The rearrangement was carried out as part of a kinetic study (ref. 9). The product was not isolated.

† This represents percent conversion (not percent yield) during a kinetic study at the temperature indicated (ref. 3). For preparative means, the rearrangement was carried out at 280–310°, but the yield was not given.

‡ The reaction was unsuccessful.

§ The yield is based on imidate.

## TABLE I—*Continued*

### Aryl N-Arylbenzimidates, N-Aroyldiarylamines, and Diarylamines

| Aryl N-Arylbenzimidate: Substituents in R″—C(OR)=NR′ R | R′ | R″ | Yield, % (Ref.) | Conditions for Rearrangement (Ref.) | N-Aroyldiarylamine R″—C(=O)—N(R)R′ Yield, % (Ref.) | Diarylamine RNHR′ Yield, % (Ref.) | References |
|---|---|---|---|---|---|---|---|
| $p$-$FC_6H_4$ | $C_6H_5$ | $C_6H_5$ | 55 | 305–310°; 1 hr. | 85 | 85 | 13 |
| $p$-$FC_6H_4$ | $p$-$FC_6H_4$ | $C_6H_5$ | 62 | 280°; 18 hr. | 80 | 96 | 12 |
| $o$-$ClC_6H_4$ | $C_6H_5$ | $C_6H_5$ | 37 (9) | 270–280°; 1 hr. (2) | 91*† | | 2, 3, 9 |
| | | | | 255°; 1.5 hr. (3) | | | |
| $o$-$ClC_6H_4$ | $o$-$ClC_6H_4$ | $C_6H_5$ | — | — | — | — | 18 |
| $o$-$ClC_6H_4$ | $o$-$CH_3C_6H_4$ | $C_6H_5$ | — | 290°; 2 hr. | — | ‡ | 19 |
| $m$-$ClC_6H_4$ | $C_6H_5$ | $C_6H_5$ | — | 255°; 1.5 hr. | 84† | | 3 |
| $m$-$ClC_6H_4$ | $m$-$ClC_6H_4$ | $C_6H_5$ | — | 290°; 2 hr. | — | — | 19 |
| $m$-$ClC_6H_4$ | $o$-$CH_3C_6H_4$ | $C_6H_5$ | — | 290°; 2 hr. | — | — | 19 |
| $p$-$ClC_6H_4$ | $C_6H_5$ | $C_6H_5$ | 57 (9) | 280–290°; 2 hr. (2) | 62*† | | 2, 3, 9 |
| | | | | 255°; 1.5 hr. (3) | | | |
| $p$-$ClC_6H_4$ | $o$-$ClC_6H_4$ | $C_6H_5$ | 90 | 300°; 2 hr. | 74 | 71 | 15 |
| $p$-$ClC_6H_4$ | $m$-$ClC_6H_4$ | $C_6H_5$ | — | 290°; 2 hr. | — | — | 19 |
| $p$-$ClC_6H_4$ | $p$-$ClC_6H_4$ | $C_6H_5$ | 64 (20) | — | 78 (20)* | 94 (20) | 9, 16, 20 |
| | | | 60 (9) | | | | |
| $p$-$ClC_6H_4$ | $p$-$BrC_6H_4$ | $C_6H_5$ | — | 290–320°; 2.5 hr. | 75 | — | 15 |
| $p$-$ClC_6H_4$ | 2,4-$Cl_2C_6H_3$ | $C_6H_5$ | 96 | 250–270°; 2 hr. | 76 | 85 | 15 |
| $o$-$BrC_6H_4$ | $o$-$BrC_6H_4$ | $C_6H_5$ | 48 | 155–260°; 2.5 hr. | 86 | 87 | 21 |
| $p$-$BrC_6H_4$ | $C_6H_5$ | $C_6H_5$ | 51 | 255° | * | | 9 |
| $p$-$BrC_6H_4$ | $p$-$BrC_6H_4$ | $C_6H_5$ | 75 | 270–290°; 2 hr. | 84 | 68 | 22 |
| $o$-$O_2NC_6H_4$ | $C_6H_5$ | $C_6H_5$ | 54 (9) | 165°; 1 hr. (3) | 40 (3) | | 1, 3, 9 |
| | | | | | ‡ (1) | | |
| $o$-$O_2NC_6H_4$ | 2-Br-4-$CH_3C_6H_3$ | $C_6H_5$ | 67 | 170–180°; 1 hr. | — | | 30 |
| $p$-$O_2NC_6H_4$ | $C_6H_5$ | $C_6H_5$ | 30 (9) | | ‡ | | 3, 9 |
| $m$-$HOC_6H_4$ | $C_6H_5$ | $C_6H_5$ | 57 | | | | 104 |
| $o$-$CH_3OC_6H_4$ | $C_6H_5$ | $C_6H_5$ | 20 (9) | 266°; 1.5 hr. | 74*† | | 3, 9 |
| $m$-$CH_3OC_6H_4$ | $C_6H_5$ | $C_6H_5$ | — | 266°; 1.5 hr. | 55† | | 3 |
| $p$-$CH_3OC_6H_4$ | $C_6H_5$ | $C_6H_5$ | 66 (9) | 266°; 1.5 hr. | 44*† | | 3, 9 |
| $o$-$CH_3C_6H_4$ | $C_6H_5$ | $C_6H_5$ | 46 | 255° | * | | 9 |

| | | | | | | | |
|---|---|---|---|---|---|---|---|
| $o\text{-}CH_3C_6H_4$ | $p\text{-}ClC_6H_4$ | $C_6H_5$ | — | — | — | — | 19 |
| $o\text{-}CH_3C_6H_4$ | $o\text{-}CH_3C_6H_4$ | $C_6H_5$ | 68 | 280–300°; 2 hr. | — | ‡ | 25 |
| $o\text{-}CH_3C_6H_4$ | $m\text{-}CH_3C_6H_4$ | $C_6H_5$ | — | 280–300°; 2 hr. | — | Satisfactory | 25 |
| $o\text{-}CH_3C_6H_4$ | $p\text{-}CH_3C_6H_4$ | $C_6H_5$ | Mod. good | 280–300°; 2 hr. | — | 62§ | 25 |
| $m\text{-}CH_3C_6H_4$ | $C_6H_5$ | $C_6H_5$ | 36 (24); 60 (9) | 280–300°; 2 hr. (24) | 85 (24)* | 68 (24) | 9, 24 |
| $m\text{-}CH_3C_6H_4$ | $o\text{-}CH_3C_6H_4$ | $C_6H_5$ | — | 280–300°; 2 hr. | Isomerized completely | Satisfactory | 25 |
| $m\text{-}CH_3C_6H_4$ | $p\text{-}CH_3C_6H_4$ | $C_6H_5$ | — | | | | 25 |
| $p\text{-}CH_3C_6H_4$ | $C_6H_5$ | $C_6H_5$ | 64 | 275° | * | | 9 |
| $p\text{-}CH_3C_6H_4$ | $o\text{-}CH_3C_6H_4$ | $C_6H_5$ | — | | | | 25 |
| $p\text{-}CH_3C_6H_4$ | $m\text{-}CH_3C_6H_4$ | $C_6H_5$ | 58 | 280–300°; 2 hr. | — | | 25 |
| $p\text{-}CH_3C_6H_4$ | $p\text{-}CH_3C_6H_4$ | $C_6H_5$ | — | 270–300°; 2 hr. | — | | 2 |
| $o\text{-}C_2H_5C_6H_4$ | $C_6H_5$ | $C_6H_5$ | — | | | | 9 |
| $p\text{-}C_2H_5C_6H_4$ | $C_6H_5$ | $C_6H_5$ | 15 | 255° | * | | 9 |
| $o\text{-}i\text{-}C_3H_7C_6H_4$ | $C_6H_5$ | $C_6H_5$ | 58 | 255° | * | | 9 |
| $p\text{-}i\text{-}C_3H_7C_6H_4$ | $C_6H_5$ | $C_6H_5$ | 39 | 255° | * | | 9 |
| $o\text{-}t\text{-}C_4H_9C_6H_4$ | $C_6H_5$ | $C_6H_5$ | 40 | 230°; 9 hr. | 40 | | 9 |
| $p\text{-}t\text{-}C_4H_9C_6H_4$ | $C_6H_5$ | $C_6H_5$ | 57 | 255° | * | | 9 |
| $p\text{-}t\text{-}C_4H_9C_6H_4$ | $p\text{-}t\text{-}C_4H_9C_6H_4$ | $C_6H_5$ | — | 300° | Almost quantitative | | 27 |
| $o\text{-}C_6H_5CH_2C_6H_4$ | $C_6H_5$ | $C_6H_5$ | 67 | 275°; 2.5 hr. | 86 | | 28 |
| $p\text{-}C_6H_5CH_2C_6H_4$ | $C_6H_5$ | $C_6H_5$ | 74 | 275°; 2.5 hr. | 83 | | 28 |
| $m\text{-}OHCC_6H_4$ | $C_6H_5$ | $C_6H_5$ | ~35 | | ‡ | | 33 |
| $p\text{-}(C_6H_5N{=}CH)C_6H_5$ | $C_6H_5$ | $C_6H_5$ | — | 270°; 40 min. | — | | 33 |
| $p\text{-}OHCC_6H_4$ | $C_6H_5$ | $C_6H_5$ | 35 | | ‡ | | 33 |
| $m\text{-}HO_2CC_6H_4$ | $C_6H_5$ | $C_6H_5$ | By hydrolysis of methyl ester | | ‡ | | 33 |
| $m\text{-}CH_3O_2CC_6H_4$ | $C_6H_5$ | $C_6H_5$ | — | | ‡ | | 33 |
| $p\text{-}CH_3O_2CC_6H_4$ | $C_6H_5$ | $C_6H_5$ | 75 (26) | 270–280°; 2 hr. (33)<br>300°; 2 hr. (26) | 81 (33)<br>84 (26) | 73 (30)¶ | 26, 33 |
| $p\text{-}CH_3O_2CC_6H_4$ | $p\text{-}ClC_6H_4$ | $C_6H_5$ | 57 | 277–290° | 90 | 61‖ | 15 |
| $p\text{-}C_2H_5O_2CC_6H_4$ | $C_6H_5$ | $C_6H_5$ | — | — | — | | 33 |

*Note:* References 106 and 107 are on p. 51.
* The rearrangement was carried out as part of a kinetic study (ref. 9). The product was not isolated.
† This represents percent conversion (not percent yield) during a kinetic study at the temperature indicated (ref. 3). For preparative means, the rearrangement was carried out at 280–310°, but the yield was not given.
‡ The reaction was unsuccessful.
§ The yield is based on imidate.
¶ The product is diphenylamine-4-carboxylic acid.
‖ The product is N-benzoyl-4′-chlorodiphenylamine-4-carboxylic acid.

## Table I—*Continued*

### Aryl N-Arylbenzimidates, N-Aroyldiarylamines, and Diarylamines

| Aryl N-Arylbenzimidate: Substituents in R″—C(OR)=NR′, R | R′ | R″ | Yield, % (Ref.) | Conditions for Rearrangement (Ref.) | N-Aroyldiarylamine R″—C(=O)—N(R)R′ Yield, % (Ref.) | Diarylamine RNHR′ Yield, % (Ref.) | References |
|---|---|---|---|---|---|---|---|
| $C_6H_5CO_2$— (ring structure) | $p$-$ClC_6H_4$ | $C_6H_5$ | 88 | 280–295° | —** | | 15 |
| $p$-$NCC_6H_4$ | $p$-$NCC_6H_4$ | $C_6H_5$ | 82 | 280–300°; 1.25 hr. | — | 76†† | 32 |
| $p$-$NCC_6H_4$ | 2-$O_2N$-4-$NCC_6H_3$ | $C_6H_5$ | 78 | | ‡ | — | 29 |
| $p$-$NCC_6H_4$ | 2-$CH_3$-4-$NCC_6H_3$ | $C_6H_5$ | 45 | Boiling Dowtherm; 1–2 hr. | — | 85 | 29 |
| $o$-$CH_3COC_6H_4$ | $C_6H_5$ | $C_6H_5$ | — | 267°; 1.5 hr. | —‡‡ | | 3 |
| $p$-$CH_3COC_6H_4$ | $C_6H_5$ | $C_6H_5$ | 45 (9) | 242°; 1.5 hr. (3) | 92*† | — | 3, 9 |
| $o$-$CH_3O_2CCH_2C_6H_4$ | $C_6H_5$ | $C_6H_5$ | 74 | 268°; 45 min. | 29 | —§§ | 35 |
| $p$-$C_6H_5COC_6H_4$ | $C_6H_5$ | $C_6H_5$ | 82 | 270–280°; 40 min. | 90 | 89 | 31 |
| 2,4-$Cl_2C_6H_3$ | $C_6H_5$ | $C_6H_5$ | — | 242°; 1.5 hr. | 93‡ | | 3 |
| 2,4-$Cl_2C_6H_3$ | $p$-$ClC_6H_4$ | $p$-$CH_3C_6H_4$ | — | 280–300°; 2.5 hr. | — | | 15 |
| 2,4-$Cl_2C_6H_3$ | $o$-$CH_3C_6H_4$ | $C_6H_5$ | — | — | — | | 19 |
| 2,4-$Cl_2C_6H_3$ | 2,4-$Cl_2C_6H_3$ | $C_6H_5$ | — | 280–300°; 2 hr. | — | ~80 | 4 |
| 3,5-$Cl_2C_6H_3$ | 3,5-$Cl_2C_6H_3$ | $C_6H_5$ | — | — | — | — | 17 |
| 4-Cl-2-$BrC_6H_3$ | $o$-$BrC_6H_4$ | $C_6H_5$ | 71 | 230–240°; 2.5 hr. | 42–55 | 71–79 | 106 |
| 2-Cl-4-$NCC_6H_3$ | $p$-$NCC_6H_4$ | $C_6H_5$ | 25 | Boiling Dowtherm; 1–2 hr. | 84 | 60 | 29 |
| 2-Cl-4-$NCC_6H_3$ | 2-$CH_3$-4-$NCC_6H_3$ | $C_6H_5$ | 36 | Boiling Dowtherm; 1–2 hr. | 80 | 73 | 29 |
| (2-Br-4-$C_6H_5$)$C_6H_3$ | 2-$BrC_6H_4$ | $C_6H_5$ | 74 | 240–255°; 2.5 hr. | 68–80 | 78 | 106 |
| 2-$O_2N$-4-$NCC_6H_3$ | $p$-$NCC_6H_4$ | $C_6H_5$ | 69 | Boiling anisole or pyridine | 99 | — | 29 |
| 2-$O_2N$-4-$NCC_6H_3$ | 2-$O_2N$-4-$NCC_6H_3$ | $C_6H_5$ | 96 | | ‡ | | 29 |
| 2-$CH_3O$-4-$NCC_6H_3$ | $p$-$NCC_6H_4$ | $C_6H_5$ | 60 | Boiling Dowtherm; 1–2 hr. | 88 | 86 | 29 |

| | | | | | | | |
|---|---|---|---|---|---|---|---|
| 2-$CH_3$-4-$NCC_6H_3$ | *p*-$NCC_6H_4$ | $C_6H_5$ | 76 | Boiling Dowtherm; 1–2 hr. | 94 | 85 | 29 |
| 2-$CH_3$-4-$NCC_6H_3$ | 2-$CH_3$-4-$NCC_6H_3$ | $C_6H_5$ | 83 | Boiling Dowtherm; 1–2 hr. | 98 | 38 | 29 |
| 2,4,6-$Cl_3C_6H_2$ | $C_6H_5$ | $C_6H_5$ | — | 220°; 1.5 hr. (3) | >91† | — (4) | 3, 4 |
| 2,4,6-$Cl_3C_6H_2$ | *o*-$ClC_6H_4$ | $C_6H_5$ | — | 250–270°; 2 hr. | — | ~80 | 4 |
| 2,4,6-$Cl_3C_6H_2$ | *p*-$ClC_6H_4$ | $C_6H_5$ | 80 | 250–270°; 2 hr. | — | ~80 | 4 |
| 2,4,6-$Cl_3C_6H_2$ | 2,4-$Cl_2C_6H_3$ | $C_6H_5$ | 88 | 250–270°; 2 hr. | 81 | 92 | 15 |
| 2,4,6-$Cl_3C_6H_2$ | 2,4,6-$Cl_3C_6H_2$ | $C_6H_5$ | — | 280–300° | — | | 3 |
| 2,4,6-$I_3C_6H_2$ | *p*-$CH_3OC_6H_4$ | $C_6H_5$ | 65 | | | | 23 |
| 2,4,6-$(NO_2)_3C_6H_2$ | $C_6H_5$ | $C_6H_5$ | — | — | ‡ | | 1 |
| 2,4,6-$(NO_2)_3C_6H_2$ | *o*-$CH_3C_6H_4$ | $C_6H_5$ | — | — | ‡ | | 1 |
| 2,6-$Cl_2$-4-$NCC_6H_2$ | *p*-$NCC_6H_4$ | $C_6H_5$ | 24 | Boiling Dowtherm; 1–2 hr. | 54 | | 29 |
| 2,6-$I_2$-4-$CH_3C_6H_2$ | *p*-$CH_3OC_6H_4$ | $C_6H_5$ | 75 | Boiling $C_6H_5OC_6H_5$; 10 min. | 56 | | 23 |
| 4,6-$I_2$-2-$CH_3C_6H_2$ | *p*-$CH_3OC_6H_4$ | $C_6H_5$ | 90 | | | | 23 |
| 2,6-$I_2$-4-$OHCC_6H_2$ | *p*-$CH_3OC_6H_4$ | $C_6H_5$ | — | — | ‡ | | 23 |
| 2,6-$I_2$-4-$CH_3O_2CC_6H_2$ | *p*-$CH_3OC_6H_4$ | $C_6H_5$ | 71 | Boiling *o*-$C_6H_4Cl_2$; 80 min. | 95 | | 23 |
| 2,6-$I_2$-4-*n*-$C_4H_9O_2CC_6H_2$ | *p*-$CH_3OC_6H_4$ | $C_6H_5$ | 92 | | | | 23 |
| 2-Br-6-$CH_3O$-4-$CH_3O_2CC_6H_2$ | 2,6-$(CH_3)_2C_6H_3$ | $C_6H_5$ | 61 | — | ‡ | | 34 |
| α-$C_{10}H_7$ | $C_6H_5$ | $C_6H_5$ | — | 266°; 1.5 hr. | 80† | | 3 |
| β-$C_{10}H_7$ | $C_6H_5$ | $C_6H_5$ | — | 266°; 1.5 hr. | 82† | | 3 |
| 8-Quinolinyl | $C_6H_5$ | $C_6H_5$ | 74 | 280°; 20 min. | 72 | | 30 |
| 8-Quinolinyl | *o*-$ClC_6H_4$ | $C_6H_5$ | 71 | — | ‡ | | 30 |
| 8-Quinolinyl | *o*-$CH_3C_6H_4$ | $C_6H_5$ | 47 | 280°; 10 min. | — | | 30 |
| 8-Quinolinyl | 2-Br-4-$CH_3C_6H_3$ | $C_6H_5$ | 62 | — | ‡ | | 30 |
| 8-Quinolinyl | 2-Br-4,6-$(CH_3)_2C_6H_2$ | $C_6H_5$ | 66 | — | ‡ | | 30 |
| 5,7-$Cl_2$-8-quinolinyl | $C_6H_5$ | $C_6H_5$ | 21 | 260°; few min. | — | | 30 |

*Note:* References 106 and 107 are on p. 51.

* The rearrangement was carried out as part of a kinetic study (ref. 9). The product was not isolated.

† This represents percent conversion (not percent yield) during a kinetic study at the temperature indicated (ref. 3). For preparative means, the rearrangement was carried out at 280–310°, but the yield was not given.

‡ The reaction was unsuccessful.

** The initial product decomposes at once to 1-menthene, benzoic acid, and 2-chloroacridone.

†† The product was obtained by pyrolysis of 4,4′-diamidino-N-benzoyldiphenylamine.

‡‡ 1,2-Diphenyl-4-quinolone was obtained.

§§ N-Phenyloxindole was obtained.

## TABLE I—*Continued*

### Aryl N-Arylbenzimidates, N-Aroyldiarylamines, and Diarylamines

| Aryl N-Arylbenzimidate: Substituents in $R''—C(OR)=NR'$ R | R′ | R″ | Yield, % (Ref.) | Conditions for Rearrangement (Ref.) | N-Aroyldiarylamine $R''—C(=O)—N(R)R'$ Yield, % (Ref.) | Diarylamine RNHR′ Yield, % (Ref.) | References |
|---|---|---|---|---|---|---|---|
| 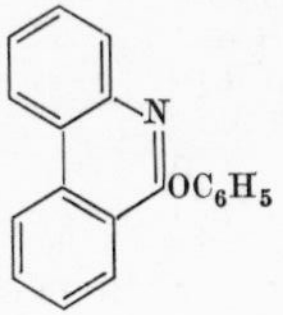 N, $OC_6H_5$ | | | 91 | 350°; 2 hr. | 60 | | 26 |
| 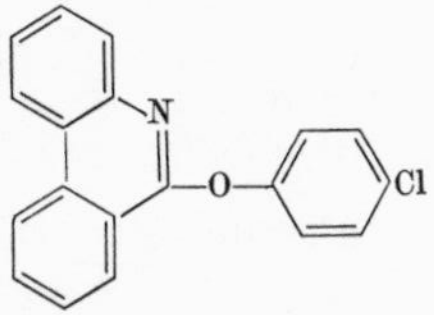 N, O, Cl | | | 93 | 350°; 2 hr. | 75 | | 26 |
| 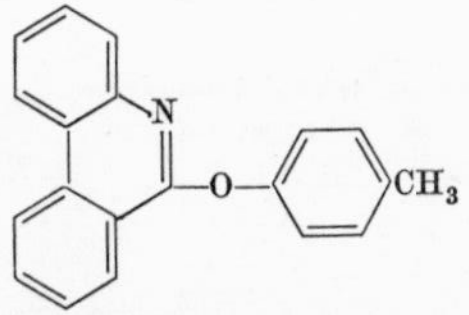 N, O, $CH_3$ | | | 81 | 360°; 2 hr. | 70 | | 26 |

*Note:* References 106 and 107 are on p. 51.

# TABLE II

## Diarylamine-2-Carboxylic Acids

| | Aryl N-Arylbenzimidate Substituents in R—(3,2,1,6)C₆H₃(CO₂CH₃)OC(C₆H₅)=NR′ | | Conditions for Rearrangement | N-Aroyldiarylamine 2-Carboxylic Ester R—C₆H₃(CO₂CH₃)—NR′—C(=O)C₆H₅ | N-Aroyldiarylamine 2-Carboxylic Acid R—C₆H₃(CO₂H)—NR′—COC₆H₅ | Diarylamine-2-carboxylic Acid R—C₆H₃(CO₂H)NHR′ | Acridone | | | |
|---|---|---|---|---|---|---|---|---|---|---|
| R | R′ | Yield, % (Ref.) | | Yield, % (Ref.) | Yield, % (Ref.) | Yield, % (Ref.)* | Substituent | Method (Ref.) | Yield, % | Refs. |
| H | $C_6H_5$ | 62 (36); 81 (15) | 270–5°; 10 min. (15) | 73 (15) | 76 (15) | 78 (36) 96 (15) | — | *b*† (15) | | 15, 36 |
| H | *o*-$ClC_6H_4$ | — | — | — | | | | | | 107 |
| H | *p*-$ClC_6H_4$ | 88 | 270–300°; few minutes | 85–91 | 76 | Almost quant. | 2-Cl | *a* *b* *c* | ? 90 87 | 15 |
| H | *o*-$CH_3O_2CCH_2C_6H_4$ | 70 | 280–295°; 12 min. | 83 | — | 28–40‡ | | | | 44 |
| H | *o*-$C_2H_5O_2CCH_2C_6H_4$ | 63 | 280–295°; 12 min. | 91 | | | | | | 44 |
| H | 2,4-$Cl_2C_6H_3$ | 72 | 260–280°; 10 min. | 65 | — | 93 | | | | 15 |
| H | 2,4-$(CH_3)_2C_6H_3$ | 56 | 275°; 10 min. | 89 | 74 | | 2,4-$(CH_3)_2$ | *a*† *b* | 71 | 15 |
| 3-Cl | *p*-$CH_3OC_6H_4$ | — | 210–215°; 70 min. | — | — | 28§ | 1-Cl-7-$OCH_3$¶ | *c* | — | 46 |

*Note:* References 106 and 107 are on p. 51.
* The yield is based on the N-benzoyl ester.
† The reaction was unsuccessful.
‡ 2′-Carboxydiphenylamine-2-acetic acid was obtained.
§ The yield is based on methyl 6-chlorosalicylate.
¶ The 9-chloroacridine was synthesized.

[a] Pyrolysis of N-benzoyl ester.
[b] Pyrolysis of N-benzoyl ester acid.
[c] Treatment of diarylamine, RNHR′, with $POCl_3$ followed by hydrolysis.

## TABLE II—*Continued*

### Diarylamine-2-Carboxylic Acids

| Aryl N-Arylbenzimidate: Substituents in R-C₆H₃(3,2-CO₂CH₃)(6,1-OC(C₆H₅)=NR′) | | | Conditions for Rearrangement | N-Aroyldiarylamine: 2-Carboxylic Ester R-C₆H₃(CO₂CH₃)(NR′-C(=O)C₆H₅) | N-Aroyldiarylamine: 2-Carboxylic Acid R-C₆H₃(CO₂H)(NR′-COC₆H₅) | Diarylamine-2-carboxylic Acid R-C₆H₃(CO₂H)(NHR′) | Acridone | | | |
|---|---|---|---|---|---|---|---|---|---|---|
| R | R′ | Yield, % (Ref.) | | Yield, % (Ref.) | Yield, % (Ref.) | Yield, % (Ref.)* | Substituent | Method (Ref.) | Yield, % | Refs. |
| 4-Cl | $p$-$CH_3OC_6H_4$ | — | 200–210°; 10 min. | — | — | — | 2-Cl-7-$OCH_3$ | *b* | Almost quant. | 15 |
| 5-$OCH_3$ | $o$-$ClC_6H_4$‖ | 86 | 270°; 10 min | — | — | — | 5-Cl-3-$OCH_3$ | *a*, *b* | — | 40 |
| 5-$OCH_3$ | $m$-$ClC_6H_4$ | 83 | — | — | | — | 1-Cl-6-$OCH_3$** | *c* — | — | 41 |
| 5-$OCH_3$ | $p$-$ClC_6H_4$‖ | 83 | 270–310° | 80 | — | 77 | 2-Cl-6-$OCH_3$ | *a*, *b*, *c* | — | 42 |
| 5-$OCH_3$ | $p$-$IC_6H_4$ | 50 | 275–290° | — | | — | 2-I-6-$OCH_3$ | *a*, *c* | — | 43 |
| 6-$CH_3$ | $C_6H_5$ | — | 260° | 92 | — | | | | | 37 |
| 6-$CH_3$ | $o$-$FC_6H_4$ | 61 | 275° | 81 | — | | | | | 34 |
| 6-$CH_3$ | $o$-$ClC_6H_4$ | 72 (36) 75 (37) | 260–270° (37) | 88 (37) | 92 (36) | — (36) | 4-Cl-5-$CH_3$ | *b* (36) | 86 | 36, 37 |
| 6-$CH_3$ | $o$-$BrC_6H_4$ | — | 280° | 80 | — | | | | | 34 |
| 6-$CH_3$ | $o$-$CH_3C_6H_4$ | 60 | 290° | — | — | | | | | 37 |
| 6-$CH_3$ | $p$-$CH_3C_6H_4$ | — | 275° | 70 | 98 | | | | | 38 |
| 6-$CH_3$ | 2,4-$Cl_2C_6H_3$ | 69 | 285–290° | 70 | — | | | | | 34 |
| 6-$CH_3$ | 2-Cl-4-$BrC_6H_3$ | — | 275° | 83 | — | | | | | 34 |
| 6-$CH_3$ | 2-Cl-4-$CH_3OC_6H_3$ | — | 285° | 82 | — | | | | | |
| 6-$CH_3$ | 2-Cl-4-$CH_3C_6H_3$ | — | 285–290° | 79 | — | | | | | 34 |
| 6-$CH_3$ | 2,4,6-$(CH_3)_3C_6H_2$ | — | — | † | | | | | | 34 |
| 6-$CH_3$ | $\alpha$-$C_{10}H_7$ | 70 | 270–300°; 45 min. | †† | | | 5-$CH_3$-3,4-benzo | *a* | 72 | 48 |
| 6-$CH_3$ | $\beta$-$C_{10}H_7$ | 63 | 360°; 0.5 hr. | †† | | | 5-$CH_3$-1,2-benzo | *a* | 38 | 48 |

| | | | | | | | | | | |
|---|---|---|---|---|---|---|---|---|---|---|
| 4,6-$Cl_2$ | $C_6H_5$ | 84 | 220° | 89 | — | | | | | 37 |
| 4,6-$Cl_2$ | 2,4-$Cl_2C_6H_3$‡‡ | 65 | 260°; 10 min. | — | | | 2,4,5,7-$Cl_4$ | *d* | 10 | 47 |
| 4,6-$Br_2$ | $C_6H_5$ | 86 | 190–200° | 80 | — | | | | | 34 |
| 4,6-$Br_2$ | *p*-$BrC_6H_4$ | 65 (15) | 270° (15) | 87 (15) | — | 90 (15) | 2,4,7-$Br_3$ | *b* | 56 (105) | 81 |
| | | | 200–210° (105) | 86–94 (105) | — | | | *c* | 91 (15) | 15, 105 |
| | | | | | | | | *d* | 87 (105) | |
| 4,6-$Br_2$ | 2,4-$Br_2C_6H_3$ | — | 260°; 5 min. | 80 | | | 2,4,5,7-$Br_4$ | *d* | 66 | 105 |
| $CO_2CH_3$; $OC{=}NC_{10}H_7$-β; $C_6H_5$ | | — | 280–320°; 15 min. | †† | | | 1,2-7,8-Dibenzo | *a* | 80 | 49 |
| $C_6H_5$; $OC{=}NC_{10}H_7$-α; $CO_2CH_3$ | | 74 | 280–300°; 0.5 hr. | †† | | | 3,4-5,6-Dibenzo | *a* | 47 | 48 |
| $CO_2CH_3$; $OC{=}NC_{10}H_7$-β; $C_6H_5$ | | 51 | 350–360° | †† | | | 1,2-6,7-Dibenzo | *a* | 54 | 48 |

*Note:* References 106 and 107 below.
* The yield is based on the N-benzoyl ester.
† The reaction was unsuccessful
|| Both the methyl and the ethyl esters were used.
** This is the probable structure of the product.
†† Pyrolysis gave the acridone directly.
‡‡ The ethyl ester was used instead of the methyl ester.

[a] Pyrolysis of N-benzoyl ester
[b] Pyrolysis of N-benzoyl acid.
[c] Treatment of diarylamine, RNHR′, with $POCl_3$ followed by hydrolysis.
[d] Treatment of the ester with sulfuric acid.



## REFERENCES


[106] Gilman and Zuech, *J. Org. Chem.*, **26,** 2013 (1961).
[107] Buchanan and Graham, *J. Chem. Soc.*, **1950,** 500.

# CHAPTER 2

## α-AMIDOALKYLATIONS AT CARBON


Harold E. Zaugg*
*Abbott Laboratories*

William B. Martin*
*Lake Forest College*


CONTENTS




* The authors are deeply indebted to Prof. Heinrich Hellmann for making available to them an unpublished review paper and bibliography on amide-aldehyde condensations written by Dr. Hans-Peter Wiedemann at the University of Tübingen. The continued interest of Prof. Hellmann during the preparation of this review is also gratefully acknowledged.

## INTRODUCTION

This chapter is concerned with reactions which lead to the formation of a new carbon-carbon bond by replacement of X from the electrophilic reagent $RCON(R')CH(R'')X$, where X is halogen, —OH, —OR, —OCOR, —NHCOR, $—NR_2$ or $—\overset{\oplus}{N}R_3$. The group R′ may be hydrogen or alkyl or, in important instances, a second acyl group, as in the corresponding derivatives of phthalimide, $o\text{-}C_6H_4(CO)_2NCH(R'')X$. In a few cases a sulfonyl group may replace the acyl group of the electrophilic reagent.

The nucleophiles that react with these reagents fall into two broad groups: aromatic compounds and aliphatic compounds containing reactive methylene or methine groups. The first may be illustrated by the phthalimidomethylation of benzene, and the second by the reaction of ethyl acetoacetate with N,N′-benzylidenebisacetamide.

$$o\text{-}C_6H_4(CO)_2NCH_2OH + C_6H_6 \xrightarrow{H_2SO_4} o\text{-}C_6H_4(CO)_2NCH_2C_6H_5 + H_2O$$

$$C_6H_5CH(NHCOCH_3)_2 + CH_3COCH_2CO_2C_2H_5 \xrightarrow{(CH_3CO)_2O}$$
$$C_6H_5CH(NHCOCH_3)CH(COCH_3)CO_2C_2H_5 + CH_3CONH_2$$

Also included in this review are the methods for preparing the electrophilic reagents. They are followed by a listing of compounds structurally related to them, and which, consequently, must be regarded as potential amidoalkylating agents. Portions of the material in this chapter have been reviewed elsewhere.[1,2]

## MECHANISMS OF THE REACTIONS

The reactions considered in this chapter include acid-catalyzed, base-catalyzed, and thermally induced processes. They, like the corresponding α-aminoalkylation reactions,[3] probably encompass a considerable portion of the mechanistic spectrum of heterolytic organic chemistry. Detailed studies are almost completely lacking. Nevertheless, some general outlines of the mechanistic possibilities can be drawn.

Although cryoscopic studies in 100% sulfuric acid have yet to be reported, it seems likely that, in very strong acids of high dielectric constant, electrophiles lacking a hydrogen atom on the nitrogen atom undergo appreciable dissociation to a carbonium-immonium ion. In the

[1] Hellmann, *Angew. Chem.*, **69**, 463 (1957).

[2] Schröter in *Houben-Weyl Methoden der Organischen Chemie*. Vol. XI/1, 4th ed., G. Thieme, Stuttgart, 1957, pp. 795–805.

[3] Hellmann and Opitz, *α-Aminoalkylierung*, Verlag Chemie, GMBH, Weinheim, 1960, pp. 64–79.

presence of weak nucleophiles (e.g., aromatic rings) bimolecular electro-

$$\mathrm{RCON(R')CH_2X} \xrightleftharpoons{\mathrm{H}^{\oplus}} \left[\begin{array}{c} \mathrm{RCO(R')N\overset{\oplus}{C}H_2} \\ \updownarrow \\ \mathrm{RCO(R')\overset{\oplus}{N}{=}CH_2} \end{array}\right] + \mathrm{HX}$$

philic displacement ($S_E2$) of protons from C—H bonds usually follows. Under the same conditions amide derivatives having a hydrogen atom attached to the nitrogen atom (R′ = H) may give the analogous cation, but other forms are more likely. Thus, when nitriles are treated with formaldehyde in strong sulfuric acid, sulfur-containing intermediates have been isolated.[4,5] Evidence indicates that these are sulfate esters which in strong acid equilibrate with their corresponding carbonium ions.[4–6] Granting the existence of these structures, identical species should, of course, be formed when the amide, $RCONHCH_2X$, is dissolved in strong

$$\mathrm{RCN} + \mathrm{CH_2O} \xrightarrow{\mathrm{H_2SO_4}} \underset{\mathrm{OSO_3H}}{\mathrm{RC{=}NCH_2X}} \xrightleftharpoons{\mathrm{H}^{\oplus}} \underset{\mathrm{OSO_3H}}{\mathrm{RC{=}N\overset{\oplus}{C}H_2}} + \mathrm{HX}$$

$$(\mathrm{X{=}OH\ or\ N{=}C(R)OSO_3H})$$

sulfuric acid. That such a common electrophilic intermediate is present is suggested by the recent finding that nitriles and formaldehyde in either strong sulfuric or strong phosphoric acid can substitute effectively for the corresponding methylolamides in the amidomethylation of aromatic compounds.[5]

In acid-catalyzed processes associated with media of low dielectric constant it is unlikely that appreciable preformed concentrations of carbonium ions are present In such processes, reaction with weak nucleophiles probably involves either tight ion pairs or incipient carbonium ions formed by an $S_N1$ process.

Reactions in neutral or basic media of amide derivatives

$$\mathrm{RCON(R')CH(R'')X}$$

lacking a hydrogen atom on the nitrogen atom exhibit characteristics of $S_N2$ processes. Thus, more nucleophilic reactants are usually required under these conditions, and reactivity toward comparable nucleophiles generally diminishes with decreasing stability of the leaving group X, i.e., halide ion $\sim R_3N > \overset{\ominus}{O}H \sim \overset{\ominus}{O}R$. It must be recognized, however,

[4] Magat, Faris, Reith, and Salisbury, *J. Am. Chem. Soc.*, **73**, 1028 (1951).

[5] Parris and Christenson, *J. Org. Chem.*, **25**, 1888 (1960).

[6] Mowry and Ringwald, *J. Am. Chem. Soc.*, **72**, 4439 (1950).

that even under these conditions the presence of proton donors can assist the removal of X by an acid-catalyzed process. These considerations also apply to similar reactions of amide derivatives with R′ = H, but with the added opportunity for a mechanism involving an azomethine intermediate. This elimination-addition scheme appears to provide the most attractive mechanism for most of the thermally induced amidoalkylation reactions.

$$\mathrm{RCONHCH(R'')X} \xrightarrow{-\mathrm{HX}} \mathrm{RCON{=}CHR''} \xrightarrow{\mathrm{HA}} \mathrm{RCONHCH(R'')A}$$

## SCOPE AND LIMITATIONS

### General Considerations

The N-methylol-amides and -imides are the most important electrophiles to be dealt with here. They are readily prepared by the reaction of formaldehyde with primary amides and imides, and their stability toward heat and acid is sufficient to permit their use under drastic reaction conditions. Furthermore, they can be employed as intermediates for the preparation of most of the other useful electrophiles.

$$\mathrm{RCONHCH_2OH} \xrightarrow{\mathrm{HX}} \mathrm{RCONHCH_2X}$$
$$\mathrm{RCONHCH_2OH} \xrightarrow{\mathrm{ROH,H^{\oplus}}} \mathrm{RCONHCH_2OR}$$
$$\mathrm{RCONHCH_2OH} \xrightarrow{\mathrm{R_2NH}} \mathrm{RCONHCH_2NR_2}$$
$$\mathrm{RCONHCH_2OH} \xrightarrow{\mathrm{RCONH_2}} \mathrm{(RCONH)_2CH_2}$$

Unlike formaldehyde, most higher aldehydes in their reactions with amides do not stop at the RCHOHNHCOR′ stage but react further to produce N,N′-alkylidenebisamides, $\mathrm{RCH(NHCOR')_2}$. Thus, whereas the amidomethylation reaction permits a relatively broad selection of electrophilic reagents, the homologous α-amidoalkylation reactions have been restricted almost exclusively to the use of N,N′-alkylidene- and N,N′-arylidene-bisamides. As the reactivities of these derivatives often are to be found at the lower end of the electrophilic scale, their usefulness is further limited by the consequent requirement of comparatively high nucleophilic reactivity for the other reactant.

### Structural Considerations

**Structure of the Electrophile.** Any structural feature tending to stabilize an incipient carbonium ion in either an $S_N1$ or an $S_N2$ process must enhance the rate of reaction with a given nucleophile. However, if ionization is not rate-limiting, the reverse is true, since stabilization of a *preformed* cation must thereby decrease its tendency to lose its charge through electrophilic attack. Therefore, in going from reaction conditions where an $S_N1$ or $S_N2$ mechanism prevails to a situation favoring the rapid pre-equilibrium production of carbonium ions (solvated or paired), one would expect to encounter an inversion in the reactivity sequence of a series of electrophiles. No quantitative work has been reported for α-amidoalkylation reagents. Nevertheless, indication that such an inversion in sequence applies to these reagents follows from a qualitative consideration of the results of several workers.

de Diesbach studied the reactions of a series of N-methylolamides in concentrated sulfuric acid at room temperature.[7] He found that amidomethylation of 1,3-dimethylanthraquinone in the 4-position succeeded with N-methyloltrichloroacetamide and N-methylolphthalimide but failed with N-methylolbenzamide. Phenanthrenequinone gave both a 2-mono- and a 2,7-di-substituted product with N-methyloltrichloroacetamide and N-methylolphthalimide, but only the monosubstitution product with N-methylolchloroacetamide. 2,4-Dimethylbenzophenone gave both a 5-mono- and a 3,5-di-substituted product with N-methylolphthalimide, but only the monosubstitution product with N-methyloltrichloroacetamide. These results enable one to construct the qualitative reactivity sequence: $o\text{-}C_6H_4(CO)_2NCH_2OH > CCl_3CONHCH_2OH > CH_2ClCONHCH_2OH > C_6H_5CONHCH_2OH$. This order is the reverse of that expected for the stabilities of the corresponding carbonium ions (i.e., delocalization of the type $RCONH\overset{\oplus}{C}H_2 \leftrightarrow RCO\overset{\oplus}{N}H{=}CH_2$ should be greatest for the benzamide derivative) and suggests that, in *concentrated sulfuric acid*, reactivity is determined by the relative electrophilicities of preformed carbonium ions.

Tawney and co-workers prepared N-methylol- and N-chloromethylmaleimide and compared their chemical properties with those of the corresponding phthalimide derivatives.[8] Although direct comparisons of their reactivities with aromatic compounds were not made, these investigators did establish that the hydroxyl group of N-methylolmaleimide was readily displaced by amines in dioxane solution and by *o*-nitrotoluene in concentrated sulfuric acid, and that N-chloromethylmaleimide readily underwent the zinc chloride-catalyzed reaction with benzene and phenol. However, unlike the corresponding phthalimide derivatives, N-methylolmaleimide would not react with hydrochloric acid to give the N-chloromethyl derivative, nor would the latter solvolyze in ethanol to give N-ethoxymethylmaleimide. Tawney and co-workers explained these striking differences by suggesting that the nitrogen atom in the maleimide,

[7] de Diesbach, *Helv. Chim. Acta*, **23**, 1232 (1940).

[8] Tawney, Snyder, Conger, Leibbrand, Stiteler, and Williams, *J. Org. Chem.*, **26**, 15 (1961).

being more electronegative than that in the phthalimide, renders stabilization of the *incipient* carbonium ion

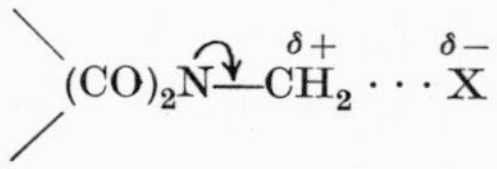

$$>(CO)_2\overset{\curvearrowright}{N}-\overset{\delta+}{C}H_2 \cdots \overset{\delta-}{X}$$

less effective in the former derivative. In an $S_N1$ process this would retard rate-limiting ionization, and in an $S_N2$ process it would make bond forming more important in the transition state and require a greater nucleophilicity of the attacking reagent.

The ultimate effect of this trend can be approached by substituting other electronegative groups directly on the methylol carbon atom. Esters of 5-hydroxy-5-hydantoincarboxylic acid gave only the corresponding amides with primary amines.[9] The hydroxyl group remained unaffected. Apparently, so many electronegative groups are attached

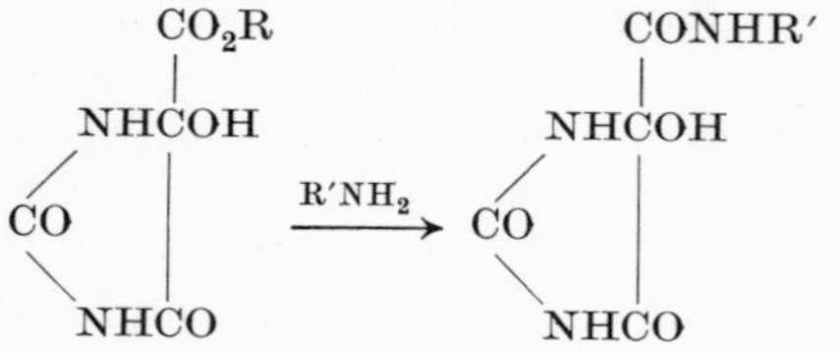

$$\text{CO}\begin{cases}\text{NHC(OH)(CO}_2\text{R)} \\ \text{NHCO}\end{cases} \xrightarrow{\text{R}'\text{NH}_2} \text{CO}\begin{cases}\text{NHC(OH)(CONHR}') \\ \text{NHCO}\end{cases}$$

to the usually reactive methylol carbon atom that insufficient development of positive charge in the transition state is available even for the attack of strongly nucleophilic amines. One might predict, however, that, if a carbonium ion could be generated from this or related hydroxyhydantoins, electrophilic reagents more reactive than anything obtainable from the simple methylolamides would be produced.

Clearly, in an electrophilic reagent that reacts by a pre-equilibrium dissociation the nature of the anionic leaving group is immaterial. The most easily accessible derivative then becomes the reagent of choice. In view of the ready availability of N-methylolamides, it is not surprising that they have been used to the virtual exclusion of other α-amidoalkyl derivatives *when strong sulfuric acid is the reaction medium.* It is only under less drastic reaction conditions, or when a more active nucleophile is involved, that the stability of the group departing from the electrophile becomes significant. This will become more apparent from a consideration of the effect of nucleophile structure on the scope of these reactions.

**Structure of the Nucleophile.** As already indicated, two general groups of nucleophiles fall within the scope of this chapter: aromatic

[9] Biltz and Lachmann, *J. Prakt. Chem.*, **113**, 309 (1926).

compounds and aliphatic compounds containing an active hydrogen atom attached to a carbon atom. The range of aromatic reactivity can best be illustrated by consideration of four examples: benzoic acid, benzene, phenol, and 2-naphthol.

As expected of such a poor nucleophile, benzoic acid requires reaction conditions ensuring attack by a strongly electrophilic species. Thus all the successful amidomethylations of benzoic acid so far reported involve reaction of a methylolamide in concentrated sulfuric acid at room temperature. Although N-methylolbenzamide gives the 3-amidomethylated product in only 8% yield under these conditions,[10] N-methylolchloroacetamide[11] and N-methylolphthalimide[12] are converted in 54% and 60% yields, respectively. Again, these results agree with expectations based

$$C_6H_5CO_2H + RCON(R')CH_2OH \xrightarrow[H_2SO_4]{\text{Concd.}} C_6H_4(CO_2H)\text{-}3\text{-}CH_2N(R')COR$$

on previous considerations of relative electrophilic reactivities in strong sulfuric acid.

Benzene is readily substituted under similar conditions. Indeed, a 32% yield of a 1,4-disubstituted product has been reported with N-methylolbenzamide in concentrated sulfuric acid.[13] Such drastic conditions, however, are not required for simple substitution. Although N-bromomethylphthalimide is inert in boiling benzene, addition of a trace of anhydrous zinc chloride suffices to catalyze an exothermic reaction leading to N-benzylphthalimide in 94% yield.[14]

$$o\text{-}C_6H_4(CO)_2NCH_2Br \xrightarrow[ZnCl_2]{C_6H_6} o\text{-}C_6H_4(CO)_2NCH_2C_6H_5$$

Phenol, which does not react with N-methylolphthalimide in boiling benzene, is readily attacked by N-bromomethylphthalimide under similar conditions even in the absence of a Lewis acid catalyst. A 51% yield of a mixture of approximately equal amounts of 2- and 4-phthalimidomethylphenol is obtained.[15] Interestingly, no 2,4-diphthalimidomethylphenol

[10] Einhorn, Bischkopff, and Szelinski, *Ann.*, **343**, 223 (1905).

[11] Einhorn and Mauermayer, *Ann.*, **343**, 282 (1905).

[12] Oda, Teramura, Tanimoto, Nomura, Suda, and Matsuda, *Bull. Inst. Chem. Res. Kyoto Univ.*, **33**, 117 (1955) [*C.A.*, **51**, 11355 (1957)].

[13] Nenitzescu and Dinulescu, *Rev. Chim. Acad. Rep. Populaire Roumaine*, **2**, 47 (1954); *Commun. Acad. Rep. Populaire Române*, **4**, 45 (1954) [*C.A.*, **50**, 15445 (1956)].

[14] H. E. Zaugg, unpublished data.

[15] Zaugg and Schaefer, *J. Org. Chem.* **28**, 2925 (1963).

is formed. This material accounts for 32% of the product when phenol is treated with N-methylolphthalimide in concentrated sulfuric acid.[15,16]

2-Naphthol represents a near approach to the ultimate in nucleophilic reactivity toward α-amidoalkylation reagents. At room temperature in an ethanol solution containing 1–2% of hydrochloric acid, it reacts with N-methylolbenzamide to give 1-benzamidomethyl-2-naphthol quantitatively in 2 hours.[10] (The more nucleophilic aromatic heterocycles, such as the polymethylpyrroles,[17] also resemble 2-naphthol in their reactivity toward various methylolamides.) 2-Naphthol is sufficiently nucleophilic to undergo smooth α-amidoalkylation with N,N′-methylene- and N,N′-arylidene-bisamides.[18] Although phenols and even phenol ethers and

$$\text{2-naphthol (OH)} + (CH_3CONH)_2CH_2 \xrightarrow[65^\circ]{POCl_3,\, CHCl_3}$$

$$\text{1-}(CH_2NHCOCH_3)\text{-2-naphthol (OH)}\ (94\%) + CH_3CONH_2$$

esters react in like manner, the diminished reactivity of the bisamides clearly restricts their usefulness in the aromatic series to substitution in activated rings.

Conceivably, aromatic nucleophilicity could be increased by employing an aromatic Grignard reagent. The orientational ambiguity often associated with the direct substitution methods also would be obviated thereby. The only example of this approach however, appears to be the reaction of phenylmagnesium bromide with N-benzoyldiphenylketimine (see p. 77).[19]

Aliphatic compounds containing methyl, methylene, or methine groups sufficiently reactive to undergo α-amidoalkylation represent a wide variety of structural types. They include cyclic and acyclic β-dicarbonyl compounds of all types, β-cyano esters, activated nitriles, nitro alkanes and β-nitro esters, certain non-aromatic heterocyclic compounds with active methine groups in the ring, heteroaromatic compounds with activated methyl groups, acetylene, and hydrocyanic acid. Most of these types possess nucleophilic reactivity in the range of that possessed by phenol in the aromatic series. Hence some of them can undergo reaction with the whole range of α-amidoalkylation reagents, including the weakly

[16] Tscherniac, Ger. pat. 134, 979 (*Chem. Zentr.*, **1902**, II, 1084).

[17] Fischer and Nenitzescu, *Ann.*, **443**, 113 (1925).

[18] Ishidate, Sekiya, and Yanaihara, *Chem. Ber.*, **93**, 2898 (1960).

[19] Ivanoff, *Doklady Akad. Nauk SSSR*, **109**, 537 (1956) [*C.A.*, **51**, 4997 (1957)].

electrophilic N,N′-bisamides. However, unlike the aromatic nucleophiles, many of the reactive methylene compounds are unstable or undergo side reactions in strong acid. For this reason and because their corresponding anions invariably exhibit enhanced nucleophilic reactivities, α-amidoalkylations of these weak aliphatic acids are usually conducted in neutral or basic media.

## α-Amidoalkylation of Aromatic Carbon Atoms

**With N-Methylol-amides and -imides (the Tscherniac-Einhorn Reaction).** In 1902 Tscherniac reported the condensation of N-methylolphthalimide with a series of aromatic compounds in concentrated or fuming sulfuric acid.[16] Subsequently Einhorn extended the reaction to a variety of readily available N-methylolamides.[10,11,20–26] The Tscherniac-Einhorn reaction,[1,2] together with its more recently developed variations (see below), bears a superficial relationship to the well-known Mannich

$$\text{RCONHCH}_2\text{OH} + \text{ArH} \xrightarrow{\text{H}_2\text{SO}_4} \text{RCONHCH}_2\text{Ar}$$

reaction.[3,27] However, the latter is usually restricted to the preparation of tertiary benzylamines, whereas amidomethylation, through hydrolysis of initial products, provides a route to primary benzylamines. Furthermore, the scope of the Mannich reaction in the aromatic series is generally restricted to phenols or to equally nucleophilic ring systems. In contrast, some of the reagents available for amidomethylation are even more electrophilic than the usual acylation reagents of the Friedel-Crafts reaction. Consequently the scope of some of the amidomethylations can be extended to aromatic systems usually considered rather inert to substitution.

The facile amidomethylation of benzoic acid has already been discussed. Although experimental details have been reported only for chlorobenzene,[13] successful amidomethylation of the other halobenzenes and of benzonitrile and benzenesulfonamide has been claimed.[28] Benzophenone reportedly is inert to N-methylolamides in concentrated sulfuric acid, but

[20] Einhorn, Ger. pat. 156,398 (*Chem. Zentr.*, **1905**, I, 55).

[21] Einhorn and Göttler, *Ber.*, **42**, 4837 (1909).

[22] Einhorn and Ladisch, *Ann.*, **343**, 264 (1905).

[23] Einhorn and Ladisch, *Ann.*, **343**, 265 (1905).

[24] Einhorn and Ladisch, *Ann.*, **343**, 277 (1905).

[25] Einhorn and Schupp, *Ann.*, **343**, 252 (1905).

[26] Einhorn and Spröngerts, *Ann.*, **361**, 161 (1908).

[27] Blicke, in Adams, *Organic Reactions*, Vol. I, John Wiley & Sons, New York, 1942, p. 303.

[28] O'Cinnéide, *Nature*, **175**, 47 (1955).

its methyl and hydroxyl derivatives are easily substituted.[7] Even nitrobenzene is subject to attack. In his original patent Tscherniac stated that a 3-substituted nitrobenzene could be obtained from N-methylolphthalimide either in concentrated sulfuric acid at 50° or in fuming sulfuric acid at room temperature.[16] More recently the isolation of 3,5-bisphthalimidomethylnitrobenzene has been reported using 20% oleum as the condensing agent.[29]

Because of its clear superiority over other readily available reagents, N-methylolphthalimide has been used most frequently in the Tscherniac-Einhorn reaction. This superiority arises from the high order of stability and reactivity of N-methylolphthalimide in strong sulfuric acid and from its ability to form crystalline readily isolable products from which, if desired, the phthalimido group can be removed easily.[30] Notwithstanding the fact that benzamides are generally more difficult to hydrolyze than trichloroacetamides or even chloroacetamides, N-methylolbenzamide is used as frequently as the methylol derivatives of the two preceding halogenated amides. Examples of the use of N-methylol derivatives of most of the commonly available amides or imides are to be found. Yet, aside from their requirement for the synthesis of specifically desired derivatives, none appears to provide any particular advantage over the generally preferred reagents. The hitherto unreported N-methyloltrifluoroacetamide, $CF_3CONHCH_2OH$ (see p. 130), however, might show some superiority over the common reagents. In strong sulfuric acid it should be even more reactive than its trichloro analog, and the extraordinary ease of alkaline hydrolysis of trifluoroacetamide derivatives[31] would make it a convenient reagent for the preparation of benzylamines.

Because they are more difficult to isolate, N-methylol derivatives of monosubstituted amides, i.e., $RCON(R')CH_2OH$, are seldom used.[32,33] However, Buc has avoided this difficulty in the case of the two lactams, 2-pyrrolidinone and $\gamma$-valerolactam, merely by using a mixture of the amide and paraformaldehyde in concentrated sulfuric acid.[34]

N,N'-Dimethylol derivatives of dicarboxamides and of urea have also been employed. For example, 1,3-dimethylolurea (**1**) and ethyl 2-furoate condense in concentrated sulfuric acid to give the symmetrically disubstituted urea **2** in 93% yield.[35] N,N'-Dimethyloloxamide,

[29] Buc, U.S. pat. 2,593,840 [*C.A.*, **46,** 6844 (1952)].

[30] Ing and Manske, *J. Chem. Soc.*, **1926,** 2348; Sheehan and Frank, *J. Am. Chem. Soc.*, **71,** 1856 (1949).

[31] Weygand and Reiher, *Chem. Ber.*, **88,** 26 (1955), and previous references.

[32] Böhme, Dick, and Driesen, *Chem. Ber.*, **94,** 1879 (1961).

[33] Böhme, Driesen, and Schünemann, *Arch. Pharm.*, **294,** 344 (1961).

[34] Buc, U.S. pat. 2,652,403 [*C.A.*, **48,** 11495 (1954)].

[35] Moldenhauer, Irion, and Marwitz, *Ann.*, **583,** 37 (1953).

$(CONHCH_2OH)_2$, reacts similarly.[35] Several reactions of 1,3-dimethylol-

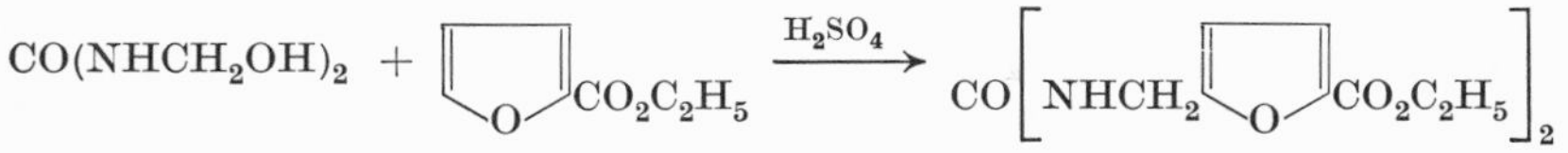

$$CO(NHCH_2OH)_2 + \text{(furan)}CO_2C_2H_5 \xrightarrow{H_2SO_4} CO\left[NHCH_2\text{(furan)}CO_2C_2H_5\right]_2$$

**1** **2**

urea have been reported in which only one of the reactive centers is attacked.[36–40] Thus, with 4-nitrophenol in sulfuric acid diluted with glacial acetic acid, the unsymmetrical urea **3** is obtained.[40] Although a

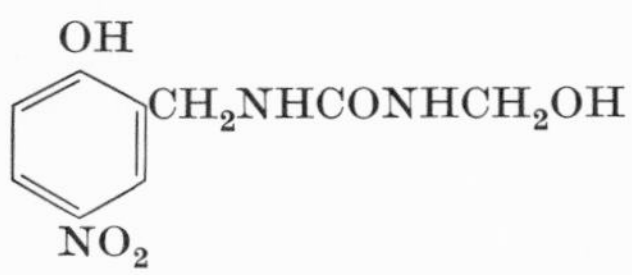

OH; $CH_2NHCONHCH_2OH$; $NO_2$ (**3**)

=O; $CON(CH_2OH)_2$ (**4**)

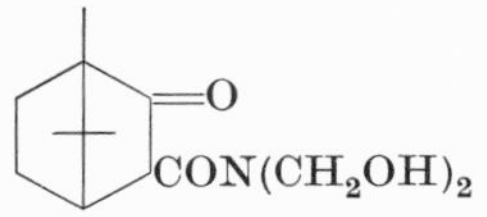

$CH_2CONHCH_2OH$ | $C(OH)CONHCH_2OH$ | $CH_2CONHCH_2OH$ (**5**)

few N,N-dimethylolamides of type **4** are known,[41,42] as is N,N′,N″-trimethylolcitramide (**5**),[43] Tscherniac-Einhorn reactions with such polyfunctional derivatives have not been reported.

A large majority of aromatic amidomethylations reported to date have been conducted either in concentrated sulfuric acid according to Tscherniac's original specifications or in the ethanolic hydrogen chloride medium used by Einhorn for substitution of the more nucleophilic aromatic systems. Despite its long history, remarkably few attempts to vary the conditions of the Tscherniac-Einhorn reaction have been reported. Even fewer qualify as reasonably systematic studies of reaction conditions. A careful investigation of the phthalimidomethylation of acetanilide has been reported by Ota, Kaneyuki, and Matsui.[44] They studied the effect of varying the sulfuric acid concentration on the reaction of equivalent quantities of acetanilide (1.6 g.) and N-methylolphthalimide (2.0 g.) at room temperature for 24 hours. In 99.5% sulfuric acid (8 ml.) only the *ortho* **6** and *para* **7** monosubstitution products were formed in 27 and 60% yields, respectively. However, in 4% oleum the yield of **7** decreased to

[36] de Diesbach, Swiss pat. 127,926–127,930 (*Chem. Zentr.*, **1929**, I, 2243).

[37] de Diesbach, Ger. pat. 507,049 (*Chem. Zentr.*, **1932**, II, 295).

[38] de Diesbach, Ger. pat. 511,210 (*Chem. Zentr.*, **1931**, II, 2514).

[39] de Diesbach, Gubser, and Spoorenberg, *Helv. Chim. Acta*, **13**, 1265 (1930).

[40] de Diesbach, Wanger, and Stockalper, *Helv. Chim. Acta*, **14**, 355 (1931).

[41] Einhorn, *Ann.*, **361**, 113 (1908).

[42] Einhorn, Ger. pat., 208,255 (*Chem. Zentr.*, **1909**, I, 1281).

[43] Einhorn and Feibelmann, *Ann.*, **361**, 140 (1908).

[44] Ota, Kaneyuki, and Matsui, *J. Chem. Soc. Japan, Pure Chem. Sect.*, **81**, 1849 (1960) [*C.A.*, **56**, 2373 (1962)].

6 % and the disubstituted product **8** was isolated in 66 % yield. Only a trace of **6** was obtained.

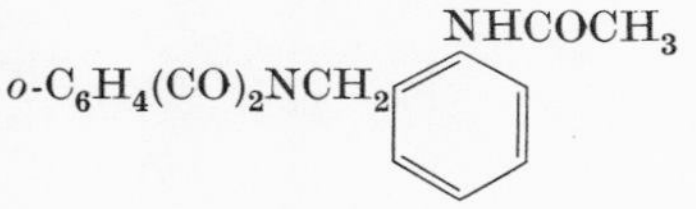

NHCOCH$_3$

$o$-C$_6$H$_4$(CO)$_2$NCH$_2$

**6**

NHCOCH$_3$

$o$-C$_6$H$_4$(CO)$_2$NCH$_2$

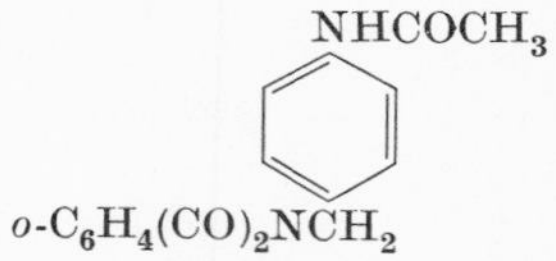

**7**

NHCOCH$_3$

$o$-C$_6$H$_4$(CO)$_2$NCH$_2$

$o$-C$_6$H$_4$(CO)$_2$NCH$_2$

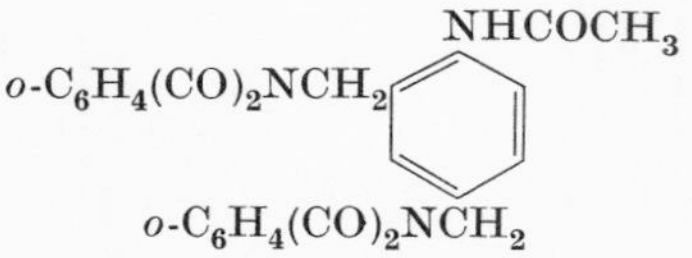

**8**

Similar results were observed with 4-methylacetanilide, but the activating effect of the methyl group in this substrate reduced the optimum sulfuric acid concentration to 97.7 %. Under these conditions a mixture of the two possible monosubstitution products was formed in 78 % yield. In 99.5 % sulfuric acid only a 20 % yield was obtained. That sulfonation as well as disubstitution played an important role in these experiments was demonstrated by the fact that, with acetanilide in 100 % sulfuric acid, sulfonation occurred to the extent of 98 % after 24 hours at room temperature, whereas, in 95 % acid under the same conditions, sulfonation proceeded to the extent of 18 %.

To minimize this undesirable property of sulfuric acid, other workers have used glacial acetic acid as a diluent.[13,40,45,46] In another instance the condensation of N-methylolchloroacetamide with phenylacetic acid was allowed to proceed in anhydrous hydrogen fluoride.[47] Other means of avoiding the use of sulfuric acid in this reaction have failed, however. They include $p$-toluenesulfonic acid in benzene and anhydrous zinc chloride in phosphorus oxychloride.

Several other media, however, have been employed successfully in condensations of N-methylolamides. Unfortunately, the aromatic substrates employed generally possessed reactive ring systems. Thus it is not possible to deduce whether any of these media could be substituted for strong sulfuric acid (or anhydrous hydrogen fluoride) under all possible conditions. The most likely candidates appear to be polyphosphoric acid, anhydrous aluminum chloride, or boron trifluoride. Polyphosphoric acid and 100% phosphoric acid alone or in glacial acetic

[45] de Diesbach, Swiss pat. 136,046 (*Chem. Zentr.*, **1930**, I, 3355).

[46] de Diesbach, Swiss pat. 139,642–139,645 (*Chem. Zentr.*, **1931**, I, 2120).

[47] Zaugg and Horrom, *J. Am. Chem. Soc.*, **80**, 4317 (1958).

acid have brought about a few relatively easy amidomethylations,[48] and 85% phosphoric acid at 70–80° has effected the condensation of N-methylolacetamide with thiophene to give the amide **9** in yields of 50–60%.[49] Phosphorus pentoxide and phosphoric acid also have been used in the substitution of phthalocyanine dyes with N-methylolphthalimide.[50]

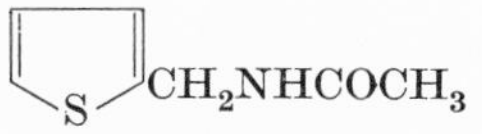

(thiophene ring, S)$CH_2NHCOCH_3$

**9**

Aluminum chloride in dry acetone or glacial acetic acid has been found to catalyze the reaction of a number of N-methylolamides with several phenols and phenol ethers in yields ranging from 60 to 73%.[51] Although less reactive aromatic compounds were not included in this study, it was noted that the aluminum chloride-catalyzed reactions seemed to proceed faster than comparable ones in sulfuric acid. With boron trifluoride in benzene at 60°, N-methylol-N-methylacetamide gave the expected product, $CH_3CON(CH_3)CH_2C_6H_5$, in 68% yield.[32]

Several sets of mild conditions have been used for carrying out aromatic alkylations with N-methylolamides, but it is doubtful whether they possess any general advantages over the ethanolic hydrogen chloride method of Einhorn. A number of di- and tri-methoxybenzene derivatives[52] and some phenolic dyes[53] have been successfully condensed with N-methylolamides by using anhydrous zinc chloride in acetic acid, but no yields were reported. Anhydrous formic acid at 50° sufficed to condense 2,4-xylenol with several N-methylolurea derivatives;[54,55] and merely heating the two N-methylolfuranamides **10** ($n = 0$ and 1) above their melting points generated the polymers **11** ($n = 0$ and 1).[35] However, heating in the absence of acid cannot be relied on to effect amidomethylation even of

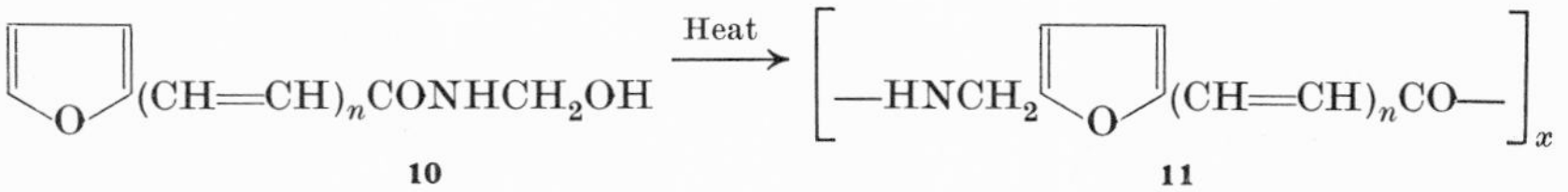

(furan ring, O)$(CH{=}CH)_nCONHCH_2OH \xrightarrow{\text{Heat}} \left[-HNCH_2\text{(furan ring, O)}(CH{=}CH)_nCO-\right]_x$

**10** **11**

[48] Arzneimittelfabrik Krewel-Leuffen, Austrian pat. 196,391 (1958) (*Chem. Zentr.*, **1960**, 16521).

[49] Hartough, *Thiophene and Its Derivatives*, Interscience Division, John Wiley & Sons, New York, 1952, p. 253.

[50] American Cyanamid Co., Brit. pat. 695,523 [*C.A.*, **48**, 1016 (1954)].

[51] Arzneimittelfabrik Krewel-Leuffen, Austrian pat. 191,878 (1957) (*Chem. Zentr.*, **1958**, 13359).

[52] Monti and Verona, *Gazz. Chim. Ital.*, **60**, 777 (1930) [*C.A.*, **25**, 1225 (1931)].

[53] Haack, U.S. pat. 2,340,528 [*C.A.*, **38**, 4385 (1944)].

[54] Zigeuner, Knierzinger, and Voglar, *Monatsh. Chem.*, **82**, 847 (1951).

[55] Zigeuner, Voglar, and Pitter, *Monatsh. Chem.*, **85**, 1196 (1954).

the most reactive aromatic systems. Thus heating the trimethylpyrrole **12** in ethanol at refluxing temperature with N-methylolchloroacetamide gave none of the expected amide. Only the dipyrrylmethane **13** was produced, presumably by reaction with formaldehyde derived in turn from cleavage of the N-methylolamide.[17]

$$\underset{\textbf{12}}{\text{2,3,5-}(CH_3)_3\text{-pyrrole (N-H)}} \xrightarrow[C_2H_5OH, 80°]{CH_2ClCONHCH_2OH} \underset{\textbf{13}}{[\text{2,3,5-}(CH_3)_3\text{-pyrrolyl (N-H)}]_2CH_2}$$

Because of the unusually high reactivity of most methylolamides in strong sulfuric acid and the consequent reduction in selectivity of substitution, many aromatic amidomethylations are complicated by the production of mixtures of isomers and polysubstitution products. For this reason many derivatives formed from substitutions involving equivocal orientations have not yet been characterized adequately. In some instances only the structure of the most readily isolable product in the mixture has been determined. In a few other instances only half-hearted attempts have been made to resolve the mixture; and in still others, particularly where the Tscherniac-Einhorn reaction has been used to introduce basic auxochrome groups into certain aromatic dye structures, little effort to purify products is apparent. However, the modern methods of isolation and characterization that are now available to the organic chemist should serve to remove what has been a formidable barrier to progress in this area.

The characterization of those aromatic amidomethylation products not readily obtainable by independent synthesis has been accomplished in many ways. Perhaps the most common is the direct permanganate[13,56] or dichromate[57] oxidation to a known carboxylic acid, $ArCH_2NHCOR \rightarrow ArCO_2H$. Another method involving two steps is hydrolysis to the benzylamine followed by conversion to a known hydroxymethyl derivative with nitrous acid, $ArCH_2NHCOR \rightarrow ArCH_2NH_2 \rightarrow ArCH_2OH$.[56] A three-step method which can be carried out with only one intervening isolation[57] involves methylation of the benzylamine and hydrogenolysis to a known methylated aromatic compound,[47]

$$ArCH_2NHCOR \rightarrow ArCH_2NH_2 \rightarrow ArCH_2N(CH_3)_2 \rightarrow ArCH_3$$

Amidomethylation occurring *ortho* to a functional group often can be diagnosed by subsequent cyclization, although completely unequivocal

[56] O'Cinnéide, *Proc. Roy. Irish Acad.*, **42B**, 359 (1935) [*C.A.*, **29**, 7326 (1935)].

[57] Truitt and Creagh, *J. Org. Chem.*, **27**, 1066 (1962).

identification may not necessarily be achieved thereby. Thus reduction of the product **14** obtained by benzamidomethylation of 4-nitroveratrole gave an amine **15** which could be cyclized to the quinazoline **16** in the presence of phosphorus oxychloride.[58]

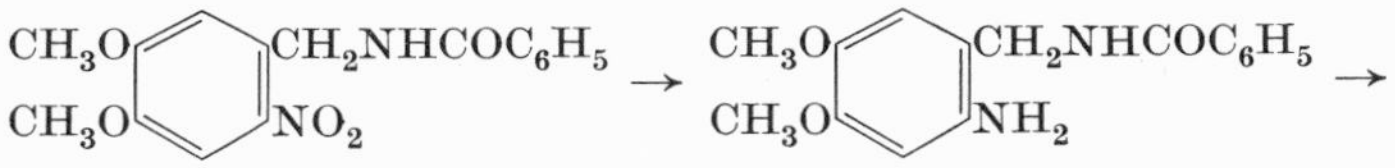

$CH_3O$, $CH_3O$, $CH_2NHCOC_6H_5$, $NO_2$ (**14**) → $CH_3O$, $CH_3O$, $CH_2NHCOC_6H_5$, $NH_2$ (**15**) →

$CH_3O$, $CH_3O$, N, N, $C_6H_5$ (**16**)

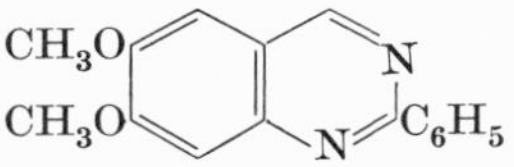

As might be expected, *m*-toluic acid (**17**) underwent substitution *ortho* to the carboxyl group. The only product isolated (in unspecified yield) from its reaction with N-methylolbenzamide in concentrated sulfuric acid was 6-methylphthalimidine (**19**) resulting from intramolecular amide interchange of the initial product **18**.[7]

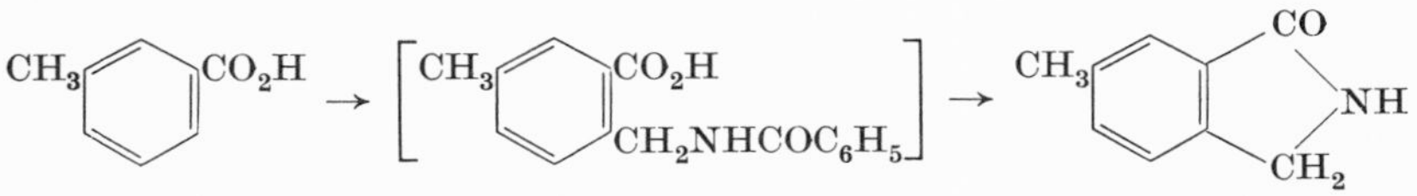

$CH_3$, $CO_2H$ (**17**) → [$CH_3$, $CO_2H$, $CH_2NHCOC_6H_5$] (**18**) → $CH_3$, CO, NH, $CH_2$ (**19**)

**With Formaldehyde and Amides or Nitriles.** Few reports of this modification of the Tscherniac-Einhorn reaction have appeared. Buc applied it to the condensation of lactams and imides with some aromatic nitro compounds, using concentrated sulfuric acid as the medium.[34] Thus 4-chloronitrobenzene with equivalent amounts of paraformaldehyde and 4-cyclohexene-1,2-dicarboximide in concentrated sulfuric acid at 65° gave the monosubstitution product **20** in 75% yield.

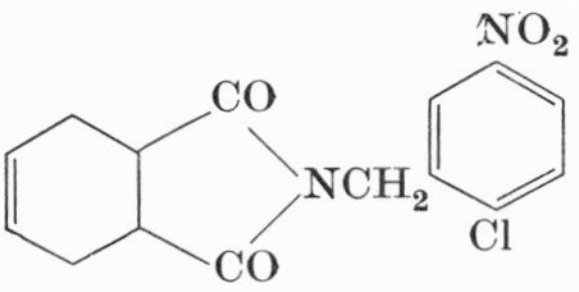

CO, CO, $NCH_2$, $NO_2$, Cl (**20**)

Parris and Christenson used a series of aliphatic and aromatic nitriles, and even hydrogen cyanide, with paraformaldehyde in sulfuric-acetic acid mixtures at temperatures varying from ambient to 90°.[5] Although yields ranged from 20 to 90% most of the aromatic compounds used were

[58] Downes and Lions, *J. Am. Chem. Soc.*, **72**, 3053 (1950).

more reactive than benzene. Bromobenzene, however, with acetonitrile and formaldehyde in concentrated sulfuric acid gave the monosubstitution product $p$-$BrC_6H_4CH_2NHCOCH_3$ in 37% yield. In the reaction of acetonitrile and formaldehyde with $m$-xylene, 85% phosphoric acid at 90° gave a better yield (66% vs. 52%) than a sulfuric-acetic acid mixture at the same temperature. Of some interest is the successful utilization of acrylonitrile. With toluene and formaldehyde in a sulfuric-acetic acid mixture at room temperature it afforded the acrylamide

$$p\text{-}CH_3C_6H_4CH_2NHCOCH{=}CH_2$$

in 89% yield.

Surprisingly, even after heating a mixture of urea, formaldehyde, and 2,4-xylenol in formic acid at 50° for 2 hours, two monomeric urea derivatives **21** and **22** could be isolated.[59]

$H_2NCONR_2$ **21** $RNHCONR_2$ **22**

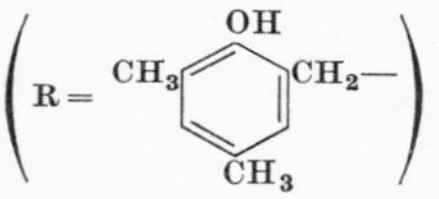

(R = 2-hydroxy-3,5-dimethylbenzyl: $CH_3$, OH, $CH_2$—, $CH_3$ on benzene ring)

Extension of this reaction to sulfonamides also has been described.[60] Sulfanilamide and its $N^4$-acetyl derivative were heated with formaldehyde and 4-methyl-2-thoiuracil in an acetic-hydrochloric acid mixture. The products were assigned structure **23** (R = H and $CH_3CO$).

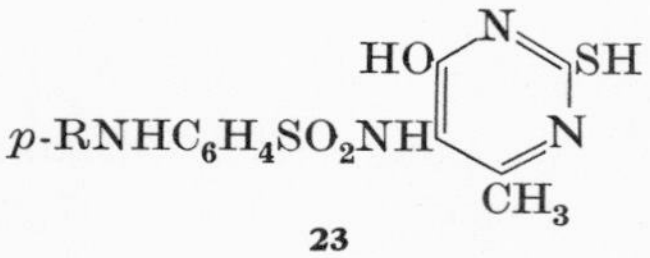

$p$-$RNHC_6H_4SO_2NH$—(pyrimidine ring bearing HO, SH, N, N, $CH_3$)

**23**

**With Ethers of N-Methylol-amides and -imides.** Most of the few reported examples of this condensation involve the diamidomethyl ethers, $(RCONHCH_2)_2O$, formed by self-condensation of corresponding methylol-amides. Thus in his original work Tscherniac was able to condense diphthalimidomethyl ether, $[o\text{-}C_6H_4(CO)_2NCH_2]_2O$, with phenol, 4-nitrophenol, and 2-nitrotoluene, using concentrated sulfuric acid.[61] Zigeuner and co-workers studied the condensations of similar ethers derived from methylolureas.[54] With anhydrous formic acid at 50°, they obtained the

[59] Zigeuner, Pitter, and Rauch, *Monatsh. Chem.*, **86**, 173 (1955).

[60] Monti and Franchi, *Gazz. Chim. Ital.*, **81**, 332 (1951) (*Chem. Zentr.*, **1952**, 7492).

[61] Tscherniac, Ger. pat. 134,980 (*Chem. Zentr.*, **1902**, II, 1164).

product **24** from the ether $(C_6H_5NHCONHCH_2)_2O$ and 2,4-xylenol in 90% yield. Under similar conditions the complex ether

$$O(CH_2NHCONHCH_2OH)_2$$

gave the disubstituted urea **25** as the main product together with small amounts of the mono- and tri-substituted ureas **26** and **22**, respectively.[62]

$$\underset{\mathbf{24}}{RNHCONHC_6H_5} \qquad \underset{\mathbf{25}}{(RNH)_2CO} \qquad \underset{\mathbf{26}}{RNHCONH_2}$$

(R = OH, $CH_3$, $CH_2$—, $CH_3$)

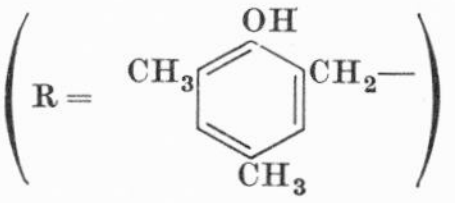

Three reactions involving unsymmetrical ethers have been described. N-Ethoxymethylphthalimide in 100% sulfuric acid at 80–90° gave monosubstituted products with benzene (5% yield) and nitrobenzene (41% yield).[63] A disubstituted product of undetermined structure was obtained when copper phthalocyanine was treated with N-methoxymethylphthalimide in 100% sulfuric acid at 100°.[64] The corresponding symmetrical diphthalimidomethyl ether at 75° in 100% acid reportedly gave only a monosubstituted derivative.[50]

**With N-Halomethyl-amides, -imides, and -carbamyl Compounds.** Since its introduction by Cherbuliez and co-workers in 1922,[65,66] this method of amidomethylation has seen comparatively little use. The original workers showed that both functions in 1,4-dichloromethyl-2,5-piperazinedione (**27**) attack naphthalene and 2-naphthol even in the absence of catalyst.[65] Merely heating the dione **27** with 2-naphthol in benzene under reflux gave **28** (Ar = 2-hydroxy-1-naphthyl) in 89% yield. Although the dichloromethyl derivative **27** was inert to benzene in carbon

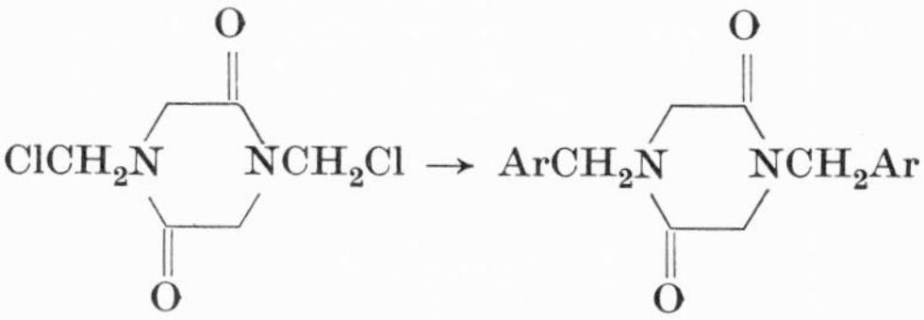

$$\underset{\mathbf{27}}{ClCH_2N\langle(CO)CH_2\rangle_2NCH_2Cl} \rightarrow \underset{\mathbf{28}}{ArCH_2N\langle(CO)CH_2\rangle_2NCH_2Ar}$$

[62] Zigeuner and Pitter, *Monatsh. Chem.*, **86**, 57 (1955).

[63] Tanimoto, Kyo, and Oda, *J. Chem. Soc. Japan, Ind. Chem. Sect.*, **65**, 1583 (1962) [*C.A.*, **59**, 505 (1963)].

[64] Rösch and Bayer, Ger. pat. 852,588 (*Chem. Zentr.*, **1953**, 8213).

[65] Cherbuliez and Feer, *Helv. Chim. Acta*, **5**, 678 (1922).

[66] Cherbuliez and Sulzer, *Helv. Chim. Acta*, **8**, 567 (1925).

disulfide, addition of a catalytic amount of anhydrous aluminum chloride gave 1,4-dibenzyl-2,5-piperazinedione (**28**, Ar = $C_6H_5$) in 43% yield.[65] More recent work has extended the method to N-substituted N-chloromethylamides, $RCON(R')CH_2Cl$.[32,67] Anhydrous aluminum chloride was the catalytic agent used in most of these studies.[32]

Many of the N-halomethylamides are difficult to isolate and purify. Both N-chloro- and N-bromo-methylphthalimide are, however, fairly stable, but reactive, crystalline solids. For this reason they have been the reagents of choice ever since the former was first used by Herzberg and Lange in 1927.[68] As an example of their high reactivity, it was noted that, when a mixture of 4-chlorophenol, N-chloromethylphthalimide, and a trace of zinc chloride was warmed on the steam bath, spontaneous heating to 130° occurred and the product, 2-phthalimidomethyl-4-chlorophenol, was formed in 70% yield.[69] Similar exothermic behavior was observed in the reaction of N-bromomethylphthalimide with benzene.[14] Indeed, this reagent proved sufficiently reactive to alkylate phenylacetic acid in the presence of zinc chloride catalyst.[47]

Utilization of the high reactivity of N-bromomethylphthalimide for the characterization of alcohols and phenols has been recommended twice.[70,71] Unfortunately, both groups of workers assigned ether structures to their phenolic derivatives. Recent work has demonstrated, however, that reaction of N-bromomethylphthalimide with phenol under their conditions (no catalyst) gave a mixture of *o*- and *p*-phthalimidomethylphenols.[15] No N-phenoxymethylphthalimide could be detected in the reaction mixture. Undoubtedly the corresponding derivatives of the other phenols (thymol and β-naphthol) reported by these workers are likewise substituted in the aromatic nucleus.[71]

The N-halomethyl-amides and -imides represent the most strongly electrophilic reagents presently available for the amidomethylation reaction. Recently Olah and co-workers have been able to isolate stable oxocarbonium salts of very strong acids, e.g., $RCO^{\oplus}\ SbCl_6^{\ominus}$.[72] These are extremely efficient aromatic acylating agents and do not require the presence of Lewis acid catalysts. This suggests that, if a salt such as $o\text{-}C_6H_4(CO)_2N\overset{\oplus}{C}H_2\ SbCl_6^{\ominus}$ could be isolated, i.e., by treatment of N-chloromethylphthalimide with antimony pentachloride, it might serve as a powerful amidomethylating reagent even in the absence of a polarizing catalyst.

[67] Gränacher and Sallmann, Ger. pat. 889,157 (*Chem. Zentr.*, **1954**, 5661).
[68] Herzberg and Lange, Ger. pat. 442,774 (*Chem. Zentr.*, **1927**, II, 505).
[69] Yamaguchi, *J. Chem. Soc. Japan, Pure Chem. Sect.*, **73**, 393 (1952) [*C.A.*, **47**, 10497 (1953)].
[70] Hopkins, *J. Am. Chem. Soc.*, **45**, 541 (1923).
[71] Mancera and Lemberger, *J. Org. Chem.*, **15**, 1253 (1950).
[72] Olah, Kuhn, Tolgyesi, and Baker, *J. Am. Chem. Soc.*, **84**, 2733 (1962).

A recently developed bifunctional reagent is chloromethylcarbamoyl chloride (**29**).[73] With β-naphthol in the absence of catalyst it gives 3,4-dihydro-2H-naphtho[1,2-*e*]-1,3-oxazin-2-one (**30**) in 25% yield. 2,4-Dichloroaniline gives the analogous nitrogen heterocycle, 3,4-dihydro-6,8-dichloro-2(1*H*)-quinazolinone (**31**). Less reactive aromatic compounds such as *m*-xylene require a zinc chloride catalyst. Chloromethyl isocyanate, $ClCH_2NCO$, readily obtainable from **29** can be used in its place in these reactions.

$$CH_2O + HNCO \longrightarrow HOCH_2NCO \xrightarrow[30^\circ]{SOCl_2} ClCONHCH_2Cl$$

**29**

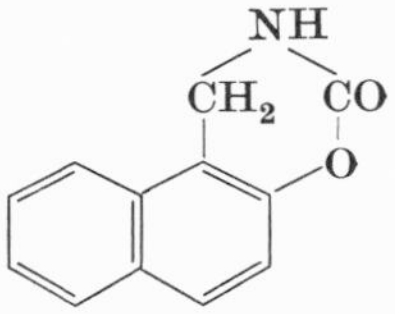

**30** 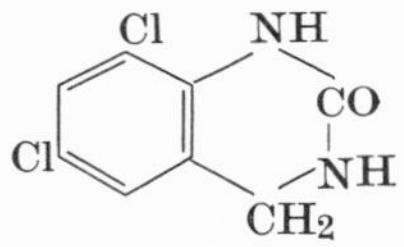 **31**

**With N,N′-Methylene-, -Alkylidene- and -Arylidenebisamides.** The diminished reactivity of these amide derivatives usually requires correspondingly increased reactivity of the nucleophiles. Hence, in the aromatic series, amidoalkylations of this type have usually been restricted to the activated systems: phenols, phenol ethers and esters, and anilides. Three methods have been used to effect condensation: heating the reactants at 190°,[74] warming them at 50° in formic acid,[55] or treating them at 65–130° with phosphorus oxychloride alone or in chloroform.[18,75–78] A single report, however, describes the extension of this reaction to benzene using 100% sulfuric acid as the condensing agent.[63]

The formic acid method is probably the least general. Whereas 1,1′-methylenebis-(3-phenylurea), $(C_6H_5NHCONH)_2CH_2$, and its *p*-tolyl analog gave with 2,4-xylenol the unsymmetrical ureas **32**,[55] neither N,N′-methylenebisbenzamide, $(C_6H_5CONH)_2CH_2$,[55] nor N,N′-methylenebisisovaleramide[54] underwent reaction under these conditions. On the

[73] Hoover, Stevenson, and Rothrock, *J. Org. Chem.*, **28**, 1825 (1963).

[74] Stefanović, Bojanović, Vandjel, Maksimović, and Mihailović, *Rec. Trav. Chim.*, **76**, 249 (1957).

[75] Ishidate, Sekiya, and Yanaihara, *Chem. Pharm. Bull.* (*Tokyo*), **8**, 1120 (1960) [*C.A.*, **57**, 9736 (1962)].

[76] Sekiya and Yanaihara, *Chem. Pharm. Bull.* (*Tokyo*), **7**, 746 (1959) [*C.A.*, **54**, 16369 (1960)].

[77] Sekiya, Yanaihara, and Masui, *Chem. Pharm. Bull.* (*Tokyo*), **9**, 945 (1961) [*C.A.*, **57**, 16459 (1962)].

[78] Sekiya, Yanaihara, and Masui, *Chem. Pharm. Bull.* (*Tokyo*), **11**, 551 (1963) [*C.A.* **59**, 8643 (1963)].

other hand, after only 1 hour at 95°, phosphorus oxychloride caused the

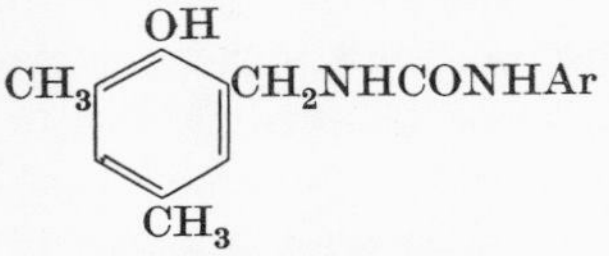

OH, $CH_3$, $CH_2NHCONHAr$, $CH_3$

**32** ($Ar = C_6H_5$ or $p$-$CH_3C_6H_4$)

OH, $CH_3$, $CH_2NHCOC_6H_5$, $CH_3$

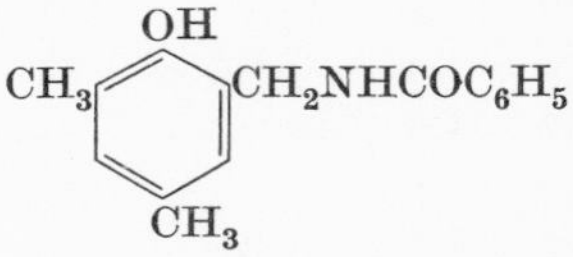

**33**

conversion of N,N′-methylenebisbenzamide and 2,4-xylenol to the expected product **33** in 93% yield.[75] Dilute mineral acid catalysis appears to be useless in these reactions. Treatment of 2-naphthol with several methylenebisamides, $(RCONH)_2CH_2$, in ethanolic hydrochloric acid led to the methylenebisnaphthol **34** instead of the expected amidomethylation products.[79]

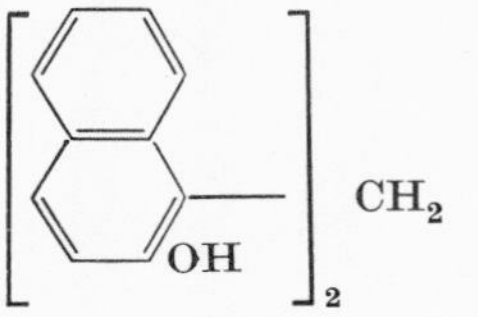

$CH_2$, OH, 2

**34**

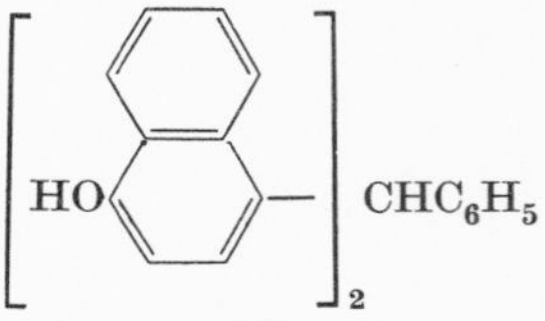

HO, $CHC_6H_5$, 2

**35**

$CH(C_6H_5)NHCOCH_3$, OH

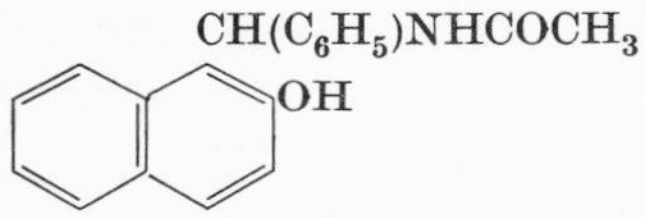

**36**

A by-product, **35,** of similar character was obtained when benzylidenebisacetamide, $(CH_3CONH)_2CHC_6H_5$, was heated 4 hours at 190° with 1-naphthol.[74] In contrast, 2-naphthol, under the same conditions, gave the α-amidoalkylation product, **36,** in 93% yield.[74] This tendency for the thermal (190°) reaction to substitute *ortho* rather than *para* to the oxygen function is further supported by the observation that benzylidenebisacetamide with 2,4-xylenol gave the *ortho* derivative **37** in 74% yield but the *para* isomer **38** from 2,6-xylenol in 30% yield.[74] Furthermore, reactions of benzylidenebisacetamide and benzylidenebisbenzamide with phenol, phenyl acetate, and phenyl benzoate (all at 190°) led to *o*-mono-

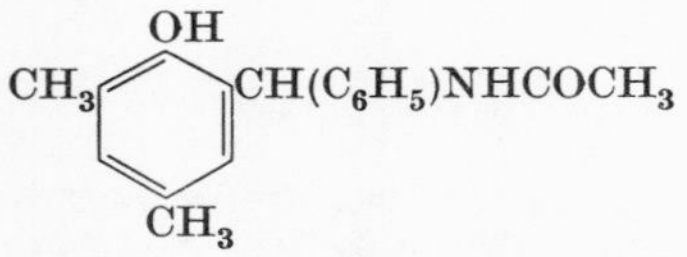

OH, $CH_3$, $CH(C_6H_5)NHCOCH_3$, $CH_3$

**37**

OH, $CH_3$, $CH_3$, $CH(C_6H_5)NHCOCH_3$

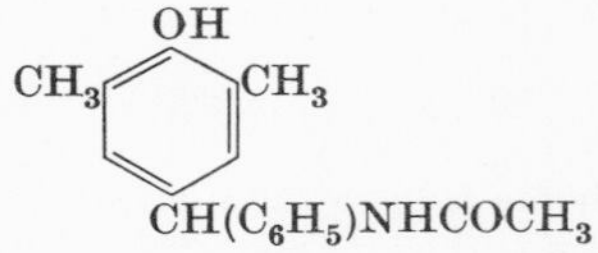

**38**

[79] Haworth, Peacock, Smith, and MacGillivray, *J. Chem. Soc.*, **1952,** 2972.

amidoalkylated derivatives as the only isolable products, albeit in poor to moderate yields (12–69%).[74]

The phosphorus oxychloride method is less selective but considerably more effective in bringing about condensation. Thus treatment of phenol and N,N′-methylenebisacetamide with phosphorus oxychloride in chloroform for 4 hours under reflux gave a mixture of equal amounts of *o*- and *p*-acetamidomethylphenol.[77] By using various reaction temperatures and times, excellent yields were obtained from a number of other aromatic compounds. These included 4-nitrophenol (85%), 4-methoxytoluene (93%), and 2-methylacetanilide (86%).[18] Other yields ranged from 71 to 98%.[18,75]

The thermal (190°) condensation method was found not to require the preformed bisamide reagent.[74] Thus heating a mixture of benzaldehyde, acetamide, and *p*-cresol at 190° for 4 hours gave the alkylated product **39** in 58% yield. The use of preformed benzylidenebisacetamide gave a

$$C_6H_5CHO + CH_3CONH_2 + p\text{-}CH_3C_6H_4OH \xrightarrow{190^\circ}$$

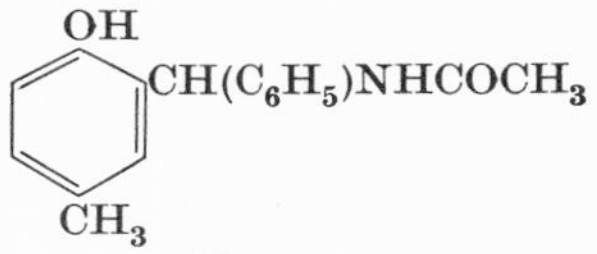

OH, $CH(C_6H_5)NHCOCH_3$, $CH_3$

**39**

lower yield, 43%. However, in most other instances the preformed reagent led to distinctly better yields.[74]

Related to these amidoalkylations is an interesting condensation reported by Pirrone in 1937.[80] He found that 8-hydroxyquinoline treated with two equivalents of benzaldehyde and one of an amide (formamide, acetamide, benzamide, or salicylamide) in warm (60–80°) ethanol or benzene gave 3-acyl derivatives of 2,4-diphenyl-2-pyrido[3,2-*h*][1,3]benzoxazine (**40**, R = H, $CH_3$, $C_6H_5$, or *o*-$HOC_6H_4$).

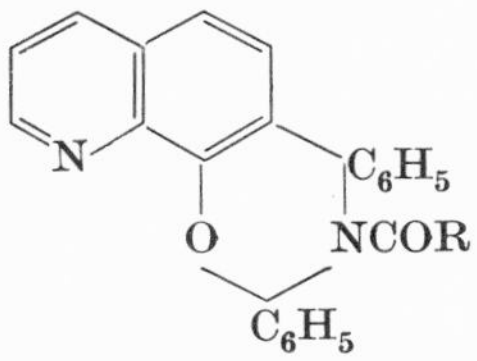

N, $C_6H_5$, O, NCOR, $C_6H_5$

**40**

The use of an unsymmetrical methylene-bisamide has been reported once.[63] Treatment of benzene with N-benzamidomethylphthalimide in

[80] Pirrone, *Gazz. Chim. Ital.*, **67**, 529 (1937) (*Chem. Zentr.*, **1938**, I, 1581) [*C.A.*, **32**, 1701 (1938)].

100% sulfuric acid at 90° gave N-benzylphthalimide in 24% yield. No N-benzylbenzamide was isolated.

**With N-Aminomethyl-amides and -imides.** Compounds of this type, e.g., $RCON(R'')CH_2NR'_2$, do not amidomethylate directly at a carbon atom of an aromatic system. Rather, they cleave as indicated $RCON(R'')$–¦–$CH_2NR_2$ and thus behave as *amino*methylating agents. A special reaction of indole is known, however, in which the end result is the same as if an amidomethylation had occurred.

When equivalent amounts of indole and N-piperidinomethylphthalimide (**41**) were heated under reflux for 9 hours in xylene containing a little powdered sodium hydroxide, 3-phthalimidomethylindole (**42**) was isolated in 48% yield.[81] When the reaction was interrupted after 1–2 hours,

Indole + $o$-$C_6H_4(CO)_2NCH_2N$(piperidine) → 3-[$CH_2N(CO)_2C_6H_4$-$o$]indole

**41** **42**

however, N-skatylpiperidine (**43**) was obtained in 50% yield. The latter compound was converted smoothly to **42** by further treatment with phthalimide. Thus the aminomethylation product **43** was formed first; but on prolonged treatment it reacted further with the phthalimide present, probably by an elimination-addition mechanism involving **44**,

3-($CH_2N$-piperidino)indole → [3-methyleneindolenine, $=CH_2$] $\xrightarrow{o\text{-}C_6H_4(CO)_2NH}$ **42**

**43** **44**

typical of alkylation reactions of gramine.

N-Piperidinomethylbenzenesulfonamide condensed with indole in the same way to give **45**, but in only 21% yield.[82] N-Dimethylaminomethylbenzamide, however, amidomethylated the nitrogen atom of indole.[83]

3-($CH_2NHSO_2C_6H_5$)indole (N–H)    1-($CH_2NHCOC_6H_5$)indole

**45** **46**

[81] Hellmann, Löschmann, and Lingens, *Chem. Ber.*, **87**, 1690 (1954).

[82] Hellmann and Teichmann, *Chem. Ber.*, **91**, 2432 (1958).

[83] Hellmann and Haas, *Chem. Ber.*, **90**, 53 (1957).

No further reaction could be induced, and the monosubstituted product **46** was isolated in 57% yield. Supposedly the methylene imide, $CH_2{=}NCOC_6H_5$, is the intermediate in this reaction and the NH group is the only function in indole sufficiently reactive to add to it. Attempts to force substitution at the 3-position of indole by using media acidic enough presumably to form the stronger electrophile $\overset{\oplus}{C}H_2NHCOC_6H_5$ led only to decomposition of the acid-sensitive substrate.[83]

**With Amidomethanesulfonic Acids.** Only two reactions of this type have been described.[63] Both involved the use of sodium benzamidomethanesulfonate (**47**) as the reagent. With benzene in 100% sulfuric acid at 95°, N-benzylbenzamide was isolated in 62% yield. With nitrobenzene under similar conditions N-(*m*-nitrobenzyl)benzamide was obtained, but in only 8% yield.

$$\underset{\mathbf{47}}{C_6H_5CONHCH_2SO_3Na} + ArH \rightarrow C_6H_5CONHCH_2Ar$$

**With N-Acylimines.** Only one example of this type of reaction has been found. Ivanoff treated N-benzoyldiphenylketimine with phenylmagnesium bromide and obtained N-benzoyltritylamine in 62% yield.[19]

$$C_6H_5CON{=}C(C_6H_5)_2 \xrightarrow{C_6H_5MgBr} C_6H_5CONHC(C_6H_5)_3$$

## α-Amidoalkylation of Aliphatic Carbon Atoms

**With N-Methylol and N-α-Alkylol Derivatives of Amides, Imides, and Carbamyl Compounds.** The extension of the Tscherniac-Einhorn reaction to the aliphatic series is limited by the side reactions that aliphatic compounds undergo in strong acid media. The nucleophiles successfully employed possess active hydrogen atoms on aliphatic carbon atoms and would be expected to exhibit a wide range of nucleophilic activity.

Several attempts to amidomethylate malonic ester with N-methylolbenzamide and N-methylolphthalimide in sulfuric acid have failed.[84,85] Monti succeeded, however, in benzamidomethylating 1-phenylpropane-1,3-dione and 1,3-diphenylpropane-1,3-dione, although no yields were given.[86] Hellman and co-workers, using the same substrates as well as pentane-2,4-dione, 5,5-dimethylcyclohexane-1,3-dione, and 1,2-diphenylpyrazolidine-3,5-dione with N-methylolphthalimide in concentrated sulfuric acid, were able to prepare the amidomethylated products in about

[84] Buc, *J. Am. Chem. Soc.*, **69**, 254 (1947).

[85] Hellmann, Aichinger, and Wiedemann, *Ann.*, **626**, 35 (1959).

[86] Monti, *Gazz. Chim. Ital.*, **60**, 39 (1930) [*C.A.*, **24**, 4013 (1930)].

80% yields.[85] The alkylated derivatives **48** and **49** are examples of the products obtained.

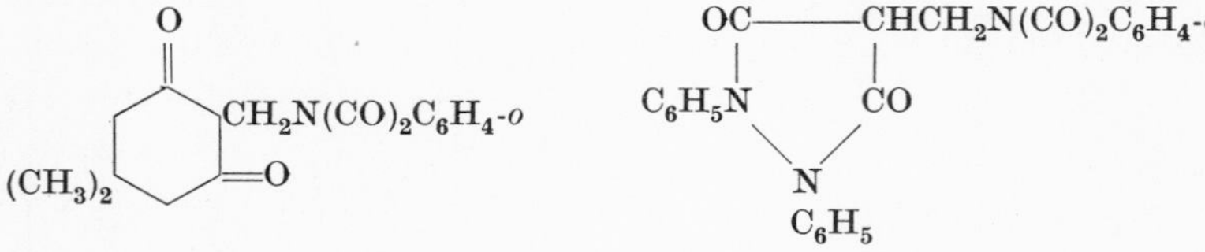

O
$CH_2N(CO)_2C_6H_4$-*o*
$(CH_3)_2$ =O

**48**

OC—$CHCH_2N(CO)_2C_6H_4$-*o*
$C_6H_5N$ CO
N
$C_6H_5$

**49**

The reaction conditions usually employed, i.e., concentrated sulfuric acid, do not seem to be generally applicable, although admittedly they have been studied only to a limited extent. The reaction of N-methylolbenzamide with cyclohexane-1,3-dione is reported to give the amidomethylated diketone **50** in 36% yield with concentrated sulfuric acid; in 20% yield with ethanol and hydrochloric acid; in 40% yield with acetic acid and zinc chloride; and in 65% yield with acetic acid and boron

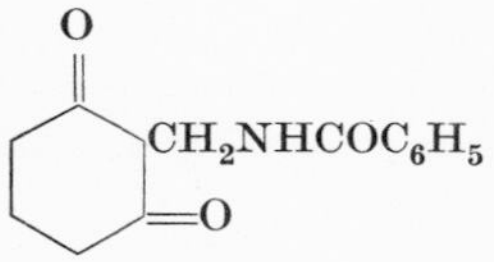

O
$CH_2NHCOC_6H_5$
=O

**50**

trifluoride etherate.[85] By using boron trifluoride etherate, a 74% yield of the ketone $C_6H_5CONHCH_2CH(COC_6H_5)_2$ was obtained from the condensation of 1,3-diphenylpropane-1,3-dione with N-methylolbenzamide. The latter reaction conditions appear to provide a serviceable alternative to the use of concentrated sulfuric acid.

A series of reactions with trinitromethane as the nucleophile was carried out successfully using N-methylolurethan and a variety of N-methylolamides.[87,88] An example is the reaction of trinitromethane with N-methylolmethacrylamide in water to give the amide

$$CH_2{=}C(CH_3)CONHCH_2C(NO_2)_3$$

in 87% yield. Yields were not given for all the reactions, but those reported were above 80%. With these compounds water often served as the solvent, and the condensation occurred readily over a wide range of *p*H (<7).

An interesting reaction has been reported in which 2-hydroxypteridine hydrate (**51**) was condensed with four different active methylene compounds (pentane-2,4-dione, ethyl acetoacetate, diethyl malonate, and

[87] Feuer and Lynch-Hart, *J. Org. Chem.*, **26**, 391 (1961).

[88] Nitroglycerin Aktiebolaget, Brit. pat. 813,477 [*C.A.*, **53**, 19884 (1959)].

ethyl cyanoacetate) in 75, 80, 90, and 90% yields, respectively, to give compounds of structure **53**.[89] Since the reactions were conducted in

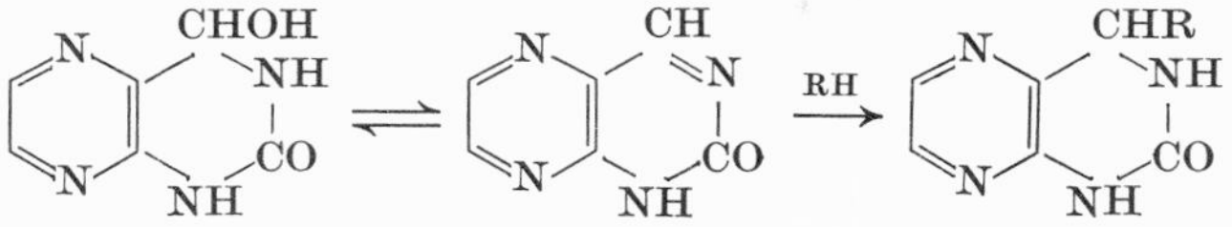

**51** ⇌ **52** $\xrightarrow{RH}$ **53**

(R = $CH_3COCHCOCH_3$, $CH_3COCHCO_2C_2H_5$, $—CH(CO_2C_2H_5)_2$, $NCCHCO_2C_2H_5$)

neutral or basic media, it is likely that they proceed by a Michael-type addition of enolate anion to the pteridine **52** rather than directly through the hydrate **51**.

Only a few reactions of N,N′-dimethylol compounds have been reported.[87] N,N′-Dimethylolfumaramide and trinitromethane in water at *p*H 0.8 gave the fumaramide **54** in 48% yield. A patent describes the reaction of trinitromethane with N,N′-dimethylolurea and with a urea-

$(O_2N)_3CCH_2NHCOCH$

‖

$HCCONHCH_2C(NO_2)_3$

**54**

formaldehyde mixture to give $CO[NHCH_2C(NO_2)_3]_2$ in 77% and 88% yields, respectively.[88] The condensation of acetylene with N,N′-dimethylolurea to give N-methylol-N′-2-propynylurea is mentioned.[90]

**With Formaldehyde or Acetaldehyde and Sulfonamides.** Alkyl- and aryl-sulfonamides with formaldehyde and alkali cyanides give products of the type $RSO_2NHCH_2CN$. Reaction of benzenesulfonamide and acetaldehyde with potassium cyanide yields the analogous derivative, $C_6H_5SO_2NHCH(CH_3)CN$.[91]

Sulfanilamide and paraformaldehyde condense at elevated temperature with 2-picoline to give the product **55**. Similar reactions occurred to give amidomethylated products from quinaldine, 9-methylacridine, 9-ethylacridine (i.e., **56**), and 2-methyl-4(3*H*)-quinazolinone (i.e., **57**).[92] Indication that amidomethylation occurred on the methyl group of 2-methyl-4(3*H*)-quinazolinone and not on the amide nitrogen atom was

[89] Albert and Howell, *J. Chem. Soc.*, **1962,** 1591.

[90] Reppe, Keyssner, and Hecht, Ger. pat. 724,759 [*C.A.*, **37,** 5733 (1943)]; Fr. pat. 839,875 (*Chem. Zentr.*, **1939,** II, 734).

[91] Reuter, Ger. pat. 847,006 [*C.A.*, **50,** 2669 (1956)].

[92] Monti and Felici, *Gazz. Chim. Ital.*, **70,** 375 (*Chem. Zentr.*, **1940,** II, 2158).

derived from the fact that no reaction took place with 4(3*H*)-quinazolinone, in which the 2-methyl group is absent. In contrast, N-methylolbenzamide

$$\underset{\mathbf{55}}{\text{2-}(RCH_2)\text{pyridine}} \qquad \underset{\mathbf{56}}{\text{9-}(RCHCH_3)\text{acridine}} \qquad \underset{\mathbf{57}}{\text{2-}(RCH_2)\text{-4(3}H\text{)-quinazolinone}}$$

($R = p\text{-}H_2NC_6H_4SO_2NHCH_2\text{—}$)

and 2-methyl-4(3*H*)-quinazolinone reportedly condense preferentially at the 3-nitrogen atom.[92a]

**With Ethers and Esters of N-Methylol- and N-α-Alkylol-amide Derivatives.** Most of the reported work in this area is concerned with the reaction of trinitromethane with N-benzoxymethyl-acrylamide and -methacrylamide, $CH_2{=}C(R)CONHCH_2OCOC_6H_5$ ($R = H, CH_3$).[93] Both polar and non-polar solvents have been used with equally favorable results. Regardless of solvent, amidomethylation with the acrylamide derivative ($R = H$) occurred with concomitant addition of trinitromethane to the carbon-carbon double bond to produce the hexanitro compound $(O_2N)_3CCH_2CH_2CONHCH_2C(NO_2)_3$. That addition to the carbon-carbon double bond was faster than the ester cleavage was indicated by the observation that, with one equivalent of trinitromethane, the principal product was the adduct $(O_2N)_3CCH_2CH_2CONHCH_2OCOC_6H_5$. Trinitromethane did not add to the methacrylamide ($R = CH_3$). The same workers examined the reaction of trinitromethane with N,N′-diacetoxymethyl- and N,N′-dibenzoxymethyl-fumaramide and reported a 2% and 47% yield, respectively, of the product **54**. Since water was the solvent for the reaction of the N,N′-diacetoxy compound, the formation of the fumaramide **54** may occur by a mechanism different from that involving the dibenzoate. The latter reaction was carried out under anhydrous conditions.

The trichloromethyl derivative **58** reacted with potassium cyanide to give the unsaturated nitrile **59** in 53% yield.[94] Elimination-substitution reactions of this type take place in a number of reactions with alkali cyanides. The reactions are described on pp. 82–83. In the same article it was reported that the unsaturated trichloromethyl derivative **60** and the

$$\underset{\mathbf{58}}{C_2H_5O_2CNHCH(OCOCH_3)CCl_3} \xrightarrow{KCN} \underset{\mathbf{59}}{C_2H_5O_2CNHC(CN){=}CCl_2}$$

[92a] Monti, Osti, and Piras, *Gazz. Chim. Ital.*, **71**, 654 (1941) (*Chem. Zentr.*, **1942**, I, 3096); Monti, *Boll. Sci. Fac. Chim. Ind. Bologna*, **1940**, 133 [*C.A.*, **34**, 7292 (1940)].

[93] Feuer and Lynch-Hart, *J. Org. Chem.*, **26**, 587 (1961).

[94] Diels and Seib, *Ber.*, **42**, 4062 (1909); Diels and Gukassianz, *ibid.*, **43**, 3314 (1910).

potassium salt of diethyl malonate gave the triethyl ester **61** in 35% yield. It was later shown, however, that **60** was actually the symmetrical

$$\underset{\mathbf{60}}{C_2H_5O_2CN{=}CHCCl_3} + K[CH(CO_2C_2H_5)_2] \rightarrow \underset{\mathbf{61}}{C_2H_5O_2CNHCH[CH(CO_2C_2H_5)_2]CCl_3}$$

ether **62**.[95] This reaction, therefore, appears to be the only one in which an ether has been utilized, albeit unknowingly, as an α-amidoalkylating agent of an aliphatic carbon atom.

$$\underset{\mathbf{62}}{(C_2H_5O_2CNHCHCCl_3)_2O}$$

**With N-Halomethyl- and N-α-Haloalkyl-amide Derivatives.** The N-halomethyl- or N-α-haloalkyl-amides are the reagents of choice for the amidoalkylation of aliphatic compounds,[33,96–101] in contrast to the aromatic series where N-alkylolamides are more commonly employed. The halomethylamides often are prepared just before use and, with little or no purification, are added to the alkali metal derivative of the active methylene compound suspended in some inert solvent. Many structural types have been utilized. A partial list would include malonic esters and their monosubstituted derivatives, cyclic and acyclic β-diketones, cyano-acetic esters, acetoacetic esters and their monosubstituted derivatives, and nitroacetic ester. Reported yields exhibit wide ranges.

No systematic investigation of the important reaction conditions is available, but the N-halomethyl derivatives of aliphatic amides generally give lower yields than do N-halomethylbenzamides and N-halomethyl-phthalimides. For example, diethyl methylmalonate condenses with N-chloromethylformamide, with N-chloromethylacetamide, with N-chloromethylbenzamide, and with N-chloromethylphthalimide in 20, 31, 77, and 81% yields, respectively.[97,100]

With N-chloro- and N-bromo-methylphthalimide, monosubstituted derivatives of active methylene compounds appear to react more smoothly than do unsubstituted ones. The amidomethylation product **63** was not obtained from N-bromomethylphthalimide and the sodium derivative of diethyl malonate in benzene. Instead, the tetra ester **64**, phthalimide,

[95] Feist, *Ber.*, **45**, 945 (1912).

[96] Böhme, Broese, Dick, Eiden, and Schünemann, *Chem. Ber.*, **92**, 1599 (1959).

[97] Böhme, Broese, and Eiden, *Chem. Ber.*, **92**, 1258 (1959).

[98] Böhme, Dick, and Driesen, *Arch. Pharm.*, **294**, 312 (1961).

[99] Böhme and Eiden, *Arch. Pharm.*, **292**, 642 (1959).

[100] Böhme and Eiden, Ger. pat. 1,025,883 [*C.A.*, **54**, 9773, (1960)].

[101] Böhme, Eiden, and Schünemann, *Arch. Pharm.*, **294**, 307 (1961).

and sodium bromide were formed.[102] This result is compatible with the

$o$-$C_6H_4(CO)_2NCH_2CH(CO_2C_2H_5)_2$ **63**

$CH_2[CH(CO_2C_2H_5)_2]_2$ **64**

observation that phthalimide is cleaved from 2-phthalimidomethylcyclohexane-1,3-dione in the presence of excess active methylene compound to produce methylenebiscyclohexane-1,3-dione **65**.[103] In both reactions it seems likely that a second molecule of the enolate ion displaces a phthalimide anion (e.g., from **63**), giving the observed products.

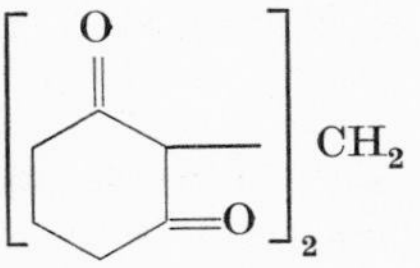

**65**

Failure to obtain the normal substitution product from N-bromomethylphthalimide and sodioacetoacetic ester also has been reported.[104] In contrast, good yields have been obtained from the reaction of N-chloromethylphthalimide with the sodium derivatives of many *monosubstituted* active methylene compounds;[97] and the amidomethylated compound **66** has been prepared in 73% yield from N-bromomethylphthalimide and the sodium salt of 3-phenylbenzofuran-2-one.[105]

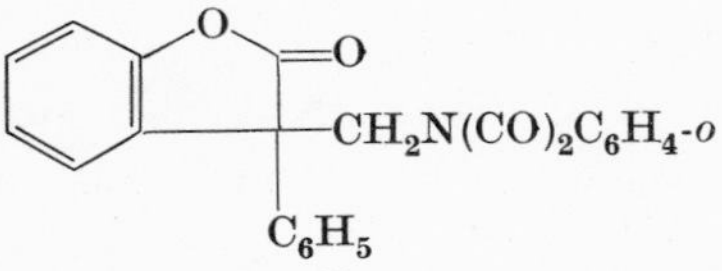

$CH_2N(CO)_2C_6H_4$-$o$, $C_6H_5$

**66**

An interesting set of reactions has been reported in which various substituted aromatic α-haloalkylamides of the type $ArCONHCHClCCl_3$ are treated in dry acetone with 2 moles of potassium cyanide to give, not the simple substitution products, but products of the type

$$ArCONHC(CN){=}CCl_2$$

**67**

[102] Vinkler and Szabó, *Magy. Kem. Folyoirat*, **56**, 209 (1950) [*C.A.*, **45**, 7984 (1951)]; published in Russian in *Acta Chim. Hung.*, **1**, 103 (1951) [*C.A.*, **46**, 2500 (1952)].

[103] Hellmann and Aichinger, *Chem. Ber.*, **92**, 2122 (1959).

[104] Pucher and Johnson, *J. Am. Chem. Soc.*, **44**, 817 (1922).

[105] H. E. Zaugg and R. W. De Net, unpublished data.

in which elimination has occurred as well.[106,107] One mole of potassium cyanide reacts rapidly, but no pure product can be isolated. Apparently the second mole of cyanide dehydrochlorinates the initially formed substitution products to the more readily isolable unsaturated nitriles **67**. The yields of the nitriles **67** are only fair, but their hydrolysis to the corresponding acids can be achieved in essentially quantitative yields.

**With N,N′-Methylenedisulfonamides and N,N′-Ethylidenebisurethan or Their Precursors.** Only a few successful reactions of this type have been described.[85,108] In general, N,N′-alkylidenebisamides fail to undergo amidomethylation with active methylene compounds.[109,110] Instead they give the corresponding alkylidene derivatives, i.e., $RCH(CHX_2)_2$, analogous to the methylenebis compounds **64** and **65.**

The condensation of cyclohexane-1,3-dione in sulfuric acid with N,N′-methylenedibenzenesulfonamide or N,N′-methylenedi-*p*-toluenesulfonamide yielded the corresponding diketones **68** in 18 and 19% yields, respectively.[85]

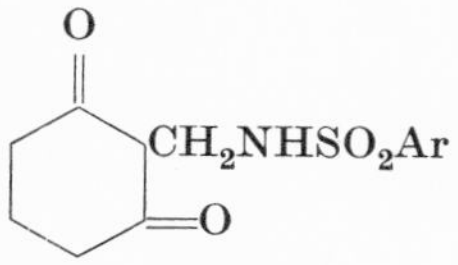

2-($CH_2NHSO_2Ar$)cyclohexane-1,3-dione (structure: six-membered ring with two C=O groups)

**68**

$C_2H_5OCONHCH(CH_3)CH(COCH_3)_2$

**69**

Ethyl carbamate, acetaldehyde, and pentane-2,4-dione were condensed to give the urethan **69.**[108] This is apparently the only reported case of amidoalkylation with an alkylidenebisamide derivative.

**With N,N′-Arylidenebisamides or Their Precursors.** The reactivity of these amide derivatives though diminished is still sufficient for condensation with most of the reactive methylene compounds. Two methods are generally used to effect the condensation: the preformed arylidenebisamide is condensed with the active methylene compound, or the aldehyde and amide (or the ammonium salt of the corresponding carboxylic acid) are treated with the active methylene compound. Both methods usually involve acetic anhydride or acetic acid-acetic anhydride mixtures as the solvent, although early workers used ethanol and hydrochloric acid.[108,111]

[106] Hirwe and Deshpande, *Proc. Indian Acad., Sci.* **13A,** 277 (1941) [*C.A.*, **35,** 6250 (1941)].

[107] Hirwe and Rana, *J. Indian Chem. Soc.*, **17,** 481 (1940) [*C.A.*, **35,** 2130 (1941)].

[108] Bianchi, *Gazz. Chim. Ital.*, **42** (I), 499 (1912) (*Chem. Zentr.*, **1912,** II, 329).

[109] Stefanović, Prekajski, and Mihailović, *Ber. Chem. Ges. Belgrade*, **22,** 113 (1957) (*Chem. Zentr.*, **1959,** 9883).

[110] Stefanović, Stefanović, and Milanović, *Ber. Chem. Ges. Belgrade*, **20,** 313 (1955) (*Chem. Zentr.*, **1957,** 6134).

[111] Bianchi and Schiff, *Gazz. Chim. Ital.*, **41** (II), 81 (1911) (*Chem. Zentr.*, **1911,** II, 1919).

The condensation of arylidenebisamides with ethyl nitroacetate clarified the reaction course. It was shown that, when a mixture of benzaldehyde, acetamide, and ethyl nitroacetate in acetic anhydride was heated, the ester **70** was obtained in 61% yield.[112] The course of the reaction was the same when N,N'-benzylidenebisacetamide was substituted for the benzaldehyde and acetamide. Under the latter conditions, however, a better yield (85%) of **70** was obtained.

$$C_6H_5CH(NHCOCH_3)_2 + O_2NCH_2CO_2C_2H_5 \xrightarrow{(CH_3CO)_2O} \underset{\mathbf{70}}{CH_3CONHCH(C_6H_5)CH(NO_2)CO_2C_2H_5} + CH_3CONH_2$$

N,N'-Benzylidenebisacetamide also condensed with nitromethane in analogous fashion, producing the nitro compound

$$CH_3CONHCH(C_6H_5)CH_2NO_2$$

in 32% yield.[113] As expected, nitromethane is less reactive than ethyl nitroacetate and requires a longer reaction time even for lower yields.

Ethyl acetoacetate has been condensed with aldehydes and amides in ethanol and hydrochloric acid[111] as well as with N,N'-arylidenebisamides in acetic anhydride.[114] Compounds of the type

$$RCONHCH(Ar)CH(COCH_3)CO_2C_2H_5$$

are obtained. Like acetoacetic ester itself, the condensation products exist in two tautomeric forms. As a consequence, they decolorize bromine water and produce a reddish brown coloration with alcoholic ferric chloride solution. Acid cleavage leads to $\beta$-aryl-$\beta$-acylaminopropionic acids, $RCONHCH(Ar)CH_2CO_2H$, in good yields, but ketone cleavage does not occur.

The reaction of arylidenebisamides with malonic esters in acetic anhydride yields the expected condensation products.[115]

$$RCONHCH(Ar)CH(CO_2R')_2$$

The reaction time appears to be important. For example, N,N'-benzylidenebisacetamide gives a 62% yield of amidoalkylated product in 3 hours, but after 9 hours only an 11% yield can be obtained. The major products isolated from these prolonged reactions are the arylidenemalonic esters $ArCH{=}C(CO_2R')_2$ resulting from the elimination of amide. Partial

[112] Stefanović and Bojanović, *J. Org. Chem.*, **17**, 816 (1952).

[113] Stefanović, Bojanović, and Sirotanović, *J. Org. Chem.*, **17**, 1110 (1952).

[114] Stefanović and Stefanović, *J. Org. Chem.*, **17**, 1114 (1952).

[115] Stefanović, Mihailović, and Stefanović, *J. Org. Chem.*, **18**, 1467 (1953).

hydrolysis of the amidoalkylated esters produces the malonic acids which decarboxylate to give β-aryl-β-acylaminopropionic acids

$$RCONHCH(Ar)CH_2CO_2H$$

in excellent yields.[115] Diethyl ethylmalonate reacts with N,N′-benzylidenebisacetamide in an analogous fashion. Although partial hydrolysis of the product leads to a half-ester of malonic acid, more severe conditions result in both decarboxylation and amide elimination to give α-ethylcinnamic acid, $C_6H_5CH{=}C(C_2H_5)CO_2H$.[115]

Reactions of cyanoacetic esters with arylidenebisamides do not take the expected course but lead to the elimination products $ArCH{=}C(CN)CO_2R$.[116] These condensations proceed best in the absence of any solvent or catalyst.

Hippuric acid as an active methylene compound condenses with arylidenebisamides in glacial acetic acid or in acetic anhydride to give, as the main product, β-aryl-α,β-diacylaminopropionic acids, **73**, in two diastereomeric forms.[117] In several reactions the corresponding azlactones **72** are formed; and in all instances small amounts (~10%) of the

$$C_6H_5CONHCH_2CO_2H \xrightarrow{(RCONH)_2CHAr} \underset{\textbf{71}}{\text{ArCH=C—CO, N–O ring, C–}C_6H_5} + \underset{\textbf{72}}{\text{RCONHCH(Ar)CH—CO, N–O ring, C–}C_6H_5} + \underset{\textbf{73}}{RCONHCH(Ar)CH(NHCOC_6H_5)CO_2H}$$

azlactones **71** of β-arylacrylic acids are obtained. The complicated configurational relationships among these products have not been completely elucidated. Hippuric acid does not react with *o*- or *p*-nitrobenzylidenebisacetamide, but the *meta* isomer gives the corresponding azlactone **71** ($Ar = m\text{-}O_2NC_6H_4$).[117] The acids **73** as well as the azlactones **72** undergo acid hydrolysis to the corresponding β-aryl-α,β-diaminopropionic acids, $ArCH(NH_2)CH(NH_2)CO_2H$.

[116] Stefanović and Nikić, *J. Org. Chem.*, **17**, 1305 (1952).

[117] Stefanović and Stefanović, *J. Org. Chem.*, **21**, 161 (1956).

One heterocyclic compound which has been reported[109] to undergo reaction in acetic anhydride with N,N′-benzylidenebisacetamide is piperazine-2,5-dione. It gives the diketone **74** in 29% yield. Other heterocyclic compounds apparently condense with N,N′-benzylidenebis-

$COCH_3$ … $CH_3CONHCH(C_6H_5)$ … N … =O … O= … N … $COCH_3$

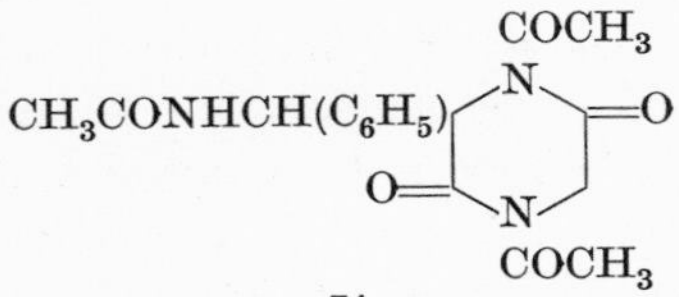

**74**

acetamide in glacial acetic acid but without leading to α-amidoalkylation products.[109,110] Thus rhodanine gives **75** in 99% yield,[110] and barbituric acid gives **76** in 91% yield.[109] The yield of the latter is lowered to 51% when the solvent is acetic anhydride.

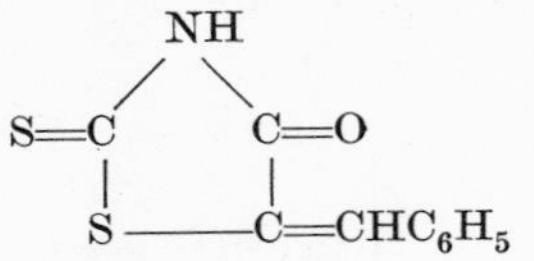

NH, S=C, C=O, S, $C{=}CHC_6H_5$

**75**

NH, O=C, C=O, HN, $C{=}CHC_6H_5$, C, O

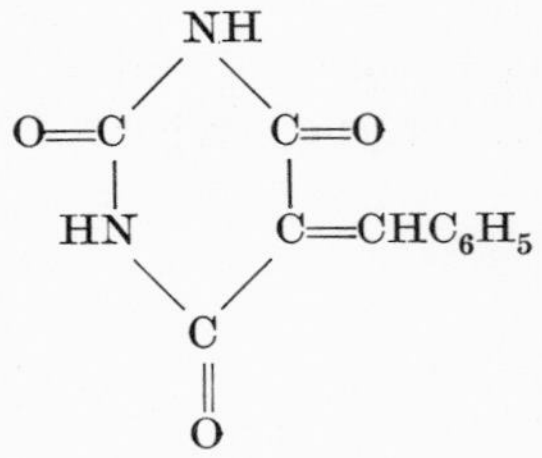

**76**

Some cyclic β-diketones are reported to condense with N,N′-benzylidenebisamides to give the corresponding 2-(α-acylaminobenzyl) derivatives in poor to moderate yields.[85] For example, 5,5-dimethylcyclohexane-1,3-dione, with N,N′-benzylidenebisacetamide, produces **77** in 12% yield. In the same manner cyclohexane-1,3-dione and N,N′-benzylidenebisbenzamide give the corresponding product **78** in 40% yield.

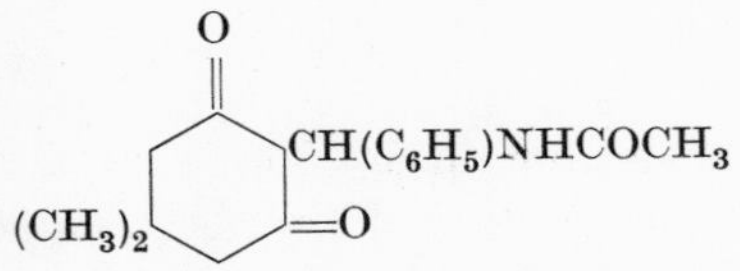

O, $CH(C_6H_5)NHCOCH_3$, $(CH_3)_2$, =O

**77**

O, $CH(C_6H_5)NHCOC_6H_5$, =O

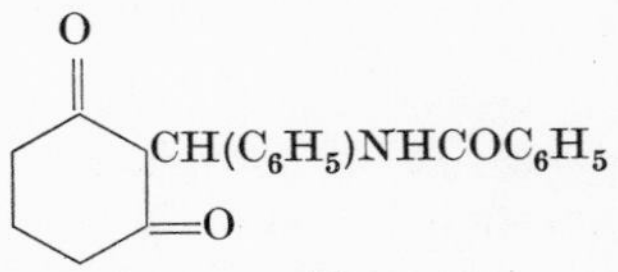

**78**

**With N-Aminomethylamides and Their Quaternary Salts.** In contrast to the aromatic systems, a number of reactive methylene compounds undergo base-catalyzed amidomethylation with N-dialkylaminomethylamides.[81,118] They include dialkyl malonates and their

[118] Hellmann and Haas, *Chem. Ber.*, **90,** 1357 (1957).

monosubstituted derivatives, ethyl acetoacetate, β-diketones, and nitrocyclohexane. For example, the last compound condenses with N-(dimethylaminomethyl)benzamide **79** in the presence of powdered sodium hydroxide in boiling toluene to give **80** in 38% yield.[118]

$$\underset{\textbf{79}}{C_6H_5CONHCH_2N(CH_3)_2} + \text{cyclo-}C_6H_{11}NO_2 \xrightarrow[C_6H_5CH_3]{NaOH} \underset{\textbf{80}}{1\text{-}O_2N\text{-}1\text{-}(CH_2NHCOC_6H_5)\text{cyclohexane}} + (CH_3)_2NH$$

Two cyano compounds have been amidoalkylated by N-diethylaminomethylbenzamide with good results.[118] Diethyl α-cyanopimelate (**81**) reacts in the expected manner to give an 85% yield of the product **82**, and ethyl α-phenylcyanoacetate undergoes analogous conversion in 74%

$$C_6H_5CONHCH_2N(C_2H_5)_2 + \underset{\textbf{81}}{NC\text{-}CH(CO_2C_2H_5)(CH_2)_4CO_2C_2H_5} \xrightarrow[C_6H_5CH_3]{NaOH} \underset{\textbf{82}}{C_6H_5CONHCH_2C(CN)(CO_2C_2H_5)(CH_2)_4CO_2C_2H_5}$$

yield. In interesting contrast, however, is the observation that compound **81** with *N-methylolbenzamide* in strong acid is converted to the N-alkylated product **83** in high yield.[84,119]

$$\underset{\textbf{83}}{C_6H_5CONHCH_2NHCOCH(CO_2C_2H_5)(CH_2)_4CO_2C_2H_5}$$

Only a few reactions of N-dialkylaminomethylphthalimides with reactive methylene compounds have been reported, and they have usually led to by-products.[81] Diethyl malonate reacts with N-dimethylaminomethylphthalimide (**84**) to yield the unstable derivative **85**. The chief products formed were phthalimide and diethyl methylenemalonate (**86**). Attempts to add phthalimide to **86** to produce **87** were not successful.

[119] English and Clapp, *J. Am. Chem. Soc.*, **67**, 2262 (1945).

On the contrary, evidence indicated that, under the reaction conditions, **87**, like **85**, undergoes elimination to give **86**.

$$o\text{-}C_6H_4(CO)_2NCH_2N(CH_3)_2 + CH_2(CO_2C_2H_5)_2 \xrightarrow{\text{NaOH}} (CH_3)_2NCH_2CH(CO_2C_2H_5)_2$$

**84** **85**

$$\downarrow -(CH_3)_2NH$$

$$o\text{-}C_6H_4(CO)_2NCH_2CH(CO_2C_2H_5)_2 \xrightarrow{-[o\text{-}C_6H_4(CO)_2NH]} CH_2{=}C(CO_2C_2H_5)_2$$

**87** **86**

The quaternary salt **88** reacted with diethyl malonate to give only a small amount of the dialkylated product **89**.[81] With sodium cyanide, however, it gave the nitrile $o$-$C_6H_4(CO)_2NCH_2CN$ in 80% yield.

$$o\text{-}C_6H_4(CO)_2NCH_2\overset{\oplus}{N}(CH_3)_3I^{\ominus} + CH_2(CO_2C_2H_5)_2 \xrightarrow[C_2H_5OH]{NaOC_2H_5} [o\text{-}C_6H_4(CO)_2NCH_2]_2C(CO_2C_2H_5)_2$$

**88** **89**

**With N-Acylimines.** Relatively few examples of this reaction have been reported. Ivanoff found that benzoyl derivatives of diaryl ketimines will add reactive nucleophilic agents such as the chloromagnesium derivative of sodium phenylacetate[19] and the lithium derivatives of acetonitrile,[120] propionitrile,[121] and butyronitrile[121] prepared *in situ* with lithium amide in liquid ammonia.

$$C_6H_5CON{=}CAr_2 \xrightarrow{C_6H_5CH(MgCl)CO_2Na} C_6H_5CONHCAr_2(C_6H_5CHCO_2H)$$

$$C_6H_5CON{=}CAr_2 \xrightarrow{RCH(Li)CN} C_6H_5CONHCAr_2(RCHCN)$$

Yields of the reactions ranged between 40 and 85%. This synthesis offers promise as a convenient entry to systems highly substituted by bulky groups.

**α-Amidoalkylation of Ethyl Acetoacetate with N,N′-Alkylidene- and N,N′-Arylidene-bisureas or Their Precursors. The Biginelli Pyrimidine Synthesis.** Biginelli treated ethyl acetoacetate with urea

[120] Ivanoff, Markov, and Dobrev, *Compt. Rend. Acad. Bulgare Sci.*, **15**, 403 (1962) [C.A., **60**, 4112 (1964)].

[121] Ivanoff, Konstantinova, and Popandova, *Compt. Rend. Acad. Bulgare Sci.*, **15**, 617 (1962) [C.A., **59**, 2699 (1963)].

and benzaldehyde and obtained 5-carbethoxy-2-keto-6-methyl-4-phenyl-1,2,3,4-tetrahydropyrimidine (**90**).[122] He extended the reaction to other

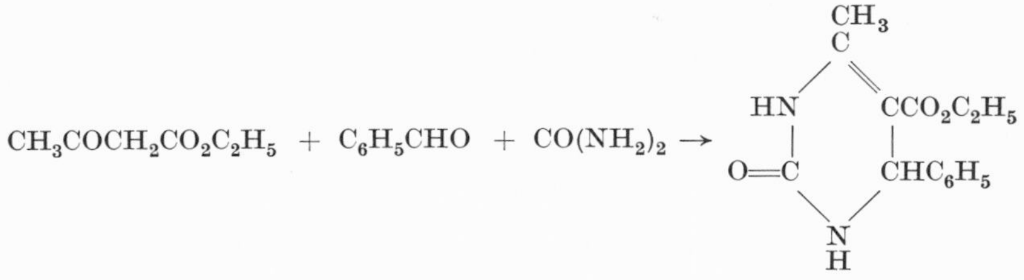

$$CH_3COCH_2CO_2C_2H_5 + C_6H_5CHO + CO(NH_2)_2 \rightarrow$$

$CH_3$, C, HN, $CCO_2C_2H_5$, O=C, $CHC_6H_5$, N, H

**90**

aldehydes and made the first contribution to an understanding of the course of the reaction by using the corresponding arylidenebisureas, $(H_2NCONH)_2CHAr$, in place of the aldehydes and urea.[122–124] Under the original conditions the reaction gave only moderate yields, but later workers, by using a 1-mole excess of ethyl acetoacetate, were able to increase the yields appreciably.[125] Using benzaldehyde, they also showed that the substituted dihydrolutidine **91** was a by-product and that the

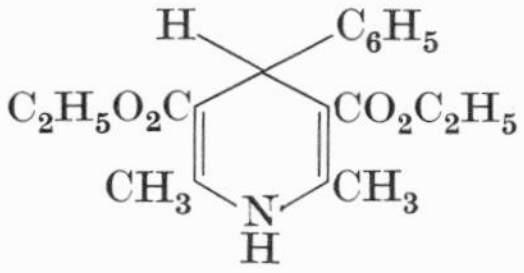

H, $C_6H_5$, $C_2H_5O_2C$, $CO_2C_2H_5$, $CH_3$, N, $CH_3$, H

**91**

course of the synthesis can be described as follows. Preformed N,N′-benzylidenebisurea reacts with ethyl acetoacetate to give **90**. The molecule of urea which is eliminated in this process condenses with a second molecule of ethyl acetoacetate to give ethyl β-ureidocrotonate, $CH_3C(NHCONH_2){=}CHCO_2C_2H_5$, which in turn reacts with benzaldehyde to give the Biginelli product **90**. However, some of the ureidocrotonate hydrolyzes to ethyl β-aminocrotonate, ammonia, and carbon dioxide. In the presence of ammonia, conditions become favorable for the Hantzsch synthesis and this accounts for the side reaction leading to the pyridine derivative **91**.

Further work confirmed these results and showed (*a*) that condensation of urea with ethyl α-benzylideneacetoacetate plays no part in the reaction; and (*b*) that under the reaction conditions the ureidocrotonate is partially

[122] Biginelli, *Ber.*, **24,** 1317 (1891).
[123] Biginelli, *Ber.*, **24,** 2962 (1891).
[124] Biginelli, *Gazz. Chim. Ital.*, **23** (I), 360 (1893); *Ber.*, **26,** 447 (1893).
[125] Hinkel and Hey, *Rec. Trav. Chim.*, **48,** 1280 (1929).

cleaved to urea and ethyl acetoacetate.[126,127] It was also demonstrated that the Biginelli synthesis is catalyzed by mineral acid, to which the yield is approximately proportional. In the absence of acid, reaction proceeds extremely slowly.

Thiourea can be substituted for urea in this process. With benzaldehyde the thiopyrimidine **92** is formed.[125]

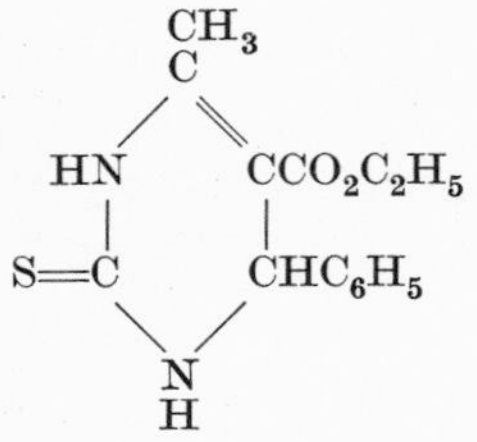

**92**

**α-Amidoalkylation of Active Methylene Compounds Other than Ethyl Acetoacetate with Urea and Aldehydes.** Some interesting work using aromatic aldehydes, urea, and β-diketones in acidic media has led to compounds of the type **93**.[128,129] Nine aromatic aldehydes were used successfully with pentane-2,4-dione ($R = R' = CH_3$), and five

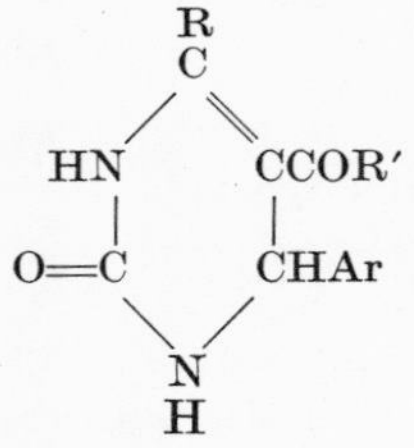

**93** 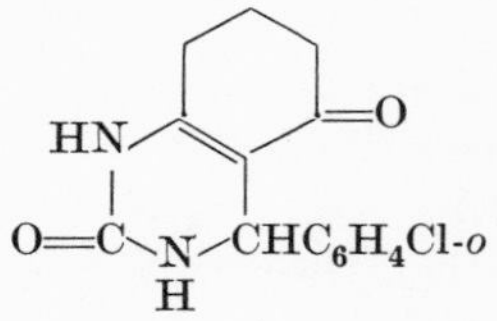 **94**

aromatic aldehydes were employed with 1-phenylbutane-1,3-dione. The yields were fair (40–90%), but the positions of the substituents in the butanedione products were not established. Propionaldehyde and urea react with pentane-2,4-dione in an analogous fashion (32% yield), but heptanal and isovaleraldehyde do not.

Cyclohexane-1,3-dione reacts with *o*-chlorobenzaldehyde and urea to give the bicyclic compound **94** in 80% yield.[130] The reaction is not general.

[126] Folkers, Harwood, and Johnson, *J. Am. Chem. Soc.*, **54**, 3751 (1932).

[127] Folkers and Johnson, *J. Am. Chem. Soc.*, **55**, 3784 (1933).

[128] Chi and Ling, *Acta Chim. Sinica*, **22**, 177 (1956); *Sci. Sinica*, **6**, 247 (1957) [*C.A.*, **52**, 396 (1958)].

[129] Chi and Wu, *Acta Chim. Sinica*, **22**, 184 (1956) [*C.A.*, **52**, 6360 (1958)].

[130] Chi and Wu, *Acta Chim. Sinica*, **22**, 188 (1956) [*C.A.*, **52**, 6360 (1958)].

*n*-Butyraldehyde, *p*-methoxybenzaldehyde, and *p*-dimethylaminobenzaldehyde all condensed preferentially with 2 molecules of the diketone to give, after intramolecular dehydration, tricyclic compounds of type **95.**

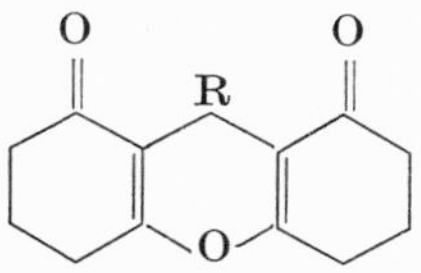

**95**

Phenylacetaldehyde did not yield the expected Biginelli product **96.** Instead the pyrimidine **97** was obtained.[131] Clearly this arises from the substitution of a molecule of phenylacetaldehyde for the usual ethyl acetoacetate. The same pyrimidine, **97,** as expected, was also formed

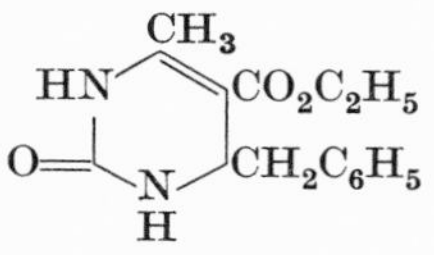

**96** 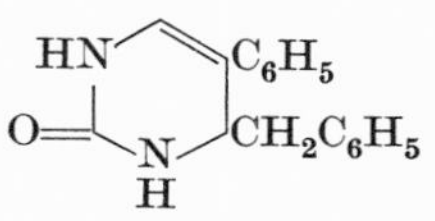 **97**

when phenylacetaldehyde and urea were condensed in the absence of acetoacetic ester.

**Amidomethylation of Potassium Cyanide with Arenesulfonamidomethanesulfonates.** Only one report concerning arenesulfonamidomethanesulfonates seems to have appeared in the literature.[132] The sulfonate **98,** prepared by heating *m*-benzenedisulfonamide with sodium hydroxymethanesulfonate, on treatment with excess aqueous

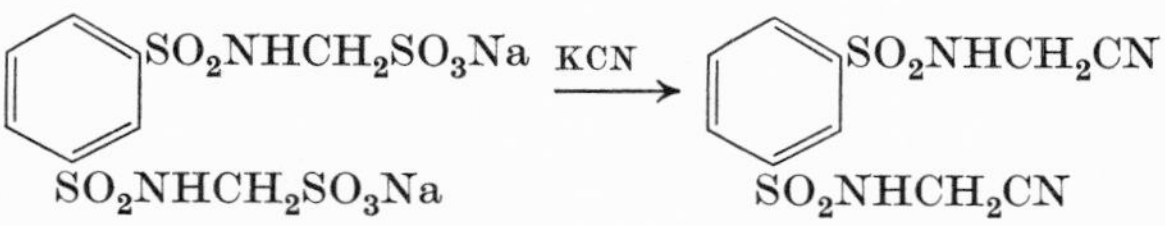

**98** $\xrightarrow{\text{KCN}}$ **99**

potassium cyanide produced the dinitrile **99** in 75% yield. The nitrile is readily hydrolyzed to the corresponding dicarboxylic acid. The reaction with benzenesulfonamide proceeds in an analogous fashion.

## Preparation of the Electrophilic Reagents

If there is any process within the scope of this review which has received adequate attention, it is the reaction of formaldehyde with cyclic and

[131] Folkers and Johnson, *J. Am. Chem. Soc.*, **55,** 3361 (1933).
[132] Knoevenagel and Lebach, *Ber.*, **37,** 4094 (1904).

acyclic mono- and poly-amides.[133] Studies of this reaction have involved the labeling of proteins and peptides,[79] the modification of nylon polymers,[134,135] the preparation of cross-linking agents for crease-proofing or softening cotton textiles,[136,137] and the development of waterproofing agents for textile materials.[138–140] Furthermore the commercial importance of urea-formaldehyde and urea-acetaldehyde[141] polymers has stimulated a number of fundamental investigations in this area.[142–149]

**N-Methylol-amides and -imides.** Amides and imides react reversibly with formaldehyde in acid, neutral, or basic media. The

$$RCONH_2 + CH_2O \rightleftharpoons RCONHCH_2OH$$

reaction is catalyzed by either acid or base,[148] but, over a relatively wide *p*H range (2–12), the activation energy of the reverse reaction remains greater than that of the forward process by a nearly constant amount (~5 kcal./mole).[144,145,148] Although the equilibrium favoring the N-methylolamide is practically unaffected by *p*H, elevated temperatures tend to favor the dissociation reaction which has the higher activation energy. For this reason the isolation of products often must be carried out at or near room temperature,[35,84,150] and purifications by recrystallization should be conducted with a minimum of heating.[151,152] In the absence of large concentrations of formaldehyde, contact with aqueous acid or alkali, as expected, causes reversal of the methylolation reaction even at room temperature.[79]

Condensation of amides with formaldehyde under neutral or basic conditions is preferred because acid catalysis often leads to further

[133] Walker, *Formaldehyde*, 2nd ed., Reinhold, New York, 1953, pp. 290–309.
[134] Cairns, Foster, Larchar, Schneider, and Schreiber, *J. Am. Chem. Soc.*, **71**, 651 (1949).
[135] Cairns, Gray, Schneider, and Schreiber, *J. Am. Chem. Soc.*, **71**, 655 (1949).
[136] Tovey, *Textile Res. J.*, **31**, 185 (1961) [*C.A.*, **55**, 11859 (1961)].
[137] Vail, Frick, Murphy, and Reid, *Am. Dyestuff Reptr.*, **50**, 19 (1961) [*C.A.*, **55**, 11860 (1961)].
[138] Weaver, Schuyten, Frick, and Reid, *J. Org. Chem.*, **16**, 1111 (1951).
[139] Sallmann and Graenacher, U.S. pat. 2,448,125 [*C.A.*, **43**, 1434 (1949)].
[140] Baldwin and Walker, U.S. pat. 2,146,392 [*C.A.*, **33**, 3495 (1939)].
[141] Scheffer, Ziechmann, and Kuntze, *Naturwiss.*, **44**, 52 (1957).
[142] Crowe and Lynch, *J. Am. Chem. Soc.*, **70**, 3795 (1948); **71**, 3731 (1949); **72**, 3622 (1950).
[143] de Jong and de Jonge, *Rec. Trav. Chim.*, **72**, 202 (1953).
[144] de Jong and de Jonge, *Rec. Trav. Chim.*, **71**, 643 (1952).
[145] Koskikallio, *Acta Chem. Scand.*, **10**, 1267 (1956).
[146] Květoň, *Chem. Listy*, **50**, 94 (1956) [*C.A.*, **50**, 4602 (1956)].
[147] Květoň, *Collection Czech. Chem. Commun.*, **21**, 593 (1956).
[148] Ugelstad and de Jonge, *Rec. Trav. Chim.*, **76**, 919 (1957).
[149] Zigeuner and Hoselmann, *Monatsh. Chem.*, **88**, 159 (1957).
[150] Weber and Tollens, *Ann.*, **299**, 340 (1898).
[151] Sachs, *Ber.*, **31**, 1225 (1898).
[152] Weber, Pott, and Tollens, *Ber.*, **30**, 2514 (1897).

transformation of the N-methylolamides to the corresponding ethers, $(RCONHCH_2)_2O$, or methylenebisamides, $(RCONH)_2CH_2$. Kinetic studies indicate that the base-catalyzed reactions occur by attack of the amide anion, $RC(=NH)O^{\ominus} \leftrightarrow RC(=O)\overset{\ominus}{N}H$, at the carbonyl carbon atom of formaldehyde.[142,148] This finding explains why reactivity toward formaldehyde under these conditions generally follows the order of increasing amide acidity, e.g.,

$$CH_3CONH_2 < C_6H_5CONH_2 < o\text{-}C_6H_4(CO)_2NH$$

Early work indicated that susceptibility to N-methylolation by formaldehyde was generally restricted to primary amides.[41] The monomethylol derivative of 1,3-dimethylurea[153] and three derivatives of the type $RCON(CH_2OH)_2$[41,42] were originally the sole exceptions to this rule. Recent work, however, has demonstrated that the reactivity of N-mono-alkylated amides toward formaldehyde is greater than had been supposed.[32,154,154a] Although the reaction may not be complete and products may not be readily isolable in crystalline form, many N-methylol-N-alkylamides can be obtained in good yields as crude oils and can be used in that form.

Unsubstituted[155,156] and N-monosubstituted sulfonamides[157] also react readily with formaldehyde to give the N-methylol derivatives. Indications are, however, that they tend to disproportionate to the methylene-bisamides and also undergo polymerization more readily than their carboxamide analogs.[156]

**N-α-Alkylolamides.** Most aliphatic and all aromatic aldehydes do not behave toward amides as formaldehyde does. Reaction usually does not stop at the N-alkylol stage, RCONHCHOHR′, but progresses further to the alkylidene- or arylidene-bisamide, $(RCONH)_2CHR'$. 1-Ethylol- and 1,3-diethylol-urea have been prepared in crude form from acetaldehyde and urea, but, aside from these, the only exceptions to the rule above are the α-halogenated aldehydes.[141] Chloral, for example, reacts readily with amides, either spontaneously or on heating with or without an acid catalyst, to give the N-alkylolamides, $RCONHCHOHCCl_3$, or "chloral-amides".[158] Sulfonamides also react readily.[159] Other aldehydes that

[153] Einhorn and Hamburger, *Ann.*, **361,** 122 (1908).
[154] Vail, Moran, and Moore, *J. Org. Chem.*, **27,** 2067 (1962).
[154a] Chupp and Speziale, *J. Org. Chem.*, **28,** 2592 (1963).
[155] Feibelmann, Ger. pat. 403,718 (*Chem. Zentr.*, **1925,** I, 440).
[156] Hug, *Bull. Soc. Chim. France*, **1934,** 990.
[157] Wiedemann and Strassberger, Ger. pat. 947,795 [*C.A.*, **53,** 2262 (1959)].
[158] Jacobson, *Ann.*, **157,** 245 (1871).
[159] Lichtenberger, Fleury, and Barette, *Bull. Soc. Chim. France*, **1955,** 669.

have been shown to undergo the same condensation are bromal, bromodichloroacetaldehyde, dichloroacetaldehyde, dibromoacetaldehyde, and 2,2,3-trichlorobutanal. Numerous examples of these alkylolamides have been reported by Feist[95,160] and by groups led by Chattaway,[161–163] Meldrum,[164–167] and Hirwe.[168–171] Through 1962, however, extension of the reaction to the presently available trifluoroacetaldehyde apparently had not been reported. Also, one might expect some of the recently accessible polyfluorinated ketones to react analogously, especially in view of a recent report showing that carbamyl derivatives of fluorinated aldehydes and ketones are stable.[73] In this instance, however, they were prepared by the action of an alcohol or an amine on an intermediate isocyanate. A number of fluoral carbamates and ureas were prepared in this way.

$$CF_3CHO + HNCO \longrightarrow OCNCHOHCF_3 \xrightarrow{RNH_2} RNHCONHCHOHCF_3$$

Of incidental interest is another recent report showing that chloral reacts with potassium cyanide or chloral cyanohydrin in aqueous alkali to give a chloral-oxazolidone derivative.[172]

$$3CCl_3CHO + KCN \xrightarrow{KOH} \begin{array}{ccc} CO & \text{—} & NCHOHCCl_3 \\ | & & | \\ Cl_3CCH & & CHCCl_3 \\ & \diagdown\ \diagup & \\ & O & \end{array}$$

As expected, these N-alkylolamides, like N-methylolamides, are unstable. On being warmed in water or heated above the melting point, they dissociate to amide and aldehyde.[171,173] In alkaline media this occurs at room temperature.[171] Hirwe and Rana have studied the effect

[160] Feist, *Ber.*, **47,** 1173 (1914).

[161] Chattaway and James, *Proc. Roy. Soc.* (*London*), **A134,** 372 (1931) [*C.A.*, **26,** 1249 (1932)].

[162] Chattaway and James, *Proc. Roy. Soc.* (*London*), **A137,** 481 (1932) [*C.A.*, **26,** 5549 (1932)].

[163] Chattaway and James, *J. Chem. Soc.*, **1934,** 109.

[164] Meldrum and Bhojraj, *J. Indian Chem. Soc.*, **13,** 185 (1936) [*C.A.*, **30,** 5940 (1936)].

[165] Meldrum and Tata, *J. Univ. Bombay*, **6,** Pt. II, 120 (1937) [*C.A.*, **32,** 3761 (1938)].

[166] Meldrum and Deodhar, *J. Indian Chem. Soc.*, **11,** 529 (1934) [*C.A.*, **29,** 136 (1935)].

[167] Meldrum and Pandya, *J. Univ. Bombay*, **6,** Pt. II, 114 (1937) [*C.A.*, **32,** 3760 (1938)].

[168] Hirwe and Deshpande, *Proc. Indian Acad. Sci.*, **13A,** 275 (1941) [*C.A.*, **35,** 6250 (1941)].

[169] Hirwe, Gavankar, and Patil, *Proc. Indian Acad. Sci.*, **11A,** 512 (1940) [*C.A.*, **34,** 7882 (1940)].

[170] Hirwe and Kulkarni, *Proc. Indian Acad. Sci.*, **13A,** 49 (1941) [*C.A.*, **35,** 5502 (1941)].

[171] Hirwe and Rana, *Ber.*, **72,** 1346 (1939).

[172] Bowman, Campbell, and Tanner, *J. Chem. Soc.*, **1963,** 692.

[173] Bischoff, *Ber.*, **7,** 628 (1874).

of ring substitution on the ease of formation and stability of the N-alkylolamides, $ArCONHCHOHCCl_3$, derived from chloral and aromatic amides.[171] Some of their findings are unusual and may be of some theoretical interest.

Cyclic N-α-alkylolamides are formed by hydride reduction of certain dicarboxylic imides. α,α-Disubstituted succinimides and glutarimides are reduced by lithium aluminum hydride at the unhindered carbonyl group.[174] Similarly, phthalimides can be reduced selectively with sodium borohydride.[175] Phthalide and *o*-hydroxymethylbenzamides often appear as by-products, however.

```
        NH                                  NH
       /  \                                /  \
     CO    CO        LiAlH4              CO    CHOH
     |     |       ---------->           |     |
C6H5C(R)---CH2      (C2H5)2O        C6H5C(R)---CH2
```

$$o\text{-}C_6H_4(CO)_2NR \xrightarrow[CH_3OH]{NaBH_4} C_6H_4\begin{matrix}CHOH\\ \rangle NR\\ CO\end{matrix} + C_6H_4\begin{matrix}CH_2OH\\ CONHR\end{matrix} + C_6H_4\begin{matrix}CH_2\\ \rangle O\\ CO\end{matrix}$$

Only one α-alkylolamide appears to have been used directly for the alkylation of a carbon atom.[89] It is 2-hydroxypteridine hydrate (**51**; see p. 79), prepared by the condensation of 2-hydroxy-4,5-diaminopyrimidine with glyoxal in aqueous solution.[176] The more common chloralamides have not been employed as α-amidoalkylating reagents. Instead they have been converted to the more reactive halides, $RCONHCHXCX_3$, with which successful alkylation is more likely.

**N-Halomethyl- and N-α-Haloalkyl-amides and -imides.** *A. N-Halomethyl Derivatives.* Most N-chloro- or N-bromo-methylamides or -imides have been prepared by treatment of the corresponding N-methylol derivatives with the reagents usually employed for making acid halides from carboxylic acids. Thionyl chloride alone,[177–180] in ether,[181,182] in chloroform,[183] or in benzene[184] provides the most generally convenient reagent for this transformation. However, phosphorus pentachloride or

[174] Tagmann, Sury, and Hoffmann, *Helv. Chim. Acta*, **37**, 185 (1954).

[175] Horii, Iwata, and Tamura, *J. Org. Chem.*, **26**, 2273 (1961).

[176] Albert, Brown, and Cheeseman, *J. Chem. Soc.*, **1951**, 474.

[177] Shostakovskii, Sidel'kovskaya, and Zelenskaya, *Bull. Acad. Sci. USSR*, **1957**, 762 (*Chem. Zentr.*, **1959**, 17131).

[178] Zinner, Hübsch, and Burmeister, *Chem. Ber.*, **90**, 2246 (1957).

[179] Zinner, Schmitt, Schritt, and Rembarz, *Chem. Ber.*, **90**, 2852 (1957).

[180] Zinner and Spangenberg, *Chem. Ber.*, **91**, 1432 (1958).

[181] Zinner and Schritt, *J. Prakt. Chem.*, [4] **15**, 72 (1962).

[182] Zinner, Zelck, and Rembarz, *J. Prakt. Chem.*, [4] **8**, 150 (1959).

[183] Stavrovskaya and Kolosova, *J. Gen. Chem. USSR* (*Engl. Transl.*), **30**, 711 (1960).

[184] Sidel'kovskaya, Zelenskaya, and Shostakovskii, *Bull. Acad. Sci., USSR, Div. Chem. Sci.* (*Engl. Transl.*), **1959**, 868 [*C.A.*, **54**, 1286 (1960)].

pentabromide alone,[185] in ether,[33,97,99,100,186] in chloroform,[66,187] in a carbon tetrachloride-acetyl chloride mixture,[65] or in dioxane[32,33] have often been used. Phosphorus trichloride or tribromide has also seen occasional service.[8,188,189]

Phosphorus oxychloride has been effective in one instance.[151] Although N-methylolphthalimide on short treatment gave only the symmetrical ether $[o\text{-}C_6H_4(CO)_2NCH_2]_2O$, prolonged heating at reflux temperature converted the ether to N-chloromethylphthalimide. N-Chloromethylphthalimide is also obtainable by warming N-ethoxymethylphthalimide[190] or N-piperidinomethylphthalimide[191] with acetyl chloride.

Certain N-methylolamides can be converted to the halides merely by heating in concentrated hydrochloric or hydrobromic acid.[104,185,192–194] This reaction is not general, for, although N-chloromethylphthalimide can be prepared in this way,[192] N-chloromethylmaleimide cannot.[8] Concentrated sulfuric acid catalyzes the reaction of N-methylolphthalimide with 48% hydrobromic acid.[70,104]

A number of N-chloromethylamides, particularly those derived from fatty acids, have been prepared directly from the amides by treatment with paraformaldehyde and dry hydrogen chloride in an inert solvent such as methylene chloride,[195] benzene,[196–200] dioxane,[201] or glacial acetic acid.[202,203] The products were not isolated but were used directly without

$$RCONH_2 + (CH_2O)_x + HCl \rightarrow RCONHCH_2Cl$$

purification. Analyses of the crude mixtures indicated conversions of the order of 20 to 30%. Crude N-chloromethylbenzenesulfonamide, $C_6H_5SO_2NHCH_2Cl$, was also prepared in this way.[201]

185 Behrend and Niemeyer, *Ann.*, **365**, 38 (1909).

186 Böhme and Eiden, *Arch. Pharm.*, **289**, 677 (1956).

187 Schlögl, Wessely, Kraupp, and Stormann, *J. Med. Pharm. Chem.*, **4**, 231 (1961) [*C.A.*, **56**, 7299 (1962)].

188 Kissinger and Ungnade, *J. Org. Chem.*, **23**, 815 (1958).

189 Sexton and Spinks, *J. Chem. Soc.*, **1948**, 1717.

190 Böhme and Müller, *Arch. Pharm.*, **296**, 54 (1963).

191 Böhme, Hartke, and Müller, *Chem. Ber.*, **96**, 595 (1963).

192 Gabriel, *Ber.*, **41**, 242 (1908).

193 Sakellarios, *J. Am. Chem. Soc.*, **70**, 2822 (1948).

194 Zinner, Herbig, and Wigert, *Chem. Ber.*, **89**, 2131 (1956).

195 Orthner, Balle, Dittus, and Wagner, U.S. pat. 2,302,885 [*C.A.*, **37**, 2389 (1943)].

196 Baldwin and Piggott, U.S. pat. 2,131,362 [*C.A.*, **32**, 9099 (1938)].

197 Pikl, U.S. pat. 2,426,790 [*C.A.*, **42**, 385 (1948)].

198 Pikl, U.S. pat. 2,331,276 [*C.A.*, **38**, 1380 (1944)].

199 Rogers, U.S. pat. 2,386,140 [*C.A.*, **40**, 1674 (1946)].

200 Wolf, U.S. pat. 2,296,412 [*C.A.*, **37**, 1278 (1943)].

201 Rust and Van Delden, U.S. pat. 2,399,603 [*C.A.*, **40**, 4229 (1946)].

202 Teramura, Niwa, and Oda, *Bull. Inst. Chem. Res., Kyoto Univ.*, **28**, 73 (1952) [*C.A.*, **47**, 3223 (1953)].

203 Teramura, Niwa, and Oda, *J. Chem. Soc. Japan, Ind. Chem. Sect.*, **57**, 405 (1954) [*C.A.*, **49**, 15734 (1955)].

Two special methods have been utilized to prepare N-chloro- and N-bromo-methylphthalimides. In one, heating phthalimidoglycyl chloride at 240° effected decarbonylation to the chloromethyl derivative;[192] in the other, N-methylphthalimide with bromine at 160–170° gave the

$$o\text{-}C_6H_4(CO)_2NCH_2COCl \xrightarrow[-CO]{240^\circ} o\text{-}C_6H_4(CO)_2NCH_2Cl$$

bromomethyl analog reportedly in quantitative yield.[151]

$$o\text{-}C_6H_4(CO)_2NCH_3 \xrightarrow[160^\circ]{Br_2} o\text{-}C_6H_4(CO)_2NCH_2Br$$

*B. N-α-Haloalkyl-amides and -imides.* Because the only α-alkylolamides that have been isolated in pure form are those derived from α-halogenated aldehydes such as chloral and bromal, most known α-haloalkylamides are derived from them. Almost invariably these have been prepared by warming the alkylolamide with phosphorus pentachloride or pentabromide in the absence of solvent.[106,160,163,204–208]

$$RCONHCHOHCCl_3 \xrightarrow{PBr_5} RCONHCHBrCCl_3 + POBr_3$$

In order to introduce a structural variation in the product, an indirect procedure has been used.[208,209] For example, reduction of the alkylolamide **100** with zinc and acetic acid gave the dichlorovinylamide **101.** Either halogen acid (HX) or halogen ($X_2$) can be added to intermediate **101** to give the α-haloalkylamides **102** and **103,** respectively.

$$\underset{\mathbf{100}}{RCONHCHOHCCl_3} \xrightarrow[CH_3CO_2H]{Zn} \underset{\mathbf{101}}{RCONHCH{=}CCl_2}$$

$$\mathbf{101} \xrightarrow{HX} \underset{\mathbf{102}}{RCONHCHXCHCl_2}$$

$$\mathbf{101} \xrightarrow{X_2} \underset{\mathbf{103}}{RCONHCHXCXCl_2}$$

[204] Willard and Hamilton, *J. Am. Chem. Soc.*, **75,** 2370 (1953).

[205] Hirwe, Gavankar, and Patil, *Proc. Indian Acad. Sci.*, **13A,** 371 (1941) [*C.A.*, **36,** 1595 (1942)].

[206] Hirwe and Rana, *J. Indian Chem. Soc.*, **16,** 677 (1939) [*C.A.*, **34,** 4731 (1940)].

[207] Yelburgi, *J. Indian Chem. Soc.*, **10,** 383 (1933) [*C.A.*, **28,** 466 (1934)].

[208] Yelburgi and Wheeler, *J. Indian Chem. Soc.*, **11,** 217 (1934) [*C.A.*, **28,** 4377 (1934)].

[209] Meldrum and Vad, *J. Indian Chem. Soc.*, **13,** 117 (1936) [*C.A.*, **30,** 4815 (1936)].

Several N-α-haloalkylphthalimides have been obtained in this way from N-vinyl-[210,211] and N-propenyl-phthalimide.[212,213] For example, addition of dry hydrogen chloride to the former gave a 97% yield of N-α-chloroethylphthalimide;[211] and addition of bromine gave the corresponding dibromide, $o\text{-}C_6H_4(CO)_2NCHBrCH_2Br$.[210] The latter compound also was

$$o\text{-}C_6H_4(CO)_2NCH{=}CH_2 \xrightarrow{\text{HCl}} o\text{-}C_6H_4(CO)_2NCHClCH_3$$

prepared by brominating N-(β-bromoethyl)phthalimide (89% yield) and N-ethylphthalimide (27% yield) with N-bromosuccinimide.[214]

Extension of the decarbonylation reaction mentioned on p. 97 to α-phthalimidobutyric acid produced an α-haloalkyl derivative. When the acid was heated with phosphorus and bromine at 100°, it gave N-(1,3-dibromopropyl)phthalimide in 52% yield.[212,215]

$$o\text{-}C_6H_4(CO)_2NCH(CO_2H)C_2H_5 \xrightarrow[100^\circ]{Br_2,\ P} o\text{-}C_6H_4(CO)_2NCHBr(CH_2)_2Br$$

Recent work has shown that acid chlorides will add to imines derived from aromatic aldehydes to give α-chloroalkylamides.[216]

$$ArCH{=}NCH_3 + RCOCl \longrightarrow RCON(CH)_3CHClAr$$

Another recent report describes the preparation of N-(1,2,2,2-tetrachloroethyl)carbamates by the treatment of the corresponding isocyanate with alcohols and phenols.[73]

$$OCNCHOHCCl_3 \xrightarrow[C_6H_5CH_3,\ 110^\circ]{SOCl_2} OCNCHClCCl_3 \xrightarrow{ROH} RO_2CNHCHClCCl_3$$

Finally, α-chloroalkyl derivatives of acetamide and of benzenesulfonamide have been prepared in crude form in solution by direct reaction of the amide, an aldehyde, and dry hydrogen chloride.[201]

$$CH_3CONH_2 + RCHO + HCl \rightarrow CH_3CONHCHClR$$

[210] Bachstez, *Ber.*, **46,** 3087 (1913).
[211] Kato, *Kogyo Kagaku Zasshi*, **59,** 1006 (1956) [*C.A.*, **52,** 10002 (1958)].
[212] Gabriel, *Ber.*, **44,** 1905 (1911).
[213] Gabriel, *Ber.*, **44,** 3084 (1911).
[214] Zaugg, *J. Am. Chem. Soc.*, **76,** 5818 (1954).
[215] Hildesheimer, *Ber.*, **43,** 2796 (1910).
[216] Böhme and Hartke, *Chem. Ber.*, **96,** 600 (1963).

**Ethers of N-Methylol- and N-α-Alkylol-amides.** *A. Ethers of N-Methylol-amides and -imides.* These substances are most simply prepared by acid-catalyzed alcoholysis of N-methylolamides. Primary alcohols react more readily than secondary,[79] and tertiary carbinols do

$$\mathrm{RCONHCH_2OH} \xrightarrow{\mathrm{R'OH,\ H^{\oplus}}} \mathrm{RCONHCH_2OR'}$$

not react at all.[217] Under similar conditions one ether can be made from another by an alkoxyl interchange process.[79,218] With mercaptans thio ethers are formed.[219] Excessive amounts of acid are to be avoided

$$\mathrm{RCONHCH_2OR'} \underset{}{\overset{\mathrm{R''OH,\ H^{\oplus}}}{\rightleftharpoons}} \mathrm{RCONHCH_2OR''}$$

because under such conditions the ethers, like N-methylol compounds, are readily converted to the corresponding methylenebisamides. A *p*H of about 3 is optimal for stopping the reaction at the ether stage.[217]

In a few instances N-methoxymethyl derivatives of urea have been obtained by direct condensation of formaldehyde with urea in methanol.[149,219] The generality of the method, however, has not been established.

Ethers of less reactive alcohols can be prepared by solvolysis of N-halomethyl-amides or -imides. Thus N-bromomethylphthalimide with triphenylcarbinol gives a 50% yield of the ether.[71]

$$o\text{-}\mathrm{C_6H_4(CO)_2NCH_2OC(C_6H_5)_3}.$$

Tertiary carbinols that undergo facile acid-catalyzed elimination, i.e., *t*-butyl alcohol, obviously cannot be etherified in this way.

The tendency for N-methylolamides to undergo self-etherification, i.e.,

$$\mathrm{2RCONHCH_2OH} \xrightarrow{-\mathrm{H_2O}} \mathrm{(RCONHCH_2)_2O}$$

varies with the structure of the amide, but etherification usually can be effected in the presence of controlled amounts of mineral acid (see above). The great ease with which ether formation occurred in some reactions, even in alkaline solution, led Einhorn to assign to the ethers the isomeric N-methylol-N,N′-methylenebisamide structure.[10,153]

$$\mathrm{RCONHCH_2N(CH_2OH)COR}$$

[217] Brace and Mantell, *J. Org. Chem.*, **26,** 5176 (1961).

[218] Zigeuner and Hoselmann, *Monatsh. Chem.*, **88,** 5 (1957).

[219] Kadowaki, *Bull. Chem. Soc. Japan*, **11,** 248 (1936) [*C.A.*, **30,** 5944 (1936)].

On the basis of chemical evidence, however, Zigeuner preferred the symmetrical ether structure.[54] Furthermore, the infrared spectra of a series of derivatives of this type prepared from long-chain alkyl amides showed no hydroxyl absorption,[217] and both the infrared and proton magnetic resonance spectra of the benzamide derivative first reported by Einhorn and co-workers[10] are consistent only for the structure $(C_6H_5CONHCH_2)_2O$.[14] Hence it must be concluded that Einhorn's original structural assignment[10,153] is in error and should be corrected along with several more recent erroneous assignments.[220–222]

2-Tetrahydropyranyloxymethyl isocyanate has been used recently to prepare ethers of N-methylolcarbamates and of N-methylolureas.[73]

$$\text{(2-tetrahydropyranyl)}OCH_2NCO \xrightarrow{ArOH} ArOCONHCH_2O\text{(2-tetrahydropyranyl)}$$

$$\text{(2-tetrahydropyranyl)}OCH_2NCO \xrightarrow{ArNH_2} ArNHCONHCH_2O\text{(2-tetrahydropyranyl)}$$

Once formed, most ethers are stable except to hydrolytic conditions. In hot aqueous solution or in cold $1N$ alkali they break down to amide, formaldehyde, and alcohol only slightly less rapidly than do the corresponding N-methylol derivatives.[79]

*B. Ethers of N-α-Alkylolamides.* Because nearly all N-α-alkylolamides are necessarily restricted to derivatives of chloral and bromal (see pp. 93–94), most of the known ethers likewise fall into the general type

$$RCONHCH(OR')CX_3$$

However, among a wide variety of examples the amides used include aliphatic, aromatic, and heterocyclic amides, a lactam, imides, carbamates, and urea derivatives.

Three methods used for the preparation of the N-methylol ethers may also be used for the synthesis of ethers of N-α-alkylolamides. They are acid-catalyzed alcoholysis of the corresponding N-α-alkylol derivative,[223] alcoholysis of the N-α-haloalkyl compound,[106,204,206,214] and direct reaction

[220] Bredereck, Gompper, Herlinger, and Woitun, *Chem. Ber.*, **93,** 2423 (1960).

[221] Chechelska, *Roczniki Chem.*, **30,** 149 (1956) [*C.A.*, **51,** 279 (1957)].

[222] Petterson and Brownell, *J. Org. Chem.*, **25,** 843 (1960).

[223] Badische Anilin- and Soda-Fabrik A.-G., Brit. pat. 779,849 [*C.A.*, **52,** 1272 (1958)]; Brit. pat. 717,287 [*C.A.*, **49,** 15976 (1955)].

of an amide derivative with an aldehyde in an acidic alcoholic medium.[224]

$$\begin{array}{l} \text{ROCONHCHOH} \\ \quad\quad\quad\quad | \\ \text{ROCONHCHOH} \end{array} \xrightarrow{\text{R'OH, H}^{\oplus}} \begin{array}{l} \text{ROCONHCHOR'} \\ \quad\quad\quad\quad | \\ \text{ROCONHCHOR'} \end{array}$$

$$\text{RCON(R')CHXR''} \xrightarrow{\text{R'''OH}} \text{RCON(R')CH(OR''')R''}$$

$$\begin{array}{l} \text{CH}_2\text{NH} \\ | \quad\quad \diagdown \\ | \quad\quad\quad \text{CO} \\ | \quad\quad \diagup \\ \text{CH}_2\text{NH} \end{array} \xrightarrow[\text{ROH, HCl}]{\text{CH}_3\text{CHO}} \begin{array}{l} \text{CH}_2\text{NCH(OR)CH}_3 \\ | \quad\quad \diagdown \\ | \quad\quad\quad \text{CO} \\ | \quad\quad \diagup \\ \text{CH}_2\text{NH} \end{array} + \begin{array}{l} \text{CH}_2\text{NCH(OR)CH}_3 \\ | \quad\quad \diagdown \\ | \quad\quad\quad \text{CO} \\ | \quad\quad \diagup \\ \text{CH}_2\text{NCH(OR)CH}_3 \end{array}$$

In addition, alkylation with dimethyl sulfate has been employed to prepare a number of methyl ethers.[101,166,205,225,226]

$$\text{RCONHCHOHR'} \xrightarrow[\text{Aq. NaOH}]{(\text{CH}_3)_2\text{SO}_4} \text{RCONHCH(OCH}_3\text{)R'}$$

A base-catalyzed ether interchange process has been utilized in several instances.[161–163] The symmetrical ether employed as the starting material is readily obtainable from the N-α-alkylol derivative by using

$$[\text{RNHCONH}\overset{|}{\text{C}}\text{H(CX}_3)]_2\text{O} \xrightarrow[\text{R'OH}]{\text{NaOR'}} 2\text{RNHCONHCH(OR')CX}_3$$

an acid anhydride or an acid chloride under Schotten-Baumann conditions (refs. 94, 95, 160, 161, 163, 166, 168, 207, 208, 226–229). Workers who first encountered this reaction believed that simple elimination had

$$2\text{RCONHCHOHCX}_3 \xrightarrow[\text{Aq. NaOH}]{(\text{CH}_3\text{CO})_2\text{O}} [\text{RCONH}\overset{|}{\text{C}}\text{H(CX}_3)]_2\text{O}$$

occurred to produce N-acylimines of type $\text{RCON{=}CHCX}_3$.[94,229,230] It remained for Feist to recognize the bimolecular nature of the dehydration and to assign the correct structure to the symmetrical ethers.[95] In acid media the dehydration reaction is largely prevented, and normal esterification of the hydroxyl group predominates (see below). In this connection it is known that certain N-acetyl ketimines spontaneously add

224 Yost, U.S. pat. 2,880,208 [*C.A.*, **54,** 569 (1960)].

225 Hirwe and Gavankar, *J. Univ. Bombay*, **6,** II, 123 (1937) [*C.A.*, **32,** 3762 (1938)].

226 Hirwe and Rana, *J. Univ. Bombay*, **7,** Pt. 3, 174 (1938) [*C.A.*, **33,** 3778 (1939)].

227 Chattaway, Kerr, and Lawrence, *J. Chem. Soc.*, **1933,** 30.

228 Hirwe and Patil, *Proc. Indian Acad. Sci.*, **13A,** 273 (1941) [*C.A.*, **35,** 6250 (1941)].

229 Moscheles, *Ber.*, **24,** 1803 (1891).

230 Hantzsch, *Ber.*, **27,** 1248 (1894).

alcohols to give ethers. In contrast, the analogous N-benzoyl ketimines do so with comparative reluctance.[231]

$$CH_3CON{=}CAr_2 + ROH \xrightarrow{\text{(Fast)}} CH_3CONHC(OR)Ar_2$$

$$C_6H_5CON{=}CAr_2 + ROH \xrightarrow{\text{(Slow)}} C_6H_5CONHC(OR)Ar_2$$

A special method has been applied to the preparation of N-α-alkoxyethyl derivatives of imides and of one sulfonamide: reaction with vinyl ethers at elevated temperatures.[232] In the presence of acid, however, the products derived from imides eliminate the elements of alcohol to give

$$o\text{-}C_6H_4(CO)_2NH + CH_2{=}CHOR \xrightarrow[\text{No acid}]{160\text{–}260°} o\text{-}C_6H_4(CO)_2NCH(OR)CH_3$$

$$p\text{-}CH_3C_6H_4SO_2NHCH_3 + CH_2{=}CHOR \xrightarrow[\text{Acid}]{60°} \underset{(84\%)}{p\text{-}CH_3C_6H_4SO_2N(CH_3)CH(OR)CH_3}$$

vinyl imides, e.g., $o\text{-}C_6H_4(CO)_2NCH{=}CH_2$. By contrast, carboxamides and primary sulfonamides generally react with vinyl ethers (with or without acid catalysis) to produce N,N′-ethylidenebisamides, $(RCONH)_2CHCH_3$ (see below).[232,233]

Another special method for preparing N-α-alkoxyethyl derivatives of carbamyl compounds consists of the treatment of α-alkoxyethylisocyanates with alcohols or amines.[234]

$$HNCO + ROCH{=}CH_2 \rightarrow$$

$$OCNCH(OR)CH_3 \begin{cases} \xrightarrow{R'OH} R'OCONHCH(OR)CH_3 \\ \xrightarrow{R'NH_2} R'NHCONHCH(OR)CH_3 \end{cases}$$

**Esters of N-Methylol- and of N-α-Alkylol-amide Derivatives.** *A. Esters of N-Methylolamides.* N-Methylolamides readily undergo acetolysis.[79] Most esters of N-methylolamides, however, have been prepared by conventional acylation methods. These include reaction of

$$RCONHCH_2OH + CH_3CO_2H \xrightarrow[\text{30 min.}]{75°} RCONHCH_2OCOCH_3$$

[231] Banfield, Brown, Davey, Davies, and Ramsey, *Australian J. Sci. Res.*, **A1**, 330 (1948) [*C.A.*, **46**, 5558 (1952)].

[232] Furukawa, Onishi, and Tsuruta, *J. Org. Chem.*, **23**, 672 (1958).

[233] Voronkov, *J. Gen. Chem. USSR (Engl. Transl.)*, **21**, 1631 (1951) [*C.A.*, **46**, 8002 (1952)].

[234] Hoover and Rothrock, *J. Org. Chem.*, **28**, 2082 (1963).

the methylolamide with an acid anhydride alone,[222,235–237] in the presence of catalytic amounts of sulfuric acid[8] or of the corresponding sodium salt;[238–241] treatment with an acid chloride or anhydride in pyridine (refs. 8, 66, 181, 194, 242, 243) or in aqueous alkali;[237,238,241] and heating with a carboxylic acid in the presence of hydrogen chloride[240] or trifluoroacetic anhydride.[8] The formyl derivative, $o\text{-}C_6H_4(CO)_2NCH_2OCHO$, was prepared (24% yield) by heating N-chloromethylphthalimide in formamide at 100°;[220] N-acetoxymethylmaleimide was obtained (43% yield) by acetolysis of N-chloromethylmaleimide at 100°;[8] and the phthalimidomethyl esters of a number of carbobenzoxyamino acids and peptides have been prepared by the treatment of N-chloromethylphthalimide with the amino acid in the presence of triethylamine.[243a]

Esters of N-methylolamides may serve as useful derivatives for isolation and purification. Thus N-methyl-N-methylolacetamide, which could not be obtained in crystalline form, was converted to the acetate $CH_3CON(CH_3)CH_2OCOCH_3$, which could be distilled under reduced pressure.[244]

A direct preparation of esters of N-methylolamides from amides would greatly increase their usefulness as amidomethylating agents. One example of this synthetic simplification is the preparation of N-acetoxymethylstearamide directly from stearamide.[138] The relatively mild conditions used in the procedure suggest that it may be generally applicable to the synthesis of other esters of this type.

*B. Esters of N-α-Alkylolamides.* Nearly all substances of this type are either acetates or benzoates of chloral- or bromal-amide derivatives, $RCONHCH(OCOR')CX_3$. They have usually been made by direct acylation of the corresponding hydroxy compound with acetyl chloride or benzoyl chloride either alone[94,168] or in the presence of pyridine.[101,166,204] Acetic anhydride with catalytic amounts of concentrated sulfuric acid generally gives the corresponding acetate (refs. 161, 162, 166, 168, 225, 228), but in the presence of aqueous sodium hydroxide it usually produces the symmetrical ether (see p. 101).[94,95,160] This tendency toward ether formation is also governed, however, by the nature of the α-alkylolamide.

[235] Pikl, U.S. pat. 2,477,346 [*C.A.*, **44,** 168 (1950)].
[236] Shipp, U.S. pat. 2,232,485 [*C.A.*, **35,** 3458 (1941)].
[237] Spirer, *Roczniki Chem.*, **28,** 455 (1954) [*C.A.*, **50,** 311 (1956)].
[238] Monti and Venturi, *Gazz. Chim. Ital.*, **76,** 365 (1946) [*C.A.*, **42,** 1261 (1948)].
[239] Zellner, Austrian pat. 176,561 [*C.A.*, **48,** 10778 (1954)].
[240] Zellner, Austrian pat. 176,562 [*C.A.*, **48,** 10778 (1954)].
[241] Zellner, Ger. pat. 927,269 (*Chem. Zentr.*, **1955,** 7980).
[242] Einhorn and Feibelmann, *Ann.*, **361,** 153 (1908).
[243] Mizuch, Kasatkin, and Gelfer, *J. Gen. Chem. USSR* (*Engl. Transl.*), **27,** 213 (1957).
[243a] Nefkens, *Nature*, **193,** 974 (1962); *Rec. Trav. Chim.*, **82,** 941 (1963).
[244] Walter, Steffen, and Heyns, *Chem. Ber.*, **94,** 2462 (1961).

Thus with acetic anhydride and aqueous sodium hydroxide the chloralamides **104** give mainly the symmetrical ethers **105**. The unmethylated analogs **106**, on the other hand, yield the corresponding acetates **107** as the major products.[226]

$$2\text{Cl}-C_6H_3(OCH_3)CONHCHOHCCl_3 \xrightarrow[\text{Aq. NaOH}]{(CH_3CO)_2O} [\text{Cl}-C_6H_3(OCH_3)CONHCHCCl_3]_2O$$

**104** → **105**

$$\text{Cl}-C_6H_3(OH)CONHCHOHCCl_3 \xrightarrow[\text{Aq. NaOH}]{(CH_3CO)_2O} \text{Cl}-C_6H_3(OCOCH_3)CONHCH(OCOCH_3)CCl_3$$

**106** → **107**

A few esters have been made by acidolysis of the corresponding N-α-chloroalkylamides.[206,211]

$$\text{RCONHCHClR}' \xrightarrow[C_6H_5CO_2H]{CH_3CO_2H \text{ or}} \text{RCONHCH(OCOR}'')\text{R}' \quad (\text{R}'' = CH_3, C_6H_5)$$

**N,N′-Methylenebis-amides and -imides.** N,N′-Methylenebisbenzamide was first prepared in 1876 by the action of methylal on benzonitrile in the presence of concentrated sulfuric acid.[245] Later this so-called

$$C_6H_5CN + CH_2(OCH_3)_2 \xrightarrow[H_2SO_4]{\text{Concd.}} (C_6H_5CONH)_2CH_2$$

"Hipparaffin" was obtained more readily by heating benzamide with formalin in acid solution.[246] Since then this method has become one of general utility, having been applied to many simple amides, some N-

$$2C_6H_5CONH_2 + CH_2O \xrightarrow{H^{\oplus}} (C_6H_5CONH)_2CH_2$$

monosubstituted amides,[247–249] acyclic[55,153,250,251] and cyclic[185,252] carbamyl compounds, and a few aromatic sulfonamides.[85,156] Although mineral acid is most often used as the catalyst, other conditions successfully employed have included heating the components alone or in acetic anhydride or glacial acetic acid, or even warming in the presence of weak

[245] Hepp and Spiess, *Ber.*, **9**, 1424 (1876).
[246] Pulvermacher, *Ber.*, **25**, 304 (1892).
[247] Breitenbach and Wolf, *Monatsh. Chem.*, **87**, 367 (1956).
[248] Reppe *et al.*, *Ann.*, **596**, 210 (1955).
[249] Sallmann and Albrecht, U.S. pat. 2,279,497 [*C.A.*, **36**, 5290 (1942)].
[250] Brian and Lamberton, *J. Chem. Soc.*, **1949**, 1633.
[251] Conrad and Hock, *Ber.*, **36**, 2206 (1903).
[252] Walker, U.S. pat. 2,418,000 [*C.A.*, **41**, 4512 (1947)].

aqueous base (dilute potassium carbonate solution). The ubiquity of this reaction is readily understood from the fact that N,N′-methylenebisamides usually represent the end products (and most stable ones) derived from the reaction of amides with formaldehyde in acid media. Thus N,N′-methylenebisamides are obtainable by acid treatment of both N-methylolamides and symmetrical ethers as well as of unsymmetrical ethers and esters of N-methylolamides.

$$RCONH_2 \xrightarrow{CH_2O} RCONHCH_2OH \begin{cases} \xrightarrow[H^{\oplus},\ -H_2O]{RCONH_2} (RCONH)_2CH_2 \\ \qquad\qquad \uparrow H^{\oplus},\ -CH_2O \\ \xrightarrow[H^{\oplus},\ -H_2O]{RCONHCH_2OH} (RCONHCH_2)_2O \end{cases}$$

An interesting specificity in catalytic requirement for one of these reactions has been observed recently.[253] Heating N-methylol-2,3-dibromoisobutyramide with aqueous hydrobromic or sulfuric acid gave the corresponding bisamide, but identical treatment with aqueous hydrochloric acid did not. Furthermore, the same transformation could be effected in the presence of bromine but not with iodine.[253,254]

$$2CH_2BrC(CH_3)BrCONHCH_2OH \xrightarrow[H_2SO_4]{HBr\ or} [CH_2BrC(CH_3)BrCONH]_2CH_2$$

Perhaps the most general method and the one giving the best yields involves treatment of an amide with an N-methylolamide in acid solution. It has the added advantage of flexibility in that unsymmetrical bisamides are obtainable in this way. These unsymmetrical derivatives can be

$$RCONH_2 + R'CONHCH_2OH \xrightarrow{H^{\oplus}} RCONHCH_2NHCOR'$$

prepared also by treatment of nitriles with N-methylolamides in strong sulfuric acid. Thus acetonitrile and N-methylolphthalimide in concentrated sulfuric acid give N-acetylaminomethylphthalimide in 93% yield.[84]

$$o\text{-}C_6H_4(CO)_2NCH_2OH + CH_3CN \xrightarrow[H_2SO_4]{Concd.} o\text{-}C_6H_4(CO)_2NCH_2NHCOCH_3$$

Under these conditions nitriles react more readily with N-methylolamides than they do with formaldehyde. A study designed to test this generalization showed that in a formic-sulfuric acid solution at 30° a mixture of

[253] Feuer and Bello, *J. Am. Chem. Soc.*, **78**, 4367 (1956).
[254] Feuer and Lynch, *J. Am. Chem. Soc.*, **75**, 5027 (1953).

*p*-chlorobenzonitrile, formaldehyde, and N-methylolbenzamide gave almost exclusively the unsymmetrical bisamide (reaction *a*).[255]

$$p\text{-ClC}_6\text{H}_4\text{CONHCH}_2\text{NHCOC}_6\text{H}_5 \xleftarrow[(a)]{\text{C}_6\text{H}_5\text{CONHCH}_2\text{OH}} p\text{-ClC}_6\text{H}_4\text{CN} \xrightarrow[(b)]{\text{CH}_2\text{O}} (p\text{-ClC}_6\text{H}_4\text{CONH})_2\text{CH}_2$$

A few special methods have been employed for making certain methylenebis-amides and -imides. N,N'-Methylenebisphthalimide has been prepared by treating potassium phthalimide with methylene iodide[256] or, preferably (48% yield), with the methiodide of N-dimethylaminomethylphthalimide.[81] Similarly, chloromethylamides[178,194] and aminomethylamides[82,257] have been used to amidomethylate imides and sulfonamides

$$o\text{-C}_6\text{H}_4(\text{CO})_2\text{NK} + o\text{-C}_6\text{H}_4(\text{CO})_2\text{NCH}_2\overset{\oplus}{\text{N}}(\text{CH}_3)_3\,\text{I}^{\ominus} \rightarrow [o\text{-C}_6\text{H}_4(\text{CO})_2\text{N}]_2\text{CH}_2$$

or their alkali metal salts.

A recently reported method for preparing N,N'-methylenebisurethans appears to be general.[258] An example is the reaction of methylal with two equivalents of methyl isocyanate to give dimethyl N,N'-dimethyl-N,N'-methylenebiscarbamate. Thioformals react analogously to give

$$\text{CH}_2(\text{OCH}_3)_2 + 2\text{CH}_3\text{NCO} \rightarrow \text{CH}_2[\text{N}(\text{CH}_3)\text{CO}_2\text{CH}_3]_2$$

methylenebisthiocarbamates. By another apparently general process, unsymmetrical combinations of amides with urethan have been obtained through the Curtius reaction of N-acylglycylhydrazides.[187,259]

$$\text{CH}_2\text{ClCONHCH}_2\text{CONHNH}_2 \xrightarrow{\text{HNO}_2}$$

$$[\text{CH}_2\text{ClCONHCH}_2\text{CON}_3] \xrightarrow{\text{C}_2\text{H}_5\text{OH}} \text{CH}_2\text{ClCONHCH}_2\text{NHCO}_2\text{C}_2\text{H}_5$$

Finally, acetylation of the Mannich-type products derived from some anilines and an oxazolidinethione has given several unsymmetrical bisamides.[260]

$$\text{C}_6\text{H}_{10}\text{(fused)}\!-\!\text{NCH}_2\text{NHAr}\;/\;\text{O}\!-\!\text{C}\!=\!\text{S} \xrightarrow[\text{Pyridine}]{\text{CH}_3\text{COCl}} \text{C}_6\text{H}_{10}\text{(fused)}\!-\!\text{NCH}_2\text{N(Ar)COCH}_3\;/\;\text{O}\!-\!\text{C}\!=\!\text{S}$$

[255] Magat and Salisbury, *J. Am. Chem. Soc.*, **73,** 1035 (1951).
[256] Neumann, *Ber.*, **23,** 1002 (1890).
[257] Hellmann and Haas, *Chem. Ber.*, **90,** 50 (1957).
[258] Brachel and Merten, *Angew. Chem.* (*Intern. Ed.*), **1,** 592 (1962).
[259] Curtius and Callan, *Ber.*, **43,** 2457 (1910).
[260] Zinner and Schritt, *J. Prakt. Chem.*, [4] **15,** 82 (1962).

N,N′-Methylenebisamides generally are stable crystalline solids. N,N′-Methylenebisbenzamide can be sublimed,[245] and N,N′-methylenebisphenylacetamide can be distilled.[261] Compounds of this type usually are stable in 1*N* sodium hydroxide at 90°,[79] but treatment with hot dilute mineral acid reconverts them to the corresponding amides. As expected, hot concentrated acid or hot alcoholic alkali hydrolyzes them to the carboxylic acids.[245,261] Of some interest is the report that methylenediamine can be prepared in solution by hydrolysis of N,N′-methylenebisformamide with cold concentrated mineral acid.[262] Einhorn had previously reported similar results from the alkaline hydrolysis of N,N′-methylenebistrichloroacetamide.[263,264]

$$CH_2(NHCHO)_2 \xrightarrow{H^\oplus} [CH_2(NH_2)_2] \xleftarrow{OH^\ominus} CH_2(NHCOCCl_3)_2$$

Treatment of methylenebisurea with dilute hydrochloric acid leads to polymeric products of the type $H_2NCONH(CH_2NHCONH)_nH$.[219,265]

**N,N′-Alkylidene- and -Arylidene-bisamides.** The first example of this type, diethyl N,N′-dichloroethylidenebiscarbamate, was prepared in 1840 by the action of chlorine on an ethanolic solution of hydrocyanic acid.[266] Its structure, however, was not determined until thirty years

$$Cl_2 + C_2H_5OH + HCN \rightarrow (C_2H_5OCONH)_2CHCHCl_2$$

$$CHCl_2CHO + 2H_2NCO_2C_2H_5 \xrightarrow{-H_2O} (C_2H_5OCONH)_2CHCHCl_2$$

later, when it was obtained by the reaction of dichloroacetaldehyde with urethan.[173,267] The latter method has been the basis for the preparation of several hundred compounds of this class. Synthetic variations include heating the components together without solvent,[268–270] in ethanol,[271,272]

[261] Hepp, *Ber.*, **10**, 1649 (1877).
[262] Knudsen, *Ber.*, **47**, 2698 (1914).
[263] Einhorn, *Ann.*, **343**, 207 (1905).
[264] Einhorn and Mauermayer, *Ann.*, **343**, 305 (1905).
[265] Staudinger and Wagner, *Makromol. Chem.*, **12**, 168 (1954) [*C.A.*, **49**, 12302 (1955)].
[266] Stenhouse, *Ann.*, **33**, 92 (1840).
[267] Bischoff, *Ber.*, **5**, 80 (1872).
[268] Bojanović, Vandjel, Mihailović, and Stefanović, *Ber. Chem. Ges. Belgrade*, **20**, 267 (1955) (*Chem. Zentr.*, **1957**, 6712).
[269] Ittyerah and Pandya, *Proc. Indian Acad. Sci.*, **15A**, 258 (1942) [*C.A.*, **37**, 2725 (1943)].
[270] Schiff, *Ann.*, **148**, 330 (1868).
[271] Nigam and Pandya, *Proc. Indian Acad. Sci.*, **29A**, 56 (1949) [*C.A.*, **43**, 6182 (1949)].
[272] Schiff, *Ann.*, **151**, 186 (1869).

water,[123,272] pyridine,[273–276] aqueous acid,[173,277,278] glacial acetic acid,[154,279] or acetic anhydride.[280–282] The last method appears to be the most generally useful. However, condensation in refluxing benzene with azeotropic removal of water has also met with moderate success.[283]

Most of the commonly available primary amides and carbamyl compounds have been utilized in reactions with aldehydes. Even a few N-monosubstituted amides have been included.[270,276] By heating benzaldehyde and 2-furaldehyde with mixtures of two different amides, mixed N,N′-benzylidene- and N,N′-furylidene-bisamides have been obtained in yields ranging from 25 to 42%.[284]

$$\text{ArCHO} + \text{RCONH}_2 + \text{R'CONH}_2 \rightarrow \text{RCONHCH(Ar)NHCOR'}$$

Diamides with suitably positioned functional groups yield cyclic bisamides. Oxamide and malonamide derivatives lead to diketotetrahydroimidazoles[285] and diketohexahydropyrimidines,[286] respectively.

$$(\text{CONH}_2)_2 + \text{C}_6\text{H}_5\text{CHO} \rightarrow \begin{array}{l}\text{CONH} \\ | \qquad \diagdown \\ | \qquad\quad \text{CHC}_6\text{H}_5 \\ | \qquad \diagup \\ \text{CONH}\end{array}$$

$$(\text{C}_2\text{H}_5)_2\text{C(CONH}_2)_2 + \text{C}_6\text{H}_5\text{CHO} \rightarrow (\text{C}_2\text{H}_5)_2\text{C}\begin{array}{c}\diagup\text{CONH}\diagdown \\ \\ \diagdown\text{CONH}\diagup\end{array}\text{CHC}_6\text{H}_5$$

Formamide generally gives poor yields with either aliphatic or aromatic

[273] Agarwal, Pandya, and Tripathi, *J. Indian Chem. Soc.*, **21**, 383 (1944) [*C.A.*, **39**, 4595 (1945)].

[274] Ittyerah and Pandya, *Proc. Indian Acad. Sci.*, **15A**, 6 (1942) [*C.A.*, **36**, 6144 (1942)].

[275] Mehra and Pandya, *Proc. Indian Acad. Sci.*, **7A**, 376 (1938) [*C.A.*, **32**, 7434 (1938)].

[276] Mehra and Pandya, *Proc. Indian Acad. Sci.*, **9A**, 508 (1939) [*C.A.*, **33**, 8589 (1939)].

[277] Bischoff, *Ber.*, **7**, 1078 (1874).

[278] Glaser and Frisch, *Arch. Pharm.*, **266**, 103 (1928).

[279] Noyes and Forman, *J. Am. Chem. Soc.*, **55**, 3493 (1933).

[280] Stefanović, Bojanović, Côrbić, and Mihailović, *Ber. Chem. Ges. Belgrade*, **22**, 29 (1957) (*Chem. Zentr.*, **1958**, 9488).

[281] Stefanović, Bojanović, and Vandjel, *Ber. Chem. Ges. Belgrade*, **18**, 579 (1953) (*Chem. Zentr.*, **1954**, 9980).

[282] Stefanović, Mihailović, Bojanović, and Vandjel, *Ber. Chem. Ges. Belgrade*, **20**, 417 (1955) (*Chem. Zentr.*, **1958**, 2710).

[283] Paulson and Mersereau, *Trans. Illinois State Acad. Sci.*, **47**, 94 (1955) [*C.A.*, **50**, 233 (1956)].

[284] Polya and Spotswood, *Rec. Trav. Chim.*, **70**, 269 (1951).

[285] Medicus, *Ann.*, **157**, 44 (1871).

[286] Burrows and Keane, *J. Chem. Soc.*, **1907**, 269.

aldehydes. One reason is the occurrence of a side reaction in which tetra-substituted pyrazines are formed.[271,287]

$$4ArCHO + 2HCONH_2 \rightarrow \text{2,3,5,6-tetraarylpyrazine}$$

Phthalimide does not react with benzaldehyde even in the presence of acetic anhydride.[85] This behavior is consistent with the fact that hydantoins undergo *acid-catalyzed* condensation with acetals (and with formaldehyde[185,252,288,289]) preferentially at the amide nitrogen atom.[290]

$$CH_2ClCH(OC_2H_5)_2 + 2\ \begin{matrix} CONH \\ | \quad\quad CO \\ CH_2NH \end{matrix} \xrightarrow[H_2SO_4]{CH_3CO_2H} \left[\begin{matrix} CONH \\ | \quad\quad CO \\ CH_2N— \end{matrix}\right]_2 CHCH_2Cl$$

In interesting contrast is the demonstration that in *neutral* solution formaldehyde reacts at the imide nitrogen atom to give the 3-methylol derivative.[187] This alteration in reaction course is consistent with the

$$\begin{matrix} CONH \\ | \quad\quad CO \\ R_2C—NH \end{matrix} + CH_2O \xrightarrow{\text{Neutral}} \begin{matrix} CONCH_2OH \\ | \quad\quad CO \\ R_2C—NH \end{matrix}$$

idea that the acid-catalyzed process involves attack at the more nucleophilic (amide) nitrogen atom by a protonated aldehyde species,[290] while the reaction in neutral solution occurs through attack of the neutral aldehyde by the anion derived from the more acidic (imide) nitrogen atom.

Sulfonamides, likewise, do not react readily with aldehydes to give bisamides. Ethylidenebis-(*p*-toluenesulfonamide), however, has been obtained (11% yield) from the acid-catalyzed reaction of *p*-toluenesulfonamide with phenyl vinyl ether.[232] This appears to be a general method for preparing the corresponding biscarboxamides as well.[232,233]

$$2\ p\text{-}CH_3C_6H_4SO_2NH_2 + CH_2{=}CHOC_6H_5 \xrightarrow{H^{\oplus}} (p\text{-}CH_3C_6H_4SO_2NH)_2CHCH_3 + C_6H_5OH$$

Simple ketones, unlike aldehydes, do not react with amides. Perfluorinated ketones, however, might be expected to behave like aldehydes and

[287] Bülow, *Ber.*, **26**, 1972 (1893).

[288] Rogers, U.S. pat. 2,404,096 [*C.A.*, **40**, 6096 (1946)].

[289] Walker, U.S. pat. 2,417,999 [*C.A.*, **41**, 4511 (1947)].

[290] Johnson and Crosby, *J. Org. Chem.*, **27**, 2077 (1962).

give bisamides, provided that the reaction does not stop at the α-alkylol stage, as it usually does with perhalogenated aldehydes, i.e., chloral. This supposition gains support from the fact that both pyruvic acid[291] and benzoylformic acid[292] give bisamides with acetamide. The pyruvic acid

$$RCOCO_2H + CH_3CONH_2 \xrightarrow[-H_2O]{Heat} (CH_3CONH)_2C(R)CO_2H$$

derivative loses one acetamide residue in hot glacial acetic acid, and α-acetamidoacrylic acid is formed.[292–294]

$$(CH_3CONH)_2C(CH_3)CO_2H \longrightarrow CH_3CONHC(CO_2H){=}CH_2 + CH_3CONH_2$$

If proper conditions are employed, even chloral and its analogs yield bisamides rather than the usual α-alkylol compounds. The conditions include the use of concentrated sulfuric acid as the condensing agent either with a nitrile[4,245,261,295] or with an amide.[296] However, a possible side

$$CCl_3CHO + 2RCN + H_2O \xrightarrow[H_2SO_4]{Concd.} (RCONH)_2CHCCl_3$$

reaction in this and similar condensations of aromatic aldehydes with carboxamides and sulfonamides must be anticipated. When a Lewis acid is used as a catalyst in the condensation of a sulfonamide with an aromatic aldehyde, an arylidenemonoamide (or sulfonylimine) is produced.[159]

$$RSO_2NH_2 + ArCHO \xrightarrow[ZnCl_2]{Heat} RSO_2N{=}CHAr$$

Of some interest is the observation that the three hydroxybenzaldehydes react with amides to give the acylimines, ArCH=NCOR, under the same conditions that lead to bisamides, $ArCH(NHCOR)_2$, from the corresponding methoxybenzaldehydes.[296a,297–299] Similarly, 5-chloro- and 3,5-dichlorosalicylaldehyde, when heated with heptanamide, benzamide, or benzenesulfonamide, give only the acylimine (or sulfonylimine) derivatives even though no Lewis acid is present.[271] However, these are exceptional cases.

[291] Bergmann and Grafe, *Z. Physiol. Chem.*, **187**, 187 (1930).
[292] Shemin and Herbst, *J. Am. Chem. Soc.*, **60**, 1954 (1938).
[293] Süs, *Ann.*, **569**, 153 (1950).
[294] Wieland, Ohnacker, and Ziegler, *Chem. Ber.*, **90**, 194 (1957).
[295] Hübner, *Ber.*, **6**, 109 (1873).
[296] Batt and Woodcock, *J. Chem. Soc.*, **1948**, 2322.
[296a] Mehra and Pandya, *Proc. Indian Acad. Sci.*, **10A**, 285 (1939) [*C.A.*, **34**, 1981 (1940)].
[297] Manzur and Pandya, *Proc. Indian Acad. Sci.*, **10A**, 282 (1939) [*C.A.*, **34**, 1980 (1940)].
[298] Mehra and Pandya, *Proc. Indian Acad. Sci.*, **10A**, 279 (1939) [*C.A.*, **34**, 1980 (1940)].
[299] Pandya and Sodhi, *Proc. Indian Acad. Sci.*, **7A**, 361 (1938) [*C.A.*, **32**, 7434 (1938)].

As a rule, alkylidene- and arylidene-bisamides resemble the methylenebisamides in their high thermal stability and in their behavior toward acids and bases.

**N-Aminomethylamides and Their Quaternary Salts.** The voluminous literature dealing with N-aminomethylamides has been reviewed recently.[3] No purpose would be served by repeating this information here. It should suffice to note that three related methods have been used to synthesize these compounds: a Mannich-type reaction of amides with formaldehyde and amines,[79,257,300–302] reaction of amines with N-methylolamides,[79,66,303] and reaction of amides with N-methylolamines or N,N′-methylenediamines.[303] The first method is, by far, the

$$\left.\begin{array}{l} RCONH_2 + CH_2O + R'_2NH \\ RCONHCH_2OH + R'_2NH \\ RCONH_2 + HOCH_2NR'_2 \end{array}\right] \rightarrow RCONHCH_2NR'_2$$

most convenient and, consequently, the most commonly used.

Three types of quaternary salts of N-aminomethylamides are known. The first is prepared by alkylation of the tertiary amine with methyl or ethyl iodide.[1,179,180,194,304] The second is obtained by treating pyridine

$$RCONHCH_2NR'_2 + R''I \rightarrow RCONHCH_2\overset{\oplus}{N}R'_2R''\overset{\ominus}{I}$$

with an N-chloromethylamide (pure[66] or prepared *in situ*[199,200,305]), or by reaction of pyridine hydrochloride in pyridine either with a preformed N-methylolamide[140,235,243] or with an amide and formaldehyde.[138,306]

$$\left.\begin{array}{l} RCONHCH_2Cl + C_5H_5N \\ RCONHCH_2OH + C_5H_5\overset{\oplus}{N}H\ Cl^{\ominus} \\ RCONH_2 + CH_2O + C_5H_5\overset{\oplus}{N}H\ Cl^{\ominus} \end{array}\right] \rightarrow RCONHCH_2\overset{\oplus}{N}C_5H_5\ Cl^{\ominus}$$

[300] Böhme, Dietz, and Leidreiter, *Arch. Pharm.*, **287**, 198 (1954).
[301] Einhorn, Ger. pat. 284,440 (*Chem. Zentr.*, **1915**, II, 108).
[302] Feldman and Wagner, *J. Org. Chem.*, **7**, 31 (1942).
[303] Weaver, Simons, and Baldwin, *J. Am. Chem. Soc.*, **66**, 222 (1944).
[304] Hellmann and Löschmann, *Chem. Ber.*, **87**, 1684 (1954).
[305] Hunt and Bradley, U.S. pat. 2,493,068 [*C.A.*, **44**, 3041 (1950)].
[306] Baldwin, Evans, and Salkeld, U.S. pat. 2,278,417 [*C.A.*, **36**, 5032 (1942)].

The third results when an aminomethylamide is heated with an acid halide.[191]

$$RCONHCH_2NR'_2 + R''COX \rightarrow (RCONHCH_2)\overset{\oplus}{N}R'_2X^{\ominus} + R''CONR'_2$$

**N-Acyl- and N-Sulfonyl-imines.** Consideration of compounds of this type will be restricted to derivatives of aldehydes and ketones lacking an α-hydrogen atom. When an α-hydrogen atom is present, tautomerization to the corresponding enamide becomes possible (see pp. 115–116).

As stated previously, reactions of amides with aromatic aldehydes generally lead to N,N′-arylidenebisamides. A number of hydroxybenzaldehyde derivatives are exceptions to this rule and give acylimines instead.[271,297–299] These have been noted on p. 110.

Several other amide-aldehyde combinations also react anomalously. These are chloroacetamide with chloral and bromal,[307] urethan with glyoxal,[308] and thiourea with chloral.[160] For the last combination, reaction occurs normally at one nitrogen atom to give the chloralurea derivative, but abnormally at the other to give the acylimine structure.

$$CH_2ClCONH_2 + CX_3CHO \xrightarrow{100^\circ} CH_2ClCON{=}CHCX_3$$

$$C_2H_5OCONH_2 + (CHO)_2 \xrightarrow{H^{\oplus}} (C_2H_5OCON{=}CH)_2$$

$$CS(NH_2)_2 + 2CCl_3CHO \longrightarrow CS\begin{cases} N{=}CHCCl_3 \\ NHCHOHCCl_3 \end{cases}$$

Bis-trichloroethylideneurea has been prepared by the elimination of 2 moles of acetic acid from the diacetate of chloralurea.[160]

$$CO[NHCH(OCOCH_3)CCl_3]_2 \xrightarrow{HCl,\ heat} CO(N{=}CHCCl_3)_2 \cdot HCl$$

Although sulfonamides react with chloral to give the corresponding N-α-alkylolamides, they condense with aromatic aldehydes to the N-sulfonylimines.[159] Zinc chloride is used as a catalyst. Interestingly, sulfonylimines are also obtained from the reaction of 2-furamide with arenesulfonyl chlorides in pyridine solution.[159]

$$RSO_2NH_2 + ArCHO \xrightarrow{ZnCl_2} RSO_2N{=}CHAr$$

$$\text{(2-furyl)}CONH_2 + ArSO_2Cl \xrightarrow{\text{Pyridine}} ArSO_2N{=}CH\text{(2-furyl)}$$

[307] Thomas and George, *Agra Univ. J. Res.*, **9**, Pt. 1, 11 (1960) [*C.A.*, **55**, 24517 (1961)].

[308] Pauly and Sauter, *Ber.*, **63**, 2063 (1930).

N-Acyl derivatives of aromatic ketimines can be prepared by acylation of the isolated imine.[231] It is usually more convenient, however, to acylate the intermediate halomagnesium derivative obtained by the action of an aromatic Grignard reagent on an aromatic nitrile.[19,120,231]

$$RCOCl + Ar_2C{=}NH \rightarrow RCON{=}CAr_2$$

$$ArCN + Ar'MgX \longrightarrow [Ar(Ar')C{=}NMgX] \xrightarrow{RCOCl} RCON{=}C(Ar)Ar'$$

**Amidomethanesulfonic Acids and Their Sodium Salts.** Few compounds of this type have been reported since Knoevenagel and Lebach first made them in 1904 by heating carboxamides or sulfonamides with aqueous sodium formaldehyde bisulfite under pressure.[132] The only

$$RSO_2NH_2 + HOCH_2OSO_2Na \xrightarrow[H_2O]{200^\circ} RSO_2NHCH_2SO_3Na$$

sulfonic *acid* of this class was prepared by the reaction of N-bromomethylphthalimide with aqueous sodium sulfite.[309]

$$o\text{-}C_6H_4(CO)_2NCH_2Br + Na_2SO_3 \rightarrow o\text{-}C_6H_4(CO)_2NCH_2SO_3H$$

## Related Compounds of Potential Utility as Electrophilic Reagents

Many compounds not yet employed as α-amidoalkylating agents bear a sufficient structural resemblance to those that have been so employed to suggest that the former might react analogously to the latter under suitable conditions. In the compilation that follows, an attempt has been made, mainly on the basis of structural analogy, to place the more likely candidates ahead of the less likely ones. For this reason the cyclic analogs, which might be expected to be most stable and least reactive, have been placed last.

**Amidomethyl Thio Ethers and Thiol Esters.** *A. From N-Methylolamides and Mercaptans*[79,219,310,311] *or Thiourea.*[312,313]

$$RCONHCH_2OH + R'SH \xrightarrow{H^\oplus} RCONHCH_2SR'$$

$$RCONHCH_2OH + CS(NH_2)_2 \xrightarrow{HCl} RCONHCH_2SC({=}NH)NH_2\cdot HCl$$

309 Balaban, *J. Chem. Soc.*, **1926,** 569.

310 Benson and Cairns, *J. Am. Chem. Soc.*, **70,** 2115 (1948).

311 Monti and Franchi, *Gazz. Chim. Ital.*, **85,** 510 (1955) [*C.A.*, **50,** 4952 (1956)].

312 Albrecht, Frei, and Sallmann, *Helv. Chim. Acta*, **24,** 233E (1941).

313 Société pour l'Industrie Chimique à Bàle, Fr. pat. 849,147 [*C.A.*, **35,** 6357 (1941)].

*B. From N-Halomethyl-amides*[190,195,314] *or -imides*[190,315] *and Mercaptans*[190,315] *or Thiourea.*[195,314]

$$o\text{-}C_6H_4(CO)_2NCH_2X + RSH \xrightarrow{(-HX)} o\text{-}C_6H_4(CO)_2NCH_2SR$$

$$RCONHCH_2Cl + CS(NH_2)_2 \longrightarrow RCONHCH_2SC({=}NH)NH_2{\cdot}HCl$$

*C. From N-Halomethyl-amides*[198] *or -imides*[8,104,316,317] *or Carbamyl Compounds*[178–180,318] *and Alkali Metal Mercaptides,*[178–180] *Thiocyanates,*[8,104,198] *Xanthates,*[317] *Dithiocarbamates,*[317] *or Dithiophosphates.*[316]

$$o\text{-}C_6H_4(NCH_2Cl)_2CX + RSNa \xrightarrow{C_2H_5OH} o\text{-}C_6H_4(NCH_2SR)_2CX$$

$$o\text{-}C_6H_4(CO)_2NCH_2X + \overset{\oplus\ominus}{MSY} \rightarrow o\text{-}C_6H_4(CO)_2NCH_2SY$$

$[X = Cl, Br; M = \overset{\oplus}{Na}, K^{\oplus}; Y = CN, —C(=S)OR, —C(=S)NR_2, —P(=S)(OR)_2]$

*D. From Dialkylaminomethylamides and Mercaptans.*[319]

$$RCONHCH_2NR'_2 + R''SH \xrightarrow[NaOH]{C_6H_5CH_3} RCONHCH_2SR'' + R'_2NH$$

$(R' = CH_3, C_2H_5;\ R = \text{alkyl, aryl})$

*E. From Amides, Aldehydes, and Thiolacetic Acid.*[320]

$$RCONH_2 + ArCHO + CH_3COSH \xrightarrow{H^{\oplus}} RCONHCH(Ar)SCOCH_3$$

$(R = CH_3, C_2H_5O;\ Ar = \text{aryl})$

*F. From Formamide and an Aminomethyl Sulfide.*[220]

$$HCONH_2 + (C_2H_5)_2NCH_2SC_6H_4CH_3\text{-}p \xrightarrow{150°} HCONHCH_2SC_6H_4CH_3\text{-}p + (C_2H_5)_2NH$$

*G. From Methyl Isocyanate and Thioformals.*[258]

$$2CH_3NCO + CH_2(SC_4H_9\text{-}n)_2 \rightarrow n\text{-}C_4H_9SCON(CH_3)CH_2SC_4H_9\text{-}n$$

[314] Société pour l'Industrie Chimique à Bàle, Fr. pat. 849,146 [*C.A.*, **35**, 6470 (1941)].
[315] Lo, *J. Org. Chem.*, **26**, 3591 (1961).
[316] Fancher, U.S. pat. 2,767,194 [*C.A.*, **51**, 3915 (1957)].
[317] Griffin and Hey, *J. Chem. Soc.*, **1952**, 3334.
[318] Kolosova and Stavrovskaya, *J. Gen. Chem. USSR* (*Engl. Transl.*), **30**, 3544 (1960).
[319] Hellmann and Haas, *Chem. Ber.*, **90**, 444 (1957).
[320] Sirotanović, *Ber. Chem. Ges. Belgrade*, **21**, 219 (1956) (*Chem. Zentr.*, **1957**, 10729).

**Miscellaneous Amidomethyl Derivatives.** *A. Amidomethanephosphonic Acids.*[321,322]

$$\mathrm{RCONHCH_2OH \xrightarrow[2.\ H_2O]{1.\ PCl_3,\ CH_3CO_2H} RCONHCH_2PO(OH)_2}$$

(R = alkyl, aryl)

*B. N-Nitromethylimides.*[188]

$$\mathrm{X(CO)_2NCH_2Br + AgNO_2 \xrightarrow{CH_3CN} X(CO)_2NCH_2NO_2 + AgBr}$$

[$X(CO)_2NH$ = succinimide, phthalimide]

*C. Amidomethylsulfones.*[190]

$$\mathrm{ArCONHCH_2SR \xrightarrow{[O]} ArCONHCH_2SO_2R}$$

**N-(α-Aminoalkyl)amides.** *A. From N-α-Haloalkylamides* (Refs. 73, 101, 106, 160, 204–206, 208, 209).

$$\mathrm{RCONHCHXCX_3 + R'_2NH \rightarrow RCONHCH(NR'_2)CX_3}$$

(R = alkyl, aryl; X = Cl, Br; R′ = H, alkyl, aryl)

*B. From Aromatic Amides, Aromatic Aldehydes, and Piperidine.*[323]

$$\mathrm{ArCONH_2 + ArCHO + HN(CH_2)_5 \xrightarrow[37^\circ]{CH_3OH} ArCONHCH(Ar)N(CH_2)_5}$$

*C. From 2-Benzoxazolone, Benzaldehyde, and Aniline.*[324]

$$\text{2-benzoxazolone (C}_6\text{H}_4\text{–NH–CO–O)} + \mathrm{C_6H_5CHO + C_6H_5NH_2 \xrightarrow[20^\circ]{CH_3OH}}$$

$$\text{C}_6\text{H}_4\text{–N(CO–O)–}\mathrm{CH(C_6H_5)NHC_6H_5}\ (76\%)$$

*D. From Acylketimines and Amines.*[231]

$$\mathrm{RCON{=}CAr_2 \xrightarrow{R'_2NH} RCONHC(NR'_2)Ar_2}$$

(R = alkyl, aryl; Ar = aryl; R′ = H, alkyl)

**Enamides.** In sufficiently acidic media, enamides should protonate to give electrophilic α-amidoalkylating species. Enamides are procurable in several ways.

$$\mathrm{RCONHCH{=}CR_2 \overset{H^{\oplus}}{\rightleftharpoons} [RCONH\overset{\oplus}{C}H{-}CHR_2]}$$

[321] Engelmann and Pikl, U.S. pat. 2,304,156 [*C.A.*, **37**, 3262 (1943)].

[322] Pikl, U.S. pat. 2,328,358 [*C.A.*, **38**, 754 (1944)].

[323] Macovski and Bachmeyer, *Ber.*, **77**, 495 (1944).

[324] Zinner and Wigert, *Chem. Ber.*, **94**, 2209 (1961).

*A. By Reduction of Chloral- or Bromal-amides.*[207–209]

$$\mathrm{RCONHCHOHCX_3} \xrightarrow[\mathrm{CH_3CO_2H}]{\mathrm{Zn}} \mathrm{RCONHCH{=}CX_2}$$

(R = alkyl, aryl; X = Cl, Br)

*B. By Dehydration of Dihaloacetalamides.*[208]

$$\mathrm{ArCONHCHOHCHCl_2} \xrightarrow[\text{or } \mathrm{(CH_3CO)_2O{-}H_2SO_4}]{\mathrm{P_2O_5}\text{, heat}} \mathrm{ArCONHCH{=}CCl_2}$$

*C. From Amides and α-Keto Acids*[291–294] *or Diarylacetaldehydes.*[324a]

$$\mathrm{RCONH_2} + \mathrm{R'_2CHCOCO_2H} \longrightarrow$$

$$\left[\begin{array}{c}\mathrm{RCONHC(OH)(CHR'_2)CO_2H} \\ \text{or} \\ \mathrm{(RCONH)_2C(CHR'_2)CO_2H}\end{array}\right] \xrightarrow[(-\mathrm{RCONH_2})]{(-\mathrm{H_2O})\text{ or}} \mathrm{RCONHC(CO_2H){=}CR'_2}$$

(R = alkyl, aryl, R′ = H, alkyl, aryl)

$$\mathrm{RCONHR'} + \mathrm{Ar_2CHCHO} \rightarrow \mathrm{RCON(R')CH{=}CAr_2}$$

(R = alkyl, aryl, alkoxy, amino; R′ = H, $CH_3$; Ar = aryl)

*D. From the Methyl Ester of N,O-Diformyl-*DL*-serine.*[325]

$$\mathrm{HCONHCH(CO_2CH_3)CH_2OCHO} \xrightarrow{150^\circ} \mathrm{HCONHC(CO_2CH_3){=}CH_2}$$

*E. By Vinylation of Amides.*[326]

$$\mathrm{RCONHR'} + \mathrm{HC{\equiv}CH} \xrightarrow[\text{Heat, pressure}]{\mathrm{KOH}} \mathrm{RCON(R')CH{=}CH_2}$$

*F. By Vinylation of Imides.*[232]

$$\mathrm{X(CO)_2NH} + \mathrm{CH_2{=}CHOR} \xrightarrow[\text{Acid}]{\text{Heat}} \mathrm{X(CO)_2NCH{=}CH_2}$$

[$X(CO)_2NH$ = succinimide, phthalimide]

*G. From Hydantoins and Aldehydes.*[327]

```
CONH                                                   CONH
|    \                                                 |    \
|     CX  +  RCHO  ---CH3CO2H or--------------->       |     CX
|    /             Pyridine,(C2H5)2NH                  |    /
CH2NH                                            RCH=C—NH
```

(X = O, S; R = aryl, alkyl)

[324a] Eiden and Nagar, *Arch. Pharm.*, **296**, 445 (1963).

[325] Heyns and Heinecke, *Z. Physiol. Chem.*, **331**, 45 (1963).

[326] Reppe, *Neue Entwicklungen auf dem Gebiet der Chemie des Acetylens und Kohlenoxyds.* Springer-Verlag, 1949, pp. 20–21; Copenhaver and Bigelow, *Acetylene and Carbon Monoxide Chemistry*, Reinhold, 1949, pp. 66–67.

[327] Ware, *Chem. Rev.*, **46**, 431–433 (1950).

**5-Hydroxyhydantoins.** *A. Alloxanic Acid Derivatives from Alloxan.*[9,328]

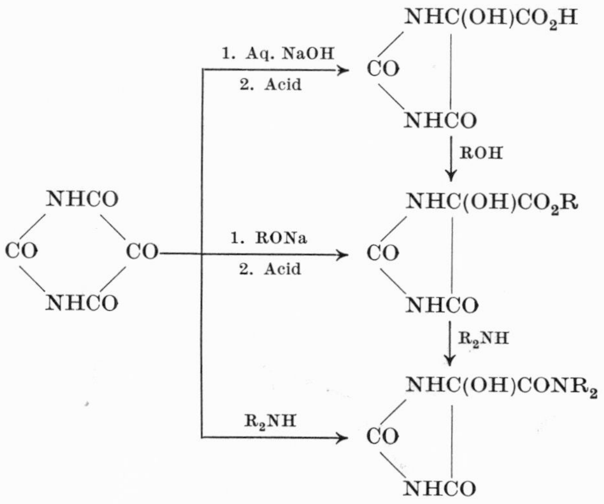


*B. 5-Hydroxyhydantoins from Alloxanic Acid and* 1,3-*Dimethylparabanic Acid.*[329]

$$\mathrm{CO}\begin{cases}\mathrm{NHC(OH)CO_2H}\\ \mathrm{NHCO}\end{cases} \xrightarrow[\text{Heat}\ (-\mathrm{CO_2})]{\mathrm{H_2O}} \mathrm{CO}\begin{cases}\mathrm{NHCHOH}\\ \mathrm{NHCO}\end{cases}$$

$$\mathrm{CO}\begin{cases}\mathrm{N(CH_3)CO}\\ \mathrm{N(CH_3)CO}\end{cases} \xrightarrow{\mathrm{Zn,\ H^{\oplus}}} \mathrm{CO}\begin{cases}\mathrm{N(CH_3)CHOH}\\ \mathrm{N(CH_3)CO}\end{cases} \longrightarrow \mathrm{CO}\begin{cases}\mathrm{N(CH_3)CHOR}\\ \mathrm{N(CH_3)CO}\end{cases}$$

$(\mathrm{R{=}CH_3,\ COC_6H_5,\ CONHC_6H_5})$

**Glycol Amides, Carbamates, and Ureas, and Derived Products.** *A. From Glyoxal and Amides or Carbamates.*[223,330]

$$2\mathrm{RCONH_2} + (\mathrm{CHO})_2 \xrightarrow{\mathrm{H_2O}} (\mathrm{RCONH\overset{|}{C}HOH})_2 \xrightarrow[\mathrm{H^{\oplus}}]{\mathrm{R'OH}} (\mathrm{RCONH\overset{|}{C}HOR'})_2$$

(R = alkyl, alkoxyl)

[328] Fisher and Day, *J. Am. Chem. Soc.*, **77**, 4894 (1955).

[329] Biltz and Kobel, *Ber.*, **54**, 1802 (1921); Biltz and Heidrich, *ibid.*, **54**, 1829 (1921).

[330] Badische Anilin and Soda-Fabrik A.-G., Fr. pat. 1,128,263 (*Chem. Zentr.*, **1959**, 1605).

*B. From Glyoxal and Diamides.*[331]

$$(CH_2)_n(NHCOR)_2 + (CHO)_2 \xrightarrow{H_2O} (CH_2)_n\begin{matrix}N(COR)—CHOH \\ | \\ N(COR)—CHOH\end{matrix}$$

($n$=1, 2; R=H, $CH_3$)

*C. From Glyoxal and Ureas.*[223,308,332–335]

$$CO(NHR)_2 + (CHO)_2 \xrightarrow{H_2O} CO\begin{matrix}N(R)CHOH \\ | \\ N(R)CHOH\end{matrix} \xrightarrow[H^{\oplus}]{R'OH} CO\begin{matrix}N(R)CHOR' \\ | \\ N(R)CHOR'\end{matrix}$$

(R=H, $CH_3$, $CH_2OH$)

*D. From Diketosuccinic Esters and Urea.*[336]

$$CO(NH_2)_2 + (COCO_2R)_2 \xrightarrow{CH_3CO_2H} CO\begin{matrix}NHC(OH)CO_2R \\ | \\ NHC(OH)CO_2R\end{matrix}$$

$$\xrightarrow[H^{\oplus}]{ROH} CO\begin{matrix}NHC(OR)CO_2R \\ | \\ NHC(OR)CO_2R\end{matrix} \qquad \xrightarrow{CH_3COCl} CO\begin{matrix}NHC(OCOCH_3)CO_2R \\ | \\ NHC(OCOCH_3)CO_2R\end{matrix} \qquad \xrightarrow{NH_3} CO\begin{matrix}NHC(OH)CONH_2 \\ | \\ NHC(OH)CONH_2\end{matrix}$$

(R=$CH_3$, $C_2H_5$)

[331] Vail, Moran, Moore, and Kullman, *J. Org. Chem.*, **27**, 2071 (1962).

[332] Badische Anilin and Soda-Fabrik A.-G., Brit. pat. 720,386 [*C.A.*, **49**, 9326 (1955)].

[333] Reibnitz, U.S. pat. 2,731,472 [*C.A.*, **50**, 13999 (1956)].

[334] Slezak, Bluestone, Magee, and Wotiz, *J. Org. Chem.*, **27**, 2181 (1962).

[335] Vail, Murphy, Frick, and Reid, *Am. Dyestuff Reptr.*, **50**, 27 (1961) [*C.A.*, **55**, 19254 (1961)].

[336] Geisenheimer and Anschütz, *Ann.*, **306**, 38 (1899).

*E. From 4,5-Diaryl-2-imidazolones.*[337–343]

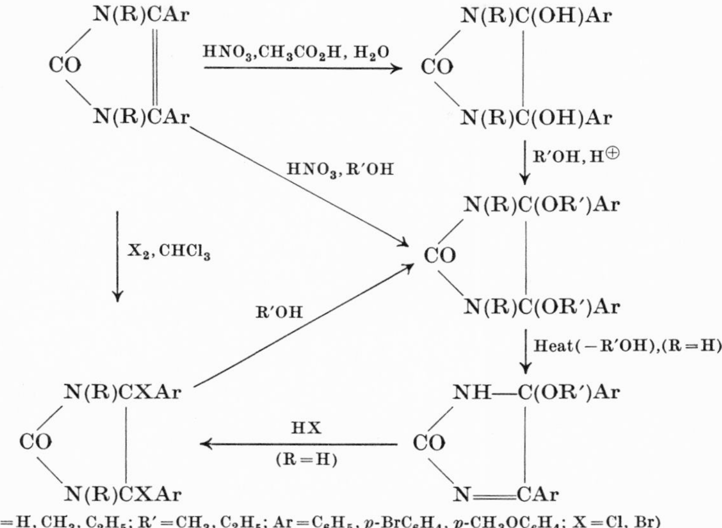


(R = H, $CH_3$, $C_2H_5$; R' = $CH_3$, $C_2H_5$; Ar = $C_6H_5$, *p*-$BrC_6H_4$, *p*-$CH_3OC_6H_4$; X = Cl, Br)

**Cyclic Esters and Carbamates of N-α-Alkylolamides.** *A. N-Acyl- and N-arenesulfonyl-5-oxazolidinones.*[344–347]

$$\mathrm{R(X)NHCH(R')CO_2H} + \mathrm{CH_2O} \xrightarrow[\mathrm{C_6H_6(-H_2O)}]{\mathrm{H}^{\oplus}} \mathrm{R(X)N{-}CHR'},\ \mathrm{CH_2{-}O{-}CO}\ \text{(ring)}$$

(R = alkyl, aryl; R' = H, alkyl; X = CO, $SO_2$)

*B. 2-Trichloromethyl-4,5-oxazolidinedione.*[160]

$$\mathrm{C_2H_5OCOCONHCHOHCCl_3} \xrightarrow[\text{2. } \mathrm{H}^{\oplus}]{\text{1. Aq. } \mathrm{Na_2CO_3}} \mathrm{NHCHCCl_3},\ \mathrm{O},\ \mathrm{COCO}\ \text{(ring)}$$

[337] Biltz, *Ber.*, **41,** 167, 1754, 1761 (1908).
[338] Biltz, *Ann.*, **368,** 262 (1909).
[339] Biltz and Behrens, *Ber.*, **43,** 1990 (1910).
[340] Biltz and Kosegarten, *Ann.*, **368,** 219, 228, 236 (1909).
[341] Biltz and Krebs, *Ann.*, **368,** 218 (1909).
[342] Biltz and Rimpel, *Ann.*, **368,** 173, 201, 206 (1909).
[343] Dunnavant, *J. Org. Chem.*, **21,** 1513 (1956).
[344] Ben-Ishai, *J. Am. Chem. Soc.*, **79,** 5736 (1957).
[345] Chemische Fabrik auf Aktien, Ger. pat. 148,669 (*Chem. Zentr.*, **1904,** I, 411).
[346] Chemische Fabrik auf Aktien, Ger. pat. 153,860 (*Chem. Zentr.*, **1904,** II, 678).
[347] Micheel and Thomas, *Chem. Ber.*, **90,** 2906 (1957).

*C. ε-Phthalimido-ε-caprolactone.*[348]

$$o\text{-}C_6H_4(CO)_2N\text{-cyclohexanone} \xrightarrow[CH_2Cl_2]{CF_3CO_3H} o\text{-}C_6H_4(CO)_2N\text{-}\epsilon\text{-caprolactone}\ (43\%)$$

*D.* 1,3,5-*Oxadiazine*-2,4-*diones.*[73]

$$OCNC(OH)RR' \xrightarrow{H_2O} \text{OC(NHCRR'O)(NHCO)}$$

($R = R' = CF_3$ or $CF_2Cl$ and $R = H$, $R' = CCl_3$)

**Cyclic Ethers and Thio Ethers of N-α-Alkylolamides.** *A.* 4-*Oxazolidinones and* 4-*Thiazolidinones.*[349–351]

$$RCH(XH)CONH_2 + R'COR'' \xrightarrow[(-H_2O)]{H^{\oplus}} \text{RCHX–C(R')(R'')–NHCO (ring)}$$

(R, R', R'' = H, $CH_3$, $C_6H_5$; X = O, S)

*B. N-Sulfonyloxazolidines.*[352]

$$RSO_2NHCH_2CH(R')OH + R''CHO \xrightarrow[C_6H_6(-H_2O)]{H^{\oplus}} \text{RSO}_2\text{N–CH}_2\text{–CHR'–O–R''CH (ring)}$$

(R, R' = H, $CH_2OH$; R'' = alkyl, aryl)

348 Smissman and Bergen, *J. Org. Chem.*, **27**, 2316 (1962).
349 Davies, Ramsey, and Stove, *J. Chem. Soc.*, **1949**, 2633.
350 Fischer, Dangschat, and Stettiner, *Ber.*, **65**, 1032 (1932).
351 Michaël and Jeanprêtre, *Ber.*, **25**, 1678 (1892).
352 Rätz, U.S. pat. 2,722,531 [*C.A.*, **50**, 3503 (1956)].

*C. N-(2-Tetrahydropyranyl)amides.*[353–355]

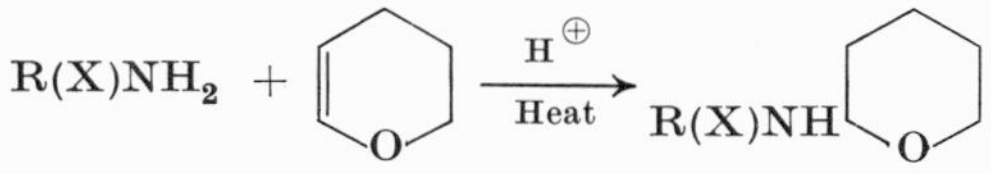

$$R(X)NH_2 + \text{(3,4-dihydro-2H-pyran)} \xrightarrow[\text{Heat}]{H^{\oplus}} R(X)NH\text{-(2-tetrahydropyranyl)}$$

(R = alkyl, aryl; X = CO, $SO_2$)

$$R'NH\text{-(2-tetrahydropyranyl)} + (RCO)_2O \longrightarrow RCON(R')\text{-(2-tetrahydropyranyl)}$$

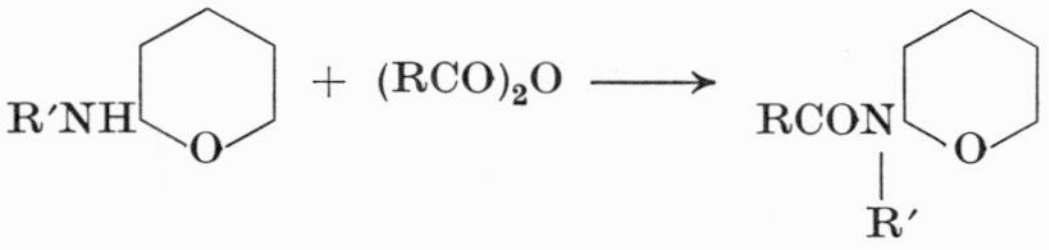

*D. N-[2,3-Dihydropyranyl-(2)]-carbamates and -carbamides.*[356]

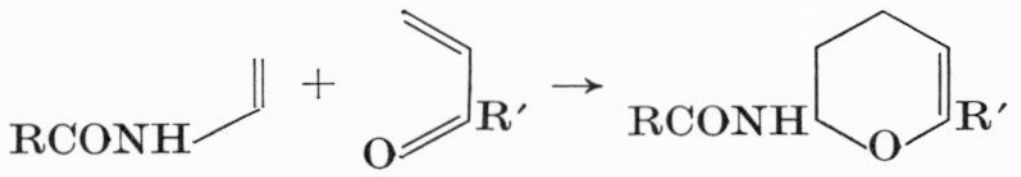

$$RCONH\text{-}CH{=}CH_2 + CH_2{=}CH\text{-}C(R'){=}O \rightarrow RCONH\text{-(2,3-dihydropyranyl-(2))-}R'$$

(R = O-alkyl, NH-alkyl; R′ = H, $CH_3$)

*E. Urons.*[357]

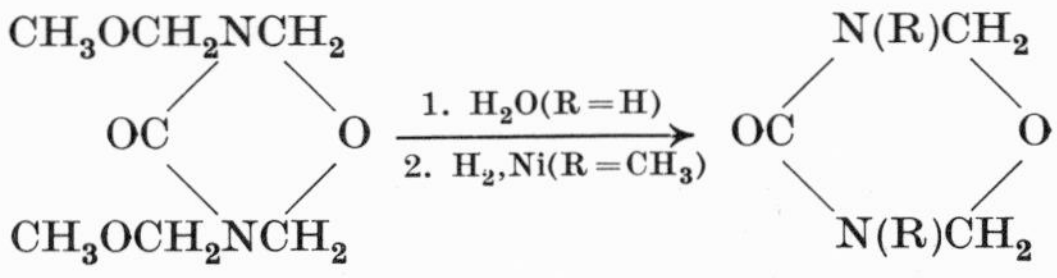

$$OC(N(CH_2OCH_3)CH_2)_2O \xrightarrow[2.\ H_2,Ni(R=CH_3)]{1.\ H_2O(R=H)} OC(N(R)CH_2)_2O$$

*F. Diacyltetrahydro-1,3,5-oxadiazines.*[305]

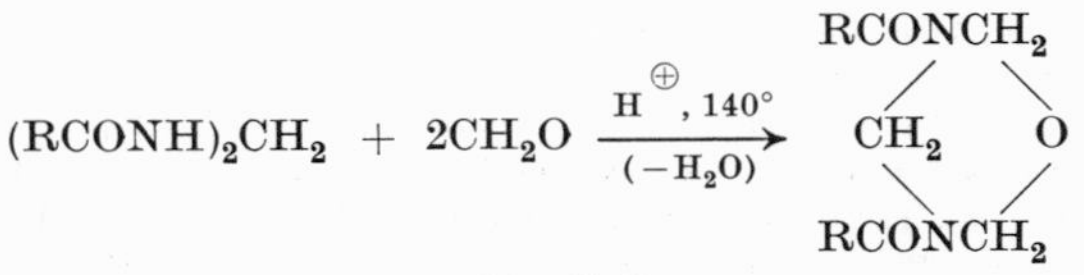

$$(RCONH)_2CH_2 + 2CH_2O \xrightarrow[(-H_2O)]{H^{\oplus},\ 140°} CH_2(N(COR)CH_2)_2O$$

(R = alkyl)

[353] Glacet and Troude, *Compt. rend.*, **253,** 681 (1961).
[354] Glacet and Overbèke, *Compt. Rend.*, **255,** 316 (1962).
[355] Speziale, Ratts, and Marco, *J. Org. Chem.*, **26,** 4311 (1961).
[356] Schulz and Hartmann, *Chem. Ber.*, **95,** 2735 (1962).
[357] Beachem, Oppelt, Cowen, Schickedantz, and Maier, *J. Org. Chem.*, **28,** 1876 (1963).

*G.* 2,3-*Dihydro*-1,3-*benzoxazin*-4-*ones*.[166,278,350,358–374]

$$X\text{-}C_6H_3(CONH_2)(OH) + RCOR' \xrightarrow[(-H_2O)]{H^{\oplus}} \text{X-2,3-dihydro-1,3-benzoxazin-4-one (2-R, 2-R')}$$

(X = H, Cl, Br, acyl; R = alkyl, aryl; R′ = H, alkyl)

*H.* 2,3-*Dihydro*-1,3-*benzthiazin*-4-*ones*.[358,375–377]

$$C_6H_4(CONH_2)(SH) + RCOR' \xrightarrow[(-H_2O)]{H^{\oplus}} \text{2,3-dihydro-1,3-benzthiazin-4-one (2-R, 2-R')}$$

(R = alkyl, aryl; R′ = H, alkyl)

$$C_6H_4(CONH_2)(SH) + C_6H_5CHCl_2 \xrightarrow{\text{Pyridine}} \text{2-}C_6H_5\text{-2,3-dihydro-1,3-benzthiazin-4-one}$$

$$C_6H_4(CO_2H)(SH) + ArCHO + CH_3NH_2 \xrightarrow[(-H_2O)]{C_6H_6} \text{3-}NCH_3\text{-2-Ar-2,3-dihydro-1,3-benzthiazin-4-one}$$

358 Böhme and Boeing, *Arch. Pharm.*, **294**, 556 (1961).
359 Hicks, *J. Chem. Soc.*, **97**, 1032 (1910).
360 Hirwe and Rana, *J. Univ. Bombay*, **8**, Pt. 3, 243 (1939) [*C.A.*, **34**, 2819 (1940)].
361 Horrom and Zaugg, *J. Am. Chem. Soc.*, **72**, 721 (1950).
362 Hughes and Titherley, *J. Chem. Soc.*, **99**, 23 (1911).
363 Irikura, Suzue, and Kasuga, *J. Pharm. Soc. Japan*, **82**, 1079 (1962).
364 Kaufmann, *Arch. Pharm.*, **265**, 226 (1927) (*Chem. Zentr.*, **1927**, II, 88).
365 Keane and Nicholls, *J. Chem. Soc.*, **91**, 264 (1907).
366 Moucka and Rögl, *Ber.*, **59**, 756 (1926).
367 Mowry, Yanko, and Ringwald, *J. Am. Chem. Soc.*, **69**, 2358 (1947).
368 Ohnacker, Brit. pat. 806,729 [*C.A.*, **53**, 14127 (1959)] (same as Ger. pat. 1,028,999).
369 Ohnacker, Ger. pat. 1,021,848 [*C.A.*, **54**, 4633 (1960)].
370 Ohnacker and Scheffler, Ger. pat. 1,063,169 [*C.A.*, **55**, 12432 (1961)].
371 Ohnacker and Scheffler, U.S. pat. 2,943,087 [*C.A.*, **54**, 24818 (1960)].
372 Rana, *J. Indian Chem. Soc.*, **19**, 299 (1942) [*C.A.*, **37**, 2361 (1943)].
373 Titherley, *J. Chem. Soc.*, **91**, 1419 (1907).
374 Titherley and Hicks, *J. Chem. Soc.*, **95**, 908 (1909); Titherley and Hughes, *ibid.*, **97**, 1368 (1910).
375 Böhme and Schmidt, *Arch. Pharm.*, **286**, 330 (1953).
376 Boudet, *Bull. Soc. Chim. France*, **1955**, 1518.
377 Loev, *J. Org. Chem.*, **28**, 2160 (1963).

## CHOICE OF EXPERIMENTAL CONDITIONS

It is difficult to make generalizations about optimum conditions for carrying out the numerous reactions included in this review. Impressions in this regard are best derived from the specific discussions of the foregoing section. A few general remarks can be made at this point, however.

### Choice of Reactants

If the end product sought from an amidoalkylation reaction is the amine, an amidoalkylation reagent with an easily removable acyl group is to be preferred. This is especially important when the reaction product may not resist treatment with the strong acid or base often necessary for acyl cleavage. Thus the phthalimido group of *p*-phthalimidomethylphenol could not be removed, either by the hydrazine method or by hydrolytic means[15] without destroying the molecule. Similarly no way was found to hydrolyze derivatives of the type $ArCH(C_6H_5)NHCOR$ (Ar = hydroxyaryl; R = $CH_3$ or $C_6H_5$) to the corresponding amines.[74] Consistent with these results is the observation that the ease of acid hydrolysis of amides of the type $ArCH_2NHCOR$ (Ar = hydroxyaryl) decreased in the order R = H, $C_2H_5$, $CH_3$, $C_6H_5$, and that the formamide derivative was especially easily hydrolyzed.[75]

To obtain the best yield of monosubstitution product from a relatively reactive substrate it is often desirable to select an amidoalkylating agent of minimal reactivity. Excessive reactivity generally leads to polysubstitution and consequent reduction in yield of the desired product.

When attempts were made to amidomethylate relatively unreactive substrates with N-methylolamides under acid-catalyzed conditions, methylenebisamides, $(RCONH)_2CH_2$, were often encountered as byproducts,[20,47] and sometimes as the only products. (N-Methylolbenzamide is especially prone to disproportionate in this way.) This result is an indication that the use of a more electrophilic reagent, e.g., N-methylolphthalimide, may be in order or that more strongly acidic conditions might be employed to force the reaction in the desired direction.

### Choice of Conditions

In most reactions, monosubstitution is favored by increasing the ratio of subtrate to reagent. In a few favorable instances either mono- or di-substitution can be effected at will, each to the virtual exclusion of the other, merely by using the reactants in the proper stoichiometric ratio. Thus equivalent quantities of *p*-toluidine and N-methylolphthalimide in concentrated sulfuric acid gave the monosubstitution product **108** in 81% yield,[44] whereas two equivalents of the phthalimide reagent gave the

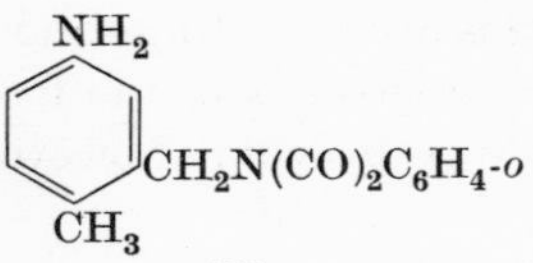

$NH_2$ / $CH_2N(CO)_2C_6H_4$-*o* / $CH_3$

**108**

*o*-$C_6H_4(CO)_2NCH_2$ / $NH_2$ / $CH_2N(CO)_2C_6H_4$-*o* / $CH_3$

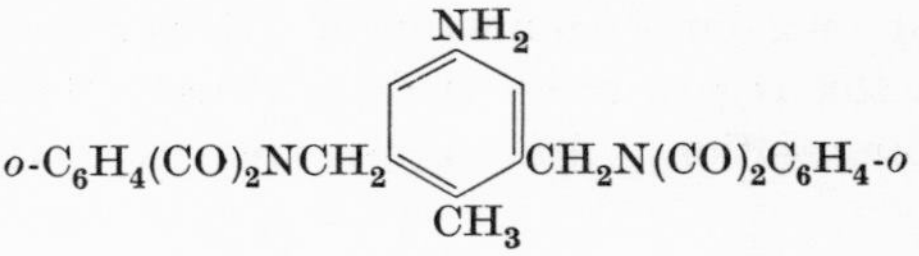

**109**

disubstituted derivative **109** in 95% yield under the same conditions.[29]

Even the concentrations of reactants can be varied to affect the yield. For example, in the reaction of *o*-chlorobenzoic acid with N-methylolchloroacetamide to give the product **110**, the 65% yield obtained was reduced to 40% when the quantity of sulfuric acid was halved.[57] The

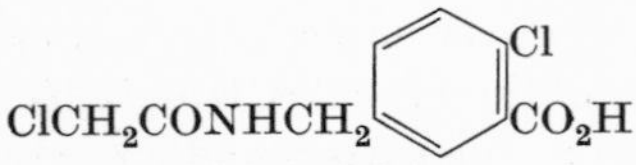

$ClCH_2CONHCH_2$ / Cl / $CO_2H$

**110**

generality of this effect of concentration on yield has not been established, but it should be regarded as a potentially important variable.

## EXPERIMENTAL PROCEDURES

### α-Amidoalkylation of Aromatic Carbon Atoms

**N-(2-Hydroxy-3,5-dinitrobenzyl)phthalimide from 2,4-Dinitrophenol and N-Methylolphthalimide in 5% Oleum.**[69] An intimate mixture of 9 g. (0.049 mole) of 2,4-dinitrophenol and 9 g. (0.051 mole) of N-methylolphthalimide is gradually added to 35 g. of 5% fuming sulfuric acid cooled in ice. After the mixture has stood at room temperature for 10 minutes with occasional stirring, it is warmed slowly and finally heated for 40 minutes on the steam bath. The cooled reaction mixture is then poured in a thin stream into 200 ml. of vigorously stirred ethanol, and the resulting mixture is boiled for 1–2 minutes. After it has cooled, the precipitate is collected and recrystallized from nitrobenzene or glacial acetic acid to give 16.8 g. (95%) of the N-arylmethylphthalimide, m.p. 210–211°.

**2-Chloro-5-(chloroacetylaminomethyl)benzoic Acid from 2-Chlorobenzoic Acid and N-Methylolchloroacetamide in Concentrated Sulfuric Acid.**[57] A mixture of 277.8 g. (1.76 moles) of 2-chlorobenzoic acid and 700 ml. of concentrated sulfuric acid is stirred and cooled to 20°. Powdered N-methylolchloroacetamide (202 g., 1.8 moles) is added over a 30-minute period while the temperature is maintained at 20°. The reaction mixture is stirred overnight, and the solution is poured over ice and allowed to stand for 2 days. The solid is collected, washed with water, dried, and triturated with diethyl ether. The solid (m.p. 126–135°) is again recrystallized from ethanol to give 300 g. (65%) of 2-chloro-5-(chloroacetylaminomethyl)benzoic acid, m.p. 144–146°. Recrystallization from dioxane and then from dimethylformamide raises the melting point to 147–148°.

**N-Benzyl-N-methylacetamide from Benzene and N-Methyl-N-methylolacetamide with Boron Trifluoride.**[32] Through a stirred suspension of 20.6 g. (0.2 mole) of N-methyl-N-methylolacetamide in 100 ml. of dry benzene is passed a slow stream of boron trifluoride while the temperature is maintained at 60° by means of a cold water bath. When the mixture is saturated and no more heat is evolved, it is cooled in an ice bath, then poured into ice water and separated. The aqueous phase is extracted with diethyl ether, and the combined benzene and ether layers are dried over anhydrous sodium sulfate. Filtration and concentration of the filtrate leaves an oil which is fractionally distilled to give 22.0 g. (68%) of N-benzyl-N-methylacetamide, b.p. 91–93°/0.2 mm., m.p. 40–41°.

**N-(2′-Methyl-5′-nitrobenzyl)-2-pyrrolidinone from 4-Nitrotoluene, 2-Pyrrolidinone, and Paraformaldehyde in Concentrated Sulfuric Acid.**[34] To 600 ml. of concentrated sulfuric acid maintained at 20° are added with stirring 274 g. (2.0 moles) of 4-nitrotoluene, 60 g. (2.0 moles) of paraformaldehyde, and 170 g. (2.0 moles) of 2-pyrrolidinone. The solution is allowed to stand at room temperature for 12–15 hours, then at 45° for 6 hours, and finally at 65° for 12 hours. It is then cooled and poured over ice, and the oily product is washed with water by decantation until free of acid. The washed oil is taken up in a minimum amount of warm carbon tetrachloride. The white crystals which deposit on cooling are collected and dried. There is obtained 174 g. (37%) of N-(2′-methyl-5′-nitrobenzyl)-2-pyrrolidinone, m.p. 94–96°.

**N,N′-Diacetyl-4,6-di(aminomethyl)-1,3-xylene from *m*-Xylene, Acetonitrile, and Formaldehyde in 85% Phosphoric Acid.**[5] In a 5-l. three-necked flask fitted with a Hershberg stirrer, condenser, and thermometer are placed 1.5 l. of 85% phosphoric acid, 360 g. (11 moles) of 91% paraformaldehyde, 530 g. (5.0 moles) of *m*-xylene, and 535 g. (13 moles) of acetonitrile. The mixture is heated with vigorous agitation to 65° whereupon a spontaneous reaction occurs. The temperature is held at 65–75° until the exothermic reaction is over, and then at 90° for 4 hours. After cooling there remains a layer of 124–135 g. of unchanged xylene. The viscous acid layer is added in a slow stream with vigorous agitation to 8 l. of ice water containing 3 l. of concentrated ammonium hydroxide. The resulting suspension is stirred overnight, filtered, and the solid washed with dilute ammonium hydroxide and dried at 75° for 24 hours. The yield of crude diamide, m.p. 225–235°, is 600–611 g. (61–66% based on xylene converted). Recrystallization from methanol gives pure N,N′-diacetyl-4,6-di-(aminomethyl)-1,3-xylene, m.p. 245–246°.

**N-(2-Hydroxy-3,5-dimethylbenzyl)-N′-phenylurea from 2,4-Xylenol and the Ether $(C_6H_5NHCONHCH_2)_2O$, in Formic Acid.**[54] A solution of 0.9 g. of 2,4-xylenol and 0.7 g. of the ether in 20 ml. of formic acid is warmed at 50° for 2 hours, poured into water, and allowed to stand for 4 hours. The product is collected and dried. The yield is 90%. Recrystallization from benzene gives the pure diarylurea, m.p. 169°.

**N-Benzylphthalimide from Benzene, N-Bromomethylphthalimide, and Zinc Chloride.**[14] A solution of 24 g. (0.1 mole) of N-bromomethylphthalimide in 50 ml. of dry benzene is treated with 1 g. of anhydrous zinc chloride and then heated under reflux for 2 hours or until the evolution of hydrogen bromide nearly ceases. The cooled red solution is poured into water, separated, and washed to neutrality with aqueous sodium bicarbonate and water. The organic layer is filtered to remove any insoluble material, and the filtrate is then concentrated. The residual solid (22.6 g., 94%, m.p. 110–113°) is recrystallized once from 100 ml. of ethanol to give 20.2 g. of N-benzylphthalimide, m.p. 113–114°.

**2- and 4-Methoxybenzylamine from Anisole and N-Chloromethylphthalimide (Zinc Chloride Catalyst).**[378] A mixture of 280 g. (2.9 moles) of anisole, 215 g. (1.1 moles) of N-chloromethylphthalimide, and 11 g. of anhydrous zinc chloride is heated for 2 hours at 120–140°. The excess anisole is removed by steam distillation, and the layer of water is decanted from the pasty residue which crystallizes after trituration with a little ethanol.

The mixture of N-(2- and 4-methoxybenzyl)phthalimide is collected and hydrolyzed directly by stirring it for 2-3 hours at room temperature with a mixture of 250 ml. of ethanol and 1.5 l. of 12% aqueous sodium hydroxide. The mixture is then concentrated under reduced pressure to two-thirds of its original volume and treated carefully with 700 ml. of concentrated hydrochloric acid. After the mixture has been heated on the steam bath for several hours, it is cooled, filtered to remove insoluble material, and the filtrate is concentrated to dryness. The residue is made alkaline with 35% sodium hydroxide solution, and the base which is released is taken up in benzene and dried over anhydrous potassium carbonate. Filtration, removal of the benzene, and distillation of the

[378] Shirakawa and Kawasaki, *J. Pharm. Soc. Japan*, **71**, 1213 (1951) [*C.A.*, **46**, 5544 (1952)].

residue under reduced pressure give a crude base, b.p. 120–140°/21 mm. Redistillation at atmospheric pressure gives 98 g. (65%) of a mixture of 2- and 4-methoxybenzylamine, b.p. 231-235°.

The mixture of amines is then added dropwise with stirring to 330 g. of 10% anhydrous ethanolic hydrogen chloride. After it has stood overnight at room temperature, the solid is collected, washed with 100 ml. of ethanol, and dried to give 73 g. (38%) of 4-methoxybenzylamine hydrochloride, m.p. 223–230°. Recrystallization from ethanol raises the melting point to 231–233°.

The alcoholic filtrate is concentrated to dryness under reduced pressure to give crude 2-methoxybenzylamine hydrochloride corresponding to a 19% yield based on the original N-chloromethylphthalimide. The crude product can be purified by solution in water and treatment with a hot aqueous solution of picric acid. The 2-methoxybenzylamine picrate, m.p. 227–228°, which is formed can be reconverted through the base, b.p. 120–122°/16 mm., to pure 2-methoxybenzylamine hydrochloride, m.p. 148–149° (from ethanol).

**4-Phthalimidomethyl-1-naphthol-8-sulfonic Acid γ-Sultone from 1,8-Naphthsultone and N-Chloromethylphthalimide with Aluminum Chloride.**[379] A stirred mixture of 62 g. (0.3 mole) of 1,8-naphthsultone, 71 g. (0.36 mole) of N-chloromethylphthalimide, and 52 g. (0.39 mole) of anhydrous aluminum chloride in 500 ml. of 1,2,4-trichlorobenzene is slowly heated to 110° over a 2-hour period and then kept at that temperature for another 2 hours and 15 minutes. The cooled reaction mixture is poured over 500 g. of ice containing 30 ml. of concentrated hydrochloric acid, and the trichlorobenzene is removed by steam distillation. The glassy residue is boiled with 300 ml. of ethyl acetate, and the resulting crystalline solid is collected and washed successively with ethyl acetate and hot water. The dried product weighs 95 g. (97%), m.p. 248–249°. Recrystallization from a chloroform-ethanol mixture and then from glacial acetic acid gives pure 4-phthalimidomethyl-1-naphthol-8-sulfonic acid γ-sultone, m.p. 252–253°.

**α-Benzoylamino-α-phenyl-*o*-cresol from Phenol and N,N′-Benzylidenebisbenzamide.**[74] A mixture of 6.6 g. (0.02 mole) of the bisamide and 1.88 g. (0.02 mole) of phenol is heated for 4 hours at 180–190° in an oil bath. The brown, hard, resinous material is dissolved in glacial acetic acid, and the product is precipitated by the addition of water. The precipitate is collected, washed with water, and dried to give 4.2 g. (69%) of a yellow powder, m.p. 118°. The powder is dissolved in the minimum amount of ethyl acetate, filtered from a small quantity of insoluble material, and recovered once again by removal of the ethyl acetate. Four

[379] Schetty, *Helv. Chim. Acta*, **31**, 1229 (1948).

recrystallizations from ethanol give pure α-benzoylamino-α-phenyl-*o*-cresol, m.p. 214°.

**2-(α-Acetylaminobenzyl)-4-nitrophenol from 4-Nitrophenol, N,N′-Benzylidenebisacetamide, and Phosphorus Oxychloride.**[18] A mixture of 3.5 g. (0.025 mole) of 4-nitrophenol, 5.76 g. (0.028 mole) of the bisamide, and 1.84 g. (0.012 mole) of phosphorus oxychloride is heated (94–96°) on the steam bath for 1 hour. The cooled mixture is decomposed with water and neutralized with sodium bicarbonate. The resulting solid is collected, washed, and dried. Recrystallization from ethanol gives 5.8 g. (81%) of pure 2-(α-acetylaminobenzyl)-4-nitrophenol, m.p. 208–209°.

## α-Amidoalkylation of Aliphatic Carbon Atoms

**2-Benzamidomethylcyclohexane-1,3-dione (50) from N-Methylolbenzamide and Cyclohexane-1,3-dione.**[85] A solution of 2.24 g. (0.02 mole) of cyclohexane-1,3-dione and 3.02 g. (0.02 mole) of N-methylolbenzamide in 20 ml. of acetic acid is treated with 1.5 g. (0.01 mole) of boron trifluoride etherate. After 20 hours a first crop of crystals is obtained (scratching may be necessary) and collected. The filtrate is mixed with water containing 2.5 g. of sodium acetate, and a second crop of colorless crystals is obtained. The combined amount is 4.0 g. of crystals, m.p. 157–160°. The crude product is digested with 100 ml. of cold 1*N* sodium hydroxide and the insoluble by-product, methylenebisbenzamide, is removed by filtration. The filtrate is acidified with concentrated hydrochloric acid, and the resulting solid is collected and recrystallized from ethanol to give 3.2 g. (65%) of pure product, m.p. 162–164°.

**N,N′-Bis-(β-trinitroethyl)fumaramide (54) from N,N′-Bis(benzoxymethyl)fumaramide and Trinitromethane.**[93] A mixture of 50 ml. of dry nitromethane, 0.38 g. (1.0 mmole) of N,N′-bis(benzoxymethyl)fumaramide, and 0.45 g. (3.0 moles) of trinitromethane is heated under reflux for 4 hours. The solvent is then removed in vacuum, and the solid residue is extracted with boiling carbon tetrachloride in order to remove the benzoic acid. The residue, which is insoluble in carbon tetrachloride, is dissolved in absolute ethanol, and water is added to the point of turbidity. On standing, 0.20 g. (47%) of product, m.p. 197° (dec.), crystallizes.

**Diethyl Methylbenzamidomethylmalonate, $C_6H_5CONHCH_2$-$C(CH_3)(CO_2C_2H_5)_2$, from N-Chloromethylbenzamide and Diethyl Methylmalonate.**[100] A suspension of 4.2 g. (0.025 mole) of N-chloromethylbenzamide and 4.7 g. (0.024 mole) of diethyl sodiomethylmalonate in 20 ml. of dry diethyl ether is heated on a steam bath under reflux for 1 hour. The slimy precipitate is removed by filtration and washed with ether. The ether fractions are combined, and the ether is removed by distillation. The viscous yellow residue is distilled under reduced pressure. Any unchanged diethyl methyl malonate distils as the oil bath is heated to 180°. The fraction distilling at 180°/0.05 mm. is collected, and it eventually crystallizes as a white, waxy substance. It is recrystallized from diisoamyl ether to give 5.8 g. (78%) of product, m.p. 67°.

**Diethyl α-Acetamidobenzylmalonate, $CH_3CONHCH(C_6H_5)$-$CH(CO_2C_2H_5)_2$, from Benzylidenebisacetamide and Diethyl Malonate.**[115] A mixture of 10.3 g. (0.05 mole) of N,N′-benzylidenebisacetamide, 8 g. (0.05 mole) of diethyl malonate, and 25 ml. of acetic anhydride is heated for 3 hours in an oil bath at 150–155°. (Prolonged heating leads to diethyl benzylidenemalonate; e.g., after heating for 9 hours the yield drops to 11%.) The acetic anhydride is removed under reduced pressure (14 mm.), and 100 ml. of water is added with stirring to the remaining viscous mass. The colorless crystals that separate are collected and washed with cold water and a little diethyl ether. The dried product, m.p. 83°, weighs 8.9 g. An additional 0.7 g. is obtained from the ether filtrate (total yield: 9.6 g., 62%). Recrystallization from 50% ethanol gives the pure product, m.p. 85°. It is readily soluble in ethanol and acetone, sparingly soluble in ether, and insoluble in water.

**Diethyl α-Benzamidomethyl-α-cyanopimelate (82) from N-(Diethylaminomethyl)benzamide and Diethyl α-Cyanopimelate.**[118] A mixture of 2.4 g. (0.01 mole) of diethyl α-cyanopimelate, 2.06 g. (0.01 mole) of N-(diethylaminomethyl)benzamide, 0.01 g. of powdered sodium hydroxide, and 30 ml. of toluene is heated at reflux under nitrogen. After 5 hours 80% of the theoretical quantity of diethylamine is evolved. The mixture is then cooled, and 100 ml. of petroleum ether is added. An oil separates which crystallizes after 3 days. This solid is recrystallized from petroleum ether to give 3.2 g. (85%) of the product **82**, m.p. 61°.

**β-Benzamido-β,β-diphenylpropionitrile from N-Benzoyldiphenylketimine and Acetonitrile.**[120] To a stirred solution of 1.49 g. (0.065 mole) of lithium amide in 200 ml. of liquid ammonia is added, during 5 minutes, a solution of 1.25 g. (0.03 mole) of acetonitrile in 10 ml. of diethyl ether. The mixture is stirred for 15 minutes and then, over another 5 minute period, is added a solution of 8.60 g. (0.03 mole) of N-benzoyldiphenylketimine [$C_6H_5CON{=}C(C_6H_5)_2$] in a mixture of benzene (100 ml.) and ether (50 ml.). After stirring for 1 hour, 3.50 g. of solid ammonium chloride is added, the ammonia is removed by distillation, and 100 ml. of water is added to the residue. The organic layer is separated, washed with water, and dried over anhydrous magnesium sulfate. Filtration and removal of the solvent by distillation give 8.40 g.

(85%) of material, m.p. 132–134°. Recrystallization from ethanol gives pure product, m.p. 134–135°.

**5-Carbethoxy-2-keto-6-methyl-4-phenyl-1,2,3,4-tetrahydropyrimidine (90) from Ethyl Acetoacetate, Benzaldehyde, and Urea.**[126] A mixture of 53 g. (0.5 mole) of benzaldehyde, 30 g. (0.5 mole) of urea, 97.5 g. (0.75 mole) of ethyl acetoacetate, 200 ml. of absolute ethanol, and 40 drops of concentrated hydrochloric acid is heated under reflux for 3 hours. The mixture is then cooled to 0°, and the crystallized pyrimidine (93.6 g.) is collected and dried at 50°. The filtrate is heated under reflux for 2 hours and then distilled until 155 ml. of ethanol is collected. On cooling the residue an additional 21.3 g. of the pyrimidine is obtained, giving a total crude yield of 114.9 g. For purification the crude product is divided into two equal portions, and each is dissolved in 800 ml. of boiling 95% ethanol. On cooling, the pyrimidine separates to give 102 g. (78%) of colorless crystals, m.p. 202–204°. The loss on recrystallization may be materially decreased by distilling the solvent to incipient crystallization. The pyrimidine dissolves slowly in ethanol, and an excess of solvent is needed to effect solution.

## Preparation of Some Electrophilic Reagents

**N-Methyloltrifluoroacetamide, $CF_3CONHCH_2OH$.**[380] Trifluoroacetamide (4.4 g., 0.039 mole), potassium carbonate (0.16 g.), and formalin (3.2 ml., 0.042 mole) are mixed. A homogeneous solution forms, and the product crystallizes almost immediately. After refrigeration overnight the product is collected and recrystallized from chloroform to yield 4.95 g. (88%) of filamentary crystals. One more recrystallization from the same solvent gives N-methyloltrifluoroacetamide, m.p. 105–105.5°, with a satisfactory elemental analysis.

**N-Methylolphthalimide.**[84] A mixture of 511 g. (3.48 moles) of phthalimide, 260 ml. of 40% formaldehyde solution, and 1.75 l. of water is heated until a clear solution results. This requires only 5 minutes after the boiling point is reached. The solution is refrigerated overnight, the product is removed by filtration, washed with ice water, and air-dried. The yield of N-methylolphthalimide, m.p. 137–141°, is 594 g. (96%), and the product is sufficiently pure for most purposes. The material should not be dried at elevated temperatures because it loses formaldehyde. Recrystallization from ethanol does not improve the melting point.

If a pure sample is needed, the following procedure may be used.[193] A solution obtained by warming 17.7 g. of N-methylolphthalimide in 30 ml.

[380] W. R. Sherman, unpublished data.

of pure pyridine is filtered, if necessary, and left to crystallize. If crystallization does not occur, seed crystals are obtained by placing a few drops of the solution in a desiccator over concentrated sulfuric acid. As soon as the first crystals appear, they are added to the solution. The pyridine complex crystallizes in long lustrous needles which are collected after cooling in an ice bath. On drying in vacuum over concentrated sulfuric acid, the crystals lose their luster and come to constant weight after 24 hours. The dried product melts at 148.5–149°, and one recrystallization from acetone gives pure N-methylolphthalimide, m.p. 149.5°.

**N-Chloromethyltrichloroacetamide.**[186] A cold stirred solution of 19.2 g. (0.1 mole) of N-methyloltrichloroacetamide[264] in 10 ml. of dry diethyl ether is treated with a suspension of 20 g. (0.096 mole) of phosphorus pentachloride in 20 ml. of dry diethyl ether. After the initial reaction and evolution of hydrogen chloride moderates, the mixture is heated for 10 minutes under reflux and allowed to stand overnight. Ice water is then added dropwise to the cooled, stirred solution. The precipitate is collected and washed with ice water. Two recrystallizations from lukewarm carbon tetrachloride give 16 g. (76%) of N-chloromethyltrichloroacetamide, m.p. 76–77°.

**N-Bromomethylphthalimide.**[104] A mixture of 80 g. (0.45 mole) of N-methylolphthalimide, 150 ml. of 48% hydrobromic acid, and 45 ml. of concentrated sulfuric acid is stirred for 2 hours at 50–60°. The crystalline product is collected and washed with water and then with dilute aqueous ammonia to remove all acid. Drying at 80° gives 75 g. (69%) of N-bromomethylphthalimide, m.p. 146–147°. Recrystallization from acetone gives a purer product, m.p. 148°.

**N-Acetoxymethylstearamide from Stearamide and Paraformaldehyde in Acetic Acid-Acetic Anhydride.**[138] A solution of 200 g. (0.71 mole) of stearamide and 25 g. (0.83 mole) of paraformaldehyde in a mixture of 200 ml. of acetic anhydride and 400 ml. of glacial acetic acid is heated at 70° for 4 hours and then cooled. The precipitate is collected and dried. Recrystallization from acetone gives N-acetoxymethylstearamide, m.p. 92–93°, in 66% yield.

**N,N′-Methylenebis-*p*-toluamide from *p*-Tolunitrile.**[4] A solution of trioxane (1.5 g., .05 mole) in *p*-tolunitrile (11.7 g., 0.1 mole) is added slowly with stirring to 38 ml. of an 85% solution of sulfuric acid in a 125-ml. three-necked flask. The temperature is maintained at 30° by means of an ice bath. After 3 hours the solution is poured into 300 ml. of ice and water. The product separates as white crystals which are collected, washed with water, and recrystallized from 95% ethanol. The yield of N,N′-methylenebis-*p*-toluamide, m.p. 209–210°, is 12.4 g. (83%).

**N,N′-Methylenebischloroacetamide from N-Methylolchloroacetamide.**[11] N-Methylolchloroacetamide (10 g., 0.081 mole) is dissolved in 25 g. of concentrated sulfuric acid with cooling. The solution is allowed to stand overnight and is then poured over ice. The precipitate is collected and the filtrate saturated with salt to obtain more of the product which is collected and combined with the main product. The yield of crude material is 95%. Recrystallization from 95% ethanol gives pure N,N′-methylenebischloroacetamide, m.p. 175°.

**N,N′Benzylidenebisacetamide.**[283] A solution of 10.6 g. (0.1 mole) of benzaldehyde and 11.8 g. (0.2 mole) of acetamide in 100 ml. of benzene is heated under reflux for 48 hours under a Soxhlet extractor containing 50 g. of anhydrous magnesium sulfate. The precipitate is filtered, washed with acetone, and recrystallized from ethanol to give a 72% yield of N,N′-benzylidenebisacetamide, m.p. 239.5°.

**N,N′-Benzylidenebisbenzamide.**[85] A mixture of freshly distilled benzaldehyde (21.2 g., 0.2 mole), benzamide (48.4 g., 0.4 mole), and freshly distilled acetic anhydride (50 ml.) is warmed until a clear solution results. Four drops of concentrated hydrochloric acid is added, and the solution is heated on a steam bath for 2 hours. The acetic anhydride is removed by distillation under reduced pressure, and the residue is taken up in hot ethanol. Cooling gives 51.7 g. (78%) of crystalline N,N′-benzylidenebisbenzamide, m.p. 231.5–232.5°.

**Stearamidomethylpyridinium Chloride, $n$-$C_{17}H_{35}CONHCH_2$-$\overset{\oplus}{N}C_5H_5Cl^{\ominus}$.**[138] Stearamide (141 g., 0.5 mole) and paraformaldehyde (20 g., 0.67 mole) are added to a solution of 0.5 mole of pyridine hydrochloride in 2 l. of pyridine. The mixture is heated at 80° for 10 hours, cooled, and filtered. An equal volume of cold acetone is added to the filtrate, and the precipitate is collected and added to the original crop. The combined solids are recrystallized from acetone to yield 160 g. (78%) of stearamidomethylpyridinum chloride, melting at about 135°. The material sinters at 92°, and the final melting point varies with the rapidity of heating, probably because of partial conversion to methylenedistearamide.

## TABULAR SURVEY

The tables fall into three groups. Tables I to VII summarize the α-amidoalkylation reactions of aromatic carbon atoms, Tables VIII to XIX those involving substitution at aliphatic carbon atoms, and Tables XX to XXXVI list the various electrophilic reagents that have been used in the α-amidoalkylation of carbon atoms.

The literature has been surveyed through 1962, but a number of more recent references have been included in both the text and the tables.

Exhaustive coverage has been attempted for the first two parts of this survey, but no such claim can be made for the third. The widely dispersed way in which these compounds appear in the literature makes omission of many examples a virtual certainty. Sufficient numbers have been listed, however, to provide a representative sampling of possible structural variation in the electrophilic reagent.

Because the N-aminomethylamides have been thoroughly reviewed recently,[3] Table XXVIII is intentionally incomplete. It is devoted exclusively to examples which were omitted from the review cited or which have been reported since its publication.

Within each table, reactions or reagents are arranged in order of increasing complexity of the amide portion of the amidoalkylating reagent. The nucleophilic substrates are generally arranged according to increasing complexity of substituent groups as listed in Heilbron's *Dictionary of Organic Compounds*.[381] Where applicable, the order aliphatic, aromatic, heterocyclic is also followed, but arylaliphatic aldehydes, acids, and amides are treated as derivatives of the aliphatic portion of the molecule. Compounds containing reactive methylene or methine groups are listed in the order nitro, aldehydo, keto, ester, polynitro, diketo, nitro ester, keto ester, diester, cyano, cyano ketone, and cyano ester.

Except where a separate column is provided for them, yields are given in parentheses after the references to which they relate.

[381] Heilbron, *Dictionary of Organic Compounds*, Oxford University Press, New York, 1953, pp. xv–xvi.

TABLE I

AMIDOMETHYLATION OF AROMATIC CARBON ATOMS WITH N-MONOMETHYLOL DERIVATIVES OF CARBOXAMIDES AND CARBAMYL COMPOUNDS

$RCONHCH_2OH + ArH \rightarrow RCONHCH_2Ar$

| Product Derived from | Position(s) of Substituent(s) | Method | References (Yield, %) |
|---|---|---|---|
| $HCONHCH_2OH$ and | | | |
| 4-Nitrophenol | *x*- | Concd. $H_2SO_4$ | 22 |
| $CH_3CONHCH_2OH$ and | | | |
| 4-Nitrophenol | *x*,*x*- | Concd. $H_2SO_4$ | 23 |
| 2-Naphthol | 1- | HCl, $C_2H_5OH$ | 96 (95) |
| Thiophene | 2- | 85 % $H_3PO_4$, 80° | 49 (50) |
| Ethyl 2,5-dimethylpyrrole-3-carboxylate | 4- | HCl, $C_2H_5OH$ | 17 |
| $CH_3CON(CH_3)CH_2OH$ and | | | |
| Benzene | 1- | $BF_3$, $C_6H_6$ | 32 (68) |
| 2-Naphthol | 1- | HCl, $C_2H_5OH$ | 32 (94) |
| $CH_3CON(C_3H_7\text{-}n)CH_2OH$ and | | | |
| 2-Naphthol | 1- | HCl, $C_2H_5OH$ | 32 |
| $CH_2ClCONHCH_2OH$ and | | | |
| Benzene | 1- | Concd. $H_2SO_4$ | 13 (88) |
| Chlorobenzene | 2- and 4- | Concd. $H_2SO_4$ | 13 (80) |
| Phenol | 4- | $AlCl_3$, $CH_3CO_2H$ | 51 |
| 2-Nitrophenol | 4- | Concd. $H_2SO_4$ | 11 (90) |
| 4-Nitrophenol | 2- | Concd. $H_2SO_4$ | 11 (90), 20 |
| Catechol | *x*- | 3 % HCl, $H_2O$ | 11 |
| Hydroquinone | *x*,*x*- | 1 % HCl, $C_2H_5OH$ | 11 (20) |
| Pyrogallol | *x*,*x*- | 1 % HCl, $C_2H_5OH$ | 11 (15) |

| | | | |
|---|---|---|---|
| 4-Nitrophenetole | 2,6- | Concd. $H_2SO_4$ | 11 (80) |
| Guaiacol | 4- | 20 % $H_2SO_4$, $C_2H_5OH$ | 11, 382 |
| | 4- | $AlCl_3$, $CH_3CO_2H$ | 51 (64) |
| Veratrole | 4- and 2,4- | $ZnCl_2$, $CH_3CO_2H$ | 52 |
| 1,2,3-Trimethoxybenzene | 4- | $ZnCl_2$, $CH_3CO_2H$ | 52 |
| 1,2,4-Trimethoxybenzene | 5- | $ZnCl_2$, $CH_3CO_2H$ | 52 |
| Acetanilide | 4- | Concd. $H_2SO_4$ | 11 (100), 20 |
| 4-Hydroxyacetanilide | 3- | $AlCl_3$, $CH_3CO_2H$ | 51 (16) |
| Acetophenetidine | *x*- | Concd. $H_2SO_4$ | 11 (70) |
| 4-($CH_3CHOHCONH$)$C_6H_4OC_2H_5$ | *x*- | Concd. $H_2SO_4$ | 11 (70) |
| Toluene | 4- | $H_2SO_4$, $CH_3CO_2H$ | 13 (75) |
| | 4- | 70 % $H_2SO_4$ | 13 (50) |
| | 2,4- | Concd. $H_2SO_4$ | 13 (47) |
| *m*-Xylene | 4- | $H_2SO_4$, $CH_3CO_2H$ | 13 (94) |
| | 4,6- | Concd. $H_2SO_4$ | 13 (35) |
| *p*-Xylene | 2- | $H_2SO_4$, $CH_3CO_2H$ | 13 (92) |
| Thymol | *x*- | 1 % HCl, $C_2H_5OH$ | 11 (20) |
| Phenylacetic acid | 4- | Concd. $H_2SO_4$ or HF | 47 (35) |
| Cinnamic acid | 3- and 4- | Concd. $H_2SO_4$ | 21 (68) |
| Benzoic acid | 3- | Concd. $H_2SO_4$ | 11 (54), 20 |
| 2-Chlorobenzoic acid | 5- | Concd. $H_2SO_4$ | 57 (65) |
| Salicylic acid | *x,x*- | Concd. $H_2SO_4$ | 11 (40), 20 |
| 3,4,5-Trimethoxybenzoic acid | 2- | $ZnCl_2$, $CH_3CO_2H$ | 52 |
| *o*-Toluic acid | 5(?)- | Concd. $H_2SO_4$ | 7 |
| *p*-Toluic acid | 3- | Concd. $H_2SO_4$ | 7 |
| Tetralin | *x*- | $H_2SO_4$, $CH_3CO_2H$ | 13 (39) |
| | *x,x*- | Concd. $H_2SO_4$ | 13 (46) |
| Naphthalene | *x*- | $H_2SO_4$, $CH_3CO_2H$ | 13 (85) |
| | *x,x,x*- | Concd. $H_2SO_4$ | 13 (28) |
| 2-Naphthol | 1- | Satd. HCl, $C_2H_5OH$ | 26 (81) |
| 2-Methoxynaphthalene | 1- | Satd. HCl, $C_2H_5OH$ | 26 (60) |

*Note:* References 382 to 537 are on pp. 266–269.

TABLE I—*Continued*

AMIDOMETHYLATION OF AROMATIC CARBON ATOMS WITH N-MONOMETHYLOL DERIVATIVES OF CARBOXAMIDES AND CARBAMYL COMPOUNDS

$RCONHCH_2OH + ArH \rightarrow RCONHCH_2Ar$

| Product Derived from | Position(s) of Substituent(s) | Method | References (Yield, %) |
|---|---|---|---|
| $CH_2ClCONHCH_2OH$ (*contd.*) and | | | |
| 1-Hydroxy-2-naphthoic acid | 4- | $H_2SO_4$, $CH_3CO_2H$ | 40 |
| 2-Hydroxy-3-naphthoic acid | 1- | $H_2SO_4$, $CH_3CO_2H$ | 40 |
| 9-Fluorenone | 2,7- | Concd. $H_2SO_4$ | 7 |
| 2-Nitro-9-fluorenone | 7- | Concd. $H_2SO_4$ | 7 |
| 9,10-Phenanthrenequinone | 2(?)- | Concd. $H_2SO_4$ | 7 |
| 7H-Benz[de]anthracen-7-one (benzanthrone) | 3,9(?)- | Concd. $H_2SO_4$ | 7 |
| Methyl 2-furoate | 5- | Concd. $H_2SO_4$ | 35 (100) |
| Ethyl 2-furoate | 5- | Concd. $H_2SO_4$ | 383 |
| | 5- | PPA,* $CH_3CO_2H$ | 48 (51) |
| 2-Thiophenecarboxylic acid | 4(?)- | Concd. $H_2SO_4$ | 56 (53) |
| Ethyl 2,4-dimethylpyrrole-3-carboxylate | 5- | HCl, $C_2H_5OH$ | 17 (good) |
| Ethyl 2,5-dimethylpyrrole-3-carboxylate | 4- | HCl, $C_2H_5OH$ | 17 (good) |
| Acridine | *x*- | Concd. $H_2SO_4$ | 384 |
| 3-Methyl-1-phenyl-5-pyrazolone | 4- | Concd. $H_2SO_4$ | 11 |
| 2,3-Dimethyl-1-phenyl-5-pyrazolone (antipyrine) | 4- | $AlCl_3$, $CH_3CO_2H$ | 51 (70) |
| | 4- | 100% $H_3PO_4$, $CH_3CO_2H$ | 48 (48) |
| $CCl_3CONHCH_2OH$ and | | | |
| 2-Hydroxybenzophenone | 3,5(?)- | Concd. $H_2SO_4$ | 7 |
| 4-Hydroxybenzophenone | 3,5(?)- | Concd. $H_2SO_4$ | 7 |
| 2,4-Dimethylbenzophenone | 5(?)- | Concd. $H_2SO_4$ | 7 |
| *o*-Toluic acid | 5(?)- | Concd. $H_2SO_4$ | 7 |
| *p*-Toluic acid | 3- | Concd. $H_2SO_4$ | 7 |
| 2,4-Dimethylbenzoic acid | 5- | Concd. $H_2SO_4$ | 7 |

| | | | |
|---|---|---|---|
| 9-Fluorenone | 2,7- | Concd. $H_2SO_4$ | 7 |
| 2-Nitro-9-fluorenone | 7- | Concd. $H_2SO_4$ | 7 |
| 2-Hydroxy-9-fluorenone | 7- | Concd. $H_2SO_4$ | 7 |
| 2-Acetamido-9-fluorenone | 7- | Concd. $H_2SO_4$ | 7 |
| 1,2-Acenaphthenequinone | 4- | Concd. $H_2SO_4$ | 7 |
| 1-Hydroxyanthraquinone | 4- | Concd. $H_2SO_4$ | 385 |
| 2-Hydroxyanthraquinone | 1- | Concd. $H_2SO_4$ | 37, 385, 386 |
| 2-Hydroxy-1-chloroanthraquinone | 3- | Concd. $H_2SO_4$ | 39 |
| 2-Hydroxy-3-chloroanthraquinone | 1- | Concd. $H_2SO_4$ | 37, 387 |
| 1,2-Dihydroxyanthraquinone | *x*- | Concd. $H_2SO_4$ | 36, 37 |
| 1,5-Dihydroxyanthraquinone | 2,4,6,8- | Concd. $H_2SO_4$ | 385 |
| 1,8-Dihydroxyanthraquinone | 2,4,5,7- | Concd. $H_2SO_4$ | 385 |
| 2,3-Dihydroxyanthraquinone | 1,4- | Concd. $H_2SO_4$ | 37, 385 |
| 1-Hydroxy-2-methylanthraquinone | 4- | Concd. $H_2SO_4$ | 37, 385 |
| 2-Hydroxy-3-methylanthraquinone | 1- | Concd. $H_2SO_4$ | 37, 385 |
| 1,3-Dimethylanthraquinone | 4- | Concd. $H_2SO_4$ | 7 |
| 2-Hydroxyanthraquinone-3-carboxylic acid | 1- | Concd. $H_2SO_4$ | 385 |
| 9,10-Phenanthrenequinone | 2(?)- and 2,7(?)- | Concd. $H_2SO_4$ | 7 |
| 2-Nitro-9,10-phenanthrenequinone | 7(?)- | Concd. $H_2SO_4$ | 7 |
| 2-Hydroxy-9,10-phenanthrenequinone | 7(?)- and 3,7(?)- | Concd. $H_2SO_4$ | 7 |
| 7H-Benz[de]anthracen-7-one (benzanthrone) | 3,9(?)- | Concd. $H_2SO_4$ | 7 |
| 3-Bromo-7H-benz[de]anthracen-7-one | 9(?)- | Concd. $H_2SO_4$ | 7 |
| 3-Nitro-7H-benz[de]anthracen-7-one | 9(?)- | Concd. $H_2SO_4$ | 7 |
| Xanthen-9-one | 2(?)- and 2,4,5,7(?)- | Concd. $H_2SO_4$ | 7 |
| $C_6H_5CH_2CONHCH_2CONHCH_2OH$ and | | | |
| *p*-Cresol | 2- | 1 % HCl, $C_2H_5OH$ | 79 |
| 2-Naphthol | 1- | HCl, $C_2H_5OH$ | 79 |
| $C_6H_5CONHCH_2CONHCH_2OH$ and | | | |
| 2-Naphthol | 1- | HCl, $C_2H_5OH$ | 79 |

*Note:* References 382 to 537 are on pp. 266–269.
*Polyphosphoric acid.

TABLE I—*Continued*

AMIDOMETHYLATION OF AROMATIC CARBON ATOMS WITH N-MONOMETHYLOL DERIVATIVES OF CARBOXAMIDES AND CARBAMYL COMPOUNDS

$$RCONHCH_2OH + ArH \rightarrow RCONHCH_2Ar$$

| Product Derived from | Position(s) of Substituent(s) | Method | References (Yield, %) |
|---|---|---|---|
| $C_6H_5CH_2CONHCH_2OH$ and | | | |
| N,N-Dimethylaniline | 4(?)- | HCl, $C_2H_5OH$ | 388 |
| *p*-Cresol | 2- | 1% HCl, $C_2H_5OH$ | 79 |
| 2-Naphthol | 1- | 1% HCl, $C_2H_5OH$ | 388 (84) |
| 6-Bromo-2-naphthol | 1- | 1% HCl, $C_2H_5OH$ | 388 (85) |
| 3,6-Dibromo-2-naphthol | 1- | HCl, $C_2H_5OH$ | 388 (80) |
| $C_6H_5CH_2CONHCH(CH_3)CONHCH_2OH$ and 2-Naphthol | 1- | HCl, $C_2H_5OH$ | 79 |
| $C_6H_5CH_2CH_2CONHCH_2OH$ and 2-Naphthol | 1- | HCl, $C_2H_5OH$ | 79 |
| N-Methylol-2-pyrrolidinone and 2-Naphthol | 1- | HCl, $C_2H_5OH$ | 33 (50) |
| $(CH_3)_2CHCH_2CONHCH_2OH$ and | | | |
| 2,4-Xylenol | 6- | | 54 |
| 3,3-Bis(*p*-hydroxyphenyl)oxindole (phenolisatin) | *x*- and *x*,*x*- | $ZnCl_2$, $CH_3CO_2H$ | 53 |
| *n*-$C_7H_{15}CONHCH_2OH$ and | | | |
| 3,3-Bis(*p*-hydroxyphenyl)oxindole (phenolisatin) | *x*- and *x*,*x*- | $ZnCl_2$, $CH_3CO_2H$ | 53 |
| 3,3-Bis(*p*-hydroxyphenyl)phthalide (phenolphthalein) | *x*,*x*- | $ZnCl_2$, $CH_3CO_2H$, 50° | 53 |

| | | | |
|---|---|---|---|
| $n$-$C_8H_{17}CONHCH_2OH$ and | | | |
| Phenol | 4(?)- | $AlCl_3$, $CH_3CO_2H$ | 51 (63) |
| Guaiacol | 4- | $AlCl_3$, $(CH_3)_2CO$ | 51 (73) |
| $n$-$C_9H_{19}CONHCH_2OH$ and | | | |
| Guaiacol | 4- | $AlCl_3$, $CH_3CO_2H$ | 51 (65) |
| | 4- | 100 % $H_3PO_4$, $(CH_3)_2CO$ | 48 (61) |
| $CH_2{=}CH(CH_2)_8CONHCH_2OH$ and | | | |
| Catechol | 4- | HCl, $C_2H_5OH$ | 389 (2) |
| Guaiacol | 4- | 20 % $H_2SO_4$, $C_2H_5OH$ | 382, 390 |
| | 4- | $AlCl_3$, $CH_3CO_2H$ | 51 (63) |
| $CH_3(CH{=}CH)_2CONHCH_2OH$ and | | | |
| Catechol | 4- | HCl, $C_2H_5OH$ | 389 (57) |
| $C_6H_5CONHCH_2OH$ and | | | |
| Benzene | 1- | $H_2SO_4$, $CH_3CO_2H$ | 13 |
| | 1,4- | Concd. $H_2SO_4$ | 13 (32) |
| Phenol | 4(?)- | $AlCl_3$, $(CH_3)_2CO$ | 51 (60) |
| | 4(?)- | 100 % $H_3PO_4$, $CH_3CO_2H$ | 48 (66) |
| 2-Nitrophenol | 4- | Concd. $H_2SO_4$ | 10 |
| 4-Nitrophenol | 2- | Concd. $H_2SO_4$ | 1,10 (90), 20 |
| | 2- | HCl, $C_2H_5OH$ | 1 (<90), 20 |
| | 2- | $ZnCl_2$, $CH_3CO_2H$ | 1 (<90), 20 |
| Catechol | $x$- | HCl, $C_2H_5OH$ | 10 (30) |
| Hydroquinone | 2- | HCl, $C_2H_5OH$ | 10 (20) |
| Pyrogallol | $x,x$- | HCl, $C_2H_5OH$ | 10 (33) |
| Guaiacol | 4- | HCl, $C_2H_5OH$ | 10 (<25), 382 (27) |
| | 4- | $AlCl_3$, $(CH_3)_2CO$ | 51 (68) |
| 4-Nitroveratrole | 5- | Concd. $H_2SO_4$ | 58 (31) |
| 1,2,3-Trimethoxybenzene | 4,5- | Concd. $H_2SO_4$ | 52 |
| | 4- and 4,5- | $ZnCl_2$, $CH_3CO_2H$ | 52 |

*Note:* References 382 to 537 are on pp. 266–269.

TABLE I—*Continued*

AMIDOMETHYLATION OF AROMATIC CARBON ATOMS WITH N-MONOMETHYLOL DERIVATIVES OF CARBOXAMIDES AND CARBAMYL COMPOUNDS

$RCONHCH_2OH + ArH \rightarrow RCONHCH_2Ar$

| Product Derived from | Position(s) of Substituent(s) | Method | References (Yield, %) |
|---|---|---|---|
| $C_6H_5CONHCH_2OH$ (*contd.*) and | | | |
| N,N-Dimethylaniline | 4- | HCl, $C_2H_5OH$ | 388 (26) |
| Acetanilide | *x*- | Concd. $H_2SO_4$ | 13 (82) |
| Toluene | 4- | $H_2SO_4$, $CH_3CO_2H$ | 13 (90) |
| *p*-Cresol | 2- | HCl, $C_2H_5OH$ | 79 |
| 2,4-Xylenol | 6- | HCl, $C_2H_5OH$ | 55 |
| Thymol | *x*- | $H_2SO_4$, $C_2H_5OH$ | 10 (62) |
| Benzoic acid | 3- | Concd. $H_2SO_4$ | 10 (8), 20 |
| *o*-Toluic acid | 5(?)- | Concd. $H_2SO_4$ | 7 |
| *p*-Toluic acid | 3- | Concd. $H_2SO_4$ | 7 |
| 4-Hydroxyphenylarsonic acid | 3- | Concd. $H_2SO_4$ | 391 |
| 2-Naphthol | 1- | HCl, $C_2H_5OH$ | 1, 10 (100), 20 |
| 6-Bromo-2-naphthol | 1- | HCl, $C_2H_5OH$ | 388 (90) |
| 3,6-Dibromo-2-naphthol | 1- | HCl, $C_2H_5OH$ | 388 |
| 2-Methoxynaphthalene | 1- | HCl, $CH_3OH$, 40° | 392 (97) |
| 2-Hydroxy-3-naphthoic acid | 1- | $H_2SO_4$, $CH_3CO_2H$ | 40 |
| Fluorene | | | 86 |
| 9-Fluorenone | 2,7- | Concd. $H_2SO_4$ | 7 |
| 1-Hydroxyanthraquinone | 4- and 2,4- | Concd. $H_2SO_4$ | 37, 385 |
| 2-Hydroxyanthraquinone | 1- | Concd. $H_2SO_4$ | 36, 37, 385 |
| 1,2-Dihydroxyanthraquinone | *x*- | Concd. $H_2SO_4$ | 37 |
| 1,8-Dihydroxyanthraquinone | 2,4,5- | Concd. $H_2SO_4$ | 385 |
| 1,3-Dimethylanthraquinone | | Concd. $H_2SO_4$ | 7 (0) |
| 2-Hydroxy-9,10-phenanthrenequinone | 7(?)- | Concd. $H_2SO_4$ | 7 |

| | | | |
|---|---|---|---|
| 7H-Benz[de]anthracen-7-one (benzanthrone) | 3- | Concd. $H_2SO_4$ | 7 |
| 2-Furoic acid | 5- | Concd. $H_2SO_4$ | 393 (30) |
| Methyl 2-furoate | 5- | Concd. $H_2SO_4$ | 35 (100) |
| Ethyl 2-furoate | 5- | Concd. $H_2SO_4$ | 383 |
| 2-Thiophenecarboxylic acid | 4- | Concd. $H_2SO_4$ | 56 |
| 6-Hydroxyquinoline | *x*- | Concd. $H_2SO_4$ | 394 |
| 8-Hydroxyquinoline | *x*- | Concd. $H_2SO_4$ | 10, 394 |
| 2-Hydroxy-6-methoxy-4-methylquinoline | *x*- | Concd. $H_2SO_4$ | 394 |
| Carbazole | | | 86 |
| Acridine | *x*- | Concd. $H_2SO_4$ | 384 |
| Rhodanine | | | 86 |
| 2,3-Dimethyl-1-phenyl-5- | 4- | $AlCl_3$, $(CH_3)_2CO$ | 51 (52) |
| pyrazolone (antipyrine) | 4- | Concd. $H_2SO_4$ | 86 |
| Picrolonic acid | | | 86 |
| 4-Methyl-2-thiouracil | 5- | Concd. $H_2SO_4$ | 395 |
| *o*-$O_2NC_6H_4CONHCH_2OH$ and | | | |
| 2-Furoic acid | 5- | Concd. $H_2SO_4$ | 393 |
| *o*-$HOC_6H_4CONHCH_2OH$ and | | | |
| Catechol | *x,x*- | HCl, $C_2H_5OH$ | 25 (50) |
| Hydroquinone | *x,x*- | HCl, $C_2H_5OH$ | 25 (50) |
| Pyrogallol | *x*- | HCl, $C_2H_5OH$ | 25 (100) |
| Phenetole | 4- | $AlCl_3$, $CH_3CO_2H$ | 51 (63) |
| Thymol | *x*- | HCl, $C_2H_5OH$ | 25 (56) |
| *p*-$CH_3C_6H_4CONHCH_2OH$ and | | | |
| 2-Furoic acid | 5- | Concd. $H_2SO_4$ | 393 |
| N-Methylol-2-furamide and | | | |
| N-Methylol-2-furamide | 2,5-Polymer | Heat, pressure | 35 |

*Note:* References 382 to 537 are on pp. 266–269.

TABLE I—*Continued*

AMIDOMETHYLATION OF AROMATIC CARBON ATOMS WITH N-MONOMETHYLOL DERIVATIVES OF CARBOXAMIDES AND CARBAMYL COMPOUNDS

$RCONHCH_2OH + ArH \rightarrow RCONHCH_2Ar$

| Product Derived from | Position(s) of Substituent(s) | Method | References (Yield, %) |
|---|---|---|---|
| N-Methylol-β-2-furanacrylamide and | | | |
| N-Methylol-β-2-furanacrylamide | 2,5-Polymer | Heat, pressure | 35 |
| $C_2H_5O_2CNHCH_2OH$ and | | | |
| 2,4-Xylenol | 6- | HCl, $C_2H_5OH$ | 218 (56) |
| $(CH_3)_2NCONHCH_2OH$ and | | | |
| 3,3-Bis-(*p*-hydroxyphenyl)oxindole (phenolisatin) | *x,x*- | $ZnCl_2$, $CH_3CO_2H$, 50° | 53 |
| $CO(NHCH_2OH)_2$† and | | | |
| Phenol | 2- | HCl | 38 (50) |
| 2-Nitrophenol | 4- | $H_2SO_4$, $CH_3CO_2H$ | 40 |
| 4-Nitrophenol | 2- | $H_2SO_4$, $CH_3CO_2H$ | 40 |
| 1-Hydroxyanthraquinone | 2- | Concd. $H_2SO_4$ | 37 |
| 1,2-Dihydroxyanthraquinone | 3- | Concd. $H_2SO_4$ | 36, 37, 39 |
| $C_6H_5NHCONHCH_2OH$ and | | | |
| 2,4-Xylenol | 6- | $HCO_2H$, 50° | 54 |
| *p*-$CH_3C_6H_4NHCONHCH_2OH$ and | | | |
| 2,4-Xylenol | 6- | $HCO_2H$, 50° | 55 |

*Note:* References 382 to 537 are on pp. 266–269.

† Reacting as a monomethylolamide. For bifunctional reactions see Table III.

Amidomethylation of Aromatic Carbon Atoms with N-Methylolimides

| Product Derived from | Position(s) of Substituent(s) | Method | References (Yield, %) |
|---|---|---|---|
| N-Methylolsuccinimide and 4-Nitroveratrole | 5- | Concd. $H_2SO_4$ | 58 (1.4) |
| N-Methylolmaleimide and 2-Nitrotoluene | 4(?)- | Concd. $H_2SO_4$ | 8 (12) |
| N-Methylolsaccharin and 4-Methyl-2-thiouracil | 5- | Concd. $H_2SO_4$ | 395 |
| N-Methylolphthalimide and | | | |
| Benzene | 1- | Concd. $H_2SO_4$ | 16 |
| Nitrobenzene | 3- | Oleum | 16 |
| | 3,5- | 20% Oleum | 29 |
| Phenol | 4- | Concd. $H_2SO_4$ | 16 |
| | 2-, 4-, and 2,4- | Concd. $H_2SO_4$ | 15 |
| 4-Nitrophenol | 2- | Concd. $H_2SO_4$ | 16 |
| 2,4-Dinitrophenol | 6- | 5% Oleum, 95° | 69 (95) |
| 4-Nitroveratrole | 5- | Concd. $H_2SO_4$ | 58 (88) |
| Aniline | 3- and 4- | 4% Oleum | 44 (>60) |
| N,N-Dimethylaniline | *x*- | Concd. $H_2SO_4$ | 16 |
| Acetanilide | 4(?)- | 95% $H_2SO_4$ | 396 (41) |
| | 2- (27%) and 4- (60%) | 99.5% $H_2SO_4$ | 44 (87) |
| | 2- (15%), 4- (33%), and 2,4- (27%) | 1.5% Oleum | 44 (75) |
| | 4- (6%) and 2,4- (66%) | 4% Oleum | 44 (72) |
| 2-Nitrotoluene | 5(?)- | Concd. $H_2SO_4$ | 16 (95) |
| 3-Nitrotoluene | *x*- | Concd. $H_2SO_4$ | 16 |
| 4-Nitrotoluene | *x*- | Concd. $H_2SO_4$ | 16 |
| | 2,6- | Concd. $H_2SO_4$ | 29 (95) |
| *o*-Toluidine | 3,5- | Concd. $H_2SO_4$ | 29 (77) |

*Note:* References 382 to 537 are on pp. 266–269.

TABLE II—*Continued*

AMIDOMETHYLATION OF AROMATIC CARBON ATOMS WITH N-METHYLOLIMIDES

| Product Derived from | Position(s) of Substituent(s) | Method | References (Yield, %) |
|---|---|---|---|
| N-Methylolphthalimide (*contd.*) and | | | |
| *p*-Toluidine | 3- | 98.3 % $H_2SO_4$ | 44 (81) |
| | 3,5- | Concd. $H_2SO_4$ | 29 (95) |
| 4-Methylacetanilide | 2- (3 %) and 3- (17 %) | 99.5 % $H_2SO_4$ | 44 (>20) |
| | 2- (23 %) and 3- (55 %) | 97.7% $H_2SO_4$ | 44 (>78) |
| 2,4-Xylidine | 3,5- | 90 % $H_2SO_4$ | 29 |
| 4-Phenylphenol | 2- | $H_2SO_4$, $C_6H_6$ | 397 (>30) |
| 4-Hydroxyacetophenone | 3- | Concd. $H_2SO_4$ | 69 (92) |
| 2-Methylbenzophenone | 3,5(?)- | Concd. $H_2SO_4$ | 7 |
| 4-Methylbenzophenone | 3(?)- | Concd. $H_2SO_4$ | 7 |
| 2,4-Dimethylbenzophenone | 5(?)- and 3,5(?)- | Concd. $H_2SO_4$ | 7 |
| Phenylacetic acid | 3- and 4- | Concd. $H_2SO_4$ | 47 (85) |
| Benzoic acid | 3- | 95 % $H_2SO_4$ | 396 (40) |
| | 3- | Concd. $H_2SO_4$ | 12 (60) |
| *o*-Toluic acid | 5(?)- | Concd. $H_2SO_4$ | 7 |
| *m*-Toluic acid | 6- | Concd. $H_2SO_4$ | 7 |
| *p*-Toluic acid | 3- | Concd. $H_2SO_4$ | 7 |
| 4-$HOC_6H_4AsO_3H$ | 3- | Concd. $H_2SO_4$ | 398 |
| 2-Naphthol | 1- | 88 % $H_2SO_4$ | 396 (53) |
| 9-Fluorenone | 2,7- | Concd. $H_2SO_4$ | 7 |
| 1-Hydroxyanthraquinone | 2,4- | Concd. $H_2SO_4$ | 37, 385 |
| 2-Hydroxyanthraquinone | 1- | Concd. $H_2SO_4$ | 36, 37, 385 |
| 2-Hydroxy-3-chloroanthraquinone | 1- | Concd. $H_2SO_4$ | 387 |
| 1,2-Dihydroxyanthraquinone | 3- | Concd. $H_2SO_4$ | 37,39 |
| 1,5-Dihydroxyanthraquinone | 4(?)- | Concd. $H_2SO_4$ | 385 |
| 1,8-Dihydroxyanthraquinone | 2,4,5,7- | Concd. $H_2SO_4$ | 385 |
| 2,3-Dihydroxyanthraquinone | 1,4- | Concd. $H_2SO_4$ | 385 |
| 1,4-Di-(4′-methylanilino)anthraquinone | 3′, 5′, 3″, 5″(?)- | 96 % $H_2SO_4$ | 399 (97) |
| 1-Amino-2-bromo-4-(4′-methylanilino)-anthraquinone | 3′(?)- and 3′, 5′ (?)- | 96 % $H_2SO_4$ | 399 (100) |

| | | | |
|---|---|---|---|
| 1,4,5,8-Tetra-(4′-methylanilino) anthraquinone | 3′, 5′, 3″, 5″, 3‴, 5‴, 3⁗, 5⁗(?)- | 98% $H_2SO_4$ | 399 (97) |
| 1-Hydroxy-2-methylanthraquinone | 4- | Concd. $H_2SO_4$ | 37, 385 |
| 2-Hydroxy-3-methylanthraquinone | 1- | Concd. $H_2SO_4$ | 37, 385 |
| 1,3-Dimethylanthraquinone | 4- | Concd. $H_2SO_4$ | 7 |
| 2-Hydroxyanthraquinone-3-carboxylic acid | 1- | Concd. $H_2SO_4$ | 385 |
| 1-(2′-Methyl-5′-chloroanilino)anthraquinone-2-carboxylic acid | 4′(?) | 96% $H_2SO_4$ | 399 (95) |
| 1-Amino-4-(4′-methylanilino)-anthraquinone-2-sulfonic acid | 3′(?)- | 96% $H_2SO_4$ | 399 |
| 9,10-Phenanthrenequinone | 2(?)- and 2,7(?)- | Concd. $H_2SO_4$ | 7 |
| 2-Hydroxy-9,10-phenanthrenequinone | 7(?)- and 3,7(?)- | Concd. $H_2SO_4$ | 7 |
| 2-Acetyl-9,10-phenanthrenequinone | 7(?)- | Concd. $H_2SO_4$ | 7 |
| 7H-Benz[de]anthracen-7-one | 3(?)- | 96% $H_2SO_4$ | 400, 401 (100) |
| (benzanthrone) | 3,9(?)- | Concd. $H_2SO_4$ | 7 |
| | 3,9(?)- | 96% $H_2SO_4$, 80° | 400 (97) |
| 3-Bromo-7H-benz[de]anthracen-7-one | 9(?)- | 96% $H_2SO_4$, 80° | 7, 400 |
| | | | 402 (100) |
| 4,4′-Bis-7H-benz[de]anthracen-7-one | 3,3′(?)- | 96% $H_2SO_4$ | 400 (95) |
| Dibenz[ah]pyrene-7,14-dione (Indanthrene Golden Yellow GK) | 12(?)- | 10% Oleum, 60° | 403 (93) |
| Dibenz[cd, jk]pyrene-6,12-dione (anthanthrone) | 10(?)- | 10% Oleum, 60° | 403 |
| Dinaphtho[1,2,3-cd : 3′,2′,1′-lm]perylene-5,10-dione (violanthrone) | 3(?)-* | 10% Oleum | 403 (100) |
| Ethyl 2-furoate | 5- | Concd. $H_2SO_4$ | 383 |
| 4-Methyl-2-thiouracil | 5- | Concd. $H_2SO_4$ | 395 |
| Quinoline | 5-(21%) and 8-(20%) | Concd. $H_2SO_4$ | 404 (41) |
| Cobalt phthalocyanine | | 100% $H_2SO_4$, 75° | 405 |
| Copper phthalocyanine | *x,x*- | 3% Oleum, 100° | 64 |
| | *x,x*- to 8*x* | $H_2SO_4$, $H_3PO_4$, or $P_2O_5$ | 50 |
| Cobalt monochlorophthalocyanine | | 100% $H_2SO_4$, 75° | 405 |

*Note:* References 382 to 537 are on pp. 266–269.

* Sulfonation also occurred in the 12(?)- position.

## TABLE III

### Amidomethylation of Aromatic Carbon Atoms with N,N′-Dimethylol Derivatives of Dicarboxamides and of Urea

$$(CH_2)_n(CONHCH_2OH)_2 + 2ArH \rightarrow (CH_2)_n(CONHCH_2Ar)_2$$

and

$$CO(NHCH_2OH)_2 + 2ArH \rightarrow CO(NHCH_2Ar)_2$$

| Product Derived from | Position(s) of Substituent(s) | Method | References (Yield, %) |
|---|---|---|---|
| $(CONHCH_2OH)_2$ and | | | |
| Methyl 2-furoate | 5- | Concd. $H_2SO_4$ | 35 (88) |
| Ethyl 2-furoate | 5- | Concd. $H_2SO_4$ | 35 (92) |
| $(CH_2)_2(CONHCH_2OH)_2$ and | | | |
| 4-Nitrophenol | 2- | Concd. $H_2SO_4$ | 20, 24 |
| 2,4-Xylenesulfonic acid | $x$- | Concd. $H_2SO_4$ | 20 |
| 2-Naphthol | 1- | HCl, $C_2H_5OH$ | 20, 24 |
| $(CH_2)_4(CONHCH_2OH)_2$ and | | | |
| Benzene | 1- | Concd. $H_2SO_4$ | 406 |
| 4-Nitrophenol | 2- | Concd. $H_2SO_4$ | 406 (80) |
| Acetanilide | 4- | Concd. $H_2SO_4$ | 406 (41) |
| 2-Methylacetanilide | $x$- | Concd. $H_2SO_4$ | 406 |
| 4-Methylacetanilide | $x$- | Concd. $H_2SO_4$ | 406 (55) |
| 2-Naphthol | 1- | Concd. $H_2SO_4$ | 406 (50) |
| 2-Acetylaminonaphthalene | $x$- | Concd. $H_2SO_4$ | 406 |
| $(CH_2)_7(CONHCH_2OH)_2$ and | | | |
| Benzene | 1- | Concd. $H_2SO_4$ | 406 (37) |
| $(CH_2)_8(CONHCH_2OH)_2$ and | | | |
| Benzene | 1- | Concd. $H_2SO_4$ | 406 (48) |

| | | | |
|---|---|---|---|
| $CO(NHCH_2OH)_2$ and | | | |
| 2-Nitrophenol | 4- | $H_2SO_4$, $CH_3CO_2H$ | 40, 45 |
| 4-Nitrophenol | 2- | $H_2SO_4$, $CH_3CO_2H$ | 40 |
| Resorcinol | 4- | $H_2SO_4$, $CH_3CO_2H$ | 40, 46 |
| 2-Nitroanisole | 4- | $H_2SO_4$, $CH_3CO_2H$ | 40 |
| 2-Nitro-4-methylphenol | 6- | $H_2SO_4$, $CH_3CO_2H$ | 40 |
| 2,4-Xylenol | 6- | $HCO_2H$, 50° | 54 |
| Salicylic acid | 5- | $H_2SO_4$, $CH_3CO_2H$ | 40, 46 |
| 1-Naphthol | 2- | $H_2SO_4$, $CH_3CO_2H$ | 40 |
| 2-Naphthol | 1- | $H_2SO_4$, $CH_3CO_2H$ | 40 |
| 1-Hydroxy-2-naphthoic acid | 4- | $H_2SO_4$, $CH_3CO_2H$ | 40, 46 |
| 2-Hydroxy-3-naphthoic acid | 1- | $H_2SO_4$, $CH_3CO_2H$ | 40, 46 |
| 2-Hydroxyanthraquinone | 1- | Concd. $H_2SO_4$ | 36, 37, 39 |
| 2-Hydroxyanthraquinone-3-carboxylic acid | 1- | Concd. $H_2SO_4$ | 37, 39 |
| Methyl 2-furoate | 5- | Concd. $H_2SO_4$ | 35 (91) |
| Ethyl 2-furoate | 5- | Concd. $H_2SO_4$ | 35 (93) |

*Note:* References 382 to 537 are on pp. 266–269.

TABLE IV

AMIDOMETHYLATION OF AROMATIC CARBON ATOMS WITH FORMALDEHYDE AND NITRILES OR AMIDES

$$RCN\ (\text{or}\ RCONH_2) + CH_2O + ArH \xrightarrow[H_3PO_4]{H_2SO_4\ \text{or}} RCONHCH_2Ar$$

| Product from Formaldehyde | Position(s) of Substituent(s) | Method | References (Yield, %) |
|---|---|---|---|
| HCN and | | | |
| *m*-Xylene | 4- | $H_2SO_4$, $CH_3CO_2H$, 90° | 5 (25) |
| $CH_3CN$ and | | | |
| Benzene | 1- | $H_2SO_4$, $CH_3CO_2H$, 90° | 5 (50) |
| Bromobenzene | 4- | Concd. $H_2SO_4$ | 5 (37) |
| Anisole | 4- | $H_2SO_4$, $CH_3CO_2H$ | 5 (28) |
| Toluene | 2- (10 %) and 4- (36 %) | $H_2SO_4$, $CH_3CO_2H$, 90° | 5 (53) |
| Ethylbenzene | 4- | $H_2SO_4$, $CH_3CO_2H$ | 5 (19) |
| Cumene | 4- | $H_2SO_4$, $CH_3CO_2H$ | 5 (75) |
| *o*-Xylene | *x*- | $H_2SO_4$, $CH_3CO_2H$ | 5 (54) |
| *m*-Xylene | 4- | $H_2SO_4$, $CH_3CO_2H$ | 5 (52) |
| | 4,6- | 85 % $H_3PO_4$, 90° | 5 (66) |
| | 4,6- | $H_2SO_4$, $CH_3CO_2H$, 90° | 5 (75) |
| *p*-Xylene | 2- (21 %) and 2,5- (55 %) | $H_2SO_4$, $CH_3CO_2H$, 90° | 5 (89) |
| 1,2,4-Trimethylbenzene | 5- | $H_2SO_4$, $CH_3CO_2H$ | 5 (67) |
| N-(2,4-Dimethylbenzyl)acetamide | 5- | $H_2SO_4$, $CH_3CO_2H$, 90° | 5 (45) |
| 1,2,4,5-Tetramethylbenzene (durene) | 3,6- | $H_2SO_4$, $CH_3CO_2H$, 85° | 5 (72) |
| Naphthalene | 1- (12 %), 1,4- (25 %) and 1,5(?)- (6 %) | $H_2SO_4$, $CH_3CO_2H$, 90° | 5 (>43) |
| $CH_2ClCN$ and | | | |
| *m*-Xylene | 4,6- | $H_2SO_4$, $CH_3CO_2H$ | 5 |
| $CH_2ClCH_2CN$ and | | | |
| *m*-Xylene | 4,6- | $H_2SO_4$, $CH_3CO_2H$ | 5 |

| | | | |
|---|---|---|---|
| $CH_2$=CHCN and | | | |
| Toluene | 4- | $H_2SO_4$, $CH_3CO_2H$ | 5 (89) |
| Cumene | 4- | $H_2SO_4$, $CH_3CO_2H$ | 5 (53) |
| *m*-Xylene | 4,6- | 85% $H_3PO_4$, 90° | 5 |
| $C_6H_5CN$ and | | | |
| *m*-Xylene | 4,6- | $H_2SO_4$, $CH_3CO_2H$ | 5 |
| 2-Pyrrolidinone and | | | |
| 4-Nitrotoluene | 2- | Concd. $H_2SO_4$ | 34 (37) |
| γ-Valerolactam and | | | |
| 4-Nitro-1,3-xylene | 6- | Concd. $H_2SO_4$ | 34 (41) |
| Succinimide and | | | |
| 4-Nitrotoluene | 2- | Concd. $H_2SO_4$ | 34 (81) |
| Copper phthalocyanine | *x*- | 95% $H_2SO_4$, 75° | 50 |
| 4-Cyclohexene-1,2-dicarboximide and | | | |
| 4-Chloronitrobenzene | 3- | Concd. $H_2SO_4$ | 34 (75) |
| Phthalimide and | | | |
| Copper phthalocyanine | *x*- to *x,x,x,x*- | $H_2SO_4$, oleum, or $P_2O_5$ | 50 |
| Urea and | | | |
| 2,4-Xylenol | $H_2NCONR_2$ and $RNHCONR_2$ (R = 2-hydroxy-3,5-dimethylbenzyl: ring bearing OH, $CH_3$, $CH_2$–, $CH_3$) | $HCO_2H$, 50° | 59 |
| Sulfanilamide and | | | |
| 4-Methyl-2-thiouracil | 5- | HCl, $CH_3CO_2H$, heat | 60 |
| N[4]-Acetylsulfanilamide and | | | |
| 4-Methyl-2-thiouracil | 5- | HCl, $CH_3CO_2H$, heat | 60 |

*Note:* References 382 to 537 are on pp. 266–269.

TABLE V

AMIDOMETHYLATION OF AROMATIC CARBON ATOMS WITH ETHERS OF N-METHYLOL-AMIDES AND -IMIDES

| Product Derived from | Position(s) of Substituent(s) | Method | References (Yield, %) |
|---|---|---|---|
| $[(CH_3)_2CHCH_2CONHCH_2]_2O$ and 2,4-Xylenol | 6- | $H_2SO_4$, $C_2H_5OH$ | 54 |
| $o$-$C_6H_4(CO)_2NCH_2OCH_3$ and Copper phthalocyanine | $x,x$- | 100% $H_2SO_4$, 100° | 64 |
| $o$-$C_6H_4(CO)_2NCH_2OC_2H_5$ and Benzene | 1- | 100% $H_2SO_4$, 90°, 1.5 hr. | 63 (5) |
| Nitrobenzene | 3- | 100% $H_2SO_4$, 80°, 1.5 hr. | 63 (41) |
| $[o$-$C_6H_4(CO)_2NCH_2]_2O$ and Phenol | $x$- and $x,x$- | Concd. $H_2SO_4$ | 61 |
| 4-Nitrophenol | 2- | Concd. $H_2SO_4$ | 61 |
| 2-Nitrotoluene | 5(?)- | Concd. $H_2SO_4$ | 61 |
| Copper phthalocyanine | $x$- | 100% $H_2SO_4$, 75° | 50 |
| $(HOCH_2NHCONHCH_2)_2O$ and 2,4-Xylenol | $RNHCONR_2$, $(RNH)_2CO$, and $RNHCONH_2$ (R = ring with OH, $CH_3$, $CH_2$–, $CH_3$) | HCl, $C_2H_5OH$, 50° | 62 |
| $(C_2H_5NHCONHCH_2)_2O$ and 2,4-Xylenol | 6- | $HCO_2H$, 50° | 54 |
| $(C_6H_5NHCONHCH_2)_2O$ and 2,4-Xylenol | 6- | $HCO_2H$, 50° | 54 (90) |

## TABLE VI

Amidomethylation of Aromatic Carbon Atoms with N-Halomethyl-amides, -imides, and -carbamyl Compounds

$RCON(R')CH_2X + ArH \rightarrow RCON(R')CH_2Ar$

| Product Derived from | Position(s) of Substituent(s) | Method | References (Yield, %) |
|---|---|---|---|
| $CH_3CON(CH_3)CH_2Cl$ and | | | |
| Benzene | 1- | $AlCl_3$, $C_6H_6$ | 32 |
| 4-Chlorophenol | 2- | $AlCl_3$ | 32 |
| Anisole | 4- | $AlCl_3$ | 32 |
| Toluene | 4- | $AlCl_3$,$C_6H_5CH_3$ | 32 (62) |
| $CH_3CON(C_3H_7\text{-}n)CH_2Cl$ and | | | |
| Benzene | 1- | $AlCl_3$, $C_6H_6$ | 32 |
| $CH_3CON(CH_2C_6H_5)CH_2Cl$ and | | | |
| Toluene | 4- | $AlCl_3$, $C_6H_5CH_3$ | 32 |
| $n\text{-}C_{17}H_{35}CON(CH_3)CH_2Cl$ and | | | |
| Phenol | | $CH_3CO_2H$, 20–60° | 67 |
| $C_6H_5CON(CH_3)CH_2Cl$ and | | | |
| Toluene | 4- | $AlCl_3$, $C_6H_5CH_3$ | 32 |
| 1,4-Dichloromethylpiperazine-2,5-dione* and | | | |
| Benzene | 1- | $AlCl_3$, $CS_2$ | 65 (43) |
| Naphthalene | *x*- | No solvent, 150° | 65 (64) |
| 2-Naphthol | 1- | $C_6H_6$, 80° | 65 (89) |
| N-Chloromethylsuccinimide and | | | |
| Benzene | 1- | $AlCl_3$, $CS_2$ | 66 |

*Note:* References 382 to 537 are on pp. 266–269.

* Bifunctional reaction.

TABLE VI—*Continued*

AMIDOMETHYLATION OF AROMATIC CARBON ATOMS WITH N-HALOMETHYL-AMIDES, -IMIDES, AND -CARBAMYL COMPOUNDS

$$RCON(R')CH_2X + ArH \rightarrow RCON(R')CH_2Ar$$

| Product Derived from | Position(s) of Substituent(s) | Method | References (Yield, %) |
|---|---|---|---|
| N-Chloromethylmaleimide and | | | |
| Benzene | 1- | $ZnCl_2$, 80° | 8 (83) |
| Phenol | 2,4- | $ZnCl_2$, $C_6H_6$, 80° | 8 (40) |
| *o*-$C_6H_4(CO)_2NCH_2Cl$ and | | | |
| Benzene | 1- | $ZnCl_2$, $C_6H_6$ | 68 |
| Phenol | 2- and 4- | $ZnCl_2$, $C_6H_6$ | 68 |
| Anisole | 4- | $ZnCl_2$, 120° | 68 (100) |
| | 2- and 4- | $ZnCl_2$, 130° | 378 (65) |
| *n*-$C_3H_7OC_6H_5$ | 4- | $ZnCl_2$, 150° | 407 |
| 4-Chlorophenol | 2- | $ZnCl_2$, 90–130° | 69 (70) |
| Guaiacol | 4- | No catalyst, 120° | 68 |
| 2-Nitrotoluene | 4- | $ZnCl_2$, 120° | 68 |
| *m*-Xylene | 4- | $ZnCl_2$, 100° | 68 |
| N-(2,4-Dimethylbenzyl)phthalimide | 5- | $ZnCl_2$, $C_6H_5NO_2$, 100° | 68 |
| Naphthalene | 1- | $ZnCl_2$, melt | 68 |
| N-(1-Naphthylmethyl)phthalimide | 4,8(?)- | $ZnCl_2$, $C_6H_5NO_2$, 120° | 68 |
| 1-Naphthol-8-sulfonic acid $\gamma$-sultone | 4- | $AlCl_3$,$C_6H_3Cl_3$, 110° | 379 (97) |
| Anthracene | 9,10- | $ZnCl_2$, $C_6H_5NO_2$, 100° | 68 |
| 7H-Benz [de] anthracen-7-one (benzanthrone) | 3(?)- | $ZnCl_2$, $C_6H_5NO_2$, 140° | 408, 409 |
| Copper phthalocyanine | 5*x*- | $AlCl_3$, piperidine, 140° | 64 |
| | *x*,*x*- | 100% $H_2SO_4$, 80° | 64 |
| Copper tetraphenylphthalocyanine | 6*x*- | $AlCl_3$, pyridine, 130° | 64 |
| | 4*x*- | Concd. $H_2SO_4$ | 64 |

| | | | |
|---|---|---|---|
| $o$-$C_6H_4(CO)_2NCH_2Br$ and | | | |
| Benzene | 1- | $ZnCl_2$, 80° | 14 (94) |
| Phenol | 2- and 4- | $C_6H_6$, no catalyst, 80° | 15 (50) |
| | 2(?)- | Heat, no solvent | 70 |
| | 4- | Heat | 71 (62) |
| Thymol | $x$- | Heat | 71 (69) |
| Phenylacetic acid | 4- | $ZnCl_2$, 120° | 47 (10) |
| 2-Naphthol | $x$- | Heat | 71 (63) |
| $ClCONHCH_2Cl$ and | | | |
| Phenol | 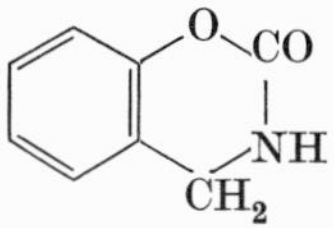 | Pentane, 40°, 3 hr. | 73 (5) |
| 2,4,Dichloroaniline | 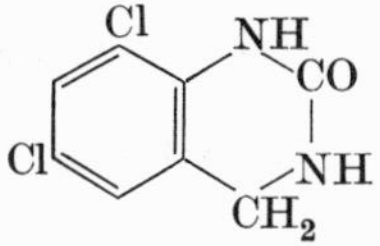 | Heptane, 25°, 16 hr. | 73 (55) |
| $m$-Xylene | 2,4-$(CH_3)_2C_6H_3CH_2NCO$ | $ZnCl_2$, $CH_2Cl_2$, 58° | 73 (35) |
| $\beta$-Naphthol | 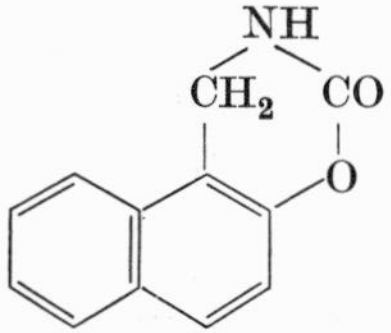 | Pentane, 25°, 24 hr. | 73 (25) |

*Note:* References 382 to 537 are on pp. 266–269.

## TABLE VII

α-AMIDOALKYLATION OF AROMATIC CARBON ATOMS WITH N,N′-METHYLENE-, -ALKYLIDENE-, AND -ARYLIDENE-BISAMIDES

$$(RCONH)_2CHR' + ArH \rightarrow RCONHCH(R')Ar + RCONH_2$$

(R′ = H, alkyl, or aryl)

| Product Derived from | Position(s) of Substituent(s) | Method | References (Yield, %) |
|---|---|---|---|
| $(HCONH)_2CH_2$ and | | | |
| 2,4-Xylenol | 6- | $POCl_3$, 95°, 1 hr. | 75 (38) |
| $(CH_3CONH)_2CH_2$ and | | | |
| Phenol | 2- (25 %) and 4- (21 %) | $POCl_3$, $CHCl_3$, 65°, 4 hr. | 76, 77 (>46) |
| 4-Nitrophenol | 2- | $POCl_3$, 95°, 1 hr. | 18 (85) |
| Anisole | 4- (46 %) and *x,x*- (26 %) | $POCl_3$, 95°, 3 hr. | 77 (72) |
| 4-Methoxytoluene | 3- | $POCl_3$, 95°, 1 hr. | 18 (93) |
| 2-Methylacetanilide | 4- | $POCl_3$, 130°, 1 hr. | 18 (86) |
| 4-Methylacetanilide | 2- | $POCl_3$, 130°, 1 hr. | 18 (71) |
| 2,4-Xylenol | 6- | $POCl_3$, 95°, 1 hr. | 75 (94) |
| Salicylic acid | 5- | $POCl_3$, 95°, 1 hr. | 76, 78 (>30) |
| 2-Naphthol | 1- | $POCl_3$, $CHCl_3$, 65° | 18 (94) |
| 8-Hydroxyquinoline | 5- | $POCl_3$, 95°, 1 hr. | 18 (84) |
| $(CH_3CH_2CONH)_2CH_2$ and | | | |
| 2,4-Xylenol | 6- | $POCl_3$, 95°, 1 hr. | 75 (89) |
| $(C_6H_5CONH)_2CH_2$ and | | | |
| Benzene | 1- | 100 % $H_2SO_4$, 95°, 6 hr. | 63 (63) |
| 2,4-Xylenol | 6- | $POCl_3$, 95°, 1 hr. | 75 (93) |
| *o*-$C_6H_4(CO)_2NCH_2NHCOC_6H_5$ and | | | |
| Benzene | 1- | 100 % $H_2SO_4$, 90°, 5.5 hr. | 63 (24)* |

| | | | |
|---|---|---|---|
| $(C_6H_5NHCONH)_2CH_2$ and | | | |
| 2,4-Xylenol | 6- | $HCO_2H$, 50° | 55 |
| ($p$-$CH_3C_6H_4NHCONH)_2CH_2$ and | | | |
| 2,4-Xylenol | 6- | $HCO_2H$, 50° | 55 |
| $(CH_3CONH)_2CHCH_3$ and | | | |
| $p$-Cresol | 2- | $POCl_3$, $C_6H_6$, 80° | 18 (93) |
| 2,4-Xylenol | 6- | $POCl_3$, $C_6H_6$, 80° | 75 (74) |
| 2-Naphthol | 1- | $POCl_3$, $CHCl_3$, 65° | 18 (91) |
| $(CH_3CONH)_2CHC_6H_5$ and | | | |
| Phenol | 2- | 190°, 4 hr. | 74 (41) |
| Phenyl acetate | 2- | 190°, 4 hr. | 74 (43) |
| 4-Nitrophenol | 2- | $POCl_3$, 95°, 1 hr. | 18 (81) |
| $p$-Cresol | 2- | 190°, 4 hr. | 74 (49) |
| 4-Methoxytoluene | 3- | $POCl_3$, 95°, 1 hr. | 18 (91) |
| 2-Methylacetanilide | 4- | $POCl_3$, 130°, 2 hr. | 18 (98) |
| 4-Methylacetanilide | | $POCl_3$ | 76 |
| 2,4-Xylenol | 6- | $POCl_3$, 95°, 1.5 hr. | 75 (98) |
| | 6- | 190°, 4 hr. | 74 (74) |
| 2,6-Xylenol | 4- | 190°, 4 hr. | 74 (30) |
| 2-Naphthol | 1- | $POCl_3$, $C_6H_6$, 80°, 1.5 hr. | 18 (98) |
| | 1- | 190°, 4 hr. | 74 (93) |
| $(C_6H_5CONH)_2CHC_6H_5$ and | | | |
| Phenol | 2- | 190°, 4 hr. | 74 (69) |
| Phenyl benzoate | 2- | 190°, 4 hr. | 74 (12) |
| 2-Naphthol | 1- | 190°, 4 hr. | 74 (93) |

*Note:* References 382 to 537 are on pp. 266–269.

* The only product obtained was N-benzylphthalimide.

## TABLE VIII

α-Amidoalkylation of Aliphatic Carbon Atoms with N-Monomethylol and N-α-Alkylol Derivatives of Amides, Imides, and Carbamyl Compounds

| Product Derived from | Method | References (Yield, %) |
|---|---|---|
| $HCONHCH_2OH$ and Trinitromethane | $H_2O$ | 88 |
| $CH_3CONHCH_2OH$ and Trinitromethane | $H_2O$, $p$H 7, 70°, 10 min. | 88 (75) |
| $CH_2ClCONHCH_2OH$ and Cyclohexane-1,3-dione | $H_2SO_4$, 3 days | 85 (0) |
| $(O_2N)_3CCH_2CH_2CONHCH_2OH$ and Trinitromethane | $CH_3OH$, 12 hr. | 87 (80) |
| $CH_2{=}CHCONHCH_2OH$ and Trinitromethane | $H_2O$, 5°, 48 hr. | 87 (5) |
| | $C_2H_5OH$, $CH(NO_2)_3$ (2 equiv.), 12 hr. | 87 (88)* |
| $CH_2{=}C(CH_3)CONHCH_2OH$ and Trinitromethane | $H_2O$, 1 day | 87 (87) |
| $C_6H_5CONHCH_2OH$ and | | |
| Trinitromethane | $H_2O$, 80° | 88 |
| Cyclohexane-1,3-dione | Concd. $H_2SO_4$, 3 days | 85 (36) |
| | $C_2H_5OH$, concd. HCl, 3 days | 85 (20) |
| | $CH_3CO_2H$, $ZnCl_2$, 3 days | 85 (40) |
| | $CH_3CO_2H$, $BF_3$-etherate, 20 hr. | 85 (65) |
| 5,5-Dimethylcyclohexane-1,3-dione | $H_2SO_4$, 25°, 3 days | 85† |
| 1-Phenylbutane-1,3-dione | $C_2H_5OH$ | 86 |
| 1,3-Diphenylpropane-1,3-dione | Concd. $H_2SO_4$, 40 hr. | 86 |
| | $CH_3CO_2H$, $BF_3$-etherate, 20 hr. | 85 (74) |
| $x$-$O_2NC_6H_4CONHCH_2OH$ and Trinitromethane | $H_2O$ | 88 |

| | | |
|---|---|---|
| $o$-$C_6H_4(CO)_2NCH_2OH$ and | | |
| Pentane-2,4-dione | Concd. $H_2SO_4$, 0°, 1 day | 85 (85) |
| 5,5-Dimethylcyclohexane-1,3-dione | $H_2SO_4$, 1 day | 85 (74) |
| 1-Phenylbutane-1,3-dione | Concd. $H_2SO_4$, 1 day | 85 (83) |
| 1,3-Diphenylpropane-1,3-dione | Concd. $H_2SO_4$, 1 day | 85 (84) |
| Ethyl acetoacetate | Concd. $H_2SO_4$, cold, 1 day | 85‡ |
| Malonic acid | $H_2SO_4$, standing overnight | 84 (0) |
| Diethyl malonate | $H_2SO_4$, 2 days | 85 (0) |
| | $H_2SO_4$, 3 days | 84 (0) |
| 1,2-Diphenylpyrazolidine-3,5-dione | Concd. $H_2SO_4$, 1 day | 85 (83) |
| 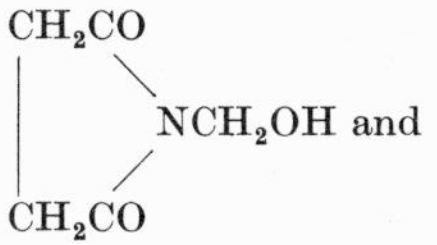 $CH_2CO$, $CH_2CO$ (ring) $NCH_2OH$ and | | |
| Cyclohexane-1,3-dione | $H_2SO_4$, 3 days | 85 (0) |
| $C_2H_5O_2CNHCH_2OH$ and | | |
| Trinitromethane | $H_2O$, 80°, short reaction time | 88 (90) |
| $CH_3NHCONHCH_2OH$ and | | |
| Trinitromethane | $H_2O$ | 88 |
| 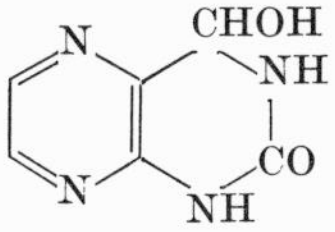 N, N, CHOH, NH, CO, NH (fused ring structure) and | | |
| Acetone | | 89 (0) |
| Pentane-2,4-dione | $H_2O$, $NaHCO_3$, 5 hr. | 89 (75) |
| Ethyl acetoacetate | $H_2O$, $K_2CO_3$, 1 day | 89 (80) |
| Diethyl malonate | $H_2O$, $K_2CO_3$, 20 hr. | 89 (90) |
| Ethyl cyanoacetate | No solvent, 100°, 4 hr. | 89 (90) |

*Note:* References 382 to 537 are on pp. 266–269.

* Trinitromethane added to the carbon-carbon double bond during the amidoalkylation. The product was the saturated material $(O_2N)_3CCH_2CH_2CONHCH_2C(NO_2)_3$.

† The product was not characterized.

‡ Mono- (33%) and bis- (16%) amidoalkylation occurred with concomitant decarboxylation.

TABLE IX

Amidomethylation of Aliphatic Carbon Atoms with Dimethylol Derivatives of Fumaramide and of Urea

| Product Derived from | Method | References (Yield, %) |
|---|---|---|
| $HOCH_2NHCOCH{=}CHCONHCH_2OH$ and | | |
| Trinitromethane | $H_2O$, *p*H 0.8, 90 min. | 87 (48) |
| $CO(NHCH_2OH)_2$ and | | |
| Trinitromethane | $H_2O$, 40°, 5 min. | 88 (77) |
| | Urea, formalin (2 equiv.), 22 hr. | 88 (88) |
| Acetylene | 10 atm., 40°, 10–20 hr. | 90 |

*Note:* References 382 to 537 are on pp. 266–269.

TABLE X

α-Amidoalkylation of Aliphatic Carbon Atoms with Formaldehyde or Acetaldehyde and Sulfonamides

| Product Derived from | Method | References |
|---|---|---|
| $CH_3(CH_2)_2CH_2SO_2NH_2$, $CH_2O$, and | | |
| Potassium cyanide | Aq. $CH_2O$, 80°, 1 hr. | 91 |
| $CH_3(CH_2)_6CH_2SO_2NH_2$, $CH_2O$, and | | |
| Potassium cyanide | Aq. $CH_2O$, 45°, 2 hr. | 91 |
| $C_6H_5SO_2NH_2$ and | | |
| $HOCH_2CN$ | 8% NaOH, 90°, 1 hr. | 91 |
| $C_6H_5SO_2NH_2$, $CH_2O$, and | | |
| Sodium cyanide | Aq. $CH_2O$, 70°, 2 hr. | 91 |
| $C_6H_5SO_2NH_2$, $CH_3CHO$, and | | |
| Potassium cyanide | $H_2O$, 95°, 2 hr. | 91 |
| *p*-$H_2NC_6H_4SO_2NH_2$, $(CH_2O)_x$, and | | |
| 2-Picoline | Paraffin oil, 130° | 92 |
| Quinaldine | Paraffin oil, 130° | 92 |
| 9-Methylacridine | Paraffin oil, 180° | 92 |
| 9-Ethylacridine | Paraffin oil, 130° | 92 |
| 2-Methyl-4(3H)-quinazolinone (structure: O, NH, N, $CH_3$) | No solvent, 160–170° | 92 |
| *p*-$CH_3CONHC_6H_4SO_2NH_2$, $(CH_2O)_x$, and | | |
| 2-Picoline | Paraffin oil, 200° | 92 |
| *p*-$CH_3C_6H_4SO_2NH_2$, $CH_2O$, and | | |
| Sodium cyanide | $H_2O$, 70°, 2 hr. | 91 |

*Note:* References 382 to 537 are on pp. 266–269.

## TABLE XI

α-AMIDOALKYLATION OF ALIPHATIC CARBON ATOMS WITH ETHERS AND ESTERS OF N-METHYLOL- AND N-α-ALKYLOL-AMIDE DERIVATIVES

| Product Derived from | Method* | References (Yield, %) |
|---|---|---|
| $(C_2H_5O_2CNHCHCCl_3)_2O$ and Diethyl malonate | $KOC_2H_5$, ether, 0° | 94 (35), 95 |
| $C_2H_5O_2CNHCH(OCOCH_3)CCl_3$ and Potassium cyanide | $H_2O$, 0°, 3–4 hr. | 94 (53)† |
| $CH_2{=}CHCONHCH_2OCOC_6H_5$ and Trinitromethane | $(CH_2Cl)_2$, 1 equiv., reflux, 0.5 hr. | 93 (4)‡ |
| | $CH_3NO_2$, 2 equiv., 101°, 15 min. | 93 (49)‡ |
| | $CH_3NO_2$, 101°, 0.5 hr. | 93 (12)‡ |
| | $(CH_2Cl)_2$, 2 equiv., 83°, 0.5 hr. | 93 (43)‡ |
| | $C_2H_5OH$, 2 equiv., 78°, 0.5 hr. | 93 (55)‡ |
| | $H_2O$, 2 equiv., 25°, 4 hr. | 93 (10)‡ |
| $CH_2{=}C(CH_3)CONHCH_2OCOC_6H_5$ and Trinitromethane | $(CH_2Cl)_2$, 25°, 12 hr. | 93 (50) |
| | $H_2O$, 25°, 0.5 hr. | 93 (64) |
| | $(CH_2Cl)_2$, 2 equiv., 25°, 3 hr. | 93 (65) |
| $CH_3CO_2CH_2NHCOCH{=}CHCONHCH_2OCOCH_3$ and Trinitromethane | $H_2O$, reflux, 25 min. | 93 (2) |
| $C_6H_5CO_2CH_2NHCOCH{=}CHCONHCH_2OCOC_6H_5$ and Trinitromethane | $CH_3NO_2$, reflux, 4 hr. | 93 (47) |

*Note:* References 382 to 537 are on pp. 266–269.

* The abbreviation "equiv." refers to the number of equivalents of trinitromethane.

† The product obtained was of the type $C_2H_5O_2CNHC(CN){=}CCl_2$.

‡ Trinitromethane added to the carbon-carbon double bond during amidoalkylation. The product was the saturated material $(O_2N)_3CCH_2CH_2CONHCH_2C(NO_2)_3$.

## TABLE XII

### α-Amidoalkylation of Aliphatic Carbon Atoms with N-Halomethyl and N-α-Haloalkyl Derivatives of Amides, Lactams, Imides, and Carbamyl Compounds

| Product Derived from | Method* | References (Yield, %) |
|---|---|---|
| $HCONHCH_2Cl$ and | | |
| Diethyl methylmalonate | Dioxane | 96, 100 (20) |
| $HCONHCHClCCl_3$ and | | |
| Diethyl malonate | $(C_2H_5)_2O$, reflux, 3–4 hr. | 160 |
| $CH_3CONHCH_2Cl$ and | | |
| Diethyl malonate | Dioxane, 50°, 1 hr. | 96, 100 (25) |
| Diethyl methylmalonate | Dioxane, 70°, 1.5 hr. | 100 (31) |
| Ethyl cyanoacetate | Dioxane, 70°, 1 hr. | 96, 100 (27) |
| $CH_3CON(CH_3)CH_2Cl$ and | | |
| Diethyl methylmalonate | Dioxane, 40°, 30 min. | 98 (68) |
| $CH_3CON(C_3H_7\text{-}n)CH_2Cl$ and | | |
| Diethyl methylmalonate | Dioxane, 100°, 1 hr. | 98 (38) |
| $CH_3CON(C_3H_7\text{-}i)CH_2Cl$ and | | |
| Diethyl methylmalonate | | 98 (28) |
| $CH_2ClCONHCH_2Cl$ and | | |
| Diethyl methylmalonate | Dioxane | 96 |
| $CCl_3CONHCH_2Cl$ and | | |
| 1-Phenylbutane-1,3-dione | Dioxane | 96 (51) |
| Diethyl ethylmalonate | $(C_2H_5)_2O$, reflux, 15 min. | 100 (62) |
| α-Phenylacetoacetonitrile | Dioxane | 96 (57) |
| 1-Chloromethyl-2-pyrrolidinone and | | |
| Diethyl methylmalonate | Dioxane-acetonitrile, 40°, 2 hr. | 33 (40) |
| N-Chloromethyl-ε-caprolactam and | | |
| Diethyl methylmalonate | Dioxane, 100°, 0.5 hr. | 33 (40) |
| $C_6H_5CONHCH_2Cl$ and | | |
| Pentane-2,4-dione | $(C_2H_5)_2O$ | 96 (60)† |
| 3-Methylpentane-2,4-dione | $(C_2H_5)_2O$, reflux, 2 hr. | 410 (88) |
| Ethyl acetoacetate | $(C_2H_5)_2O$, reflux, 0.5 hr. | 96, 100 (57) |
| Ethyl α-methylacetoacetate | $(C_2H_5)_2O$, reflux, 0.5 hr. | 96, 100 (50) |
| Ethyl α-isopropylacetoacetate | $(C_2H_5)_2O$, reflux, 0.5 hr. | 96, 100 (50) |
| 2-Carbethoxycyclohexanone | Dioxane, 100°, 3 hr. | 96 (84) |
| Dimethyl malonate | $(C_2H_5)_2O$ | 96 (81)† |
| Diethyl methylmalonate | $(C_2H_5)_2O$, reflux, 1 hr. | 96, 100 (77) |
| Diethyl ethylmalonate | $(C_2H_5)_2O$, reflux, 0.5 hr. | 96 (72), 100 (72) |
| Ethyl α-cyanopropionate | $(C_2H_5)_2O$, reflux, 1 hr. | 96 (48), 100 (48) |
| 1-Phenylbutane-1,3-dione | $(C_2H_5)_2O$, reflux, 0.5 hr. | 96 (51), 100 |
| α-Phenylacetoacetonitrile | $(C_2H_5)_2O$, reflux, 1.5 hr. | 96 (79), 100 (51) |
| $C_6H_5CON(CH_3)CH_2Cl$ and | | |
| Ethyl methylacetoacetate | Dioxane, 100°, 1 hr. | 98 (28) |
| Diethyl methylmalonate | Dioxane, 100°, 1 hr. | 98 (37) |
| $C_6H_5CONHCHClCCl_3$ and | | |
| 1-Phenylbutane-1,3-dione | Dioxane, 50°, 3 hr. | 101 (73) |

*Note:* References 382 to 537 are on pp. 266–269.

* Unless otherwise noted, the sodium salt of the active methylene compound was prepared before the alkylating agent was added.

† The bisamidoalkylated product was obtained.

TABLE XII—*Continued*

α-AMIDOALKYLATION OF ALIPHATIC CARBON ATOMS WITH N-HALOMETHYL AND N-α-HALOALKYL DERIVATIVES OF AMIDES, LACTAMS, IMIDES, AND CARBAMYL COMPOUNDS

| Product Derived from | Method* | References (Yield, %) |
|---|---|---|
| 2,4-$Cl_2C_6H_3CONHCH_2Cl$ and | | |
| Diethyl methylmalonate | Dioxane | 96 |
| 2-$CH_3O$-5-$ClC_6H_3CONHCHClCCl_3$ and | | |
| Potassium cyanide | $CH_3COCH_3$, 0.5 hr. | 107 (ca. 39)‡ |
| 2-$CH_3O$-3,5-$Cl_2C_6H_2CONHCHClCCl_3$ and | | |
| Potassium cyanide | $CH_3COCH_3$, 0.5 hr. | 107 (ca. 39)‡ |
| 2-$CH_3O$-5-$BrC_6H_3CONHCHClCCl_3$ and | | |
| Potassium cyanide | $CH_3COCH_3$, 0.5 hr. | 107 (ca. 39)‡ |
| 2-$CH_3O$-3,5-$Br_2C_6H_2CONHCHClCCl_3$ and | | |
| Potassium cyanide | $CH_3COCH_3$, 0.5 hr. | 107 (ca. 39)‡ |
| *m*-$CH_3C_6H_4CONHCHClCCl_3$ and | | |
| Potassium cyanide | $CH_3COCH_3$, 0.5 hr. | 106‡ |
| *p*-$CH_3C_6H_4CONHCHClCCl_3$ and | | |
| Potassium cyanide | $CH_3COCH_3$, 0.5 hr. | 106‡ |
| *o*-$C_6H_4(CO)_2NCH_2Cl$ and | | |
| Diphenylacetaldehyde | $(C_2H_5)_2O$ | 97 |
| 2-Methyl-1-phenylbutane-1,3-dione | $CH_3COCH_3$, reflux, 30 min. | 97 (54) |
| Cyclohexane-1,3-dione | $CH_3OH/C_6H_6$, reflux | 85 (20) |
| 1,1,4-Trimethylcyclohexane-3,5-dione | $C_2H_5OH$, reflux, 30 min. | 97 (96) |
| Ethyl α-methylacetoacetate | $(C_2H_5)_2O$ | 97 (63) |
| Diethyl methylmalonate | Dioxane | 97 (47) |
| | $(C_2H_5)_2O$ | 97 (81) |
| Diethyl ethylmalonate | $(C_2H_5)_2O$ | 97 (67) |
| Diethyl acetamidomalonate | $C_2H_5OH$ | 97 (60) |
| Diethyl phenylmalonate | $(C_2H_5)_2O$ | 97 (83) |
| α-Phenylacetoacetonitrile | $C_2H_5OH/CH_3COCH_3$, reflux 30 min. | 97 (72) |
| *o*-$C_6H_4(CO)_2NCH_2Br$ and | | |
| 3-Phenyl-2-benzofuranone | $(C_2H_5)_2O$, reflux, 2–3 hr. | 105 (73) |
| Ethyl acetoacetate | $C_2H_5OH$ | 104 (0) |
| Ethyl 4,4-diethoxyacetoacetate | $C_2H_5OH$, reflux, 3 hr. | 104 (0) |
| Diethyl malonate | $C_6H_6$ | 102 (0) |
| Potassium cyanide | $CH_3COCH_3$, reflux, 3 hr. | 104 (0) |
| N-Chloromethylsaccharin and | | |
| Ethyl methylacetoacetate | $(C_2H_5)_2O$, reflux, 3 hr. | 99 (56) |
| Diethyl methylmalonate | $(C_2H_5)_2O$, reflux, 4 hr. | 99 (59) |
| Ethyl methylcyanoacetate | $(C_2H_5)_2O$, reflux, 2 hr. | 99 (61) |
| $C_2H_5O_2CNHCHClCCl_3$ and | | |
| Ethyl acetoacetate | $(C_2H_5)_2O$, reflux, 6–7 hr. | 160 |

*Note:* References 382 to 537 are on pp. 266–269.

* Unless otherwise noted, the sodium salt of the active methylene compound was prepared before the alkylating agent was added.

‡ The product was of the type $ArCONHC(CN){=}CCl_2$.

## TABLE XIII

α-Amidoalkylation of Aliphatic Carbon Atoms with N,N′-Methylenedisulfonamides and N,N′-Ethylidene-bisurethan or Their Precursors

Method A: $(ArSO_2NH)_2CH_2 + HC\lt \rightarrow ArSO_2NHCH_2C\lt + ArSO_2NH_2$

Method B: $C_2H_5O_2CNH_2 + CH_3CHO + HC\lt \rightarrow C_2H_5O_2CNHCH(CH_3)C\lt$

| Product Derived from | Method | References (Yield, %) |
|---|---|---|
| $(C_6H_5SO_2NH)_2CH_2$ and Cyclohexane-1,3-dione | A; $H_2SO_4$, 25°, 2 days | 85 (18) |
| $(p\text{-}CH_3C_6H_4SO_2NH)_2CH_2$ and Cyclohexane-1,3-dione | A; $H_2SO_4$, 25°, 2 days | 85 (19) |
| $C_2H_5O_2CNH_2$, $CH_3CHO$, and Pentane-2,4-dione | B; $C_2H_5OH$, HCl, 24 hr. | 108 (30) |

*Note:* References 382 to 537 are on pp. 266–269.

## TABLE XIV

### α-Amidoalkylation of Aliphatic Carbon Atoms with N,N′-Arylidenebisamides or their Precursors

Method A: $(RCONH)_2CHAr + HC\lessgtr \rightarrow RCONHCH(Ar)C\lessgtr + RCONH_2$

Method B: $RCONH_2 + ArCHO + HC\lessgtr \rightarrow RCONHCH(Ar)C\lessgtr$

| Product Derived from | Method | References (Yield, %) |
|---|---|---|
| Acetamide, benzaldehyde, and | | |
| Nitromethane | A; $(CH_3CO)_2O$, 100°, 24 hr. | 113 (32) |
| Cyclohexane-1,3-dione | A; $(CH_3CO)_2O$, 130°, 3 hr. | 85 (0) |
| 5,5-Dimethylcyclohexane-1,3-dione | A; $(CH_3CO)_2O$, 130°, 3 hr. | 85 (12) |
| Ethyl nitroacetate | A; $(CH_3CO)_2O$, 100°, 7 hr. | 112 (85) |
| | B; $(CH_3CO)_2O$, 100°, 7 hr. | 112 (61) |
| Ethyl acetoacetate | A; $(CH_3CO)_2O$, 120°, 5 hr. | 114 (72) |
| Dimethyl malonate | A; $(CH_3CO)_2O$, HCl, 155°, 3 hr. | 85 (62) |
| Diethyl malonate | A; $(CH_3CO)_2O$, 155°, 3 hr. | 115 (62) |
| | A; 180°, 9 hr. | 115 (11) |
| Diethyl ethylmalonate | A; $(CH_3CO)_2O$, 155°, 5 hr. | 115 (76) |
| Hippuric acid | A; $CH_3CO_2H$, $(CH_3CO)_2O$, 110° | 117 (61)* |

*Note:* References 382 to 537 are on pp. 266–269.

* The product was

$CH_3CONHCH(C_6H_5)CH—C{=}O$ (ring: CH–N=C($C_6H_5$)–O–C=O)

All other alkylations of hippuric acid gave the expected products.

TABLE XIV—*Continued*

α-Amidoalkylation of Aliphatic Carbon Atoms with N,N′-Arylidenebisamides or Their Precursors

Method A: $(RCONH)_2CHAr + HC\!\!\lessgtr \rightarrow RCONHCH(Ar)C\!\!\lessgtr + RCONH_2$

Method B: $RCONH_2 + ArCHO + HC\!\!\lessgtr \rightarrow RCONHCH(Ar)C\!\!\lessgtr$

| Product Derived from | Method | References (Yield, %) |
|---|---|---|
| Acetamide, benzaldehyde (*contd.*) and | | |
| $CH_3CHCO$–NH–CS–S (ring: $CH_3CHCO$, NH, CS, S) | A; $(CH_3CO)_2O$, 150°, 2 hr. | 411 (93) |
| Piperazine-2,5-dione | A; $(CH_3CO)_2O$, 165°, 12 hr. | 109 (29)† |
| Acetamide, *o*-nitrobenzaldehyde, and | | |
| Ethyl nitroacetate | B; $(CH_3CO)_2O$, 100°, 7 hr. | 112 (12) |
| Hippuric acid | A; $CH_3CO_2H$, $(CH_3CO)_2O$ | 117 (0) |
| Acetamide, *p*-nitrobenzaldehyde, and | | |
| Hippuric acid | A; $CH_3CO_2H$, $(CH_3CO)_2O$ | 117 (0) |
| Acetamide, *o*-methoxybenzaldehyde, and | | |
| Hippuric acid | A; $CH_3CO_2H$, $(CH_3CO)_2O$, heat, 1 hr. | 117 (57)‡ |
| Acetamide, *p*-methoxybenzaldehyde, and | | |
| Diethyl malonate | A; $(CH_3CO)_2O$, 155°, 3 hr. | 115 (45) |
| Hippuric acid | A; $CH_3CO_2H$, $(CH_3CO)_2O$, heat, 1 hr. | 117 (69)‡ |
| Acetamide, veratraldehyde, and | | |
| Ethyl nitroacetate | A; $(CH_3CO)_2O$, 100°, 7 hr. | 112 (44) |
| | B; $(CH_3CO)_2O$, 100°, 7 hr. | 112 (41) |
| Ethyl acetoacetate | A; $(CH_3CO)_2O$, 130°, 5 hr. | 114 (75) |

| | | |
|---|---|---|
| Acetamide, piperonal, and | | |
| Ethyl nitroacetate | B; $(CH_3CO)_2O$, 100°, 8.5 hr. | 112 (49) |
| Ethyl acetoacetate | A; $(CH_3CO)_2O$, 120°, 5 hr. | 114 (50) |
| Diethyl malonate | A; $(CH_3CO)_2O$, 155°, 3 hr. | 115 (53) |
| Hippuric acid | A; $CH_3CO_2H$, $(CH_3CO)_2O$, heat, 1.5 hr. | 117 (50)‡ |
| Acetamide, O-acetylvanillin, and | | |
| Ethyl nitroacetate | A; $(CH_3CO)_2O$, 100°, 7 hr. | 112 (43) |
| | B; $(CH_3CO)_2O$, 100°, 7 hr. | 112 (40) |
| Acetamide, *p*-tolualdehyde, and | | |
| Hippuric acid | A; $CH_3CO_2H$, $(CH_3CO)_2O$, heat, 1.5 hr. | 117 (55)‡ |
| Benzamide, benzaldehyde, and | | |
| Cyclohexane-1,3-dione | A; $(CH_3CO)_2O$, 140°, 4 hr. | 85 (40) |
| 5,5-Dimethylcyclohexane-1,3-dione | A; $(CH_3CO)_2O$, 130°, 3 hr. | 85 (0) |
| Ethyl nitroacetate | A; $(CH_3CO)_2O$, 135°, 2 hr. | 85 (47) |
| Dimethyl malonate | A; $(CH_3CO)_2O$, 155°, 3 hr. | 85 (52) |
| Benzamide, *o*-nitrobenzaldehyde, and | | |
| Cyclohexane-1,3-dione | A; $(CH_3CO)_2O$, 130°, 3 hr. | 85 (0) |

*Note:* References 382 to 537 are on pp. 266–269.

† The product was

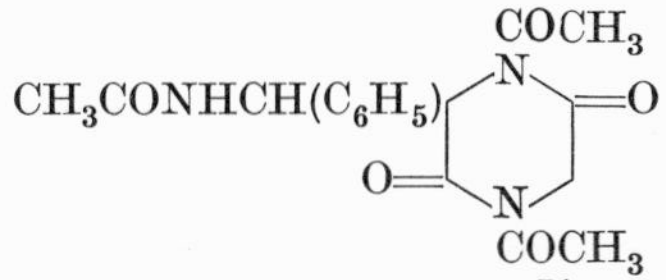

COCH$_3$
N
$CH_3CONHCH(C_6H_5)$ =O
O=
N
COCH$_3$
**74**

‡ This yield is based on the combined weight of the stereoisomeric products $CH_3CONHCH(Ar)CH(CO_2H)NHCOC_6H_5$ (**73**).

TABLE XIV—*Continued*

α-Amidoalkylation of Aliphatic Carbon Atoms with N,N′-Arylidenebisamides or Their Precursors

Method A: $(RCONH)_2CHAr + HC{<} \rightarrow RCONHCH(Ar)C{<} + RCONH_2$

Method B: $RCONH_2 + ArCHO + HC{<} \rightarrow RCONHCH(Ar)C{<}$

| Product Derived from | Method | References (Yield, %) |
|---|---|---|
| Ethyl carbamate, acetaldehyde, and | | |
| Pentane-2,4-dione | B; HCl, $C_2H_5OH$ | 108 |
| Ethyl carbamate, cinnamaldehyde, and | | |
| Pentane-2,4-dione | B; HCl, $C_2H_5OH$ | 111 |
| Ethyl acetoacetate | B; HCl, $C_2H_5OH$ | 111 |
| Ethyl carbamate, benzaldehyde, and | | |
| Pentane-2,4-dione | B; HCl, $C_2H_5OH$ | 111 |
| Ethyl acetoacetate | B; HCl, $C_2H_5OH$ | 111 |
| Ethyl benzoylacetate | B; HCl, $C_2H_5OH$ | 111 |
| Ethyl carbamate, salicylaldehyde, and | | |
| Pentane-2,4-dione | B; HCl, $C_2H_5OH$ | 111 |
| Ethyl carbamate, *p*-methoxybenzaldehyde, and | | |
| Pentane-2,4-dione | B; HCl, $C_2H_5OH$ | 111 |

*Note:* References 382 to 537 are on pp. 266–269.

TABLE XV

AMIDOMETHYLATION OF ALIPHATIC CARBON ATOMS WITH N-AMINOMETHYLAMIDES AND THEIR QUATERNARY SALTS

| Product Derived from | Method | References (Yield, %) |
|---|---|---|
| $C_6H_5CONHCH_2N(CH_3)_2$ and | | |
| Nitrocyclohexane | Toluene, NaOH, reflux, 50 hr. | 118 (38) |
| Dimethyl acetamidomalonate | Toluene, NaOH, reflux, 2.5 hr. | 118 (87) |
| Diethyl methylmalonate | Toluene, NaOH, reflux, 5.5 hr. | 118 (48) |
| Diethyl benzylmalonate | Toluene, NaOH, reflux, 11.5 hr. | 118 (68) |
| Diethyl phenylmalonate | Toluene, NaOH, reflux, 1 hr. | 118 (100) |
| $C_6H_5CONHCH_2N(C_2H_5)_2$ and | | |
| 1,3-Diphenylpropane-1,3-dione | Toluene, $N(C_4H_9\text{-}n)_3$, reflux, 30 hr. | 118 (3) |
| Ethyl α-ethylacetoacetate | Toluene, NaOH, reflux, 30 hr. | 118 (38) |
| Dimethyl malonate | Toluene, NaOH, reflux, 3 hr. | 118 (68) |
| Diethyl ethylmalonate | Toluene, NaOH, reflux, 7 hr. | 118 (44) |
| $C_2H_5O_2C(CH_2)_4CH(CN)CO_2C_2H_5$ | Toluene, NaOH, reflux, 5 hr. | 118 (85) |
| Ethyl α-phenylcyanoacetate | Toluene, NaOH, reflux, 8 hr. | 118 (74) |
| $o\text{-}C_6H_4(CO)_2NCH_2N(CH_3)_2$ and | | |
| Diethyl malonate | No solvent, NaOH, 100°, 5–15 hr. | 81 (0) |
| Diethyl acetamidomalonate | Xylene, reflux | 1* |
| $o\text{-}C_6H_4(CO)_2NCH_2N$ (piperidino) and | | |
| Diethyl formamidomalonate | Xylene, 150°, 4 hr. | 81 (0) |
| $o\text{-}C_6H_4(CO)_2NCH_2\overset{\oplus}{N}(CH_3)_3\, I^{\ominus}$ and | | |
| Cyclohexane-1,3-dione | $CH_3OH$, Na, reflux, 6 hr. | 85 (0) |
| Diethyl malonate | $C_2H_5OH$, $NaOC_2H_5$, reflux, 7 hr. | 81† |
| Diethyl formamidomalonate | $C_2H_5OH$, $NaOC_2H_5$, reflux, 5 hr. | 81 (88) |
| Sodium cyanide | $HCON(CH_3)_2$, NaOH, reflux, 3–4 hr. | 81 (76) |
| | $HCON(CH_3)_2$, reflux, 1 hr. | 412 (81) |

*Note:* References 382 to 537 are on pp. 266–269.

* Monodecarboxylation occurred during amidomethylation. The product isolated was $o\text{-}C_6H_4(CO)_2NCH_2CH(NHCOCH_3)CO_2C_2H_5$.

† The bisamidomethylated derivative was obtained in low yield as the sole product.

## TABLE XVI

### α-Amidoalkylation of Aliphatic Carbon Atoms with Acylimines

$$C_6H_5CON{=}CAr_2 + HC\overset{/}{\underset{\backslash}{-}} \rightarrow C_6H_5CONHC(Ar)_2C\overset{/}{\underset{\backslash}{-}}$$

| Product Derived from | Method | References (Yield, %) |
|---|---|---|
| $C_6H_5CON{=}C(C_6H_5)_2$ and | | |
| Acetonitrile | $LiNH_2$, liq. $NH_3$ | 120 (85) |
| Propionitrile | $LiNH_2$, liq. $NH_3$ | 121 (50) |
| Butyronitrile | $LiNH_2$, liq. $NH_3$ | 121 (53) |
| $C_6H_5CH_2CO_2Na$ | $i$-$C_3H_7MgCl$, $C_6H_6$/ether, reflux | 19 (49) |
| $C_6H_5CON{=}C(C_6H_5)C_6H_4OCH_3$-$p$ and | | |
| $C_6H_5CH_2CO_2Na$ | $i$-$C_3H_7MgCl$, $C_6H_6$/ether, reflux | 19 (53) |
| $C_6H_5CON{=}C(C_6H_5)$(1-naphthyl) and | | |
| Acetonitrile | $LiNH_2$, liq. $NH_3$ | 120 (59) |
| Propionitrile | $LiNH_2$, liq. $NH_3$ | 121 (36) |
| Butyronitrile | $LiNH_2$, liq. $NH_3$ | 121 (40)* |

*Note:* References 382 to 537 are on pp. 266–269.

* Two diastereoisomers were isolated; one in a 37% and the other in a 3% yield.

## TABLE XVII

α-AMIDOALKYLATION OF ETHYL ACETOACETATE WITH N,N′-ALKYLIDENE- AND N,N′-ARYLIDENE-BISUREAS OR THEIR PRECURSORS. THE BIGINELLI PYRIMIDINE SYNTHESIS

Method A: $CH_3COCH_2CO_2C_2H_5 + (H_2NCONH)_2CHR \longrightarrow$

Method B: $CH_3COCH_2CO_2C_2H_5 + CO(NH_2)_2 + RCHO \rightarrow$

(ring: HN–C($CH_3$)=C($CO_2C_2H_5$)–CHR–NH–C(=O)–, i.e. O=C, HN, $CCO_2C_2H_5$, CHR, N–H)

| Product Derived from Urea and | Method | References (Yield, %) |
|---|---|---|
| Formaldehyde | A; HCl, $CH_3CO_2H$, 120° | 126 (18) |
| | A; 120°, 5–8 hr. | 124 |
| Acetaldehyde | B; HCl, $C_2H_5OH$, 80°, 3 hr. | 126 (26) |
| | B; HCl, dioxane, 100°, 2 hr. | 126 (47) |
| | B; 100°, 2–3 hr. | 124 |
| Heptanal | B; HCl, $C_2H_5OH$, 80°, 5.5 hr. | 126 (7) |
| β-Phenylpropionaldehyde | B; $CH_3CO_2H$, 25°, 450 hr. | 126 (13) |
| Cinnamaldehyde | A; $C_2H_5OH$, 80° | 123 |
| | B; $C_2H_5OH$, 80°, 2 hr. | 122 |
| | B; $CH_3CO_2H$, 25° | 126 (71) |

*Note:* References 382 to 537 are on pp. 266–269.

TABLE XVII—*Continued*

α-Amidoalkylation of Ethyl Acetoacetate with N,N′-Alkylidene- and N,N′-Arylidene-bisureas or Their Precursors. The Biginelli Pyrimidine Synthesis

Method A: $CH_3COCH_2CO_2C_2H_5 + (H_2NCONH)_2CHR \longrightarrow$ (ring product below)

Method B: $CH_3COCH_2CO_2C_2H_5 + CO(NH_2)_2 + RCHO \rightarrow$ (ring product below)

Product ring: HN–C($CH_3$)=C($CO_2C_2H_5$)–CHR–NH–C(=O)– (ring closed back to HN)

| Product Derived from Urea and | Method | References (Yield, %) |
|---|---|---|
| Benzaldehyde | A; $C_2H_5OH$, 80°, 5–6 hr. | 122, 125 |
| | B; $C_2H_5OH$, 80°, 2 hr. | 122, 125 (56) |
| | B; HCl, $C_2H_5OH$, 80°, 3 hr. | 126 (78) |
| | B; 120°, 6 hr. | 125 (60) |
| | B; $CS(NH_2)_2$, $C_2H_5OH$, reflux | 125* |
| *m*-Nitrobenzaldehyde | B; HCl, $C_2H_5OH$, 80°, 6 hr. | 124, 126 (56) |
| *p*-Nitrobenzaldehyde | B; HCl, $C_2H_5OH$, 80°, 3 hr. | 126 (31) |
| | B; HCl, $CH_3CO_2H$, 120°, 3 hr. | 126 (58) |
| Salicylaldehyde | A, B; $C_2H_5OH$, 80°, 2 hr. | 122, 123 |
| | B; HCl, $C_2H_5OH$, 80°, 3 hr. | 126 (19) |

| | | |
|---|---|---|
| *p*-Hydroxybenzaldehyde | B; HCl, $C_2H_5OH$, 80°, 3 hr. | 126 (66) |
| 3,5-Diiodo-4-hydroxybenzaldehyde | B; HCl, $CH_3CO_2H$, 120°, 2.25 hr. | 126 (11) |
| *p*-Anisaldehyde | B; HCl, $C_2H_5OH$, 80°, 3 hr. | 126 (61) |
| 2,4-Dihydroxybenzaldehyde | B; $CH_3CO_2H$, 25°, 120 hr. | 126 (34) |
| Vanillin | B; HCl, $C_2H_5OH$, 80°, 3 hr. | 126 (42) |
| Veratraldehyde | B; HCl, $C_2H_5OH$, 80°, 2 hr. | 126 (47) |
| Piperonal | B; HCl, $C_2H_5OH$, 80°, 3 hr. | 126 (49) |
| 2,4,6-Trimethoxybenzaldehyde | B; HCl, $C_2H_5OH$, 80°, 2 hr. | 126 (38) |
| *p*-Isopropylbenzaldehyde | A, B; $C_2H_5OH$, 80° | 123 |
| 2-Furaldehyde | A; HCl, $C_2H_5OH$, 80° | 124 |
| | B; $C_2H_5OH$, 80°, 2 hr. | 122, 123, 126 (36) |

*Note:* References 382 to 537 are on pp. 266–269.

* Thiourea replaced urea to give the corresponding product

```
        CH3
        C
      /   \\
   HN       CCO2C2H5
   |        |
 S=C        CHC6H5
      \   /
       NH
```

## TABLE XVIII

α-Amidoalkylation of Active Methylene Compounds Other than Ethyl Acetoacetate with Urea and Aldehydes. The Biginelli Pyrimidine Synthesis

$$\mathrm{RC(=O)CH_2C(=O)R'} + \mathrm{CO(NH_2)_2} + \mathrm{R''CHO} \rightarrow$$

```
          R
          C         O
       /    \\      ||
    HN        C—C—R'
    |         |
  O=C        CHR''
       \    /
         N
         H
```

| Product Derived from Urea | Method | References (Yield, %) |
|---|---|---|
| Pentane-2,4-dione and | | |
| Propionaldehyde | HCl, $C_2H_5OH$, reflux, 5 hr. | 128 (32) |
| Isovaleraldehyde | HCl, $C_2H_5OH$, reflux, 4 hr. | 128 (0) |
| Heptanal | HCl, $C_2H_5OH$, reflux, 4 hr. | 128 (0) |
| Cinnamaldehyde | $CH_3CO_2H$, 100°, 3 hr. | 128 (92) |
| Benzaldehyde | HCl, $C_2H_5OH$, reflux, 3.5 hr. | 128 (55) |
| *m*-Nitrobenzaldehyde | HCl, $C_2H_5OH$, reflux, 3 hr. | 128 (70) |
| Salicylaldehyde | HCl, $C_2H_5OH$, reflux, 2 hr. | 128 (49) |
| *o*-Methoxybenzaldehyde | HCl, $C_2H_5OH$, reflux, 3.5 hr. | 128 (40) |
| 3,4-Dimethoxybenzaldehyde | HCl, $C_2H_5OH$, reflux, 1 hr. | 128 (44) |
| *p*-$(CH_3)_2NC_6H_4CHO$ | HCl, $C_2H_5OH$, reflux, 3 hr. | 128 (40) |
| 4-Hydroxy-3-methylbenzaldehyde | HCl, $C_2H_5OH$, reflux, 3.5 hr. | 128 (65) |
| 2-Furaldehyde | $CH_3CO_2H$, 100°, 2 hr. | 128 (89) |

| | | |
|---|---|---|
| Cyclohexane-1,3-dione and | | |
| *n*-Butyraldehyde | HCl, $C_2H_5OH$, reflux, 2.5 hr. | 130 (0) |
| *o*-Chlorobenzaldehyde | HCl, $C_2H_5OH$, reflux, 3.5 hr. | 130 (80) |
| *p*-Methoxybenzaldehyde | HCl, $C_2H_5OH$, reflux, 3 hr. | 130 (0) |
| *p*-$(CH_3)_2NC_6H_4CHO$ | HCl, $C_2H_5OH$, reflux, 3 hr. | 130 (0) |
| 1-Phenylbutane-1,3-dione and | | |
| Benzaldehyde | HCl, $C_2H_5OH$, reflux, 3 hr. | 129 (51)* |
| *o*-Chlorobenzaldehyde | HCl, $C_2H_5OH$, reflux, 2 hr. | 129 (68)* |
| *m*-Nitrobenzaldehyde | HCl, $C_2H_5OH$, reflux, 2 hr. | 129 (42)* |
| *p*-Hydroxybenzaldehyde | HCl, $C_2H_5OH$, reflux, 2 hr. | 129 (53)* |
| *p*-Methoxybenzaldehyde | HCl, $C_2H_5OH$, reflux, 2 hr. | 129 (50)* |
| Ethyl benzoylacetate and | | |
| Acetaldehyde | HCl, $CH_3CO_2H$, 100°, 24 hr. | 413 |

*Note:* References 382 to 537 are on pp. 266–269.
* The relative positions of R and R′ in the product were not established.

## TABLE XIX

AMIDOMETHYLATION OF POTASSIUM CYANIDE WITH ARENESULFONAMIDOMETHANESULFONATES

$$ArSO_2NHCH_2SO_3Na + KCN \rightarrow ArSO_2NHCH_2CN + NaKSO_3$$

| Product Derived from | Method | References (Yield, %) |
|---|---|---|
| $C_6H_5SO_2NHCH_2SO_3Na$ | $H_2O$ | 132 |
| *m*-$C_6H_4(SO_2NHCH_2SO_3Na)_2$ | $H_2O$ | 132 (75) |

*Note:* References 382 to 537 are on pp. 266–269.

## TABLE XX

N-MONOMETHYLOL DERIVATIVES OF CARBOXAMIDES, IMIDES, AND SULFONAMIDES

| N-Methylol Derivative | Method* | References (Yield, %) |
|---|---|---|
| *Of Aliphatic Amides* | | |
| $HCONHCH_2OH$ | Formalin, base | 22 |
| | 20% aq. $CH_2O$, 60° | 154 |
| | $(CH_2O)_x$, heat | 414 |
| $HCON(CH_3)CH_2OH$ | 20% aq. $CH_2O$, 60° | 154 |
| $CH_3CONHCH_2OH$ | Formalin, base | 23 |
| | 20% aq. $CH_2O$, 60° | 154 |
| | $(CH_2O)_x$, heat | 96, 414 |
| | $(CH_2O)_x$, base | 217, 415 |
| $CH_3CON(CH_3)CH_2OH$ | 20% aq. $CH_2O$, 60° | 154 |
| | $(CH_2O)_x$, 120–145° | 32 (good), 244 |
| $CH_3CON(C_3H_7\text{-}n)CH_2OH$ | $(CH_2O)_x$, base, 110° | 32 |
| $CH_3CON(C_3H_7\text{-}i)CH_2OH$ | $(CH_2O)_x$, base, 110° | 32 |
| $CH_3CON(CH_2C_6H_5)CH_2OH$ | $(CH_2O)_x$, base, 110° | 32 |
| $CH_3CON(CH_2NHCOCH_3)CH_2OH$ | 20% aq. $CH_2O$, 60° | 154 |
| $CH_3CON(CH_2CH_2NHCOCH_3)CH_2OH$ | 20% aq. $CH_2O$, 60° | 154 |
| $CH_3CON(CH_2CH_2CH_2NHCOCH_3)CH_2OH$ | 20% aq. $CH_2O$, 60° | 154 |
| $CH_2ClCONHCH_2OH$ | Formalin, base | 11 (100), 52 |
| | Formalin, acid | 416 |
| $CCl_3CONHCH_2OH$ | Formalin, base | 264 (100) |
| | Formalin, acid | 416 |
| $CH_2BrCONHCH_2OH$ | Formalin, acid | 416, 417 |
| $CH_2ICONHCH_2OH$ | Formalin, acid | 416, 417 |
| $o\text{-}CH_3OC_6H_4OCH_2CONHCH_2OH$ | Formalin, base | 418 |
| $CH_3CH{=}CH\text{-}C_6H_3(OCH_3)\text{-}OCH_2CONHCH_2OH$ (ring structure) | $RCON(CH_2OH)_2$, $H_2O$ | 42 |

| Product | Reactants* | Refs. (Yield, %) |
|---|---|---|
| 2-$OCH_3$-4-($CH_2$=$CHCH_2$)-$C_6H_3OCH_2CONHCH_2OH$ | $RCON(CH_2OH)_2$, $H_2O$ | 42 |
| $C_6H_5CONHCH_2CONHCH_2OH$ | Formalin, base | 79 |
| $C_6H_5CH_2CONHCH_2CONHCH_2OH$ | Formalin, base | 79 (86) |
| $C_5H_5\overset{\oplus}{N}CH_2CONHCH_2OH\ Cl^{\ominus}$ | $CH_2ClCONHCH_2OH$, pyridine | 312 (30) |
| $C_6H_5CH_2CONHCH_2OH$ | Formalin, base | 388 (83) |
| $C_6H_5CHOHCONHCH_2OH$ | Formalin, base | 43 |
| $CH_3CH_2CONHCH_2OH$ | Formalin, base | 153 (90) |
| $CH_3CHBrCONHCH_2OH$ | Formalin, acid | 416, 417 |
| $CH_2BrCHBrCONHCH_2OH$ | $CH_2$=$CHCONHCH_2OH$, $Br_2$ | 253 |
| $CH_3CHOHCONHCH_2OH$ | Formalin, base | 43 |
| $C_6H_5CH_2CONHCH(CH_3)CONHCH_2OH$ | Formalin, base | 79 (87) |
| $C_6H_5CH_2CH_2CONHCH_2OH$ | Formalin, base | 79 |
| $CH_2CO$–$NCH_2OH$–$CH_2CH_2$ (cyclic) | Formalin, base, 85° | 177, 184 (85) |
| | $(CH_2O)_x$, 120° | 33 (80) |
| $(CH_3)_2CHCONHCH_2OH$ | Formalin, base | 153 |
| $CH_2BrCBr(CH_3)CONHCH_2OH$ | Formalin, base | 253 (51) |
| $(CH_3)_2CHCH_2CONHCH_2OH$ | Formalin, base | 54, 418, 419 (65) |
| $CH_2CH_2CO$–$NCH_2OH$–$CH_2CH_2CH_2$ (cyclic) | Formalin, base, 85° | 184 (70) |
| | $(CH_2O)_x$, base, 140° | 310 (67) |
| | $(CH_2O)_x$, 110° | 33 |
| | Formaldehyde | 420 |
| $(C_2H_5)_2CHCONHCH_2OH$ | Formalin, base | 419 (80) |
| $(n\text{-}C_3H_7)_2CHCONHCH_2OH$ | Formalin, base | 153 (82) |

*Note:* References 382 to 537 are on pp. 266–269.

* Unless otherwise indicated, the amide or imide is an unnamed reactant.

TABLE XX—*Continued*

N-Monomethylol Derivatives of Carboxamides, Imides, and Sulfonamides

| N-Methylol Derivative | Method* | References (Yield, %) |
|---|---|---|
| *Of Aliphatic Amides* (*contd.*) | | |
| $n$-$C_{11}H_{23}CONHCH_2OH$ | Formalin, base, heat | 222 (88) |
| $CF_3(CH_2)_{11}CONHCH_2OH$ | $(CH_2O)_x$, base, $C_6H_6$, 50° | 217 (100) |
| $n$-$C_{17}H_{35}CONHCH_2OH$ | Formalin, base, 70° | 138 (83), 139, 140, 312 |
| | $(CH_2O)_x$, 60° | 217 (85) |
| $n$-$C_{17}H_{35}CON(CH_3)CH_2OH$ | $(CH_2O)_x$, heat | 139 |
| $n$-$C_{21}H_{43}CONHCH_2OH$ | Formalin, base | 235 |
| $CH_2{=}CHCONHCH_2OH$ | $(CH_2O)_x$, base | 254 (70), 421 (87), 422, 423 |
| $o$-$ClC_6H_4CH{=}CHCONHCH_2OH$ | $(CH_2O)_x$, base | 312 (91) |
| (2-furyl)$CH{=}CHCONHCH_2OH$ | $(CH_2O)_x$, $NaOC_2H_5$, $CCl_4$ | 35 (82) |
| $CH_2{=}C(CH_3)CONHCH_2OH$ | $(CH_2O)_x$, base | 254 (70), 421 (83), 424 (64) |
| $(CH_3)_2C{=}CHCONHCH_2OH$ | $(CH_2O)_x$, $NaOC_2H_5$, $CCl_4$ | 422 (77) |
| $CH_2{=}CH(CH_2)_8CONHCH_2OH$ | Formalin, base | 382 (96) |
| $CH_3(CH_2)_7CH{=}CH(CH_2)_7CONHCH_2OH$ | $(CH_2O)_x$, heat | 139 |
| $CH_3(CH{=}CH)_2CONHCH_2OH$ | Formalin, base | 389 (78), 423 |
| *Of Aromatic Amides* | | |
| $C_6H_5CONHCH_2OH$ | Formalin, base | 1, 10 (100), 52, 79, 88 |
| | Formalin, acid | 263, 425 (64) |
| $C_6H_5CON(CH_3)CH_2OH$ | $(CH_2O)_x$, base, 110° | 32 |
| 2,4-$Cl_2C_6H_3CONHCH_2OH$ | Formalin, base | 96 (75) |
| $o$-$O_2NC_6H_4CONHCH_2OH$ | Formalin, base | 221 |
| $m$-$O_2NC_6H_4CONHCH_2OH$ | Formalin, base | 221 |

| | | |
|---|---|---|
| $p$-$O_2NC_6H_4CONHCH_2OH$ | Formalin, base | 221 (56) |
| | $(CH_2O)_x$, heat | 139 |
| $o$-$HOC_6H_4CONHCH_2OH$ | Formalin, base | 25 (93), 418 |
| | Formalin, acid | 25 (62), 425 |
| $p$-$HOC_6H_4CONHCH_2OH$ | Formalin, base | 418 |
| $p$-$CH_3C_6H_4CONHCH_2OH$ | Formalin, base | 418 |
| *Of Heteroaromatic Amides* | | |
|  O ring, $CONHCH_2OH$ | $(CH_2O)_x$, base | 35 |
|  N ring, $CONHCH_2OH$ | Formalin, base, heat | 426 |
|  $CH_3$, N ring, $CONHCH_2OH$ | Formalin, base | 426 |
|  N ring, $CONHCH_2OH$ | Formalin, base | 426 |
| 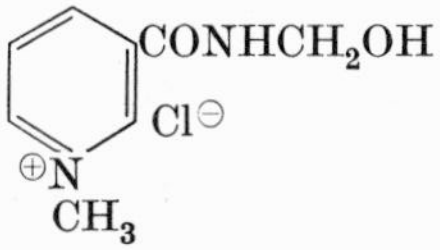 $CONHCH_2OH$, $Cl^{\ominus}$, $^{\oplus}N$, $CH_3$ | Formalin, heat | 427 (69) |
| | $RCONHCH_2OH$, $CH_3Cl$ | 427 (78) |
| 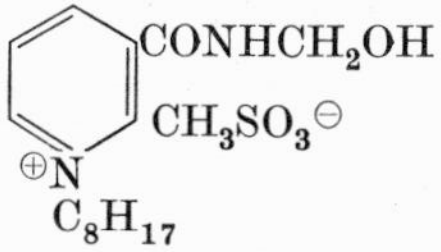 $CONHCH_2OH$, $CH_3SO_3^{\ominus}$, $^{\oplus}N$, $C_8H_{17}$ | Formalin, 60° | 427 (59) |
| | $RCONHCH_2OH$, $C_8H_{17}OSO_2CH_3$ | 427 (61) |

*Note:* References 382 to 537 are on pp. 266–269.

* Unless otherwise indicated, the amide or imide is an unnamed reactant.

TABLE XX—*Continued*

N-MONOMETHYLOL DERIVATIVES OF CARBOXAMIDES, IMIDES, AND SULFONAMIDES

| N-Methylol Derivative | Method* | References (Yield, %) |
|---|---|---|
| *Of Heteroaromatic Amides (contd.)* | | |
| $CONHCH_2OH$, $Br^{\ominus}$, $^{\oplus}N$, $C_{12}H_{25}n$ | $RCONHCH_2OH$, $n$-$C_{12}H_{25}Br$ | 427 (68) |
| $CONHCH_2OH$, $Br^{\ominus}$, $^{\oplus}N$, $C_{16}H_{33}$-$n$ | Formalin, 30° | 427 (66) |
| | $RCONHCH_2OH$, $n$-$C_{16}H_{33}Br$ | 427 (79) |
| $CONHCH_2OH$, $Cl^{\ominus}$, $^{\oplus}N$, $CH_2CO_2H$ | Formalin, 50° | 427 (73) |
| | $RCONHCH_2OH$, $CH_2ClCO_2H$ | 427 |
| $CONHCH_2OH$, $Cl^{\ominus}$, $^{\oplus}N$, $CH_2CONH_2$ | Formalin, 40° | 427 (80) |
| | $RCONHCH_2OH$, $CH_2ClCONH_2$ | 427 (84) |
| $CONHCH_2OH$, $CO_2H$, N | Formalin, base, 80° | 428 |
| $HO_2C$, $CONHCH_2OH$, N | Formalin, base, 80° | 428 (97) |

| | | |
|---|---|---|
| 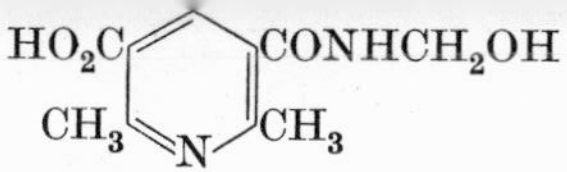 $HO_2C$, $CONHCH_2OH$, $CH_3$, N, $CH_3$ (pyridine) | Formalin, base, 100° | 428 (95) |
| 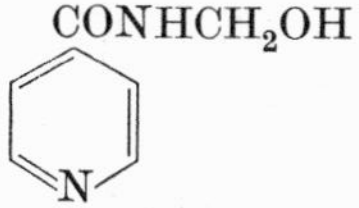 $CONHCH_2OH$ (pyridine, N) | Formalin, $NaHCO_3$, 50°<br>Formalin, 100° | 429 (80)<br>430 (66) |
|  $CONHCH_2OH$ (quinoline, N) | Formalin, base, heat | 426 |
| 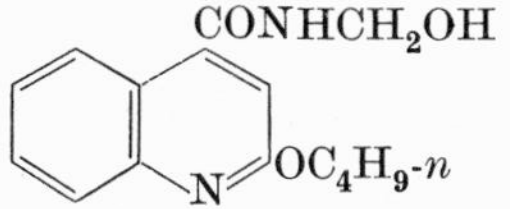 $CONHCH_2OH$, $OC_4H_9$-*n* (quinoline, N) | Formalin, base, heat | 426 |
| 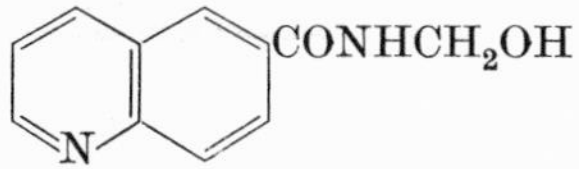 $CONHCH_2OH$ (quinoline, N) | Formalin, base | 242 |
| *Of Imides* | | |
| 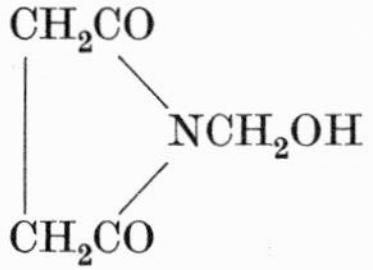 $CH_2CO$, $NCH_2OH$, $CH_2CO$ | Formalin, base | 66 (100) |
| 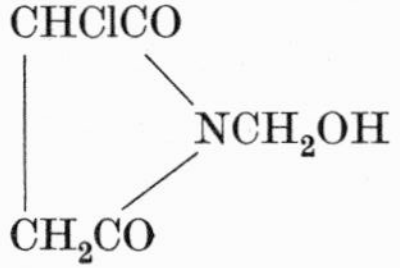 $CHClCO$, $NCH_2OH$, $CH_2CO$ | N-Methylolmaleimide, HCl | 8 |

*Note:* References 382 to 537 are on pp. 266–269.

* Unless otherwise indicated, the amide or imide is an unnamed reactant.

TABLE XX—*Continued*

N-Monomethylol Derivatives of Carboxamides, Imides, and Sulfonamides

| N-Methylol Derivative | Method* | References (Yield, %) |
|---|---|---|
| *Of Imides (contd.)* | | |
| $C_6H_5CONHCHCO$ / $CH_2CO$ (ring) $>NCH_2OH$ | Formalin | 431 (92) |
| $CHCO$ ‖ $CHCO$ $>NCH_2OH$ | Formalin, base, 35° | 8 (75), 432 |
| (bicyclic, $CH_2$ bridge) $<CO, CO>NCH_2OH$ | | 433 |
| $o$-$C_6H_4(CO)_2NCH_2OH$ | Formalin, heat | 1, 84 (96), 434, 435 (90) |
| | 10–15% aq. $CH_2O$, heat | 104 (excellent) |
| $NO_2$-$C_6H_3(CO)_2NCH_2OH$ (3-nitro) | Formalin, 100° | 436 |
| $O_2N$-$C_6H_3(CO)_2NCH_2OH$ (4-nitro) | Formalin, 100° | 436 |
| $C_6H_4(SO_2)(CO)NCH_2OH$ | Formalin, 100° | 99 (69) |
| | Formalin, heat | 182 (81), 437 |

| | | |
|---|---|---|
| 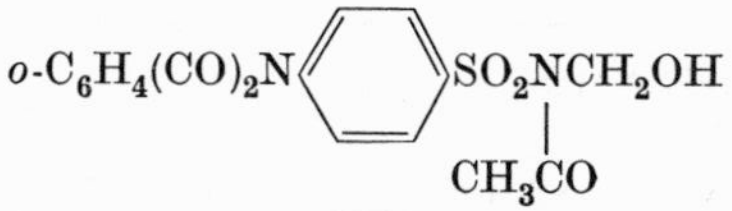 $o\text{-}C_6H_4(CO)_2N\text{-}C_6H_4\text{-}SO_2N(COCH_3)CH_2OH$ | Formalin, base | 157 |
| 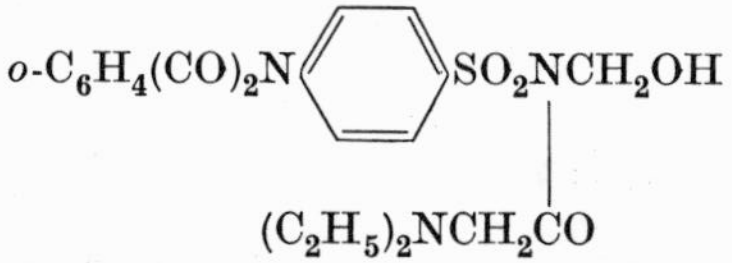 $o\text{-}C_6H_4(CO)_2N\text{-}C_6H_4\text{-}SO_2N(COCH_2N(C_2H_5)_2)CH_2OH$ | Formalin, base | 157 |
| *Of Sulfonamides* | | |
| $C_6H_5SO_2NHCH_2OH$ | Formalin, base | 155 (60), 156 |
| $p\text{-}CH_3C_6H_4SO_2NHCH_2OH$ | Formalin, base | 155 (93), 156 (60), 438 |
| 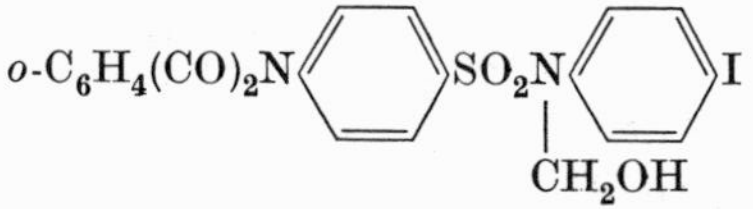 $o\text{-}C_6H_4(CO)_2N\text{-}C_6H_4\text{-}SO_2N(CH_2OH)\text{-}C_6H_4\text{-}I$ | Formalin, base | 157 |
| 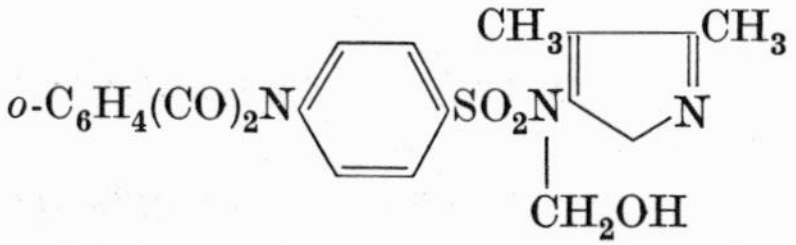 $o\text{-}C_6H_4(CO)_2N\text{-}C_6H_4\text{-}SO_2N(CH_2OH)$-(dimethylpyrazolinyl, $CH_3$, $CH_3$) | Formalin, base | 157 |

*Note:* References 382 to 537 are on pp. 266–269.

* Unless otherwise indicated, the amide or imide is an unnamed reactant.

## TABLE XX—*Continued*

### N-Monomethylol Derivatives of Carboxamides, Imides, and Sulfonamides

| N-Methylol Derivatives | Method* | References (Yield, %) |
|---|---|---|
| *Of Sulfonamides* (*contd.*) | | |
| 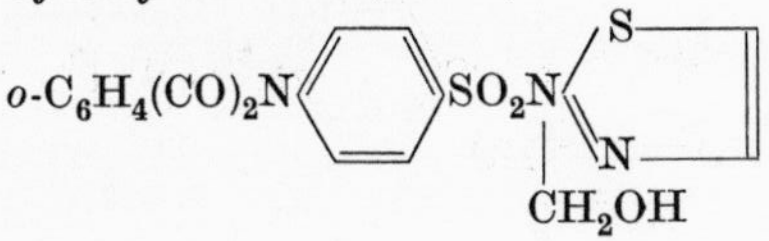 $o\text{-}C_6H_4(CO)_2N\text{-}C_6H_4\text{-}SO_2N(CH_2OH)\text{-}$[thiazole ring: S, N] | Formalin, base | 157 |
| 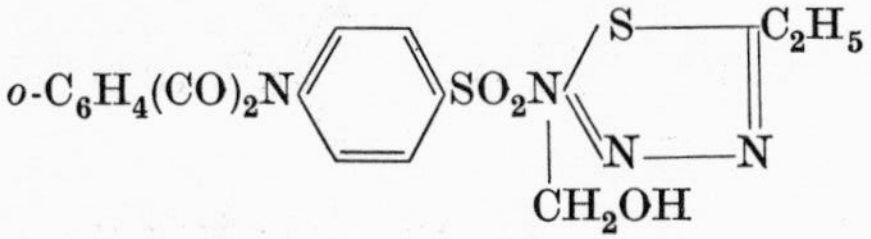 $o\text{-}C_6H_4(CO)_2N\text{-}C_6H_4\text{-}SO_2N(CH_2OH)\text{-}$[thiadiazole ring: S, N—N, $C_2H_5$] | Formalin, base | 157 |
| 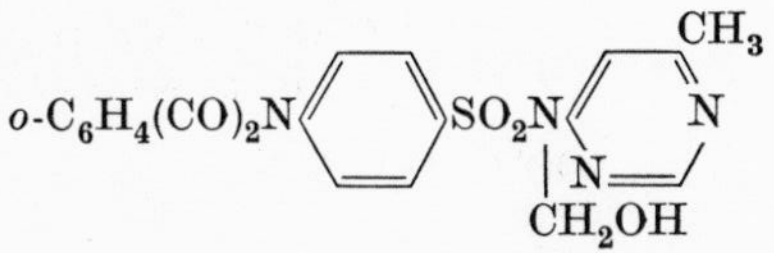 $o\text{-}C_6H_4(CO)_2N\text{-}C_6H_4\text{-}SO_2N(CH_2OH)\text{-}$[pyrimidine ring: N, N, $CH_3$] | Formalin, base | 157 |
| 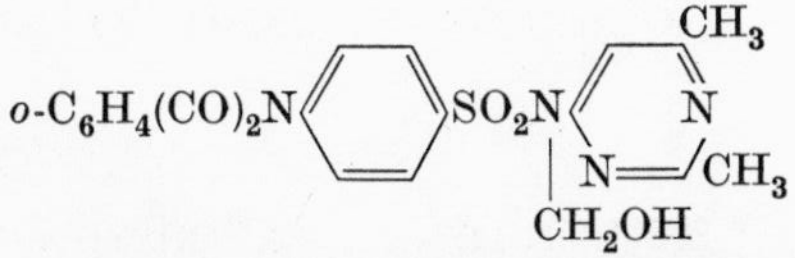 $o\text{-}C_6H_4(CO)_2N\text{-}C_6H_4\text{-}SO_2N(CH_2OH)\text{-}$[pyrimidine ring: N, N, $CH_3$, $CH_3$] | Formalin, base | 157 |

*Note:* References 382 to 537 are on pp. 266–269.

* Unless otherwise indicated, the amide or imide is an unnamed reactant.

## TABLE XXI

POLY-N-METHYLOL DERIVATIVES OF MONO- AND POLY-CARBOXAMIDES

| N-Methylol Derivatives | Method* | References (Yield, %) |
|---|---|---|
| *Of Monocarboxamides* | | |
| =O, $CON(CH_2OH)_2$ | Formalin, base | 41 (100) |
| $CH_3CH{=}CH$–(ring)–$OCH_3$, $OCH_2CON(CH_2OH)_2$ | Formalin, base | 42 |
| $CH_2{=}CHCH_2$–(ring)–$OCH_3$, $OCH_2CON(CH_2OH)_2$ | Formalin, base | 42 |
| *Of Polycarboxamides* | | |
| $(CONHCH_2OH)_2$ | Formalin, base | 35 (47) |
| $(C_2H_5)_2C(CONHCH_2OH)_2$ | Formalin, base | 263 (100) |
| $(CH_2)_2(CONHCH_2OH)_2$ | Formalin, base | 24 (80) |
| $CHOHCONHCH_2OH$–$CHOHCONHCH_2OH$ | Formalin, base | 43 (70) |
| $(CH_2)_4(CONHCH_2OH)_2$ | Formalin, base | 406 (71) |
| $(CH_2)_6(CONHCH_2OH)_2$ | Formalin, base | 406 (50) |
| $(CH_2)_7(CONHCH_2OH)_2$ | Formalin, base | 406 (60) |

*Note:* References 382 to 537 are on pp. 266–269.

* The amide is an unnamed reactant.

TABLE XXI—*Continued*

POLY-N-METHYLOL DERIVATIVES OF MONO- AND POLY-CARBOXAMIDES

| N-Methylol Derivatives | Method* | References (Yield, %) |
|---|---|---|
| *Of Polycarboxamides (contd.)* | | |
| $(CH_2)_8(CONHCH_2OH)_2$ | Formalin, base | 406 (81) |
| $HOC(CONHCH_2OH)(CH_2CONHCH_2OH)_2$ (structure: central C bearing HO, $CONHCH_2OH$ and two $CH_2CONHCH_2OH$ groups) | Formalin, base | 43 (20) |
| $HOCH_2N\langle CH_2CO \rangle_2 NCH_2OH$ (cyclic: 1,4-bis(hydroxymethyl)-2,5-dioxopiperazine) | Formalin | 1, 65 (90) |
| | Formalin, base | 79 |
| | Formalin, acid | 79 (0) |
| $CH_2[N(CHO)CH_2OH]_2$ | 20% aq. $CH_2O$, 60° | 154 |
| $[CH_2N(CHO)CH_2OH]_2$ | 20% aq. $CH_2O$, 60° | 154 |
| $CH_3CHN(CHO)CH_2OH$ / $CH_2N(CHO)CH_2OH$ (i.e., $CH_3CH[N(CHO)CH_2OH]CH_2N(CHO)CH_2OH$) | 20% aq. $CH_2O$, 60° | 154 |
| $CH_2[CH_2N(CHO)CH_2OH]_2$ | 20% aq. $CH_2O$, 60° | 154 |
| $HOCH[CH_2N(CHO)CH_2OH]_2$ | 20% aq. $CH_2O$, 60° | 154 |

| | | |
|---|---|---|
| $HCON[CH_2CH_2CH_2N(CHO)CH_2OH]_2$ | 20% aq. $CH_2O$, 60° | 154 |
| $HCCONHCH_2OH$ ‖ $HCCONHCH_2OH$ | | 423 |
| $HOCH_2NHCOCH$ ‖ $HCCONHCH_2OH$ | Formalin, acid | 254 (63) |
| | Formalin, base | 254 (49) |
| $C_6H_4(OCH_2CONHCH_2OH)_2$ (1,2-) | Formalin, base | 439 |
| $C_6H_4(OCH_2CONHCH_2OH)_2$ (1,3-) | Formalin, base | 439 |
| 3,5-$(HOCH_2NHCO)_2C_5H_3N$ | Formalin, base 100° | 428 (50) |
| 2,6-dimethyl-3,5-$(HOCH_2NHCO)_2C_5HN$ | Formalin, $NaHCO_3$, 100° | 428 |

*Note:* References 382 to 537 are on pp. 266–269.

* The amide is an unnamed reactant.

## TABLE XXII

N-Mono- and N,N′-Di-methylol Derivatives of Carbamyl Compounds

| N-Methylol Derivatives | Method* | References (Yield, %) |
|---|---|---|
| $C_2H_5O_2CNHCH_2OH$ | Formalin, acid | 153 (60) |
| $n$-$C_{22}H_{45}O_2CNHCH_2OH$ | $(CH_2O)_x$, pyridine, 90° | 235 |
| $C_6H_5CH_2NHCOCH_2N(CO_2CH_2C_6H_5)CH_2OH$ | $C_6H_5CH_2O_2CN\text{-}CH_2CO\text{-}O\text{-}CH_2$ (ring), $C_6H_5CH_2NH_2$ | 344 (91) |
| $H_2NCONHCH_2OH$ | Formalin, base | 153, 440–443 |
| | Formalin, $p$H 8.5 | 144 (20) |
| | $(CH_2O)_x$, base | 444 |
| $CO(NHCH_2OH)_2$ | Formalin, base | 153, 442, 445 (90) |
| | $(CH_2O)_x$, base, 15° | 444 (100), 446 |
| | Formalin | 440, 443 |
| $O(CH_2NHCONHCH_2OH)_2$ | $CO(NHCH_2OH)_2$, base | 62 |
| $CH_3NHCON(CH_3)CH_2OH$ | Formalin, base | 153 (71) |
| $(CH_3)_2NCONHCH_2OH$ | Formalin, base | 153 (75) |
| $C_6H_5NHCONHCH_2OH$ | $(CH_2O)_x$, base | 54, 55, 62 |
| $p$-$CH_3C_6H_4NHCONHCH_2OH$ | $(CH_2O)_x$, base | 55 |
| $H_2NCONHCH_2NHCONHCH_2OH$ | Formalin, base | 219 (10) |
| $CH_2(NHCONHCH_2OH)_2$ | Formalin, base | 219 (20) |
| $H_2NCSNHCH_2OH$ | Formalin, base | 265 (96) |
| $CS(NHCH_2OH)_2$ | Formalin, base | 265 (100) |

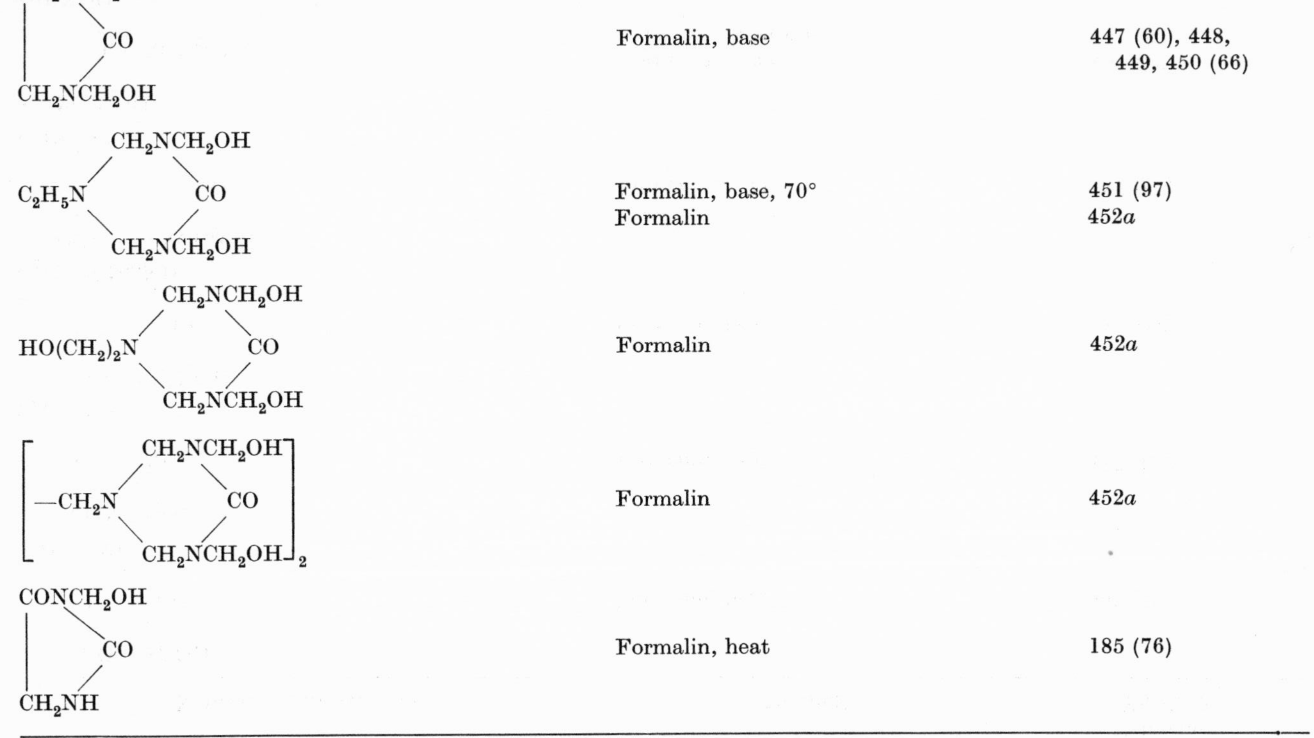

| Product | Conditions | Refs. (Yield %) |
|---|---|---|
| $CH_2N(CH_2OH)COCH_2N(CH_2OH)$ (ring: $CH_2NCH_2OH$ / CO / $CH_2NCH_2OH$) | Formalin, base | 447 (60), 448, 449, 450 (66) |
| $C_2H_5N$ ring with $CH_2NCH_2OH$, CO, $CH_2NCH_2OH$ | Formalin, base, 70°<br>Formalin | 451 (97)<br>452*a* |
| $HO(CH_2)_2N$ ring with $CH_2NCH_2OH$, CO, $CH_2NCH_2OH$ | Formalin | 452*a* |
| [—$CH_2N$ ring with $CH_2NCH_2OH$, CO, $CH_2NCH_2OH$]$_2$ | Formalin | 452*a* |
| $CONCH_2OH$ / CO / $CH_2NH$ (ring) | Formalin, heat | 185 (76) |

*Note:* References 382 to 537 are on pp. 266–269.

* Unless otherwise indicated, the carbamyl compound is an unnamed reactant.

TABLE XXII—*Continued*

N-Mono- and N,N′-Di-methylol Derivatives of Carbamyl Compounds

| N-Methylol Derivatives | Method* | References (Yield, %) |
|---|---|---|
| $CONCH_2OH$ / CO / $(CH_3)_2CNH$ (ring: $CONCH_2OH$—CO—NH—$C(CH_3)_2$) | Formalin, heat | 453 (99) |
| $CONCH_2OH$ / CO / $C_6H_5CHNH$ (ring) | Formalin, heat | 187 (62) |
| $OC$—$NCH_2OH$ / CO / $C_6H_5C(CH_3)NH$ (ring) | Formalin, heat | 187 (66) |
| $OC$—$NCH_2OH$ / CO / $C_6H_5C(C_2H_5)NH$ (ring) | Formalin, heat | 187 (67) |
| $CONCH_2OH$ / $C{=}NCH_2OH$ / $CH_2NCH_3$ (ring) | Formalin, creatine<br>Formalin, creatinine | 454 (90)<br>454 (80) |

| Product | Conditions | Refs. (Yield, %) |
| --- | --- | --- |
| 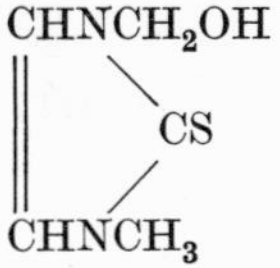 $CHNCH_2OH$ / ‖ CS / $CHNCH_3$ (ring) | Formalin, base | 455 (72) |
| 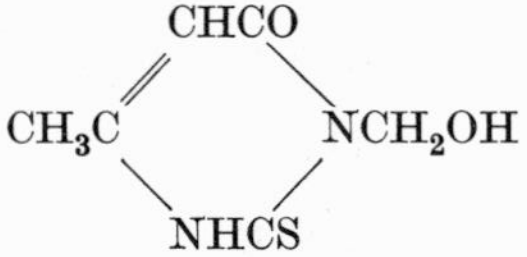 $CH_3C$=CHCO–$NCH_2OH$–NHCS (ring) | Formalin, heat | 239 |
| 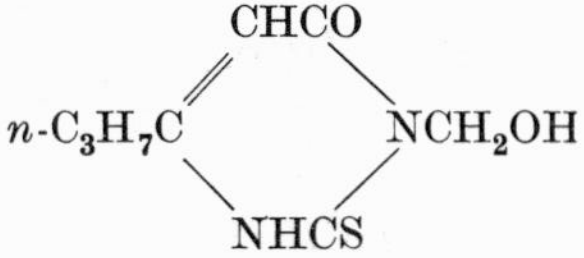 $n$-$C_3H_7C$=CHCO–$NCH_2OH$–NHCS (ring) | Formalin, heat | 239 |
| 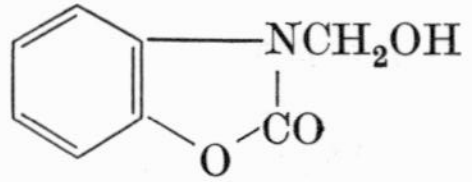 Benzo-fused ring: –$NCH_2OH$, –CO, –O– | Formalin, 100° | 194 (82) |
| 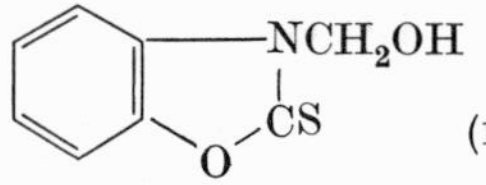 Benzo-fused ring: –$NCH_2OH$, –CS, –O– (DL-*trans*) | Formalin, heat | 181 (75) |
| 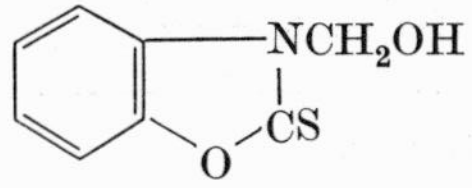 Benzo-fused ring: –$NCH_2OH$, –CS, –O– | Formalin, 100° | 178 (70) |

*Note:* References 382 to 537 are on pp. 266–269.

* Unless otherwise indicated, the carbamyl compound is an unnamed reactant.

TABLE XXII—*Continued*

N-MONO- AND N,N′-DI-METHYLOL DERIVATIVES OF CARBAMYL COMPOUNDS

| N-Methylol Derivatives | Method* | References (Yield, %) |
|---|---|---|
| $C_6H_4$ with $NCH_2OH$, $CS$ and $S$ in a ring | Formalin, heat | 183 (96), 189 (84) |
| $C_6H_4$ with $NCH_2OH$, $CO$ and $NCH_2N(CH_3)_2$ in a ring | Formalin, $(CH_3)_2NH$ | 180 (54) |
| $C_6H_4$ with $NCH_2OH$, $CO$ and $NCH_2N(C_2H_5)_2$ in a ring | Formalin, $(C_2H_5)_2NH$ | 180 (30) |
| $C_6H_4$ with $NCH_2OH$, $CO$ and $NCH_2OH$ in a ring | Formalin, 100° | 180 (80), 238, 456 |
| $C_2H_5OC_6H_3$ with $NCH_2OH$, $CO$ and $NCH_2OH$ in a ring | Formalin, 100° | 237 |

| Structure | Conditions | Refs. (Yield %) |
|---|---|---|
| $CH_3$-benzene ring fused to a ring of $NCH_2OH$, $CO$, $NCH_2OH$ | Formalin, heat | 456 (71) |
| Benzene ring fused to a ring of $NCH_2OH$, $CS$, $NH$ | Formalin, base | 457 (77) |
| Benzene ring fused to a ring of $NCH_2OH$, $CS$, $NCH_2OH$ | Formalin, 100° | 179 (83), 238 |
| $C_2H_5O$-benzene ring fused to a ring of $NCH_2OH$, $CS$, $NCH_2OH$ | Formalin, 100° | 237 |
| Naphthalene ring system fused to a ring of $NCH_2OH$, $CS$, $NH$ | | 239 |

*Note:* References 382 to 537 are on pp. 266–269.
* Unless otherwise indicated, the carbamyl compound is an unnamed reactant.

TABLE XXII—*Continued*

N-Mono- and N,N′-Di-methylol Derivatives of Carbamyl Compounds

| N-Methylol Derivatives | Method* | References (Yield, %) |
|---|---|---|
| (naphthalene-2,3-diyl) $NCH_2OH$ / CS / $NCH_2OH$ | | 239, 457 |
| $HOCH_2N$—CS—$NCH_2OH$ (naphthalene-1,2-diyl) | Formalin, 100° | 241 |
| S ring —$NCH_2OH$ / CO / —$NCH_2OH$, $(CH_2)_4CO_2H$ | $(CH_2O)_x$, base, 50° | 458 (45) |

*Note:* References 382 to 537 are on pp. 266–269.

* Unless otherwise indicated, the carbamyl compound is an unnamed reactant.

## TABLE XXIII

α-Hydroxyalkyl Derivatives of Carboxamides, Carbamyl Compounds, and Sulfonamides

$$RCONH_2 + R'CHO \rightarrow RCONHCHOHR'$$

(R′ = haloalkyl)

| α-Alkylolamide Derivative | References (Yield, %) | α-Alkylolamide Derivative | References (Yield, %) |
|---|---|---|---|
| *Of Aliphatic Monoamides* | | $\begin{array}{l} CH_2CO \\ \vert \\ CH_2CH_2 \end{array} \!> NCHOHCCl_3$ | 101 (60), 467 |
| $HCONHCHOHCCl_3$ | 95, (100), 459 | $n$-$C_4H_9CONHCHOHCCl_3$ | 164, 460 (78) |
| $HCONHCHOHCBr_3$ | 207 | $n$-$C_4H_9CONHCHOHCCl_2CHClCH_3$ | 460 (69) |
| $HCONHCHOHCCl_2CHClCH_3$ | 463 | $i$-$C_4H_9CONHCHOHCCl_3$ | 460 (41) |
| $C_2H_5O_2CCONHCHOHCCl_3$ | 160, 163, 229 | $i$-$C_4H_9CONHCHOHCBr_3$ | 207 |
| $CH_3CONHCHOHCHCl_2$* | 208 | $i$-$C_4H_9CONHCHOHCCl_2CHClCH_3$ | 460 (14) |
| $CH_3CONHCHOHCCl_3$ | 95, 158, 460 (80), 461, 462 | $(CH_3)_2CHCHBrCONHCHOHCCl_3$ | 468 |
| $CH_3CONHCHOHCBr_3$ | 463 | $n$-$C_5H_{11}CONHCHOHCCl_3$ | 164, 460 (69) |
| $CH_3CONHCHOHCCl_2CHClCH_3$ | 460, 463 (70), 464 | $n$-$C_5H_{11}CONHCHOHCBr_3$ | 207 |
| $CH_2ClCONHCHOHCCl_3$ | 465 (50), 466 | $n$-$C_5H_{11}CONHCHOHCCl_2CHClCH_3$ | 460 (62) |
| $CH_2ClCONHCHOHCCl_2CHClCH_3$ | 460 (43) | $(C_2H_5)_2CHCONHCHOHCCl_3$ | 460 (78) |
| $C_6H_5CH_2CONHCHOHCCl_3$ | 164, 466 (77) | $(C_2H_5)_2CHCONHCHOHCCl_2CHClCH_3$ | 460 (42) |
| $C_6H_5CH_2CONHCHOHCBr_3$ | 166 | $n$-$C_6H_{13}CONHCHOHCCl_3$ | 164, 460 (81) |
| $C_6H_5CH_2CONHCHOHCCl_2CHClCH_3$ | 466 (37) | $n$-$C_6H_{13}CONHCHOHCBr_3$ | 207 |
| $NCCH_2CONHCHOHCCl_3$ | 163 | $n$-$C_6H_{13}CONHCHOHCCl_2CHClCH_3$ | 460 (72) |
| $C_2H_5CONHCHOHCHCl_2$* | 208 | $n$-$C_7H_{15}CONHCHOHCCl_3$ | 460 (100) |
| $C_2H_5CONHCHOHCCl_3$ | 164, 460 (76) | $n$-$C_7H_{15}CONHCHOHCBr_3$ | 207 |
| $C_2H_5CONHCHOHCBr_3$ | 207 | $n$-$C_7H_{15}CONHCHOHCCl_2CHClCH_3$ | 460 (66) |
| $C_2H_5CONHCHOHCCl_2CHClCH_3$ | 460 (65) | $n$-$C_4H_9CH(C_2H_5)CONHCHOHCCl_3$ | 460 (88) |
| $n$-$C_3H_7CONHCHOHCCl_3$ | 460 (79) | $n$-$C_8H_{17}CONHCHOHCCl_3$ | 460 (76) |
| $n$-$C_3H_7CONHCHOHCBr_3$ | 166 | $n$-$C_8H_{17}CONHCHOHCCl_2CHClCH_3$ | 460 (57) |
| $n$-$C_3H_7CONHCHOHCCl_2CHClCH_3$ | 460 (66) | $n$-$C_9H_{19}CONHCHOHCCl_3$ | 460 (59) |
| $C_2H_5CH(C_6H_5)CONHCHOHCCl_3$ | 466 (28) | $n$-$C_9H_{19}CONHCHOHCBr_3$ | 207 |
| $i$-$C_3H_7CONHCHOHCCl_3$ | 164 | $n$-$C_9H_{19}CONHCHOHCCl_2CHClCH_3$ | 460 (43) |
| $i$-$C_3H_7CONHCHOHCBr_3$ | 207 | | |
| $i$-$C_3H_7CONHCHOHCCl_2CHClCH_3$ | 466 (75) | | |

*Note:* References 382 to 537 are on pp. 266–269.

* This substance was prepared by an indirect method.

TABLE XXIII—*Continued*

α-Hydroxyalkyl Derivatives of Carboxamides, Carbamyl Compounds, and Sulfonamides

$RCONH_2 + R'CHO \rightarrow RCONHCHOHR'$

($R'$=haloalkyl)

| α-Alkylolamide Derivative | References (Yield, %) |
|---|---|
| *Of Aliphatic Monoamides (contd.)* | |
| $n\text{-}C_{15}H_{31}CONHCHOHCCl_3$ | 469 |
| $CH_2{=}CHCONHCHOHCCl_3$ | 101 (66) |
| (furan ring, O)$CH{=}CHCONHCHOHCCl_3$ | 204 (30) |
| OC——$NCHOHCCl_3$<br>$Cl_3CCH$ $CHCCl_3$<br>O | 172 |
| *Of Aromatic Amides* | |
| $C_6H_5CONHCHOHCHCl_2$* | 208 |
| $C_6H_5CONHCHOHCHBr_2$* | 208 |
| $C_6H_5CONHCHOHCCl_3$ | 95, 171 158, 47 |
| $C_6H_5CONHCHOHCBrCl_2$* | 208 |
| $C_6H_5CONHCHOHCBr_3$ | 166, 208 |
| $C_6H_5CONHCHOHCCl_2CHClCH_3$ | 463, 464 |
| $o\text{-}HOC_6H_4CONHCHOHCCl_3$ | 364 |
| $o\text{-}HOC_6H_4CONHCHOHCBr_3$ | 166 |
| $m\text{-}HOC_6H_4CONHCHOHCCl_3$ | 170 (0) |
| $p\text{-}HOC_6H_4CONHCHOHCCl_3$ | 171 (0) |
| 2-HO-3-$ClC_6H_3CONHCHOHCCl_3$ | 171 (81), 470*a* |
| 2-HO-5-$ClC_6H_3CONHCHOHCCl_3$ | 171 (65) |
| 2-HO-3,5-$Cl_2C_6H_2CONHCHOHCCl_3$ | 171 (82) |
| 2-HO-3-$BrC_6H_3CONHCHOHCCl_3$ | 169 |
| 2-HO-5-$BrC_6H_3CONHCHOHCCl_3$ | 169 |
| 2-HO-3,5-$Br_2C_6H_2CONHCHOHCCl_3$ | 169 |
| 2-HO-3-$O_2NC_6H_3CONHCHOHCCl_3$ | 169 (77) |
| 2-HO-5-Br-3-$O_2NC_6H_2CONHCHOHCCl_3$ | 169 (96) |
| 2-HO-3,5-$(O_2N)_2C_6H_2CONHCHOHCCl_3$ | 169 (85) |
| 2-HO-5-$CH_3CONHC_6H_3CONHCHOHCCl_3$ | 372 (68) |
| $o\text{-}CH_3OC_6H_4CONHCHOHCCl_3$ | 166, 171 (good) |
| $m\text{-}CH_3OC_6H_4CONHCHOHCCl_3$ | 171 (good) |
| $p\text{-}CH_3OC_6H_4CONHCHOHCCl_3$ | 171 (good) |
| 2-$CH_3O$-3-$ClC_6H_3CONHCHOHCCl_3$ | 171 (82) |
| 2-$CH_3O$-5-$ClC_6H_3CONHCHOHCCl_3$ | 171 (79) |
| 2-$CH_3O$-3,5-$Cl_2C_6H_2CONHCHOHCCl_3$ | 171 (79) |
| 2-$CH_3O$-2-$BrC_6H_3CONHCHOHCCl_3$ | 169 |
| 2-$CH_3O$-3,5-$Br_2C_6H_2CONHCHOHCCl_3$ | 169 |
| 2-$CH_3O$-3-$O_2NC_6H_3CONHCHOHCCl_3$ | 169 |
| 2-$CH_3O$-5-$O_2NC_6H_3CONHCHOHCCl_3$ | 169 |
| 2-$CH_3O$-3,5-$(O_2N)_2C_6H_2CONHCHOHCCl_3$ | 169 |
| $o\text{-}CH_3CO_2C_6H_4CONHCHOHCCl_3$ | 364 |
| $o\text{-}CH_3CONHC_6H_4CONHCHOHCCl_3$ | 170 (90) |
| $m\text{-}CH_3CONHC_6H_4CONHCHOHCCl_3$ | 170 (85) |
| $p\text{-}CH_3CONHC_6H_4CONHCHOHCCl_3$ | 170 (90) |
| 2-$CH_3CONH$-5-$BrC_6H_3CONHCHOHCCl_3$ | 170 (85) |
| $o\text{-}C_6H_5CONHC_6H_4CONHCHOHCCl_3$ | 170 |
| $m\text{-}C_6H_5CONHC_6H_4CONHCHOHCCl_3$ | 170 (90) |
| $p\text{-}C_6H_5CONHC_6H_4CONHCHOHCCl_3$ | 170 (60) |
| 2-$C_6H_5CONH$-5-$BrC_6H_3CONHCHOHCCl_3$ | 170 (75) |
| $o\text{-}CH_3C_6H_4CONHCHOHCCl_3$ | 168 |
| $m\text{-}CH_3C_6H_4CONHCHOHCCl_3$ | 168 |
| $p\text{-}CH_3C_6H_4CONHCHOHCCl_3$ | 166, 168 |
| $Cl_3C$ (O, O ring fused to benzene ring) $CONHCHOHCCl_3$ | 165 |

| *Of Heteroaromatic Amides* | |
|---|---|
| [furan ring, O]$CONHCHOHCCl_3$ | 204 (86) |
| Br[furan ring, O]$CONHCHOHCCl_3$ | 204 (84) |
| $_2$ON[furan ring, O]$CONHCHOHCCl_3$ | 204 (64) |
| $CH_3$[furan ring, O]$CONHCHOHCCl_3$ | 204 (74) |
| $(CH_3)_3C$[furan ring, O]$CONHCHOHCCl_3$ | 204 (91) |
| [pyridine ring, N]$CONHCHOHCCl_3$ | 471 |
| *Of Dicarboxamides* | |
| $(CONHCHOHCCl_3)_2$ | 163 (31) |
| $C_6H_5NHCOCONHCHOHCCl_3$ | 163 |
| $CH_2(CONHCHOHCCl_3)_2$ | 166 |
| $C_2H_5CH(CONHCHOHCCl_3)_2$ | 163 |
| $(C_2H_5)_2C(CONH_2)CONHCHOHCCl_3$ | 286 |
| $C_6H_5NHCOCH(C_2H_5)CONHCHOHCCl_3$ | 163 |
| $CCl_3CHOHCH(CONHCHOHCCl_3)_2$ | 163 |
| $C_6H_5NHCOCH(CHOHCCl_3)CONHCHOHCCl_3$ | 163 |
| $(CH_2CONHCHOHCCl_3)_2$ | 166 |

| *Of Lactams* | |
|---|---|
| NH–CO–$C(C_6H_5)(CH_2CH_2N(C_2H_5)_2)$–$CH_2$–$CH_2$–CHOH (ring) | 174 |
| NH–CO–$C(C_6H_5)$([pyridine ring, N])–$CH_2$–CHOH (ring) | 174 (30) |
| [benzene ring] CHOH >NH CO | 175 (66) |
| [benzene ring] CHOH $>NCH_3$ CO | 175 (56) |
| [benzene ring] CHOH $>NC_6H_5$ CO | 175 (23) |

*Note:* References 382 to 537 are on pp. 266–269.
* This substance was prepared by an indirect method.

## TABLE XXIII—*Continued*

α-HYDROXYALKYL DERIVATIVES OF CARBOXAMIDES, CARBAMYL COMPOUNDS, AND SULFONAMIDES

$RCONH_2 + R'CHO \rightarrow RCONHCHOHR'$

($R'$ = haloalkyl)

| α-Alkylolamide Derivative | References (Yield, %) |
|---|---|
| *Of Lactams (contd.)* | |
| 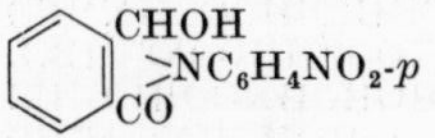 $C_6H_4$(CHOH)(CO)>$NC_6H_4NO_2$-*p* | 175 (81) |
| 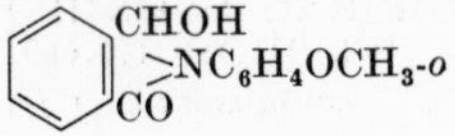 $C_6H_4$(CHOH)(CO)>$NC_6H_4OCH_3$-*o* | 175 (27) |
| 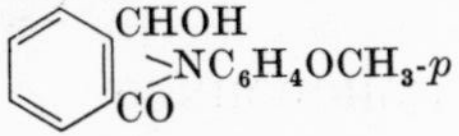 $C_6H_4$(CHOH)(CO)>$NC_6H_4OCH_3$-*p* | 175 (18) |
| 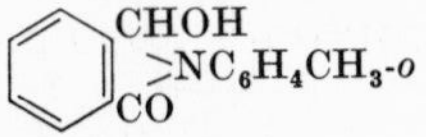 $C_6H_4$(CHOH)(CO)>$NC_6H_4CH_3$-*o* | 175 (23) |
| 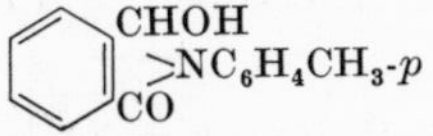 $C_6H_4$(CHOH)(CO)>$NC_6H_4CH_3$-*p* | 175 (40) |
| 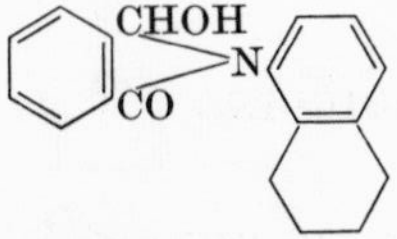 $C_6H_4$(CHOH)(CO)>N–(tetrahydronaphthyl) | 175 (37) |
| 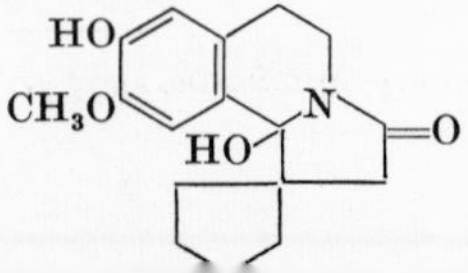 HO, $CH_3O$, HO, N, =O (tricyclic lactam) | 472 |
| 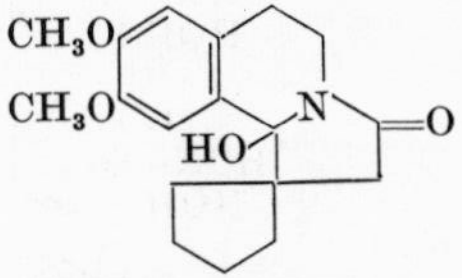 $CH_3O$, $CH_3O$, HO, N, =O (tricyclic lactam) | 472 |
| *Of Carbamyl Compounds* | |
| $CH_3O_2CNHCHOHCCl_3$ | 95 (100), 167 |
| $C_2H_5O_2CNHCHOHCCl_3$ | 94 (100), 95, 167, 173, 229 |
| $C_2H_5O_2CNHCHOHCBr_3$ | 173 |
| $C_2H_5O_2CNHCHOHCCl_2CHClCH_3$ | 173 |
| $C_2H_5O_2CNHC(OH)(CF_2Cl)_2$ | 73 |
| *n*-$C_3H_7O_2CNHCHOHCCl_3$ | 167 (96) |
| *i*-$C_4H_9O_2CNHCHOHCCl_3$ | 167 |
| *i*-$C_5H_{11}O_2CNHCHOHCCl_3$ | 95 |
| Menthyl-$O_2CNHCHOHCCl_3$ | 95 |
| Bornyl-$O_2CNHCHOHCCl_3$ | 160 |
| $H_2NCONHCHOHCCl_3$ | 158, 161, 162, 473 (66) |
| $CO(NHCHOHCCl_3)_2$ | 73, 158, 160, 161, 162, 473 (71) |
| CO($NHCHOHCCl_3$)($NHCHOHCBr_3$) | 162, 474 |
| CO($NHCHOHCCl_3$)($NHCH(OCH_3)CCl_3$) | 161 |

*Of Carbamyl Compounds (contd.)*

| Compound | Refs. |
|---|---|
| $CO(NHCHOHCCl_3)(NHCH(OC_2H_5)CCl_3)$ | 161 |
| $CO(NHCHOHCCl_3)(NHCH(OC_3H_7\text{-}n)CCl_3)$ | 161 |
| $CO(NHCHOHCCl_3)(NHCH(OCH_3)CBr_3)$ | 162 |
| $CO(NHCHOHCCl_3)(NHCH(OC_2H_5)CBr_3)$ | 162 |
| $CH_2(NHCONHCHOHCCl_3)_2$ | 475 |
| $CH_3NHCONHCHOHCCl_3$ | 163 |
| $CH_3NHCONHCHOHCBr_3$ | 163 |
| $C_2H_5NHCONHCHOHCCl_3$ | 163 |
| $(CH_3)_2NCONHCHOHCCl_3$ | 163 |
| $(CH_3)_2NCONHCHOHCBr_3$ | 163 |
| $(C_2H_5)_2NCONHCHOHCCl_3$ | 163 |
| $C_6H_5NHCONHCHOHCF_3$ | 73 |
| $C_6H_5NHCONHCHOH(CF_2)_3CHF_2$ | 73 |
| $C_6H_5NHCONHCHOHCCl_3$ | 73, 161 (83) |
| $C_6H_5NHCONHCHOHCBr_3$ | 163 |
| $C_6H_5NHCONHC(OH)(CF_2Cl)_2$ | 73 |
| $p\text{-}ClC_6H_4NHCONHCHOHCCl_3$ | 73 |
| $m\text{-}O_2NC_6H_4NHCONHCHOHCCl_3$ | 227 |
| $p\text{-}O_2NC_6H_4NHCONHCHOHCCl_3$ | 227 |
| $o\text{-}CH_3C_6H_4NHCONHCHOHCCl_3$ | 227 |
| $p\text{-}CH_3C_6H_4NHCONHCHOHCCl_3$ | 227 |
| $(C_6H_5)_2NCONHCHOHCCl_3$ | 163 |
| $(C_6H_5)_2NCONHCHOHCBr_3$ | 163 |
| $CS(NHCHOHCCl_3)_2$ | 160 |
| $CCl_3CH{=}NC({=}S)NHCHOHCCl_3$ | 160 |
| Pyrazine ring fused with CHOH–NH–CO–NH (N, CHOH, NH, CO, NH, N) | 176 (80) |

*Of Sulfonamides*

| Compound | Refs. |
|---|---|
| $C_6H_5CH_2SO_2NHCHOHCCl_3$ | 159 |
| $C_6H_5SO_2NHCHOHCCl_3$ | 159 |
| $p\text{-}ClC_6H_4SO_2NHCHOHCCl_3$ | 159 |
| $p\text{-}O_2NC_6H_4SO_2NHCHOHCCl_3$ | 159 |
| $p\text{-}CH_3OC_6H_4SO_2NHCHOHCCl_3$ | 159 |
| $o\text{-}CH_3C_6H_4SO_2NHCHOHCCl_3$ | 159 |
| $m\text{-}CH_3C_6H_4SO_2NHCHOHCCl_3$ | 159 |
| $p\text{-}CH_3C_6H_4SO_2NHCHOHCCl_3$ | 159 |
| $2,4\text{-}(CH_3)_2C_6H_3SO_2NHCHOHCCl_3$ | 159 |
| $2,4,6\text{-}(CH_3)_3C_6H_2SO_2NHCHOHCCl_3$ | 159 |
| Naphthyl-$SO_2NHCHOHCCl_3$ | 159 |

*Note:* References 382 to 537 are on pp. 266–269.

# TABLE XXIV

## Ethers of N-Methylol Derivatives of Amides, Lactams, Imides, and Carbamyl Compounds

| N-Methylol Ether | References (Yield, %) |
|---|---|
| *Amides* | |
| $HCONHCH_2OCH_3$ | 415 (11) |
| $[HCON(CH_3)CH_2]_2O$ | 476 (>64) |
| $CH_3CONHCH_2OCH_3$ | 415 (90–94) |
| $CH_2ClCONHCH_2OCH_2C_6H_5$ | 312 |
| $C_6H_5CH_2CONHCH_2CONHCH_2OCH_3$ | 79 |
| $C_6H_5CH_2CONHCH_2CONHCH_2OC_2H_5$ | 79 |
| $C_6H_5CH_2CONHCH_2CONHCH_2OC_3H_7\text{-}n$ | 79 |
| $C_6H_5CH_2CONHCH_2CONHCH_2OC_4H_9\text{-}i$ | 79 |
| $C_6H_5CH_2CONHCH_2CONHCH_2OCH_2C_6H_5$ | 79 |
| $C_6H_5CH_2CONHCH_2CONHCH_2OCH_2CHCO_2C_2H_5$<br>\|<br>$NHCOC_6H_5$ | 79 (27) |
| $C_6H_5CONHCH_2CONHCH_2OCH_3$ | 79 |
| $C_6H_5CONHCH_2CONHCH_2OC_2H_5$ | 79 |
| $C_6H_5CONHCH_2CONHCH_2OCH_2C_6H_5$ | 79 |
| $C_5H_5\overset{\oplus}{N}CH_2CONHCH_2OCH_2C_6H_5\ Cl^{\ominus}$ | 312 |
| $C_6H_5CH_2CONHCH_2OC_2H_5$ | 79 |
| $C_6H_5CH_2CONHCH_2OC_4H_9\text{-}n$ | 79 |
| $C_6H_5CH_2CONHCH_2OCH_2C_6H_5$ | 388 (73) |
| $C_6H_5CH_2CONHCH_2OCH_2CHCO_2C_2H_5$<br>\|<br>$NHCOC_6H_5$ | 79 |
| $(C_2H_5CONHCH_2)_2O$ | 153 (55) |
| $C_6H_5CH_2CONHCH(CH_3)CONHCH_2OC_2H_5$ | 79 |
| $C_6H_5CH_2CONHCH(CH_3)CONHCH_2OCH_2C_6H_5$ | 79 |
| $(i\text{-}C_4H_9CONHCH_2)_2O$ | 54 |
| $(n\text{-}C_{11}H_{23}CONHCH_2)_2O$ | 217 (70), 222 |
| $n\text{-}C_{17}H_{35}CONHCH_2OCH_3$ | 138 (85), 217, 478 |
| $n\text{-}C_{17}H_{35}CONHCH_2OC_2H_5$ | 138 (84), 217 |
| $n\text{-}C_{17}H_{35}CONHCH_2OC_3H_7\text{-}n$ | 217 (70) |
| $n\text{-}C_{17}H_{35}CONHCH_2OC_3H_7\text{-}i$ | 138 (85), 217 |
| $n\text{-}C_{17}H_{35}CONHCH_2OC_4H_9\text{-}n$ | 138, 217 |
| $n\text{-}C_{17}H_{35}CONHCH_2OC_8H_{17}$ | 138 (56), 478, 479 |
| $n\text{-}C_{17}H_{35}CONHCH_2OCH_2C_6H_5$ | 478 |
| $n\text{-}C_{17}H_{35}CONHCH_2OC_6H_5$ | 138 (50), 217 |
| $(n\text{-}C_{17}H_{35}CONHCH_2)_2O$ | 217 (95) |
| $n\text{-}C_{17}H_{35}CONHCH_2OCH_2CO_2H$ | 479 |
| $n\text{-}C_{17}H_{35}CONHCH_2OCH(CH_3)CO_2H$ | 479 |
| $n\text{-}C_{17}H_{35}CONHCH_2OCH_2PO(OH)_2$ | 479 |
| $n\text{-}C_{21}H_{43}CONHCH_2OC_2H_5$ | 235 |
| $(n\text{-}C_{21}H_{43}CONHCH_2)_2O$ | 217 (90) |
| $CH_2{=}CHCONHCH_2OCH_3$ | 423 (70), 477, 480 |
| $CH_2{=}CHCONHCH_2OC_4H_9$ | 477 |
| $(CH_2{=}CHCONHCH_2)_2O$ | 422 (55) |
| $CH_2{=}C(CH_3)CONHCH_2OCH_3$ | 477 (90), 480 (80) |
| $CH_2{=}C(CH_3)CONHCH_2OC_2H_5$ | 477 |
| $CH_2{=}C(CH_3)CONHCH_2OC_3H_7$ | 477 |
| $CH_2{=}C(CH_3)CONHCH_2OCH(CH_2Cl)_2$ | 477 |
| $CH_2{=}C(CH_3)CONHCH_2OC_4H_9$ | 477 |
| $CH_2{=}C(CH_3)CONHCH_2OC_5H_{11}$ | 477 |
| $CH_2{=}C(CH_3)CONHCH_2OC_6H_{13}$ | 477 |
| $CH_2{=}C(CH_3)CONHCH_2OC_{12}H_{25}$ | 477 |
| $CH_2{=}C(CH_3)CONHCH_2OC_{18}H_{37}$ | 477 |

| Compound | References (yield) |
|---|---|
| *Amides (contd.)* | |
| $CH_2{=}C(CH_3)CONHCH_2O$–(cyclohexyl ring) | 477 |
| $CH_2{=}C(CH_3)CONHCH_2OCH_2CH{=}CH_2$ | 477 |
| $CH_2{=}C(CH_3)CONHCH_2OCH_2C_6H_5$ | 477 |
| $[CH_2{=}C(CH_3)CONHCH_2]_2O$ | 424 (43) |
| $CH_3(CH{=}CH)_2CONHCH_2OCH_3$ | 423 (64), 480 |
| $o$-$ClC_6H_4CH{=}CHCONHCH_2OCH_2C_6H_5$ | 312 |
| $CHCONHCH_2OCH_3$ ‖ $CHCONHCH_2OCH_3$ | 423 |
| $C_6H_5CONHCH_2OC_4H_9$-$n$ | 79 |
| $C_6H_5CONHCH_2OC_8H_{17}$ | 478, 479 |
| $C_6H_5CONHCH_2OCH_2C_6H_5$ | 96 (54) |
| $(C_6H_5CONHCH_2)_2O$ | 10, 14 (80), 220 (77) |
| $(o$-$O_2NC_6H_4CONHCH_2)_2O$ | 221 (50) |
| (4-pyridyl)$CONHCH_2OCH_3$ | 430 (43) |
| (4-pyridyl)$CONHCH_2OC_2H_5$ | 430 |
| (4-pyridyl)$CONHCH_2OC_3H_7$-$n$ | 430 |
| (4-pyridyl)$CONHCH_2OC_3H_7$-$i$ | 430 |
| (4-pyridyl)$CONHCH_2OC_4H_9$-$n$ | 430 |
| *Lactams* | |
| $CH_2CO$, $CH_2CH_2$ (ring) $>NCH_2OCH_3$ | 481 (64) |
| $CH_2CO$, $CH_2CH_2$ (ring) $>NCH_2OC_2H_5$ | 33 (77), 481 (58) |
| $CH_2CO$, $CH_2CH_2$ (ring) $>NCH_2OC_4H_9$-$n$ | 481 (58) |
| $CH_2CH_2CO$, $CH_2CH_2CH_2$ (ring) $>NCH_2OC_4H_9$-$n$ | 481 (59) |
| $CH_3OCH_2N(COCH_2)_2NCH_2OCH_3$ (piperazinedione ring) | 65 (86) |
| $C_2H_5OCH_2N(COCH_2)_2NCH_2OC_2H_5$ (piperazinedione ring) | 65 |

*Note:* References 382 to 537 are on pp. 266–269.

## TABLE XXIV—*Continued*

### Ethers of N-Methylol Derivatives of Amides, Lactams, Imides, and Carbamyl Compounds

| N-Methylol Ether | References (Yield, %) |
|---|---|
| *Imides* | |
| $CH_2CO$ / $CH_2CO$ ring, $\text{>}NCH_2OCH_3$ | 188 |
| $CH_2CO$ / $CH_2CO$ ring, $\text{>}NCH_2OC_2H_5$ | 66 |
| $[CH_2CO$ / $CH_2CO$ ring, $\text{>}NCH_2]_2O$ | 85 (17) |
| $[CHCO$ = $CHCO$ ring, $\text{>}NCH_2]_2O$ | 8 (41) |
| $o\text{-}C_6H_4(CO)_2NCH_2OCH_3$ | 70, 71, 151, 188 |
| $o\text{-}C_6H_4(CO)_2NCH_2OC_2H_5$ | 70, 71 (92), 104 (good), 151 |
| $o\text{-}C_6H_4(CO)_2NCH_2OC_3H_7\text{-}n$ | 70 |
| $o\text{-}C_6H_4(CO)_2NCH_2OC_3H_7\text{-}i$ | 70 |
| $o\text{-}C_6H_4(CO)_2NCH_2OC_4H_9\text{-}n$ | 71 (78) |
| $o\text{-}C_6H_4(CO)_2NCH_2OC_4H_9\text{-}i$ | 71 (48) |
| $o\text{-}C_6H_4(CO)_2NCH_2OC_{14}H_{29}\text{-}n$ | 71 (83) |
| $o\text{-}C_6H_4(CO)_2NCH_2OC_{16}H_{33}\text{-}n$ | 71 (66) |
| $o\text{-}C_6H_4(CO)_2NCH_2OC_{18}H_{37}\text{-}n$ | 71 (97) |
| $o\text{-}C_6H_4(CO)_2NCH_2OC_6H_{11}\text{-}cyclo$ | 71 (94) |
| $o\text{-}C_6H_4(CO)_2NCH_2OCH_2CH_2C_6H_5$ | 71 (86) |
| $o\text{-}C_6H_4(CO)_2NCH_2OCH_2CH(C_6H_5)_2$ | 71 (65) |
| $o\text{-}C_6H_4(CO)_2NCH_2OC(C_6H_5)_3$ | 71 (50) |
| $o\text{-}C_6H_4(CO)_2NCH_2OCH_2$ — $o\text{-}C_6H_4(CO)_2NCH_2OCHCH_3$ | 71 (50) |
| $o\text{-}C_6H_4(CO)_2NCH_2OCH_2$ — $o\text{-}C_6H_4(CO)_2NCH_2OCH$ — $o\text{-}C_6H_4(CO)_2NCH_2OCH_2$ | 70 |
| $[o\text{-}C_6H_4(CO)_2NCH_2]_2O$ | 61, 151, 220, 263 |
| $o\text{-}C_6H_4(CO)_2NCH_2OCH(CH_3)CO_2H$ | 70 |
| *Carbamyl Compounds* | |
| $C_2H_5O_2CNHCH_2OCH_3$ | 218 (95) |
| $C_2H_5O_2CNHCH_2OC_2H_5$ | 218 (82) |
| $C_2H_5O_2CNHCH_2OC_3H_7\text{-}n$ | 218 (96) |
| $(C_2H_5O_2CNHCH_2)_2O$ | 153, 218 |
| $n\text{-}C_{18}H_{37}O_2CNHCH_2OCH_2CO_2H$ | 479 |
| $2,4\text{-}Cl_2C_6H_3O_2CNHCH_2O$-(tetrahydropyran-2-yl) | 73 |
| $1\text{-}C_{10}H_7O_2CNHCH_2O$-(tetrahydropyran-2-yl) | 73 |

*Carbamyl Compounds (contd.)*

| Compound | Refs. |
|---|---|
| 2-$C_{10}H_7O_2CNHCH_2O$–(ring, O) | 73 |
| $H_2NCONHCH_2OCH_3$ | 62, 219 (98), 441 |
| $H_2NCONHCH_2OC_2H_5$ | 219 (55) |
| $CO(NHCH_2OCH_3)_2$ | 62, 149, 219, 446, 482 (86), 483 |
| $CO(NHCH_2OC_2H_5)_2$ | 219 (81), 446 |
| $CO(NHCH_2OC_3H_7\text{-}n)_2$ | 219 (85), 446 |
| $CO(NHCH_2OC_3H_7\text{-}i)_2$ | 482 |
| $CO(NHCH_2OC_4H_9\text{-}n)_2$ | 219 (50), 446 |
| $CO(NHCH_2OC_4H_9\text{-}i)_2$ | 482 (60) |
| $CO(NHCH_2OC_5H_{11}\text{-}n)_2$ | 219 (46), 446 |
| $CO(NHCH_2OC_6H_{13}\text{-}n)_2$ | 446 |
| $CO(NHCH_2OC_7H_{15}\text{-}n)_2$ | 446 |
| $CO(NHCH_2OC_8H_{17}\text{-}n)_2$ | 446 |
| $CO(NHCH_2OC_9H_{19}\text{-}n)_2$ | 446 |
| $CO(NHCH_2OCH_2C_6H_5)_2$ | 219 (40) |
| $(H_2NCONHCH_2)_2O$ | 146, 147, 149 |
| $CH_2(NHCONHCH_2OCH_3)_2$ | 62, 219 (63), 483 |
| $(C_2H_5NHCONHCH_2)_2O$ | 54, 153 |
| $CH_3NHCON(CH_3)CH_2OCH_3$ | 357, 484 |
| $(CH_3)_2NCONHCH_2OCH_3$ | 219 (53) |
| $C_6H_5NHCONHCH_2OCH_3$ | 55 (83) |
| $C_6H_5NHCONHCH_2OC_2H_5$ | 55 (86) |
| $C_6H_5NHCONHCH_2OC_3H_7\text{-}i$ | 55 (84) |
| $C_6H_5NHCONHCH_2OC_4H_9\text{-}n$ | 55 (86) |
| $(C_6H_5NHCONHCH_2)_2O$ | 54, 55 |
| $p\text{-}ClC_6H_4NHCONHCH_2OCH(CH_3)OC_4H_9\text{-}n$ | 73 |
| $p\text{-}ClC_6H_4NHCONHCH_2O$–(ring, O) | 73 |
| $p\text{-}CH_3C_6H_4NHCONHCH_2OC_4H_9\text{-}n$ | 55 (83) |
| $(p\text{-}CH_3C_6H_4NHCONHCH_2)_2O$ | 55 |
| $CH_2NCH_2OCH_3$<br>CO<br>$CH_2NCH_2OCH_3$ | 447 (66), 450 (66) |
| $CH_2NCH_2OC_2H_5$<br>CO<br>$CH_2NCH_2OC_2H_5$ | 447 (40) |
| $CH_2NCH_2OC_4H_9\text{-}n$<br>CO<br>$CH_2NCH_2OC_4H_9\text{-}n$ | 450 |
| $CH_2NCH_2OCH_3$<br>O CO<br>$CH_2NCH_3$ | 219 (69), 484 |
| $CH_2NCH_2OCH_3$<br>O CO<br>$CH_2NC_2H_5$ | 219 (92) |

*Note:* References 382 to 537 are on pp. 266–269.

TABLE XXIV—*Continued*

ETHERS OF N-METHYLOL DERIVATIVES OF AMIDES, LACTAMS, IMIDES, AND CARBAMYL COMPOUNDS

| N-Methylol Ether | References (Yield, %) | N-Methylol Ether | References (Yield, %) |
|---|---|---|---|
| *Carbamyl Compounds (contd.)* | | | |
| $CH_2NCH_2OCH_3$<br>O CO<br>$CH_2NCH_2OCH_3$ | 62, 219 (66), 357, 484 | $C_6H_4$ $NCH_2OCH_3$<br>CO<br>$NCH_2OCH_3$ | 180 (41) |
| $CH_2NCH_2OCH_3$<br>$CH_3N$ CO<br>$CH_2NCH_2OCH_3$ | 485 (64) | $C_6H_4$ $NCH_2OC_2H_5$<br>CO<br>$NCH_2OC_2H_5$ | 180 (82) |
| $CH_2NCH_2OCH_3$<br>$i\text{-}C_4H_9N$ CO<br>$CH_2NCH_2OCH_3$ | 485 (30) | $C_6H_4$ $NCH_2OCH_3$<br>CS<br>$NCH_2OCH_3$ | 240, 457 |
| $CH_2NCH_2OCH_3$<br>$HO(CH_2)_2N$ CO<br>$CH_2NCH_2OCH_3$ | 485 | $C_6H_4$ $NCH_2OC_2H_5$<br>CS<br>$NCH_2OC_2H_5$ | 179 (87) |
| $[CONCH_2—$<br>CO<br>$CH_2NH]_2$ O | 185 (92) | $C_6H_4$ $NCH_2OCH_2C_6H_5$<br>CS<br>$NCH_2OCH_2C_6H_5$ | 240 |

*Note:* References 382 to 537 are on pp. 266–269.

## TABLE XXV

ETHERS OF N-α-ALKYLOL DERIVATIVES OF AMIDES, LACTAMS, IMIDES, CARBAMYL COMPOUNDS, AND SULFONAMIDES

| N-α-Alkylol Ether | References (Yield, %) |
|---|---|
| *Aliphatic Amides* | |
| $HCONHCH(OCH_3)CCl_3$ | 95 |
| $(HCONHCHCCl_3)_2O$ | 95, 229 |
| $(HCONHCHCBr_3)_2O$ | 207 |
| $CH_3CONHCH(OCH_3)CCl_3$ | 95 |
| $(CH_3CONHCHCCl_3)_2O$ | 95, 229 |
| $(CH_3CONHCHCBr_3)_2O$ | 207 |
| $(CH_3CONHCHOCH_3)_2$ | 223 |
| $CH_3CONHC(OCH_3)(C_6H_5)_2$ | 231 (>70) |
| $CH_3CONHC(OC_2H_5)(C_6H_5)_2$ | 231 (>70) |
| $CH_3CONHC(OC_3H_7\text{-}n)(C_6H_5)_2$ | 231 (>70) |
| $CH_3CONHC(OC_4H_9\text{-}n)(C_6H_5)_2$ | 231 (>70) |
| $CH_3CONHC(OC_5H_{11}\text{-}n)(C_6H_5)_2$ | 231 (>70) |
| $OC_2H_5$ / $NHCOCH_3$ | 231 |
| $CH_3CON(CH_3)CH(OCH_3)C_6H_5$ | 216 (62) |
| $CH_2ClCONHC(OCH_3)(C_6H_5)_2$ | 231 |
| $CH_2ClCONHC(OC_2H_5)(C_6H_5)_2$ | 231 |
| $CH_2BrCONHC(OC_2H_5)(C_6H_5)_2$ | 231 |
| $CH_2ICONHC(OC_2H_5)(C_6H_5)_2$ | 231 |
| $C_6H_5CH_2CONHCH(OCH_3)CBr_3$ | 166 |
| $(C_6H_5CH_2CONHCHCBr_3)_2O$ | 166 |
| $C_2H_5CONHCH(OCH_3)CBr_3$ | 207 |
| $(C_2H_5CONHCHCBr_3)_2O$ | 207 |
| $C_2H_5CONHC(OC_2H_5)(C_6H_5)_2$ | 231 |
| $C_2H_5CONHC(OC_4H_9\text{-}n)(C_6H_5)_2$ | 231 |
| $CH_2ClCH_2CONHC(OC_2H_5)(C_6H_5)_2$ | 231 |
| $n\text{-}C_3H_7CONHCH(OCH_3)CCl_3$ | 166 |
| $(n\text{-}C_3H_7CONHCHCCl_3)_2O$ | 166 |
| $n\text{-}C_3H_7CONHCH(OCH_3)CBr_3$ | 207 |
| $(n\text{-}C_3H_7CONHCHCBr_3)_2O$ | 207 |
| $i\text{-}C_3H_7CONHCH(OCH_3)CBr_3$ | 207 |
| $(i\text{-}C_3H_7CONHCHCBr_3)_2O$ | 207 |
| $n\text{-}C_4H_9CONHCH(OCH_3)CBr_3$ | 207 |
| $(n\text{-}C_4H_9CONHCHCBr_3)_2O$ | 207 |
| $i\text{-}C_4H_9CONHCH(OCH_3)CBr_3$ | 207 |
| $(i\text{-}C_4H_9CONHCHCBr_3)_2O$ | 207 |
| $n\text{-}C_5H_{11}CONHCH(OCH_3)CBr_3$ | 207 |
| $n\text{-}C_6H_{13}CONHCH(OCH_3)CBr_3$ | 207 |

*Note:* References 382 to 537 are on pp. 266–269.

## TABLE XXV—*Continued*

### Ethers of N-α-Alkylol Derivatives of Amides, Lactams, Imides, Carbamyl Compounds, and Sulfonamides

| N-α-Alkylol Ether | References (Yield, %) |
|---|---|
| *Aliphatic Amides (contd.)* | |
| C | |
| $n$-$C_6H_{13}CONHCHCBr_3)_2O$ | 207 |
| $( -C_8H_{17}CONHCH(OCH_3)CBr_3$ | 207 |
| $n_2H_5O_2CCONHCH(OCH_3)CCl_3$ | 160 |
| $C_2H_5O_2CCONHCH(OC_2H_5)CCl_3$ | 160 |
| $CONHCH(OC_2H_5)CCl_3]_2$ | 163 |
| *Aromatic Amides* | |
| $C_6H_5CONHCH(OCH_3)CHCl_2$ | 208 |
| $(C_6H_5CONHCHCHCl_2)_2O$ | 208 |
| $C_6H_5CONHCH(OCH_3)CHBr_2$ | 208 |
| $(C_6H_5CONHCHCHBr_2)_2O$ | 208 |
| $C_6H_5CONHCH(OCH_3)CCl_3$ | 95 |
| $C_6H_5CONHCH(OC_2H_5)CCl_3$ | 95 |
| $(C_6H_5CONHCHCCl_3)_2O$ | 95, 229 |
| $C_6H_5CONHCH(OCH_3)CBr_3$ | 166 |
| $(C_6H_5CONHCHCBr_3)_2O$ | 166 |
| $C_6H_5CONHC(OC_2H_5)(C_6H_5)_2$ | 231 |
| $C_6H_5CONHC(OC_4H_9$-$n)(C_6H_5)_2$ | 231 |
| 2-HO-3,5-$Br_2C_6H_2CONHCH(OCH_3)CCl_3$ | 228 |
| 2-HO-3-$O_2NC_6H_3CONHCH(OCH_3)CCl_3$ | 225 |
| 2-HO-5-Br-3-$O_2NC_6H_2CONHCH(OCH_3)CCl_3$ | 225 |
| 2-HO-3,5-$(O_2N)_2C_6H_2CONHCH(OCH_3)CCl_3$ | 225 |
| 2-$CH_3OC_6H_4CONHCH(OCH_3)CBr_3$ | 166 |
| (2-$CH_3OC_6H_4CONHCHCBr_3)_2O$ | 166 |
| 2-$CH_3O$-3-$ClC_6H_3CONHCH(OCH_3)CCl_3$ | 226 |
| 2-$CH_3O$-5-$ClC_6H_3CONHCH(OCH_3)CCl_3$ | 206, 226 |
| 2-$CH_3O$-5-$ClC_6H_3CONHCH(OC_2H_5)CCl_3$ | 206 |
| 2-$CH_3O$-5-$ClC_6H_3CONHCH(OC_6H_5)CCl_3$ | 206 |
| (2-$CH_3O$-5-$ClC_6H_3CONHCHCCl_3)_2O$ | 226 |
| 2-$CH_3O$-3,5-$Cl_2C_6H_2CONHCH(OCH_3)CCl_3$ | 206, 226 |
| 2-$CH_3O$-3,5-$Cl_2C_6H_2CONHCH(OC_6H_5)CCl_3$ | 206 |
| (2-$CH_3O$-3,5-$Cl_2C_6H_2CONHCHCCl_3)_2O$ | 226 |
| 2-$CH_3O$-5-$BrC_6H_3CONHCH(OCH_3)CCl_3$ | 228 |
| 2-$CH_3O$-5-$BrC_6H_3CONHCH(OC_2H_5)CCl_3$ | 205 |
| 2-$CH_3O$-5-$BrC_6H_3CONHCH(OC_6H_5)CCl_3$ | 205 |
| (2-$CH_3O$-5-$BrC_6H_3CONHCHCCl_3)_2O$ | 228 |
| 2-$CH_3O$-3,5-$Br_2C_6H_2CONHCH(OCH_3)CCl_3$ | 228 |
| (2-$CH_3O$-3,5-$Br_2C_6H_2CONHCHCCl_3)_2O$ | 228 |
| 2-$CH_3O$-3-$O_2NC_6H_3CONHCH(OCH_3)CCl_3$ | 205, 225 |
| (2-$CH_3O$-3-$O_2NC_6H_3CONHCHCCl_3)_2O$ | 225 |
| 2-$CH_3O$-5-$O_2NC_6H_3CONHCH(OCH_3)CCl_3$ | 205, 225 |
| 2-$CH_3O$-5-$O_2NC_6H_3CONHCH(OC_2H_5)CCl_3$ | 205 |
| (2-$CH_3O$-5-$O_2NC_6H_3CONHCHCCl_3)_2O$ | 225 |
| $o$-$CH_3C_6H_4CONHCH(OCH_3)CCl_3$ | 106, 168 |
| $o$-$CH_3C_6H_4CONHCH(OC_2H_5)CCl_3$ | 106 |
| $o$-$CH_3C_6H_4CONHCH(OC_6H_5)CCl_3$ | 106 |
| ($o$-$CH_3C_6H_4CONHCHCCl_3)_2O$ | 168 |
| $m$-$CH_3C_6H_4CONHCH(OCH_3)CCl_3$ | 106 |
| $m$-$CH_3C_6H_4CONHCH(OC_2H_5)CCl_3$ | 106 |
| $m$-$CH_3C_6H_4CONHCH(OC_6H_5)CCl_3$ | 106 |
| $m$-$CH_3C_6H_4CONHCH(OC_6H_4CH_3$-$o)CCl_3$ | 106 |

| Compound | Refs. (Yield, %) |
|---|---|
| *Aromatic Amides (contd.)* | |
| $(m\text{-}CH_3C_6H_4CONHCHCCl_3)_2O$ | 168 |
| $p\text{-}CH_3C_6H_4CONHCH(OCH_3)CCl_3$ | 106 |
| $p\text{-}CH_3C_6H_4CONHCH(OC_2H_5)CCl_3$ | 106 |
| $p\text{-}CH_3C_6H_4CONHCH(OC_6H_5)CCl_3$ | 106 |
| $(p\text{-}CH_3C_6H_4CONHCHCCl_3)_2O$ | 168 |
| (2-furyl)$CONHCH(OCH_3)CCl_3$ | 204 (80) |
| (2-furyl)$CONHCH(OC_2H_5)CCl_3$ | 204 (70) |
| Or(ring, B)$CONHCH(OCH_3)CCl_3$ | 204 (92) |
| $O_2N$(2-furyl)$CONHCH(OCH_3)CCl_3$ | 204 (57) |
| $(CH_3)_3C$(2-furyl)$CONHCH(OCH_3)CCl_3$ | 204 (100) |
| *Lactams and Imides* | |
| $CH_2CO$—$NCH(OCH_3)CCl_3$—$CH_2CH_2$ (ring) | 101 (50) |
| $CH_2CO$—$NCH(OC_2H_5)CH_3$—$CH_2CO$ (ring) | 232 (93) |
| $CH_2CO$—$NCH(OC_4H_9\text{-}n)CH_3$—$CH_2CO$ (ring) | 232 (95) |
| $CH_2CO$—$NCH(OC_6H_5)CH_3$—$CH_2CO$ (ring) | 232 (79) |
| $CH_2(CH_2CO)_2NCH(OC_4H_9\text{-}n)CH_3$ (ring) | 232 (91) |
| $o\text{-}C_6H_4(CO)_2NCH(OC_2H_5)CH_3$ | 232 (82) |
| $o\text{-}C_6H_4(CO)_2NCH(OC_6H_5)CH_3$ | 232 (41) |
| $o\text{-}C_6H_4(CO)_2NCH(OC_2H_5)CH_2Br$ | 214 (63) |
| *Carbamyl Compounds* | |
| $CH_3O_2CNHCH(OCH_3)CCl_3$ | 73, 95 |
| $(CH_3O_2CNHCHCCl_3)_2O$ | 95 |
| $C_2H_5O_2CNHCH(OC_4H_9\text{-}n)CH_3$ | 234 (79) |
| $C_2H_5O_2CNHCH(OCH_3)CCl_3$ | 94 |
| $C_2H_5O_2CNHCH(OC_2H_5)CCl_3$ | 94 |
| $(C_2H_5O_2CNHCHCCl_3)_2O$ | 94, 95, 229, 230 |
| $C_2H_5O_2CNHCH(OCH_2CH{=}CH_2)CH(NHCO_2C_2H_5)_2$ | 223 |
| $C_2H_5O_2CNHCH(OCH_2C_6H_5)CH(NHCO_2C_2H_5)_2$ | 223 |
| $(C_2H_5O_2CNHCHOCH_3)_2$ | 223 |
| $C_2H_5O_2CN(CH_3)CH(OCH_3)C_6H_5$ | 216 (62) |
| $n\text{-}C_4H_9O_2CNHCH(OC_2H_5)CH(NHCO_2C_4H_9\text{-}n)_2$ | 223 (good) |

*Note:* References 382 to 537 are on pp. 266–269.

TABLE XXV—*Continued*

ETHERS OF N-α-ALKYLOL DERIVATIVES OF AMIDES, LACTAMS, IMIDES, CARBAMYL COMPOUNDS, AND SULFONAMIDES

| N-α-Alkylol Ether | References (Yield, %) |
|---|---|
| *Carbamyl Compounds (contd.)* | |
| $(i\text{-}C_5H_{11}O_2CNHCHCCl_3)_2O$ | 95 |
| $1\text{-}C_{10}H_7O_2CNHCH(OCH_3)CCl_3$ | 73 |
| $2\text{-}C_{10}H_7O_2CNHCH(OCH_3)CCl_3$ | 73 |
| $H_2NCONHCH(OCH_3)CCl_3$ | 161 |
| $H_2NCONHCH(OC_2H_5)CCl_3$ | 161 |
| $H_2NCONHCH(OC_3H_7\text{-}n)CCl_3$ | 161 |
| $H_2NCONHCH(OC_4H_9\text{-}n)CCl_3$ | 161 |
| $(H_2NCONHCHCCl_3)_2O$ | 161, 473 |
| $H_2NCONHCH(OCH_3)CBr_3$ | 162 |
| $H_2NCONHCH(OC_2H_5)CBr_3$ | 162 |
| $H_2NCONHCH(OC_3H_7\text{-}n)CBr_3$ | 162 |
| $(H_2NCONHCHCBr_3)_2O$ | 162 |
| $(ClNHCONClCHCCl_3)_2O$ | 161 |
| $CH_3NHCONHCH(OC_2H_5)CCl_3$ | 163 |
| $(CH_3NHCONHCHCCl_3)_2O$ | 163 |
| $CH_3NHCONHCH(OC_2H_5)CBr_3$ | 163 |
| $(CH_3NHCONHCHCBr_3)_2O$ | 163 |
| $C_2H_5NHCONHCH(OCH_3)CCl_3$ | 163 |
| $C_2H_5NHCONHCH(OC_2H_5)CCl_3$ | 163 |
| $C_2H_5NHCONHCH(OC_3H_7\text{-}n)CCl_3$ | 163 |
| $C_2H_5NHCONHCH(OC_4H_9\text{-}n)CCl_3$ | 163 |
| $(C_2H_5NHCONHCHCCl_3)_2O$ | 163 |
| $C_6H_5NHCONHCH(OC_4H_9\text{-}n)CH_3$ | 234 |
| $C_6H_5NHCONHCH(OCH_3)CCl_3$ | 161 |
| $C_6H_5NHCONHCH(OC_2H_5)CCl_3$ | 161 |
| $(C_6H_5NHCONHCHCCl_3)_2O$ | 161 |
| $C_6H_5NHCONHCH(OCH_3)CBr_3$ | 163 |
| $C_6H_5NHCONHCH(OC_2H_5)CBr_3$ | 163 |
| $(C_6H_5NHCONHCHCBr_3)_2O$ | 163 |
| $p\text{-}ClC_6H_4NHCONHCH(OC_4H_9\text{-}n)CH_3$ | 234 (60) |
| $m\text{-}O_2NC_6H_4NHCONHCH(OCH_3)CCl_3$ | 227 |
| $m\text{-}O_2NC_6H_4NHCONHCH(OC_2H_5)CCl_3$ | 227 |
| $(m\text{-}O_2NC_6H_4NHCONHCHCCl_3)_2O$ | 227 |
| $p\text{-}O_2NC_6H_4NHCONHCH(OCH_3)CCl_3$ | 227 |
| $p\text{-}O_2NC_6H_4NHCONHCH(OC_2H_5)CCl_3$ | 227 |
| $(p\text{-}O_2NC_6H_4NHCONHCHCCl_3)_2O$ | 227 |
| $o\text{-}CH_3C_6H_4NHCONHCH(OCH_3)CCl_3$ | 227 |
| $o\text{-}CH_3C_6H_4NHCONHCH(OC_2H_5)CCl_3$ | 227 |
| $(o\text{-}CH_3C_6H_4NHCONHCHCCl_3)_2O$ | 227 |
| $p\text{-}CH_3C_6H_4NHCONHCH(OCH_3)CCl_3$ | 227 |
| $p\text{-}CH_3C_6H_4NHCONHCH(OC_2H_5)CCl_3$ | 227 |

| Compound | Refs. |
|---|---|
| *Carbamyl Compounds (contd.)* | |
| $(p\text{-}CH_3C_6H_4NHCONHCHCCl_3)_2O$ | 227 |
| $CH_3CONHCONHCH(OCH_3)CCl_3$ | 161 |
| $CH_3CONHCONHCH(OC_2H_5)CCl_3$ | 161 |
| $(CH_3CONHCONHCHCCl_3)_2O$ | 161 |
| $CO[NHCH(OCH_3)CCl_3]_2$ | 161 |
| $CO[NHCH(OC_2H_5)CCl_3]_2$ | 160, 161 |
| $CO[NHCH(OC_3H_7\text{-}n)CCl_3]_2$ | 161 |
| $CO[NHCH(OCH_3)CBr_3]_2$ | 162 |
| $CO[NHCH(OC_2H_5)CBr_3]_2$ | 162 |
| $CO(NHCH(OCH_3)CCl_3)(NHCH(OC_2H_5)CBr_3)$ | 162 |
| $CO(NHCH(OC_2H_5)CCl_3)(NHCH(OCH_3)CBr_3)$ | 162 |
| 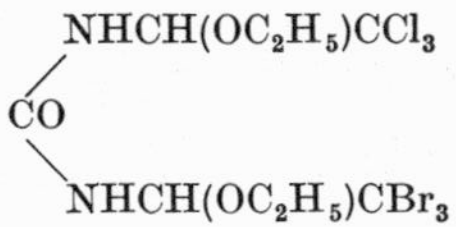 $CO(NHCH(OC_2H_5)CCl_3)(NHCH(OC_2H_5)CBr_3)$ | 162 |
| $[CCl_3CH(OCOCH_3)NHCONHCHCCl_3]_2O$ | 160 |
| $CCl_3CH(OCOC_6H_5)NHCONHCHCCl_3$ — O — $CCl_3CHOHNHCONHCHCCl_3$ | 160 |
| $CH_2NCH(OCH_3)CH_3$, CO, $CH_2NH$ (ring) | 224 |
| $CH_2NCH(OC_3H_7\text{-}i)CH_3$, CO, $CH_2NH$ (ring) | 224 |
| $CH_2NCH(OCH_3)CH_3$, CO, $CH_2NCH(OCH_3)CH_3$ (ring) | 224 |
| 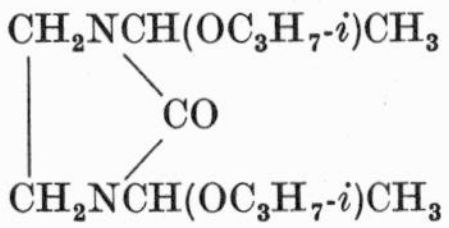 $CH_2NCH(OC_3H_7\text{-}i)CH_3$, CO, $CH_2NCH(OC_3H_7\text{-}i)CH_3$ (ring) | 224 |
| *Sulfonamides* | |
| $p\text{-}CH_3C_6H_4SO_2N(CH_3)CH(OC_4H_9\text{-}n)CH_3$ | 232 (84) |

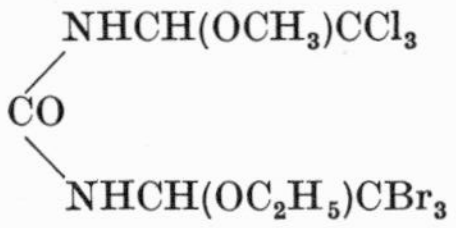

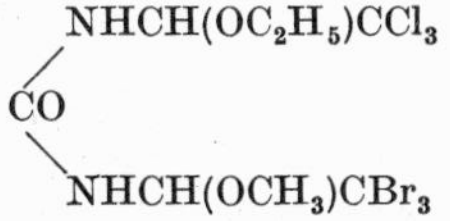

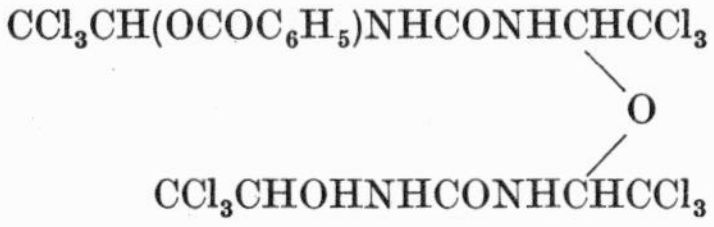

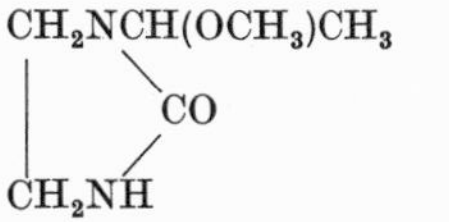

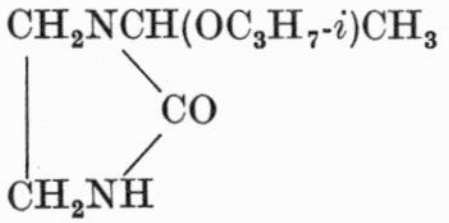

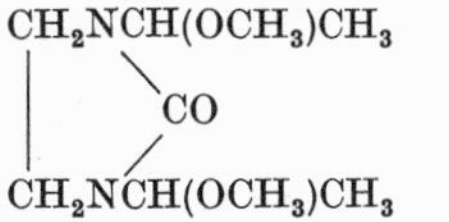

*Note:* References 382 to 537 are on pp. 266–269.

# TABLE XXVI

## Esters of N-Methylol Derivatives of Amides, Lactams, Imides, and Carbamyl Compounds

| N-Methylol Ester | References (Yield, %) |
|---|---|
| *Amides and Lactams* | 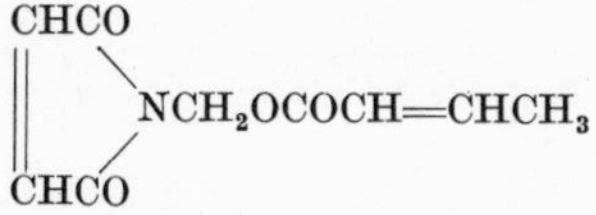 |
| $CH_3CON(CH_3)CH_2OCOCH_3$ | 244 (21) |
| $C_6H_5CH_2CONHCH_2OCOCH_3$ | 79 (60) |
| $C_6H_5(CH_2)_2CONHCH_2OCOCH_3$ | 79 |
| $n$-$C_{11}H_{23}CONHCH_2OCOCH_3$ | 222 |
| $n$-$C_{17}H_{35}CONHCH_2OCOCH_3$ | 138 (66), 236 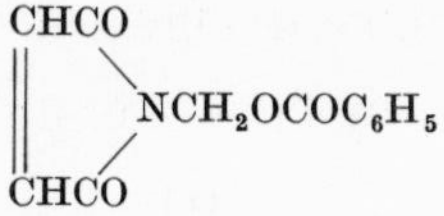 |
| $n$-$C_{21}H_{43}CONHCH_2OCOCH_3$ | 235 |
| $CH_2{=}C(CH_3)CONHCH_2OCOCH_3$ | 477 (61) |
| $CH_2{=}C(CH_3)CONHCH_2OCOC_2H_5$ | 477 |
| $CON(CH_2OCOC_6H_5)_2$<br>$=O$ | 242 |
| O<br>$C_6H_5CO_2CH_2N$ $NCH_2OCOC_6H_5$<br>O | 65 (50) |
| *Imides* | |
| $CH_2CO$<br>$NCH_2OCOCH_3$<br>$CH_2CO$ | 66 (75) |
| $CHCO$<br>$NCH_2OCOCH_3$<br>$CHCO$ | 8 (93) |
| $CHCO$<br>$NCH_2OCOCH{=}CHCH_3$<br>$CHCO$ | 8 (34) |
| $CHCO$<br>$NCH_2OCOC_6H_5$<br>$CHCO$ | 8 (66) |
| $CHCO$<br>$NCH_2OCONHC_6H_5$<br>$CHCO$ | 8 (21) |
| $o$-$C_6H_4(CO)_2NCH_2OCOH$ | 220 (24) |
| $o$-$C_6H_4(CO)_2NCH_2OCOC_6H_5$ | 243 |
| *Carbamyl Compounds* | |
| $NCH_2OCOCH_3$<br>O—CS (DL-*trans*) | 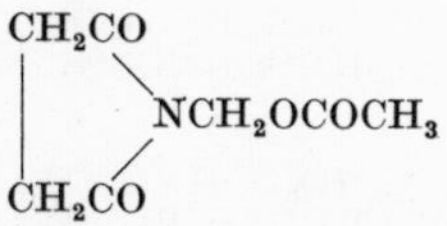 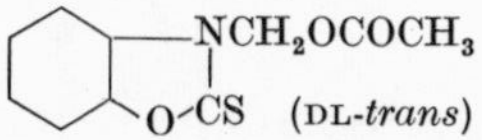 181 (65) |
| 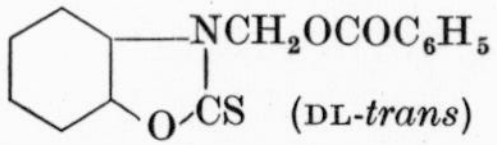 $NCH_2OCOC_6H_5$<br>O—CS (DL-*trans*) | 181 (55) |
| $NCH_2OCONHC_6H_5$<br>O—CS (DL-*trans*) | 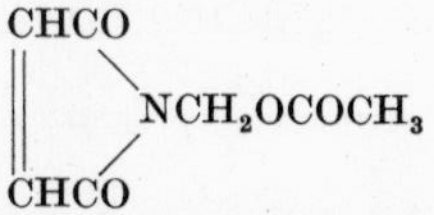 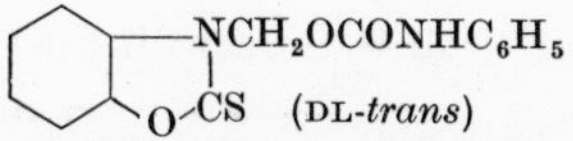 181 (65) |

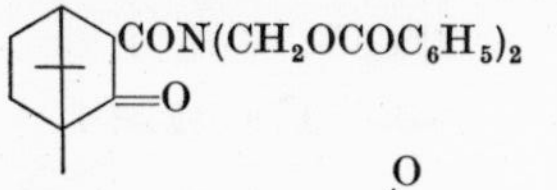

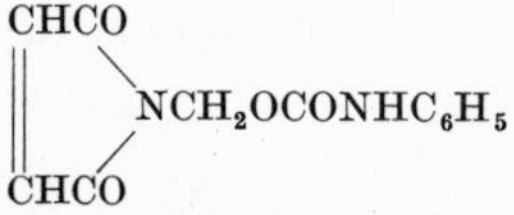

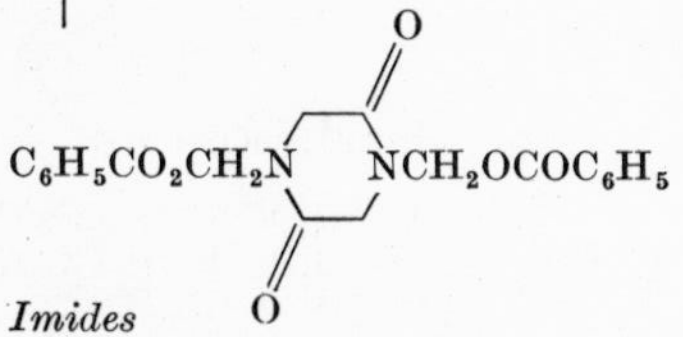

| Compound | Refs. | Compound | Refs. |
|---|---|---|---|
| $C_6H_4$—$NCH_2OCOCH_3$, O—CO | 194 (75) | $C_2H_5O$—$C_6H_3$, CO, $NCH_2OCOC_6H_5$ | 237 |
| $C_6H_4$—$NCH_2OCOCH_3$, O—CS | 178 (56) | $C_6H_4$, $NCH_2OCOCH_3$, CS, $NCH_2OCOCH_3$ | 238 |
| $C_6H_4$—$NCH_2OCOC_6H_5$, O—CS | 178 (56) | $C_6H_4$, $NCH_2OCOC_2H_5$, CS, $NCH_2OCOC_2H_5$ | 240 |
| CHCO, $CH_3C$, $NCH_2OCOCH_3$, NHCS | 239 | $C_6H_4$, $NCH_2OCOC_3H_7$-$n$, CS, $NCH_2OCOC_3H_7$-$n$ | 240, 457 |
| CHCO, $CH_3C$, $NCH_2OCOC_6H_5$, NHCS | 239 | $C_6H_4$, $NCH_2OCOC_6H_5$, CS, $NCH_2OCOC_6H_5$ | 238 |
| $C_6H_4$, $NCH_2OCOCH_3$, CO, $NCH_2OCOCH_3$ | 238 | $C_6H_4$, $NCH_2OCOC_6H_4CO_2H$-$o$, CS, $NCH_2OH$ | 240 |
| $C_6H_4$, $NCH_2OCOC_6H_5$, CO, $NCH_2OCOC_6H_5$ | 238 | $C_2H_5O$—$C_6H_3$, $NCH_2OCOCH_3$, CS, $NCH_2OCOCH_3$ | 237 |
| $C_2H_5O$—$C_6H_3$, $NCH_2OCOCH_3$, CO, $NCH_2OCOCH_3$ | 237 | $n$-$C_3H_7CO_2CH_2N$—CS, $NCH_2OCOC_3H_7$-$n$ (naphthalene) | 241 |

*Note:* References 382 to 537 are on pp. 266–269.

## TABLE XXVII

### Esters of N-α-Alkylol Derivatives of Amides, Lactams, Imides, Carbamyl Compounds, and Sulfonamides

| N-α-Alkylol Ester | References (Yield, %) |
|---|---|
| *Aliphatic Amides* | |
| $HCONHCH(OCOCH_3)CBr_3$ | 207 |
| $HCONHCH(OCOC_6H_5)CBr_3$ | 207 |
| $CH_3CONHCH(OCOC_6H_5)CCl_3$ | 166 |
| $CH_3CONHCH(OCOCH_3)CBr_3$ | 166 |
| $CH_3CONHCH(OCOC_6H_5)CBr_3$ | 207 |
| $C_6H_5CH_2CONHCH(OCOCH_3)CCl_3$ | 166 |
| $C_6H_5CH_2CONHCH(OCOC_6H_5)CCl_3$ | 166 |
| $C_6H_5CH_2CONHCH(OCOCH_3)CBr_3$ | 166 |
| $C_6H_5CH_2CONHCH(OCOC_6H_5)CBr_3$ | 166 |
| $NCCH_2CONHCH(OCOCH_3)CCl_3$ | 163 |
| $C_2H_5CONHCH(OCOCH_3)CBr_3$ | 207 |
| $C_2H_5CONHCH(OCOC_6H_5)CBr_3$ | 207 |
| *n*-$C_3H_7CONHCH(OCOCH_3)CCl_3$ | 166 |
| *n*-$C_3H_7CONHCH(OCOC_6H_5)CCl_3$ | 166 |
| *n*-$C_3H_7CONHCH(OCOCH_3)CBr_3$ | 207 |
| *n*-$C_3H_7CONHCH(OCOC_6H_5)CBr_3$ | 207 |
| *i*-$C_3H_7CONHCH(OCOCH_3)CBr_3$ | 207 |
| *i*-$C_3H_7CONHCH(OCOC_6H_5)CBr_3$ | 207 |
| *i*-$C_4H_9CONHCH(OCOCH_3)CBr_3$ | 207 |
| *i*-$C_4H_9CONHCH(OCOC_6H_5)CBr_3$ | 207 |
| *n*-$C_5H_{11}CONHCH(OCOCH_3)CBr_3$ | 207 |
| *n*-$C_5H_{11}CONHCH(OCOC_6H_5)CBr_3$ | 207 |
| *n*-$C_6H_{13}CONHCH(OCOCH_3)CBr_3$ | 207 |
| *n*-$C_6H_{13}CONHCH(OCOC_6H_5)CBr_3$ | 207 |
| *n*-$C_7H_{15}CONHCH(OCOCH_3)CBr_3$ | 207 |
| *n*-$C_7H_{15}CONHCH(OCOC_6H_5)CBr_3$ | 207 |
| *n*-$C_9H_{19}CONHCH(OCOCH_3)CBr_3$ | 207 |
| *n*-$C_9H_{19}CONHCH(OCOC_6H_5)CBr_3$ | 207 |
| $CCl_3CHOHCONHCH(OCOC_6H_5)CCl_3$ | 172 |
| *Amides of Dicarboxylic Acids* | |
| $C_2H_5O_2CCONHCH(OCOCH_3)CCl_3$ | 160 |
| $C_6H_5NHCOCONHCH(OCOCH_3)CCl_3$ | 163 |
| $[CONHCH(OCOCH_3)CCl_3]_2$ | 163 |
| $C_2H_5CH(CONHC_6H_5)CONHCH(OCOCH_3)CCl_3$ | 163 |
| $C_2H_5CH[CONHCH(OCOCH_3)CCl_3]_2$ | 163 |
| $CCl_3CH(OCOCH_3)CH(CONHC_6H_5)CONHCH$-$(OCOCH_3)CCl_3$ | 163 |
| *Aromatic Amides* | |
| $C_6H_5CONHCH(OCOCH_3)CHCl_2$ | 208 |
| $C_6H_5CONHCH(OCOC_6H_5)CHCl_2$ | 208 |
| $C_6H_5CONHCH(OCOCH_3)CHBr_2$ | 208 |
| $C_6H_5CONHCH(OCOCH_3)CCl_3$ | 95, 166 |
| $C_6H_5CONHCH(OCOC_6H_5)CCl_3$ | 95 |
| $C_6H_5CONHCH(OCOCH_3)CBr_3$ | 166 |
| $C_6H_5CONHCH(OCOC_6H_5)CBr_3$ | 166 |
| 2-$CH_3CO_2C_6H_4CONHCH(OCOCH_3)CCl_3$ | 166 |
| 2-$CH_3CO_2$-3-$ClC_6H_3CONHCH(OCOCH_3)CCl_3$ | 226 |
| 2-$CH_3CO_2$-5-$ClC_6H_3CONHCH(OCOCH_3)CCl_3$ | 226 |
| 2-$CH_3CO_2$-3,5-$Cl_2C_6H_2CONHCH(OCOCH_3)CCl_3$ | 226 |
| 2-$CH_3CO_2$-3-$BrC_6H_3CONHCH(OCOCH_3)CCl_3$ | 228 |
| 2-$CH_3CO_2$-5-$BrC_6H_3CONHCH(OCOCH_3)CCl_3$ | 228 |
| 2-$CH_3CO_2$-3,5-$Br_2C_6H_2CONHCH(OCOCH_3)CCl_3$ | 228 |
| 2-$CH_3CO_2$-3-$O_2NC_6H_3CONHCH(OCOCH_3)CCl_3$ | 225 |
| 2-$CH_3CO_2$-3-$O_2N$-5-$BrC_6H_2CONHCH(OCOCH_3)CCl_3$ | 225 |
| 2-$C_6H_5CO_2$-3,5-$Cl_2C_6H_2CONHCH(OCOC_6H_5)CCl_3$ | 226 |
| 2-$C_6H_5CO_2$-5-$BrC_6H_3CONHCH(OCOC_6H_5)CCl_3$ | 228 |
| 2-$C_6H_5CO_2$-3,5-$(O_2N)_2C_6H_2CONHCH(OCOC_6H_5)CCl_3$ | 225 |
| 2-$CH_3OC_6H_4CONHCH(OCOCH_3)CBr_3$ | 166 |
| 2-$CH_3O$-3-$ClC_6H_3CONHCH(OCOCH_3)CCl_3$ | 226 |
| 2-$CH_3O$-5-$ClC_6H_3CONHCH(OCOCH_3)CCl_3$ | 226 |
| 2-$CH_3O$-5-$ClC_6H_3CONHCH(OCOC_6H_5)CCl_3$ | 206 |
| 2-$CH_3O$-3,5-$Cl_2C_6H_2CONHCH(OCOCH_3)CCl_3$ | 226 |
| 2-$CH_3O$-5-$BrC_6H_3CONHCH(OCOCH_3)CCl_3$ | 228 |
| 2-$CH_3O$-5-$BrC_6H_3CONHCH(OCOC_6H_5)CCl_3$ | 228 |
| 2-$CH_3O$-3,5-$Br_2C_6H_2CONHCH(OCOCH_3)CCl_3$ | 228 |
| 2-$CH_3O$-3,5-$Br_2C_6H_2CONHCH(OCOC_6H_5)CCl_3$ | 228 |

| *Aromatic Amides (contd.)* | |
|---|---|
| $2\text{-}CH_3O\text{-}3\text{-}O_2NC_6H_3CONHCH(OCOCH_3)CCl_3$ | 225 |
| $2\text{-}CH_3O\text{-}5\text{-}O_2NC_6H_3CONHCH(OCOCH_3)CCl_3$ | 225 |
| $2\text{-}CH_3O\text{-}5\text{-}O_2NC_6H_3CONHCH(OCOC_6H_5)CCl_3$ | 225 |
| $2\text{-}CH_3O\text{-}3,5\text{-}(O_2N)_2C_6H_2CONHCH(OCOCH_3)CCl_3$ | 225 |
| $o\text{-}CH_3C_6H_4CONHCH(OCOCH_3)CCl_3$ | 168 |
| $o\text{-}CH_3C_6H_4CONHCH(OCOC_6H_5)CCl_3$ | 168 |
| $m\text{-}CH_3C_6H_4CONHCH(OCOCH_3)CCl_3$ | 168 |
| $m\text{-}CH_3C_6H_4CONHCH(OCOC_6H_5)CCl_3$ | 168 |
| $p\text{-}CH_3C_6H_4CONHCH(OCOCH_3)CCl_3$ | 168 |
| (2-furyl)$CONHCH(OCOC_6H_5)CCl_3$ | 204 (60) |
| (5-Br-2-furyl)$CONHCH(OCOC_6H_5)CCl_3$ | 204 (43) |
| *Lactams and Imides* | |
| $CH_2CO$ / $CH_2CH_2$ > $NCH(OCOCH_3)CCl_3$ | 101 (60) |
| $CH_2CO$ / $CH_2CH_2$ > $NCH(OCOC_6H_5)CCl_3$ | 101 (56) |
| $o\text{-}C_6H_4(CO)_2NCH(OCOCH_3)CH_3$ | 211 (96) |

| *Carbamyl Compounds* | |
|---|---|
| $C_2H_5O_2CNHCH(OCOCH_3)CCl_3$ | 94 (80) |
| $CCl_3CH{=}NCSNHCH(OCOCH_3)CCl_3$ | 160 |
| $CO[NHCH(OCOCH_3)CCl_3]_2$ | 161 |
| $CO[NHCH(OCOCH_3)CBr_3]_2$ | 162 |
| $CH_3CONHCONHCH(OCOCH_3)CCl_3$ | 161 |
| $CH_3CON(CH_3)CONHCH(OCOCH_3)CCl_3$ | 163 |
| $CH_3CON(CH_3)CONHCH(OCOCH_3)CBr_3$ | 163 |
| $CH_3CON(C_2H_5)CONHCH(OCOCH_3)CCl_3$ | 163 |
| $CH_3CON(C_6H_5)CONHCH(OCOCH_3)CBr_3$ | 163 |
| $CH_3CON(C_6H_4NO_2\text{-}m)CONHCH(OCOCH_3)CCl_3$ | 227 |
| $CH_3CON(C_6H_4NO_2\text{-}p)CONHCH(OCOCH_3)CCl_3$ | 227 |
| $CH_3CON(C_6H_4CH_3\text{-}o)CONHCH(OCOCH_3)CCl_3$ | 227 |
| $CH_3CON(C_6H_4CH_3\text{-}p)CONHCH(OCOCH_3)CCl_3$ | 227 |
| *Sulfonamides* | |
| $C_6H_5CH_2SO_2NHCH(OCOCH_3)CCl_3$ | 159 |
| $C_6H_5SO_2NHCH(OCOCH_3)CCl_3$ | 159 |
| $p\text{-}ClC_6H_4SO_2NHCH(OCOCH_3)CCl_3$ | 159 |
| $p\text{-}O_2NC_6H_4SO_2NHCH(OCOCH_3)CCl_3$ | 159 |
| $p\text{-}CH_3OC_6H_4SO_2NHCH(OCOCH_3)CCl_3$ | 159 |
| $o\text{-}CH_3C_6H_4SO_2NHCH(OCOCH_3)CCl_3$ | 159 |
| $m\text{-}CH_3C_6H_4SO_2NHCH(OCOCH_3)CCl_3$ | 159 |
| $p\text{-}CH_3C_6H_4SO_2NHCH(OCOCH_3)CCl_3$ | 159 |
| $2,4\text{-}(CH_3)_2C_6H_3SO_2NHCH(OCOCH_3)CCl_3$ | 159 |
| $2,4,6\text{-}(CH_3)_3C_6H_2SO_2NHCH(OCOCH_3)CCl_3$ | 159 |
| (2-naphthyl)$SO_2NHCH(OCOCH_3)CCl_3$ | 159 |

*Note:* References 382 to 537 are on pp. 266–269.

## TABLE XXVIII

### N-Halomethyl and N-α-Haloalkyl Derivatives of Amides, Lactams, Imides, Carbamyl Compounds, and Sulfonamides

$$RCON(R')CH(R'')OH \rightarrow RCON(R')CH(R'')X$$

| N-α-Haloalkyl Derivatives | Method* | References (Yield, %) |
|---|---|---|
| *Amides* | | |
| $HCONHCH_2Cl$ | $PCl_5$, dioxane | 96 |
| $HCONHCHClCCl_3$ | $PCl_5$, $(C_2H_5)_2O$ | 101 (76) |
| | $PCl_5$, heat | 160 |
| $HCONHCHClCBr_3$ | $PCl_5$, heat | 207 |
| $CH_3CONHCH_2Cl$ | $PCl_5$, dioxane | 96 |
| | $CH_3CONH_2$, $(CH_2O)_x$, HCl, $CH_3CO_2H$ | 202 (60), 203 (60) |
| | $CH_3CONH_2$, $(CH_2O)_x$, HCl, dioxane | 201 |
| $CH_3CONHCH_2Br$ | $PBr_5$, dioxane | 96 |
| $CH_3CONHCHClCH_3$ | $CH_3CONH_2$, $CH_3CHO$, HCl, dioxane | 201 |
| $CH_3CONHCHClCCl_3$ | $PCl_5$, heat | 209 |
| | $CH_3CONHCH{=}CCl_2$, $Cl_2$, $CHCl_3$ | 209 |
| $CH_3CONHCHClCBr_3$ | $PCl_5$, heat | 207 |
| $CH_3CONHCHCl(CH_2)_2CH_3$ | $CH_3CONH_2$, $n$-$C_3H_7CHO$, HCl, dioxane | 201 |
| $CH_3CON(CH_3)CH_2Cl$ | $PCl_5$, dioxane, 30–40° | 32 |
| $CH_3CON(C_3H_7\text{-}n)CH_2Cl$ | $PCl_5$, dioxane, 30–40° | 32 |
| $CH_3CON(C_3H_7\text{-}i)CH_2Cl$ | $PCl_5$, dioxane, 30–40° | 32 (good) |
| $CH_3CON(CH_2C_6H_5)CH_2Cl$ | $PCl_5$, dioxane, 30–40° | 32 |
| $CH_3CON(C_{11}H_{23}\text{-}n)CH_2Cl$ | $CH_3CONHR$, $(CH_2O)_x$, HCl, $C_6H_6$ | 196 |
| $CH_3CON(CH_3)CHClC_6H_5$ | $CH_3COCl$, $CH_3N{=}CHC_6H_5$ | 216 |
| $CH_2ClCONHCH_2Cl$ | $PCl_5$, $(C_2H_5)_2O$/dioxane | 96 |
| $CH_2ClCONHCHClCBr_3$ | $PCl_5$, heat | 207 |
| $CCl_3CONHCH_2Cl$ | $PCl_5$, $(C_2H_5)_2O$ | 100, 186 (76) |
| $C_2H_5CONHCH_2Cl$ | $PCl_5$, dioxane | 96 |

| | | |
|---|---|---|
| $C_2H_5CONHCH_2Br$ | $PBr_5$, dioxane | 96 |
| $C_2H_5CONHCHClCBr_3$ | $PCl_5$, heat | 207 |
| $n$-$C_3H_7CONHCHClCBr_3$ | $PCl_5$, heat | 207 |
| $n$-$C_{11}H_{23}CON(C_6H_{11}$-*cyclo*$)CH_2Cl$ | $RCONHR'$, $(CH_2O)_x$, HCl, $C_6H_6$ | 196 |
| $n$-$C_{17}H_{35}CONHCH_2Cl$ | $RCONH_2$, $(CH_2O)_x$, HCl, $CH_2Cl_2$ | 195 |
| | $RCONH_2$, $(CH_2O)_x$, HCl, $C_6H_6$ | 200 |
| $n$-$C_{17}H_{35}CON(CH_3)CH_2Cl$ | $RCONHCH_3$, $(CH_2O)_x$, HCl, $C_6H_6$ | 196 |
| $n$-$C_{17}H_{35}CON(CH_3)CH_2Br$ | $RCONHCH_3$, $(CH_2O)_x$, HBr, $C_6H_6$ | 196 |
| $(n$-$C_{17}H_{35}CONCH_2Cl)_2CH_2$ [$CH_2$ bridging two $n$-$C_{17}H_{35}CONCH_2Cl$ units] | $(RCONH)_2CH_2$, $(CH_2O)_x$, HCl, $C_6H_6$ | 198, 199 |
| $n$-$C_{21}H_{43}CONHCH_2Cl$ | $RCONH_2$, $(CH_2O)_x$, HCl, $C_6H_6$ | 197 |
| $C_6H_5CONHCH_2Cl$ | $PCl_5$, $(C_2H_5)_2O$ | 96 (78), 100 |
| | $C_6H_5CONH_2$, $(CH_2O)_x$, HCl, $CH_3CO_2H$ | 202 (62) |
| $C_6H_5CONHCH_2Br$ | $PBr_5$, $(C_2H_5)_2O$ | 96 (80) |
| $C_6H_5CONHCHClCHCl_2$ | $C_6H_5CONHCH{=}CCl_2$, HCl | 208 |
| $C_6H_5CONHCHBrCHCl_2$ | $C_6H_5CONHCH{=}CCl_2$, HBr | 208 |
| $C_6H_5CONHCHClCCl_3$ | $PCl_5$, $(C_2H_5)_2O$/dioxane | 101 (72) |
| $C_6H_5CON(CH_3)CH_2Cl$ | $PCl_5$, dioxane, 30–40° | 32 |
| $C_6H_5CON(CH_3)CHClC_6H_5$ | $C_6H_5COCl$, $CH_3N{=}CHC_6H_5$ | 216 (55) |
| 2,4-$Cl_2C_6H_3CONHCH_2Cl$ | $PCl_5$, dioxane | 96 |
| 2-$CH_3O$-5-$ClC_6H_3CONHCHClCCl_3$ | $PCl_5$, heat | 206 (94) |
| 2-$CH_3O$-3,5-$Cl_2C_6H_2CONHCHClCCl_3$ | $PCl_5$, heat | 206 |
| 2-$CH_3O$-5-$BrC_6H_3CONHCHClCCl_3$ | $PCl_5$, heat | 205 |
| 2-$CH_3O$-5-$BrC_6H_3CONHCHBrCCl_3$ | $PBr_5$, heat | 205 |
| 2-$CH_3O$-3,5-$Br_2C_6H_2CONHCHClCCl_3$ | $PCl_5$, heat | 205 |
| 2-$CH_3O$-5-$O_2NC_6H_3CONHCHClCCl_3$ | $PCl_5$, heat | 205 |
| $o$-$CH_3C_6H_4CONHCHClCCl_3$ | $PCl_5$, heat | 106 |

*Note:* References 382 to 537 are on pp. 266–269.

* Unless otherwise indicated, the N-α-hydroxyalkyl derivative is an unnamed reactant.

TABLE XXVIII—*Continued*

N-Halomethyl and N-α-Haloalkyl Derivatives of Amides, Lactams, Imides, Carbamyl Compounds, and Sulfonamides

$RCON(R')CH(R'')OH \rightarrow RCON(R')CH(R'')X$

| N-α-Haloalkyl Derivatives | Method* | References (Yield, %) |
|---|---|---|
| *Amides (contd.)* | | |
| $m$-$CH_3C_6H_4CONHCHClCCl_3$ | $PCl_5$, heat | 106 |
| $p$-$CH_3C_6H_4CONHCHClCCl_3$ | $PCl_5$, heat | 106 |
| $C_2H_5O_2CNHCHClCCl_3$ | $PCl_5$, heat | 160 |
| $C_2H_5O_2CNHCHBrCCl_3$ | $PBr_5$, heat | 160 |
| $(CONHCHClCCl_3)_2$ | $PCl_5$, heat | 163 |
| (furan ring, O) $CONHCHClCCl_3$ | $PCl_5$, heat | 204 |
| $O_2N$ (furan ring, O) $CONHCHClCCl_3$ | $PCl_5$, heat | 204 |
| *Lactams* | | |
| $CH_2CO$<br>$NCH_2Cl$<br>$CH_2CH_2$ | $SOCl_2$<br>$PCl_5$, $(C_2H_5)_2O$/dioxane | 177 (85), 184 (87)<br>33 |
| $CH_2CO$<br>$NCHClCCl_3$<br>$CH_2CH_2$ | $PCl_5$, $(C_2H_5)_2O$/dioxane | 101 (99) |

| Compound | Conditions | Refs. (Yield, %) |
|---|---|---|
| $CH_2CH_2CO-N(CH_2Cl)-CH_2CH_2CH_2$ (ring) | $SOCl_2$, $C_6H_6$ | 184 (70) |
| | $PCl_5$, $(C_2H_5)_2O$/dioxane | 33 |
| 1,4-Bis(chloromethyl)-2,5-piperazinedione ($ClCH_2N$, $NCH_2Cl$, two C=O) | $PCl_5$, $CH_3COCl$, $CCl_4$ | 1, 65 (91) |
| *Imides* | | |
| $CH_2CO-N(CH_2Cl)-COCH_2$ (ring) | $PCl_5$, $CHCl_3$ | 66 (75) |
| $CH_2CO-N(CH_2Br)-COCH_2$ (ring) | $PBr_3$, $C_6H_6$ | 188 (91) |
| | $PBr_5$, $CHCl_3$ | 66 |
| $CHClCO-N(CH_2Cl)-COCH_2$ (ring) | N-Chloromethylmaleimide, HCl | 8 |

*Note:* References 382 to 537 are on pp. 266–269.
* Unless otherwise indicated, the N-α-hydroxyalkyl derivative is an unnamed reactant.

TABLE XXVIII—*Continued*

N-HALOMETHYL AND N-α-HALOALKYL DERIVATIVES OF AMIDES, LACTAMS, IMIDES, CARBAMYL COMPOUNDS, AND SULFONAMIDES

$RCON(R')CH(R'')OH \rightarrow RCON(R')CH(R'')X$

| N-α-Haloalkyl Derivatives | Method* | References (Yield, %) |
|---|---|---|
| *Imides* (*contd.*) | | |
| CHCO=CHCO (cyclic) $\rangle NCH_2Cl$ | $PCl_3$, $(CH_3)_2CO$ | 8 (81) |
| CHCO=CHCO (cyclic) $\rangle NCH_2Br$ | $PBr_3$, $(CH_3)_2CO$ | 8 (52) |
| $o$-$C_6H_4(CO)_2NCH_2Cl$ | $PCl_5$, $(C_2H_5)_2O$ | 97 (82) |
| | $o$-$C_6H_4(CO)_2NH$, $(CH_2O)_x$, HCl, $CH_3CO_2H$ | 202 (70) |
| | Concd. HCl, heat | 192, 193 |
| | $o$-$C_6H_4(CO)_2NCH_2COCl$, heat (−CO) | 192 |
| | $o$-$C_6H_4(CO)_2NCH_2OC_2H_5$, $CH_3COCl$ | 190 (90) |
| | $o$-$C_6H_4(CO)_2NCH_2N$(piperidino), $CH_3COCl$ | 191 (81) |

| | | |
|---|---|---|
| $o\text{-}C_6H_4(CO)_2NCH_2Br$ | $PBr_3$, $C_6H_6$ | 188 (82) |
| | $o\text{-}C_6H_4(CO)_2NCH_3$, $Br_2$, 170° | 151 (100) |
| | Concd. HBr, heat | 104 (69), 192, 193 |
| | Concd. HBr, $H_2SO_4$, heat | 70 (65) |
| | $o\text{-}C_6H_4(CO)_2NCH_2N\langle\;\rangle O$, $CH_3COBr$ | 191 (86) |
| $o\text{-}C_6H_4(CO)_2NCH_2I$ | Concd. HI, heat | 192, 193 |
| $o\text{-}C_6H_4(CO)_2NCHClCH_3$ | $o\text{-}C_6H_4(CO)_2NCH{=}CH_2$, HCl, $CHCl_3$ | 211 (97) |
| $o\text{-}C_6H_4(CO)_2NCHBrCH_2Br$ | $o\text{-}C_6H_4(CO)_2NCH_2CH_2Br$, NBS† | 214 (89) |
| | $o\text{-}C_6H_4(CO)_2NCH{=}CH_2$, $Br_2$ | 210 |
| $o\text{-}C_6H_4(CO)_2NCHBrCHBrCH_3$ | $o\text{-}C_6H_4(CO)_2NCH{=}CHCH_3$, $Br_2$ | 212 |
| $o\text{-}C_6H_4(CO)_2NCHBr(CH_2)_2Br$ | $o\text{-}C_6H_4(CO)_2NCH(CO_2H)CH_2CH_3$, $Br_2$, P | 212, 215 (52) |
| $o\text{-}C_6H_4(CO)_2NC(CH_3)BrCH_2Br$ | $o\text{-}C_6H_4(CO)_2NC(CH_3){=}CH_2$, $Br_2$ | 213 |
| $C_6H_4\langle^{CO}_{SO_2}\rangle NCH_2Cl$ | $PCl_5$, $(C_2H_5)_2O$ | 99 (65) |
| | $SOCl_2$, $(C_2H_5)_2O$ | 182 (54) |
| *Carbamyl Compounds* | | |
| $ClCONHCH_2Cl$ | $OCNCH_2OH$, $SOCl_2$ | 73 (68) |
| $ClCONHCHClCCl_3$ | $OCNCHOHCCl_3$, $SOCl_2$ | 73 |
| $CH_3OCONHCHClCCl_3$ | $OCNCHClCCl_3$, $CH_3OH$ | 73 |
| $n\text{-}C_{22}H_{45}OCONHCH_2Cl$ | $RO_2CNH_2$, $(CH_2O)_x$, HCl, $C_6H_6$ | 197 |
| $CH_3O_2CN(C_4H_9\text{-}n)CH_2Cl$ | $CH_3O_2CNHR$, $(CH_2O)_x$, HCl, $C_6H_6$ | 196 |
| $CH_3O_2CN(C_{11}H_{23}\text{-}n)CH_2Cl$ | $CH_3O_2CNHR$, $(CH_2O)_x$, HCl, $C_6H_6$ | 196 |
| $CH_3O_2CN(C_{17}H_{35}\text{-}n)CH_2Cl$ | $CH_3O_2CNHR$, $(CH_2O)_x$, HCl, $C_6H_6$ | 196 |
| $C_2H_5O_2CN(CH_3)CHClC_6H_5$ | $C_2H_5O_2CCl$, $CH_3N{=}CHC_6H_5$ | 216 |
| $2,4\text{-}Cl_2C_6H_3O_2CNHCHClCCl_3$ | $OCNCHClCCl_3$, $2,4\text{-}Cl_2C_6H_3OH$ | 73 |
| $1\text{-}C_{10}H_7O_2CNHCHClCCl_3$ | $OCNCHClCCl_3$, $1\text{-}C_{10}H_7OH$ | 73 |

*Note:* References 382 to 537 are on pp. 266–269.

* Unless otherwise indicated, the N-α-hydroxyalkyl derivative is an unnamed reactant.

† N-Bromosuccinimide.

TABLE XXVIII—*Continued*

N-Halomethyl and N-α-Haloalkyl Derivatives of Amides, Lactams, Imides, Carbamyl Compounds, and Sulfonamides

$RCON(R')CH(R'')OH \rightarrow RCON(R')CH(R'')X$

| N-α-Haloalkyl Derivatives | Method* | References (Yield, %) |
|---|---|---|
| *Carbamyl Compounds (contd.)* | | |
| $2\text{-}C_{10}H_7O_2CNHCHClCCl_3$ | $OCNCHClCCl_3$, $2\text{-}C_{10}H_7OH$ | 73 |
| 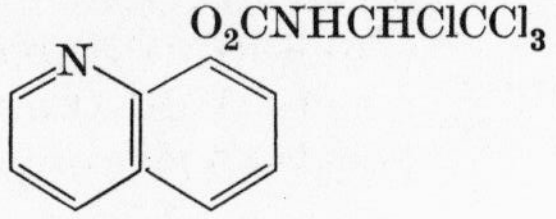 8-Quinolyl–$O_2CNHCHClCCl_3$ (quinoline structure) | $OCNCHClCCl_3$, 8-hydroxyquinoline (structure) 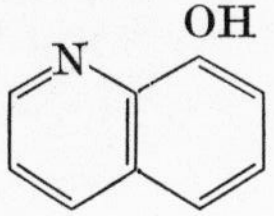 | 73 |
| $CS(NHCHClCCl_3)_2$ | $PCl_5$, heat | 160 |
| 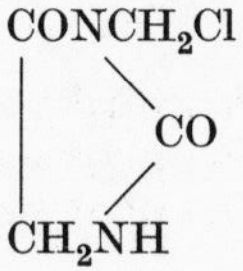 $CONCH_2Cl$—CO—$CH_2NH$ (cyclic structure) | $PCl_5$, heat or concd. HCl | 185 (65) |
| 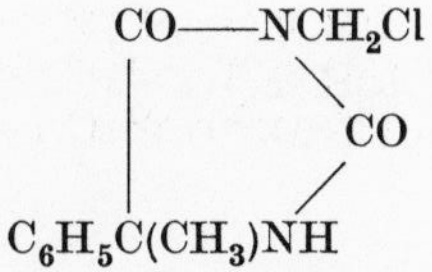 CO—$NCH_2Cl$—CO—$C_6H_5C(CH_3)NH$ (cyclic structure) | $PCl_5$, $CHCl_3$ | 187 (60) |
| 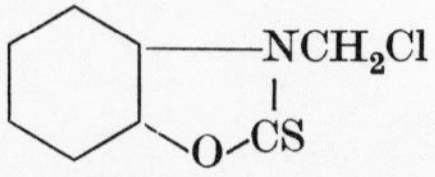 Cyclohexane-fused $NCH_2Cl$, O—CS (cyclic structure) (DL-*trans*) | $SOCl_2$, $(C_2H_5)_2O$ | 181 (53) |

| | | |
|---|---|---|
| 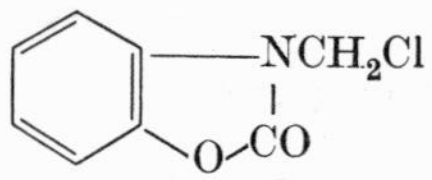 $NCH_2Cl$, O–CO | Concd. HCl, heat | 194 (91) |
| 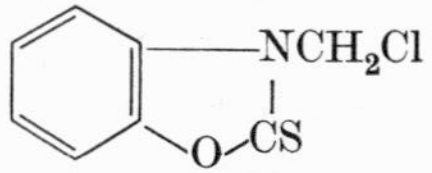 $NCH_2Cl$, O–CS | $SOCl_2$ | 178 (60) |
| 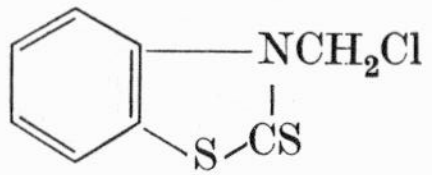 $NCH_2Cl$, S–CS | $SOCl_2$, $CHCl_3$ | 183 (55) |
| | $PCl_3$, heat | 189 |
| 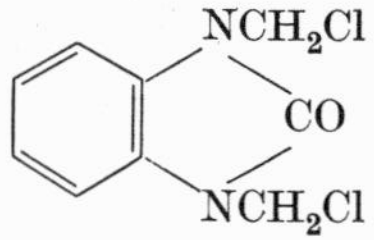 $NCH_2Cl$, CO, $NCH_2Cl$ | $SOCl_2$ | 180 (59) |
| 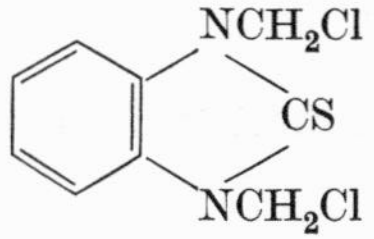 $NCH_2Cl$, CS, $NCH_2Cl$ | $SOCl_2$, heat | 179 (55) |
| *Sulfonamides* | | |
| $C_6H_5SO_2NHCH_2Cl$ | $C_6H_5SO_2NH_2$, $(CH_2O)_x$, HCl, dioxane | 201 |
| $C_6H_5SO_2NHCHClCH_3$ | $C_6H_5SO_2NH_2$, $CH_3CHO$, HCl, dioxane | 201 |
| $C_6H_5SO_2NHCHCl(CH_2)_2CH_3$ | $C_6H_5SO_2NH_2$, $n$-$C_3H_7CHO$, HCl, dioxane | 201 |

*Note:* References 382 to 537 are on pp. 266–269.

* Unless otherwise indicated, the N-α-hydroxyalkyl derivative is an unnamed reactant.

TABLE XXIX

N-AMINOMETHYL DERIVATIVES OF AMIDES, LACTAMS, IMIDES, AND CARBAMYL COMPOUNDS*

$$RCONH_2 + CH_2O + R'_2NH \rightarrow RCONHCH_2NR'_2$$

| N-Aminomethyl Derivative | References (Yield, %) | N-Aminomethyl Derivative | References (Yield, %) |
|---|---|---|---|
| 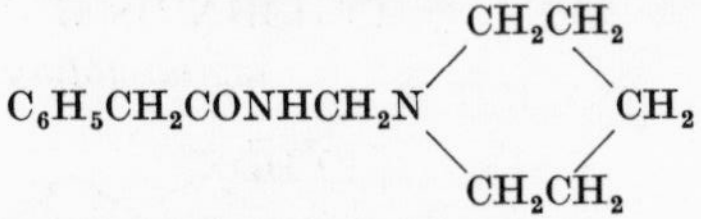 $C_6H_5CH_2CONHCH_2N$ $CH_2CH_2$ $CH_2$ $CH_2CH_2$ | 388 (78) | 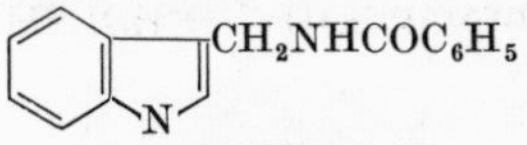 $CH_2NHCOC_6H_5$ N | 83 |
| $CH_2{=}CHCONHCH_2N(CH_3)_2$ | 477 | 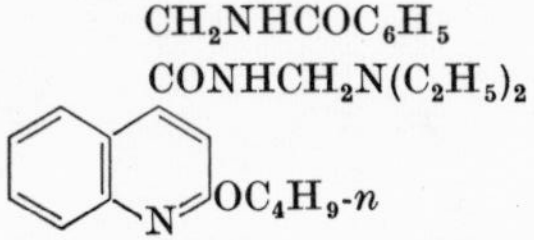 $CH_2NHCOC_6H_5$ $CONHCH_2N(C_2H_5)_2$ N $OC_4H_9$-$n$ | 426 |
| 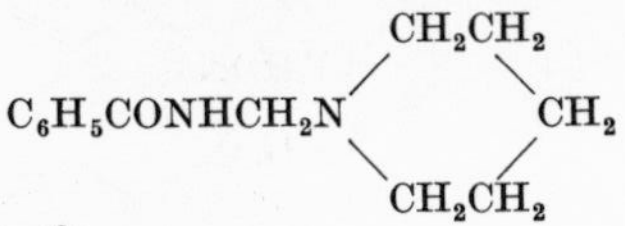 $C_6H_5CONHCH_2N$ $CH_2CH_2$ $CH_2$ $CH_2CH_2$ | 96 (72) | 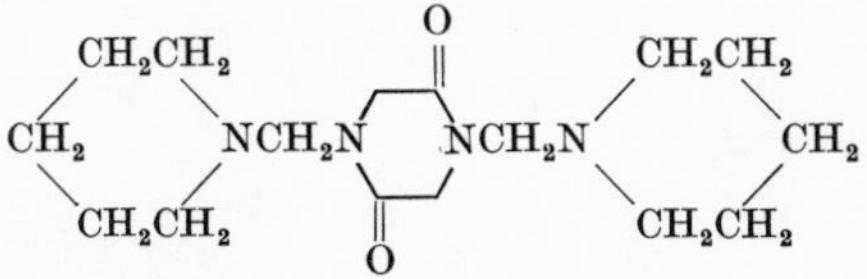 O $CH_2CH_2$ $CH_2$ $NCH_2N$ $NCH_2N$ $CH_2CH_2$ $CH_2$ $CH_2CH_2$ O $CH_2CH_2$ | 65 (65) |
| 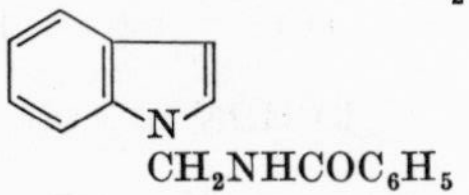 N $CH_2NHCOC_6H_5$ | 83 (57) | 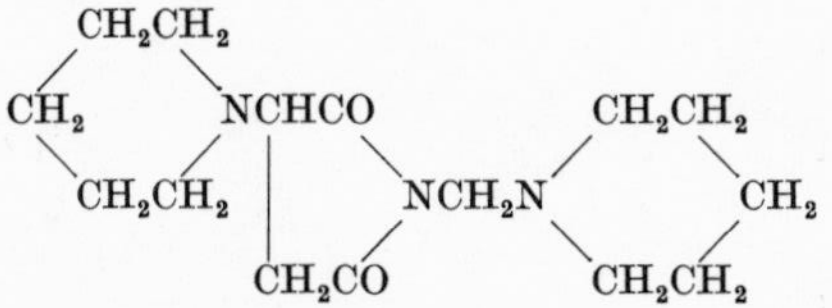 $CH_2CH_2$ $CH_2$ NCHCO $CH_2CH_2$ $CH_2CH_2$ $NCH_2N$ $CH_2$ $CH_2CO$ $CH_2CH_2$ | 8 (49) |
| 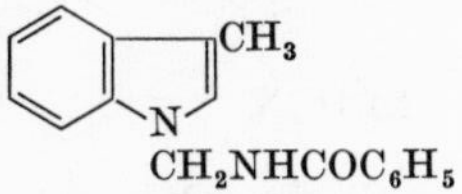 $CH_3$ N $CH_2NHCOC_6H_5$ | 83 (74) | From $\alpha$-methylacrylamide and | |
| 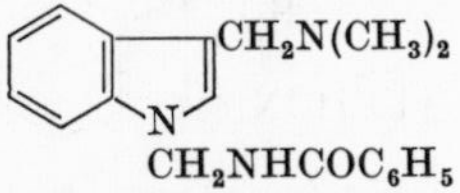 | | Dimethylamine | 477 |
| $CH_2N(CH_3)_2$ N $CH_2NHCOC_6H_5$ | 83 | Di-$n$-propylamine | 477 |
| | | Di-$n$-butylamine | 477 |
| 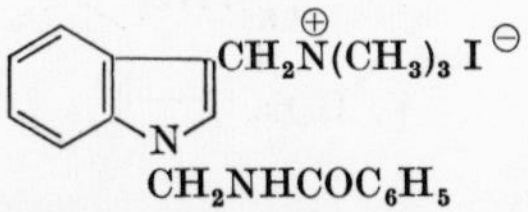 $CH_2\overset{\oplus}{N}(CH_3)_3\ I^{\ominus}$ N $CH_2NHCOC_6H_5$ | 83 | Morpholine | 477 (75) |
| | | $CH_3NHCH_2CH_2OH$ | 477 |
| | | $HN(CH_2CH_2OH)_2$ | 477 |

| Reactants | References |
|---|---|
| From succinimide and | |
| Dimethylamine | 486 (75) |
| Piperidine | 66, 188 |
| Morpholine | 486 (60) |
| From maleimide and | |
| Diethylamine | 432 |
| Di-*n*-butylamine | 8 (74), 432 |
| Piperidine | 8 (82), 432 |
| Aniline | 8 (88), 432 |
| From phthalimide and | |
| Dimethylamine | 1 (42), 412 |
| Piperidine | 188 |
| Aniline | 434 |
| From 3-(and 4-)Nitrophthalimide and | |
| Aniline | 436 |
| *o*-Chloroaniline | 436 |
| *m*-Chloroaniline | 436 |
| *p*-Chloroaniline | 436 |
| 2,4-Dichloroaniline | 436 |
| *o*-Bromoaniline | 436 |
| *m*-Bromoaniline | 436 |
| *p*-Bromoaniline | 436 |
| *o*-Nitroaniline | 436 |
| *m*-Nitroaniline | 436 |
| *p*-Nitroaniline | 436 |
| 4-Chloro-2-nitroaniline | 436 |
| 2-Chloro-4-nitroaniline | 436 |
| *o*-Anisidine | 436 |
| *p*-Anisidine | 436 |
| *o*-Toluidine | 436 |
| *m*-Toluidine | 436 |
| *p*-Toluidine | 436 |
| 2-Naphthylamine | 436 |
| From saccharin and | |
| Dimethylamine | 182 (12) |
| Diethylamine | 182 (87) |
| Di-*n*-propylamine | 182 (39) |
| Di-*n*-butylamine | 182 (33) |
| Piperidine | 182 (65) |
| Morpholine | 182 (41) |
| $CONCH_2N(CH_2CH_2)_2CH_2$ / $CO$ / $CH_2NCOCH_3$ (cyclic) | 487 (98) |
| $CONCH_2N(CH_2CH_2)_2O$ / $CO$ / $CH_2NCOCH_3$ (cyclic) | 487 (91) |
| From (hexahydrobenzoxazolethione: ring NH, O–CS) (DL-*trans*) and | |
| Diethylamine | 260 (66) |
| Piperidine | 260 (86) |
| Morpholine | 260 (86) |
| Aniline | 260 (64) |
| *p*-Chloroaniline | 260 (81) |
| *p*-Bromoaniline | 260 (73) |
| *m*-Nitroaniline | 260 (81) |
| *o*-Anisidine | 260 (75) |
| *p*-Anisidine | 260 (78) |
| *p*-Phenetidine | 260 (75) |
| *o*-Toluidine | 260 (83) |
| *p*-Toluidine | 260 (72) |
| 2,3-Xylidine | 260 (76) |

*Note:* References 382 to 537 are on pp. 266–269.

* For an exhaustive compilation of compounds of this class through 1958 see ref. 3

TABLE XXIX—*Continued*

N-Aminomethyl Derivatives of Amides, Lactams, Imides, and Carbamyl Compounds*

$$RCONH_2 + CH_2O + R'_2NH \rightarrow RCONHCH_2NR'_2$$

| N-Aminomethyl Derivative | References (Yield, %) | N-Aminomethyl Derivative | References (Yield, %) |
|---|---|---|---|
| From 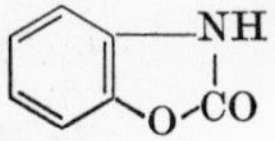 (NH, O–CO) and | | 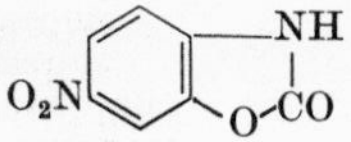 From $O_2N$ (NH, O–CO) and | |
| Aniline | 324 (84) | | |
| *o*-Chloroaniline | 324 (73) | Piperidine | 488 (85) |
| *m*-Chloroaniline | 324 (82) | Morpholine | 488 (77) |
| *p*-Chloroaniline | 324 (86) | | |
| 2,4-Dichloroaniline | 324 (38) | From 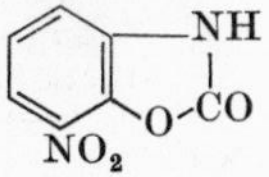 (NH, O–CO, $NO_2$) and | |
| *m*-Bromoaniline | 324 (88) | | |
| *p*-Bromoaniline | 324 (81) | | |
| 3,4-Dibromoaniline | 324 (60) | | |
| *o*-Anisidine | 324 (85) | Morpholine | 488 (81) |
| *p*-Anisidine | 324 (87) | | |
| *p*-Phenetidine | 324 (88) | 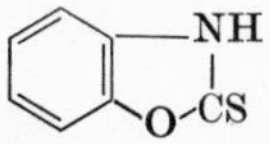 | |
| *o*-Phenylenediamine (*bis* cpd.) | 324 (96) | From (NH, O–CS) and | |
| *p*-Phenylenediamine (*bis* cpd.) | 324 (96) | | |
| N-Acetyl-*p*-phenylenediamine | 324 (84) | | |
| *o*-Toluidine | 324 (81) | Diethylamine | 183 (96), 489 (63) |
| *m*-Toluidine | 324 (83) | | |
| *p*-Toluidine | 324 (87) | Piperidine | 183 (98), 189, 318 |
| 3,4-Dimethylaniline | 324 (56) | | |
| 2,4,5-Trimethylaniline | 324 (70) | Morpholine | 183 (85), 318 |
| 2,4,6-Trimethylaniline | 324 (59) | Aniline | 183 (88) |
| 2-Naphthylamine | 324 (86) | | |
| Benzidine (*bis* cpd.) | 324 (97) | | |
| 2-Aminopyridine | 324 (41) | 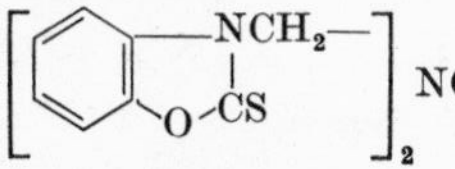 [(NCH$_2$—, O–CS)]$_2$ $NC_6H_{11}$ | 183 (54) |
| From 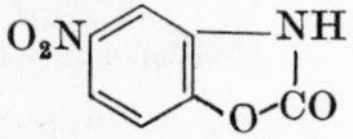 $O_2N$ (NH, O–CO) and | | | |
| Piperidine | 488 (80) | 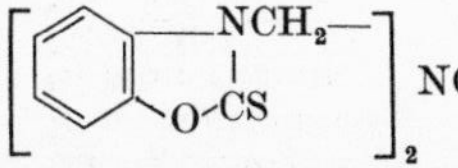 [(NCH$_2$—, O–CS)]$_2$ $NCH_2CH_2OH$ | 183 (58) |
| Morpholine | 488 (77) | | |

*Note:* References 382 to 537 are on pp. 266–269.

# TABLE XXX

## N-Quaternary Aminomethyl Derivatives of Amides, Imides, and Carbamyl Compounds

$RCONHCH_2—\overset{\oplus}{N}\lt \quad X^{\ominus}$

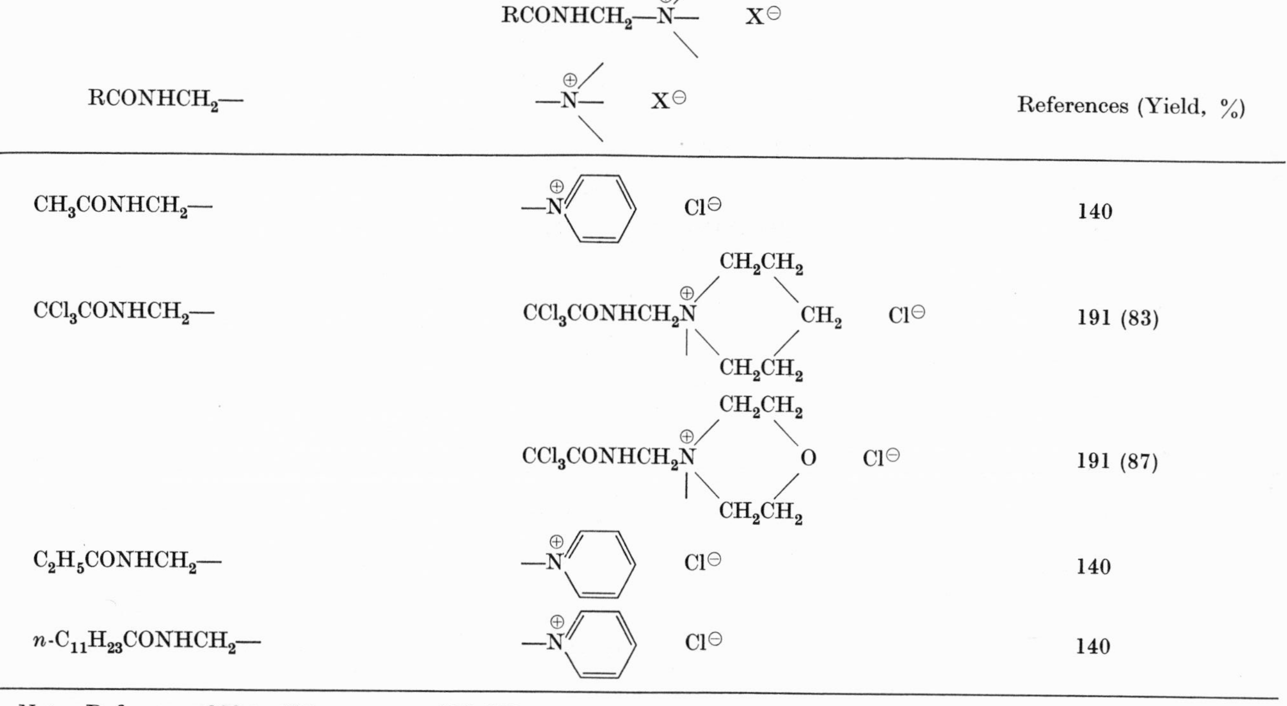

| $RCONHCH_2—$ | $—\overset{\oplus}{N}\lt \quad X^{\ominus}$ | References (Yield, %) |
|---|---|---|
| $CH_3CONHCH_2—$ | $—\overset{\oplus}{N}$(pyridinium) $Cl^{\ominus}$ | 140 |
| $CCl_3CONHCH_2—$ | $CCl_3CONHCH_2\overset{\oplus}{N}$(CH₃)(CH₂CH₂)₂CH₂ [piperidinium] $Cl^{\ominus}$ | 191 (83) |
| | $CCl_3CONHCH_2\overset{\oplus}{N}$(CH₃)(CH₂CH₂)₂O [morpholinium] $Cl^{\ominus}$ | 191 (87) |
| $C_2H_5CONHCH_2—$ | $—\overset{\oplus}{N}$(pyridinium) $Cl^{\ominus}$ | 140 |
| $n$-$C_{11}H_{23}CONHCH_2—$ | $—\overset{\oplus}{N}$(pyridinium) $Cl^{\ominus}$ | 140 |

*Note:* References 382 to 537 are on pp. 266–269.

TABLE XXX—*Continued*

N-Quaternary Aminomethyl Derivatives of Amides, Imides, and Carbamyl Compounds

$RCONHCH_2—\overset{\oplus}{N}{<}\ \ X^{\ominus}$

| $RCONHCH_2—$ | $—\overset{\oplus}{N}{<}\ \ X^{\ominus}$ | References (Yield, %) |
|---|---|---|
| $n$-$C_{17}H_{35}CONHCH_2—$ | $—\overset{\oplus}{N}C_5H_5$ (pyridinium) $Cl^{\ominus}$ | 138 (78), 140, 200, 243 |
| | $—\overset{\oplus}{N}C_5H_5$ (pyridinium) $NO_3^{\ominus}$ | 140 |
| | $—\overset{\oplus}{N}C_5H_5$ (pyridinium) $HSO_4^{\ominus}$ | 140, 306 |
| | $—\overset{\oplus}{N}C_5H_5$ (pyridinium) $HO_2CCO_2^{\ominus}$ | 140 |
| | $—\overset{\oplus}{N}C_5H_5$ (pyridinium) $p$-$CH_3C_6H_4SO_2^{\ominus}$ | 140 |
| | $—\overset{\oplus}{N}C_9H_7$ (quinolinium) $Cl^{\ominus}$ | 140 |
| $n$-$C_{21}H_{43}CONHCH_2—$ | $—\overset{\oplus}{N}C_5H_5$ (pyridinium) $Cl^{\ominus}$ | 235 |

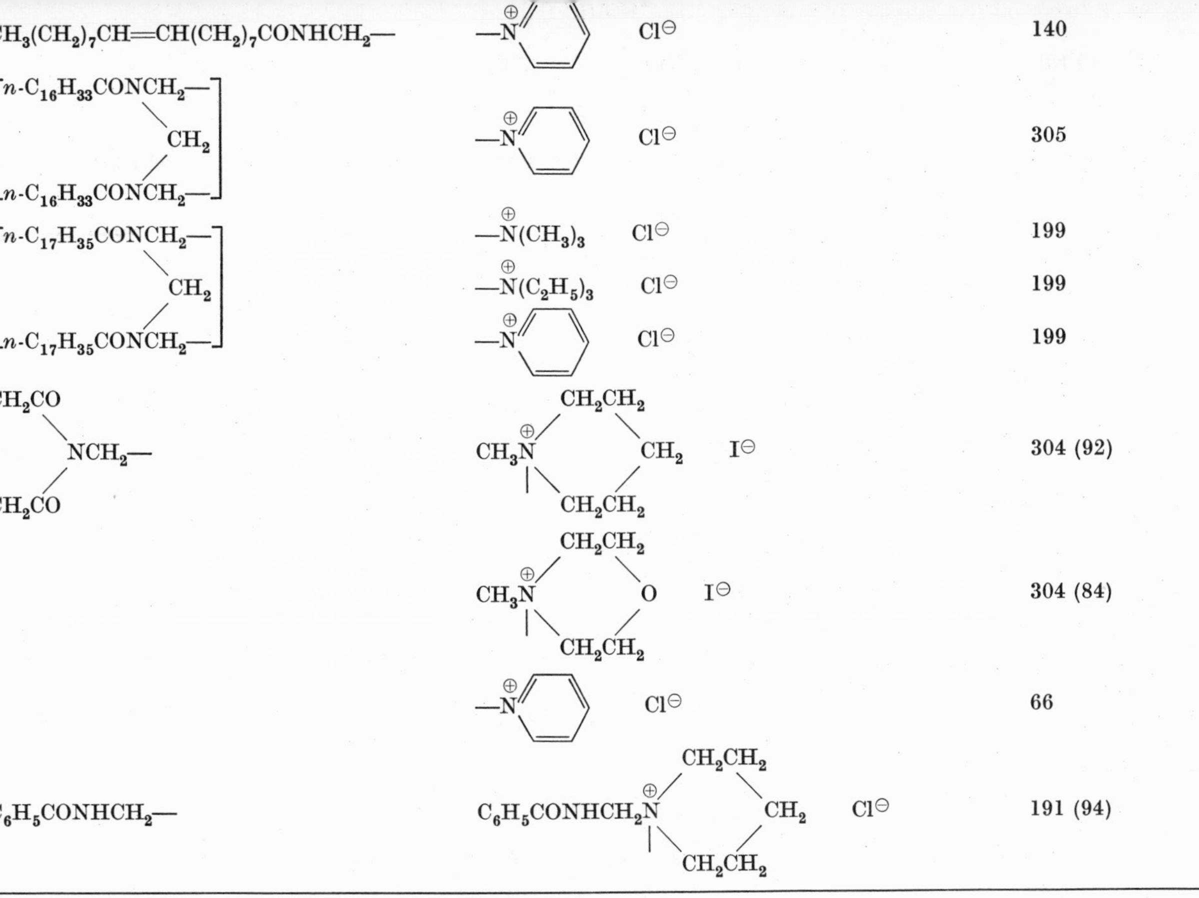

| | | Refs. |
|---|---|---|
| $CH_3(CH_2)_7CH{=}CH(CH_2)_7CONHCH_2$— | $—\overset{\oplus}{N}C_5H_5$ (pyridinium) $Cl^{\ominus}$ | 140 |
| $[n\text{-}C_{16}H_{33}CONCH_2—]_2CH_2$ | $—\overset{\oplus}{N}C_5H_5$ (pyridinium) $Cl^{\ominus}$ | 305 |
| $[n\text{-}C_{17}H_{35}CONCH_2—]_2CH_2$ | $—\overset{\oplus}{N}(CH_3)_3$ $Cl^{\ominus}$ | 199 |
| | $—\overset{\oplus}{N}(C_2H_5)_3$ $Cl^{\ominus}$ | 199 |
| | $—\overset{\oplus}{N}C_5H_5$ (pyridinium) $Cl^{\ominus}$ | 199 |
| $(CH_2CO)_2NCH_2$— | $CH_3\overset{\oplus}{N}$ (ring: $CH_2CH_2$, $CH_2$, $CH_2CH_2$) $I^{\ominus}$ | 304 (92) |
| | $CH_3\overset{\oplus}{N}$ (ring: $CH_2CH_2$, O, $CH_2CH_2$) $I^{\ominus}$ | 304 (84) |
| | $—\overset{\oplus}{N}C_5H_5$ (pyridinium) $Cl^{\ominus}$ | 66 |
| $C_6H_5CONHCH_2$— | $C_6H_5CONHCH_2\overset{\oplus}{N}$ (ring: $CH_2CH_2$, $CH_2$, $CH_2CH_2$) $Cl^{\ominus}$ | 191 (94) |

*Note:* References 382 to 537 are on pp. 266–269.

TABLE XXX—*Continued*

N-Quaternary Aminomethyl Derivatives of Amides, Imides, and Carbamyl Compounds

$RCONHCH_2—\overset{\oplus}{N}(—)_3 \quad X^{\ominus}$

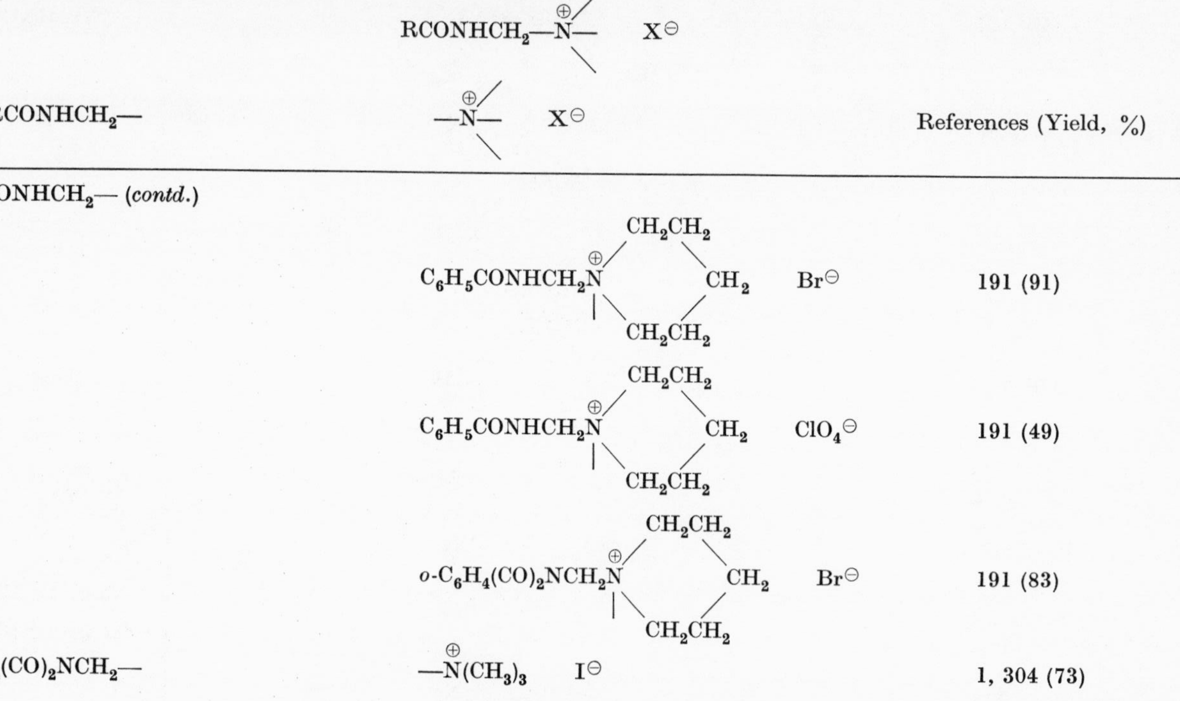

| $RCONHCH_2—$ | $—\overset{\oplus}{N}(—)_3 \quad X^{\ominus}$ | References (Yield, %) |
|---|---|---|
| $C_6H_5CONHCH_2—$ (*contd.*) | | |
| | $C_6H_5CONHCH_2\overset{\oplus}{N}(CH_3)$ in ring with $CH_2CH_2$, $CH_2$, $CH_2CH_2$ (piperidinium)  $Br^{\ominus}$ | 191 (91) |
| | $C_6H_5CONHCH_2\overset{\oplus}{N}(CH_3)$ in ring with $CH_2CH_2$, $CH_2$, $CH_2CH_2$ (piperidinium)  $ClO_4^{\ominus}$ | 191 (49) |
| | $o\text{-}C_6H_4(CO)_2NCH_2\overset{\oplus}{N}(CH_3)$ in ring with $CH_2CH_2$, $CH_2$, $CH_2CH_2$ (piperidinium)  $Br^{\ominus}$ | 191 (83) |
| $o\text{-}C_6H_4(CO)_2NCH_2—$ | $—\overset{\oplus}{N}(CH_3)_3 \quad I^{\ominus}$ | 1, 304 (73) |
| | $CH_3\overset{\oplus}{N}$ in ring with $CH_2CH_2$, $CH_2$, $CH_2CH_2$ (piperidinium)  $I^{\ominus}$ | 304 (96) |

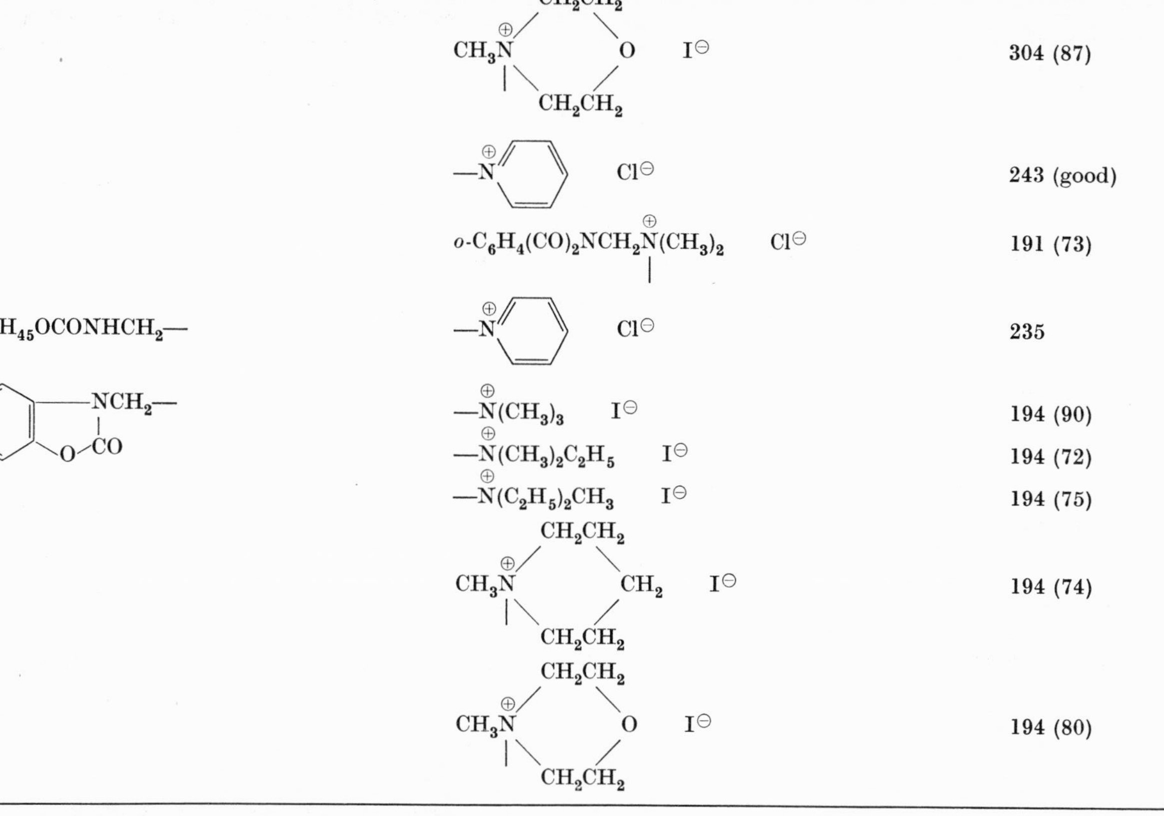

| | | |
|---|---|---|
| | $I^{\ominus}$ | 304 (87) |
| | $Cl^{\ominus}$ | 243 (good) |
| | $o\text{-}C_6H_4(CO)_2NCH_2\overset{\oplus}{N}(CH_3)_2$ $Cl^{\ominus}$ | 191 (73) |
| $C_{22}H_{45}OCONHCH_2$— | $Cl^{\ominus}$ | 235 |
| | —$\overset{\oplus}{N}(CH_3)_3$ $I^{\ominus}$ | 194 (90) |
| | —$\overset{\oplus}{N}(CH_3)_2C_2H_5$ $I^{\ominus}$ | 194 (72) |
| | —$\overset{\oplus}{N}(C_2H_5)_2CH_3$ $I^{\ominus}$ | 194 (75) |
| | $I^{\ominus}$ | 194 (74) |
| | $I^{\ominus}$ | 194 (80) |

*Note:* References 382 to 537 are on pp. 266–269.

TABLE XXX—*Continued*

N-Quaternary Aminomethyl Derivatives of Amides, Imides, and Carbamyl Compounds

$$RCONHCH_2-\overset{\oplus}{N}{<}\ \ X^{\ominus}$$

| $RCONHCH_2-$ | $-\overset{\oplus}{N}{<}\ \ X^{\ominus}$ | References (Yield, %) |
|---|---|---|
| 1,3-bis(methylene)benzimidazolone ($C_6H_4$, NCH$_2$—, CO, NCH$_2$—) | $-\overset{\oplus}{N}(CH_3)_3\ \ I^{\ominus}$ | 180 (34) |
| | $CH_3\overset{\oplus}{N}{<}(CH_2CH_2)_2CH_2\ \ I^{\ominus}$ | 180 (33) |
| | $CH_3\overset{\oplus}{N}{<}(CH_2CH_2)_2O\ \ I^{\ominus}$ | 180 (27) |
| 1,3-bis(methylene)benzimidazolethione ($C_6H_4$, NCH$_2$—, CS, NCH$_2$—) | $-\overset{\oplus}{N}(CH_3)_3\ \ I^{\ominus}$ | 179 (88) |

*Note:* References 382 to 537 are on pp. 266–269.

## TABLE XXXI

N-Acyl- and N-Sulfonyl-imines

| Imine Derivative | References (Yield, %) |
|---|---|
| *Amides* | |
| $HCON{=}CHC_6H_4OH\text{-}p$ | 297 (60) |
| $CH_3CON{=}CHC_6H_4OH\text{-}p$ | 297 (90) |
| $CH_3CON{=}C(C_6H_5)_2$ | 231 (70) |
| $CH_3CON{=}C[C_6H_4N(CH_3)_2\text{-}p]_2$ | 231, 490 (77) |
| $CH_3CON{=}C(C_{10}H_7\text{-}1)C_6H_5$ | 231 |
| $CH_3CON{=}C(C_{10}H_7\text{-}2)C_6H_5$ | 231 |
| $CH_3CON{=}C(C_{10}H_7\text{-}1)_2$ | 231 |
| $CH_3CON{=}$ 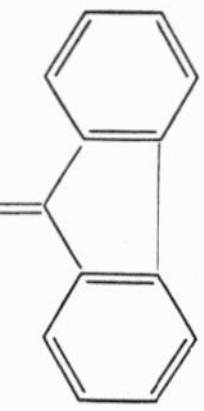 | 231 |
| $CH_3CON{=}$ 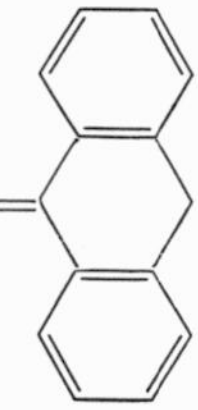 | 231, 491 (59), 492 |
| $CH_2ClCON{=}CHCCl_3$ | 307 |
| $CH_2ClCON{=}CHCCl_3$ | 307 |
| $CH_2ClCON{=}C(C_6H_5)_2$ | 231 |
| $C_6H_5CH_2CON{=}CHC_6H_4OH\text{-}o$ | 299 |
| $C_6H_5CH_2CON{=}CHC_6H_4OH\text{-}m$ | 298 (40) |
| $C_6H_5CH_2CON{=}CHC_6H_4OH\text{-}p$ | 297 (80) |
| $C_2H_5CON{=}CHC_6H_4OH\text{-}o$ | 299 |
| $C_2H_5CON{=}CHC_6H_4OH\text{-}m$ | 298 (27) |
| $C_2H_5CON{=}CHC_6H_4OH\text{-}p$ | 297 (83) |
| $C_2H_5CON{=}C(C_6H_5)_2$ | 231 (71) |
| $C_2H_5CON{=}C(C_6H_5)C_6H_4CH_3\text{-}o$ | 231 |
| $n\text{-}C_3H_7CON{=}C(C_6H_5)_2$ | 231 (65) |
| $n\text{-}C_6H_{13}CON{=}CHC_6H_3OH\text{-}2\text{-}Cl\text{-}5$ | 271 |

*Note:* References 382 to 537 are on pp. 266–269.

TABLE XXXI—*Continued*

N-Acyl- and N-Sulfonyl-imines

| Imine Derivative | References (Yield, %) |
|---|---|
| *Amides (contd.)* | |
| $C_6H_5CON{=}CHC_6H_4OH\text{-}o$ | 299 |
| $C_6H_5CON{=}CHC_6H_4OH\text{-}m$ | 298 (58) |
| $C_6H_5CON{=}CHC_6H_4OH\text{-}p$ | 297 (81) |
| $C_6H_5CON{=}CHC_6H_3OH\text{-}2\text{-}Cl\text{-}5$ | 271 |
| $C_6H_5CON{=}C(C_6H_5)_2$ | 19 (70), 231 (70), 493 (35) |
| $C_6H_5CON{=}C(C_6H_5)C_6H_4OCH_3\text{-}p$ | 19 (48) |
| $C_6H_5CON{=}C[C_6H_4N(CH_3)_2\text{-}p]_2$ | 231 |
| $C_6H_5CON{=}C(C_{10}H_7\text{-}1)C_6H_5$ | 120 (59) |
| $C_6H_5CON{=}$(fluorenylidene, structure) | 493 |
| *Carbamyl Compounds* | |
| $(C_2H_5O_2CN{=}CH)_2$ | 308 |
| $C_2H_5O_2CN{=}C(C_6H_5)_2$ | 231 |
| $C_2H_5O_2CN{=}C(C_6H_5)C_6H_4CH_3\text{-}o$ | 231 (66) |
| $CO(N{=}CHCCl_3)(NHCH(OC_2H_5)CCl_3)$ | 160 |
| $CO(N{=}CHCCl_3)_2$ HCl | 160 |
| $CS(N{=}CHCCl_3)(NHCHOHCCl_3)$ | 160 |
| *Sulfonamides* | |
| $C_6H_5SO_2N{=}CHC_6H_4NO_2\text{-}p$ | 159 (50) |
| $C_6H_5SO_2N{=}CHC_6H_2(NO_2)_3\text{-}2,4,6$ | 159 (65) |
| $C_6H_5SO_2N{=}CHC_6H_3OH\text{-}2\text{-}Cl\text{-}5$ | 271 |

*Note:* References 382 to 537 are on pp. 266–269.

TABLE XXXI—*Continued*

N-ACYL- AND N-SULFONYL-IMINES

| Imine Derivative | References (Yield, %) |
|---|---|
| *Sulfonamides (contd.)* | |
| $C_6H_5SO_2N{=}CHC_6H_2OH$-2-$Cl_2$-3,5 | 271 |
| $C_6H_5SO_2N{=}CHC_6H_4N(CH_3)_2$-*p* | 159 |
| $C_6H_5SO_2N{=}CH$-(2-furyl) | 159 (70) |
| $C_6H_5SO_2N{=}C[C_6H_4N(CH_3)_2$-*p*$]_2$ | 490 (65) |
| *o*-$CH_3C_6H_4SO_2N{=}CHC_6H_4NO_2$-*p* | 159 |
| *o*-$CH_3C_6H_4SO_2N{=}CHC_6H_2(NO_2)_3$-2,4,6 | 159 |
| *o*-$CH_3C_6H_4SO_2N{=}CHC_6H_4N(CH_3)_2$-*p* | 159 |
| *o*-$CH_3C_6H_4SO_2N{=}CH$-(2-furyl) | 159 (50) |
| *p*-$CH_3C_6H_4SO_2N{=}CHC_6H_4NO_2$-*p* | 159 |
| *p*-$CH_3C_6H_4SO_2N{=}CHC_6H_2(NO_2)_3$-2,4,6 | 159 |
| *p*-$CH_3C_6H_4SO_2N{=}CHC_6H_4N(CH_3)_2$-*p* | 159 |
| *p*-$CH_3C_6H_4SO_2N{=}CH$-(2-furyl) | 159 (20) |
| 2-$C_{10}H_7SO_2N{=}CH$-(2-furyl) | 159 (70) |

*Note:* References 382 to 537 are on pp. 266–269.

TABLE XXXII

SYMMETRICAL N,N′-METHYLENE-*bis* DERIVATIVES OF AMIDES, LACTAMS, THIOAMIDES, IMIDES, CARBAMYL COMPOUNDS, AND SULFONAMIDES

| N,N′-Methylene-*bis* Derivative | Method | References (Yield, %) |
|---|---|---|
| *Amides* | | |
| $(HCONH)_2CH_2$ | Amide, $(CH_2O)_3$ | 154 (32), 262, 494 |
| | Amide, formalin, $(CH_3CO)_2O$ | 281 (21) |
| | Amide, hexamine | 495 (53–83) |
| $(CH_3CONH)_2CH_2$ | Amide, formalin, $(CH_3CO)_2O$ | 281 (76) |
| | Amide, formalin, $CH_3CO_2H$ | 154 (16), 279 (54) |
| | Amide, formalin, acid | 246, 465 |
| | Amide, $(CH_2O)_3$ | 494 (good) |
| | Amide, $(CH_2O)_x$ | 250 (62) |
| | Amide, hexamine, base | 495 (good) |
| | RCN, $CH_2O$, HCl, $H_2O$, 300° | 496 |
| | RCN, 33% $CH_2O$, concd. HCl | 497 |
| $(CH_2ClCONH)_2CH_2$ | $RCONHCH_2OH$, acid | 11 (95), 465 |
| $(CHCl_2CONH)_2CH_2$ | Amide, $RCONHCH_2OH$, acid | 465 |
| $(CCl_3CONH)_2CH_2$ | Amide, $RCONHCH_2OH$, acid | 264 |
| | $RCONHCH_2OH$, acid | 264 (95), 465 |
| $(CH_2BrCONH)_2CH_2$ | Amide, $RCONHCH_2OH$, acid | 465 |
| $(CH_2ICONH)_2CH_2$ | $RCONHCH_2OH$, acid | 11, 465 |
| $(NCSCH_2CONH)_2CH_2$ | $(CH_2ClCONH)_2CH_2$, KSCN, $C_2H_5OH$ | 465 |
| $[(C_2H_5)_2NCH_2CONH]_2CH_2$ | $RCONHCH_2OH$, acid | 153, 465 |
| $(C_6H_5CH_2CONHCH_2CONH)_2CH_2$ | $RCONHCH_2OH$, acid | 79 |
| | $RCONHCH_2OR$, acid | 79 |
| | Amide, formalin, acid | 79 |
| $(C_6H_5CONHCH_2CONH)_2CH_2$ | $RCONHCH_2OH$, acid | 79 |
| | Amide, formalin, acid | 79 |

| | | |
|---|---|---|
| $(CH_2{=}CHCH_2CONH)_2CH_2$ | RCN, $(CH_2O)_x$, 75% $H_2SO_4$ | 4 (18) |
| $(CH_3CH{=}CHCH_2CONH)_2CH_2$ | RCN, $(CH_2O)_x$, 90% $H_2SO_4$ | 4 (9) |
| $(C_6H_5CH_2CONH)_2CH_2$ | Amide, formalin, acid | 388 (88) |
| | $RCONHCH_2OR'$, acid | 79 |
| | RCN, $(CH_2O)_x$, concd. $H_2SO_4$ | 4 (30), 261 |
| $(C_2H_5O_2CCH_2CONH)_2CH_2$ | RCN, $(CH_2O)_x$, $C_6H_{12}$, $H_2SO_4$ | 4 (5) |
| $(H_2NCOCH_2CONH)_2CH_2$ | Amide, hexamine, $HCON(CH_3)_2$, heat | 495 (83) |
| $(NCCH_2CONH)_2CH_2$ | Amide, formaldehyde, acid | 465 |
| $(C_2H_5CONH)_2CH_2$ | Amide, $RCONHCH_2OH$, acid | 153 (79) |
| | $RCONHCH_2OH$, heat | 153 (low) |
| | $RCONHCH_2OH$, acid | 153 (40) |
| | Amide, formalin, acid | 153 (75) |
| $(CH_3CHClCONH)_2CH_2$ | Amide, formaldehyde, acid | 465 |
| $(CH_2ClCH_2CONH)_2CH_2$ | RCN, 33% $CH_2O$, concd. HCl | 497 |
| | $CH_2{=}CHCN$, $CH_2O$, HCl, $H_2O$, 400° | 465, 498 |
| $(CH_2BrCH_2CONH)_2CH_2$ | RCN, 37% $CH_2O$, concd. $H_2SO_4$ | 499 (79) |
| $(CH_2ICH_2CONH)_2CH_2$ | Amide, $(CH_2O)_x$, HCl | 499 (91) |
| $(CH_2ClCHClCONH)_2CH_2$ | RCN, 53% $CH_2O$, HCl, 300° | 500 |
| $(CH_2BrCHBrCONH)_2CH_2$ | $RCONHCH_2OH$, acid | 253 (55) |
| | RCN, 35% $CH_2O$, HCl, 380° | 500 |
| $[C_6H_5CH_2CONHCH(CH_3)CONH]_2CH_2$ | $RCONHCH_2OH$, acid | 79 |
| | Amide, formalin, acid | 79 |
| $[HO_2C(CH_2)_2CONH]_2CH_2$ | $RCONHCH_2OH$, acid | 79 |
| $[H_2NCO(CH_2)_2CONH]_2CH_2$ | Amide, $(CH_2O)_x$, acid | 501 |
| | Amide, hexamine, heat | 502 |
| $(n\text{-}C_3H_7CONH)_2CH_2$ | RCN, $(CH_2O)_x$, 85% $H_2SO_4$ | 4 (40) |
| $[CH_2BrCBr(CH_3)CONH]_2CH_2$ | $RCONHCH_2OH$, HBr or $H_2SO_4$ | 253 (94) |
| | $CH_2{=}C(CH_3)CONHCH_2OH$, $Br_2$ | 253 (40) |
| | $CH_2BrCBr(CH_3)CONHCH_2OH$, $Br_2$ | 253 (86) |

*Note:* References 382 to 537 are on pp. 266–269.

TABLE XXXII—*Continued*

SYMMETRICAL N,N′-METHYLENE-*bis* DERIVATIVES OF AMIDES, LACTAMS, THIOAMIDES, IMIDES, CARBAMYL COMPOUNDS, AND SULFONAMIDES

| N,N′-Methylene-*bis* Derivative | Method | References (Yield, %) |
|---|---|---|
| *Amides* (*contd.*) | | |
| $(i\text{-}C_4H_9CONH)_2CH_2$ | Amide, $RCONHCH_2OH$, acid | 419 (85) |
| | $RCONHCH_2OH$, heat | 419 (73) |
| $[(C_2H_5)_2CHCONH]_2CH_2$ | Amide, $RCONHCH_2OH$, acid | 153 (67) |
| $(n\text{-}C_7H_{15}CONH)_2CH_2$ | RCN, $(CH_2O)_x$, acid | 503 (good) |
| $(n\text{-}C_{11}H_{23}CONH)_2CH_2$ | Amide, formalin, acid | 217 (70), 222 **(57)** |
| | RCOCl, hexamine | 504 (20) |
| | RCN, $(CH_2O)_x$, $BF_3$, $CH_3CO_2H$ | 503 (good) |
| $[CF_3(CH_2)_{11}CONH]_2CH_2$ | $RCONHCH_2OH$, acid, $CH_3COC_2H_5$ | 217 (poor) |
| $(n\text{-}C_{15}H_{31}CONH)_2CH_2$ | RCN, $(CH_2O)_x$, 90 % $H_2SO_4$ | 503 (good) |
| $(n\text{-}C_{17}H_{35}CONH)_2CH_2$ | RCN, $(CH_2O)_x$, 90 % $H_2SO_4$ | 503 (100) |
| | $(RCONHCH_2)_2O$, acid | 217 (100) |
| | $RCONHCH_2N^{\oplus}$(pyridinium) $Cl^{\ominus}$, acid | 243 (99) |
| $(n\text{-}C_{21}H_{43}CONH)_2CH_2$ | $RCONHCH_2OH$, acid | 235 |
| $(CH_2{=}CHCONH)_2CH_2$ | RCN, $(CH_2O)_x$, 85% $H_2SO_4$ | 4 (86) |
| | Amide, $(CH_2O)_x$, acid | 254 (89), 505 (50) |
| | $RCONHCH_2OH$, acid | 254 (67) |
| $(CH_2{=}CHCH_2CONH)_2CH_2$ | RCN, formalin, concd. $H_2SO_4$ | 499 (58) |
| $[CH_2{=}C(CH_3)CONH]_2CH_2$ | Amide, $RCONHCH_2OH$, acid | 254 (80) |
| | $(RCONHCH_2)_2O$, acid, 60° | 424 (81) |
| | $RCONHCH_2OH$, acid | 254 (67) |
| | Amide, $(CHO_2)_x$, acid | 254 (70), 505 (81) |
| $[(CH_3)_2C{=}CHCONH]_2CH_2$ | $RCONHCH_2OH$, HCl, $CCl_4$ | 422 (65) |

| Product | Reagents | Refs. (Yield %) |
|---|---|---|
| $(C_6H_5CONH)_2CH_2$ | $RCONHCH_2OH$, acid | 10, 79, 132 |
| | $RCONHCH_2OR'$, heat | 79 |
| | $RCONHCH_2OR'$, acid | 79 |
| | Amide, formalin, acid | 10 (100), 246, 388 |
| | Amide, hexamine, base | 495 (good) |
| | RCN, $(CH_2O)_x$, 85–98% $H_2SO_4$ | 4 (90), 245, 506, 507 |
| | Amide, $RCONHCH_2SO_3Na$, $NaOC_2H_5$, 200° | 508 (84) |
| | RCOCl, hexamine, base | 509 |
| | $(C_6H_5CONH)_2Hg$, $(CH_2S)_3$ | 246 |
| $(o\text{-}ClC_6H_4CONH)_2CH_2$ | RCN, 33% $CH_2O$, concd. HBr | 465, 497 |
| $(p\text{-}ClC_6H_4CONH)_2CH_2$ | RCN, $(CH_2O)_x$, $H_2SO_4$, $CHCl_3$ | 4 (75) |
| $(o\text{-}O_2NC_6H_4CONH)_2CH_2$ | $(RCONHCH_2)_2O$, heat | 221 |
| $(m\text{-}O_2NC_6H_4CONH)_2CH_2$ | Amide, formalin, acid | 221 |
| $(p\text{-}O_2NC_6H_4CONH)_2CH_2$ | Amide, formalin, acid | 221 |
| $(o\text{-}HOC_6H_4CONH)_2CH_2$ | Amide, $RCONHCH_2OH$, acid | 25 (100) |
| $(p\text{-}CH_3OC_6H_4CONH)_2CH_2$ | $RCONHCH_2OH$, acid | 132 (100) |
| $(p\text{-}CH_3C_6H_4CONH)_2CH_2$ | RCN, $(CH_2O)_x$, 85% $H_2SO_4$ | 4 (83) |
| $(p\text{-}CH_3O_2CC_6H_4CONH)_2CH_2$ | RCN, $(CH_2O)_x$, $H_2SO_4$, $CHCl_3$ | 4 (50), 510 (84) |
| $(p\text{-}HO_2CC_6H_4CONH)_2CH_2$ | Methyl ester, 5% NaOH, 100° | 510 (95) |
| [4-pyridyl-CONH]$_2CH_2$ | $RCONHCH_2OH$, 200° | 430 |
| *Lactams, Thioamides, and Imides* | | |
| [$CH_2CO$–N–$CH_2CH_2$ (2-pyrrolidinon-1-yl)]$_2CH_2$ | Lactam, formalin, acid | 247, 248 |

*Note:* References 382 to 537 are on pp. 266–269.

TABLE XXXII—*Continued*

SYMMETRICAL N,N′-METHYLENE-*bis* DERIVATIVES OF AMIDES, LACTAMS, THIOAMIDES, IMIDES, CARBAMYL COMPOUNDS, AND SULFONAMIDES

| N,N′-Methylene-*bis* Derivative | Method | References (Yield, %) |
|---|---|---|
| *Lactams, Thioamides, and Imides (contd.)* | | |
| $(CH_3CSNH)_2CH_2$ | $(CH_3CONH)_2CH_2$, $P_2S_5$ | 246, 465 |
| $[o\text{-}C_6H_4(CO)_2N]_2CH_2$ | $o\text{-}C_6H_4(CN)_2$, $(CH_2O)_x$, $H_2SO_4$, $CHCl_3$ | 4 (20) |
| | $o\text{-}C_6H_4(CO)_2NK$, $o\text{-}C_6H_4(CO)_2NCH_2\overset{\oplus}{N}(CH_3)_3\ I^{\ominus}$ | 81 (48) |
| | $o\text{-}C_6H_4(CO)_2NK$, $CH_2I_2$ | 256 |
| | $o\text{-}C_6H_4(CO)_2NH$, hexamine, heat | 502 |
| | $o\text{-}C_6H_4(CO)_2NCH_2N(CH_3)_2$, 100° | 220 (low) |
| $[C_6H_4(CO)(SO_2)N]_2CH_2$ (saccharin ring) | Imide, formalin, acid | 511 (low) |
| *Carbamyl Compounds* | | |
| $(C_2H_5O_2CNH)_2CH_2$ | $C_2H_5O_2CNHCH_2OH$, acid | 153 |
| | $C_2H_5O_2CNH_2$, formalin, acid | 250 (85), 251 |
| $(CH_3O_2CNCH_3)_2CH_2$ | $CH_3NCO$, $CH_2(OCH_3)_2$ | 258 |
| $(C_2H_5O_2CNCH_3)_2CH_2$ | $C_2H_5O_2CNHCH_3$, formalin, acid | 250 (fair), 251 |
| $(C_6H_5CH_2O_2CNCH_2CO_2H)_2CH_2$ | $RO_2CNHCH_2CO_2C_2H_5$, $(CH_2O)_x$, acid, $C_6H_6$ | 344 (65) |
| $[n\text{-}C_4H_9OC({=}S)NCH_3]_2CH_2$ | $CH_3NCO$, $CH_2(SC_4H_9\text{-}n)_2$ | 258 |
| $(H_2NCONH)_2CH_2$ | $H_2NCONHCH_2OH$, acid | 146, 147 |
| | $CO(NH_2)_2$, formalin, acid | 146, 147, 219 (17) |
| $(CH_3NHCONH)_2CH_2$ | $CH_3NHCONH_2$, formalin, acid | 219 (74), 483 |
| $(C_2H_5NHCONH)_2CH_2$ | $C_2H_5NHCONH_2$, formalin, acid | 153 |

| Product | Reactants | Ref. (Yield %) |
|---|---|---|
| $(CH_3NHCONCH_3)_2CH_2$ | $CO(NHCH_3)_2$, formalin, acid | 153 |
| $[(CH_3)_2NCONH]_2CH_2$ | $(CH_3)_2NCONH_2$, formalin, acid | 153 |
| $(C_6H_5NHCONH)_2CH_2$ | $(H_2NCONH)_2CH_2$, $C_6H_5NHCONH_2$ | 475 (60) |
| $(p\text{-}CH_3C_6H_4NHCONH)_2CH_2$ | $p\text{-}CH_3C_6H_4NHCONH_2$, formalin, acid | 55 |
| [(CHCO=CHCO)NCONH]$_2CH_2$ | $NCONH_2$, $(CH_2O)_x$, concd. $H_2SO_4$ | 511*a* (22) |
| $(H_2NCSNH)_2CH_2$ | $CS(NH_2)_2$, formalin, acid | 265 |
| $(CH_3NHCSNCH_3)_2CH_2$ | $CS(NHCH_3)_2$, formalin, acid | 512 |
| [CONH–CO–$CH_2N$—]$_2CH_2$ | Hydantoin, formalin, acid | 185 (42) |
| | Hydantoin, $(CH_2O)_x$, concd. HCl | 252 (87) |
| | Hydantoin, formalin, $H_2SO_4$, $CH_3CO_2H$ | 290 (97) |
| [CONH–CO–$n\text{-}C_3H_7CHN$—]$_2CH_2$ | *n*-Propylhydantoin, formalin, $ZnCl_2$, concd. HCl | 289 (54) |
| [CONH–CO–$i\text{-}C_3H_7CHN$—]$_2CH_2$ | *i*-Propylhydantoin, aq. $CH_2O$, $ZnCl_2$, concd. HCl | 289 (64) |
| | *i*-Propylhydantoin, formalin, $H_2SO_4$, $CH_3CO_2H$ | 290 (60) |

*Note:* References 382 to 537 are on pp. 266–269.

TABLE XXXII—*Continued*

SYMMETRICAL N,N′-METHYLENE-*bis* DERIVATIVES OF AMIDES, LACTAMS, THIOAMIDES, IMIDES, CARBAMYL COMPOUNDS, AND SULFONAMIDES

| N,N′-Methylene-*bis* Derivative | Method | References (Yield, %) |
|---|---|---|
| *Carbamyl Compounds (contd.)* | | |
| $[(CH_3)_2C\text{—}N\text{—}$ (ring: CONH, CO) $]_2CH_2$ | Dimethylhydantoin, aq. $CH_2O$, HCl | 252 (96) |
| | Dimethylhydantoin, formalin, $ZnCl_2$, concd. HCl | 289 (81) |
| $[(CH_3)_2C\text{—}N\text{—}$ (ring: CONCl, CO) $]_2CH_2$ | Methylenebisdimethylhydantoin, $Cl_2$ | 288 (92) |
| $[C_2H_5C(CH_3)N\text{—}$ (ring: OC—NH, CO) $]_2CH_2$ | Methylethylhydantoin, formalin, $ZnCl_2$, concd. HCl | 289 (82) |
| $[i\text{-}C_4H_9C(CH_3)N\text{—}$ (ring: OC—NH, CO) $]_2CH_2$ | Methyl-*i*-butylhydantoin, formalin, $ZnCl_2$, concd. HCl | 289 (67) |

| | | |
|---|---|---|
| $[C_6H_4(N\text{—})(O\text{-}CO)]_2CH_2$ | 3-Chloromethyl-2-benzoxazolone, 2-benzoxazolone | 194 (60) |
| $[C_6H_4(N\text{—})(S\text{-}CS)]_2CH_2$ | $C_6H_4(S\text{-}CS)NCH_2S\text{-}C(=N)S\text{-}C_6H_4$ | 318 (100) |
| | $I_2$, 200° | |
| *Sulfonamides* | | |
| $(CH_2ClSO_2NH)_2CH_2$ | Amide, formaldehyde, acid | 465 |
| $(C_6H_5SO_2NH)_2CH_2$ | Amide, formalin, $K_2CO_3$ | 85 (70) |
| | $C_6H_5SO_2NHCH_2OH$, $C_6H_6$, heat | 156 |
| $(o\text{-}CH_3C_6H_4SO_2NH)_2CH_2$ | Amide, formalin, $K_2CO_3$ | 156 |
| $(p\text{-}CH_3C_6H_4SO_2NH)_2CH_2$ | Amide, formalin, $K_2CO_3$ | 85 (80) |
| | $ArSO_2NHCH_2OH$, $C_6H_6$, heat | 156 |
| $(p\text{-}CH_3C_6H_4SO_2NCH_3)_2CH_2$ | Amide, $(CH_2O)_x$, HCl | 156 (47) |
| $(p\text{-}CH_3C_6H_4SO_2NC_6H_5)_2CH_2$ | Amide, $(CH_2O)_x$, HCl | 156 |

*Note:* References 382 to 537 are on pp. 266–269.

## TABLE XXXIII

UNSYMMETRICAL N,N′-METHYLENE-*bis* DERIVATIVES OF AMIDES, LACTAMS, IMIDES, CARBAMYL COMPOUNDS, SULFONAMIDES, AND SULFAMYL COMPOUNDS

| N,N′-Methylene-*bis* Derivative | Method | References (Yield, %) |
|---|---|---|
| *Amides* | | |
| $HO_2CCH_2CONH$–$CH_2$–$NHCOCH_2Cl$ ($HO_2CCH_2CONH$, $CH_2ClCONH$ joined by $CH_2$) | $CH_2ClCONHCH_2OH$, $CH_2(CO_2H)CONH_2$, acid | 465 |
| $(C_2H_5)_2NCH_2CONH$–$CH_2$–$NHCOCH(C_2H_5)_2$ ($(C_2H_5)_2NCH_2CONH$, $(C_2H_5)_2CHCONH$ joined by $CH_2$) | $RCONHCH_2OH$, $(C_2H_5)_2NCH_2CONH_2$, concd. $H_2SO_4$ | 153 |
| $n$-$C_{17}H_{35}CONH$–$CH_2$–$NHCO(CH_2)_2S$-$n$-$C_{14}H_{29}$ ($n$-$C_{17}H_{35}CONH$, $n$-$C_{14}H_{29}S(CH_2)_2CONH$ joined by $CH_2$) | $CH_2{=}CHCONH$–$CH_2$–$NHCO$-$n$-$C_{17}H_{35}$ ($CH_2{=}CHCONH$, $n$-$C_{17}H_{35}CONH$ joined by $CH_2$), $n$-$C_{14}H_{29}SH$ | 217 |
| $n$-$C_{17}H_{35}CONH$–$CH_2$–$NHCOCH_2SO_3Na$ ($n$-$C_{17}H_{35}CONH$, $NaO_3SCH_2CONH$ joined by $CH_2$) | $CH_2(SO_3Na)CONHCH_2OH$, $RCONH_2$, $CH_3CO_2H$, 100° | 249 |

| | | |
|---|---|---|
| $n\text{-}C_{17}H_{35}CONCH_3$<br>\\<br>$CH_2$<br>/<br>$NaO_3SCH_2CONH$ | $CH_2(SO_3Na)CONHCH_2OH$, $RCONHCH_3$, $CH_3CO_2H$, 100° | 249 |
| $CH_2{=}CHCONH$<br>\\<br>$CH_2$<br>/<br>$CH_3CONH$ | $CH_3CONHCH_2OH$, $CH_2{=}CHCONH_2$, acid, 30–50° | 217 (69) |
| $CH_2{=}CHCONH$<br>\\<br>$CH_2$<br>/<br>$n\text{-}C_5H_{11}CONH$ | $RCONHCH_2OH$, $CH_2{=}CHCONH_2$, acid, 50–55° | 217 (85) |
| $CH_2{=}CHCONH$<br>\\<br>$CH_2$<br>/<br>$n\text{-}C_7H_{15}CONH$ | $RCONHCH_2OH$, $CH_2{=}CHCONH_2$, acid, 50–55° | 217 (81) |
| $CH_2{=}CHCONH$<br>\\<br>$CH_2$<br>/<br>$n\text{-}C_{11}H_{23}CONH$ | $RCONHCH_2OH$, $CH_2{=}CHCONH_2$, acid, 50–55° | 217 (85) |
| $CH_2{=}CHCONH$<br>\\<br>$CH_2$<br>/<br>$n\text{-}C_{17}H_{35}CONH$ | $RCONHCH_2OH$, $CH_2{=}CHCONH_2$, acid, 50–55° | 217 (87) |

*Note:* References 382 to 537 are on pp. 266–269.

TABLE XXXIII—*Continued*

UNSYMMETRICAL N,N′-METHYLENE-*bis* DERIVATIVES OF AMIDES, LACTAMS, IMIDES, CARBAMYL COMPOUNDS, SULFONAMIDES, AND SULFAMYL COMPOUNDS

| N,N′-Methylene-*bis* Derivative | Method | References (Yield, %) |
|---|---|---|
| *Amides (contd.)* | | |
| $CH_2{=}C(CH_3)CONH$<br>$CH_2$<br>$n\text{-}C_{17}H_{35}CONH$ | $RCONHCH_2OH$, $CH_2{=}C(CH_3)CONH_2$, acid, 50–55° | 217 (81) |
| $CH_2{=}C(CH_3)CONH$<br>$CH_2$<br>$CH_2{=}CHCONH$ | $RCONHCH_2OH$, $CH_2{=}C(CH_3)CONH_2$, acid, 80° | 513 (63) |
| $C_6H_5CONH$<br>$CH_2$<br>$CH_3CONH$ | $(C_6H_5CONH)_2CH_2$, $(CH_3CONH)_2CH_2$, 270° | 514 (5) |
| $C_6H_5CONH$<br>$CH_2$<br>$CH_2ClCONH$ | $CH_2ClCONHCH_2OH$, $C_6H_5CONH_2$, acid | 465 |
| $C_6H_5CONH$<br>$CH_2$<br>$C_2H_5O_2C(CH_2)_4CH(CO_2C_2H_5)CONH$ | $C_6H_5CONHCH_2OH$, concd. $H_2SO_4$, $C_2H_5O_2C(CH_2)_4CH(CO_2C_2H_5)CN$ | 84, 119 (91) |

| Product | Reactants and Conditions | Refs. (Yield, %) |
|---|---|---|
| $C_6H_5CONH\text{—}CH_2\text{—}NHCO\text{-}n\text{-}C_{17}H_{35}$ ($C_6H_5CONH$, $CH_2$, $n\text{-}C_{17}H_{35}CONH$) | $RCONHCH_2OH$, $C_6H_5CONH_2$, HCl, dioxane | 249 |
| $C_6H_5CONH\text{—}CH_2\text{—}NHCOCH{=}CH_2$ ($C_6H_5CONH$, $CH_2$, $CH_2{=}CHCONH$) | $C_6H_5CONHCH_2OH$, $CH_2{=}CHCN$, 96 % $H_2SO_4$, 30° | 255 (60) |
| $C_6H_5CONH\text{—}CH_2\text{—}NHCOC_6H_4OH\text{-}o$ ($C_6H_5CONH$, $CH_2$, $o\text{-}HOC_6H_4CONH$) | $ArCONHCH_2OH$, $C_6H_5CONH_2$, acid | 25 (88) |
| *Lactams and Imides* | | |
| (pyrrolidinone ring, C=O) $NCH_2NHCOC(CH_3){=}CH_2$ | Lactam, $RCONHCH_2OCH_3$, $NaOC_2H_5$, 130–170° | 477 |
| (caprolactam ring, C=O) $NCH_2NHCOC(CH_3){=}CH_2$ | Lactam, $RCONHCH_2OCH_3$, $NaOC_2H_5$, 130–170° | 477 |

*Note:* References 382 to 537 are on pp. 266–269.

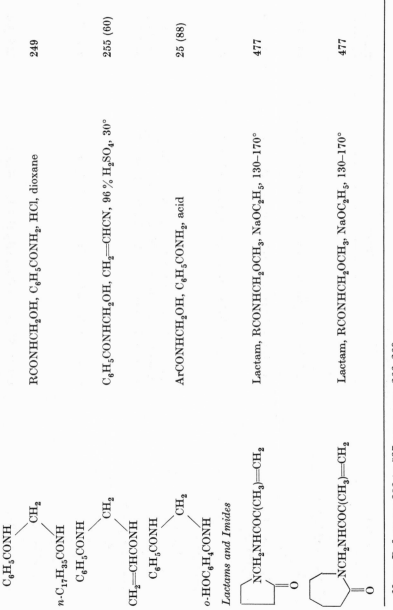

## TABLE XXXIII—*Continued*

UNSYMMETRICAL N,N′-METHYLENE-*bis* DERIVATIVES OF AMIDES, LACTAMS, IMIDES, CARBAMYL COMPOUNDS, SULFONAMIDES, AND SULFAMYL COMPOUNDS

| N,N′-Methylene-*bis* Derivative | Method | References (Yield, %) |
|---|---|---|
| *Lactams and Imides (contd.)* | | |
| $o$-$C_6H_4$(CO)(CO)N$CH_2NHCOC_6H_5$ (cyclic: benzene ring with NCH₂NHCOC₆H₅, CO, CO) | $C_6H_5CONHCH_2N(CH_3)_2$, isatin, NaOH, $C_6H_5CH_3$, heat | 257 (75) |
| $o$-$C_6H_4(CO)_2NCH_2NHCOCH_3$ | $o$-$C_6H_4(CO)_2NCH_2OH$, $CH_3CN$, concd. $H_2SO_4$, 80° | 84 (83) |
| $o$-$C_6H_4(CO)_2N$—$CH_2$—$NHCOCH_2CO_2H$ ($o$-$C_6H_4(CO)_2N$ / $CH_2$ / $HO_2CCH_2CONH$) | $o$-$C_6H_4(CO)_2NCH_2OH$, $NCCH_2CO_2H$, concd. $H_2SO_4$, 45° | 84 (91) |
| $o$-$C_6H_4(CO)_2NCH_2NHCOC(CH_3){=}CH_2$ | Imide, $RCONHCH_2OCH_3$, $NaOC_2H_5$, 130–170° | 477 (92) |
| $o$-$C_6H_4(CO)_2NCH_2NHCOC_6H_5$ | Imide, $C_6H_5CONHCH_2SO_3Na$, $NaOC_2H_5$, 200° | 508 (51) |
| | Imide, $C_6H_5CONHCH_2N(CH_3)_2$, NaOH, $C_6H_5CH_3$ | 257 (68) |
| | $o$-$C_6H_4(CO)_2NCH_2OH$, $C_6H_5CN$, concd. $H_2SO_4$, 80° | 84 (77) |
| *Carbamyl Compounds* | | |
| $C_2H_5O_2CNH$—$CH_2$—$NHCOCH_2Cl$ ($C_2H_5O_2CNH$ / $CH_2$ / $CH_2ClCONH$) | $CH_2ClCONHCH_2CONHNH_2$, $HNO_2$, $C_2H_5OH$ | 259 (37), 465 |
| $C_2H_5O_2CNH$—$CH_2$—$NHCOCH_2Br$ ($C_2H_5O_2CNH$ / $CH_2$ / $CH_2BrCONH$) | $CH_2BrCONHCH_2CONHNH_2$, $HNO_2$, $C_2H_5OH$ | 259 (35) |

| Product | Reagents | Ref. (Yield, %) |
|---|---|---|
| $C_2H_5O_2CNH$—$CH_2$—$CH_2ICONH$ | $CH_2ICONHCH_2CON_3$, $C_2H_5OH$ | 259 (94) |
| CO——$NCH_2NHCO_2C_2H_5$; CO; $C_6H_5C(C_2H_5)NH$ (ring) | CO——$NCH_2CON_3$; CO, $C_2H_5OH$; $C_6H_5C(C_2H_5)NH$ (ring) | 187 (70) |
| $NCH_2N(C_6H_5)COCH_3$; O—CS (DL-*trans*) | $NCH_2NHAr$; O—CS, $CH_3COCl$, pyridine | 260 (82) |
| $NCH_2N(C_6H_4OCH_3\text{-}p)COCH_3$; O—CS (DL-*trans*) | $NCH_2NHAr$; O—CS, $CH_3COCl$, pyridine | 260 (87) |
| $NCH_2N(C_6H_4OC_2H_5\text{-}p)COCH_3$; O—CS (DL-*trans*) | $NCH_2NHAr$; O—CS, $CH_3COCl$, pyridine | 260 (86) |
| $NCH_2N(C_6H_4CH_3\text{-}p)COCH_3$; O—CS (DL-*trans*) | $NCH_2NHAr$; O—CS, $CH_3COCl$, pyridine | 260 (78) |
| N—$CH_2N$ (CO, $O_2S$); O—CS | $NCH_2Cl$; O—CS, saccharin | 178 (48) |

*Note:* References 382 to 537 are on pp. 266–269.

# TABLE XXXIII—*Continued*

## Unsymmetrical N,N′-Methylene-*bis* Derivatives of Amides, Lactams, Imides, Carbamyl Compounds, Sulfonamides, and Sulfamyl Compounds

| N,N′-Methylene-*bis* Derivative | Method | References (Yield, %) |
|---|---|---|
| *Carbamyl Compounds (contd.)* | | |
| N—CH$_2$—N (O—CS; OC—O) | NCH$_2$Cl (O—CS), $\overset{\ominus}{\text{N}}$ Na$^{\oplus}$ (O—CO) | 178 (60) |
| *Sulfonamides and Sulfamyl Compounds* | | |
| $C_6H_5SO_2NH$—CH$_2$—$C_6H_5CONH$ | $C_6H_5SO_2NHCH_2OH$, $C_6H_5CONH_2$, xylene | 82 (11) |
| | $C_6H_5CONHCH_2OH$, $C_6H_5SO_2NH_2$, NaOH, $C_6H_5CH_3$ | 257 (86) |
| $p$-$CH_3C_6H_4SO_2NCH_3$—CH$_2$—$C_6H_5CONH$ | $ArSO_2NHCH_3$, $C_6H_5CONHCH_2N(CH_3)_2$, NaOH, $C_6H_5CH_3$ | 257 (70) |
| $HO_3SNH$—CH$_2$—$C_6H_5CONH$ | $C_6H_5CONH_2$, $H_2NSO_3H$, $CH_2O$, 99% $H_2SO_4$, 70° | 515 (20) |

*Note:* References 382 to 537 are on pp. 266–269.

## TABLE XXXIV

Symmetrical N,N′-Alkylidene-*bis* and N,N′-Arylidene-*bis* Derivatives of Amides, Lactams, Imides, Carbamyl Compounds, and Sulfonamides

$$2RCONH_2 + R'CHO \rightarrow (RCONH)_2CHR'$$

| Derivative of | Method* | Yield, % | References |
|---|---|---|---|
| *Aliphatic Amides* | | | |
| Formamide and | | | |
| Acetaldehyde | $(CH_3CO)_2O$, heat | 4 | 281 |
| β-Phenylpropionaldehyde | No solvent, heat | 0 | 516 |
| Heptanal | Heat alone, or in $(CH_3CO)_2O$, or in pyridine | 0 | 517 |
| Cinnamaldehyde | $(CH_3CO)_2O$, heat | 28 | 281 |
| Benzaldehyde | $(CH_3CO)_2O$, heat | 56 | 281 |
| | No solvent, heat | Poor | 287, 518 |
| *o*-Chlorobenzaldehyde | No solvent, heat | Poor | 518 |
| *o*-Nitrobenzaldehyde | Pyridine | 40 | 274 |
| | No solvent, heat | 40 | 274 |
| *m*-Nitrobenzaldehyde | No solvent, heat | 67 | 269 |
| *p*-Nitrobenzaldehyde | $H_2O$, acid | 68 | 269 |
| 3,5-Dichlorosalicylaldehyde | No solvent, heat | — | 271 |
| *o*-Anisaldehyde | No solvent, heat | 0 | 296*a* |
| *m*-Anisaldehyde | No solvent, heat | 0 | 296*a* |
| *p*-Anisaldehyde | $(CH_3CO)_2O$, heat | 25 | 281 |
| | No solvent, heat | 0 | 296*a* |
| Veratraldehyde | $(CH_3CO)_2O$, heat | 20 | 281 |
| O-Acetylvanillin | $(CH_3CO)_2O$, heat | 21 | 281 |
| Piperonal | No solvent, heat | 0 | 519 |
| 6-Nitropiperonal | No solvent, heat | 0 | 520 |
| *p*-$(CH_3)_2NC_6H_4CHO$ | $(CH_3CO)_2O$, heat | 18 | 281 |
| *m*-Tolualdehyde | Pyridine | 48 | 273 |

*Note:* References 382 to 537 are on pp. 266–269.

* Except when otherwise indicated, the amide and aldehyde are unnamed reactants in this column.

TABLE XXXIV—*Continued*

SYMMETRICAL N,N′-ALKYLIDENE-*bis* AND N,N′-ARYLIDENE-*bis* DERIVATIVES OF AMIDES, LACTAMS, IMIDES, CARBAMYL COMPOUNDS, AND SULFONAMIDES

$2RCONH_2 + R'CHO \rightarrow (RCONH)_2CHR'$

| Derivative of | Method* | Yield, % | References |
|---|---|---|---|
| *Aliphatic Amides (contd.)* | | | |
| Formamide (*contd.*) and | | | |
| *p*-Tolualdehyde | Pyridine | <50 | 276 |
| | No solvent, heat | 0 | 276 |
| 1-Naphthaldehyde | $(CH_3CO)_2O$, heat | 51 | 282 |
| Acetamide and | | | |
| Acetaldehyde | $(CH_3CO)_2O$, heat | 66 | 281 |
| | $CH_3CO_2H$, heat | 38, 44 | 154, 279 |
| | No solvent, heat | — | 521 |
| | $C_2H_5OCH{=}CH_2$, heat, acid | 82 | 233 |
| | $CH_3CHO$, $CH_3CN$, HCl, $H_2O$, 300[c] | — | 496 |
| Chloral | $CCl_3CHO$, $CH_3CN$, concd. $H_2SO_4$ | — | 261, 295 |
| Propionaldehyde | $(CH_3CO)_2O$, heat | 62 | 281 |
| | $CH_3CO_2H$, heat | 35, 7 | 154, 279 |
| | $C_6H_6$, heat | 48 | 283 |
| | Pyridine, heat | 11 | 452 |
| *β*-Phenylpropionaldehyde | Pyridine | 9 | 516 |
| | No solvent, heat | 26 | 516 |
| *n*-Butyraldehyde | $(CH_3CO)_2O$, heat | 76 | 281 |
| | $CH_3CO_2H$, heat | 11 | 279 |
| | $C_6H_6$, heat | 49 | 283 |
| Isovaleraldehyde | $(CH_3CO)_2O$, heat | 47 | 281 |
| | $CH_3CO_2H$, heat | 26 | 279 |
| Heptanal | $CH_3CO_2H$, heat | 6 | 279 |
| | No solvent, heat | — | 517 |
| Cyclohexanecarboxaldehyde | $(CH_3CO)_2O$, heat | 54 | 280 |

| | | | |
|---|---|---|---|
| Cinnamaldehyde | $(CH_3CO)_2O$, heat | 47 | 281 |
| | $C_6H_6$, heat | 33 | 283 |
| | Pyridine | 15 | 275 |
| | No solvent, heat | 52 | 275 |
| Benzaldehyde | $(CH_3CO)_2O$, $H_2SO_4$, heat | 40 | 85 |
| | $(CH_3CO)_2O$, heat | 71 | 281 |
| | $CH_3CO_2H$, heat | 17 | 154 |
| | $CH_3CO_2H$, heat | 48 | 279 |
| | $C_6H_6$, heat | 72 | 283 |
| | No solvent, heat | 49 | 284 |
| | No solvent, heat | — | 287, 518, 522, 523 |
| *o*-Chlorobenzaldehyde | No solvent, heat | Good | 518 |
| *o*-Nitrobenzaldehyde | $(CH_3CO)_2O$, heat | 67 | 281 |
| | $H_2O$, acid | — | 278 |
| | Pyridine | 48 | 274 |
| | No solvent, heat | 48 | 274 |
| *m*-Nitrobenzaldehyde | $(CH_3CO)_2O$, heat | 74 | 281 |
| | $C_6H_6$, heat | 44 | 283 |
| | $H_2O$, acid | — | 278 |
| | No solvent, heat | 60 | 269 |
| *p*-Nitrobenzaldehyde | $(CH_3CO)_2O$, heat | 61 | 281 |
| | No solvent, heat | 54 | 269 |
| Salicylaldehyde | $C_6H_6$, heat | Poor | 283 |
| 5-Chlorosalicylaldehyde | No solvent, heat | 53 | 271 |
| 3,5-Dichlorosalicylaldehyde | No solvent, heat | — | 271 |
| *o*-Anisaldehyde | No solvent, heat | 40–60 | 296*a* |
| *m*-Anisaldehyde | No solvent, heat | 40–60 | 296*a* |
| *p*-Anisaldehyde | $C_6H_6$, heat | 84 | 283 |
| | No solvent, heat | 40–60 | 296*a*, 524 |

*Note:* References 382 to 537 are on pp. 266–269.

* Except when otherwise indicated, the amide and aldehyde are unnamed reactants in this column.

TABLE XXXIV—*Continued*

SYMMETRICAL N,N′-ALKYLIDENE-*bis* AND N,N′-ARYLIDENE-*bis* DERIVATIVES OF AMIDES, LACTAMS, IMIDES, CARBAMYL COMPOUNDS, AND SULFONAMIDES

$$2RCONH_2 + R'CHO \rightarrow (RCONH)_2CHR'$$

| Derivative of | Method* | Yield, % | References |
|---|---|---|---|
| *Aliphatic Amides* (*contd.*) | | | |
| Acetamide (*contd.*) and | | | |
| Veratraldehyde | $(CH_3CO)_2O$, heat | 52 | 112, 281 |
| O-Acetylvanillin | $(CH_3CO)_2O$, heat | 64 | 112, 281 |
| Piperonal | $(CH_3CO)_2O$, heat | 43 | 281 |
| | $C_6H_6$, heat | 43 | 283 |
| | No solvent, heat | 42 | 519 |
| 6-Bromopiperonal | No solvent, heat | 29 | 525 |
| 6-Nitropiperonal | No solvent, heat | 61 | 520 |
| *p*-$(CH_3)_2NC_6H_4CHO$ | $(CH_3CO)_2O$, heat | 30 | 281 |
| *m*-Tolualdehyde | Pyridine | 43 | 273 |
| *p*-Tolualdehyde | $(CH_3CO)_2O$, heat | 65 | 281 |
| | Pyridine or heat alone | $<50$ | 276 |
| Cumaldehyde | $(CH_3CO)_2O$, heat | 51 | 281 |
| 1-Naphthaldehyde | $(CH_3CO)_2O$, heat | 63 | 282 |
| 2-Naphthaldehyde | $(CH_3CO)_2O$, heat | 36 | 282 |
| 9-Anthraldehyde | $(CH_3CO)_2O$, heat | 65 | 282 |
| 1-Pyrenecarboxaldehyde | $(CH_3CO)_2O$, heat | 80 | 282 |
| 2-Furaldehyde | $(CH_3CO)_2O$, heat | 69 | 281 |
| | $C_6H_6$, heat | 67 | 283 |
| | No solvent, heat | 51 | 284 |
| 2-Thiophenecarboxaldehyde | $(CH_3CO)_2O$, heat | 74 | 280, 282 |
| 2-Pyrrolecarboxaldehyde | $(CH_3CO)_2O$, heat | 0 | 280 |
| Picolinaldehyde | No solvent, heat | 0 | 280 |
| Nicotinaldehyde | No solvent, heat | 68 | 280 |
| Isonicotinaldehyde | No solvent, heat | 0 | 280 |
| 3-Indolecarboxaldehyde | $(CH_3CO)_2O$, heat | 71 | 280 |

| | | | |
|---|---|---|---|
| $HO_2C(CH_2)_2CHO$ | $HO_2C(CH_2)_2COCO_2H$, heat ($-CO_2$) | 25 | 292 |
| Pyruvic acid | Heat in vacuum or $CH_3CN$, $CH_3COCO_2H$, $H_2SO_4$ | — | 291 |
| Benzoylformic acid | Heat in vacuum | 55 | 292 |
| Acetanilide and | | | |
| *p*-Tolualdehyde | Pyridine or heat alone | <50 | 276 |
| Fluoroacetamide and | | | |
| *o*-Chlorobenzaldehyde | $(CH_3CO)_2O$, heat | — | 465 |
| Chloroacetamide and | | | |
| Acetaldehyde | No solvent, $H_2SO_4$, heat | — | 465 |
| Phenylacetaldehyde | $CH_3CO_2H$, $H_2SO_4$ | — | 465 |
| Propionaldehyde | $CH_3CO_2H$, heat | — | 465 |
| $CH_3SCH_2CH_2CHO$ | $CH_3CO_2H$, $H_2SO_4$ | — | 465 |
| Butyraldehyde | $CH_3CO_2H$, heat | — | 465 |
| Isobutyraldehyde | $CH_3CO_2H$, $H_2SO_4$ | — | 465 |
| Valeraldehyde | $CH_3CO_2H$, $H_2SO_4$ | — | 465 |
| Heptanal | $CH_3CO_2H$, $H_2SO_4$ | — | 465 |
| 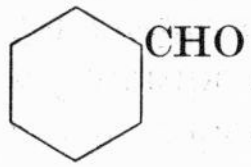 Cyclohexanecarboxaldehyde (structure: CHO) | $CH_3CO_2H$, $H_2SO_4$ | — | 465 |
| 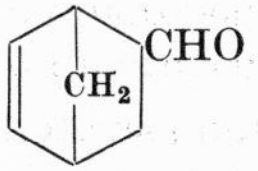 Bicyclic aldehyde (structure: CHO, $CH_2$) | $CH_3CO_2H$, $H_2SO_4$ | — | 465 |
| Crotonaldehyde | $CH_3CO_2H$, $H_2SO_4$ | — | 465 |
| Cinnamaldehyde | $(CH_3CO)_2O$, heat | — | 465 |
| Benzaldehyde | $(CH_3CO)_2O$, HCl, heat | 78 | 85 |
| | $CH_3CO_2H$, heat | — | 465 |

*Note:* References 382 to 537 are on pp. 266–269.

* Except when otherwise indicated, the amide and aldehyde are unnamed reactants in this column.

TABLE XXXIV—*Continued*

SYMMETRICAL N,N′-ALKYLIDENE-*bis* AND N,N′-ARYLIDENE-*bis* DERIVATIVES OF AMIDES, LACTAMS, IMIDES, CARBAMYL COMPOUNDS, AND SULFONAMIDES

$2RCONH_2 + R'CHO \rightarrow (RCONH)_2CHR'$

| Derivative of | Method* | Yield, % | References |
|---|---|---|---|
| *Aliphatic Amides* (*contd.*) | | | |
| Chloroacetamide (*cont.*) and | | | |
| *o*-Fluorobenzaldehyde | $(CH_3CO)_2O$, heat | — | 465 |
| *o*-Chlorobenzaldehyde | $(CH_3CO)_2O$, heat | — | 465 |
| *m*-Chlorobenzaldehyde | $(CH_3CO)_2O$, heat | — | 465 |
| *p*-Chlorobenzaldehyde | $(CH_3CO)_2O$, heat | — | 465 |
| 2,4-Dichlorobenzaldehyde | $(CH_3CO)_2O$, heat | — | 465 |
| 2,5-Dichlorobenzaldehyde | $(CH_3CO)_2O$, heat | — | 465 |
| 2,6-Dichlorobenzaldehyde | $CH_3CO_2H$, $ZnCl_2$, heat | — | 465 |
| *o*-Bromobenzaldehyde | $(CH_3CO)_2O$, heat | — | 465 |
| *o*-Nitrobenzaldehyde | $CH_3CO_2H$, $ZnCl_2$, heat | — | 465 |
| | No solvent, heat | — | 307 |
| *m*-Nitrobenzaldehyde | No solvent, heat | — | 307 |
| *p*-Nitrobenzaldehyde | $(CH_3CO)_2O$, heat | — | 465 |
| | No solvent, heat | — | 307 |
| 2-Chloro-5-nitrobenzaldehyde | $(CH_3CO)_2O$, heat | — | 465 |
| *o*-Acetoxybenzaldehyde | $CH_3CO_2H$, heat | — | 465 |
| *p*-Acetoxybenzaldehyde | $(CH_3CO)_2O$, heat | — | 465 |
| *o*-Anisaldehyde | $(CH_3CO)_2O$, heat | — | 465 |
| *p*-Anisaldehyde | $CH_3CO_2H$, heat | — | 465 |
| *o*-$HO_2CCH_2OC_6H_4CHO$ | $CH_3CO_2H$, heat | — | 465 |
| Piperonal | No solvent, heat | — | 307 |
| *p*-$(CH_3)_2NC_6H_4CHO$ | $(CH_3CO)_2O$, heat | — | 465 |
| *p*-$CH_3CONHC_6H_4CHO$ | $CH_3CO_2H$, heat | — | 465 |
| *o*-$NCC_6H_4CHO$ | $(CH_3CO)_2O$, heat | — | 465 |

| | | | |
|---|---|---|---|
| *o*-Tolualdehyde | $(CH_3CO)_2O$, heat | — | 465 |
| *m*-Tolualdehyde | No solvent, heat | — | 307 |
| *p*-Tolualdehyde | No solvent, heat | — | 307 |
| 3-Cl-4-$CH_3C_6H_3CHO$ | $CH_3CO_2H$, heat | — | 465 |
| *p*-$C_6H_4(CHO)_2$ (*bis* cpd.) | $CH_3CO_2H$, heat | — | 465 |
| 1-Naphthaldehyde | $(CH_3CO)_2O$, heat | — | 465 |
| 5-Nitro-2-furaldehyde | $(CH_3CO)_2O$, heat | — | 465 |
| 2-Thiophenecarboxaldehyde | $(CH_3CO)_2O$, heat | — | 465 |
| | No solvent, heat | — | 307 |
| (benzo[b]thiophene-2-CHO) | $(CH_3CO)_2O$, heat | — | 465 |
| (1-$COCH_3$-indole-3-CHO) | $(CH_3CO)_2O$, heat | — | 465 |
| Pyruvic acid | Heat, in vacuum | — | 291, 465 |
| Dichloroacetamide and | | | |
| *o*-Chlorobenzaldehyde | $(CH_3CO)_2O$, heat | — | 465 |
| Pyruvic acid | Heat, in vacuum | — | 291 |
| Trichloroacetamide and | | | |
| Benzaldehyde | Heat alone or in $(CH_3CO)_2O$ | 0 | 85 |
| Bromoacetamide and | | | |
| *o*-Chlorobenzaldehyde | $(CH_3CO)_2O$, heat | — | 465 |
| Iodoacetamide and | | | |
| *o*-Chlorobenzaldehyde | *o*-$ClC_6H_4CH(NHCOCH_2Cl)_2$, KI, $CH_3OH$ | — | 465 |

*Note:* References 382 to 537 are on pp. 266–269.

* Except when otherwise indicated, the amide and aldehyde are unnamed reactants in this column.

TABLE XXXIV—*Continued*

SYMMETRICAL N,N′-ALKYLIDENE-*bis* AND N,N′-ARYLIDENE-*bis* DERIVATIVES OF AMIDES, LACTAMS, IMIDES, CARBAMYL COMPOUNDS, AND SULFONAMIDES

$2RCONH_2 + R'CHO \rightarrow (RCONH)_2CHR'$

| Derivative of | Method* | Yield, % | References |
|---|---|---|---|
| *Aliphatic Amides* (*contd.*) | | | |
| $(C_2H_5)_2NCH_2CONH_2$ and *o*-Chlorobenzaldehyde | *o*-$ClC_6H_4CH(NHCOCH_2Cl)_2$, $(C_2H_5)_2NH$ | — | 465 |
| $N_3CH_2CONH_2$ and *o*-Chlorobenzaldehyde | *o*-$ClC_6H_4CH(NHCOCH_2Cl)_2$, $NaN_3$, $CH_3OH$ | — | 465 |
| Phenylacetamide and | | | |
| Acetaldehyde | No solvent, heat, acid | — | 526 |
| | *n*-$C_4H_9OCH{=}CH_2$, heat, acid | 81 | 233 |
| | $CH_3CHO$, $C_6H_5CH_2CN$, concd. HCl | — | 527 |
| Chloral | $CCl_3CHO$, $C_6H_5CH_2CN$, concd. $H_2SO_4$ | — | 261 |
| Cinnamaldehyde | Pyridine or heat alone | 18 | 275 |
| *o*-Chlorobenzaldehyde | No solvent, heat | Good | 518 |
| *m*-Nitrobenzaldehyde | No solvent, heat | 97 | 269 |
| *p*-Nitrobenzaldehyde | No solvent, heat | 97 | 269 |
| 5-Chlorosalicylaldehyde | No solvent, heat | 35 | 528 |
| 3,5-Dichlorosalicylaldehyde | No solvent, heat | 86 | 528 |
| *o*-Anisaldehyde | No solvent, heat | 40–60 | 296*a* |
| *m*-Anisaldehyde | No solvent, heat | 40–60 | 296*a* |
| *p*-Anisaldehyde | No solvent, heat | 40–60 | 296*a* |
| Piperonal | No solvent, heat | 54 | 519 |
| 6-Nitropiperonal | No solvent, heat | 10 | 520 |
| *m*-Tolualdehyde | Pyridine | 48 | 273 |

| | | | |
|---|---|---|---|
| Propionamide and | | | |
| β-Phenylpropionaldehyde | No solvent, heat | 29 | 516 |
| Heptanal | No solvent, heat | — | 517 |
| Cinnamaldehyde | No solvent, heat | 51 | 275 |
| Benzaldehyde | No solvent, heat | 45 | 284, 518 |
| *o*-Chlorobenzaldehyde | No solvent, heat | Good | 518 |
| *o*-Nitrobenzaldehyde | Pyridine or heat alone | 49 | 274 |
| *m*-Nitrobenzaldehyde | No solvent, heat | 67 | 269 |
| *p*-Nitrobenzaldehyde | No solvent, heat | 54 | 269 |
| 5-Chlorosalicylaldehyde | No solvent, heat | — | 271 |
| 3,5-Dichlorosalicylaldehyde | No solvent, heat | — | 271 |
| *o*-Anisaldehyde | No solvent, heat | 40–60 | 296*a* |
| *m*-Anisaldehyde | No solvent, heat | 40–60 | 296*a* |
| *p*-Anisaldehyde | No solvent, heat | 40–60 | 296*a* |
| Piperonal | No solvent, heat | 36 | 519 |
| 6-Bromopiperonal | No solvent, heat | 32 | 525 |
| 6-Nitropiperonal | No solvent, heat | 15 | 529 |
| *m*-Tolualdehyde | Pyridine | 36 | 273 |
| *p*-Tolualdehyde | Pyridine or heat alone | 50 | 276 |
| 2-Furaldehyde | No solvent, heat | 46 | 284 |
| $CH_2ClCH_2CONH_2$ and | | | |
| Acetaldehyde | $CH_3CHO$, $CH_2ClCH_2CN$, HCl, $H_2O$, 300° | — | 465, 496, 500 |
| Benzaldehyde | $CH_3CO_2H$, heat | — | 465 |
| $CH_2ClCHClCONH_2$ and | | | |
| Acetaldehyde | $CH_3CHO$, $CH_2ClCHClCN$, HCl, $H_2O$, 250° | — | 496, 500 |
| $CH_2ClCCl_2CONH_2$ and | | | |
| Acetaldehyde | $CH_3CHO$, $CH_2ClCCl_2CN$, HCl, $H_2O$, 250° | — | 500 |
| Benzaldehyde | $C_6H_5CHO$, $CH_2ClCCl_2CN$, HCl, $H_2O$, 380° | — | 465, 500 |

*Note:* References 382 to 537 are on pp. 266–269.

* Except when otherwise indicated, the amide and aldehyde are unnamed reactants in this column.

TABLE XXXIV—*Continued*

SYMMETRICAL N,N′-ALKYLIDENE-*bis* AND N,N′-ARYLIDENE-*bis* DERIVATIVES OF AMIDES, LACTAMS, IMIDES, CARBAMYL COMPOUNDS, AND SULFONAMIDES

$2RCONH_2 + R'CHO \rightarrow (RCONH)_2CHR'$

| Derivative of | Method* | Yield, % | References |
|---|---|---|---|
| *Aliphatic Amides (contd.)* | | | |
| $CH_2BrCH_2CONH_2$ and | | | |
| Acetaldehyde | $CH_3CHO$, $CH_2BrCH_2CN$, HCl, $H_2O$, heat | — | 500 |
| $CH_2BrCHBrCONH_2$ and | | | |
| Acetaldehyde | $CH_3CHO$, $CH_2BrCHBrCN$, HCl, $H_2O$, 380° | — | 465, 500 |
| Butyramide and | | | |
| Chloral | $CCl_3CHO$, RCN, 96% $H_2SO_4$ | 67 | 4 |
| β-Phenylpropionaldehyde | No solvent, heat | 22 | 516 |
| Butyraldehyde | RCHO, RCN, 96% $H_2SO_4$ | 19 | 4 |
| Heptanal | Heat alone, or in $(CH_3CO)_2O$, or in pyridine | 0 | 517 |
| Benzaldehyde | No solvent, heat | Good | 518, 523 |
| *o*-Chlorobenzaldehyde | No solvent, heat | Good | 518 |
| *o*-Nitrobenzaldehyde | No solvent, heat | 65 | 274 |
| *m*-Nitrobenzaldehyde | No solvent, heat | 84 | 269 |
| *p*-Nitrobenzaldehyde | No solvent, heat | 72 | 269 |
| 5-Chlorosalicylaldehyde | No solvent, heat | — | 271 |
| 3,5-Dichlorosalicylaldehyde | No solvent, heat | — | 271 |
| Piperonal | No solvent, heat | 55 | 519 |
| 6-Nitropiperonal | No solvent, heat | 57 | 520 |
| *m*-Tolualdehyde | Pyridine | 39 | 273 |
| $HO_2C(CH_2)_4CONH_2$ and | | | |
| Acetaldehyde | $CH_3CHO$, $HO_2C(CH_2)_4CN$, concd. HCl | — | 497 |
| | $CH_3CHO$, $HO_2C(CH_2)_4CN$, HCl, $H_2O$, 250° | — | 496, 500 |

| | | | |
|---|---|---|---|
| Heptanamide and | | | |
| $\beta$-Phenylpropionaldehyde | No solvent, heat | 40 | 516 |
| Heptanal | Heat alone, or in $(CH_3CO)_2O$, or in pyridine | 0 | 517 |
| Cinnamaldehyde | No solvent, heat | 41 | 530 |
| Benzaldehyde | No solvent, heat | Good | 518 |
| *o*-Chlorobenzaldehyde | No solvent, heat | Good | 518 |
| *o*-Nitrobenzaldehyde | No solvent, heat | 100 | 274 |
| *m*-Nitrobenzaldehyde | No solvent, heat | 97 | 269 |
| *p*-Nitrobenzaldehyde | No solvent, heat | 97 | 269 |
| 3,5-Dichlorosalicylaldehyde | No solvent, heat | — | 271 |
| Piperonal | No solvent, heat | 77 | 519 |
| 6-Bromopiperonal | No solvent, heat | 36 | 525 |
| 6-Nitropiperonal | No solvent, heat | 55 | 520 |
| *m*-Tolualdehyde | Pyridine or heat alone | <40 | 273 |
| Cyclohexanecarboxamide and | | | |
| Acetaldehyde | $(CH_3CO)_2O$, heat | 13 | 268 |
| Benzaldehyde | $(CH_3CO)_2O$, heat | 73 | 268 |
| Acrylamide and | | | |
| Octanal | Concd. HCl, heat | | 505 |
| Benzaldehyde | Concd. HCl, heat | | 505 |
| Cinnamamide and | | | |
| Heptanal | No solvent, heat | — | 517 |
| Benzaldehyde | No solvent, heat | 62 | 284, 518 |
| 5-Chlorosalicylaldehyde | No solvent, heat | 72 | 528 |
| 3,5-Dichlorosalicylaldehyde | No solvent, heat | 90 | 528 |
| Piperonal | No solvent, heat | 82 | 519 |
| 6-Nitropiperonal | No solvent, heat | 83 | 520 |

*Note:* References 382 to 537 are on pp. 266–269.

* Except when otherwise indicated, the amide and aldehyde are unnamed reactants in this column.

TABLE XXXIV—*Continued*

SYMMETRICAL N,N′-ALKYLIDENE-*bis* AND N,N′-ARYLIDENE-*bis* DERIVATIVES OF AMIDES, LACTAMS, IMIDES, CARBAMYL COMPOUNDS, AND SULFONAMIDES

$$2RCONH_2 + R'CHO \rightarrow (RCONH)_2CHR'$$

| Derivative of | Method* | Yield, % | References |
|---|---|---|---|
| *Aliphatic Amides* (*contd.*) | | | |
| α-Methylcinnamamide and | | | |
| Benzaldehyde | No solvent, heat | 48 | 284 |
| $(CONH_2)_2$ and | | | |
| Benzaldehyde | $C_2H_5O_2CCONH_2$, $C_6H_5CHO$, 150° | — | 285 |
| | (Product: cyclic CONH / CONH bridged by $CHC_6H_5$) | | |
| $(C_2H_5)_2C(CONH_2)_2$ and | | | |
| Cinnamaldehyde | No solvent, acid, 150° | — | 286 |
| Benzaldehyde | No solvent, acid, 150° | — | 286 |
| | (Products: $(C_2H_5)_2C$ and CHR joined by two CONH bridges, ring) | | |
| $(CH_2CONHC_6H_5)_2$ and | | | |
| Heptanal | No solvent, 110° (product not characterized) | — | 270 |

| | | | |
|---|---|---|---|
| *Aromatic Amides* | | | |
| Benzamide and | | | |
| Acetaldehyde | $(CH_3CO)_2O$, heat | 48 | 281 |
| | $H_2O$, acid | — | 531 |
| | $n$-$C_4H_9OCH{=}CH_2$, $C_6H_5CONH_2$, heat | 63, 80 | 232, 233 |
| | $CH_3CHO$, $C_6H_5CN$, concd. $H_2SO_4$ | — | 245, 527 |
| | $CH_3CHO$, $C_6H_5CN$, HCl, $H_2O$, 300° | — | 496 |
| | $CH_3CHO$, $NH_4OH$, $C_6H_5COCl$ | — | 532 |
| Chloral | $[C_6H_5CONHCH(CCl_3)_2]_2O$, 140°/40 mm. | — | 95 |
| | $CCl_3CHO$, $C_6H_5CN$, concd. $H_2SO_4$ | 37 | 245, 261, 296 |
| | $CCl_3CHO$, $C_6H_5CONH_2$, concd. $H_2SO_4$ | 67 | 296 |
| Bromal | $CBr_3CHO$, $C_6H_5CN$, concd. $H_2SO_4$ | — | 245 |
| β-Phenylpropionaldehyde | No solvent, heat | 40 | 275, 516 |
| $CH_3CHClCCl_2CHO$ | RCHO, $C_6H_5CN$, concd. $H_2SO_4$ | — | 245 |
| Valeraldehyde | $n$-$C_4H_9CHO$, $C_6H_5CN$, concd. $H_2SO_4$ | — | 245 |
| Heptanal | No solvent, heat | — | 285, 517 |
| Cyclohexanecarboxaldehyde | $(CH_3CO)_2O$, heat | 69 | 280 |
| Cinnamaldehyde | No solvent, heat | 55 | 275 |
| Benzaldehyde | $(CH_3CO)_2O$, heat | 43 | 281 |
| | $(CH_3CO)_2O$, heat | 78 | 85 |
| | No solvent, heat | 27 | 284 |
| | No solvent, heat | — | 518, 522, 523, 533 |
| $o$-Chlorobenzaldehyde | No solvent, heat | Good | 518 |
| $o$-Nitrobenzaldehyde | $H_2O$, acid | — | 278 |
| | No solvent, heat | 54 | 274 |
| $m$-Nitrobenzaldehyde | $H_2O$, acid | — | 278 |
| | No solvent, heat | 80 | 269 |

*Note:* References 382 to 537 are on pp. 266–269.

* Except when otherwise indicated, the amide and aldehyde are unnamed reactants in this column.

TABLE XXXIV—*Continued*

SYMMETRICAL N,N′-ALKYLIDENE-*bis* AND N,N′-ARYLIDENE-*bis* DERIVATIVES OF AMIDES, LACTAMS, IMIDES, CARBAMYL COMPOUNDS, AND SULFONAMIDES

$2RCONH_2 + R'CHO \rightarrow (RCONH)_2CHR'$

| Derivative of | Method* | Yield, % | References |
|---|---|---|---|
| *Aromatic Amides (contd.)* | | | |
| Benzamide (*contd.*) and | | | |
| *p*-Nitrobenzaldehyde | No solvent, heat | 80 | 269 |
| 5-Chlorosalicylaldehyde | Pyridine | — | 271 |
| 3,5-Dichlorosalicylaldehyde | No solvent, heat | — | 271 |
| *o*-Anisaldehyde | No solvent, heat | 40–60 | 296*a* |
| *m*-Anisaldehyde | No solvent, heat | 40–60 | 296*a* |
| *p*-Anisaldehyde | No solvent, heat | 40–60 | 296*a*, 524 |
| Piperonal | No solvent, heat | 64 | 519 |
| 6-Bromopiperonal | No solvent, heat | 4 | 525 |
| 6-Nitropiperonal | No solvent, heat | 25 | 520 |
| *m*-Tolualdehyde | Pyridine | 37 | 273 |
| *p*-Tolualdehyde | Pyridine or heat alone | 50 | 276 |
| 1-Naphthaldehyde | $(CH_3CO)_2O$, heat | 43 | 282 |
| 2-Naphthaldehyde | $(CH_3CO)_2O$, heat | 59 | 282 |
| 9-Anthraldehyde | $(CH_3CO)_2O$, heat | 38 | 282 |
| 1-Pyrenecarboxaldehyde | $(CH_3CO)_2O$, heat | 75 | 282 |
| 2-Furaldehyde | No solvent, heat | 40 | 284 |
| 2-Thiophenecarboxaldehyde | $(CH_3CO)_2O$, heat | 68 | 280, 282 |
| 2-Pyrrolecarboxaldehyde | $(CH_3CO)_2O$, heat | 0 | 280 |
| Picolinaldehyde | No solvent, heat | 99 | 280 |
| Nicotinaldehyde | No solvent, heat | 73 | 280 |
| Isonicotinaldehyde | No solvent, heat | 27 | 280 |
| 3-Indolecarboxaldehyde | $(CH_3CO)_2O$, heat | 60 | 280 |
| Benzanilide and | | | |
| Heptanal | No solvent, 110° | — | 270 |

| | | | |
|---|---|---|---|
| *o*-Chlorobenzamide and | | | |
| Acetaldehyde | $CH_3CHO$, ArCN, concd. HBr | — | 497 |
| | $CH_3CHO$, ArCN, HCl, $H_2O$, 280° | — | 500 |
| *p*-Chlorobenzamide and | | | |
| Chloral | $CCl_3CHO$, ArCN, 95 % $H_2SO_4$ | — | 296 |
| *o*-Nitrobenzamide and | | | |
| Benzaldehyde | $(CH_3CO)_2O$, HCl, heat | 46 | 85 |
| *x*-Nitrobenzamide and | | | |
| Heptanal | No solvent, heat | — | 285 |
| *o*-Anisamide and | | | |
| Benzaldehyde | $H_2O$, acid | — | 365 |
| *o*-Toluamide and | | | |
| *o*-Nitrobenzaldehyde | No solvent, heat | 47 | 274 |
| 2-$CH_3$-5-$O_2NC_6H_3CONH_2$ and | | | |
| Benzaldehyde | No solvent, heat | — | 534 |
| *m*-Nitrobenzaldehyde | No solvent, heat | — | 534 |
| *p*-Nitrobenzaldehyde | No solvent, heat | — | 534 |
| Piperonal | No solvent, heat | — | 534 |
| *p*-Toluamide and | | | |
| Chloral | $CCl_3CHO$, ArCN, 95 % $H_2SO_4$ | 40 | 296 |
| 1-Naphthamide and | | | |
| Acetaldehyde | $(CH_3CO)_2O$, heat | 58 | 268 |
| Benzaldehyde | No solvent, heat | 93 | 268 |
| Nicotinamide and | | | |
| Acetaldehyde | No solvent, heat | 9 | 268 |
| Benzaldehyde | No solvent, heat | 66 | 268 |

*Note:* References 382 to 537 are on pp. 266–269.

* Except when otherwise indicated, the amide and aldehyde are unnamed reactants in this column.

TABLE XXXIV—*Continued*

SYMMETRICAL N,N′-ALKYLIDENE-*bis* AND N,N′-ARYLIDENE-*bis* DERIVATIVES OF AMIDES, LACTAMS, IMIDES, CARBAMYL COMPOUNDS, AND SULFONAMIDES

$$2RCONH_2 + R'CHO \rightarrow (RCONH)_2CHR'$$

| Derivative of | Method* | Yield, % | References |
|---|---|---|---|
| *Lactams and Imides* | | | |
| 2-Pyrrolidinone and | | | |
| Acetaldehyde | $H_2O$, acid | Good | 247 |
| Phthalimide and | | | |
| Benzaldehyde | $(CH_3CO)_2O$, heat | 0 | 85 |
| *Carbamyl Compounds* | | | |
| Ethyl carbamate and | | | |
| Acetaldehyde | $H_2O$, acid | 78 | 173, 250, 531 |
| Chloroacetaldehyde | $H_2O$, acid | — | 173 |
| | $C_2H_5OH$, HCN, $Cl_2$ | — | 266, 267 |
| Dichloroacetaldehyde | $H_2O$, acid | — | 173 |
| | $C_2H_5OH$, HCN, $Cl_2$ | — | 266, 267 |
| Chloral | $NH_2CO_2C_2H_5$, $(C_2H_5O_2CNH\overset{\vert}{C}HCCl_3)_2O$ | 38 | 94 |
| | $NH_2CO_2C_2H_5$, $Cl_2$ | 20 | 94 |
| | $NH_2CO_2C_2H_5$, $CCl_3CHO$, acid | — | 94 |
| | $[C_2H_5O_2CNH\overset{\vert}{C}H(CCl_3)]_2O$, $PCl_5$ | — | 95 |
| Bromoacetaldehyde | $NH_2CO_2C_2H_5$, $Br_2$ | — | 230 |
| Dibromoacetaldehyde | $NH_2CO_2C_2H_5$, $Br_2$ | — | 230 |
| Bromal | $H_2O$, acid | — | 173 |
| | $C_2H_5OH$, HCN, $Br_2$ | — | 267 |
| | $NH_2CO_2C_2H_5$, $Br_2$ | 27 | 535 |
| | $NH_2CO_2C_2H_5$, $CBr_3CHO$, acid | — | 535 |

| | | | |
|---|---|---|---|
| Valeraldehyde | $H_2O$, acid | — | 173 |
| Cinnamaldehyde | $H_2O$, acid | — | 277 |
| Glyoxylic acid | Heat alone | — | 230 |
| Benzaldehyde | $H_2O$, acid | — | 173 |
| | $NaOC_2H_5$ | — | 230 |
| *o*-Nitrobenzaldehyde | $H_2O$, acid | — | 278 |
| *m*-Nitrobenzaldehyde | $H_2O$, acid | — | 278 |
| *p*-Anisaldehyde | $H_2O$, acid | — | 277 |
| Piperonal | $H_2O$, acid | 98 | 529 |
| 2-Furaldehyde | $H_2O$, acid | — | 277 |
| *n*-Propyl carbamate and | | | |
| Acetaldehyde | $H_2O$, acid | — | 277 |
| Benzaldehyde | $H_2O$, acid | — | 277 |
| $NH_2CS(OC_2H_5)$ and | | | |
| Valeraldehyde | $H_2O$, acid | — | 277 |
| Urea and† | | | |
| Chloral | $H_2NCONHCHOHCCl_3$, urea, $(CH_3CO)_2O$ | Poor | 473 |
| | Heat alone | — | 536 |
| $CH_3CHClCCl_2CHO$ | Heat alone | — | 536 |
| Valeraldehyde | $H_2O$ alone | — | 272 |
| Heptanal | $C_2H_5OH$ alone | — | 272 |
| Acrolein | $H_2O$ alone | — | 272 |
| Cinnamamide | $H_2O$ alone | — | 123, 272 |
| Benzaldehyde | $C_2H_5OH$ alone | — | 272 |
| *x*-Nitrobenzaldehyde | $C_2H_5OH$ alone | — | 272 |
| Salicylaldehyde | $C_2H_5OH$ alone | — | 272 |

*Note:* References 382 to 537 are on pp. 266–269.

* Except when otherwise indicated, the amide and aldehyde are unnamed reactants in this column.

† The products are of the type $(H_2NCONH)_2CHR$.

TABLE XXXIV—*Continued*

SYMMETRICAL N,N′-ALKYLIDENE-*bis* AND N,N′-ARYLIDENE-*bis* DERIVATIVES OF AMIDES, LACTAMS, IMIDES, CARBAMYL COMPOUNDS, AND SULFONAMIDES

$2RCONH_2 + R'CHO \rightarrow (RCONH)_2CHR'$

| Derivative of | Method* | Yield, % | References |
|---|---|---|---|
| *Carbamyl Compounds* (*contd.*) | | | |
| Urea (*contd.*) and† | | | |
| 5-Chlorosalicylaldehyde | $C_2H_5OH$ alone | — | 271 |
| 3,5-Dichlorosalicylaldehyde | $C_2H_5OH$ alone | — | 271 |
| *p*-Anisaldehyde | $C_2H_5OH$ alone | — | 272 |
| *o*-Ethoxybenzaldehyde | $C_2H_5OH$ alone | — | 272 |
| Cumaldehyde | $C_2H_5OH$ alone | — | 123 |
| Hydantoin and | | | |
| Acetaldehyde | Hydantoin, $CH_2(OC_2H_5)_2$, $CH_3CO_2H$, $H_2SO_4$, heat | 96 | 290 |
| Chloroacetaldehyde | Hydantoin, $CH_2ClCH(OC_2H_5)_2$, $CH_3CO_2H$, $H_2SO_4$, heat | 70 | 290 |
| *Sulfonamides* | | | |
| *p*-Toluenesulfonamide and | | | |
| Acetaldehyde | $ArSO_2NH_2$, $C_6H_5OCH{=}CH_2$, acid | 11 | 232 |

*Note:* References 382 to 537 are on pp. 266–269.

* Except when otherwise indicated, the amide and aldehyde are unnamed reactants in this column.

† The products are of the type $(H_2NCONH)_2CHR$.

## TABLE XXXV

### UNSYMMETRICAL N,N′-ARYLIDENEBISAMIDES[284]

$$RCONH_2 + R'CONH_2 + ArCHO \xrightarrow[\text{No solvent}]{\text{Heat}} RCONHCH(Ar)NHCOR'$$

| R | R′ | Ar | Yield, % |
|---|---|---|---|
| $CH_3$ | $C_2H_5$ | $C_6H_5$ | 30 |
| $CH_3$ | $C_6H_5CH{=}CH$ | $C_6H_5$ | 39 |
| $CH_3$ | $C_6H_5CH{=}C(CH_3)$ | $C_6H_5$ | 38 |
| $CH_3$ | $C_6H_5$ | $C_6H_5$ | 42 |
| $CH_3$ | $C_2H_5$ | 2-Furyl | 25 |
| $CH_3$ | $C_6H_5CH{=}CH$ | 2-Furyl | 26 |
| $C_2H_5$ | $C_6H_5CH{=}CH$ | $C_6H_5$ | 33 |
| $C_2H_5$ | $C_6H_5CH{=}C(CH_3)$ | $C_6H_5$ | 30 |
| $C_2H_5$ | $C_6H_5$ | $C_6H_5$ | 38 |
| $C_6H_5CH{=}CH$ | $C_6H_5CH{=}C(CH_3)$ | $C_6H_5$ | 25 |
| $C_6H_5CH{=}CH$ | $C_6H_5$ | $C_6H_5$ | 29 |
| $C_6H_5CH{=}C(CH_3)$ | $C_6H_5$ | $C_6H_5$ | 26 |

*Note:* References 382 to 537 are on pp. 266–269.

## TABLE XXXVI

### AMIDOMETHANESULFONIC ACIDS AND SALTS

| Sulfonic Acid Derivative | References (Yield, %) |
|---|---|
| $CH_3(CH_2)_7CH{=}CH(CH_2)_7CONHCH_2SO_3Na$ | 537 |
| $C_6H_5CONHCH_2SO_3Na$ | 132 |
| $p\text{-}CH_3OC_6H_4CONHCH_2SO_3Na$ | 132 (60) |
| $o\text{-}C_6H_4(CO)_2NCH_2SO_3H$ | 309 (13) |
| $C_6H_5SO_2NHCH_2SO_3Na$ | 132 |
| $C_6H_4(SO_2NHCH_2SO_3Na)_2$ (benzene ring bearing two $SO_2NHCH_2SO_3Na$ groups) | 132 |

*Note:* References 382 to 537 are on pp. 266–269.

## REFERENCES

[382] Jones and Pyman, *J. Chem. Soc.*, **1925,** 2588.
[383] Marini, *Gazz. Chim. Ital.*, **69,** 340 (*Chem. Zentr.*, **1939,** II, 3279).
[384] Monti, *Gazz. Chim. Ital.*, **63,** 724 (1933) (*Chem. Zentr.*, **1934,** I, 1327).
[385] de Diesbach and Gubser, *Helv. Chim. Acta*, **11,** 1098 (1928).
[386] de Diesbach, Swiss pat., 124,526 (*Chem. Zentr.*, **1929,** I, 144).
[387] de Diesbach, Gubser, and Lempen, *Helv. Chim. Acta*, **13,** 120 (1930).
[388] Haworth, MacGillivray, and Peacock, *J. Chem. Soc.*, **1950,** 1493.
[389] Haworth and Lamberton, *J. Chem. Soc.*, **1946,** 1003.
[390] Bradner and Sherrill, U.S. pat. 1,503,631 [*C.A.*, **19,** 658 (1925)].
[391] I. G. Farbenindustrie A-G., Swiss pat. 137,210 (1930) [same as Brit. pat. 318,315 (*Chem. Zentr.*, **1930,** I, 584)].
[392] Snyder and Brewster, *J. Am. Chem. Soc.*, **71,** 1058 (1949).
[393] O'Cinnéide, *Proc. Roy. Irish Acad.*, **49B,** 143 (1943) [*C.A.*, **38,** 1230 (1944)].
[394] Monti and Verona, *Gazz. Chim. Ital.*, **62,** 878 (1932) (*Chem. Zentr.*, **1933,** I, 1133).
[395] Monti and Franchi, *Gazz. Chim. Ital.*, **81,** 191 (1951) (*Chem. Zentr.*, **1952,** 1650).
[396] Maki, Ishida, Satake, and Oda, *J. Chem. Soc. Japan, Ind. Chem. Sect.*, **57,** 44 (1954) [*C.A.*, **49,** 10907 (1955)].
[397] Burckhalter, Tendick, Jones, Holcomb, and Rawlins, *J. Am. Chem. Soc.*, **68,** 1894 (1946).
[398] I. G. Farbenindustrie A.-G., Swiss pat. 141,211 (1930) [same as Brit. pat. 318,315 (*Chem. Zentr.*, **1930,** I, 584)].
[399] Randall and Renfrew, U.S. pat. 2,533,178 [*C.A.*, **45,** 2220 (1951)].
[400] Randall and Buc, U.S. pat. 2,536,984 [*C.A.*, **45,** 4053 (1951)].
[401] Randall and Buc, U.S. pat. 2,584,368 [*C.A.*, **46,** 4808 (1952)].
[402] Randall and Buc, U.S. pat. 2,584,367 [*C.A.*, **46,** 9609 (1952)].
[403] Randall and Buc, U.S. pat. 2,535, 987 [*C.A.*, **45,** 3607 (1951)].
[404] Adachi and Ota, *Ann. Rept. Fac. Pharm. Kanazawa Univ.*, **7,** 10 (1957) [*C.A.*, **52,** 9148 (1958)].
[405] General Aniline and Film Corp., Brit. pat., 779,324 [*C.A.*, **51,** 17186 (1957)].
[406] Arbuzov and Livshits, *Izv. Akad. Nauk SSSR Otd. Khim. Nauk*, **1946,** 391 [*C.A.*, **42,** 6335 (1948)].
[407] Profft, Runge, and Jumar, *J. Prakt. Chem.*, [4] **1,** 79 (1954).
[408] Reddelien and Lange, Ger. pat. 511,951 [*C.A.*, **25,** 1262 (1931)].
[409] Bayer, U.S. pat. 1,946,829 [*C.A.*, **28,** 2374 (1934)].
[410] Böhme, Broese, and Driesen, *Arch. Pharm.*, **292,** 677 (1959).
[411] Stefanović, Stojiljković, and Stefanović, *Tetrahedron*, **18,** 413 (1962).
[412] Atkinson, *J. Chem. Soc.*, **1954,** 1329.
[413] Folkers and Johnson, *J. Am. Chem. Soc.*, **55,** 1140 (1933).
[414] Kalle and Co., A. G., Ger. pat. 164,610 (*Chem. Zentr.*, **1905,** II, 1751).
[415] Chwala, *Monatsh. Chem.*, **78,** 172 (1948).
[416] Einhorn, Ger. pat. 162,395 (*Chem. Zentr.*, **1905,** II, 728).
[417] Einhorn, and Ladisch, *Ann.*, **343,** 279 (1905).
[418] Einhorn, Ger. pat. 157,355 (*Chem. Zentr.*, **1905,** I, 57).
[419] Einhorn and Spröngerts, *Ann.*, **343,** 267 (1905).
[420] Tokura and Oda, *Bull. Inst. Phys. Chem. Res.* (*Tokyo*) *Chem. Ed.*, **24,** 14 (1947) [*C.A.*, **43,** 2176 (1949)].
[421] Feuer and Lynch-Hart, U.S. pat. 2,760,977 [*C.A.*, **51,** 2851 (1957)].
[422] Arbuzova and Mosevich, *J. Gen. Chem. USSR* (*Engl. Transl.*), **31,** 2819 (1961) [*C.A.*, **56,** 11441 (1962)].
[423] Müller and Holdschmidt, Ger. pat. 1,002,326 [*C.A.*, **53,**14001 (1959)].
[424] Arbuzova, Ushakov, Plotkina, Efremova, and Ulezlo, *J. Gen. Chem. USSR* (*Engl. Transl.*), **28,** 1322 (1958) [*C.A.*, **52,** 19923 (1958)].
[425] Einhorn, Ger. pat. 158,088 (*Chem. Zentr.*, **1905,** I, 573).
[426] Graf, *J. Prakt. Chem.*, [2] **138,** 292 (1933).

[427] Suter, Stocker, and Zutter, U.S. pat. 2,660,582 [*C.A.*, **48,** 8271 (1954)].
[428] Suter, Habicht, and Kundig, Can. pat. 511,544 (1955).
[429] Offe and Siefken, Ger. pat. 919,351 [*C.A.*, **52,** 14706 (1958)].
[430] Martin, Habicht, and Zutter, U.S. pat. 2,741,620 [*C.A.*, **51,** 7445 (1957)].
[431] Cherbuliez and Chambers, *Helv. Chim. Acta.*, **8,** 395 (1925).
[432] Tawney, U.S. pat. 2,743,260 [*C.A.*, **50,** 11716 (1956)].
[433] Drábek, *Chem. Zvesti*, **12,** 29 (1958) [*C.A.*, **52,** 13705 (1958)].
[434] Sachs, *Ber.*, **31,** 3230 (1898).
[435] Winstead and Heine, *J. Am. Chem. Soc.*, **77,** 1913 (1955).
[436] Vasa and Kshatriya, *J. Indian Chem. Soc.*, **36,** 527 (1959) [*C.A.*, **54,** 9837 (1960)].
[437] Maselli, *Gazz. Chim. Ital.*, **30** (II), 31 (1900) (*Chem. Zentr.*, **1900,** II, 629).
[438] Schroy, U.S. pat. 2,361,322 [*C.A.*, **39,** 2228 (1945)].
[439] Einhorn and Göttler, *Ann.*, **361,** 148 (1908).
[440] Dixon, *J. Chem. Soc.*, **113,** 238 (1918).
[441] Dunn, *J. Chem. Soc.*, **1957,** 1446.
[442] Einhorn and Hamburger, *Ber.*, **41,** 24 (1908).
[443] Scheibler, Trostler, and Scholz, *Angew. Chem.*, **41,** 1305 (1928).
[444] Walter and Gewing, *Kolloid-Beih.*, **34,** 163 (1931); Walter, Ger. pat. 519,322 (*Chem. Zentr.*, **1928,** II, 1383) [*C.A.*, **26,** 5183 (1932)].
[445] Pollopas Ltd., Ger. pat. 504,863 (*Chem. Zentr.*, **1930,** II, 2050).
[446] Alexander, Carter, and Earland, *Melliand Textilber.*, **32,** 238 (1951).
[447] Imperial Chemical Industries Ltd., British pat. 577,735 [*C.A.*, **41,** 2254 (1947)].
[448] Farlow, U.S. pat. 2,422,400 [*C.A.*, **41,** 6279 (1947)].
[449] Wayland, U.S. pat. 2,825,732 [*C.A.*, **52,** 14666 (1958)].
[450] Hoover and Vaala, U.S. pat. 2,373,136 [*C.A.*, **39,** 4763 (1945)].
[451] Oshima, Japan. pat. 5576 (1959) [*C.A.*, **54,** 14285 (1960)].
[452] Reich, *Monatsh. Chem.*, **25,** 933 (1904).
[452a] Reid, Frick, Reinhardt, and Arceneaux, *Am. Dyestuff Reptr.*, **48,** 81 (1959) [*C.A.*, **53,** 8637 (1959)].
[453] Walker, Brit. pat. 564,424 [*C.A.*, **40,** 3300 (1946)].
[454] Jaffe, *Ber.*, **35,** 2896 (1902).
[455] Arnold and Vogelsang, Ger. pat. 1,016,267 [*C.A.*, **53,** 22019 (1959)].
[456] July, Knott, and Pollak, U.S. pat. 2,732,316 [*C.A.*, **50,** 7640 (1956)].
[457] Zellner, Austrian pat. 176,560 [*C.A.*, **48,** 10777 (1954)].
[458] Jansen and Stokes, *J. Chem. Soc.*, **1962,** 4909.
[459] Chemische Fabrik auf Aktien, Ger. pat. 50,586 (1889); Friedlaender, *Fortschr. Theerfarbenfabrikation*, **2,** 524 (1887–1890).
[460] La Rocca, Leonard, and Weaver, *J. Org. Chem.*, **16,** 47 (1951).
[461] Schiff, *Ber.*, **10,** 168 (1877).
[462] Wallach, *Ber.*, **5,** 254 (1872).
[463] Schiff and Tassinari, *Ber.*, **10,** 1783 (1877).
[464] Pinner, *Ann.*, **179,** 40 (1875).
[465] Schraufstätter and Gönnert, *Z. Naturforsch.*, **17b,** 505 (1962) [*C.A.*, **57,** 13666 (1962)].
[466] La Rocca and Byrum, *J. Am. Pharm. Assoc.*, **43,** 63 (1954).
[467] Schroeder and Aman, Ger. pat. D.A.S. 1,078,576 (*Chem. Zentr.*, **1960,** 16521).
[468] Chemische Fabrik Gedeon Richter, Ger. pat. 234,741 [*C.A.*, **5,** 2911 (1911)].
[469] Sulzberger, Ger. pat. 198,715 [*C.A.*, **2,** 2715 (1908)].
[470] Pinner and Klein, *Ber.*, **11,** 10 (1878).
[470a] Hirwe, Rana, and Gavankar, *Proc. Indian Acad. Sci.*, **8A,** 208 (1938) [*C.A.*, **33,** 2124 (1939)].
[471] Hill, U.S. pat. 2,755,283 [*C.A.*, **50,** 15032 (1956)].
[472] Mondon, *Tetrahedron*, **19,** 911 (1963).
[473] Coppin and Titherley, *J. Chem. Soc.*, **105,** 32 (1914).
[474] Kalle and Co., in Biebrich, Ger. pat. 128,462 (*Chem. Zentr.*, **1902,** I, 547).
[475] Zigeuner, Pitter, Berger, and Rauch, *Monatsh. Chem.* **86,** 165 (1955).
[476] Robson and Reinhart, *J. Am. Chem. Soc.*, **77,** 2453 (1955).

[477] Müller, Dinges, and Graulich, *Macromol. Chem.*, **57,** 27 (1962) [*C.A.*, **58,** 3511 (1963)].
[478] Engelmann and Pikl, U.S. pat. 2,361,185 [*C.A.*, **39,** 2214 (1945)].
[479] Engelmann and Pikl, U.S. pat. 2,313,741 [*C.A.*, **37,** 5168 (1943)].
[480] Farbenfabriken Bayer A.-G., Brit. pat. 780,284 [*C.A.*, **52,** 2053 (1958)].
[481] Shostakovskii, Sidel'kovskaya, Rogova, Kolodkin, and Ibragimov, *Bull. Acad. Sci. USSR, Phys. Ser.* (*Engl. Transl.*), **1961,** 1026 (*Chem. Zentr.*, **1962,** 17763).
[482] Sorenson, U.S. pat. 2,201,927 [*C.A.*, **34,** 6301 (1940)].
[483] de Jong, and de Jonge *Rec. Trav. Chim.*, **72,** 169 (1953).
[484] Becher and Griffel, *Chem. Ber.*, **91,** 2032 (1958).
[485] Burke, U.S. pat. 2,321,989 [*C.A.*, **37,** 6908 (1943)].
[486] Böhme, Bohn, Köhler, and Roehr, *Ann*, **664,** 130 (1963).
[487] Coker and Fields, *J. Org. Chem.*, **27,** 2226 (1962).
[488] Zinner, Herbig, Wistup, and Wigert, *Chem. Ber.*, **92,** 407 (1959).
[489] Ayad, McCall, Neale, and Jackman, *J. Chem. Soc.*, **1962,** 2070.
[490] Semper, *Ann.*, **381,** 234 (1911).
[491] Kliegl, *Ber.*, **43,** 2488 (1910).
[492] Goldmann, *Ber.*, **23,** 2522 (1890).
[493] Reddelien and Danilof, *Ber.*, **54,** 3132 (1921).
[494] Kalle and Co., A.-G., Ger. pat. 164,611 (*Chem. Zentr.*, **1905,** II, 1751).
[495] Sauer and Bruni, *J. Am. Chem. Soc.*, **77,** 2559 (1955).
[496] Heisenberg, Kleine, and Lotz, Ger. pat. 891,986 (*Chem. Zentr.*, **1954,** 9393).
[497] Vereinigte Glanzstoff-Fabriken A.-G., Brit. pat. 726,937 (*Chem. Zentr.*, **1956,** 9868).
[498] Heisenberg, Kleine, and Lotz, Ger. pat. 886,906 (*Chem. Zentr.*, **1954,** 4740).
[499] Tomcufcik, Willson, and Vogel, U.S. pat. 3,085,940 (1963) (same as Brit. pat. 905,186) [*C.A.*, **58,** 454 (1963)].
[500] Vereinigte Glanzstoff-Fabriken A.-G., Brit. pat. 726,933 [*C.A.*, **50,** 5019 (1956)].
[501] Bechert, *J. Prakt. Chem.*, **50,** 3 (1894).
[502] Passerini, *Gazz. Chim. Ital.*, **53,** 338 (*Chem. Zentr.*, **1923,** III, 1010).
[503] Mowry, U.S. pat. 2,534,204 [*C.A.*, **45,** 3867 (1951)].
[504] Cannepin and Parisot, *Compt. Rend.*, **239,** 180 (1954).
[505] Lundberg, U.S. pat. 2,475,846 [*C.A.*, **44,** 873 (1950)].
[506] Kraut and Schwartz, *Ann.*, **223,** 46 (1884).
[507] Kraut, *Ann.*, **258,** 109 (1890).
[508] Oda, Tanimoto, Nomura, Nishimura, and Kyô, *J. Chem. Soc. Japan, Ind. Chem. Sect.* **60,** 18 (*Chem. Zentr.*, **1957,** 5252).
[509] Duden and Scharff, *Ann.*, **288,** 247 (1895).
[510] Yoda, *Kobunshi Kagaku*, **19,** 613 (1962) [*C.A.*, **59,** 6304 (1963)].
[511] Eckenroth and Koerppen, *Ber.*, **30,** 1265 (1897).
[511a] Tawney and Kelly, U.S. pat. 2,745,841 (1956) [*C.A.*, **51,** 1256 (1957)].
[512] Staudinger and Niessen, *Makromol. Chem.*, **15,** 75 (1955) [*C.A.*, **50,** 3224 (1956)].
[513] Tomcufcik, Willson, and Vogel, U.S. pat. 3,085,941 (1963) (same as Brit. pat. 905,186) [*C.A.*, **58,** 454 1963)].
[514] Oda, Nomura, and Tanimoto, *Bull. Inst. Chem. Res. Kyoto Univ.*, **32,** 231 (1954) [*C.A.*, **50,** 7112 (1956)].
[515] Tanimoto and Oda, *J. Chem. Soc. Japan, Ind. Chem. Sect.*, **56,** 942 (1953) [*C.A.*, **49,** 6878 (1955)].
[516] Pandya and Bhandari, *J. Indian Chem. Soc.*, **24,** 185 (1947) [*C.A.*, **43,** 2600 (1949)].
[517] Pandya and Sodhi, *Proc. Indian Acad. Sci.*, **27A,** 196 (1948) [*C.A.*, **44,** 4415 (1950)].
[518] Bhatnagar and Pandya, *Proc. Indian Acad. Sci.*, **24,** 487 (1946) [*C.A.*, **41,** 3774 (1947)].
[519] Pandya and Varghese, *Proc. Indian Acad. Sci.*, **14A,** 18 (1941) [*C.A.*, **36,** 1308 (1942)].
[520] Pandya and Varghese, *Proc. Indian Acad. Sci.*, **14A,** 25 (1941) [*C.A.*, **36,** 1308 (1942)].
[521] Tavildarov, *J. Russ. Chem. Soc.*, **4,** 142 (1872); *Ber.*, **5,** 477 (1872).
[522] Chattaway and Swinton, *J. Chem. Soc.*, **101,** 1206 (1912).
[523] Roth, *Ann.*, **154,** 72 (1870).
[524] Schuster, *Ann.*, **154,** 80 (1870).

[525] Pandya and Sethi, *J. Indian Chem. Soc.*, **25,** 145 (1948) [*C.A.*, **44,** 3465 (1950)].
[526] Bernthsen, *Ann.*, **184,** 318 (1877).
[527] Henle and Schupp, *Ber.*, **38,** 1369 (1905).
[528] Nigam and Pandya, *Proc. Indian Acad. Sci.*, **29A,** 364 (1949) [*C.A.*, **44,** 1454 (1950)].
[529] Bianchi, *Gazz. Chim. Ital.*, **43** (I), 237 (1913) (*Chem. Zentr.*, **1913,** I, 1962).
[530] Pandya and Bhandari, *J. Indian Chem. Soc.*, **24,** 209 (1947) [*C.A.*, **43,** 2600 (1949)].
[531] Nencki, *Ber.*, **7,** 158 (1874).
[532] Limpricht, *Ann.*, **99,** 119 (1856).
[533] Hoffmann and Meyer, *Ber.*, **25,** 209 (1892).
[534] Macovski and Bachmeyer, *Ber.*, **77,** 487 (1944).
[535] Diels and Ochs, *Ber.*, **40,** 4571 (1907).
[536] Pinner and Lifschütz, *Ber.*, **20,** 2345 (1887).
[537] Yamashita and Yoshizaki, U.S. pat. 2,313,695 [*C.A.*, **37,** 5168 (1943)].

# CHAPTER 3

# THE WITTIG REACTION*

ADALBERT MAERCKER†
*University of Heidelberg*

## CONTENTS



* Translated from the German by Dr. E. Ciganek, Central Research Department, E. I. du Pont de Nemours & Co., Wilmington, Delaware.

† The author is indebted to Prof. Dr. G. Wittig for many helpful suggestions, as well as for permission to use unpublished material.

## INTRODUCTION

In 1953 Wittig and Geissler[1] found that reaction of benzophenone with methylenetriphenylphosphorane gave 1,1-diphenylethylene and triphenylphosphine oxide in almost quantitative yield; the phosphorane had been prepared from triphenylmethylphosphonium bromide and phenyllithium. This discovery led in the following years to the development of

$$[(C_6H_5)_3\overset{\oplus}{P}—CH_3]Br^{\ominus} + C_6H_5Li \rightarrow (C_6H_5)_3P{=}CH_2 + C_6H_6 + LiBr$$

$$(C_6H_5)_3P{=}CH_2 + (C_6H_5)_2C{=}O \rightarrow (C_6H_5)_2C{=}CH_2 + (C_6H_5)_3P{=}O$$

[1] Wittig and Geissler, *Ann.*, **580**, 44 (1953).

a new method for the synthesis of olefins[2–6] which, under the name Wittig reaction, soon attained importance in preparative organic chemistry.

One advantage of this new method is that the carbonyl group is replaced specifically by a carbon-carbon double bond without the formation of isomeric olefins. In contrast, the older method of converting carbonyl

$$(R'CH_2)(R''CH_2)C{=}O + (C_6H_5)_3P{=}CHR''' \rightarrow (R'CH_2)(R''CH_2)C{=}CHR'''$$

compounds to olefins using the Grignard reaction followed by dehydration of the resulting carbinol usually gives a mixture of isomeric olefins.

Another advantage of the Wittig reaction is that it is carried out in alkaline medium and usually under very mild conditions. Consequently it is the only method available for the preparation of sensitive olefins such as carotenoids, methylene steroids, and other natural products.

As early as 1919 Staudinger and Meyer[7,8] showed that the reaction of diphenylmethylenetriphenylphosphorane (**1**) with phenyl isocyanate gave N-phenyldiphenylketeneimine; but this discovery did not lead to the

$$\underset{\mathbf{1}}{(C_6H_5)_3P{=}C(C_6H_5)_2} + C_6H_5N{=}C{=}O \rightarrow C_6H_5N{=}C{=}C(C_6H_5)_2$$

discovery of a new synthesis of olefins. One reason is the unusual stability of the phosphorane **1**, which could not be made to react with normal carbonyl compounds. It has been found that, even with diphenylketene, reaction will occur only under forcing conditions (at 140° in benzene under pressure).[9–11]

$$\underset{\mathbf{1}}{(C_6H_5)_3P{=}C(C_6H_5)_2} + (C_6H_5)_2C{=}C{=}O \rightarrow (C_6H_5)_2C{=}C{=}C(C_6H_5)_2$$

A second reason is that the phosphorane **1** had been prepared by the thermal decomposition of the phosphazine **2**, which in turn had been

[2] Wittig and Schöllkopf, *Chem. Ber.*, **87,** 1318 (1954).
[3] Wittig and W. Haag, *Chem. Ber.*, **88,** 1654 (1955).
[4] Wittig, *Angew. Chem.*, **68,** 505 (1956).
[5] Wittig, *Experientia*, **12,** 41 (1956).
[6] Wittig, *Festschr. Arthur Stoll,* **1957,** 48 [*C.A.*, **52,** 15414c (1958)].
[7] Staudinger and Meyer, *Helv. Chim. Acta.*, **2,** 619 (1919).
[8] Staudinger and Meyer, *Helv. Chim. Acta.*, **2,** 635 (1919).
[9] G. Lüscher, Doctoral Dissertation, Eidg. Techn. Hochschule, Zürich, 1922.
[10] Meyer, *Chem. Ber.*, **89,** 842 (1956).
[11] Meyer, *Helv. Chim. Acta*, **40,** 1052 (1957).

obtained from triphenylphosphine and diphenyldiazomethane. This

$$(C_6H_5)_3P + \overset{\ominus}{N}{=}\overset{\oplus}{N}{=}C(C_6H_5)_2 \rightarrow$$

$$\underset{\mathbf{2}}{(C_6H_5)_3P{=}N{-}N{=}C(C_6H_5)_2} \xrightarrow[190^\circ]{-N_2} \underset{\mathbf{1}}{(C_6H_5)_3P{=}C(C_6H_5)_2}$$

roundabout method could not be extended to the preparation of other phosphoranes since, in the pyrolysis of phosphazines, ketazines are usually formed instead.[7] It was not until the development of modern techniques in organometallic chemistry that reactive methylene phosphoranes became easily accessible and led to routes to olefins in good yields from a variety of aldehydes and ketones.

This chapter includes the more important literature to the end of 1962 and a few selected articles published in 1963. Other reviews have been published elsewhere.[12–17a]

## ALKYLIDENE TRIPHENYLPHOSPHORANES

### Structure and Properties of Ylides

The existence of pentaphenylphosphorane[18] shows that, unlike nitrogen, phosphorus is capable of being pentacovalent, since it can expand its valency shell to 10 electrons by inclusion of *d* orbitals. Alkylidene phosphoranes can therefore be considered as resonance hybrids of two limiting structures, the ylide form **3*a*** and the ylene form **3*b***. An in-

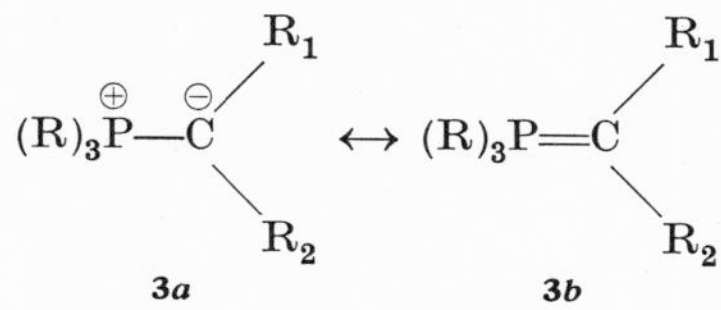

**3*a*** **3*b***

dication that alkylidene phosphoranes possess some ylene character was obtained from a kinetic study which showed that tetramethylphosphonium iodide is converted to the ylide much faster than is tetramethylammonium iodide, despite the fact that the protons in the phosphonium salt are under the influence of weaker repulsive Coulomb forces than those

[12] Levisalles, *Bull. Soc. Chim. France*, **1958,** 1021.
[13] Schöllkopf, *Angew. Chem.*, **71,** 260 (1959).
[14] Pelc, *Chem. Listy*, **53,** 177 (1959) [*C.A.*, **53,** 7965a (1959)].
[15] Kostka, *Wiadomosci Chemi*, **12,** 521 (1958).
[16] Trippett, *Advan. Org. Chem.*, **1,** 83 (1960).
[17] Yanovskaya, *Usp. Khim.*, **30,** 813 (1961) [*C.A.*, **56,** 1323a (1962)].
[17a] Trippett, *Quart. Rev.*, **17,** 406 (1963).
[18] Wittig and Rieber, *Ann.*, **562,** 187 (1949).

in the ammonium salt (bond distances: P—C, 1.87 Å; N—C, 1.47 Å).[19] This greater tendency for the formation of phosphorus ylides reflects stabilization of the transition state by *p-d* orbital overlap.

The reactivity of alkylidene phosphoranes is determined by the distribution of the negative charge in the molecule, which in turn depends on the nature of the substituents $R_1$ and $R_2$ in the alkylidene portion as well as of the groups R on phosphorus. Thus the nucleophilic character of the phosphorane is decreased and the stability of the phosphorane increased if the lone electron pair on the α-carbon atom of form **3*a*** is delocalized into groups $R_1$ and $R_2$. Generally speaking, electron-withdrawing substituents $R_1$ and $R_2$ will stabilize the negative charge and consequently reduce the reactivity of the ylide. Methylenetriphenylphosphorane (**3*a*, 3*b***, $R_1 = R_2 =$ H), where there is no such interaction, is an extremely reactive and unstable phosphorane of high nucleophilicity.

The groups R on phosphorus also influence the reactivity of the alkylidene phosphoranes since they may be capable of increasing or decreasing the *d*-orbital resonance with a consequent change in the relative importance of form **3*b*** in the resonance hybrid. Decreased *d*-orbital resonance would result in greater importance of the ylide form **3*a*** and therefore increased reactivity of the phosphorane. Investigations of the ultraviolet spectra of polyphenyl derivatives of phosphorus led to the conclusion that the formation of double bonds with inclusion of *d* orbitals is possible only if the central atom is positively charged in the single-bonded structure.[20–22] Applied to the alkylidene phosphoranes, this would mean that the ylene form **3*b*** would be the more important the larger the formal charge on phosphorus. Electron-withdrawing groups R on phosphorus will, other things being equal, increase the *d*-orbital resonance and therefore favor the ylene form **3*b***, whereas electron-releasing groups will increase the importance of the ylide form **3*a***. These hypotheses are supported by investigations of transition metal complexes in which trivalent phosphorus occurs as a ligand and forms $\pi$ as well as $\sigma$ bonds with the metal. It has been shown that the capability for the formation of $\pi$ bonds decreases in the series: $PF_3 > PCl_3 > P(OCH_3)_3 > P(C_3H_7\text{-}n)_3$, i.e., in the same sequence as the positive inductive effect (+I effect) of substituents on phosphorus.[23]

Subsequently, studies were made of the ultraviolet and visible spectra of a series of planar complexes of the structure *trans*-[L, piperidine $PtCl_2$], where the ligand L was an aliphatic amine, phosphine, etc. Again

[19] Doering and Hoffmann, *J. Am. Chem. Soc.*, **77**, 521 (1955).

[20] Jaffé, *J. Phys. Chem.*, **58**, 185 (1954).

[21] Jaffé, *J. Chem. Phys.*, **22**, 1430 (1954).

[22] Jaffé and Freedman, *J. Am. Chem. Soc.*, **74**, 1069 (1952).

[23] Chatt and Williams, *J. Chem. Soc.*, **1951**, 3061.

the extent of the phosphorus *d*-orbital participation in the interaction with the metal increases with increasing +I effect of the groups on phosphorus: $L = P(C_3H_7\text{-}n)_3 < L = P(OCH_3)_3$.[24] Finally, studies of the carbonyl stretching frequency in the infrared spectra of nickel dicarbonyl-diphosphines $[Ni(CO)_2(PR_3)_2]$[25] and of complexes of the structure $(PR_3)_3Cr(CO)_3$[26] showed that the *d*-orbital participation in the metal phosphorus bond decreases in the series $PCl_3 > P(OC_6H_5)_3 > P(OC_2H_5)_3 > P(C_6H_5)_3 > P(C_4H_9\text{-}n)_3$, i.e., again following the +I effect. Similar results using $^{31}P$ nuclear magnetic resonance spectroscopy were obtained.[27]

It would therefore be expected that alkylidene trialkylphosphoranes, in which the formal positive charge on phosphorus is reduced by the inductive effect of the alkyl groups, would under otherwise equal conditions be more reactive than alkylidene triphenylphosphoranes in which the +I effect of the phenyl groups has the opposite effect. This is the case. Carbomethoxymethylenetricyclohexylphosphorane **(4)**, for instance, is a stronger base than carbomethoxymethylenetriphenylphosphorane **(5)**, and, unlike the latter, is sensitive to water.[28] The same explanation

$$(C_6H_{11})_3P{=}CHCO_2CH_3 \qquad (C_6H_5)_3P{=}CHCO_2CH_3$$

**4** **5**

rationalizes the increased stability of fluorenylidenetriphenylphosphorane toward hydrolysis as compared to that of fluorenylidene trialkylphosphoranes.[29] In general, the conjugate acid of a phosphorane should be the stronger the more important the ylene form **3*b*** (and consequently the more important the *d*-orbital resonance); i.e., a more acidic phosphonium salt will be related to a less reactive phosphorane, and vice versa. It is possible to estimate the reactivity of an ylide from the $pK_a$ value of its conjugate acid.[29]

The discussion of the mechanism of the Wittig reaction will show, however, that alkylidene trialkylphosphoranes are superior to alkylidene triphenylphosphoranes only in certain special cases.

The Wittig reagents may be divided into two groups according to their reactivity. The first and larger group includes the alkylidene phosphoranes of low stability and high reactivity, whereas the second group comprises the highly stable unreactive resonance-stabilized alkylidene phosphoranes.

[24] Chatt, Gamlen, and Orgel, *J. Chem. Soc.*, **1959,** 1047.
[25] Meriwether and Fiene, *J. Am. Chem. Soc.*, **81,** 4200 (1959).
[26] Abel, Bennett, and Wilkinson, *J. Chem. Soc.*, **1959,** 2323.
[27] Meriwether and Leto, *J. Am. Chem. Soc.*, **83,** 3192 (1961).
[28] Bestmann and Kratzer, *Chem. Ber.*, **95,** 1894 (1962).
[29] Johnson and LaCount, *Tetrahedron*, **9,** 130 (1960).

Phosphoranes containing substituents in the alkylidene group that have little effect on the carbanion character of the molecule are found in the first group. These Wittig reagents are markedly nucleophilic and react readily at low temperatures with carbonyl and other polar groups.

The second group consists of phosphoranes with electron-withdrawing substituents in the alkylidene portion which decrease the nucleophilicity to a certain extent and, in some cases, completely prevent the attack on a carbonyl group. Unlike the phosphoranes of the first group, they are essentially stable toward hydrolysis. For instance, the acylmethylene triphenylphosphoranes[30,31] are colorless crystalline compounds that are hydrolyzed only at elevated temperatures. This stability is probably due to resonance between the limiting structures **6*a*** to **6*c*** in which the carbonyl group is included. This is reflected in the shift of the carbonyl band in the infrared spectrum to 6.5–6.7 $\mu$.[32] In spite of this resonance stabili-

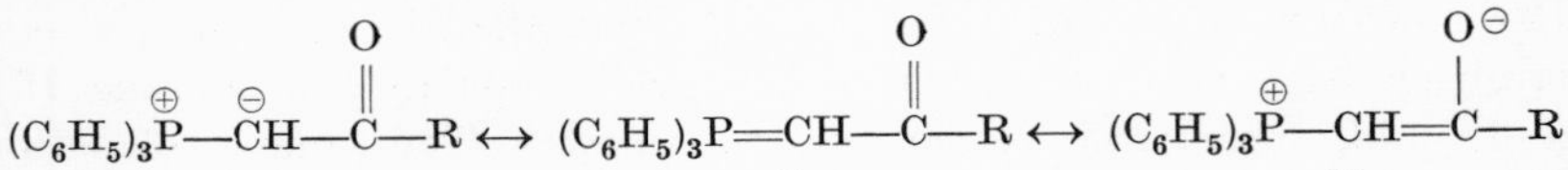

$$\underset{6a}{(C_6H_5)_3\overset{\oplus}{P}-\overset{\ominus}{C}H-\overset{O}{\overset{\|}{C}}-R} \leftrightarrow \underset{6b}{(C_6H_5)_3P{=}CH-\overset{O}{\overset{\|}{C}}-R} \leftrightarrow \underset{6c}{(C_6H_5)_3\overset{\oplus}{P}-CH{=}\overset{O^{\ominus}}{\overset{|}{C}}-R}$$

zation the acylmethylene triphenylphosphoranes undergo the Wittig reaction with reactive carbonyl compounds such as benzaldehyde, but not with cyclohexanone under the same conditions.[31] Similarly formylmethylenetriphenylphosphorane (**6,** R = H), a stable compound of melting point 187°, reacts readily with aldehydes in boiling benzene but not with ketones.[33]

In the synthesis of olefins having electron-withdrawing groups on the α-carbon atom the reagent of choice is often the anion derived from a phosphonate ester. For example,

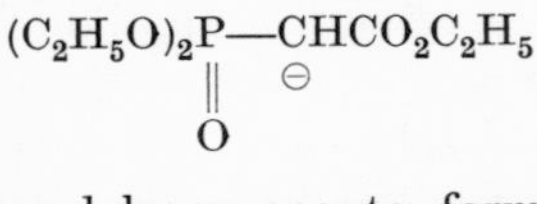

$$(C_2H_5O)_2\underset{\underset{O}{\|}}{P}-\overset{\ominus}{C}HCO_2C_2H_5$$

will react readily with cyclohexanone to form ethyl cyclohexylideneacetate. Reagents of this type are discussed on pp. 379–382.

The preparatively important carbalkoxymethylene triphenylphosphoranes **7*a*** will also react with ketones but only under forcing conditions.[34–37] However, the phosphorane **7*a*** reacts readily with cyclic

[30] Michaelis and Köhler, *Ber.*, **32,** 1566 (1899).

[31] Ramirez and Dershowitz, *J. Org. Chem.*, **22,** 41 (1957).

[32] Bestmann and Arnason, *Chem. Ber.*, **95,** 1513 (1962).

[33] Trippett and Walker, *Chem. Ind.* (*London*), **1960,** 202.

[34] Sugasawa and Matsuo, *Chem. Pharm. Bull.* (*Tokyo*), **8,** 819 (1960) [*C.A.*, **55,** 20901a (1961)].

[35] Fodor and Tömösközi, *Tetrahedron Letters*, **1961,** 579.

[36] Openshaw and Whittaker, *Proc. Chem. Soc.*, **1961,** 454.

[37] Trippett and Walker, *Chem. Ind.* (*London*), **1961,** 990.

and methyl ketones in the presence of benzoic acid as a catalyst.[38] Reaction with aldehydes, on the other hand, occurs readily.[3,39]

$$(C_6H_5)_3P{=}CHCO_2R \quad (\textbf{7a}) \qquad (C_6H_5)_3P{=}CHCONH_2 \quad (\textbf{7b}) \qquad (C_6H_5)_3P{=}CHCN \quad (\textbf{7c})$$

Carbalkoxymethylene triphenylphosphoranes (**7*a***) are stable crystalline compounds which can be kept for a long time without decomposition. Participation of the ester group in charge delocalization is evident from the shift of the carbonyl band in the infrared to 6.15 $\mu$,[40] as well as from the stability of the compounds. Carbamidomethylenetriphenylphosphorane (**7*b***)[41] and cyanomethylenetriphenylphosphorane (**7*c***)[41–43] also belong to this group of stable phosphoranes.

Replacement of the second hydrogen atom in the methylene group by an electron-delocalizing substituent results in further decrease of the reactivity of the phosphoranes to such an extent that they may become unreactive even toward aldehydes. Examples of such phosphoranes are diphenylmethylenetriphenylphosphorane (**1**),[8] 4-nitrobenzhydrylenetriphenylphosphorane (**8**),[44] and compounds of type **9**,[45] whose ultraviolet spectra show that the negative charge is extensively delocalized into the two groups X and Y.

$$(C_6H_5)_3P{=}C(C_6H_5)_2 \quad (\textbf{1}) \qquad (C_6H_5)_3P{=}C(C_6H_4NO_2\text{-}p)(C_6H_5) \quad (\textbf{8}) \qquad (C_6H_5)_3\overset{\oplus}{P}{-}\overset{\ominus}{C}(X)(Y) \quad (\textbf{9})$$

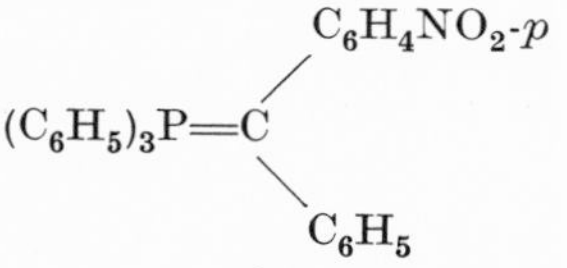

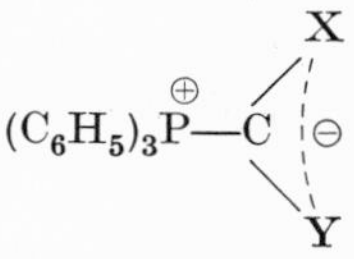

(X, Y = $CO_2R$, CN, $SO_2C_6H_5$, $COCH_3$)

The methylene carbon atom may also be made part of a quasi-aromatic system with a resulting decrease in the carbanion character of the ylide. Cyclopentadienylidenetriphenylphosphorane (**10**),[46,47] for instance, is a yellow crystalline compound, m.p. 229–231°, stable to prolonged heating with concentrated aqueous or ethanolic potassium hydroxide. It does not react with benzophenone, fluorenone, or cyclohexanone on heating in

[38] Rüchardt, Eichler, and Panse, *Angew. Chem.*, **75**, 858 (1963).

[39] Isler, Gutmann, Montavon, Rüegg, and Zeller, *Helv. Chim. Acta*, **40**, 1242 (1957).

[40] Bestmann and Schulz, *Chem. Ber.*, **95**, 2921 (1962).

[41] Trippett and Walker, *J. Chem. Soc.*, **1959**, 3874.

[42] Novikov and Shvekhgeĭmer, *Izv. Akad. Nauk SSSR, Otd. Khim. Nauk*, **1960**, 2061 [*C.A.*, **55**, 13353g (1961)].

[43] Schiemenz and Engelhard, *Chem. Ber.*, **94**, 578 (1961).

[44] Drefahl, Plötner, and Scholz, *Z. Chem.*, **1**, 93 (1961).

[45] Horner and Oediger, *Chem. Ber.*, **91**, 437 (1958).

[46] Ramirez and Levy, *J. Org. Chem.*, **21**, 488 (1956).

[47] Ramirez and Levy, *J. Am. Chem. Soc.*, **79**, 67 (1957).

diethyl ether, chloroform, ethanol, or tetrahydrofuran for 120 hours. Reaction with benzaldehyde appears to occur, but no pure products could be isolated.

$$(C_6H_5)_3P{=}C_5H_4 \leftrightarrow (C_6H_5)_3\overset{\oplus}{P}{-}C_5H_4^{\ominus}$$

**10*a*** **10*b***

Fluorenylidenetriphenylphosphorane (**11**)[48,49] is considerably more reactive; it is hydrolyzed by ethanolic sodium hydroxide, and it undergoes the Wittig reaction with a number of aldehydes to give olefins. The dipole moment of 7.09 D points to the importance of resonance form **11*b*** in the hybrid. Replacement of the phenyl groups on phosphorus by

$$C_{13}H_8{=}P(C_6H_5)_3 \leftrightarrow C_{13}H_8^{\ominus}{-}\overset{\oplus}{P}(C_6H_5)_3$$

**11*a*** **11*b***

alkyl groups in **11**[29,50] results in a marked increase in reactivity. Thus the hydrolysis of fluorenylidenetrimethylphosphorane[50,51] occurs readily under the influence of atmospheric moisture, and fluorenylidenetributylphosphorane[29] may even be made to react with a number of ketones.

## Preparation of Alkylidene Phosphoranes

The Wittig reagents are usually prepared by the action of bases on the easily accessible triphenylalkylphosphonium halides. Formation of

$$(C_6H_5)_3\overset{\oplus}{P}{-}CH(R_1)(R_2) + B^{\ominus} \rightleftharpoons (C_6H_5)_3P{=}C(R_1)(R_2) + HB$$

alkylidene phosphoranes by removal of a proton from the salt under the influence of base is a reversible reaction. The choice of conditions depends entirely on the nature of the desired ylide. The air- and moisture-sensitive phosphoranes of the first group have to be prepared in an anhydrous medium under an atmosphere of inert gas, employing

[48] Pinck and Hilbert, *J. Am. Chem. Soc.*, **69**, 723 (1947).
[49] Johnson, *J. Org. Chem.*, **24**, 282 (1959).
[50] Johnson and LaCount, *Chem. Ind.* (*London*), **1959**, 52.
[51] Wittig and Laib, *Ann.*, **580**, 57 (1953).

organometallic compounds as proton acceptors. In one method[2] an ether solution of phenyllithium or butyllithium is added under nitrogen to a suspension of 1 equivalent of the phosphonium salt in ether, tetrahydrofuran, or other suitable solvent. The reaction usually occurs in the cold, and formation of the alkylidene phosphoranes can be followed by the appearance of an intense yellow or red coloration.

A variation of this method consists in preparing the ylide in liquid ammonia with sodium amide as the base and then replacing ammonia by diethyl ether or tetrahydrofuran.[52] Sodium amide has also been used in other solvents, such as benzene, diethyl ether, or tetrahydrofuran.[53]

Alkylidene phosphoranes that do not react with a carboxamide function can also be prepared in dimethylformamide as the solvent; in these reactions, sodium acetylide has proved to be particularly useful as a base.[54,55] Since many phosphonium salts can be readily prepared in dimethylformamide, the Wittig reaction can be carried out without isolation of the quaternary salt.

Another method for the preparation of alkylidene phosphoranes employs alkali metal alkoxides as proton acceptors, usually in the corresponding alcohol as solvent.[3] The simplicity of this method is gaining more and more attention for it. The alkoxide procedure permits preparation of unstable ylides directly in the presence of carbonyl compounds, thus minimizing side reactions.

Since, unlike the reactive phosphoranes of the first group, the resonance-stabilized ylides of the second group are stable toward hydrolysis, they can be prepared by the action of alkali metal hydroxides on an aqueous solution of the phosphonium salt.[39,56] The phosphorane usually precipitates in crystalline form and can be dried in air. With sufficiently acidic phosphonium salts, weaker bases can also be used. For instance, fluorenylidenetriphenylphosphorane **(11)**[48] has been prepared by the action of ammonium hydroxide on the corresponding phosphonium salt, and *p*-nitrobenzylidenetriphenylphosphorane **(12)**[57] has been prepared

$P(C_6H_5)_3$

**11**

$(C_6H_5)_3P{=}CHC_6H_4NO_2\text{-}p$

**12**

[52] Wittig, Eggers, and Duffner, *Ann.*, **619,** 10 (1958).

[53] Wittig and Pommer, Ger. pat. 1,003,730 (to BASF) [*C.A.*, **53,** 16063c (1959)].

[54] Wittig and Pommer, Ger. pat. 954,247 (to BASF) [*C.A.*, **53,** 2279e (1959)].

[55] Pommer, Wittig, and Sarnecki, Ger. pat. 1,026,745 (to BASF) [*C.A.*, **54,** 11074f (1960)].

[56] Gerecke, Ryser, and Zeller, Ger. pat. 1,125,922 (to Hoffman-La Roche), 1958; corresponds to U.S. pat. 2,912,467 [*C.A.*, **54,** 2254e (1960)].

[57] Kröhnke, *Chem. Ber.*, **83,** 291 (1950).

using sodium carbonate as the base. Carbomethoxymethyltriphenylphosphonium bromide also reacts with sodium carbonate, whereas sodium hydroxide must be used in order to convert the less acidic carbomethoxymethyltricyclohexylphosphonium bromide to its ylide.[28] Triethylamine[58–60] and pyridine[58,60] have also been used for the preparation of resonance-stabilized alkylidene phosphoranes.

Other methods for the preparation of alkylidene phosphoranes from phosphonium salts are known. Hydrogen bromide can be eliminated from triphenylbenzylphosphonium bromide with metallic sodium.[61]

$$[(C_6H_5)_3\overset{\oplus}{P}—CH_2C_6H_5]Br^{\ominus} + Na \rightarrow (C_6H_5)_3P{=}CHC_6H_5 + [H]$$

Ethylidenetriphenylphosphorane is obtained in very good yield by the action of methylsulfinyl carbanion on triphenylethylphosphonium bromide in dimethyl sulfoxide.[62,63] Finally it should be mentioned that

$$CH_3—S(O)CH_2^{\ominus}\ Na^{\oplus} + (C_6H_5)_3\overset{\oplus}{P}C_2H_5Br^{\ominus} \rightarrow (C_6H_5)_3P{=}CHCH_3$$

Coffmann and Marvel[64] were the first to prepare ylides by the organometallic route by allowing trityl sodium to react with triphenylalkylphosphonium salts.

The mechanism of ylide formation by the interaction of organolithium compounds with phosphonium salts was the subject of an interesting study.[65] The organic base not only removes a proton from the α-carbon atom (path A) but also adds to a certain extent to phosphorus, forming a presumably pentavalent intermediate which then collapses into alkylidenephosphorane and a hydrocarbon (path B). The formation of benzene,

$$(C_6H_5)_3\overset{\oplus}{P}—CH_3Br^{\ominus} \xrightarrow[CH_3Li]{A} (C_6H_5)_3P{=}CH_2 + CH_4$$

$$\downarrow B\ CH_3Li$$

$$(C_6H_5)_3P(CH_3)—CH_3 \longrightarrow (C_6H_5)_2P(CH_3){=}CH_2 + C_6H_6$$

[58] Trippett and Walker, *J. Chem. Soc.*, **1961,** 1266.

[59] Märkl, *Chem. Ber.*, **95,** 3003 (1962).

[60] Denney and Ross, *J. Org. Chem.* **27,** 998 (1962).

[61] Parfent'ev and Shamshurin, *J. Gen. Chem. USSR (Engl. Transl.)*, **9,** 865 (1939) [*C.A.*, **34,** 392 (1940)].

[62] Corey and Chaykovsky, *J. Am. Chem. Soc.*, **84,** 866 (1962).

[63] Greenwald, Chaykovsky, and Corey, *J. Org. Chem.*, **28,** 1128 (1963).

[64] Coffmann and Marvel, *J. Am. Chem. Soc.*, **51,** 3496 (1929).

[65] Seyferth, Heeren, and Hughes, *J. Am. Chem. Soc.*, **84,** 1764 (1962).

obtained in 20% yield, could take place only by path B. Similarly, interaction of triphenylmethylphosphonium bromide with *p*-deuteriophenyllithium gave benzene in addition to deuteriobenzene.

$$(C_6H_5)_3\overset{\oplus}{P}{-}CH_3Br^{\ominus} + LiC_6H_4D \xrightarrow{A} (C_6H_5)_3P{=}CH_2 + C_6H_5D$$

$$\downarrow B$$

$$(C_6H_5)_3P(C_6H_4D){-}CH_3 \longrightarrow (C_6H_5)_3P{=}CH_2 \text{ or } (C_6H_5)_2P(C_6H_4D){=}CH_2 + C_6H_6$$

The reaction of tetraphenylphosphonium bromide with methyllithium must have proceeded via path B exclusively, since benzene was obtained in quantitative yield. Path A is ruled out in this reaction since the α-carbon atom does not carry a hydrogen atom.

$$(C_6H_5)_4\overset{\oplus}{P}Br^{\ominus} + CH_3Li \xrightarrow{B} (C_6H_5)_3P{=}CH_2 + C_6H_6$$

These results indicate that phenyllithium rather than butyllithium will be preferred for the organometallic preparation of ylides, since butyllithium may give rise to the formation of butylidene phosphoranes in addition to the desired ylide. It must be stressed, however, that

$$(C_6H_5)_3\overset{\oplus}{P}{-}CHR_1R_2\ Br^{\ominus} + C_4H_9Li \xrightarrow{A} (C_6H_5)_3P{=}CR_1R_2 + C_4H_{10}$$

$$\downarrow B$$

$$(C_6H_5)_3P(C_4H_9){-}CHR_1R_2 \longrightarrow (C_6H_5)_2P(C_4H_9){=}CR_1R_2 \text{ or } (C_6H_5)_2P(=CHC_3H_7){-}CHR_1R_2$$

certain reactions proceed by path A exclusively. Thus the reaction of benzyltriphenylphosphonium bromide with methyllithium apparently does not go by way of a pentavalent intermediate, since benzene was not a product.[65]

A side reaction involving attack of the base directly on phosphorus can also occur in the alkoxide method, but only at elevated temperatures.[66,67]

$$(C_6H_5)_3\overset{\oplus}{P}\text{—}CH_2C_6H_5\ Cl^{\ominus} + C_2H_5O^{\ominus} \rightarrow (C_6H_5)_3P(\text{—}CH_2C_6H_5)(\text{—}O\text{—}C_2H_5) \rightarrow (C_6H_5)_3PO + C_6H_5C_3H_7$$

The main reaction is the reversible removal of a proton from the α-carbon atom. The phosphorane so obtained is in equilibrium with the alcohol, as shown by the formation of tritium-labeled ylides on addition of $C_2H_5OT$.[68] The same method was used to prove the existence of a

$$(C_6H_5)_3P\text{=}CHR \underset{-C_2H_5OT}{\overset{+C_2H_5OT}{\rightleftharpoons}} (C_6H_5)_3\overset{\oplus}{P}\text{—}CHRT\ \ C_2H_5O^{\ominus} \underset{+C_2H_5OH}{\overset{-C_2H_5OH}{\rightleftharpoons}} (C_6H_5)_3P\text{=}CTR$$

resonance interaction in the allylidene phosphoranes in which the tritium label was found on both the α- and the γ-carbon atoms.

$$R_3P\text{=}CHCH\text{=}CHC_6H_5 \leftrightarrow R_3\overset{\oplus}{P}\text{—}\overset{\ominus}{C}HCH\text{=}CHC_6H_5 \leftrightarrow R_3\overset{\oplus}{P}\text{—}CH\text{=}CH\overset{\ominus}{C}HC_6H_5$$

$$\downarrow \begin{matrix} +C_2H_5OT \\ -C_2H_5OH \end{matrix}$$

$$R_3P\text{=}C(T)\text{—}CH\text{=}CHC_6H_5 + R_3P\text{=}CH\text{—}CH\text{=}C(T)C_6H_5$$

As far as is known, these ylides react in the Wittig reaction exclusively on the α-carbon atom, but the resonance effect is evident in some cases

[66] Friedrich and Henning, *Chem. Ber.*, **92**, 2944 (1959).

[67] Hey and Ingold, *J. Chem. Soc.*, **1933**, 531.

[68] Bestmann, Kratzer, and Simon, *Chem. Ber.*, **95**, 2750 (1962).

by the *cis-trans* isomerization of the partial allylic double bond which occurs more or less readily at room temperature. Thus the ylene **13** reacts with cyclohexanone at −25° with the almost exclusive formation of the expected *cis*-diene **14,** whereas at +40° a 20% yield of the *trans* isomer **15** is also obtained.[69] In order to get pure isomers, it is therefore

O

H

CH$_2$OH

+ $(C_6H_5)_3P{=}C$

C=

H

**13**

−25°

40°

H

CH$_2$OH

=C

C=

H

**14**

H

=C

C=

H

CH$_2$OH

+ **14**

**15**

necessary to work at low temperatures. This requirement poses no problem in view of the high reactivity of the allylidene phosphoranes.

Hydroxyl groups in the ylide may be protected by conversion to acetals or tetrahydropyranyl ethers.[70–72] Example **13** shows that this is not absolutely necessary provided sufficient base is present to convert the alcohol to alkoxide.

Side reactions in the preparation of alkylidene phosphoranes from phosphonium salts are very rare and can usually be avoided by suitable choice of conditions. In fact, complications are to be expected only if the ylide carries a substituent in the $\beta$-position to the phosphorus atom which

[69] Harrison and Lythgoe, *J. Chem. Soc.*, **1958,** 843.

[70] Inhoffen, Brückner, and Hess, *Chem. Ber.*, **88,** 1850 (1955).

[71] Bohlmann, Bornowski, and Herbst, *Chem. Ber.*, **93,** 1931 (1960).

[72] Bohlmann and Ruhnke, *Chem. Ber.*, **93,** 1945 (1960).

is prone to nucleophilic displacement, or if a proton in the $\beta$-position is so acidic that Hofmann elimination will be favored over ylide formation. Thus the ylides **16** and **17**[73] decompose very readily, as evidenced by the rapid disappearance of the color.

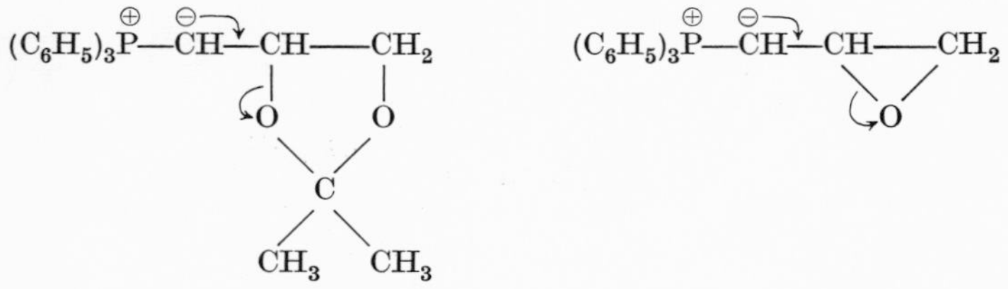

$$(C_6H_5)_3\overset{\oplus}{P}—\overset{\ominus}{C}H—CH—CH_2 \text{ (O—C(CH}_3)_2\text{—O ketal ring)}$$

**16**

$$(C_6H_5)_3\overset{\oplus}{P}—\overset{\ominus}{C}H—CH—CH_2 \text{ (epoxide, O)}$$

**17**

Halogen atoms also may undergo an intramolecular nucleophilic displacement, which in special cases leads to cyclic products.[74,75]

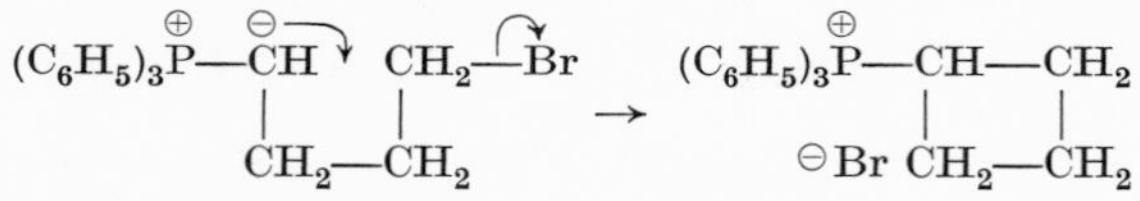

$$(C_6H_5)_3\overset{\oplus}{P}—\overset{\ominus}{C}H(CH_2—CH_2—CH_2—Br) \rightarrow (C_6H_5)_3\overset{\oplus}{P}—\underset{|}{C}H—CH_2\ \ \overset{\ominus}{Br}\ CH_2—CH_2$$

Attempts to prepare a heterocyclic phosphorus ylene from the phosphonium salt **18** were not successful; Hofmann elimination with ring opening occurred instead.[76] On the other hand, certain other cyclic

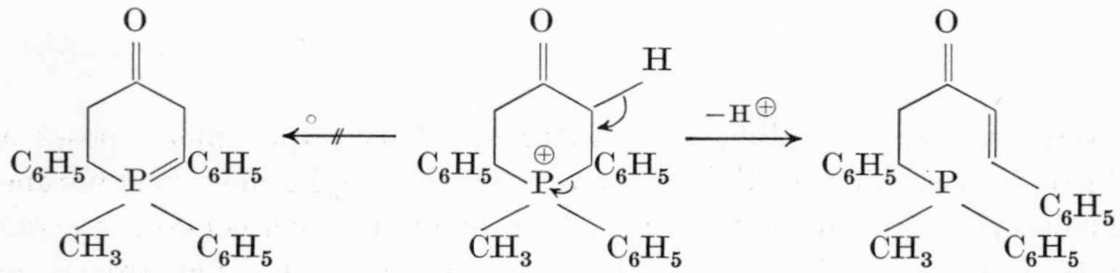

$$\text{C}_6\text{H}_5,\ \text{C}_6\text{H}_5\text{-substituted P(CH}_3\text{)(C}_6\text{H}_5\text{) ylide (4-oxo ring)} \xleftarrow{\ \ /\!/\ \ } \mathbf{18} \xrightarrow{-H^{\oplus}} \text{ring-opened P(CH}_3\text{)(C}_6\text{H}_5\text{)(C}_6\text{H}_5\text{) enone with }C_6H_5$$

**18**

alkylidene phosphoranes can be prepared.[77]

The bis-phosphonium salt **19** also undergoes Hofmann elimination under the influence of phenyllithium in ether with the formation of triphenylvinylphosphonium bromide and triphenylphosphine, the mono ylide **20**

[73] Bohlmann and Herbst, *Chem. Ber.*, **92**, 1319 (1959).

[74] Mondon, *Ann.*, **603**, 115 (1957).

[75] Bestmann and Häberlein, *Z. Naturforsch.*, **17b**, 787 (1962).

[76] Welcher and Day, *J. Org. Chem.*, **27**, 1824 (1962).

[77] Märkl, *Angew. Chem.*, **75**, 168 (1963).

being a probable intermediate.[52]

$$(C_6H_5)_3\overset{\oplus}{P}—CH_2CH_2—\overset{\oplus}{P}(C_6H_5)_3 \ \ Br^{\ominus} \ \ Br^{\ominus} \xrightarrow{C_6H_5Li}$$

**19**

$$(C_6H_5)_3\overset{\oplus}{P}—\overset{\ominus}{C}H—CH_2—\overset{\oplus}{P}(C_6H_5)_3 \ Br^{\ominus} \longrightarrow (C_6H_5)_3\overset{\oplus}{P}—CH{=}CH_2 \ Br^{\ominus} + (C_6H_5)_3P$$

**20**

The ylide **22,** a vinylog of **20,** also eliminates triphenylphosphine by a process involving the electrons of the central double bond.[78,79]

$$(C_6H_5)_3\overset{\oplus}{P}—CH_2—CH{=}CH—CH_2—\overset{\oplus}{P}(C_6H_5)_3 \ \ Br^{\ominus} \ \ Br^{\ominus}$$

**21**

$$\big\downarrow \ C_6H_5Li \text{ or } Na_2CO_3$$

$$(C_6H_5)_3\overset{\oplus}{P}—\overset{\ominus}{C}H—CH{=}CH—CH_2—\overset{\oplus}{P}(C_6H_5)_3 \ Br^{\ominus}$$

**22**

$$\big\downarrow$$

$$(C_6H_5)_3\overset{\oplus}{P}—CH{=}CH—CH{=}CH_2 \ Br^{\ominus} + (C_6H_5)_3P$$

If such unstable alkylidene phosphoranes are encountered in the course of a Wittig reaction, it is necessary to carry out the condensation with carbonyl compounds very rapidly and, if possible, *in statu nascendi.* Best suited for this purpose is the alkoxide method which permits the addition of aldehydes or ketones before the ylide is prepared. Thus, if the bis-phosphonium salt **21** is treated with lithium ethoxide in ethanol in the presence of aldehydes,[79a] decomposition of the mono ylide **22** is prevented and the normal products of the Wittig reaction are obtained.

[78] H. Burger, Doctoral Dissertation, Universität Tübingen, 1958.

[79] Ford and Wilson, *J. Org. Chem.*, **26,** 1433 (1961).

[79a] Heitman, Sperna Weiland, and Huisman, *Koninkl. Ned. Akad. Wetenschap., Proc., Ser. B*, **64,** 165 (1961) [*C.A.*, **55,** 17562f (1961)].

Bifunctional ylides with several central double bonds or with a central triple bond **(23, 24)** can be prepared without complication by the organometallic method, since the initially formed mono ylides are more stable than the mono ylide **22.**[80]

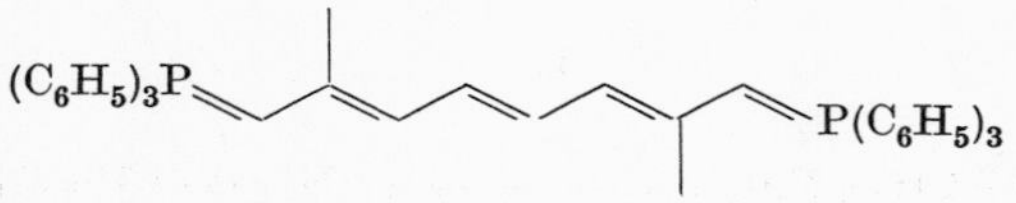

$(C_6H_5)_3P$ ... $P(C_6H_5)_3$

**23**

$(C_6H_5)_3P$ ... $-C{\equiv}C-$ ... $P(C_6H_5)_3$

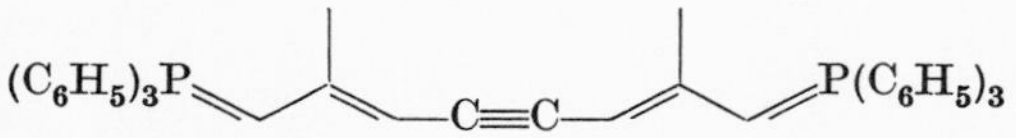

**24**

The interaction of bases with methylene bis(triphenylphosphonium bromide) **(25)**[81] leads to interesting ylides in which the two positively charged phosphorus atoms are separated by only one carbon atom. On addition of aqueous sodium carbonate there is first obtained the colorless, resonance-stabilized mono ylide **26,** which is then converted under the influence of metallic potassium in diglycol dimethyl ether to the stable yellow hexaphenylcarbodiphosphorane **27.**

$$(C_6H_5)_3\overset{\oplus}{P}-CH_2-\overset{\oplus}{P}(C_6H_5)_3$$

**25**

$$\downarrow Na_2CO_3$$

$$(C_6H_5)_3\overset{\oplus}{P}-\overset{\ominus}{C}H-\overset{\oplus}{P}(C_6H_5)_3 \leftrightarrow (C_6H_5)_3P{=}CH-\overset{\oplus}{P}(C_6H_5)_3 \leftrightarrow (C_6H_5)_3\overset{\oplus}{P}-CH{=}P(C_6H_5)_3$$

**26**

$$\downarrow K$$

$$(C_6H_5)_3P{=}C{=}P(C_6H_5)_3$$

$$\updownarrow$$

$$(C_6H_5)_3\overset{\oplus}{P}-\overset{\ominus\ominus}{C}-\overset{\oplus}{P}(C_6H_5)_3, \quad \text{etc.}$$

**27**

[80] Surmatis and Ofner, *J. Org. Chem.*, **26,** 1171 (1961).

[81] Ramirez, Desai, Hansen, and McKelvie, *J. Am. Chem. Soc.*, **83,** 3539 (1961).

The same mono ylide **26** was obtained by the elimination of bromine from dibromomethyltriphenylphosphonium bromide with triphenylphosphine.[82]

$$(C_6H_5)_3\overset{\oplus}{P}{-}CHBr_2\ Br^{\ominus} + (C_6H_5)_3P \rightleftharpoons (C_6H_5)_3P{=}CHBr + (C_6H_5)_3PBr_2$$

$$\xrightarrow{(C_6H_5)_3P} (C_6H_5)_3P{=}CH{-}\overset{\oplus}{P}(C_6H_5)_3\ Br^{\ominus}\quad \mathbf{26}$$

In the preparation of alkylidene phosphoranes from α-haloalkylphosphonium salts and organometallic compounds the possibility of halogen-metal interchange reactions must also be taken into account.[83,84]

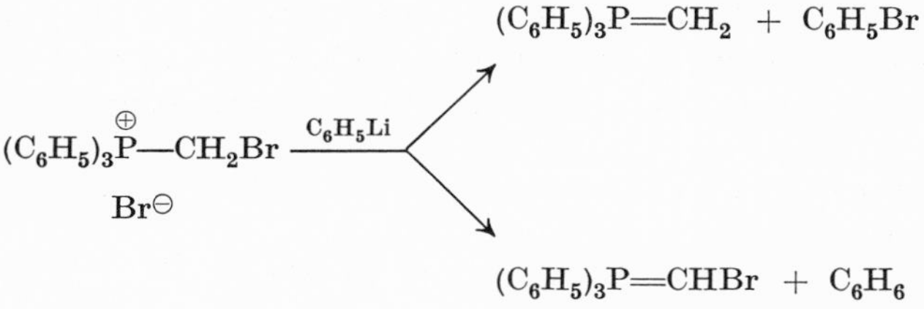


Bromomethyltriphenylphosphonium bromide will react with butyllithium in ether with the exclusive formation of methylenetriphenylphosphorane, whereas with lithium piperidide no halogen-metal interchange takes place.[84]

$$(C_6H_5)_3P{=}CHBr \xleftarrow{C_5H_{10}NLi} (C_6H_5)_3\overset{\oplus}{P}{-}CH_2Br\ Br^{\ominus} \xrightarrow{C_4H_9Li} (C_6H_5)_3P{=}CH_2$$

The interaction of bases with chloromethyltriphenylphosphonium halides leads to the formation of chloromethylenetriphenylphosphorane **(28)**. Under the further influence of bases this compound undergoes an interesting rearrangement in which a chloride ion is lost and a phenyl group migrates from phosphorus to the methylene carbon atom.[85,86]

[82] Ramirez, Desai, and McKelvie, *J. Am. Chem. Soc.*, **84**, 1745 (1962).
[83] Seyferth, Heeren, and Grim, *J. Org. Chem.*, **26**, 4783 (1961).
[84] Köbrich, *Angew. Chem.*, **74**, 33 (1962).
[85] Schlosser, *Angew. Chem.*, **74**, 291 (1962).
[86] Hellmann and Bader, *Tetrahedron Letters*, **1961**, 724.

Although the mechanism is not entirely clear, it may be formulated as in the accompanying equations.[85]

$$(C_6H_5)_3P{=}CHCl \;(\mathbf{28}) + R^{\ominus} \rightarrow (C_6H_5)_3P(R){-}\overset{\ominus}{C}HCl \xrightarrow{-Cl^{\ominus}} (C_6H_5)_2P(R){=}CHC_6H_5 \xrightarrow{\text{When } R=OH} (C_6H_5)_2P(=O){-}CH_2C_6H_5$$

$(R = C_6H_5,\ C_4H_9,\ OH)$

Methylene phosphoranes in which one hydrogen atom of the methylene group is replaced by an alkoxide group decompose in an even more complicated way. The autodecomposition of *n*-butoxymethylenetriphenylphosphorane **(29)** gave, in addition to triphenylphosphine as the main product, and depending on the conditions, varying amounts of 1,2-di-*n*-butoxyethylene, di-*n*-butoxymethane, 1-butene, and *n*-butanol.[87]

$$(C_6H_5)_3P{=}CHOC_4H_9\text{-}n \;(\mathbf{29}) \xrightarrow[\text{20 hr.}]{\text{Ether, 45°}} \begin{cases} \underset{(64\%)}{(C_6H_5)_3P} + \underset{(16\%)}{n\text{-}C_4H_9OCH{=}CHOC_4H_9\text{-}n} + \\ \underset{(2.4\%)}{CH_2(OC_4H_9\text{-}n)_2} + \underset{(1.5\%)}{C_2H_5CH{=}CH_2} + \underset{(12\%)}{n\text{-}C_4H_9OH} \end{cases}$$

The mechanism has not been elucidated, but it has been suggested that the phosphorane **29** decomposes initially into triphenylphosphine and *n*-butoxycarbene. *t*-Butoxymethylenetriphenylphosphorane **(30)** and diphenoxymethylenetriphenylphosphorane **(31)** decompose in an essentially analogous manner.[87]

$$\underset{\mathbf{30}}{(C_6H_5)_3P{=}CHOC(CH_3)_3} \qquad \underset{\mathbf{31}}{(C_6H_5)_3P{=}C(OC_6H_5)_2}$$

Even these phosphoranes, however, may be used in a Wittig reaction by operating at a sufficiently low temperature so that their rate of decomposition is slow.

So far only methods for the preparation of alkylidene phosphoranes which use phosphonium salts as starting materials have been mentioned. There are, however, other theoretically very interesting ways of preparation which do not start with phosphonium salts. The first alkylidene

[87] Wittig and Böll, *Chem. Ber.*, **95**, 2526 (1962).

phosphorane **1** was obtained by the thermal decomposition of the phosphazine **2**.[7,8] The applicability of this method remained limited, however,

$$(C_6H_5)_3P + (C_6H_5)_2CN_2 \longrightarrow (C_6H_5)_3P{=}N{-}N{=}C(C_6H_5)_2 \xrightarrow{190^\circ} (C_6H_5)_3P{=}C(C_6H_5)_2$$

**2** **1**

since thermal elimination of nitrogen usually occurs at temperatures at which the phosphoranes are unstable. Even the similarly resonance-stabilized fluorenylidenetriphenylphosphorane **(11)** could not be prepared by this route.[88] The action of diazomethane on triphenylphosphine in ethereal solution gives formaldehyde triphenylphosphazine **(32)**,[3] which at elevated temperatures decomposes into the starting materials.

$$(C_6H_5)_3P + CH_2N_2 \rightleftharpoons (C_6H_5)_3P{=}N{-}N{=}CH_2$$

**32**

The preparation of methylenetriphenylphosphorane from diazomethane did, however, succeed under the catalytic influence of metal salts such as cuprous chloride[89] in the presence of triphenylphosphine.[90]

$$(C_6H_5)_3P + CH_2N_2 \xrightarrow{Cu^{\oplus}} (C_6H_5)_3P{=}CH_2 + N_2$$

Similarly other diazo compounds such as phenyldiazomethane, diazoacetic ester, or diazoacetophenone have been converted to the corresponding ylides. Since the ylides can undergo side reactions with excess diazo compound, it is best to generate them in the presence of carbonyl compounds if they are to be used for the preparation of olefins.

The addition of carbenes to triphenylphosphine, simultaneously discovered by Wittig,[91,92] Speziale,[93,94] Seyferth,[95,96] and their co-workers, also leads to alkylidene phosphoranes which were used for the syntheses of monohalo- and 1,1-dihalo-olefins. The action of *n*-butyllithium on a solution of triphenylphosphine in methylene chloride at $-60^\circ$, for instance, leads to chloromethylenetriphenylphosphorane.[91,92,95,96]

$$(C_6H_5)_3P + CH_2Cl_2 \xrightarrow{n\text{-}C_4H_9Li} (C_6H_5)_3P{=}CHCl$$

[88] Horner and Lingnau, *Ann.*, **591,** 135 (1955).
[89] Wittig and Schwarzenbach, *Angew. Chem.*, **71,** 652 (1959); *Ann.*, **650,** 1 (1961).
[90] Wittig and Schlosser, *Tetrahedron*, **18,** 1023 (1962).
[91] Wittig and Schlosser, *Angew. Chem.*, **72,** 324 (1960).
[92] Wittig and Schlosser, *Chem. Ber.*, **94,** 1373 (1961).
[93] Speziale, Marco, and Ratts, *J. Am. Chem. Soc.*, **82,** 1260 (1960).
[94] Speziale and Ratts, *J. Am. Chem. Soc.*, **84,** 854 (1962).
[95] Seyferth, Grim, and Read, *J. Am. Chem. Soc.*, **82,** 1510 (1960).
[96] Seyferth, Grim, and Read, *J. Am. Chem. Soc.*, **83,** 1617 (1961).

In an analogous manner dihalomethylene triphenylphosphoranes were prepared from triphenylphosphine, haloform, and potassium *t*-butoxide in heptane.[93,94] The preparation of difluoromethylenetriphenylphosphorane[97] by a similar method could not be reproduced.[94]

$$(C_6H_5)_3P + CHX_3 \xrightarrow[(X=Cl,\ Br)]{KOC_4H_9\text{-}t} (C_6H_5)_3P{=}CX_2$$

Dihalomethylene triphenylphosphoranes can also be prepared in good yields by the action of triphenylphosphine on tetrahalomethanes. Heating a mixture of triphenylphosphine, carbon tetrachloride, and benzophenone to 60° for 4 hours, for instance, gave 1,1-diphenyl-2,2-dichloroethylene **(33)** in 78% yield.[82,98]

$$2(C_6H_5)_3P + CCl_4 \rightarrow (C_6H_5)_3P{=}CCl_2 + (C_6H_5)_3PCl_2$$

$$(C_6H_5)_3P{=}CCl_2 + (C_6H_5)_2C{=}O \rightarrow \underset{\mathbf{33}}{(C_6H_5)_2C{=}CCl_2} + (C_6H_5)_3PO$$

$$2(C_6H_5)_3P + BrCCl_3 \rightarrow (C_6H_5)_3P{=}CCl_2 + (C_6H_5)_3PBrCl$$

$$2(C_6H_5)_3P + Cl_2CF_2 \rightarrow (C_6H_5)_3P{=}CF_2 + (C_6H_5)_3PCl_2$$

$$2(C_6H_5)_3P + CBr_4 \rightarrow (C_6H_5)_3P{=}CBr_2 + (C_6H_5)_3PBr_2$$

Similarly, tetrabromomethane reacts with triphenylphosphine in methylene chloride to give dibromomethylenetriphenylphosphorane, which was trapped with benzaldehyde to give $\beta,\beta$-dibromostyrene in 84% yield.[82]

$$(C_6H_5)_3P{=}CBr_2 + C_6H_5CHO \rightarrow C_6H_5CH{=}CBr_2 + (C_6H_5)_3PO$$

Benzoquinone and triphenylphosphine readily combine to give the yellow-green resonance-stabilized phosphorane **34**.[99]

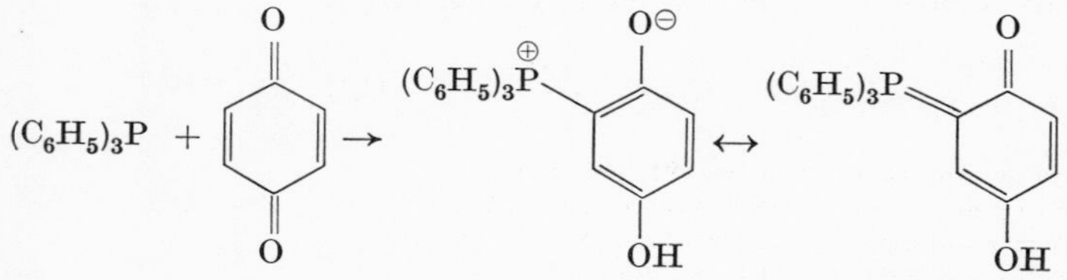


**34**

[97] Franzen, *Angew. Chem.*, **72**, 566 (1960).
[98] Rabinowitz and Marcus, *J. Am. Chem. Soc.*, **84**, 1312 (1962).
[99] Ramirez and Dershowitz, *J. Am. Chem. Soc.*, **78**, 5614 (1956).

Another method for the preparation of resonance-stabilized alkylidene phosphoranes uses triphenylphosphine dichloride as a starting material.[45]

$$(C_6H_5)_3PCl_2 + CH_2(X)(Y) \xrightarrow{(C_2H_5)_3N} (C_6H_5)_3P{=}C(X)(Y)$$

$(X, Y = CO_2R, CN, COCH_3, SO_2C_6H_5)$

Finally, the phosphocyanine dye **35**,[100] a resonance-stabilized phosphorane, may be considered to be a substituted vinylog of the mono ylide **26.**

$$2(C_6H_5)_3\overset{\oplus}{P}{-}CH_2CO_2C_2H_5\ Cl^{\ominus} + (C_2H_5O)_2CHCH_2CH(OC_2H_5)_2$$

$$\downarrow C_5H_5N,\ NaClO_4$$

$$(C_6H_5)_3P{=}\underset{CO_2C_2H_5}{C}{-}CH{=}CH{-}CH{=}\underset{CO_2C_2H_5}{C}{-}\overset{\oplus}{P}(C_6H_5)_3\ \ ClO_4^{\ominus}$$

**35**

## Reactions of Alkylidene Phosphoranes

Some of the reactions of the Wittig reagents reflect their markedly basic properties. All ylides react with acids to form phosphonium salts.[1]

$$(C_6H_5)_3P{=}CH_2 + HCl \rightarrow (C_6H_5)_3\overset{\oplus}{P}{-}CH_3\ Cl^{\ominus}$$

Similarly, hydrolysis[1] of unstable alkylidene phosphoranes gives phosphonium hydroxides which usually decompose irreversibly into a phosphine oxide and a hydrocarbon.[101] The most electronegative group is always removed from phosphorus.

$$(C_6H_5)_3P{=}CH_2 \xrightarrow{H_2O} (C_6H_5)_3\overset{\oplus}{P}{-}CH_3\ HO^{\ominus} \longrightarrow (C_6H_5)_2\overset{O}{\overset{\|}{P}}CH_3 + C_6H_6$$

[100] Kukhtin, Kazymov, and Voskoboeva, *Dokl. Akad. Nauk SSSR*, **140**, 601 (1961).

[101] Fenton and Ingold, *J. Chem. Soc.*, **1929**, 2342.

In order to hydrolyze resonance-stabilized alkylidene phosphoranes it is necessary to apply elevated temperatures and to use aqueous or alcoholic solutions of alkali metal hydroxides, as shown in the example of fluorenylidenetriphenylphosphorane **(11)**.[49]

$$\text{Fluorenylidene}{=}P(C_6H_5)_3 \ (\mathbf{11}) \xrightarrow{NaOH/C_2H_5OH} \text{Fluorene} + (C_6H_5)_3PO$$

Hydrolysis of acylalkylidene triphenylphosphoranes[31,32] gives ketones; of carbomethoxyalkylidene triphenylphosphoranes,[40] carboxylic acids. In

$$(C_6H_5)_3P{=}C(R)COR' \xrightarrow[HO^{\ominus}]{H_2O} (C_6H_5)_3PO + RCH_2COR'$$

$$(C_6H_5)_3P{=}C(R)CO_2CH_3 \xrightarrow[HO^{\ominus}]{2H_2O} (C_6H_5)_3PO + RCH_2CO_2H + CH_3OH$$

certain cases, phosphonium hydroxides are quite stable. Hexaphenylcarbodiphosphorane **(27)**, for instance, dissolves readily in water to give a strong diacidic base **(36)** that can be titrated with hydrochloric acid. The elimination of benzene with the formation of a new ylene **37** which is stabilized by a phosphoryl group occurs only very slowly.[81]

$$\underset{\mathbf{27}}{(C_6H_5)_3P{=}C{=}P(C_6H_5)_3} \rightarrow \underset{\mathbf{36}}{(C_6H_5)_3P{=}CH{-}\overset{\oplus}{P}(C_6H_5)_3 \ OH^{\ominus}}$$

$$\mathbf{36} \xrightarrow[-H_2O]{+2H^{\oplus}} (C_6H_5)_3\overset{\oplus}{P}{-}CH_2{-}\overset{\oplus}{P}(C_6H_5)_2$$

$$\mathbf{36} \rightarrow \underset{\mathbf{37}}{(C_6H_5)_3P{=}CH{-}P(O)(C_6H_5)_2} + C_6H_6$$

The addition of alkyl halides to alkylidene phosphoranes leads to phosphonium halides.[51,102]

$$(CH_3)_3P{=}CH_2 + CH_3I \rightarrow (CH_3)_3\overset{\oplus}{P}{-}CH_2CH_3 \quad I^{\ominus}$$

$$\text{9-fluorenylidene-}P(CH_3)_3 + CH_3I \rightarrow \text{9-methyl-9-fluorenyl-}\overset{\oplus}{P}(CH_3)_3 \; I^{\ominus}$$

Similarly, trialkyloxonium fluoborates give the corresponding phosphonium fluoborates.[102,103]

$$(C_6H_5)_3P{=}CHC_6H_5 + (C_2H_5)_3\overset{\oplus}{O}\overset{\ominus}{B}F_4 \rightarrow (C_6H_5)_3\overset{\oplus}{P}{-}CH(C_2H_5)C_6H_5 \quad BF_4^{\ominus}$$

Resonance-stabilized ylides are not necessarily attacked on the α-carbon atom. Acylmethylene triphenylphosphoranes, for instance, are alkylated on oxygen.[31]

$$(C_6H_5)_3P{=}CH{-}C({=}O)R + C_2H_5I \rightarrow (C_6H_5)_3\overset{\oplus}{P}{-}CH{=}C(OC_2H_5)R \quad I^{\ominus}$$

Carbomethoxymethylenetriphenylphosphorane, on the other hand, is alkylated on carbon.[40,104]

$$(C_6H_5)_3P{=}CHCO_2CH_3 + RX \rightarrow (C_6H_5)_3\overset{\oplus}{P}{-}CH(R)CO_2CH_3 \quad X^{\ominus}$$

An electron-withdrawing group, R, will increase the acidity of the newly formed phosphonium salt relative to that of the unsubstituted one, and the salt therefore reacts with excess starting ylide by *trans* ylidation with

[102] Wittig and Rieber, *Ann.*, **562,** 177 (1949).

[103] Märkl, *Tetrahedron Letters*, **1962,** 1027.

[104] Bestmann and Schulz, *Tetrahedron Letters*, **4,** 5 (1960).

formation of a new phosphorane.[16,105]

$$\underset{X^{\ominus}}{(C_6H_5)_3\overset{\oplus}{P}}—\underset{R}{\underset{|}{CHCO_2CH_3}} + (C_6H_5)_3P{=}CHCO_2CH_3 \rightleftharpoons (C_6H_5)_3P{=}\underset{R}{\underset{|}{C}}CO_2CH_3 + \underset{X^{\ominus}}{(C_6H_5)_3\overset{\oplus}{P}CH_2CO_2CH_3}$$

The substituted carbomethoxymethylene phosphoranes yield, on hydrolysis, carboxylic acids in which the carbon chain of the alkyl halide used for the alkylation has been extended by two carbon atoms. The result is similar to that of a malonic ester synthesis.

If phenacyl bromide or other α-bromo ketones are used as alkylating agents, the initially formed phosphonium salt undergoes a Hofmann elimination instead of *trans* ylidation, with the formation of α,β-unsaturated γ-ketonic esters.[106]

$$\underset{Br^{\ominus}}{(C_6H_5)_3\overset{\oplus}{P}}—\underset{H—CH—COC_6H_5}{\underset{|}{CH}}—CO_2CH_3 + (C_6H_5)_3P{=}CHCO_2CH_3 \longrightarrow (C_6H_5)_3P + \begin{matrix}CHCO_2CH_3\\ \|\\ CHCOC_6H_5\end{matrix} + \underset{Br^{\ominus}}{(C_6H_5)_3\overset{\oplus}{P}}—CH_2CO_2CH_3$$

Similarly the reaction of phenacyl bromide with benzoylmethylenetriphenylphosphorane gives mainly *trans*-dibenzoylethylene **(38)**; the formation of some *trans*-tribenzoylcyclopropane **(39)** in this reaction led to the proposal of a carbene mechanism.[107,108]

$$(C_6H_5)_3P{=}CHCOC_6H_5 + C_6H_5COCH_2Br \rightarrow [C_6H_5COCH{:}] + \underset{Br^{\ominus}}{(C_6H_5)_3\overset{\oplus}{P}CH_2COC_6H_5}$$

$$\underset{\mathbf{38}}{C_6H_5COCH{=}CHCOC_6H_5} \xrightarrow{[C_6H_5COCH:]} \underset{\mathbf{39}}{\text{1,2,3-tri}(COC_6H_5)\text{cyclopropane}}$$

[105] Bestmann, *Chem. Ber.*, **95**, 58 (1962).

[106] Bestmann and Schulz, *Angew. Chem.*, **73**, 620 (1961).

[107] Siemiatycki and Strzelecka, *Compt. Rend.*, **250**, 3489 (1960).

[108] Strzelecka, Simalty-Siemiatycki, and Prévost, *Compt. Rend.*, **253**, 491 (1961).

The postulation of carbene intermediates is not necessary since, as will be shown later, ylides are capable of adding to activated double bonds with the formation of cyclopropane derivatives. Phosphoranes containing halogen atoms can undergo an intramolecular carbon alkylation[74,109] and may thus be used for the synthesis of cyclic compounds such as phenanthrene.[109]

$$CH_2Br \quad CH{=}P(C_6H_5)_3 \longrightarrow CH{-}CH \quad H \quad \overset{\oplus}{P}(C_6H_5)_3 \quad Br^{\ominus} \xrightarrow{NaOC_2H_5} \text{phenanthrene}$$

Alkylidene phosphoranes react readily with acid chlorides. The initially formed phosphonium salts undergo *trans* ylidation very easily owing to the strongly electron-withdrawing effect of the acyl group.[32,41,58,110]

$$(C_6H_5)_3P{=}CHR + R'COCl \rightarrow (C_6H_5)_3\overset{\oplus}{P}{-}CH(COR')R \quad Cl^{\ominus}$$

$$\xrightarrow{(C_6H_5)_3P=CHR} (C_6H_5)_3P{=}C(COR')R + (C_6H_5)_3\overset{\oplus}{P}CH_2R \quad Cl^{\ominus}$$

Strongly basic ylides may eliminate hydrogen chloride from acid chlorides containing activated α-hydrogen atoms (e.g., phenylacetyl chloride). The ketenes so formed react with the alkylidene phosphoranes to give allenes.[32]

Chloroformic esters and alkylidene triphenylphosphoranes react to give carbalkoxymethylene triphenylphosphoranes.[111]

$$2(C_6H_5)_3P{=}CHR + ClCO_2CH_3 \rightarrow$$

$$(C_6H_5)_3P{=}CRCO_2CH_3 + (C_6H_5)_3\overset{\oplus}{P}{-}CH_2R \quad Cl^{\ominus}$$

109 Bestmann and Häberlein, *Z. Naturforsch.*, **17b**, 787 (1962).

110 Bestmann, *Tetrahedron Letters*, **4**, 7 (1960).

111 Bestmann and Schulz, *Angew. Chem.*, **73**, 27 (1961).

In a reaction similar to that of acid chlorides, carboxylic esters react with alkylidene phosphoranes with the formation of phosphonium alkoxides.[2]

$$(C_6H_5)_3P{=}CH_2 + C_6H_5CO_2C_2H_5 \rightarrow (C_6H_5)_3\overset{\oplus}{P}{-}CH_2COC_6H_5 \quad {}^{\ominus}OC_2H_5$$

$$\rightleftharpoons (C_6H_5)_3P{=}CHCOC_6H_5 + C_2H_5OH$$

Using thiocarboxylic acid S-ethyl esters, a convenient method for the preparation of acylalkylidene triphenylphosphoranes was developed.[32,112] Removal of the volatile mercaptan shifts the equilibrium from the initially formed phosphonium ethyl mercaptide completely to the side of the corresponding ylene. Compared with the acylation with acid

$$(C_6H_5)_3P{=}CHR + R'COSC_2H_5 \rightarrow (C_6H_5)_3\overset{\oplus}{P}{-}CHR(COR') \quad C_2H_5S^{\ominus}$$

$$\rightleftharpoons (C_6H_5)_3P{=}CRCOR' + C_2H_5SH$$

chlorides, this method has the advantage that the components may be used in a 1 : 1 ratio since deprotonation of the initially formed phosphonium salt is carried out by mercaptide ion and not by an excess of the starting ylene. In addition, the possibility of ketene formation is reduced, and, consequently, the yields are usually considerably higher.

With formic esters, which may be considered to contain both an aldehyde and an ester function, phosphoranes react to give different products depending on the conditions used. Adding an ylide to an excess of ethyl formate results in the normal ester reaction products, in this case formylethylidenetriphenylphosphorane.[33,58]

$$(C_6H_5)_3P{=}CHR + HCO_2C_2H_5 \rightarrow$$

$$(C_6H_5)_3\overset{\oplus}{P}{-}CHR(CHO) \quad C_2H_5O^{\ominus} \rightleftharpoons (C_6H_5)_3P{=}CR(CHO) + C_2H_5OH$$

[112] Bestmann and Arnason, *Tetrahedron Letters*, **1961**, 455.

Inverse addition, on the other hand, yields the products of a Wittig reaction, namely, an enol ether and triphenylphosphine oxide.[113]

$$(C_6H_5)_3P{=}C_6H_{10} + HCO_2C_2H_5 \rightarrow C_6H_{10}{=}CHOC_2H_5 + (C_6H_5)_3PO$$

An explanation for these differing results may possibly be found in a solvation of the positively charged phosphorus by excess formic ester. In the first case, such a solvation might preclude the Wittig reaction by making a nucleophilic attack of the oxygen on phosphorus impossible; the zwitterion **40,** which is probably initially formed, therefore collapses in a different way. Support for this explanation is found in the fact that

$$(C_6H_5)_3PO + \begin{array}{c}CHR\\ \| \\ CHOC_2H_5\end{array} \leftarrow \underset{\mathbf{40}}{\begin{array}{l}(C_6H_5)_3\overset{\oplus}{P}\text{—}CHR\\ \qquad\quad | \\ {}^{\ominus}O\text{—}CHOC_2H_5\end{array}} \rightarrow \begin{array}{l}(C_6H_5)_3\overset{\oplus}{P}\text{——}CHR\\ \qquad\qquad\quad | \\ C_2H_5O^{\ominus} \quad CHO\end{array}$$

lithium bromide, which is known to shield phosphorus by complex formation,[114] increases the yields in the above-mentioned reaction of alkylidene phosphoranes with normal esters[32] in which a side reaction involving attack of the oxygen on phosphorus must also be expected.

The imidazolides, which are related to acid chlorides and esters, have been found to be particularly useful in the preparation of acylalkylidene triphenylphosphoranes.[115]

$$2(C_6H_5)_3P{=}CHR + R'COIm \rightarrow (C_6H_5)_3P{=}CRCOR' + (C_6H_5)_3\overset{\oplus}{P}\text{—}CH_2R \quad Im^{\ominus}$$

$$\left(Im = \text{N-imidazolyl}\right)$$

Using N-formylimidazole, the corresponding formylalkylidene triphenylphosphoranes **(41)** are obtained; with N,N′-carbonyldiimidazole, carboimidazolidoalkylidene triphenylphosphoranes **(42)**[116] are formed.

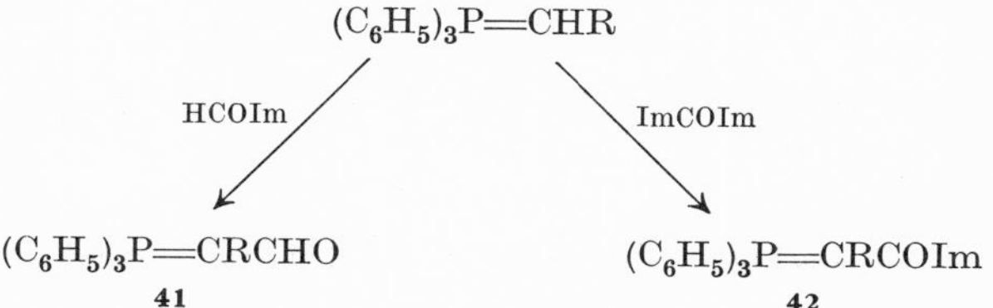


[113] Pommer and Wittig, Ger. pat. 1,047,763 (to BASF) (*Chem. Zentr.*, **1959,** 13577).

[114] Bergelson and Shemyakin, *Tetrahedron*, **19,** 149 (1963).

[115] Bestmann, Sommer, and Staab, *Angew. Chem.*, **74,** 293 (1962).

[116] Staab and Sommer, *Angew. Chem.*, **74,** 294 (1962).

The interaction of halogens with resonance-stabilized ylides, such as carbomethoxymethylenetriphenylphosphorane and acylmethylene triphenylphosphoranes, gives halogenated alkylidene phosphoranes.[59,60,117]

$$(C_6H_5)_3P{=}CHCOR + X_2 \longrightarrow (C_6H_5)_3\overset{\oplus}{P}{-}CHXCOR\ X^{\ominus} \xrightarrow{-HX} (C_6H_5)_3P{=}CXCOR$$

$(X = Cl, Br, I;\ R = CH_3O, C_2H_5O, CH_3, C_6H_5)$

The use of iodobenzene dichloride[59,117] or *t*-butyl hypochlorite[60] in place of chlorine has been found to be advantageous. Elimination of hydrogen halide is effected either by *trans* ylidation using excess phosphorane or by addition of bases such as sodium hydroxide, triethylamine, or pyridine.

The reaction with halogens is not limited to resonance-stabilized alkylidene phosphoranes, as is shown by the reaction of dibromomethylenetriphenylphosphorane with bromine, which leads to a product that cannot eliminate hydrogen halide.[82]

$$(C_6H_5)_3P{=}CBr_2 + Br_2 \rightarrow (C_6H_5)_3\overset{\oplus}{P}{-}CBr_3\ Br^{\ominus}$$

Hexaphenylcarbodiphosphorane **(27)** also reacts readily with bromine in methylene chloride.[81]

$$\underset{\mathbf{27}}{(C_6H_5)_3P{=}C{=}P(C_6H_5)_3} + Br_2 \rightarrow (C_6H_5)_3P{=}\underset{Br}{\underset{|}{C}}{-}\overset{\oplus}{P}(C_6H_5)_3\ Br^{\ominus}$$

Interaction of acylated carbomethoxymethylene triphenylphosphoranes with phosphorus pentachloride or Vilsmeier reagents, on the other hand, gives phosphonium salts halogenated in the $\beta$-position to the phosphorus atom.[118] Hydrolysis of these products gives alkyne carboxylic acids.

$$(C_6H_5)_3\overset{\oplus}{P}{-}\underset{\overset{\|}{^{\ominus}O{-}C{-}R}}{C}{-}CO_2CH_3 \xrightarrow{PCl_5} (C_6H_5)_3\overset{\oplus}{P}{-}\underset{\overset{\|}{Cl{-}C{-}R}}{C}{-}CO_2CH_3\ Cl^{\ominus} \xrightarrow[NaOH]{H_2O}$$

$$(C_6H_5)_3PO + RC{\equiv}CCO_2H + CH_3OH + HCl$$

Phosphonium salts substituted by phosphorus on the $\alpha$-carbon atom are obtained in the reaction of alkylidene triphenylphosphoranes with

[117] Märkl, *Chem. Ber.*, **94**, 2996 (1961).

[118] Märkl, *Angew. Chem.*, **74**, 217 (1962).

phenyldibromophosphine, diphenylbromophosphine,[119] or triphenylphosphine dibromide.[82]

Metalloidal, tin, and mercury halides undergo an interesting nucleophilic displacement of halide ion under the influence of ylides.[120–122] The products are phosphonium salts substituted on the α-carbon atom by metalloids, tin, or mercury.

$$(C_6H_5)_3P{=}CH_2 + (CH_3)_3SiBr \rightarrow (C_6H_5)_3\overset{\oplus}{P}CH_2Si(CH_3)_3 \quad Br^{\ominus}$$

Alkylidene phosphoranes are also capable of cleaving silicon-silicon bonds; for instance, methylenetriphenylphosphorane reacts with octaphenylcyclotetrasilane with opening of the ring.[123]

Metals may be used as reducing agents for ylides. Thus zinc in acetic acid converts acylmethylene triphenylphosphoranes to ketones and triphenylphosphine.[58]

$$(C_6H_5)_3P{=}CHCOC_6H_5 \xrightarrow[CH_3CO_2H]{Zn} (C_6H_5)_3P + CH_3COC_6H_5$$

The same result is obtained with Raney nickel, but triphenylphosphine is unstable toward Raney nickel and cannot be isolated.[124]

$$\text{9-fluorenylidene}{=}P(C_6H_5)_3 \xrightarrow{Ni} \text{fluorene}$$

Reduction of ylides with lithium aluminum hydride takes a different course. One phenyl group is removed from phosphorus with the formation of benzene.[125]

$$(C_6H_5)_3P{=}CHCOC_6H_5 \xrightarrow[2.\ H_2O]{1.\ LiAlH_4} (C_6H_5)_2PCH_2COC_6H_5 + C_6H_6$$

The reactive alkylidene phosphoranes are easily oxidized. They are so sensitive to oxygen that their preparation has to be carried out in an inert atmosphere. Oxidation leads, initially, to triphenylphosphine oxide and a carbonyl compound; the latter undergoes a Wittig reaction with

[119] Seyferth and Brändle, *J. Am. Chem. Soc.*, **83,** 2055 (1961).
[120] Grim and Seyferth, *Chem. Ind. (London)*, **1959,** 849.
[121] Seyferth, *Angew. Chem.*, **72,** 36 (1960).
[122] Seyferth and Grim, *J. Am. Chem. Soc.*, **83,** 1610 (1961).
[123] Gilman and Tomasi, *J. Org. Chem.*, **27,** 3647 (1962).
[124] Schönberg, Brosowski, and Singer, *Chem. Ber.*, **95,** 2984 (1962).
[125] Saunders and Burchman, *Tetrahedron Letters*, **1,** 8 (1959).

unoxidized ylide to form the symmetrical olefin in which both halves come from the alkylidene phosphorane.[126,127]

$$(C_6H_5)_3P{=}CRR' + O_2 \rightarrow (C_6H_5)_3PO + RR'C{=}O$$

$$(C_6H_5)_3P{=}CRR' + RR'C{=}O \rightarrow RR'C{=}CRR'$$

By this method, vitamin A was converted to β-carotene **(44)** via axerophthylenetriphenylphosphorane **(43)**.[128]

$$2\ \textbf{43} \xrightarrow{O_2} \textbf{44}$$

**43**

**44**

Oxidation of bifunctional ylides may lead to ring closure[127] as, for instance, in a synthesis of phenanthrene. Peracetic acid can be used

$$\text{biphenyl-2,2'-bis}(CH{=}P(C_6H_5)_3) + O_2 \rightarrow \text{phenanthrene}$$

instead of oxygen as the oxidizing agent.[129] This reagent is also capable of oxidizing the resonance-stabilized ylides such as acylalkylidene and carbalkoxyalkylidene triphenylphosphoranes.

[126] Bestmann, *Angew. Chem.*, **72,** 34 (1960).

[127] Bestmann and Kratzer, *Angew. Chem.*, **74,** 494 (1962).

[128] Bestmann and Kratzer, *Angew. Chem.*, **73,** 757 (1961).

[129] Denney and Smith, *J. Am. Chem. Soc.*, **82,** 2396 (1960).

Staudinger found that, in analogy to its oxidation, diphenylmethylenetriphenylphosphorane reacts with elemental sulfur to give triphenylphosphine sulfide and thiobenzophenone. Because of the low reactivity of the phosphorane, a further reaction did not take place.[8]

$$(C_6H_5)_3P{=}C(C_6H_5)_2 + 2S \rightarrow (C_6H_5)_3PS + (C_6H_5)_2C{=}S$$

In their capacity as nucleophiles, alkylidene phosphoranes can add to activated double bonds. Depending on the nature of the substituent, the initially formed zwitterion **45** can stabilize itself in three different ways.

$$(C_6H_5)_3P{=}CHR_1 + R_2R_3C{=}CR_4R_5 \rightarrow (C_6H_5)_3\overset{\oplus}{P}{-}CH(R_1){-}C(R_2)(R_3){-}\overset{\ominus}{C}(R_4)(R_5)$$

**45**

$$\textbf{46} \text{ (cyclopropane: } R_1, H;\ R_2, R_3;\ R_4, R_5) \xleftarrow{A} (C_6H_5)_3\overset{\oplus}{P}{-}CH(R_1){-}C(R_2)(R_3){-}\overset{\ominus}{C}(R_4)(R_5) \xrightarrow{B} (C_6H_5)_3P{=}C(R_1){-}C(R_2)(R_3){-}CH(R_4)(R_5)$$

**46** **45** **47**

$$\downarrow C$$

$$(C_6H_5)_3\overset{\oplus}{P}{-}CH(R_1){-}C(R_3){=}C(R_4)(R_5) \quad R_2^{\ominus}$$

**48**

$$\downarrow$$

$$(C_6H_5)_3P{=}C(R_1){-}C(R_3){=}C(R_4)(R_5)$$

**49**

Formation of a cyclopropane derivative **46**[130,131] by elimination of triphenylphosphine (path A) is preferred if $R_1$ is a group which is not electron-withdrawing, e.g., hydrogen or alkyl.[132] This is illustrated by

[130] Mechoulam and Sondheimer, *J. Am. Chem. Soc.*, **80,** 4386 (1958).

[131] Freeman, *Chem. Ind.* (*London*), **1959,** 1254.

[132] Bestmann and Seng, *Angew. Chem.*, **74,** 154 (1962).

the reaction of 9-*n*-butylidenefluorene with *n*-butylidenetriphenylphosphorane, which yields the spiro compound **50**.[130]

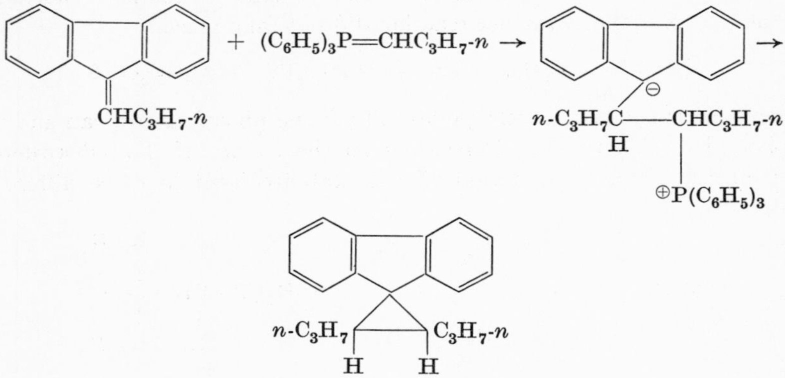

$$\text{9-(}CHC_3H_7\text{-}n\text{)fluorene} + (C_6H_5)_3P{=}CHC_3H_7\text{-}n \rightarrow \text{fluorenyl anion: } n\text{-}C_3H_7CH{-}CHC_3H_7\text{-}n\text{ with } {}^{\oplus}P(C_6H_5)_3 \rightarrow$$

$$\text{spiro[cyclopropane-fluorene] with } n\text{-}C_3H_7\text{, H and } C_3H_7\text{-}n\text{, H}$$

**50**

Michael addition of the alkylidene phosphorane to the double bond to form the new ylide **47** (path B) occurs preferentially if $R_1$ is an electron-withdrawing group capable of resonance interaction.[132]

$$(C_6H_5)_3P{=}CHCO_2CH_3 + C_6H_5COCH{=}CHCO_2CH_3 \rightarrow$$

$$(C_6H_5)_3P{=}C(CO_2CH_3){-}CH(COC_6H_5)CH_2CO_2CH_3$$

Path C is possible only if $R_2$ or $R_3$ are substituents capable of forming stable anions such as ethoxide or cyanide.[133]

$$(C_6H_5)_3P{=}CHCN + C_2H_5OCH{=}C(CO_2C_2H_5)_2 \rightarrow$$

$$(C_6H_5)_3P{=}C(CN){-}CH{=}C(CO_2C_2H_5)_2 + C_2H_5OH$$

As expected, the Wittig reagents add to the electrophilic derivatives of trivalent boron, such as boron hydrides,[134,135] boron halides,[120,121,136] and triphenylboron.[136–138]

[133] Trippett, *J. Chem. Soc.*, **1962,** 4733.
[134] Hawthorne, *J. Am. Chem. Soc.*, **80,** 3480 (1958).
[135] Hawthorne, *J. Am. Chem. Soc.*, **83,** 367 (1961).
[136] Seyferth and Grim, *J. Am. Chem. Soc.*, **83,** 1613 (1961).
[137] Wittig, *Angew. Chem.*, **70,** 67 (1958).
[138] P. Duffner, Doctoral Dissertation, Universität Tübingen, 1957.

$$2(C_6H_5)_3P{=}CHR + B_2H_6 \rightarrow 2(C_6H_5)_3\overset{\oplus}{P}{-}CHR\overset{\ominus}{B}H_3$$

$$(C_6H_5)_3P{=}CH_2 + (CH_3)_3\overset{\oplus}{N}{-}\overset{\ominus}{B}H_3 \rightarrow (C_6H_5)_3\overset{\oplus}{P}{-}CH_2\overset{\ominus}{B}H_3$$

$$\uparrow LiAlH_4$$

$$(C_6H_5)_3P{=}CH_2 + BF_3 \rightarrow (C_6H_5)_3\overset{\oplus}{P}{-}CH_2\overset{\ominus}{B}F_3$$

$$\downarrow C_6H_5MgX$$

$$(C_6H_5)_3P{=}CH_2 + B(C_6H_5)_3 \rightarrow (C_6H_5)_3\overset{\oplus}{P}{-}CH_2\overset{\ominus}{B}(C_6H_5)_3$$

The reactions of alkylidene phosphoranes with a number of diazo compounds and diazonium salts are also of interest. Even the relatively unreactive cyclopentadienylidenetriphenylphosphorane **(10)** is readily attacked by benzenediazonium chloride. Strikingly, addition does not occur on the ylide carbon but in the 2-position, giving rise to the longest possible conjugated system.[139–141] The other resonance-stabilized Wittig

$P(C_6H_5)_3$ $\overset{\oplus}{P}(C_6H_5)_3$ $\leftrightarrow$ $\ominus$ $+ \; C_6H_5N_2^{\oplus}Cl^{\ominus} \rightarrow$

**10**

$\overset{\oplus}{P}(C_6H_5)_3$ H $N{=}NC_6H_5 \rightarrow$ $P(C_6H_5)_3$ $-N{=}NC_6H_5$

reagents are attacked by diazonium salts in the normal fashion on the α-carbon atom.[142,143] The arylazoalkylphosphonium salts **51** so obtained can be converted with bases to arylazoalkylidene triphenylphosphoranes **(52)**.

$$(C_6H_5)_3P{=}CHR + C_6H_5N_2^{\oplus}X^{\ominus} \rightarrow (C_6H_5)_3\overset{\oplus}{P}{-}CHR{-}N{=}N{-}C_6H_5 \; X^{\ominus} \xrightarrow{\text{Base}}$$

**51**

$$(C_6H_5)_3P{=}CR{-}N{=}N{-}C_6H_5$$

**52**

[139] Ramirez and Levy, *J. Org. Chem.*, **21,** 1333 (1956).
[140] Ramirez and Levy, *J. Am. Chem. Soc.*, **79,** 6167 (1957).
[141] Ramirez and Levy, *J. Org. Chem.* **23,** 2035 (1958).
[142] Märkl, *Tetrahedron Letters*, **1961,** 807.
[143] Märkl, *Z. Naturforsch.*, **17b,** 782 (1962).

Strongly basic ylides (e.g., $R = C_6H_5$) furnish arylazoalkylidene phosphoranes **(52)** by *trans* ylidation without the addition of bases. Further reaction leads to the formation of bis-arylazoalkylphosphonium salts **(53).**

$$\mathbf{52} + C_6H_5N_2^{\oplus}X^{\ominus} \rightarrow (C_6H_5)_3\overset{\oplus}{P}-C(R)(N{=}N{-}C_6H_5)_2 \quad {}^{\ominus}X$$

**53**

If $R = CO_2H$, subsequent elimination of carbon dioxide and hydrogen halide leads to bis-azoarylmethylene triphenylphosphoranes.[143]

From benzylidenetriphenylphosphorane and aliphatic diazo compounds of structure **54**, mixed azines **55** are formed.[144]

$$(C_6H_5)_3P{=}CHC_6H_5 + R'CO-\overset{\ominus}{C}(R)-\overset{\oplus}{N}_2 \rightarrow$$

**54**

$$(C_6H_5)_3\overset{\oplus}{P}-CHC_6H_5,\ {}^{\ominus}N-N{=}CR-COR' \rightarrow C_6H_5CH{=}N-N{=}C(R)COR'$$

**55**

The same phosphorane reacts with phenyldiazomethane in a similar fashion to give benzalazine **(56)**.[90]

$$(C_6H_5)_3P{=}CHC_6H_5 + C_6H_5CHN_2 \rightarrow$$

$$(C_6H_5)_3\overset{\oplus}{P}-CHC_6H_5,\ {}^{\ominus}N-N{=}CHC_6H_5 \rightarrow C_6H_5CH{=}N-N{=}CHC_6H_5$$

**56**

An unexpected result was obtained in the interaction of diazoacetophenone with benzoylmethylenetriphenylphosphorane to give the heterocyclic product **57**. The only fact pertaining to the mechanism of this reaction is that the benzylidene group of the pyran derivative **57** is derived from the diazo compound.[145,146]

$$2(C_6H_5)_3P{=}CHCOC_6H_5 + \overset{*}{C}_6H_5COCHN_2 \rightarrow$$

$$\text{4-}(\overset{*}{C}HC_6H_5{=})\text{-2,6-}(C_6H_5)_2\text{-pyran} + 2(C_6H_5)_3PO$$

**57**

[144] Märkl, *Tetrahedron Letters*, **1961**, 811.

[145] Strzelecka, Simalty-Siemiatycki, and Prévost, *Compt. Rend.*, **254**, 696 (1962).

[146] Strzelecka, *Compt. Rend.*, **255**, 731 (1962).

The reaction of benzylidenetriphenylphosphorane with phenyl azide resembles that with diazo compounds.[147] The triphenylphosphine, which is formed in addition to benzylideneaniline **(58)**, reacts with excess phenyl azide to furnish tetraphenylphosphine imide **(59)**.

$$(C_6H_5)_3P{=}CHC_6H_5 + C_6H_5N_3 \longrightarrow (C_6H_5)_3\overset{\oplus}{P}{-}CHC_6H_5\ (\text{with } {}^{\ominus}NC_6H_5 \text{ on C}) \xrightarrow{C_6H_5N_3}$$

$$\underset{\mathbf{58}}{C_6H_5CH{=}NC_6H_5} + \underset{\mathbf{59}}{(C_6H_5)_3P{=}NC_6H_5}$$

Of the further reactions of alkylidene phosphoranes, only those with carbonium and nitrilium salts will be mentioned.[103] In both cases nucleophilic attack by the ylide occurs on the carbon atom with the initial formation of substituted phosphonium salts, which may react further with bases.

The preparatively most important aspect of the alkylidene phosphoranes is without doubt their reaction with carbonyl compounds to form olefins. The details of the Wittig reaction are discussed in the next section.

## MECHANISM AND STEREOCHEMISTRY

### Mechanism

Definitive kinetic studies of reactions of unstabilized alkylidene phosphoranes have not yet been made. In fact, it is not possible at present to make a final statement about the mechanism of the Wittig reaction. Using the available facts, it is possible, however, to outline the path of this complex reaction.

Olefin formation from alkylidene triphenylphosphoranes and carbonyl compounds occurs by way of the intermediates shown in the accompanying

$$(C_6H_5)_3P{=}CR_1R_2 + O{=}CR_3R_4 \xrightarrow{A} \underset{\mathbf{60}}{(C_6H_5)_3\overset{\oplus}{P}{-}CR_1R_2{-}CR_3R_4{-}O^{\ominus}} \xrightarrow{B}$$

$$\underset{\mathbf{61}}{(C_6H_5)_3P{-}CR_1R_2{-}CR_3R_4{-}O \text{ (four-membered ring, P–O bonded)}} \xrightarrow{C} (C_6H_5)_3PO + R_1R_2C{=}CR_3R_4$$

[147] Hoffmann, *Chem. Ber.*, **95**, 2563 (1962).

formulation. In the first step, nucleophilic addition of the alkylidene phosphorane in its ylide form to the polarized carbonyl group gives the phosphonium betaine **60**. As a consequence of the great affinity of phosphorus for oxygen and the possibility of expanding the valence shell of phosphorus to 10 electrons, a P—O bond is formed next, giving rise to the four-membered ring compound **61**, which then collapses into triphenylphosphine oxide and an olefin.

Experimental proof for the formation of a zwitterion in step A of the reaction was obtained from the interaction of methylenetriphenylphosphorane with benzaldehyde.[2] The zwitterion **62** is stable at room temperature and may be characterized as its hydrobromide **63**. The

$$(C_6H_5)_3P{=}CH_2 + C_6H_5CHO \rightarrow$$

$$\underset{\mathbf{62}}{(C_6H_5)_3\overset{\oplus}{P}CH_2\underset{|\atop O^{\ominus}}{C}HC_6H_5} \underset{\text{Base}}{\overset{\text{Acid}}{\rightleftharpoons}} \underset{\mathbf{63}}{(C_6H_5)_3\overset{\oplus}{P}\underset{Br^{\ominus}}{}{-}CH_2\underset{|\atop OH}{C}HC_6H_5}$$

$$\downarrow$$

$$(C_6H_5)_3PO + C_6H_5CH{=}CH_2$$

action of phenyllithium on the $\beta$-hydroxyphosphonium salt **63** regenerates **62**, which must be heated to 65° for an extended period in order to effect decomposition into triphenylphosphine oxide and styrene.

Initial formation of **62** has also been postulated for the reaction of triphenylphosphine with styrene oxide.[3] Since ring opening of the epoxide requires a temperature of 165°, the reaction proceeds directly to styrene and triphenylphosphine oxide.

$$(C_6H_5)_3P + C_6H_5\underbrace{CH{-\!-}CH_2}_{O} \xrightarrow{165^\circ}$$

$$\underset{\mathbf{62}}{(C_6H_5)_3\overset{\oplus}{P}CH_2\underset{|\atop O^{\ominus}}{C}HC_6H_5} \rightarrow (C_6H_5)_3PO + C_6H_5CH{=}CH_2$$

This is a general method for the conversion of epoxides to olefins. For example, cinnamic ester is obtained in 82% yield from phenylglycidic ester.

$$(C_6H_5)_3P + C_6H_5\underbrace{CH{-\!-}CH}_{O}CO_2R \rightarrow (C_6H_5)_3PO + C_6H_5CH{=}CHCO_2R$$

A stereochemical investigation of this deoxygenation reaction confirmed the mechanism proposed.[148] It was found that tributylphosphine reacts with *cis*-2-butene epoxide to form mainly *trans*-2-butene, whereas *trans*-2-butene epoxide gives mainly *cis*-2-butene.

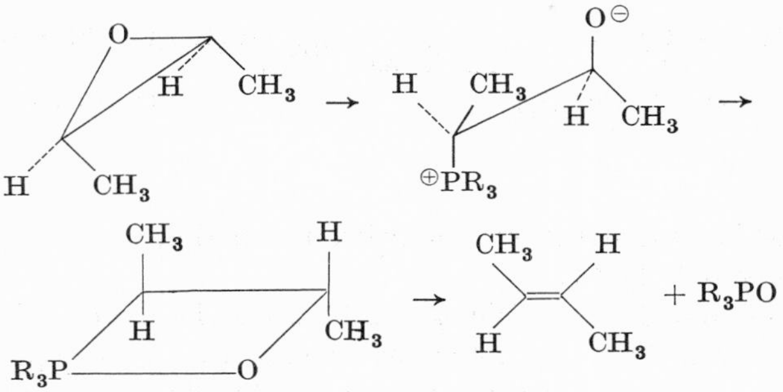

$$\text{cis-2-butene epoxide (H, CH}_3\text{; H, CH}_3) \rightarrow \text{H, CH}_3\text{C}(\overset{\oplus}{\text{P}}\text{R}_3)\text{—C(H)(CH}_3)\text{O}^{\ominus} \rightarrow$$

$$\text{R}_3\text{P—O four-membered intermediate (CH}_3\text{, H; H, CH}_3) \rightarrow \text{(CH}_3\text{)(H)C=C(H)(CH}_3\text{)} + \text{R}_3\text{PO}$$

An ionic intermediate has also been proposed for the reaction of cyclic carbonates with phosphines.[149]

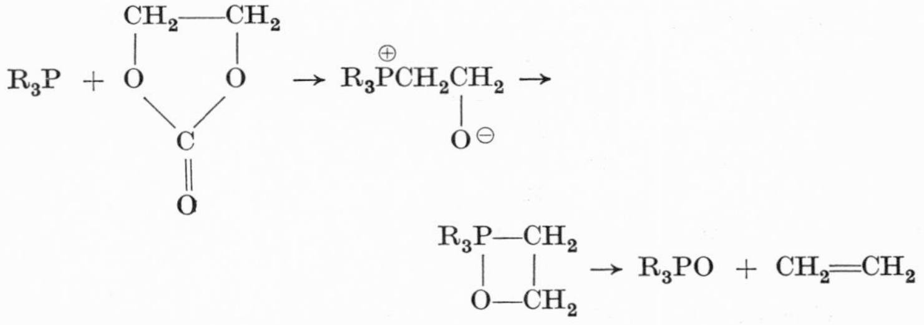

$$\text{R}_3\text{P} + \text{CH}_2\text{—CH}_2\text{ (O—C(=O)—O ring)} \rightarrow \text{R}_3\overset{\oplus}{\text{P}}\text{CH}_2\text{CH}_2\text{O}^{\ominus} \rightarrow$$

$$\begin{matrix}\text{R}_3\text{P—CH}_2\\ |\quad\ \ |\\ \text{O—CH}_2\end{matrix} \rightarrow \text{R}_3\text{PO} + \text{CH}_2\text{=CH}_2$$

The isolation of unusually stable intermediates in a Wittig reaction has been reported recently.[150] Interaction of diphenylketene with isopropylidenetriphenylphosphorane gave a pale yellow crystalline compound, **64,** which cleaved to triphenylphosphine oxide and 1,1-dimethyl-3,3-diphenylallene only at temperatures above its melting point (140°). Hydrogen

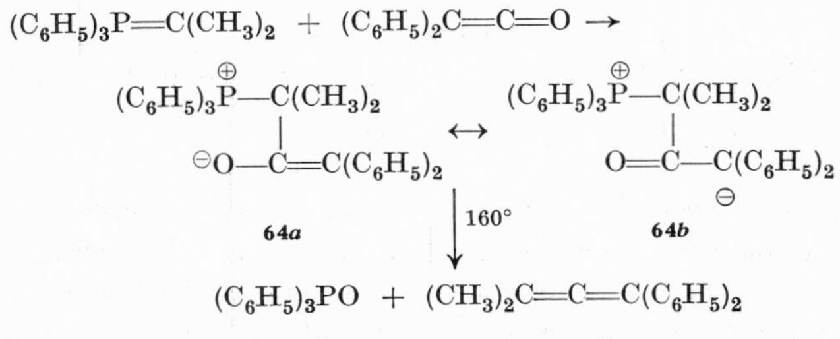

$$(C_6H_5)_3P{=}C(CH_3)_2 + (C_6H_5)_2C{=}C{=}O \rightarrow$$

$$\begin{matrix}(C_6H_5)_3\overset{\oplus}{P}\text{—}C(CH_3)_2\\ |\\ {}^{\ominus}O\text{—}C{=}C(C_6H_5)_2\end{matrix} \leftrightarrow \begin{matrix}(C_6H_5)_3\overset{\oplus}{P}\text{—}C(CH_3)_2\\ |\\ O{=}C\text{—}\overset{\ominus}{C}(C_6H_5)_2\end{matrix}$$

***64a*** ***64b***

$$\downarrow 160°$$

$$(C_6H_5)_3PO + (CH_3)_2C{=}C{=}C(C_6H_5)_2$$

148 Boskin and Denney, *Chem. Ind.* (*London*), **1959,** 330.
149 Keough and Grayson, *J. Org. Chem.*, **27,** 1817 (1962).
150 Wittig and Haag, *Chem. Ber.*, **96,** 1535 (1963).

bromide and methyl iodide attack compound **64** on carbon. In view of the relatively small dipole moment of 4.34 D, it cannot be said with certainty

$$\underset{\substack{|\\ O=C-C(C_6H_5)_2\\ |\\ CH_3}}{\overset{I^{\ominus}}{(C_6H_5)_3\overset{\ominus}{P}-C(CH_3)_2}} \xleftarrow{CH_3I} \mathbf{64} \xrightarrow{HBr} \underset{\substack{|\\ O=CCH(C_6H_5)_2}}{\overset{Br^{\ominus}}{(C_6H_5)_3\overset{\oplus}{P}-C(CH_3)_2}}$$

whether **64** has the structure of a zwitterion or that of a four-membered cyclic compound like **61**.

The final step C of the Wittig reaction, *cis* elimination of triphenylphosphine oxide by way of the four-membered ring compound **61**, is formally related to the decomposition of a phosphonium alkoxide into a phosphine oxide and a hydrocarbon.[67]

$$\underset{\mathbf{60}}{\begin{matrix} (C_6H_5)_3\overset{\oplus}{P}-C(R_1)(R_2) \\ | \\ {}^{\ominus}O-C(R_3)(R_4) \end{matrix}} \xrightarrow{B} \underset{\mathbf{61}}{\begin{matrix} (C_6H_5)_3P-C(R_1)(R_2) \\ |\qquad\quad | \\ O-C(R_3)(R_4) \end{matrix}} \xrightarrow{C} (C_6H_5)_3PO + \begin{matrix} R_1 & & R_3 \\ & C=C & \\ R_2 & & R_4 \end{matrix}$$

$$\underset{{}^{\ominus}OC_2H_5}{(CH_3)_3\overset{\oplus}{P}-CH_3} \rightleftharpoons \begin{matrix} (CH_3)_3P-CH_3 \\ | \\ O-C_2H_5 \end{matrix} \rightarrow (CH_3)_3PO + C_3H_8$$

Since it has never been observed that step C is the slowest and, therefore, the rate-determining step in a Wittig reaction, it cannot be decided at the present moment whether the four-membered ring compound **61** with a pentavalent phosphorus atom is actually an intermediate or a transition state. Depending on the reactants, however, either step A or B may become rate-determining. Studies of the interaction of fluorenylidene phosphoranes with a number of substituted carbonyl compounds showed that, with resonance-stabilized phosphoranes, the first step, formation of a betaine, is rate-determining.[29,49,50] Introduction of electron-withdrawing substituents into the benzaldehyde molecule resulted in an increased rate of reaction with fluorenylidenetriphenylphosphorane **(11)**. The reverse was true with electron-releasing substituents. No reaction

$$\text{Fluorenylidene}{=}P(C_6H_5)_3 \ (\mathbf{11}) + R_1R_2C{=}O \rightarrow \text{Fluorenylidene}{=}CR_1R_2 + (C_6H_5)_3PO$$

| $R_1R_2C{=}O$ | Yield, % |
|---|---|
| $p$-$O_2NC_6H_4CHO$ | 96 |
| $p$-$ClC_6H_4CHO$ | 93 |
| $C_6H_5CHO$ | 84 |
| $p$-$CH_3OC_6H_4CHO$ | 37 |
| $p$-$(CH_3)_2NC_6H_4CHO$ | 0 |

was observed with ketones such as acetone, benzophenone, 4,4′-dinitrobenzophenone, and fluorenone, but 2,4,7-trinitrofluorenone and the phosphorane **11** gave the corresponding olefin in quantitative yield.

A characteristic feature of resonance-stabilized phosphoranes, for which addition of the ylide to the carbonyl group is the critical step, is that replacement of the phenyl groups by alkyl groups facilitates formation of olefins. This is the result of the fact that electron-releasing groups on phosphorus result in a larger contribution to the ground state of the more reactive ylide form. Replacement of the phenyl groups in **11** by *n*-butyl groups does, indeed, give rise to a considerably more reactive phosphorane,[29] which under otherwise equal conditions gives almost quantitative yields of olefins with all the benzaldehyde derivatives mentioned above. In addition, it will react with a number of ketones which are completely unreactive toward the triphenyl derivative **11**. Thus good yields of olefins are obtained with 4,4′-dinitrobenzophenone as well as with *m*- and *p*-nitroacetophenone; the unsubstituted ketones will not react. Replacement of phenyl groups on phosphorus by alkyl groups such as methyl[58,151] or cyclohexyl[28] has a favorable effect on the yields of the Wittig reaction of other resonance-stabilized alkylidene phosphoranes

[151] Trippett and Walker, *Chem. Ind.* (*London*), **1960**, 933.

as well. In no case was it possible to isolate betaine intermediates such as **60** or **62**. Addition of hydrogen bromide gave only starting materials and end products,[29] a result which again supports the assumption that A is the slowest step, followed rapidly by steps B and C.

Recent kinetic studies with resonance-stabilized alkylidene phosphoranes indicate that the over-all reaction is best described as a slow, reversible formation of the betaine (rate-controlling) with rapid decomposition of the betaine into phosphine oxide and olefin.[152–154]

A completely different case is the interaction of carbonyl compounds with the reactive alkylidene phosphoranes, which constitute the majority of the Wittig reagents.[155] Here the addition of the ylides to the carbonyl compounds takes place within a few minutes, whereas the subsequent decomposition of the betaines into phosphine oxide and olefins often requires prolonged standing at room temperature or heating for a number of hours. Step B, decomposition of the betaine, is therefore rate-determining. Consequently electron-releasing groups on phosphorus, despite their facilitation of step A, will impede the subsequent decomposition via the four-membered cyclic intermediate, since the phosphorus is less able to accept the anionic betaine oxygen. Thus interaction of benzophenone with methylenetrimethylphosphorane led only to the betaine **65**, which could be characterized as the hydroiodide,[102] whereas methylenetriphenylphosphorane[1] under the same conditions gave an almost quantitative yield of olefin. Prolonged heating of the betaine **65** in tetrahydrofuran was

$$(CH_3)_3P{=}CH_2 + (C_6H_5)_2CO \longrightarrow (CH_3)_3\overset{\oplus}{P}{-}CH_2C(C_6H_5)_2(O^{\ominus}) \xrightarrow{HI}$$

**65**

$$(CH_3)_3\overset{\oplus}{P}{-}CH_2C(C_6H_5)_2(OH) \quad I^{\ominus}$$

required in order to effect decomposition into trimethylphosphine oxide and 1,1-diphenylethylene, but even then the yield was less than 40%.[58,155]

A number of other methylene phosphoranes (**66*a*–*e***) containing electron-releasing groups on phosphorus have also been investigated.[155] In agreement with the assumption that electron-releasing groups on phosphorus

[152] Speziale and Ratts, *J. Am. Chem. Soc.*, **85**, 2790 (1963).
[153] Speziale and Bissing, *J. Am. Chem. Soc.*, **85**, 1888 (1963).
[154] Speziale and Bissing, *J. Am. Chem. Soc.*, **85**, 3878 (1963).
[155] Wittig, Weigmann, and Schlosser, *Chem. Ber.*, **94**, 676 (1961).

considerably decrease the activation energy for step A, it was observed that betaine formation was very rapid for each compound. The subsequent decomposition of the betaine by steps B and C, however, was more difficult in all examples and in some did not occur at all.

$(p\text{-}CH_3C_6H_4)_3P{=}CH_2$ **66*a***

$(p\text{-}CH_3OC_6H_4)_3P{=}CH_2$ **66*b***

$(o\text{-}CH_3OC_6H_4)_3P{=}CH_2$ **66*c***

$(C_5H_{10}N)_3P{=}CH_2$ (piperidino) **66*d***

$(OC_4H_8N)_3P{=}CH_2$ (morpholino) **66*e***

Since the polarity of the carbonyl group is of little consequence when the second step of the Wittig reaction is rate-determining, differently substituted aldehydes or ketones will usually give olefins in about the same yields. It is striking, furthermore, that the reactive phosphoranes, unlike the resonance-stabilized phosphoranes, react more readily with benzophenone than with benzaldehyde. Nucleophilic attack on benzaldehyde will in each case be easier than on the less reactive benzophenone and, consequently, reaction with the aldehyde will be preferred by the phosphoranes of low reactivity where the first step of the Wittig reaction requires the larger activation energy. However, introduction of phenyl groups on the $\beta$-carbon atom will facilitate the decomposition of the betaine into phosphine oxide and olefin (i.e., the rate-determining step with the reactive phosphoranes), and this will result in a more rapid collapse of the benzophenone adduct as compared with the benzaldehyde adduct. In general, it may be said that, whenever reaction of an ylide with benzophenone is faster than with benzaldehyde, the second step of the reaction, the decomposition of the betaine, is rate-determining. This explains why a number of resonance-stabilized alkylidene phosphoranes will react exclusively with aldehydes but not with benzophenone, whereas some of the reactive phosphoranes will interact readily with benzophenone but not with benzaldehyde. It also makes clear why betaines can usually be isolated only when benzaldehyde is used as the carbonyl component, whereas the benzophenone adducts as a rule are much less stable. Only from the methylene phosphoranes **66*d*** and **66*e*** was it possible to isolate adducts with benzophenone, since decomposition of the betaine is rendered extremely difficult as a consequence of the high electron density on phosphorus;[155] but even these adducts are less stable than the corresponding benzaldehyde adducts, which cannot be decomposed into olefin and phosphine oxide. The marked stability of the ketene adduct **64** is probably due to its stabilization by resonance.

## Stereochemistry

Experiments with the optically active phosphonium salts **67** showed that the Wittig reaction takes place with retention of configuration on phosphorus.[156,157]

$$\underset{dextro\ \mathbf{67}}{CH_3-\overset{\oplus}{\underset{I^{\ominus}\ \ CH_2C_6H_5}{\overset{CH_2CH_3}{P}}}-C_6H_5} \xrightarrow[2.\ C_6H_5CHO]{1.\ C_6H_5Li} CH_3-\overset{\oplus}{\overset{CH_2CH_3}{P}}(-C_6H_5)(-CHC_6H_5-CHC_6H_5-O^{\ominus}) \xrightarrow{\text{Retention}}$$

$$\underset{dextro\ \mathbf{68}}{CH_3-\overset{CH_2CH_3}{\underset{O}{\overset{|}{\underset{\|}{P}}}}-C_6H_5} + C_6H_5CH{=}CHC_6H_5$$

The same phosphine oxide, *dextro* **68,** is formed with inversion from the phosphonium salt, *levo* **67,** under the influence of alkali.

$$\begin{array}{ccc} dextro\ \text{salt} & \xrightarrow[\text{(Inversion)}]{OH^{\ominus}} & levo\ \text{oxide} \\ \text{Wittig (Retention)}\downarrow & & \uparrow\text{Wittig (Retention)} \\ dextro\ \text{oxide} & \xleftarrow[\text{(Inversion)}]{OH^{\ominus}} & levo\ \text{salt} \end{array}$$

If the ylide and carbonyl components are unsymmetrically substituted, a mixture of *cis* and *trans* olefins is usually obtained. As a rule, the *trans* olefin predominates, as illustrated by the reaction of benzylidenetriphenylphosphorane with benzaldehyde which leads to a mixture containing 70% *trans*- and 30% *cis*-stilbene.[2] However, exclusive

$$(C_6H_5)_3P{=}CHC_6H_5 + C_6H_5CHO \rightarrow \underset{(70\%)}{\overset{H\qquad C_6H_5}{\underset{C_6H_5\qquad H}{C{=}C}}} + \underset{(30\%)}{\overset{C_6H_5\qquad C_6H_5}{\underset{H\qquad H}{C{=}C}}}$$

formation of one isomer has been occasionally observed. Whereas steric reasons may be advanced for the preferred formation of the *trans* compound,[158] no explanation has been provided for the occasional

[156] Bladé-Font, McEwen, and Vander Werf, *J. Am. Chem. Soc.*, **82,** 2646 (1960).
[157] Bladé-Font, Vander Werf, and McEwen, *J. Am. Chem. Soc.*, **82,** 2396 (1960).
[158] Inhoffen, Brückner, Domagk, and Erdmann, *Chem. Ber.*, **88,** 1415 (1955).

predominance of the *cis* isomer.[159–161] Strikingly, the *trans* isomer is always formed predominantly or exclusively if resonance-stabilized alkylidene phosphoranes are used. Thus almost pure *trans* olefins were obtained by using cyanomethylenetriphenylphosphorane[42,43] and carbomethoxymethylenetriphenylphosphorane[39,162,163] as well as their derivatives.[40,111,117,164] The observation that carbomethoxyethylidenetriphenylphosphorane **(69)** gives the pure *trans* compound with acrolein led to the systematic investigation of this reaction.[164]

$$\underset{\mathbf{69}}{(C_6H_5)_3P{=}C(CH_3)CO_2CH_3} + CH_2{=}CHCHO \rightarrow CH_2{=}CHCH{=}C(CH_3)CO_2CH_3$$

By means of the Wittig reaction, methyl 2-methyl-2-butenoate was prepared in two different ways. The ylide **69** and acetaldehyde furnished almost exclusively methyl tiglate **(71)**, i.e., the *trans* isomer. Methyl pyruvate and ethylidenetriphenylphosphorane gave a mixture of isomers that contained 32% of the *cis* compound, methyl angelate **(73)**.[164] In

$$\underset{\mathbf{70}}{\text{betaine: } CH_3(H)C(O^{\ominus}){-}C(CH_3)(CO_2CH_3)P^{\oplus}(C_6H_5)_3} \rightarrow \underset{\mathbf{71}\ (96.5\%)}{(CH_3)HC{=}C(CH_3)CO_2CH_3\ (trans)}$$

$$\updownarrow$$

$$CH_3CHO + \underset{\mathbf{69}}{(C_6H_5)_3P{=}C(CH_3)CO_2CH_3}$$

$$\updownarrow$$

$$\underset{\mathbf{72}}{\text{betaine: } CH_3(H)C(O^{\ominus}){-}C(CO_2CH_3)(CH_3)P^{\oplus}(C_6H_5)_3} \rightarrow \underset{\mathbf{73}\ (3.5\%)}{(CH_3)HC{=}C(CO_2CH_3)CH_3\ (cis)}$$

[159] Wailes, *Chem. Ind. (London)*, **1958,** 1086.

[160] Bohlmann, Inhoffen, and Herbst, *Chem. Ber.*, **90,** 1661 (1957).

[161] Truscheit and Eiter, *Ann.*, **658,** 65 (1962).

[162] Novikov and Shvekhgeĭmer, *Izv. Akad. Nauk. SSSR, Otd. Khim. Nauk*, **1960,** 673 [*C.A.*, **54,** 22474h (1960)].

[163] Kucherov, Kovalev, Kogan, and Yanovskaya, *Dokl. Akad. Nauk SSSR*, **138,** 1115 (1961) [*C.A.*, **55,** 24560i (1961)].

[164] House and Rasmusson, *J. Org. Chem.*, **26,** 4278 (1961).

$$\underset{\mathbf{74}}{(C_6H_5)_3\overset{\oplus}{P}\text{—}C(CH_3)(H)\text{—}C(CH_3)(CO_2CH_3)\text{—}\overset{\ominus}{O}} \rightarrow \underset{\mathbf{71}\ (68\%)}{(CH_3)HC{=}C(CH_3)CO_2CH_3}$$

$$\uparrow$$

$$CH_3COCO_2CH_3 + (C_6H_5)_3P{=}CHCH_3$$

$$\downarrow$$

$$\underset{\mathbf{75}}{(C_6H_5)_3\overset{\oplus}{P}\text{—}C(CH_3)(H)\text{—}C(CO_2CH_3)(CH_3)\text{—}\overset{\ominus}{O}} \rightarrow \underset{\mathbf{73}\ (32\%)}{(CH_3)HC{=}C(CO_2CH_3)CH_3}$$

order to rationalize the observed isomer distributions, one may assume that the resonance-stabilized phosphorane **69** is in equilibrium with the two betaines **70** and **72.** Here the first and, for the unreactive phosphoranes, slowest step will probably be reversible, so that the reaction may proceed predominantly by way of the sterically less hindered betaine **70,** thus leading to the *trans* olefin. While such an interpretation of the isomer distribution from the stabilized ylide **69** seems reasonably satisfactory, the products formed from the reactive ethylidenephosphorane are harder to rationalize. Decomposition of the betaines derived from reactive phosphoranes to starting materials was originally thought not to occur.[155] More recent studies, however, provide convincing evidence that benzaldehyde is formed from a simple betaine under mild conditions.[165]

$$(C_6H_5)_3\overset{\oplus}{P}\text{—}CH_2\text{—}CH(C_6H_5)\text{—}\overset{\ominus}{O} \rightleftharpoons (C_6H_5)_3P{=}CH_2 + O{=}C(H)C_6H_5$$

Thus it seems that a full understanding of the isomer distributions outlined above must await further research.

Another example of the differences in isomer distributions in the Wittig reaction is the preparation of 4-nitro-4′-methoxystilbene **(78)** by two

[165] Fliszár, Hudson, and Salvadori, *Helv. Chim. Acta*, **46,** 1580 (1963).

methods.[166] Reaction of the resonance-stabilized *p*-nitrobenzylidenephosphorane **(76)** with the relatively unreactive anisaldehyde leads to the

$$(C_6H_5)_3P{=}CHC_6H_4NO_2\text{-}p \;+\; p\text{-}CH_3OC_6H_4CHO$$

**76**

↓ 89% *trans* only

$$p\text{-}CH_3OC_6H_4CH{=}CHC_6H_4NO_2\text{-}p$$

**78**

↑ 89% *cis* and *trans* (1:1)

$$(C_6H_5)_3P{=}CHC_6H_4OCH_3\text{-}p \;+\; p\text{-}O_2NC_6H_4CHO$$

**77**

exclusive formation of the *trans* compound **78.** The less stable, more reactive *p*-methoxybenzylidenephosphorane **(77),** on the other hand, reacts with the extremely reactive *p*-nitrobenzaldehyde to give a mixture of the *cis*- and *trans*-stilbenes **78** in the ratio of 1:1.

In the reactions of resonance-stabilized alkylidene phosphoranes with carbonyl compounds, the isomer ratio may be shifted further in favor of the *trans* products if the stability of the initially formed betaines ***79a*** and ***79b*** is increased, thus preventing further reaction before equilibrium has been reached. This was accomplished by substituting cyclohexyl groups for the phenyl groups on phosphorus.[28] The phosphorus is thus rendered less electrophilic, and the conversion of **79** to the end products by way of the four-membered ring compound **80** becomes more difficult. If the rate

R″, H / C—C / H, R′ / $O^{\ominus}$ $^{\oplus}PR_3$ (**79*a***) → R″, H / C—C / H, R′ / O—$PR_3$ (**80*a***) → R″, H / C=C / H, R′ (*trans*)

⇅

$$R''CHO + R_3P{=}CHR'$$

⇅

R″, R′ / C—C / H, H / $O^{\ominus}$ $^{\oplus}PR_3$ (**79*b***) → R″, R′ / C—C / H, H / O—$PR_3$ (**80*b***) → R″, R′ / C=C / H, H (*cis*)

[166] Ketchan, Jambotkar, and Martinelli, *J. Org. Chem.*, **27**, 4666 (1962).

of conversion of **79** to **80** is so rapid that the equilibrium ***79a*** $\rightleftharpoons$ ***79b*** cannot be established, considerable amounts of *cis* olefins will be formed in addition to the *trans* compounds. If, on the other hand, establishment of the equilibrium is possible, the betaine ***79a*** leading to the *trans* isomer will be energetically favored to a large extent.

An effort to account for the predominance of *trans* compounds from resonance-stabilized phosphoranes has been made by postulating a nucleophilic attack on the phosphorus atom by the oxygen of the carbonyl component as the first step in such reactions.[114] Other studies, however, indicate that the primary step is probably nucleophilic attack by the ylide carbon atom on the carbonyl carbon atom.[167,168] In addition, substitution of phenyl groups on the phosphorus atom by cyclohexyl groups[28] leads not only to an increased yield of *trans* olefins but also to an increase of the total yield. Electron-releasing groups on phosphorus would be expected to impede rather than to facilitate the nucleophilic attack of the carbonyl oxygen on the phosphorus atom.

Studies have also been reported which indicate that reactive phosphoranes with aldehydes tend to give increased amounts of *cis* isomers in the products when the reaction is carried out in the presence of Lewis bases.[114,169–172]

## SCOPE AND LIMITATIONS

In the few years since the discovery of the Wittig reaction, many olefins have been synthesized by this method. The reaction is not limited to simple alkyl- or aryl-substituted ethylene derivatives but is also applicable to the synthesis of $\alpha,\beta$-unsaturated carbonyl compounds and carboxylic esters as well as vinyl halides and vinyl ethers. The large number of natural products prepared by the Wittig reaction speaks for the importance that this olefin synthesis has attained in a short period.

### Alkyl-Substituted Olefins

The Wittig reaction is especially valuable for the introduction of exocyclic double bonds and is the only method for converting a cyclic

[167] Goetz, Nerdel, and Michaelis, *Naturwiss.*, **14,** 496 (1963).

[168] Speziale and Ratts, *J. Org. Chem.*, **28,** 465 (1963).

[169] Bergel'son, Vaver, Barsukov, and Shemyakin, *Dokl. Akad. Nauk SSSR,* **143,** 111 (1962) [*C.A.*, **57,** 7298e (1962)].

[170] Bergel'son, Vaver, Kovtun, Senyavina, and Shemyakin, *Zh. Obshch. Khim.*, **32,** 1802 (1962) [*C.A.*, **58,** 4415g (1963)].

[171] Bergel'son, Vaver, and Shemyakin, *Izv. Akad. Nauk SSSR, Otd. Khim. Nauk,* **1960,** 1900 [*C.A.*, **55,** 14294e (1961)].

[172] Bergel'son, Vaver, and Shemyakin, *Izv. Akad. Nauk SSSR, Otd. Khim. Nauk,* **1961,** 729 [*C.A.*, **55,** 22196c (1961)].

ketone to the corresponding exocyclic olefin. The Grignard method is well known to give practically only the endocyclic isomer (Saytzeff rule). For example, cyclohexanone and methylenetriphenylphosphorane give methylenecyclohexane.[2]

$$\text{C}_6\text{H}_{10}\text{=O} + (C_6H_5)_3P{=}CH_2 \rightarrow \text{C}_6\text{H}_{10}\text{=CH}_2$$

Similarly a whole series of methylenesteroids has been synthesized, and a methylene group has been successfully introduced into the vitamin $D_2$ skeleton. Recently methylenecycloheptane and methylenecyclo-öctane have been prepared by this method,[173] as were derivatives of methylenedihydronaphthalene and methylenetetralin.[174] 1-Methylene-2,2-dimethyltetralin **(81)**, for instance, was obtained in 83% yield.

$$\text{2,2-dimethyl-1-tetralone} + (C_6H_5)_3P{=}CH_2 \rightarrow \text{1-methylene-2,2-dimethyltetralin } (\mathbf{81})$$

Catalytic hydrogenation of the methylene compounds permits the conversion $>C{=}O \rightarrow >CHCH_3$.[175–177]

The synthesis of 1,2-disubstituted ethylene derivatives from aldehydes and monosubstituted methylene triphenylphosphoranes can be effected in two ways.

$$RCHO + (C_6H_5)_3P{=}CHR' \searrow$$
$$RCH{=}CHR'$$
$$R'CHO + (C_6H_5)_3P{=}CHR \nearrow$$

As a rule, mixtures of *cis* and *trans* olefins are formed in these reactions, for instance, in the synthesis of 1,1,1-triphenyl-3-pentene **(82)** from β,β,β-triphenylpropionaldehyde and ethylidenetriphenylphosphorane.[178]

$$(C_6H_5)_3CCH_2CHO + (C_6H_5)_3P{=}CHCH_3 \xrightarrow{58\%} (C_6H_5)_3CCH_2CH{=}CHCH_3 \quad \mathbf{82}$$

Unsaturated aldehydes can also be used. Propargylaldehyde, for instance, reacts with *n*-dodecylidenetriphenylphosphorane to give pentadec-3-en-1-yne **(83)**.[159] The strikingly large amount of *cis* compound

[173] Schriesheim, Müller, and Rowe, *J. Am. Chem. Soc.*, **84**, 3164 (1962).

[174] Wittig, Reppe, and Eicher, *Ann.*, **643**, 47 (1961).

[175] Chadha and Rapoport, *J. Am. Chem. Soc.*, **79**, 5730 (1957).

[176] Büchi and MacLeod, *J. Am. Chem. Soc.*, **84**, 3205 (1962).

[177] DeGraw and Bonner, *Tetrahedron*, **18**, 1311 (1962).

[178] Wittig and Wittenberg, *Ann.*, **606**, 1 (1957).

(80%) in the isomer mixture **83** is probably due to the presence of lithium bromide in the solution.

$$HC{\equiv}CCHO + (C_6H_5)_3P{=}CHC_{11}H_{23}\text{-}n \rightarrow HC{\equiv}CCH{=}CHC_{11}H_{23}\text{-}n$$

**83**

Both groups, R and R′, may be unsaturated, as illustrated by the synthesis of hexa-3,5-dien-1-yne **(84)** from propargylaldehyde and allylidenetriphenylphosphorane.[179]

$$HC{\equiv}CCHO + (C_6H_5)_3P{=}CHCH{=}CH_2 \rightarrow HC{\equiv}CCH{=}CHCH{=}CH_2$$

**84**

A series of compounds containing alternating double and triple bonds was synthesized in this way;[180] for instance, the $C_{20}$ hydrocarbon containing alternating pairs of ethylenic and acetylenic linkages.

$$(CH_3)_3C(C{\equiv}C)_2CHO + (C_6H_5)_3P{=}CHCH{=}CH(C{\equiv}C)_2C(CH_3)_3 \xrightarrow{19\%} (CH_3)_3C(C{\equiv}C)_2(CH{=}CH)_2(C{\equiv}C)_2C(CH_3)_3$$

The interaction of bifunctional carbonyl compounds with 2 equivalents of phosphorane can also be used for the preparation of polyenes, as shown in the synthesis of the dimethylpolyene **85**.[181]

$$OCHCH{=}CH(C{\equiv}C)_2CH{=}CHCHO + 2(C_6H_5)_3P{=}CH(CH{=}CH)_2CH_3 \longrightarrow CH_3(CH{=}CH)_4(C{\equiv}C)_2(CH{=}CH)_4CH_3 \xrightarrow{H_2} CH_3(CH{=}CH)_{10}CH_3$$

**85**

This method has proved especially valuable in the synthesis of carotenoids. Conversely, bifunctional phosphoranes can be used. Thus $\beta$-carotene **(88)** was prepared from $\beta$-ionylideneacetaldehyde **(86)** and the bis-ylide **87**.[80]

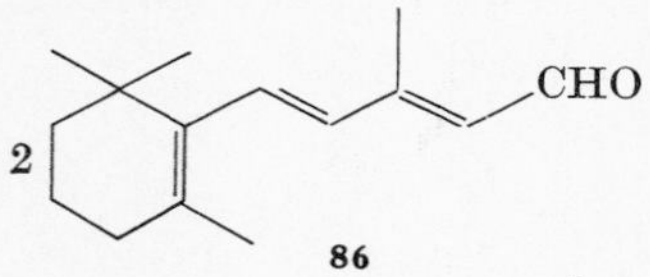

2 [β-ionylideneacetaldehyde, CHO] **86** + $(C_6H_5)_3P$=[C10 polyene]=$P(C_6H_5)_3$ **87** $\xrightarrow{51\%}$ **88**

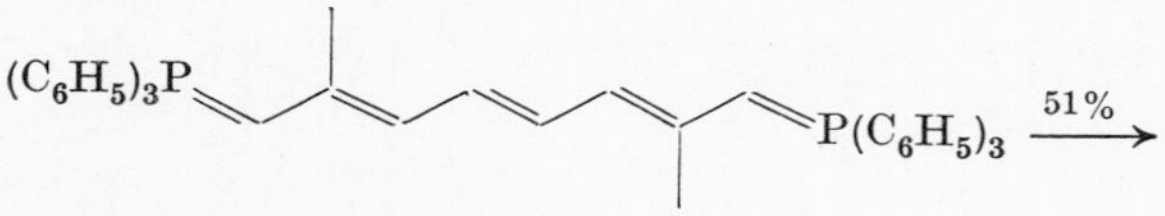

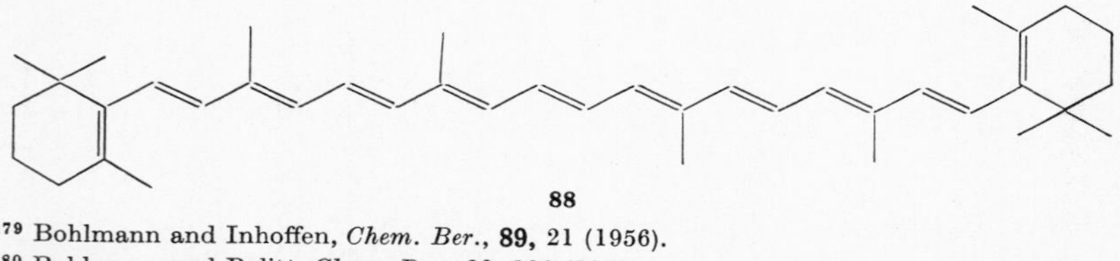

[179] Bohlmann and Inhoffen, *Chem. Ber.*, **89,** 21 (1956).

[180] Bohlmann and Politt, *Chem. Ber.*, **90,** 130 (1957).

[181] Bohlmann and Mannhardt, *Chem. Ber.*, **89,** 1307 (1956).

Tritium-labeled olefins have been prepared by the action of tritium-labeled alkylidene phosphoranes with aldehydes.[68]

$$\mathrm{RCHO} + (\mathrm{C_6H_5})_3\mathrm{P{=}CTR'} \rightarrow \mathrm{RCH{=}CTR'}$$

1,1,2-Trisubstituted ethylene derivatives may be prepared by two different paths.

$$\mathrm{RCHO} + (\mathrm{C_6H_5})_3\mathrm{P{=}C(R')(R'')} \searrow$$

$$\mathrm{RCH{=}C(R')(R'')}$$

$$\mathrm{R'(R'')C{=}O} + (\mathrm{C_6H_5})_3\mathrm{P{=}CHR} \nearrow$$

By using the first path, 1,1,1-triphenyl-4-methyl-3-pentene **(89)** was obtained in 68% yield starting from $\beta,\beta,\beta$-triphenylpropionaldehyde.[178]

$$(\mathrm{C_6H_5})_3\mathrm{CCH_2CHO} + (\mathrm{C_6H_5})_3\mathrm{P{=}C(CH_3)_2} \rightarrow \underset{\mathbf{89}}{(\mathrm{C_6H_5})_3\mathrm{CCH_2CH{=}C(CH_3)_2}}$$

An illustration of the second method, which starts from ketones, is the preparation of a number of dimethylheptadienes (geraniolenes) and methylhexenes such as 2-methyl-2-hexene **(90)** from acetone and *n*-butylidenetriphenylphosphorane.[182] This method has been applied most

$$(\mathrm{CH_3})_2\mathrm{C{=}O} + (\mathrm{C_6H_5})_3\mathrm{P{=}CHC_3H_7}\text{-}n \xrightarrow{20\%} \underset{\mathbf{90}}{(\mathrm{CH_3})_2\mathrm{C{=}CHC_3H_7}\text{-}n}$$

frequently in the area of natural products chemistry. Bifunctional carbonyl compounds or bifunctional ylides (squalene synthesis) have been used for this purpose.

Starting from cyclic ketones, the corresponding exocyclic olefins are obtained, as shown by the preparation of dicylohexylideneëthane **(91)** from cyclohexanone.[183,184]

$$\mathrm{C_6H_{10}{=}O} + (\mathrm{C_6H_5})_3\mathrm{P{=}CHCH{=}C_6H_{10}} \rightarrow \underset{\mathbf{91}}{\mathrm{C_6H_{10}{=}CHCH{=}C_6H_{10}}}$$

[182] Ansell and Thomas, *J. Chem. Soc.*, **1961,** 539.

[183] Harrison, Lythgoe, and Trippett, *J. Chem. Soc.*, **1955,** 4016.

[184] Harrison, Lythgoe, and Trippett, *Chem. Ind.* (*London*), **1955,** 507.

Substituted cyclohexanes have been prepared by variation of either the carbonyl[158,183,184] or the ylide component.[69] In the first case the *trans*

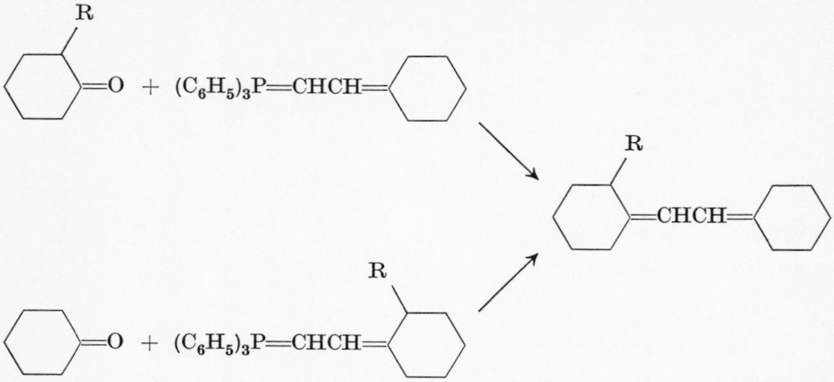

R

=O + $(C_6H_5)_3P$=CHCH=

R

=CHCH=

R

=O + $(C_6H_5)_3P$=CHCH=

compound, R = $CH_2N(CH_3)_2$, was obtained exclusively. In the second case, starting from the *cis* ylide, the *cis* compound, R = $CH_2OH$ or OH, was obtained predominantly; increasing the temperature resulted in increased formation of the *trans* olefin. When the substituent R was Br, subsequent elimination of hydrogen bromide under the influence of the ylide was observed. The triene so obtained did not contain any exocyclic double bonds.[16]

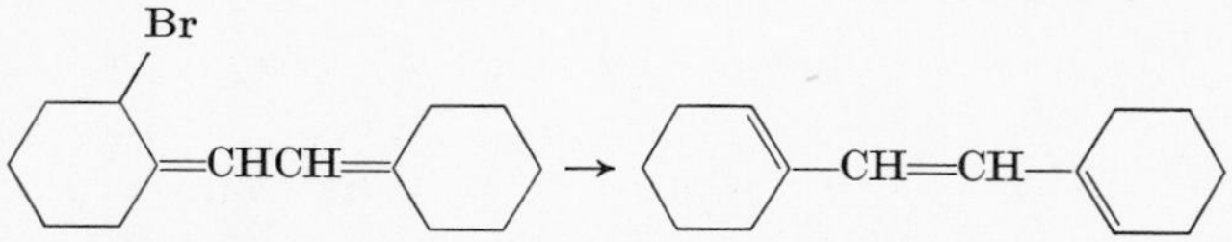

Br

=CHCH= → —CH=CH—

With benzylidenetriphenylphosphorane, cyclohexanone gives benzylidenecyclohexane in 60% yield.[3] The same compound is obtained in 70% yield from the deep red cyclohexylidenetriphenylphosphorane and benzaldehyde.[185]

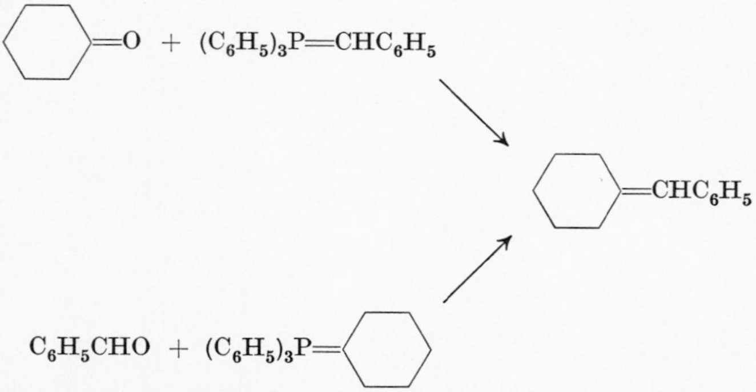

=O + $(C_6H_5)_3P{=}CHC_6H_5$

$=CHC_6H_5$

$C_6H_5CHO$ + $(C_6H_5)_3P$=

[185] U. Schöllkopf, Doctoral Dissertation, Universität Tübingen, 1956.

Cyclopentylidenetriphenylphosphorane, another ylide containing an exocyclic double bond, is also intensely colored. It reacts with benzaldehyde to give benzylidenecyclopentane.[185]

$$C_6H_5CHO + (C_6H_5)_3P{=}C_5H_8 \xrightarrow{65\%} C_6H_5CH{=}C_5H_8$$

## Aryl-Substituted Olefins

Styrene, the simplest monoaryl derivative of ethylene, has been obtained in 67% yield in the reaction of benzaldehyde with methylenetriphenylphosphorane.[2]

$$C_6H_5CHO + (C_6H_5)_3P{=}CH_2 \rightarrow C_6H_5CH{=}CH_2$$

Similarly, 9-vinylanthracene is obtained from 9-anthraldehyde.[186]

$$\text{9-anthraldehyde (CHO)} + (C_6H_5)_3P{=}CH_2 \rightarrow \text{9-vinylanthracene (CH=CH}_2\text{)}$$

Vinyl aromatics are also accessible starting from the corresponding arylidene phosphoranes and formaldehyde,[187] as illustrated by the preparation of 4,4′-divinylbiphenyl in 80% yield. This route is especially

$$2CH_2O + (C_6H_5)_3P{=}CH\text{–}C_6H_4\text{–}C_6H_4\text{–}CH{=}P(C_6H_5)_3 \rightarrow CH_2{=}CH\text{–}C_6H_4\text{–}C_6H_4\text{–}CH{=}CH_2$$

useful when the aromatic halogen compound is more accessible than the carbonyl compound. Aliphatic vinyl compounds may be prepared in a similar fashion.[188,189]

From *o*-phthalaldehyde and methylenetriphenylphosphorane, *o*-divinylbenzene is formed in 75% yield.[52]

$$o\text{-}C_6H_4(CHO)_2 + 2(C_6H_5)_3P{=}CH_2 \rightarrow o\text{-}C_6H_4(CH{=}CH_2)_2$$

[186] Hawkins, *J. Chem. Soc.*, **1957,** 3858.

[187] Drefahl, Plötner, and Rudolph, *Chem. Ber.*, **93,** 998 (1960).

[188] Hauser, Miles, and Butler, 142nd Meeting, Am. Chem. Soc., Atlantic City, N.J., Sept., 1962, Abstracts, p. 59Q.

[189] Hauser, Brooks, Miles, Raymond, and Butler, *J. Org. Chem.*, **28,** 372 (1963).

Starting from the bifunctional ylides **92** and **94,** 1,2-benzocyclohepta-1,3,6-triene **(93)** and 1,2-benzocycloöcta-1,3,7-triene **(95),** respectively, are obtained.[52]

$$\text{o-}C_6H_4(CHO)_2 + \begin{matrix}(C_6H_5)_3P{=}CH \\ \quad\diagdown \\ \qquad CH_2 \\ \quad\diagup \\ (C_6H_5)_3P{=}CH\end{matrix} \xrightarrow{28\%} \textbf{93}$$

**92** **93**

$$\text{o-}C_6H_4(CHO)_2 + \begin{matrix}(C_6H_5)_3P{=}CH \\ \quad\diagdown \\ \qquad CH_2 \\ \qquad | \\ \qquad CH_2 \\ \quad\diagup \\ (C_6H_5)_3P{=}CH\end{matrix} \xrightarrow{18\%} \textbf{95}$$

**94** **95**

3-Benzoxepin **(96)** was prepared in an analogous manner.[190]

$$\text{o-}C_6H_4(CHO)_2 + \begin{matrix}(C_6H_5)_3P{=}CH \\ \quad\diagdown \\ \qquad O \\ \quad\diagup \\ (C_6H_5)_3P{=}CH\end{matrix} \rightarrow \textbf{96}$$

**96**

1,1-Diaryl olefins are particularly easy to obtain by the interaction of methylenetriphenylphosphorane and aromatic ketones.[1] Benzophenone

$$\begin{matrix}Ar \\ \quad\diagdown \\ \qquad C{=}O \\ \quad\diagup \\ Ar'\end{matrix} + (C_6H_5)_3P{=}CH_2 \rightarrow \begin{matrix}Ar \\ \quad\diagdown \\ \qquad C{=}CH_2 \\ \quad\diagup \\ Ar'\end{matrix}$$

gave 1,1-diphenylethylene in 84% yield.[1] For preparative purposes, however, the Wittig reaction with aromatic ketones is of minor importance

[190] Dimroth and Pohl, *Angew. Chem.*, **73,** 436 (1961).

because aromatic ketones can usually be converted to the corresponding olefins without difficulty by the Grignard method, since an isomerization of the double bond cannot take place. There are reactions, however, in which the Grignard method fails to give the desired product. Thus the reaction of 2,2′-dibenzoylbiphenyl **(97)** with methyllithium followed by hydrolysis and dehydration gives the cyclic hydrocarbon **99** rather than 2,2′-distyrylbiphenyl **(98)**, whereas the latter can be prepared in 85% yield by the Wittig reaction.[191]

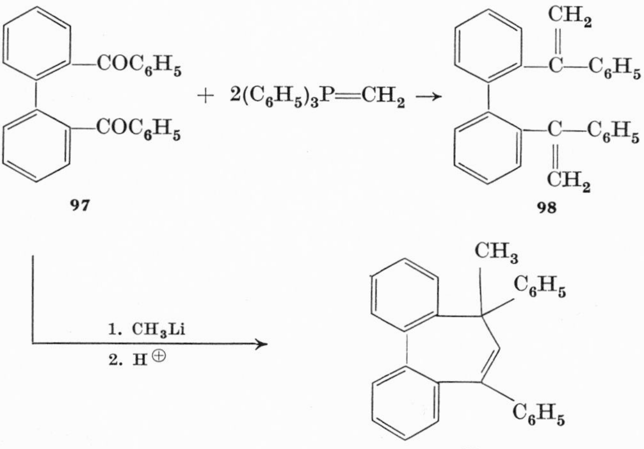

$COC_6H_5$ / $COC_6H_5$ + $2(C_6H_5)_3P{=}CH_2$ → $CH_2{=}C{-}C_6H_5$ / $C{-}C_6H_5{=}CH_2$

**97** **98**

1. $CH_3Li$
2. $H^{\oplus}$

$CH_3$ $C_6H_5$ $C_6H_5$

**99**

The preparation of olefins from nitro-substituted carbonyl compounds is not possible by the Grignard method because the reagent attacks nitro groups. This complication can usually be avoided by using an alkylidene phosphorane as in the accompanying example.[2]

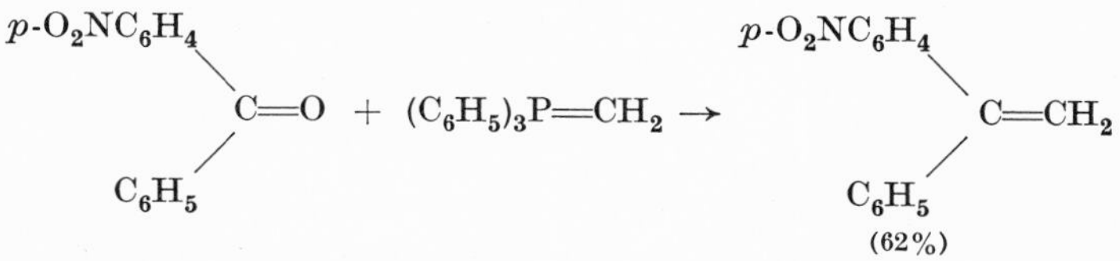

$$p\text{-}O_2NC_6H_4(C_6H_5)C{=}O + (C_6H_5)_3P{=}CH_2 \rightarrow p\text{-}O_2NC_6H_4(C_6H_5)C{=}CH_2 \quad (62\%)$$

Only extremely nucleophilic alkylidene phosphoranes containing electron-releasing groups on phosphorus, such as **100**, attack nitro groups. The ylide **100** has been reported to react with nitrobenzophenone to give

[191] Wittig and Stilz, *Ann.*, **598**, 93 (1956).

dark-colored products; olefin formation could not be observed.[192] Normally, however, olefins are formed very readily from nitro ketones and nitro aldehydes, and yields of over 90% are not unusual.[3,29,49,50,166,193]

$$p\text{-}(CH_3)_2NC_6H_4\text{—}\underset{\displaystyle CH_3}{\overset{\displaystyle C_6H_5}{\underset{|}{\overset{|}{P}}}}\text{=}CH_2$$

**100**

1,2-Diaryl olefins or stilbenes are formed in the interaction of aromatic aldehydes with aryl-substituted methylene triphenylphosphoranes. Mixtures of *cis* and *trans* olefins are obtained, the ratio in this synthesis of

$$ArCHO + (C_6H_5)_3P\text{=}CHAr' \searrow$$
$$ArCH\text{=}CHAr'$$
$$Ar'CHO + (C_6H_5)_3P\text{=}CHAr \nearrow$$

$$C_6H_5CHO + (C_6H_5)_3P\text{=}CHC_6H_5 \rightarrow C_6H_5CH\text{=}CHC_6H_5 \quad (82\%)$$

stilbene being 70% *trans* and 30% *cis*.[2] By using the alkoxide method, the ratio was found to be 47:53, the increased amount of the *cis* isomer being ascribed to increased polarity of the solvent.[3] The different isomer ratios obtained in the synthesis of 4-nitro-4′-methoxystilbene by two different ways (Ar and Ar′ exchanged) have been mentioned previously (p. 314).[166]

Reaction of 2-methoxy-3-methylbenzaldehyde with benzylidenetriphenylphosphorane gives 2-methoxy-3-methylstilbene.[194] A large

$$\text{2-}OCH_3\text{-3-}CH_3C_6H_3\text{—}CHO + (C_6H_5)_3P\text{=}CHC_6H_5 \xrightarrow{56\%} \text{2-}OCH_3\text{-3-}CH_3C_6H_3\text{—}CH\text{=}CHC_6H_5$$

number of other stilbene derivatives has been prepared in a similar manner. The alkoxide method proved to be particularly useful for this purpose.[66,195,196]

[192] Trippett and Walker, *J. Chem. Soc.*, **1961**, 2130.
[193] Campbell and McDonald, *J. Org. Chem.*, **24**, 1246 (1959).
[194] Wessely, Zbiral, and Lahrmann, *Chem. Ber.*, **92**, 2141 (1959).
[195] Drefahl and Plötner, *Chem. Ber.*, **93**, 990 (1960).
[196] Drefahl and Plötner, *Chem. Ber.*, **94**, 907 (1961).

The preparation of symmetrical and unsymmetrical 1,2-diarylethylenes containing larger aromatic groups has also been reported.[197] 1,2-Bis-(3-pyrenyl)ethylene (**101**) is formed in 91% yield.

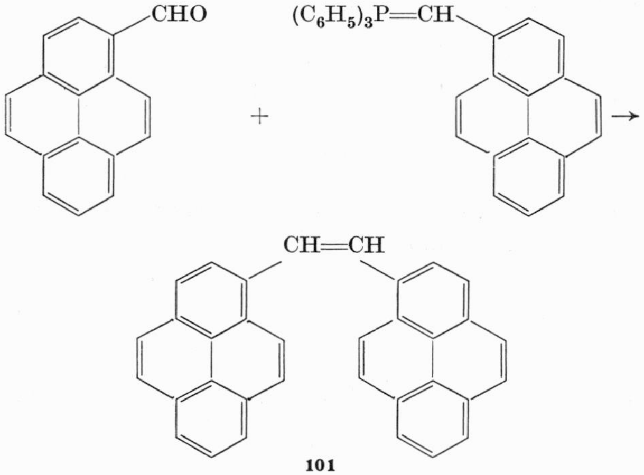

CHO + $(C_6H_5)_3P{=}CH$ →

CH=CH

**101**

Stilbazole derivatives can be prepared from pyridine aldehydes.[198]

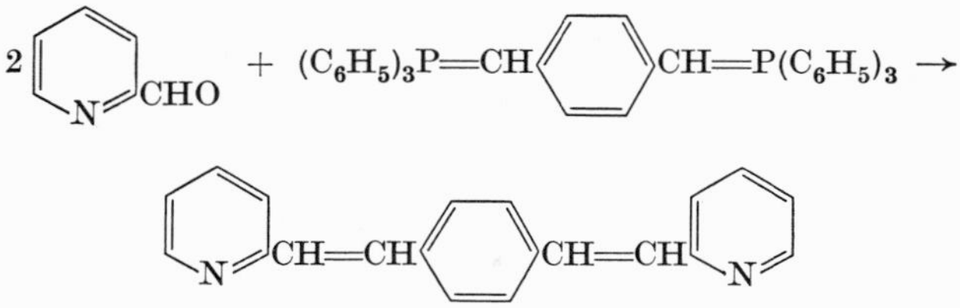

2 CHO + $(C_6H_5)_3P{=}CH$ $CH{=}P(C_6H_5)_3$ →

CH=CH CH=CH

(56%)

Distyrylbenzenes can be prepared similarly.[66,193]

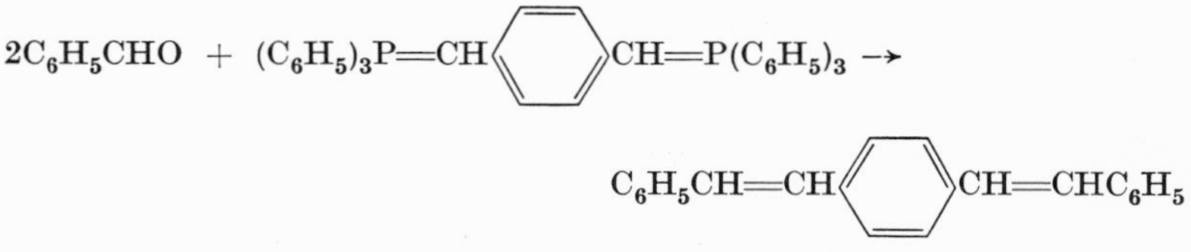

$2C_6H_5CHO$ + $(C_6H_5)_3P{=}CH$ $CH{=}P(C_6H_5)_3$ →

$C_6H_5CH{=}CH$ $CH{=}CHC_6H_5$

(88%)

[197] Geerts and Martin, *Bull. Soc. Chim. Belges*, **69**, 563 (1960) [*C.A.*, **55**, 14410c (1961)].

[198] Drefahl, Plötner, and Buchner, *Chem. Ber.*, **94**, 1824 (1961).

These compounds, which are of commercial interest as fluorescent brightening agents, can also be synthesized from terephthalaldehyde.[66,193,199–201]

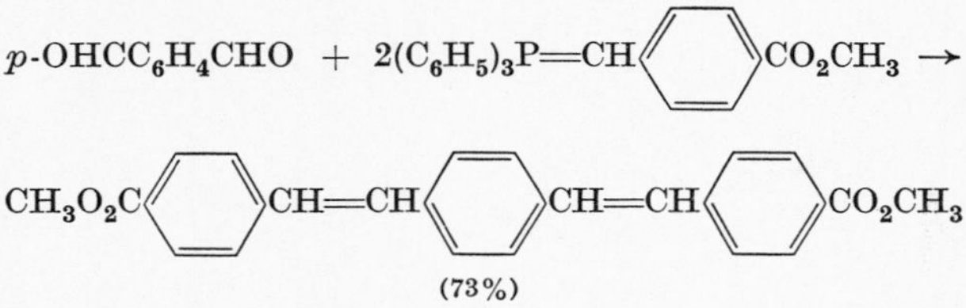

$$p\text{-}OHCC_6H_4CHO + 2(C_6H_5)_3P{=}CH\text{-}C_6H_4\text{-}CO_2CH_3 \rightarrow$$

$$CH_3O_2C\text{-}C_6H_4\text{-}CH{=}CH\text{-}C_6H_4\text{-}CH{=}CH\text{-}C_6H_4\text{-}CO_2CH_3$$

(73%)

Longer chains with alternating benzene rings and ethylene groups can be prepared from stilbenedialdehyde.[196]

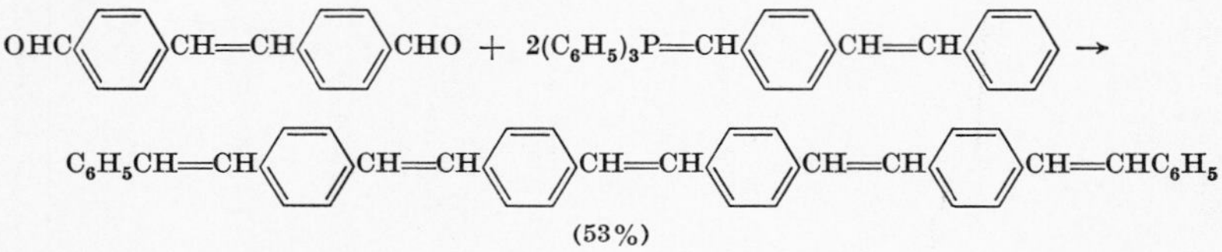

$$OHC\text{-}C_6H_4\text{-}CH{=}CH\text{-}C_6H_4\text{-}CHO + 2(C_6H_5)_3P{=}CH\text{-}C_6H_4\text{-}CH{=}CH\text{-}C_6H_5 \rightarrow$$

$$C_6H_5CH{=}CH\text{-}C_6H_4\text{-}CH{=}CH\text{-}C_6H_4\text{-}CH{=}CH\text{-}C_6H_4\text{-}CH{=}CH\text{-}C_6H_4\text{-}CH{=}CHC_6H_5$$

(53%)

1,3-Distyrylbenzenes[66] and 1,2-distyrylbenzenes[201a] have also been synthesized. Starting from the 1,2-distyrylbenzene **102,** the macrocyclic compound **103** can be prepared.[201a]

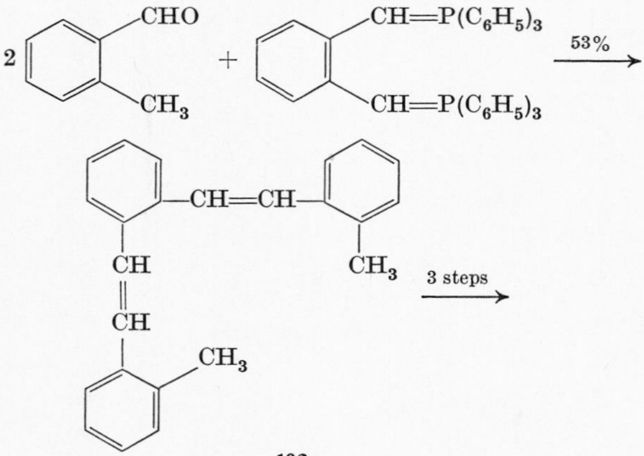

$$2\ o\text{-}CH_3C_6H_4CHO + o\text{-}C_6H_4[CH{=}P(C_6H_5)_3]_2 \xrightarrow{53\%}$$

$$o\text{-}C_6H_4[CH{=}CH\text{-}C_6H_4\text{-}CH_3\text{-}o]_2 \xrightarrow{\text{3 steps}}$$

**102**

[199] Stilz, Pommer, Wolff, and Fessmann, Fr. pat. 1,266,688 (to BASF), 1959.

[200] Stilz, Pommer, Gehm, Schmidt, Mertens, Hehl, and Grunwald, Belg. pat. 590,794 (to BASF), 1959.

[201] Pommer, Siebel, Schwen, and Stilz, Belg. pat. 593,216 (to BASF), 1960.

[201a] Griffin, Martin, and Douglas, *J. Org. Chem.*, **27,** 1627 (1962).

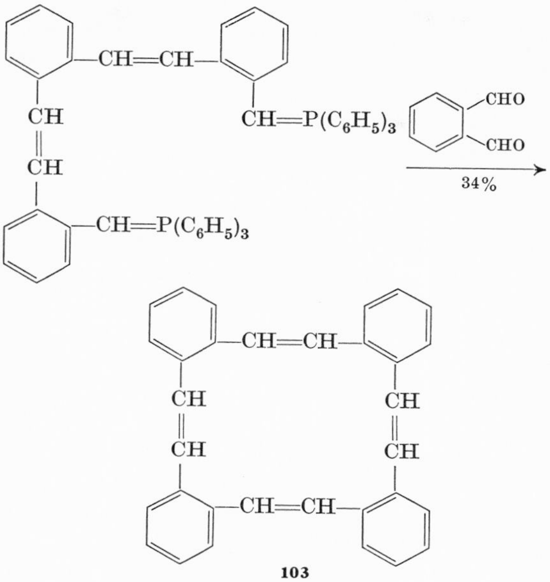

$CH{=}CH$ ; $CH{=}P(C_6H_5)_3$ ; $CH{=}P(C_6H_5)_3$ ; CHO, CHO; 34%

**103**

Cyclic compounds may also be prepared by intramolecular Wittig reactions. Thus the alkylidene phosphorane **104,** which contains a carbonyl group, cyclizes to 1-phenylcyclopentene.[202]

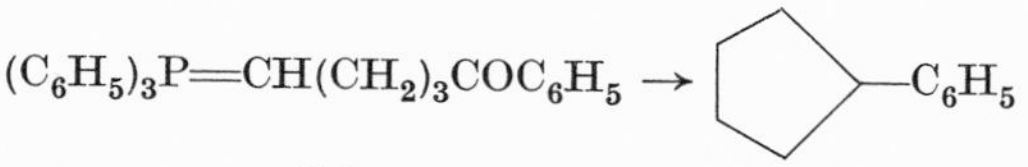

$$(C_6H_5)_3P{=}CH(CH_2)_3COC_6H_5 \rightarrow \text{1-phenylcyclopentene } (-C_6H_5)$$

**104** (24%)

2-Benzoylethylidenetriphenylphosphorane **(105),** on the other hand, does not react intramolecularly to give 1-phenylcyclopropene. Instead an intermolecular condensation of two molecules of the phosphorane leads to 1,4-diphenyl-1,4-cyclohexadiene.[203]

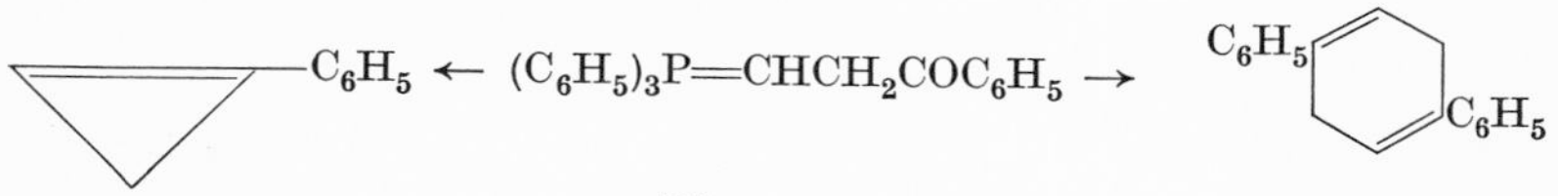

$$\text{1-phenylcyclopropene } (-C_6H_5) \leftarrow (C_6H_5)_3P{=}CHCH_2COC_6H_5 \rightarrow \text{1,4-diphenyl-1,4-cyclohexadiene } (C_6H_5, C_6H_5)$$

**105** (12%)

Aromatic compounds may also be prepared by using intramolecular Wittig reactions. The vinylog of benzoylmethylenetriphenylphosphorane

[202] Bieber and Eisman, *J. Org. Chem.*, **27**, 678 (1962).

[203] Griffin and Witschard, *J. Org. Chem.*, **27**, 3334 (1962).

**107,** prepared from the pyrylium salt **106,** cyclizes to symmetrical triphenylbenzene.[204]

$C_6H_5$ $C_6H_5$ $C_6H_5$ O $\oplus$ $BF_4^{\ominus}$ **106** $+ \; 2(C_6H_5)_3P{=}CH_2 \rightarrow$

$C_6H_5$ $C_6H_5$ O $C_6H_5$ CH $\|$ $P(C_6H_5)_3$ **107** $+ \; (C_6H_5)_3\overset{\oplus}{P}{-}CH_3 \; BF_4^{\ominus}$

$\downarrow$

$C_6H_5$ $C_6H_5$ $C_6H_5$ (59%) $+ \; (C_6H_5)_3PO$

Cyclizations may also occur in the ylide prior to condensation with a carbonyl component.[205]

$$(C_6H_5)_3\overset{\oplus}{P}{-}\overset{\ominus}{C}H(CH_2)_4CO_2C_2H_5 \longrightarrow$$

$(C_6H_5)_3\overset{\oplus}{P}$— O$^{\ominus}$ $\xrightarrow{C_6H_5CHO}$ $C_6H_5CH{=}$ O

A cyclization has also been reported in which a rearrangement occurred in the alkylidene group of the phosphorane.[206]

CHO $O(CH_2)_2\overset{\ominus}{C}H{-}\overset{\oplus}{P}(C_6H_5)_3$

CH=CH $CH_2$ O—$CH_2$ (8%) + O $CH_3$ (87%)

[204] Märkl, *Angew. Chem.*, **74,** 696 (1962).

[205] House and Babad, *J. Org. Chem.*, **28,** 90 (1963).

[206] Schweizer and Schepers, *Tetrahedron Letters*, **1963,** 979.

If there is no possibility for intramolecular ring closure, an intermolecular reaction may take place exclusively. *p*-Formylbenzylidenetriphenylphosphorane **(108)**, for instance, polymerizes spontaneously to give a lemon-yellow compound **109** which melts above 360°.[207]

$$(C_6H_5)_3P{=}CH{-}C_6H_4{-}CHO \rightarrow OHC{-}\left(C_6H_4{-}CH{=}CH\right)_n{-}C_6H_4{-}CHO$$

**108** **109** ($n = 11$)

In a similar fashion, the colorless poly-*m*-xylylidene derivative **111**, m.p. 180°, is obtained from the ylide **110**.[208]

$$(C_6H_5)_3P{=}CH{-}C_6H_3(CHO)(OCH_3) \rightarrow$$

**110**

$$OHC{-}\left(C_6H_3(OCH_3){-}CH{=}CH\right)_n{-}C_6H_3(OCH_3){-}CHO$$

**111**

Polymerizations also can be effected by interaction of bifunctional carbonyl compounds with bis-ylides. Terephthalaldehyde reacts quantitatively with the bifunctional phosphorane **112** to give a yellow poly-*p*-xylylidene which, except for its degree of polymerization, is probably identical with the product **109** mentioned above.[208]

$$OHC{-}C_6H_4{-}CHO + (C_6H_5)_3P{=}CH{-}C_6H_4{-}CH{=}P(C_6H_5)_3 \rightarrow \mathbf{109}$$

**112** ($n = 9$)

By means of the Wittig reaction, a large number of aryl-substituted butadiene derivatives and polyenes has been prepared. 1-Phenylbutadiene was obtained in two different ways.[2] Starting from cinnamaldehyde

$$C_6H_5CH{=}CHCHO + (C_6H_5)_3P{=}CH_2 \xrightarrow{A} C_6H_5CH{=}CHCH{=}CH_2$$

$$C_6H_5CHO + (C_6H_5)_3P{=}CHCH{=}CH_2 \xrightarrow{B} C_6H_5CH{=}CHCH{=}CH_2$$

[207] A. Haag, Doctoral Dissertation, Universität Heidelberg, 1962.

[208] McDonald and Campbell, *J. Am. Chem. Soc.*, **82**, 4669 (1960).

(path A), no isomerization of the double bond took place and the *trans* isomer was obtained exclusively in 69% yield. Path B, on the other hand, gave 58% of a mixture containing 55% *trans*- and 45% *cis*-phenylbutadiene.

1,1-Diphenylbutadiene was prepared from $\beta$-phenylcinnamaldehyde by using path A.[138]

$$(C_6H_5)_2C{=}CHCHO + (C_6H_5)_3P{=}CH_2 \rightarrow \underset{(54\%)}{(C_6H_5)_2C{=}CHCH{=}CH_2}$$

1,4-Diarylbutadiene derivatives may also be obtained by two different methods.[66,209–211]

$$C_6H_5CH{=}CHCHO + (C_6H_5)_3P{=}CHC_6H_5 \xrightarrow{84\%} C_6H_5CH{=}CHCH{=}CHC_6H_5$$

$$C_6H_5CHO + (C_6H_5)_3P{=}CHCH{=}CHC_6H_5 \xrightarrow{63\%} C_6H_5CH{=}CHCH{=}CHC_6H_5$$

Paths A and B have been used for the synthesis of a large number of unsymmetrically substituted 1,4-diarylbutadienes.[210,211] The symmetrically substituted compounds may also be prepared by a third method in which glyoxal is the starting carbonyl compound.[211]

$$\begin{matrix}CHO\\ |\\ CHO\end{matrix} + 2(C_6H_5)_3P{=}CH\text{-}C_6H_4\text{-}CH{=}CH\text{-}C_6H_5 \rightarrow$$

$$\underset{(40\%)}{C_6H_5CH{=}CH\text{-}C_6H_4\text{-}CH{=}CHCH{=}CH\text{-}C_6H_4\text{-}CH{=}CHC_6H_5}$$

A number of 1,1,4-triarylbutadienes was prepared from the alkylidene phosphorane **113**.[211] For example,

$$C_6H_5CHO + \underset{\mathbf{113}}{(C_6H_5)_3P{=}CHCH{=}C(C_6H_5)_2} \rightarrow C_6H_5CH{=}CHCH{=}C(C_6H_5)_2$$

The same diene is formed in 64% yield by the action of cinnamylidenetriphenylphosphorane **(114)** on benzophenone.[138]

$$(C_6H_5)_2C{=}O + \underset{\mathbf{114}}{(C_6H_5)_3P{=}CHCH{=}CHC_6H_5} \rightarrow C_6H_5CH{=}CH{-}CH{=}C(C_6H_5)_2$$

[209] Wittig and Pommer, Ger. Pat. 971,986 (to BASF) (*Chem. Zentr.*, **1959,** 16097).
[210] McDonald and Campbell, *J. Org. Chem.*, **24,** 1969 (1959).
[211] Drefahl, Plötner, Hartrodt, and Kühmstedt, *Chem. Ber.*, **93,** 1799 (1960).

*p*-Bis-(4-arylbutadienyl)benzenes have been prepared by starting from bifunctional ylides[66,210] and from bifunctional carbonyl compounds.[66,211] The synthesis of *p*-bis-4-phenylbutadienylbenzene **(115)** is an example of the first route,[210] and the synthesis of *p*-bis(4,4-diphenylbutadienyl)-

$$2C_6H_5CH{=}CHCHO + (C_6H_5)_3P{=}CH\text{-}C_6H_4\text{-}CH{=}P(C_6H_5)_3 \xrightarrow{(88\%)}$$

**112**

$$C_6H_5CH{=}CHCH{=}CH\text{-}C_6H_4\text{-}CH{=}CHCH{=}CHC_6H_5$$

**115**

benzene **(116)** is an example of the second.[211]

$$OHC\text{-}C_6H_4\text{-}CHO + 2(C_6H_5)_3P{=}CHCH{=}C(C_6H_5)_2 \xrightarrow{(57\%)}$$

$$(C_6H_5)_2C{=}CHCH{=}CH\text{-}C_6H_4\text{-}CH{=}CHCH{=}C(C_6H_5)_2$$

**116**

The synthesis of higher diarylpolyenes is effected in basically the same way, starting either from bifunctional ylides or from dicarbonyl compounds as illustrated by the synthesis of 1,10-diphenyldecapentaene **(117)**[209,212] and its dimethyl derivative **118**[209] by two different methods.

$$2C_6H_5CH{=}CHCHO + (C_6H_5)_3P{=}CHCH{=}CHCH{=}P(C_6H_5)_3 \rightarrow$$

$$C_6H_5(CH{=}CH)_5C_6H_5$$

**117**

$$OHC\text{-}C(CH_3){=}CH\text{-}CH{=}CH\text{-}CH{=}CH\text{-}CH{=}C(CH_3)\text{-}CHO + 2(C_6H_5)_3P{=}CHC_6H_5 \rightarrow$$

$$C_6H_5CH{=}CH\text{-}C(CH_3){=}CH\text{-}CH{=}CH\text{-}CH{=}CH\text{-}CH{=}C(CH_3)\text{-}CH{=}CH\text{-}C_6H_5$$

**118**

Finally, the preparation of the cross-conjugated chromophore systems **119** has been reported.[213]

$$C_6H_5(CH{=}CH)_nCO(CH{=}CH)_{n'}C_6H_5 + (C_6H_5)_3P{=}CH(CH{=}CH)_{n''}C_6H_5 \rightarrow$$

$$C_6H_5(CH{=}CH)_nC(CH{=}CH)_{n'}C_6H_5$$
$$\|$$
$$CH(CH{=}CH)_{n''}C_6H_5$$

**119** $(n=1, 2; n'=0, 1, 2; n''=0, 1)$

[212] Heitman, Sperna Weiland, and Huisman, *Konikl. Ned. Akad. Wetenschap., Proc.*, *Ser. B.*, **64,** 165 (1961) [*C.A.*, **55,** 17562f (1961)].

[213] Bohlmann, *Chem. Ber.*, **89,** 2191 (1956).

Allenes can also be prepared by the Wittig reaction. Again, two methods may be used for the preparation of tetraphenylallene. Path A

$$(C_6H_5)_2C{=}C{=}O + (C_6H_5)_3P{=}C(C_6H_5)_2 \xrightarrow{A} (C_6H_5)_2C{=}C{=}C(C_6H_5)_2$$

$$(C_6H_5)_2C{=}O + (C_6H_5)_3P{=}C{=}C(C_6H_5)_2 \xrightarrow{B} (C_6H_5)_2C{=}C{=}C(C_6H_5)_2$$

uses diphenylketene and diphenylmethylenetriphenylphosphorane.[9] Path B, which employs milder conditions, gives tetraphenylallene in 54% yield.[123] This was the first use of a vinylidene phosphorane in a Wittig reaction. Using path A, alkyl-substituted allenes such as 1,1-dimethyl-3,3-diphenylallene **(120)** may be prepared by dry distillation of the preformed, exceedingly stable betaine **64.**[150]

$$\underset{\mathbf{64}}{(C_6H_5)_3\overset{\oplus}{P}{-}C(CH_3)_2{-}C(O^{\ominus}){=}C(C_6H_5)_2} \xrightarrow[64\%]{160^\circ} \underset{\mathbf{120}}{(CH_3)_2C{=}C{=}C(C_6H_5)_2}$$

## Unsaturated Carbonyl Compounds

**Aldehydes.** The introduction of an aldehyde group may be effected by the reaction between an alkylidene phosphorane and a dicarbonyl compound in which one carbonyl group has been protected by acetal formation. Subsequent treatment with acid converts the acetal to the aldehyde.[214,215]

$P(C_6H_5)_3$ + O, CH, O—CH$_2$, O—CH$_2$ →

CH, O—CH$_2$, O—CH$_2$ $\xrightarrow{H^{\oplus}}$ CHO

[214] Isler, Rüegg, Montavon, and Zeller, Ger. pat. 1,021,361 (to Hoffmann-LaRoche) [*C.A.*, **53,** 19882c (1959)].

[215] Isler, Montavon, Rüegg, and Zeller, U.S. pat. 2,819,312 (to Hoffmann-LaRoche) [*C.A.*, **52,** 11915g (1958)].

Alternatively, a monocarbonyl compound may be made to react with an alkylidene phosphorane containing an acetal group which is subsequently converted to the aldehyde by treatment with acid.[216] Thus the $\beta$-$C_{14}$ aldehyde **121** was converted in 78% yield to the $\beta$-$C_{19}$ aldehyde **122.**

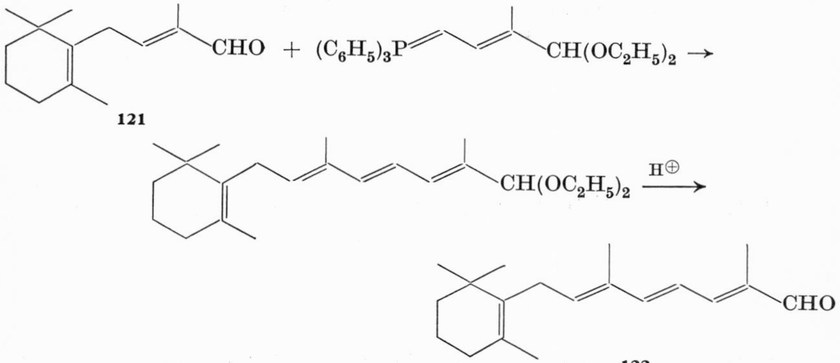

$$\textbf{121}\text{-CHO} + (C_6H_5)_3P{=}CH{-}CH{=}C(CH_3){-}CH(OC_2H_5)_2 \rightarrow \ldots CH(OC_2H_5)_2 \xrightarrow{H^{\oplus}} \ldots CHO\ \textbf{122}$$

**121**

**122**

The carbonyl group in resonance-stabilized formylalkylidenetriphenylphosphoranes, for example, **123**, need not be protected. Reaction with aldehydes gives $\alpha,\beta$-unsaturated aldehydes in good yields.[33,58,116]

$$(C_6H_5)_3P{=}CHCHO\ \textbf{(123)} \xrightarrow{n\text{-}C_6H_{13}CHO} n\text{-}C_6H_{13}CH{=}CHCHO\ (81\%)$$

$$(C_6H_5)_3P{=}CHCHO\ \textbf{(123)} \xrightarrow{C_6H_5CHO} C_6H_5CH{=}CHCHO\ (60\%)$$

The $\beta,\gamma$-unsaturated aldehyde **125** has been prepared from the enol ester **124**, the ester group being subsequently removed by saponification.[217]

$$C_6H_5CO_2CH{=}C(CH_3)CHO\ \textbf{(124)} + (C_6H_5)_3P{=}CHCO_2C_2H_5 \rightarrow$$

$$C_6H_5CO_2CH{=}C(CH_3)CH{=}CHCO_2C_2H_5 \xrightarrow{H^{\oplus}} OHCCH(CH_3)CH{=}CHCO_2H\ \textbf{(125)}$$

[216] Makin, *Dokl. Akad. Nauk SSSR*, **138,** 387 (1961) [*C.A.*, **55,** 20989 (1961)].

[217] Kucherov, Yanovskaya, and Kovalev, *Dokl. Akad. Nauk SSSR*, **133,** 370 (1960) [*C.A.*, **54,** 24699h (1960)].

**Ketones.** Unless there is the possibility of a resonance interaction with the ylene double bond, keto groups in alkylidene phosphoranes must be protected, preferably by conversion to a ketal, as illustrated by the synthesis of 7-fluoro-6-methylhept-5-en-2-one **(126)**.[218]

$$CH_2FCOCH_3 + (C_6H_5)_3P{=}CHCH_2CH_2{-}C(OCH_2CH_2O){-}CH_3 \rightarrow$$

$$CH_2FC(CH_3){=}CH(CH_2)_2{-}C(OCH_2CH_2O){-}CH_3 \xrightarrow{H^{\oplus}} CH_2FC(CH_3){=}CHCH_2CH_2COCH_3$$

**126**

Acylmethylene phosphoranes, on the other hand, are stable and may be used directly in the synthesis of $\alpha,\beta$-unsaturated ketones.[31]

$$C_6H_5CHO + (C_6H_5)_3P{=}CHCOR \rightarrow C_6H_5CH{=}CHCOR$$

$(R = CH_3, C_6H_5)$

$$p\text{-}O_2NC_6H_4CHO + (C_6H_5)_3P{=}CHCOCH_3 \rightarrow p\text{-}O_2NC_6H_4CH{=}CHCOCH_3$$

(92%) (Ref. 42)

Reactions of ketones with acylmethylene triphenylphosphoranes, as with all other resonance-stabilized ylides, take place less readily.[31] An exception is the formation of the unsaturated fluoroketone **127** in 93% yield.[219]

$$(CF_3)_2C{=}O + (C_6H_5)_3P{=}CHCOCH_3 \rightarrow (CF_3)_2C{=}CHCOCH_3$$

**127**

$\alpha,\beta$-Unsaturated ketones substituted on the $\alpha$-position are accessible in a similar way, as illustrated by the preparation of the ketone **128**.[32]

$$C_6H_5CHO + (C_6H_5)_3P{=}C(CH_3)COCH{=}CHC_6H_5 \rightarrow$$

$$C_6H_5CH{=}C(CH_3)COCH{=}CHC_6H_5$$

**128** (81%)

$\alpha$-Halo-$\alpha,\beta$-unsaturated ketones have also been prepared by this method; forcing conditions are required.[59,60] The highest yields were obtained by using excess aldehyde as the solvent. No reaction was

[218] Machleidt, Hartmann, Wessendorf, and Grell, *Angew. Chem.*, **74**, 505 (1962).
[219] Plakhova and Gambaryan, *Izv. Akad. Nauk SSSR, Otd. Khim. Nauk*, **1962**, 681.

observed when α-haloacylmethylene triphenylphosphoranes were heated with ketones in benzene for several hours.[59]

$$C_6H_5CHO + (C_6H_5)_3P{=}C(Cl)COC_6H_5 \xrightarrow[2\ days]{100°} C_6H_5CH{=}C(Cl)COC_6H_5$$

(Quant.) (Ref. 60)

**Carboxylic Esters and Other Acid Derivatives.** The interaction of carbalkoxymethylene triphenylphosphoranes with carbonyl compounds leads to α,β-unsaturated carboxylic esters. Aldehydes react very readily, as illustrated by the preparation of ethyl cinnamate.[3] Analogously,

$$C_6H_5CHO + (C_6H_5)_3P{=}CHCO_2C_2H_5 \rightarrow C_6H_5CH{=}CHCO_2C_2H_5$$

(77%)

good yields of nitrocinnamic esters were obtained from *o*-, *m*-, and *p*-nitrobenzaldehyde.[162]

The successful reaction of carbethoxymethylenetriphenylphosphorane with heterocyclic aldehydes such as furfural,[220] pyridine aldehydes,[34] and 8-xanthinecarboxaldehyde[221] has also been reported.

Aliphatic aldehydes are also easily converted to α,β-unsaturated carboxylic esters, as illustrated by the preparation of 1,9-bis(methoxycarbonyl)decene **(129)**.[161]

$$CH_3O_2C(CH_2)_8CHO + (C_6H_5)_3P{=}CHCO_2CH_3 \rightarrow CH_3O_2C(CH_2)_8CH{=}CHCO_2CH_3$$

**129** (95%)

Reactions of ketones with carbalkoxymethylene triphenylphosphoranes occur, as a rule, much less readily. The ease of reaction of fluoroacetone is an exception.[218]

$$(CH_2F)(CH_3)C{=}O + (C_6H_5)_3P{=}CHCO_2C_2H_5 \rightarrow (CH_2F)(CH_3)C{=}CHCO_2C_2H_5$$

Aromatic or aliphatic ketones, however, will give satisfactory yields in their reactions with carbalkoxymethylene triphenylphosphoranes only

[220] Kucherov, Kovalev, Nazarova, and Yanovskaya, *Izv. Akad. Nauk SSSR, Otd. Khim. Nauk*, **1960**, 1512 [*C.A.*, **55**, 1420b (1961)].

[221] Bredereck and Föhlisch, *Chem. Ber.*, **95**, 414 (1962).

after reaction times of several days at room temperature[34] or several hours at 100–170°.[35] Thus the ketone **130** reacts slowly in ethanol at 20° with carbethoxymethylenetriphenylphosphorane to give the ester **131** in 90% yield after several days.[34]

$$4\text{-}C_5H_4N\text{-}C(CH_3){=}O + (C_6H_5)_3P{=}CHCO_2C_2H_5 \rightarrow 4\text{-}C_5H_4N\text{-}C(CH_3){=}CHCO_2C_2H_5$$

**130** **131**

Heating the same ylide with acetophenone without a solvent gives ethyl β-methylcinnamate in 58% yield.[35]

$$C_6H_5C(CH_3){=}O + (C_6H_5)_3P{=}CHCO_2C_2H_5 \xrightarrow[10\ \text{hr.}]{170°} C_6H_5C(CH_3){=}CHCO_2C_2H_5$$

The preparation of esters containing exocyclic double bonds from cyclic ketones also requires forcing conditions.[34,35] Ethyl cyclohexylideneacetate **(132)** is formed in 44% yield in ethanol solution at room temperature within 1 week;[34] heating the reactants without solvent to 170° for 10 hours gives a yield of 60%.[35]

$$C_6H_{10}{=}O + (C_6H_5)_3P{=}CHCO_2C_2H_5 \rightarrow C_6H_{10}{=}CHCO_2C_2H_5$$

**132**

Substituents in the α-position to the carbonyl group further reduce the reactivity of ketones. Thus 61% of the ketone was recovered when a mixture of 2-methylcyclohexanone and carbethoxymethylenetriphenylphosphorane was kept at room temperature for 12 days; the yield of ethyl 2-methylcyclohexylideneacetate **(133)** was only 7%.[34]

$$2\text{-}CH_3C_6H_9{=}O + (C_6H_5)_3P{=}CHCO_2C_2H_5 \rightarrow 2\text{-}CH_3C_6H_9{=}CHCO_2C_2H_5$$

**133**

In general, it is advantageous to use high temperatures and little or no solvent when sterically hindered ketones are employed.[36] Thus, in the reaction of 2-methylcyclohexanone mentioned above, the yield of the ester **133** was increased to 30% by heating the reactants without solvent to 150°.[36] Under the same conditions, the tricyclic compound **134** could be obtained in 58% yield.[36]

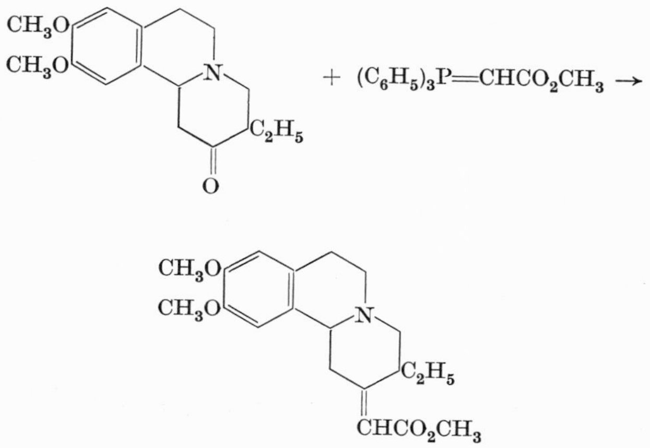

$+ \ (C_6H_5)_3P{=}CHCO_2CH_3 \rightarrow$

**134**

Reactions like these take place much more readily in the presence of benzoic acid as a catalyst.[38]

Substitution of phenyl groups on phosphorus by electron-releasing groups also results, under otherwise equal conditions, in an increase of the yields.[28,58,222–226]

In the reactions of dicarbonyl compounds with 1 mole of carbomethoxymethylenetriphenylphosphorane, it is usually possible to isolate the compound resulting from conversion of only one carbonyl group to an olefin.[227–229] Thus the monoketone **136** is obtained in 93% yield by the

[222] Trippett and Walker, *Chem. Ind. (London)*, **1960**, 933.
[223] Horner, Hoffmann, and Wippel, *Chem. Ber.*, **91**, 61 (1958).
[224] Horner, Hoffmann, Wippel, and Klahre, *Chem. Ber.*, **92**, 2499 (1959).
[225] Horner, Hoffmann, Klink, Ertel, and Toscano, *Chem. Ber.*, **95**, 581 (1962).
[226] Wadsworth, Jr., and Emmons, *J. Am. Chem. Soc.*, **83**, 1733 (1961).
[227] Cava and Pohl, *J. Am. Chem. Soc.*, **82**, 5242 (1960).
[228] Eiter, *Angew. Chem.*, **73**, 619 (1961).
[229] Eiter, *Ann.*, **658**, 91 (1962).

interaction of benzocyclobutenedione (**135**) with 1 mole of ylide, whereas with 2 moles of ylide the final product **137** is formed in 85% yield.[227]

$$\textbf{135} + (C_6H_5)_3P{=}CHCO_2CH_3 \rightarrow \textbf{136}\ (=O,\ =CHCO_2CH_3) \quad \text{or} \quad \textbf{137}\ (=CHCO_2CH_3,\ =CHCO_2CH_3)$$

Unsaturated dialdehydes such as **138** lead to polyene dicarboxylic esters, as illustrated by the synthesis of the ester **139**.[230] Methylbixin,

$$OHC\text{–polyene–}CHO\ (\textbf{138}) + 2(C_6H_5)_3P{=}CHCO_2CH_3 \rightarrow CH_3O_2C\text{–polyene–}CO_2CH_3$$

**139** (72%)

crocetin dimethyl ester[39,231–235] and a large number of other polyene dicarboxylic esters[163,236] have been prepared in this way.

In compounds containing both a keto and an aldehyde group, the latter reacts preferentially. If conversion of the keto group to an olefin is to be effected, the aldehyde group must be protected by acetal formation, as illustrated by the reaction of the carbonyl compound **140** with carbomethoxymethylenetriphenylphosphorane.[237]

[230] Buchta and Andree, *Ann.*, **640**, 29 (1961).
[231] Isler, Ryser, and Zeller, Belg. pat. 559,699 (to Hoffmann-La Roche), 1957.
[232] Buchta and Andree, *Naturwiss.*, **46**, 74 (1959).
[233] Buchta and Andree, *Naturwiss.*, **46**, 75 (1959).
[234] Buchta and Andree, *Chem. Ber.*, **92**, 3111 (1959).
[235] Buchta and Andree, *Chem. Ber.*, **93**, 1349 (1960).
[236] Guex, Isler, Rüegg, and Ryser, Belg. pat. 573,097 (to Hoffmann-La Roche), 1958.
[237] Serratosa, *Tetrahedron*, **16**, 185 (1961).

In the ketone **140**, a hydroxyl group has also been protected by conversion to its tetrahydropyranyl ether. Two neighboring hydroxyl groups, as in

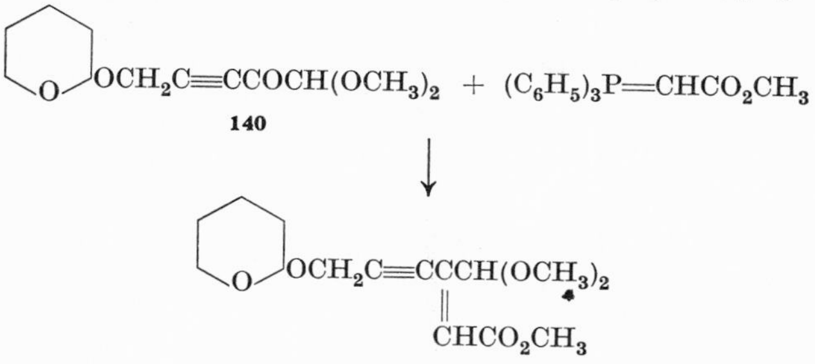

$$\text{THP-}OCH_2C{\equiv}CCOCH(OCH_3)_2 + (C_6H_5)_3P{=}CHCO_2CH_3$$

**140**

$$\downarrow$$

$$\text{THP-}OCH_2C{\equiv}CC({=}CHCO_2CH_3)CH(OCH_3)_2$$

(60%)

sugars, may also be blocked by ketal formation, as shown by the interaction of the dimethylketal of D-glyceraldehyde with carbethoxymethylenetriphenylphosphorane to form the ester **141**.[238]

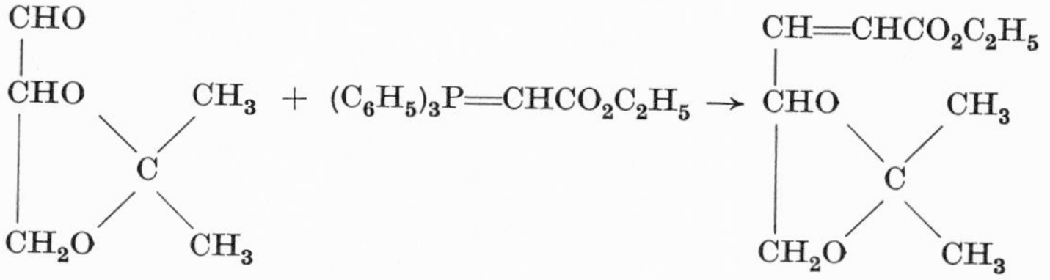

$$\text{CHO—CHO(—CH}_2\text{O—)C(CH}_3)_2 + (C_6H_5)_3P{=}CHCO_2C_2H_5 \rightarrow \text{CH}{=}\text{CHCO}_2\text{C}_2\text{H}_5\text{—CHO(—CH}_2\text{O—)C(CH}_3)_2$$

**141** (75%)

Polyene aldehydes react with carbalkoxymethylene triphenylphosphoranes to form polyene carboxylic esters which contain one more double bond than the aldehyde. For $n = 1$ through 4, the yields are

$$CH_3(CH{=}CH)_nCHO + (C_6H_5)_3P{=}CHCO_2C_2H_5 \rightarrow CH_3(CH{=}CH)_{n+1}CO_2C_2H_5$$

higher than 80%.[220]

Polyene aldehydes containing triple bonds may also be employed in this reaction,[239] which has been used extensively for the synthesis of

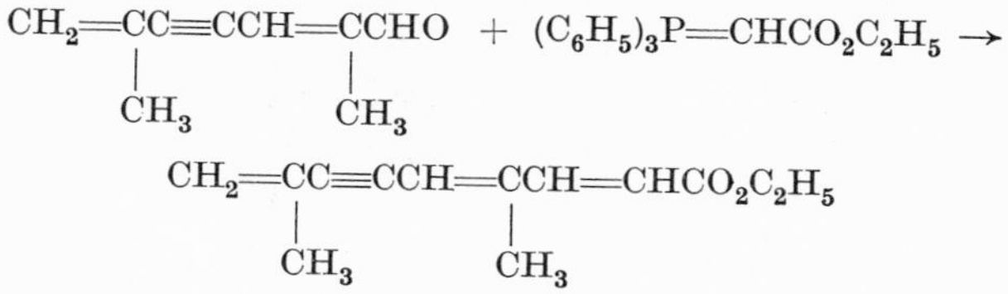

$$CH_2{=}C(CH_3)C{\equiv}CCH{=}C(CH_3)CHO + (C_6H_5)_3P{=}CHCO_2C_2H_5 \rightarrow$$

$$CH_2{=}C(CH_3)C{\equiv}CCH{=}C(CH_3)CH{=}CHCO_2C_2H_5$$

(65%)

carboxylic esters in the carotene series.

[238] Kuhn and Brossmer, *Angew. Chem.*, **74**, 252 (1962).

[239] Yanovskaya, Kucherov, and Kovalev, *Izv. Akad. Nauk SSSR, Otd. Khim. Nauk*, **1962**, 674 [*C.A.*, **57**, 16379a (1962)].

α,β-Unsaturated carboxylic esters may also be obtained from a glyoxylic ester or its vinylogs, as illustrated by the preparation of ethyl β-ionylideneacetate **(142)**.[240]

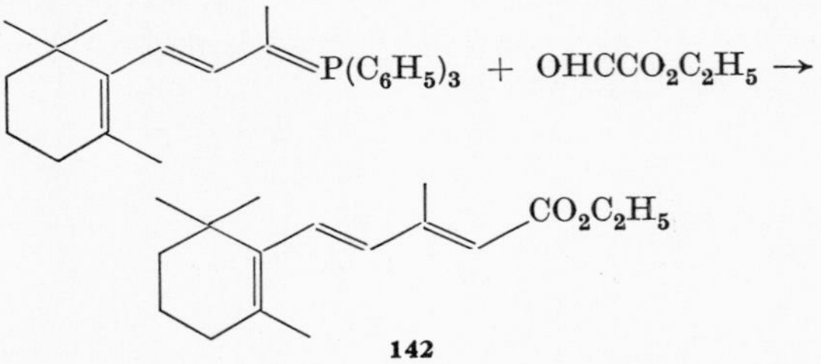

$$\text{(ionylidene ylide)}{=}P(C_6H_5)_3 + OHCCO_2C_2H_5 \rightarrow$$

$CO_2C_2H_5$

**142**

In order to prevent side reactions involving the ester group, it has been proposed to add the ylide solutions to the aldehyde and not vice versa. This procedure furnished the unsaturated ester **143** in 46% yield.[241]

$$n\text{-}C_3H_7(C{\equiv}C)_2CH_2CH_2CH{=}P(C_6H_5)_3 + OHCCH{=}CHCO_2CH_3 \rightarrow$$
$$n\text{-}C_3H_7(C{\equiv}C)_2CH_2CH_2(CH{=}CH)_2CO_2CH_3$$
**143**

Instead of a vinylog of glyoxylic ester, a vinylog of a carbalkoxymethylenephosphorane may also be used as illustrated by the reaction of substituted benzaldehydes with the phosphorane **144**.[242] The phosphorane **144** is easily accessible from methyl γ-bromocrotonate. A

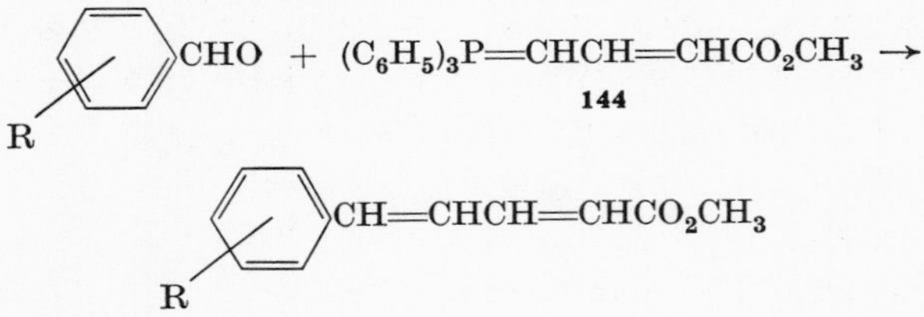

$$R\text{-}C_6H_4CHO + (C_6H_5)_3P{=}CHCH{=}CHCO_2CH_3 \rightarrow$$
**144**

$$R\text{-}C_6H_4CH{=}CHCH{=}CHCO_2CH_3$$

comparison of the results obtained by this reaction with those obtained in the Reformatsky reaction showed that the Wittig reaction is definitely to be preferred, especially with benzaldehydes containing nitro, dimethylamino, or chloro substituents. Even when comparable yields are obtained by both methods, the Wittig reaction has the advantages of convenience and speed. By using ultraviolet spectroscopy it was shown that the reaction was practically complete after 5 minutes. The only exception occurs in reactions using 2,4,6-trimethoxybenzaldehyde for which the

240 Pommer and Sarnecki, Ger. pat. 1,068,706 (to BASF) [*C.A.*, **56**, 512a (1962)].

241 Bohlmann and Inhoffen, *Chem. Ber.*, **89**, 1276 (1956).

242 Bohlmann, *Chem. Ber.*, **90**, 1519 (1957).

Reformatsky reaction is superior. It is of interest to note that aldehydes containing *o*-hydroxyl groups can also be employed in the Wittig reaction.

The reaction between aldehydes and carbalkoxy ylides containing α-alkyl substituents leads to α-branched α,β-unsaturated carboxylic esters. Thus cinnamaldehyde and the phosphorane **145** give the α-substituted ester **146**.[40,111] The vinylogous ylide **147** furnishes the corresponding hexatrienecarboxylic ester **148**.[232,235] By employing this

$$C_6H_5CH{=}CHCHO + (C_6H_5)_3P{=}C(CH_2C_6H_5)CO_2CH_3 \rightarrow C_6H_5CH{=}CHCH{=}C(CH_2C_6H_5)CO_2CH_3$$

**145** **146** (83%)

$$C_6H_5CH{=}CHCHO + (C_6H_5)_3P{=}CHCH{=}C(CH_3)CO_2CH_3 \rightarrow$$

**147**

$$C_6H_5CH{=}CHCH{=}CHCH{=}C(CH_3)CO_2CH_3$$

**148** (70%)

method and starting from polyene dialdehydes, branched polyene dicarboxylic esters have been prepared.[39]

Whereas in all these reactions using resonance-stabilized phosphoranes the *trans* olefin is formed predominantly, interaction of reactive ylides with α-keto esters leading to α-branched α,β-unsaturated esters always yields a considerable amount of the *cis* isomers.[164]

$$RCHO + (C_6H_5)_3P{=}C(R')CO_2CH_3 \rightarrow RCH{=}C(R')CO_2CH_3$$

$$RCH{=}P(C_6H_5)_3 + O{=}C(R')CO_2CH_3 \rightarrow RCH{=}C(R')CO_2CH_3$$

Similarly, α-halo-α,β-unsaturated carboxylic esters[60,117] are prepared from halocarbomethoxymethylene triphenylphosphoranes and aldehydes, as illustrated by the synthesis of the ester **149**.[117]

$$C_6H_5CH{=}CHCHO + (C_6H_5)_3P{=}C(Br)CO_2CH_3 \rightarrow C_6H_5CH{=}CHCH{=}C(Br)CO_2CH_3$$

**149** (Quant.)

α-Halo-α,β-unsaturated carboxylic esters that easily eliminate hydrogen halide furnish substituted propiolic acids on saponification.[117]

$$C_6H_5CHO + (C_6H_5)_3P{=}C(Br)CO_2CH_3 \rightarrow C_6H_5CH{=}C(Br)CO_2CH_3 \xrightarrow{OH^\ominus} C_6H_5C{\equiv}CCO_2H \quad (82\%)$$

Esters of ω-hydroxy-α,β-unsaturated carboxylic acids **(150)** are formed by the interaction of 2-hydroxytetrahydropyran with carbalkoxyalkylidene triphenylphosphoranes.[114] The *trans* isomer is formed almost exclusively.

$$\text{2-hydroxytetrahydropyran} + (C_6H_5)_3P{=}C(R)CO_2R' \rightarrow \underset{\mathbf{150}}{HO(CH_2)_4CH{=}C(R)CO_2R'}$$

*cis*-Carboxylic esters are obtained only if there is no possibility for resonance interaction between the ester group and the ylene double bond. By using dimethylformamide as the solvent and preferably in the presence of iodide ion, the *cis* compounds are formed almost exclusively.[114,170,171] Pelargonaldehyde and the ylide **151,** for instance, give ethyl oleate **(152)** stereospecifically in 73% yield.[114,170] Probably as a

$$CH_3(CH_2)_7CHO + \underset{\mathbf{151}}{(C_6H_5)_3P{=}CH(CH_2)_7CO_2C_2H_5} \xrightarrow[I^\ominus]{DMF}$$

$$\underset{\mathbf{152}}{\textit{cis-}CH_3(CH_2)_7CH{=}CH(CH_2)_7CO_2C_2H_5}$$

consequence of the solvation of this ylide in the polar solvent, the ester group in the reactive phosphorane **151** remains essentially unattacked. In other circumstances an ester group remote from the ylide portion of the molecule may, however, be attacked.[205]

Ethyl α-eleostearate (**153**) was also synthesized from the ylide **151**.[114]

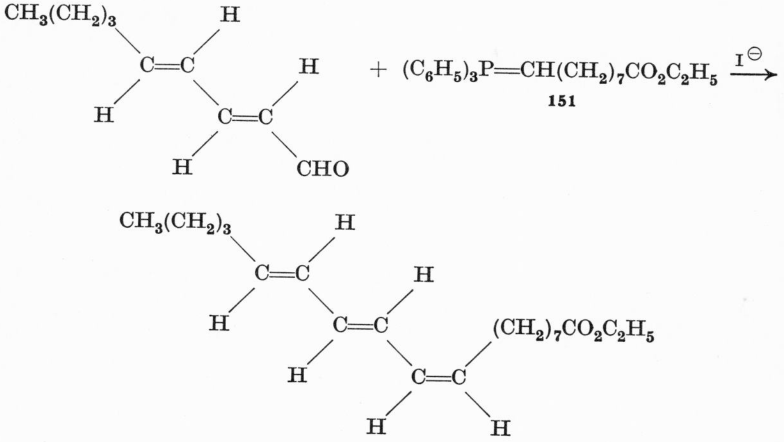

$$CH_3(CH_2)_3CH{=}CH{-}CH{=}CH{-}CHO + (C_6H_5)_3P{=}CH(CH_2)_7CO_2C_2H_5 \xrightarrow{I^{\ominus}}$$

**151**

$$CH_3(CH_2)_3CH{=}CH{-}CH{=}CH{-}CH{=}CH(CH_2)_7CO_2C_2H_5$$

**153**

Again, this type of compound may be prepared by two different ways, as illustrated by the synthesis of the biologically important acids of the *cis-cis*-divinylmethane type.[114]

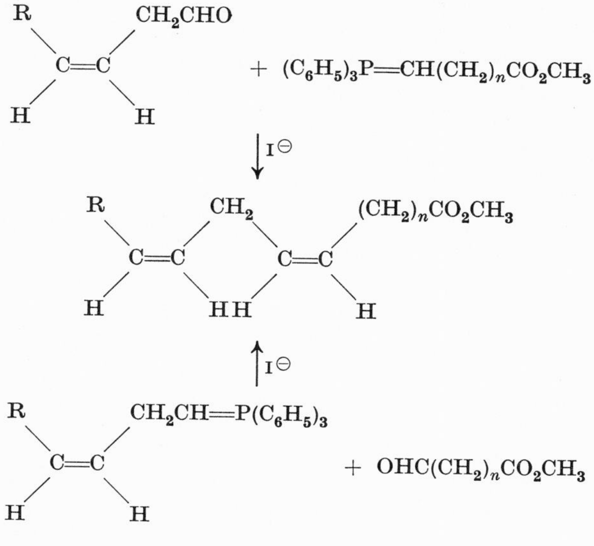

$$RCH{=}CHCH_2CHO + (C_6H_5)_3P{=}CH(CH_2)_nCO_2CH_3$$

$$\downarrow I^{\ominus}$$

$$RCH{=}CHCH_2CH{=}CH(CH_2)_nCO_2CH_3$$

$$\uparrow I^{\ominus}$$

$$RCH{=}CHCH_2CH{=}P(C_6H_5)_3 + OHC(CH_2)_nCO_2CH_3$$

The reaction between ketones and alkylidene phosphoranes of the type **154** is not stereospecific, but saturated branched fatty acids such as

**155** can be obtained in good yield by saponification followed by catalytic hydrogenation.[114]

$$CH_3(CH_2)_m\overset{\overset{CH_3}{|}}{C}{=}O + (C_6H_5)_3P{=}CH(CH_2)_nCO_2C_2H_5 \rightarrow$$

**154**

$$CH_3(CH_2)_m\overset{\overset{CH_3}{|}}{C}{=}CH(CH_2)_nCO_2C_2H_5 \rightarrow CH_3(CH_2)_m\overset{\overset{CH_3}{|}}{C}H(CH_2)_{n+1}CO_2H$$

**155**

Unsaturated amides and nitriles have also been synthesized. Thus carbamidomethylenetriphenylphosphorane **(156)** reacts with crotonaldehyde to give the amide of sorbic acid,[243] and with benzaldehyde to give cinnamamide.[58]

$$(C_6H_5)_3P{=}CHCONH_2 \xrightarrow{CH_3CH{=}CHCHO} CH_3CH{=}CHCH{=}CHCONH_2$$

$$(C_6H_5)_3P{=}CHCONH_2 \xrightarrow{C_6H_5CHO} C_6H_5CH{=}CHCONH_2$$

**156**

The interaction of cyanomethylenetriphenylphosphorane **(157)** with benzaldehyde and its derivatives gives the corresponding cinnamonitriles in good yields.[42,43,58] *p*-Nitrocinnamonitrile, for instance, is obtained from *p*-nitrobenzaldehyde in 74% yield.[42]

$$p\text{-}O_2NC_6H_4CHO + (C_6H_5)_3P{=}CHCN \rightarrow p\text{-}O_2NC_6H_4CH{=}CHCN$$

**157**

As illustrated by the synthesis of the vitamin $A_2$ carbonitrile **(158)**, vinylogous cyanomethylene triphenylphosphoranes can also be employed as starting materials.[244]

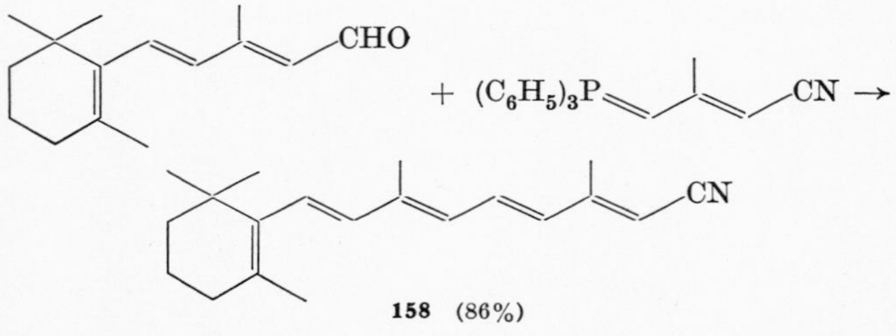

[Structure: trimethylcyclohexenyl dienal (CHO) + $(C_6H_5)_3P{=}$CH–C(CH₃)=CH–CN → vitamin A₂-type tetraene nitrile (CN)]

**158** (86%)

[243] Wittig and Pommer, Ger. pat. 943,648 (to BASF) [*C.A.*, **52,** 16292d (1958)].

[244] Eiter, Oediger, and Truscheit, Ger. pat. 1,110,633 (to Farbenf. Bayer) [*C.A.*, **56,** 3522c (1962)].

## Vinyl Halides

Monohaloölefins are formed in the reaction of halomethylene triphenylphosphoranes with aldehydes and ketones.[91,92] Benzaldehyde and benzophenone react with chloromethylenetriphenylphosphorane to give the corresponding vinyl halides in 67% yield.[92]

$$(C_6H_5)_3P{=}CHCl + C_6H_5CHO \rightarrow C_6H_5CH{=}CHCl$$
$$(C_6H_5)_3P{=}CHCl + (C_6H_5)_2CO \rightarrow (C_6H_5)_2C{=}CHCl$$

Aliphatic ketones can also be converted to vinyl chlorides.[95,96] If the

$$\text{cyclohexanone} + (C_6H_5)_3P{=}CHCl \rightarrow \text{cyclohexylidene}{=}CHCl \quad (80\%)$$

ylides are generated from triphenylphosphine and carbenes, the yields on the average decrease to 30%.[96]

The preparation of vinyl bromides has also been reported. The reaction with $\beta$-ionone, for instance, proceeds in 70% yield.[84]

$$\beta\text{-ionone} + (C_6H_5)_3P{=}CHBr \rightarrow \text{(ionylidene)}C{=}CH(Br)$$

1,1-Dihaloölefins are obtained from dihalomethylene triphenylphosphoranes.[82,93,94,98] The reactions with dichloromethylenetriphenylphosphorane are illustrative.

$$(C_6H_5)_3P{=}CCl_2 \xrightarrow{p\text{-}O_2NC_6H_4CHO} p\text{-}O_2NC_6H_4CH{=}CCl_2 \quad (83\%)$$
$$(C_6H_5)_3P{=}CCl_2 \xrightarrow{C_6H_5CH{=}CHCHO} C_6H_5CH{=}CHCH{=}CCl_2 \quad (77\%)$$

The preparation of mixed haloölefins is exemplified by the reaction of fluorochloromethylenetriphenylphosphorane with benzophenone.[94]

$$(C_6H_5)_2C{=}O + (C_6H_5)_3P{=}C(Cl)(F) \rightarrow (C_6H_5)_2C{=}C(Cl)(F) \quad (40\%)$$

Since carbon tetrachloride[98] and carbon tetrabromide[82] react directly with triphenylphosphine with the formation of dihalomethylene triphenylphosphoranes, 1,1-dihaloölefins can be prepared in one step by carrying out this reaction in the presence of carbonyl compounds.

$$CBr_4 + 2(C_6H_5)_3P \longrightarrow (C_6H_5)_3P{=}CBr_2 + (C_6H_5)_3PBr_2 \xrightarrow{C_6H_5CHO} C_6H_5CH{=}CBr_2 \; (84\%)$$

## Vinyl Ethers

The synthesis of vinyl ethers is important because they can be converted to aldehydes by saponification. The original procedure consisted of slow addition of formic ester to the alkylidene phosphorane followed by acid hydrolysis of the initially formed enol ether.[113] Hexahydrobenzaldehyde was obtained in moderate yield in this way, which is a general method for

$$(C_6H_5)_3P{=}C_6H_{10} + HCO_2C_2H_5 \rightarrow C_6H_{10}{=}CHOC_2H_5 \rightarrow C_6H_{10}(H)CHO$$

the conversion of a halogen compound to an aldehyde containing one more carbon atom. Since, however, the formic ester is partially cleaved by the ylide into carbon monoxide and alcohol, the yields are usually not very good.[245]

A better synthesis of vinyl ethers involves the reaction of a carbonyl compound with an alkoxymethylene triphenylphosphorane.[92,246–248] For example, benzophenone reacts with methoxymethylenetriphenylphosphorane to give 1,1-diphenylvinyl methyl ether **(159)** in 83% yield;[247] treatment with acid gives diphenylacetaldehyde.

$$(C_6H_5)_2C{=}O + (C_6H_5)_3P{=}CHOCH_3 \rightarrow \underset{\mathbf{159}}{(C_6H_5)_2C{=}CHOCH_3} \rightarrow (C_6H_5)_2CHCHO$$

The over-all result is the conversion of a carbonyl compound to an aldehyde containing one more carbon atom. The best reagent for the

$$>C{=}O \rightarrow >CHCHO$$

[245] W. Böll, Doctoral Dissertation, Universität Heidelberg, 1961.

[246] Levine, *J. Am. Chem. Soc.*, **80**, 6150 (1958).

[247] Wittig and Knauss, *Angew. Chem.*, **71**, 127 (1959).

[248] Wittig, Böll, and Krück, *Chem. Ber.*, **95**, 2514 (1962).

occasionally difficult hydrolysis of the enol ether is perchloric acid in diethyl ether.[246]

Cyclic ketones may also be subjected to this reaction. Cyclohexanone is converted to methoxymethylenecyclohexane which yields hexahydrobenzaldehyde on heating with ethereal perchloric acid.[248]

$$\text{cyclohexanone} + (C_6H_5)_3P{=}CHOCH_3 \rightarrow \underset{(71\%)}{\text{C}_6\text{H}_{10}{=}CHOCH_3} \rightarrow \underset{(84\%)}{\text{C}_6\text{H}_{11}CHO}$$

Alkoxymethylene triphenylphosphoranes are relatively unstable at room temperature. It is therefore advantageous to prepare them at as low a temperature as practicable and to add the carbonyl compound before warming the reaction mixture to room temperature.[87] The alkoxide method is of advantage with carbonyl compounds that are stable to alkoxides, as in the synthesis of **160.**[92]

$$o\text{-}C_6H_4(CHO)_2 + 2(C_6H_5)_3P{=}CHOCH_3 \rightarrow \underset{\textbf{160}\ (55\%)}{o\text{-}C_6H_4(CH{=}CHOCH_3)_2}$$

$n$-Butoxymethylenetriphenylphosphorane is less stable than methoxymethylenetriphenylphosphorane.[87] Low temperatures are therefore particularly important in the synthesis of butyl vinyl ethers. When diethyl ketone was allowed to react with $n$-butoxymethylenetriphenylphosphorane at $-40°$ and the reaction mixture was then left at room temperature for several hours, 1-$n$-butoxy-2-ethylbutene **(161)** was obtained in 73% yield.[248]

$$(C_2H_5)_2C{=}O + (C_6H_5)_3P{=}CHOC_4H_9\text{-}n \rightarrow \underset{\textbf{161}}{(C_2H_5)_2C{=}CHOC_4H_9\text{-}n} \rightarrow (C_2H_5)_2CHCHO$$

Hydrolysis of the ether with perchloric acid furnished diethylacetaldehyde in 70% yield. Cyclic ketones react analogously.[248]

$$\text{cyclohexanone} + (C_6H_5)_3P{=}CHOC_4H_9\text{-}n \rightarrow \underset{(74\%)}{\text{C}_6\text{H}_{10}{=}CHOC_4H_9\text{-}n}$$

Aryloxymethylene triphenylphosphoranes are more stable than alkoxymethylene triphenylphosphoranes. Vinyl aryl ethers are therefore particularly accessible. For example, *p*-methylphenoxymethylenecyclohexane **(163)** is obtained in 82% yield by allowing *p*-methylphenoxymethylenetriphenylphosphorane **(162)** to interact with cyclohexanone in ether at room temperature for 2 hours.[248]

$$\text{cyclohexanone} + (C_6H_5)_3P{=}CHOC_6H_4CH_3\text{-}p \rightarrow \text{cyclohexylidene}{=}CHOC_6H_4CH_3\text{-}p$$

**162** **163**

*β*-(*p*-Methylphenoxy)styrene **(164)**, obtained similarly in 75% yield from benzaldehyde, could be readily hydrolyzed with perchloric acid[246] to give phenylacetaldehyde in 73% yield.[248]

$$\underset{\mathbf{164}}{C_6H_5CH{=}CHOC_6H_4CH_3\text{-}p} \rightarrow C_6H_5CH_2CHO$$

In the same way, diethylacetaldehyde is prepared from diethyl ketone by way of diethylvinyl *p*-tolyl ether **(165)**.[248]

$$(C_2H_5)_2CO + \underset{\mathbf{162}}{(C_6H_5)_3P{=}CHOC_6H_4CH_3\text{-}p} \xrightarrow{72\%}$$

$$\underset{\mathbf{165}}{(C_2H_5)_2C{=}CHOC_6H_4CH_3\text{-}p} \xrightarrow{80\%} (C_2H_5)_2CHCHO$$

Vinyl phenyl ethers may be obtained from phenoxymethylenetriphenylphosphorane **(166)**.[248]

$$(C_6H_5)_2CO + \underset{\mathbf{166}}{(C_6H_5)_3P{=}CHOC_6H_5} \rightarrow \underset{(65\%)}{(C_6H_5)_2C{=}CHOC_6H_5}$$

Vinyl thio ethers have also been prepared by this method.[92] They are hydrolyzed to aldehydes much less readily.

$$(C_6H_5)_2CO + (C_6H_5)_3P{=}CHSCH_3 \rightarrow \underset{(84\%)}{(C_6H_5)_2C{=}CHSCH_3}$$

## Analytical Applications

The alkylidene phosphorane **167** has been suggested as a specific reagent for the characterization of aldehydes in the presence of ketones.[249–251] In ethanol-water (95:5) at 75° it reacts readily with all aldehydes

[249] Siemiatycki, *Compt. Rend.*, **248**, 817 (1959).

[250] Simalty-Siemiatycki, Carretto, and Malbec, *Bull. Soc. Chim. France*, **1962**, 125.

[251] Simalty-Siemiatycki, Malbec, and Carretto, *Bull. Soc. Chim. France*, **1962**, 129.

investigated but not with ketones. The reactions are complete when the yellow color of the ylide disappears. In pure ethanol, ketones react to a

$$(C_6H_5)_3P{=}CH\text{—}C_6H_4\text{—}SO_2\text{—}C_6H_4\text{—}Br + RCHO \rightarrow RCH{=}CH\text{—}C_6H_4\text{—}SO_2\text{—}C_6H_4\text{—}Br$$

**167**

limited extent (5–10%); on addition of 5% water, only aldehydes react and the yields are between 50% and 90%. The Wittig reagent **167** may therefore be used for the selective characterization and identification of aldehydes.

## Side Reactions

Fortunately, side reactions occur only rarely and can usually be suppressed by suitable choice of conditions. The handling of unstable alkylidene phosphoranes has already been mentioned. Reactions are either carried out at low temperatures, or the alkoxide method which permits interaction of the carbonyl compound with the alkylidene phosphorane *in statu nascendi* is used.

The strongly basic alkylidene phosphoranes can also function as proton acceptors to remove a proton from the α-position of carbonyl compounds. This reaction is particularly pronounced with easily enolizable ketones such as cyclohexanone and cyclopentanone. If the reaction between the ylide and the carbonyl compound is sterically hindered, enolization becomes the main reaction.[185]

$$\text{C}_6\text{H}_{10}{=}P(C_6H_5)_3 + \text{C}_6\text{H}_{10}{=}O \rightarrow \text{C}_6\text{H}_{10}(H)\overset{\oplus}{P}(C_6H_5)_3 + \text{cyclohexanone enolate } (O{=}C\text{–}C^{\ominus} \leftrightarrow O^{\ominus}\text{–}C{=}C)$$

Methoxymethylenetriphenylphosphorane converts cyclopentanone partially to 2-cyclopentylidenecyclopentanone **(168)**.[248]

$$(C_6H_5)_3P{=}CHOCH_3 + \text{C}_5\text{H}_8{=}O \rightarrow (C_6H_5)_3\overset{\oplus}{P}CH_2OCH_3 + \text{cyclopentanone enolate}^{\ominus}$$

$$\text{cyclopentanone enolate}^{\ominus} + \text{cyclopentanone} \rightarrow \text{2-(1-oxidocyclopentyl)cyclopentanone } (O^{\ominus}) \rightarrow \text{2-cyclopentylidenecyclopentanone}$$

**168**

Analogous condensations have been observed to a lesser extent with acetone, which is partially converted to mesityl oxide **(169)** by Wittig reagents.[182] Because decreasing temperatures favor the Wittig reaction

$$(C_6H_5)_3P{=}CH_2 + (CH_3)_2CO \longrightarrow (C_6H_5)_3\overset{\oplus}{P}CH_3 + CH_3CO\overset{\ominus}{C}H_2 \xrightarrow{(CH_3)_2CO}$$

$$CH_3COCH_2C(CH_3)_2(O^{\ominus}) \longrightarrow \underset{\mathbf{169}}{CH_3COCH{=}C(CH_3)_2}$$

over aldol condensations, reactions with enolizable ketones are best done at low temperatures.[248]

The elimination of acids from reactants or products under the influence of Wittig reagents has also been observed. An example is the elimination of hydrogen bromide from the diene **170.**[16]

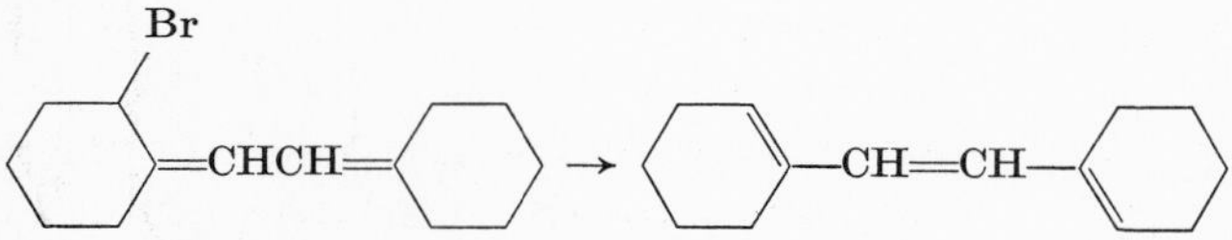

**170**

In the reaction of methylenetriphenylphosphorane with the keto steroid **171,** the expected product is accompanied by the dienone **172** formed by elimination of acetic acid.[252]

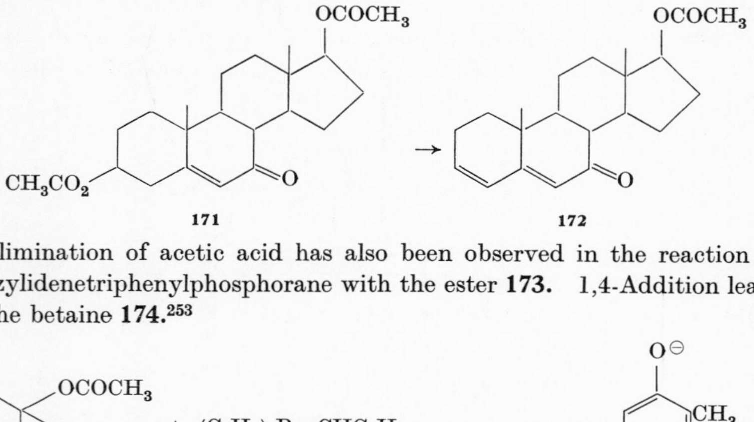

**171** **172**

Elimination of acetic acid has also been observed in the reaction of benzylidenetriphenylphosphorane with the ester **173.** 1,4-Addition leads to the betaine **174.**[253]

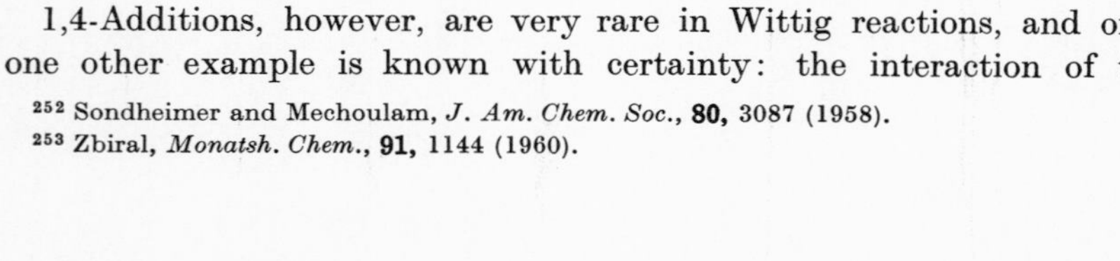

$$\mathbf{173} + (C_6H_5)_3P{=}CHC_6H_5 \rightarrow \mathbf{174}$$

1,4-Additions, however, are very rare in Wittig reactions, and only one other example is known with certainty: the interaction of the

[252] Sondheimer and Mechoulam, *J. Am. Chem. Soc.*, **80,** 3087 (1958).

[253] Zbiral, *Monatsh. Chem.*, **91,** 1144 (1960).

alkylidenephosphorane **175** with methylenecyclohexanone, for which the following mechanism has been assumed.[158]

$$+ \ (C_6H_5)_3P{=}CHCH{=} \rightarrow$$

**175**

$$CH \quad CH{-}CH{=} \quad H \quad P(C_6H_5)_3 \quad O$$

$$+ \ (C_6H_5)_3PO$$

The formation of 1,6-diphenylhexa-1,3,5-triene **(178)** as a side product in the reaction of ketone **176** with the phosphorane **177** has also been explained on the basis of a 1,4-addition.[213]

$$\underset{\mathbf{176}}{C_6H_5CH{=}CHCH{=}CHCOC_6H_5} + \underset{\mathbf{177}}{(C_6H_5)_3P{=}CHCH{=}CHC_6H_5} \rightarrow$$

$$C_6H_5CH{=}CHCH \quad CH{=}CC_6H_5 \quad O \quad CH{-}P(C_6H_5)_3 \quad C_6H_5CH{=}CH \rightarrow$$

$$C_6H_5C{\equiv}CH + \underset{\mathbf{178}}{C_6H_5(CH{=}CH)_3C_6H_5} + (C_6H_5)_3PO$$

Phenylacetylene could not be detected. An alternative path that has been suggested involves reaction of the ylide **177** with the starting phosphonium salt followed by a Hofmann elimination.[12]

$$\underset{\oplus P(C_6H_5)_3}{C_6H_5CH{=}CH{-}CH_2} + \underset{\oplus P(C_6H_5)_3 \quad \mathbf{177}}{{}^{\ominus}CH{-}CH{=}CHC_6H_5} \rightarrow$$

$$C_6H_5CH{=}CH{-}\overset{H}{CH}{-}\underset{\oplus P(C_6H_5)_3}{CH}{-}CH{=}CHC_6H_5 \rightarrow$$

$$C_6H_5(CH{=}CH)_3C_6H_5 + (C_6H_5)_3P$$

The reaction of methylenetriphenylphosphorane with the sterically hindered mesityl styryl ketone **(179)** may also be considered as a 1,4-addition. In this reaction, triphenylphosphine rather than triphenylphosphine oxide is eliminated and the cyclopropane **180** is formed.[131]

$$C_6H_5CH{=}CHC({=}O)\text{—}C_6H_2(CH_3)_3 + (C_6H_5)_3P{=}CH_2 \rightarrow$$

$$C_6H_5CH(CH_2\overset{\oplus}{P}(C_6H_5)_3)\text{—}CH{=}C(O^{\ominus})\text{—}C_6H_2(CH_3)_3$$

$$\swarrow$$

$$C_6H_5CH\text{—}CHC(\text{—}CH_2\text{—})(C{=}O)\text{—}C_6H_2(CH_3)_3$$

**180** (52%)

Thus addition to an activated double bond takes place in preference to the Wittig reaction. It is therefore necessary, when the Wittig reaction is used to prepare compounds containing an activated double bond, to avoid an excess of phosphorane in order to prevent subsequent addition of the ylide to the double bond.[130]

As mentioned previously, α-bromoketones such as phenacyl bromide do not undergo Wittig reactions with alkylidene phosphoranes.[106–108]

In compounds that contain both ester and carbonyl groups, the latter react preferentially with Wittig reagents as long as the ylide is not present in excess. The ylide should therefore be added slowly to the carbonyl component.[241]

If an excess of alkylidene phosphorane is used in reactions with compounds such as acetylated steroid keto alcohols,[252,254] the ester group is removed by the ylide by the following process, and the reaction product may have to be reacetylated.

$$(C_6H_5)_3P{=}CH_2 + CH_3CO_2R \rightarrow (C_6H_5)_3\overset{\oplus}{P}\text{—}CH_2\text{—}C(O^{\ominus})(CH_3)\text{—}OR \rightarrow$$

$$(C_6H_5)_3\overset{\oplus}{P}\text{—}CH_2COCH_3 + RO^{\ominus}$$

[254] Sondheimer and Mechoulam, *J. Am. Chem. Soc.*, **79**, 5029 (1957).

A similar side reaction occurred to a lesser extent in the interaction of allylidenetriphenylphosphorane **(181)** with N-methylformanilide. Formation of 1-methylanilino-1,3-butadiene **(182)** is accompanied by elimination of methylanilide anion.[255]

$$C_6H_5N(CH_3)CHO + (C_6H_5)_3P{=}CHCH{=}CH_2 \;(\mathbf{181}) \rightarrow$$

$$(C_6H_5)_3\overset{\oplus}{P}\text{—}CH(CH{=}CH_2)\text{—}CH(\overset{\ominus}{O})\text{—}N(CH_3)C_6H_5 \rightarrow C_6H_5N(CH_3)CH{=}CHCH{=}CH_2 \;(\mathbf{182})$$

$$\downarrow$$

$$(C_6H_5)_3\overset{\oplus}{P}\text{—}CH(CHO)CH{=}CH_2 + \overset{\ominus}{N}(CH_3)\text{—}C_6H_5$$

An abnormal decomposition of the betaine has also been observed in the reaction of *n*-butoxymethylenetriphenylphosphorane with butyraldehyde, in which formation of the expected 1-*n*-butoxy-1-pentene **(183)** is accompanied by a hydride shift which gives mostly 1-*n*-butoxypentan-2-one **(184)**.[87]

$$n\text{-}C_3H_7CHO + (C_6H_5)_3P{=}CHOC_4H_9\text{-}n \longrightarrow$$

$$n\text{-}C_3H_7\text{—}C(H)(\overset{\ominus}{O})\text{—}CH(\overset{\oplus}{P}(C_6H_5)_3)OC_4H_9\text{-}n \xrightarrow{12\%} n\text{-}C_3H_7CH{=}CHOC_4H_9\text{-}n \;(\mathbf{183})$$

$$\downarrow 26\%$$

$$n\text{-}C_3H_7C({=}O)CH_2OC_4H_9\text{-}n \;(\mathbf{184})$$

If the oxygen in the betaine may interact with a silicon or a phosphorus atom, attack on silicon occurs preferentially as in the accompanying reaction sequence.[123] The ylide **185** so formed reacts with excess

[255] Wittig and Sommer, *Ann.*, **594**, 1 (1955).

benzophenone in a normal Wittig reaction to give tetraphenylallene **(186)**, which was isolated in 28% yield.

$$(C_6H_5)_3P{=}CH{-}Si(CH_3)_3 + (C_6H_5)_2CO \rightarrow (C_6H_5)_3\overset{\oplus}{P}{-}CH{-}Si(CH_3)_3 \rightarrow$$
$$(C_6H_5)_2C{-}O^{\ominus}$$

$$(C_6H_5)_3\overset{\oplus}{P}{-}CH{=}C(C_6H_5)_2 + (CH_3)_3SiO^{\ominus} \rightarrow$$
$$(C_6H_5)_3P{=}C{=}C(C_6H_5)_2 + (CH_3)_3SiOH$$
**185**

$$\downarrow (C_6H_5)_2CO$$

$$(C_6H_5)_2C{=}C{=}C(C_6H_5)_2$$
**186**

The decisive influence on the course of a Wittig reaction exerted by the halide ion in the phosphonium salt is understandable in terms of the fact that different halide ions can shield the ylide phosphorus atom by complex formation to a different extent.[114] Thus methylenecyclopentane could not be obtained starting from triphenylmethylphosphonium iodide, whereas the reaction occurred readily when the corresponding bromide was used as the starting material.[256]

Finally, it should be mentioned that alkylidene phosphoranes may react with *t*-butyl alcohol to form isobutylene by elimination of water.[94]

$$(C_6H_5)_3P{=}CCl_2 + (CH_3)_3COH \rightarrow (C_6H_5)_3PO + CH_2Cl_2 + (CH_3)_2C{=}CH_2$$

The use of *t*-butyl alcohol as a solvent for Wittig reactions should therefore be avoided.

## THE SYNTHESIS OF NATURAL PRODUCTS

In no field has the Wittig reàction attained importance as rapidly as in the area of natural products chemistry. Three classes of compounds, especially, have become much more readily accessible. These are the naturally occurring polyacetylenic compounds; the carotenoids, including vitamin A; and methylene steroids and vitamin D. The Wittig reaction has also been applied to the synthesis of a number of other classes of natural products.

### Polyacetylenic Compounds

In recent years it has been found that compounds containing conjugated double and triple bonds occur widely in plants. The synthesis of such unsaturated carbon chains was previously very difficult because no

[256] Collins and Hammond, *J. Org. Chem.*, **25**, 1434 (1960).

suitable method existed for the introduction of a double bond in a precisely defined position. It is therefore not surprising that the Wittig reaction found very early application in the synthesis of naturally occurring polyacetylenic compounds. Bohlmann and collaborators, especially, provided proof for the usefulness of the new method in this area.

The first step in the synthesis of oenanthetol **(188)**, found in hemlock water dropwort (*Oenanthe crocata*) consisted of the preparation of dodeca-3,5-dien-1-yne **(187)** in 43% yield by means of a Wittig reaction.[257] Oxidative coupling of the hydrocarbon **187** with pent-2-en-4-yn-1-ol in the presence of cuprous chloride gave oenanthetol **(188)** in 14.1% yield in addition to the two symmetrical polyenynes.

$$n\text{-}C_6H_{13}CH{=}CHCH{=}P(C_6H_5)_3 + HC{\equiv}CCHO$$

$$\downarrow$$

$$n\text{-}C_6H_{13}CH{=}CHCH{=}CHC{\equiv}CH$$

**187**

$$\downarrow \; Cu_2Cl_2/O_2 + HC{\equiv}CCH{=}CHCH_2OH$$

$$n\text{-}C_6H_{13}(CH{=}CH)_2(C{\equiv}C)_2CH{=}CHCH_2OH$$

**188**

The same method was used for the synthesis of cicutol **(190)**, isolated from European water hemlock (*Cicuta virosa*). Oxidative coupling of dodeca-3,5,7-trien-1-yne **(189)**, obtained in 60% yield by a Wittig reaction, gave a mixture of three hydrocarbons from which cicutol **(190)** could be isolated by chromatography.[258] Small amounts of 10-*cis*-cicutol

$$n\text{-}C_4H_9CH{=}CHCH{=}P(C_6H_5)_3 + HC{\equiv}CCH{=}CHCHO$$

$$\downarrow$$

$$n\text{-}C_4H_9CH{=}CHCH{=}CHCH{=}CHC{\equiv}CH$$

**189**

$$\downarrow \; Cu_2Cl_2/O_2 + HC{\equiv}CCH_2CH_2CH_2OH$$

$$n\text{-}C_4H_9(CH{=}CH)_3(C{\equiv}C)_2CH_2CH_2CH_2OH$$

**190**

also formed were converted into the all-*trans* form with iodine in petroleum ether.

[257] Bohlmann and Viehe, *Chem. Ber.*, **88**, 1245 (1955).
[258] Bohlmann and Viehe, *Chem. Ber.*, **88**, 1347 (1955).

Trideca-1,3,5,11-tetraene-7,9-diyne **(191)**, found in a number of coreopsis species, was synthesized by two different methods.[259] Path A

$$CH_3CH{=}CH(C{\equiv}C)_2CH{=}CHCH{=}P(C_6H_5)_3 + CH_2{=}CHCHO$$

$$\downarrow A$$

$$\underset{\textbf{191}}{CH_3CH{=}CH(C{\equiv}C)_2(CH{=}CH)_2CH{=}CH_2}$$

$$\uparrow B$$

$$CH_3CH{=}CH(C{\equiv}C)_2CH{=}CHCHO + (C_6H_5)_3P{=}CHCH{=}CH_2$$

gave only poor yields as compared with about a 50% yield via path B.

The synthesis of aethusanol B **(194)**, found in fool's-parsley (*Aethusa cynapium L.*), also started with a Wittig reaction.[260] Propargyl aldehyde and the phosphorane **192** furnished the intermediate **193** which with

$$\underset{\textbf{192}}{CH_3CH_2CH{=}CHCH{=}P(C_6H_5)_3} + HC{\equiv}CCHO \rightarrow$$

$$\underset{\textbf{193}}{CH_3CH_2CH{=}CHCH{=}CHC{\equiv}CH} \xrightarrow{BrC{\equiv}CCH{=}CHCH_2OH}$$

$$\underset{\textbf{194}}{CH_3CH_2(CH{=}CH)_2(C{\equiv}C)_2CH{=}CHCH_2OH}$$

5-bromopent-2-en-4-yn-1-ol gave aethusanol B **(194)**, identical with one of the three *Aethusa* polyynes.

Hydroxyl groups may also be introduced in the initial stage of a synthesis by using a suitably substituted alkylidene phosphorane, as illustrated by the following two examples. The "chamomilla ester" **195** found in the German camomile (*Matricaria chamomilla L.*) and in *Matricaria discoidea DC.* was prepared by the following method.[71]

$$CH_3(C{\equiv}C)_2(CH{=}CH)_2CHO + (C_6H_5)_3P{=}CHCH_2CH_2O{-}\text{(tetrahydropyran-2-yl)} \rightarrow$$

$$CH_3(C{\equiv}C)_2(CH{=}CH)_3CH_2CH_2OH \rightarrow$$

$$\underset{\textbf{195}}{CH_3(C{\equiv}C)_2(CH{=}CH)_3CH_2CH_2OCOCH_3}$$

Similarly, matricarianal **(196)** was converted to "centaur $X_2$" **(197)**, the principal polyyne of the cornflower (*Centaurea cyanus L.*).[72]

[259] Bohlmann and Mannhardt, *Chem. Ber.*, **88**, 1330 (1955).

[260] Bohlmann, Arndt, Bornowski, and Herbst, *Chem. Ber.*, **93**, 981 (1960).

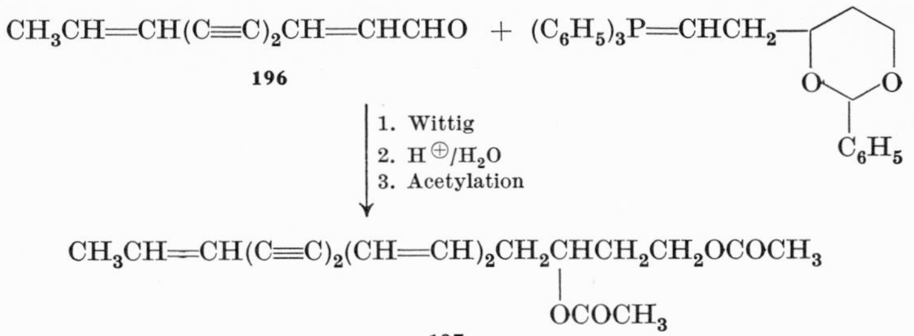

$CH_3CH{=}CH(C{\equiv}C)_2CH{=}CHCHO$ + $(C_6H_5)_3P{=}CHCH_2$–(2-phenyl-1,3-dioxan-5-yl)

**196**

1. Wittig
2. $H^{\oplus}/H_2O$
3. Acetylation

$CH_3CH{=}CH(C{\equiv}C)_2(CH{=}CH)_2CH_2CH(OCOCH_3)CH_2CH_2OCOCH_3$

**197**

The polyyne hydrocarbons **198** and **199** were also synthesized by the Wittig reaction. The former is found in common mugwort (*Artemisia vulgaris*) and in the cornflower (*Centaurea cyanus L.*), the latter in a number of coreopsis species.[160,261]

$CH_3(C{\equiv}C)_3(CH{=}CH)_2(CH_2)_4CH{=}CH_2$ **198**

$C_6H_5(C{\equiv}C)_3(CH{=}CH)_2CH_3$ **199**

## Carotenoids and Vitamin A

The synthesis of carotenoids and vitamin A using the Wittig reaction has received particular attention.[262,263] The synthesis of unsymmetrical carotenoids such as α- and γ-carotene is best carried out using only two components, a polyene aldehyde and an alkylidene phosphorane. Thus γ-carotene **(200)** can be prepared according to the scheme $C_{30} + C_{10} = C_{40}$.[264]

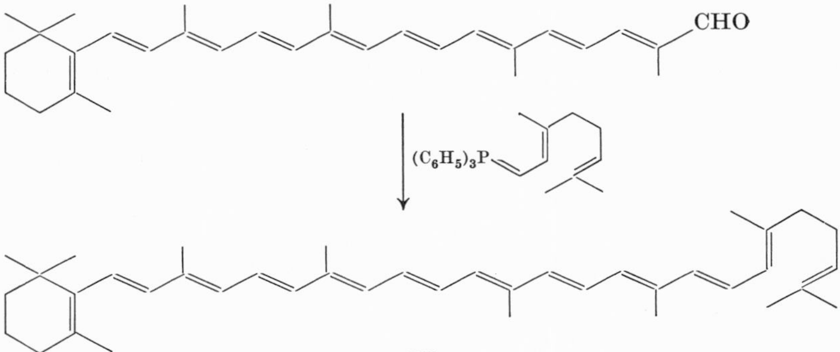

CHO

$(C_6H_5)_3P$

**200**

Another possibility is the combination of a $C_{25}$-aldehyde with a $C_{15}$-phosphorane according to the scheme $C_{25} + C_{15} = C_{40}$. Often the

[261] Bohlmann, *Chem. Ber.*, **88**, 1755 (1955).
[262] Pommer, *Angew. Chem.*, **72**, 811 (1960).
[263] Pommer, *Angew. Chem.*, **72**, 911 (1960).
[264] Rüegg, Schwieter, Ryser, Schudel, and Isler, *Helv. Chim. Acta*, **44**, 985 (1961).

15,15′-dehydro derivative is prepared first, then partially hydrogenated and isomerized to the all-*trans*-γ-carotene **(200)**.[264]

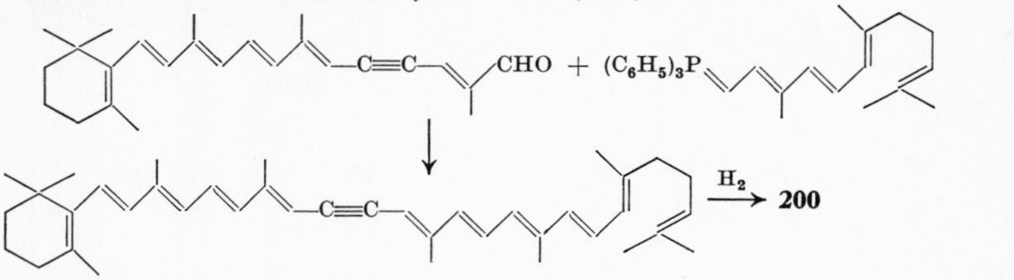

$\xrightarrow{H_2}$ **200**

7′,8′-Dihydro-γ-carotene (β-zeacarotene), found in corn, was synthesized by the same method.[265] 3′,4′-Dehydro-γ-carotene (torulin, **201**), isolated from *Torula rubra*, on the other hand, was prepared according to the scheme $C_{35} + C_5 = C_{40}$.[265]

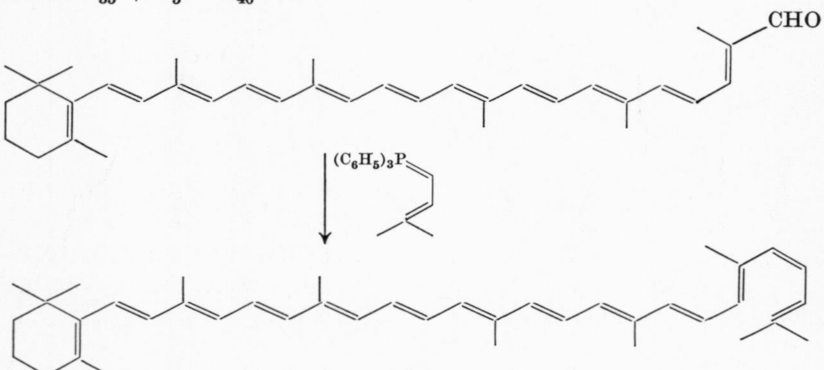

**201**

The synthesis of α-carotene **(202)** followed the scheme $C_{25} + C_{15} = C_{40}$.[264]

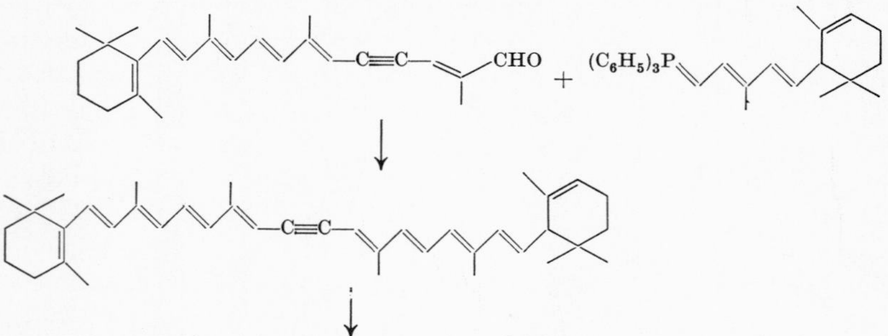

**202**

[265] Rüegg, Schwieter, Ryser, Schudel, and Isler, *Helv. Chim. Acta*, **44**, 994 (1961).

Symmetrical carotenoids such as β-carotene and lycopene may, of course, be prepared by the same method.[264] β-Carotene **(44)** has also been obtained by the scheme $C_{20} + C_{20} = C_{40}$, starting from vitamin A aldehyde **(203)**.[266]

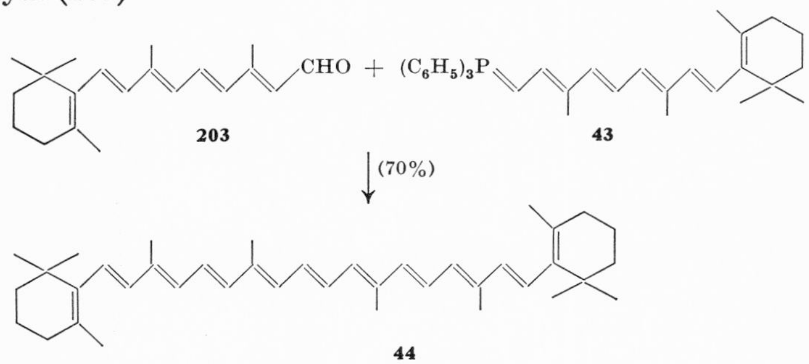

CHO + $(C_6H_5)_3P$

**203** **43**

(70%)

**44**

In place of axerophthylenetriphenylphosphorane **(43)**, the isomeric ylide **204,** which is in equilibrium with **43,** may also be used.[267]

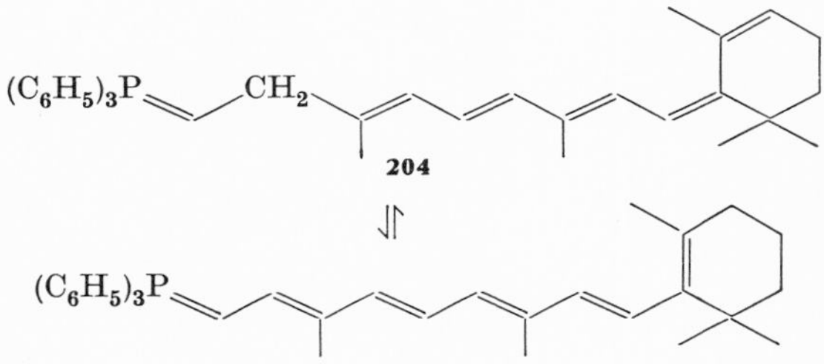

$(C_6H_5)_3P$ $CH_2$

**204**

$(C_6H_5)_3P$

**43**

The synthesis of symmetrical carotenoids, however, is better carried out by the interaction of dialdehydes with 2 moles of an alkylidene phosphorane or of a bifunctional ylide with 2 moles of an aldehyde. Thus β-carotene **(44)** is obtained in good yield according to the scheme $C_{15} + C_{10} + C_{15} = C_{40}$ by either path A[54,268–270] or path B[80,271] (equations on p. 360). Using path B but starting from vitamin A aldehyde, decapreno-β-carotene is obtained in 38% yield.[80,271]

The scheme $C_{13} + C_{14} + C_{13} = C_{40}$ illustrates another possible combination.[209,240,272] In each case, mixtures of the *cis* and *trans* isomers are

[266] Pommer and Sarnecki, Ger. pat. 1,068,709 (to BASF) [*C.A.*, **55,** 13472i (1961)].
[267] Stern, U.S. pat. 2,945,069 (to Eastman Kodak) [*C.A.*, **55,** 608e (1961)].
[268] Pommer and Sarnecki, Ger. pat. 1,068,703 (to BASF) [*C.A.*, **55,** 13473i (1961)].
[269] Pommer and Sarnecki, Ger. pat. 1,068,705 (to BASF) [*C.A.*, **56,** 1487f (1962)].
[270] Pommer and Sarnecki, Ger. pat. 1,068,710 (to BASF) [*C.A.*, **55,** 12446g (1961)].
[271] Surmatis, Ger. pat. 1,105,869 (to Hoffmann-LaRoche), 1959.
[272] Pommer and Sarnecki, Ger. pat. 1,068,704 (to BASF) [*C.A.*, **55,** 13473f (1961)].

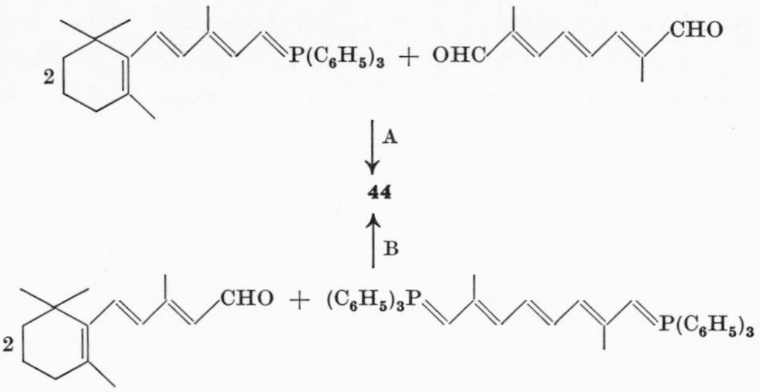

2 ... $P(C_6H_5)_3$ + OHC ... CHO

↓ A

**44**

↑ B

2 ... CHO + $(C_6H_5)_3P$ ... $P(C_6H_5)_3$

obtained. The mixtures can be converted to the all-*trans* form by heating with iodine. The isomerization to the all-*trans* compound can also be effected by nitric oxide.[272]

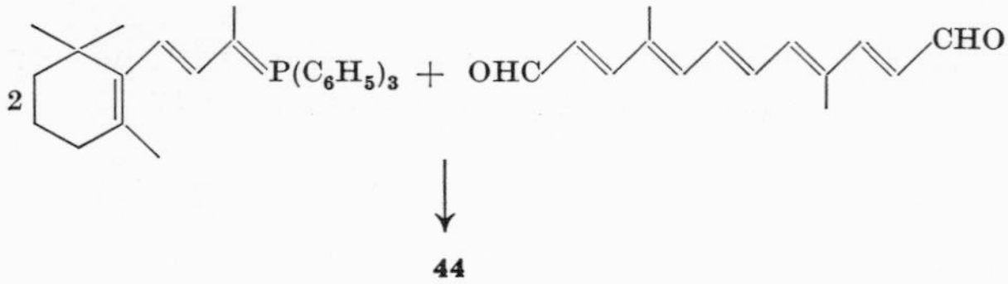

2 ... $P(C_6H_5)_3$ + OHC ... CHO

↓

**44**

Another possibility for the synthesis of symmetrical carotenoids is expressed by the scheme $C_{10} + C_{20} + C_{10} = C_{40}$. β-Carotene **(44)** can thus be obtained from crocetin dialdehyde **(206)** and 2 moles of β-cyclogeranylidenetriphenylphosphorane **(205)**.[273,274] Good yields were obtained only by using dimethylformamide as the solvent.[274]

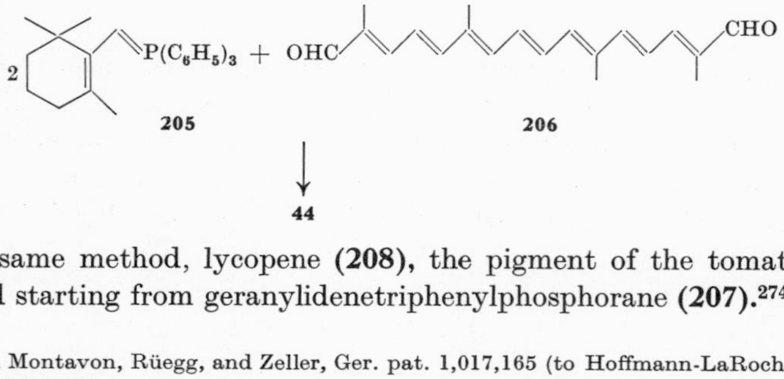

2 ... $P(C_6H_5)_3$ + OHC ... CHO

**205** **206**

↓

**44**

By the same method, lycopene **(208)**, the pigment of the tomato, was prepared starting from geranylidenetriphenylphosphorane **(207)**.[274–277]

[273] Isler, Montavon, Rüegg, and Zeller, Ger. pat. 1,017,165 (to Hoffmann-LaRoche) [*C.A.*, **53**, 18982b (1959)].

[274] Pommer and Sarnecki, Ger. pat. 1,068,707 (to BASF) [*C.A.*, **55**, 10499a (1961)].

[275] Isler, Gutmann, Lindlar, Montavon, Rüegg, Ryser, and Zeller, *Helv. Chim. Acta*, **39**, 463 (1956).

[276] Isler, Montavon, Rüegg, and Zeller, Ger. pat. 1,038,033 (to Hoffmann-LaRoche) [*C.A.*, **54**, 17456h (1960)].

[277] Isler, Montavon, Rüegg, and Zeller, U.S. pat. 2,842,599 (to Hoffmann-LaRoche) [*C.A.*, **53**, 2279h (1959)].

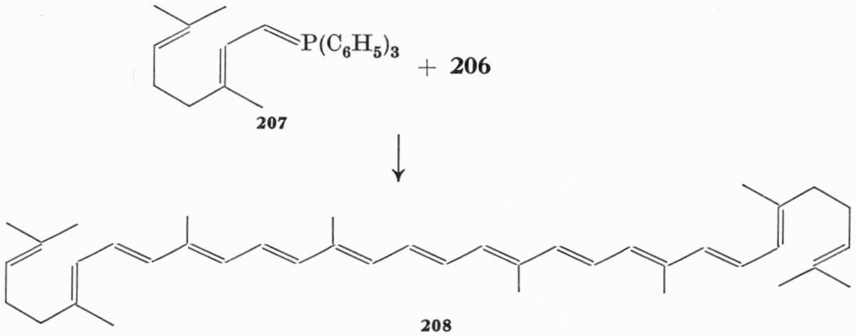

$P(C_6H_5)_3$ + **206**

**207**

↓

**208**

Spirilloxanthene,[278] dehydrolycopene, and 1,1′-dihydroxy-1,2,1′,2′-tetrahydrolycopene[279] were synthesized by essentially the same method.

All these possible synthetic routes may, of course, be applied to the preparation of 15,15′-dehydrocarotenoids. Partial hydrogenation of these compounds leads to the 15,15′-*cis*-carotenoids, which are converted to the all-*trans*-carotenoids on recrystallization or heating in an inert solvent.[280]

The synthesis of vitamin A and its derivatives is carried out by starting from monofunctional carbonyl compounds and ylides. Vitamin A methyl ether **(209)** is thus obtained according to the scheme $C_{18} + C_2 = C_{20}$.[281]

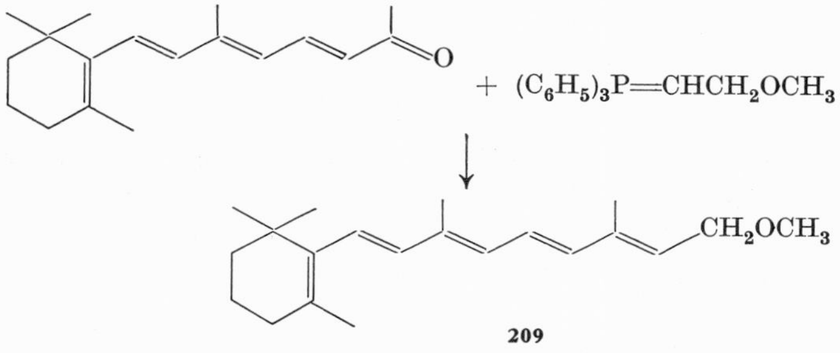

O + $(C_6H_5)_3P{=}CHCH_2OCH_3$

↓

$CH_2OCH_3$

**209**

The low yield, 10%, is probably due to the instability of the ylide **210**.[138]

$$(C_6H_5)_3P{=}CH{-}CH_2{-}OCH_3 \rightarrow (C_6H_5)_3\overset{\oplus}{P}{-}CH{=}CH_2 + CH_3\overset{\ominus}{O}$$

**210**

The vitamin A skeleton may be obtained by generating any one of the four double bonds in the side chain by combination of suitable aldehydes

[278] Surmatis and Ofner, 142nd Meeting, Am. Chem. Soc., Atlantic City, N.J., Sept. 1962, Abstracts, p. 48Q.

[279] Surmatis and Ofner, 142nd Meeting, Am. Chem. Soc., Atlantic City, N.J., Sept. 1962, Abstracts, p. 50Q.

[280] Brit. pat. 793,236 (to Hoffmann-LaRoche) [*C.A.*, **54,** 627g (1960)].

[281] Isler, Montavon, Rüegg, and Zeller, Ger. pat. 1,017,163 (to Hoffmann-LaRoche) [*C.A.*, **53,** 18982c (1959)].

and phosphoranes. The vitamin A acid methyl ester **(212*a*)**, for example, was prepared from β-ionylideneacetaldehyde **(211)** according to the scheme $C_{15} + C_5 = C_{20}$.[55,282]

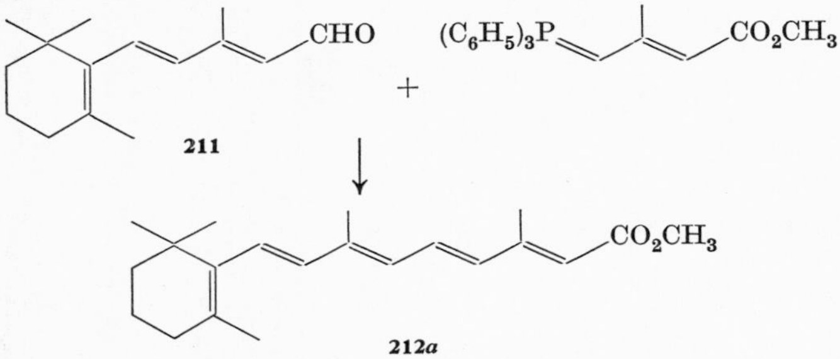

CHO $(C_6H_5)_3P$ $CO_2CH_3$

**211**

$CO_2CH_3$

**212*a***

Better yields are obtained by the reaction between the phosphorane **213** and β-formylcrotonic ester **(214)**.[270,283–286]

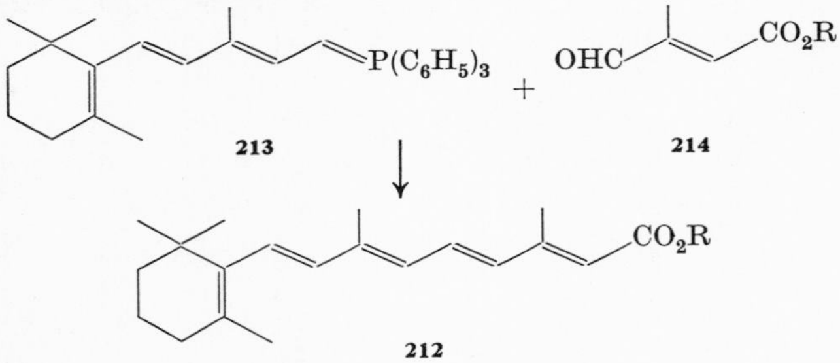

$P(C_6H_5)_3$ OHC $CO_2R$

**213** **214**

$CO_2R$

**212**

The esters of vitamin A acid **(212)** are readily saponified to vitamin A acid **(216)**. The latter can be obtained directly from β-formylcrotonic acid **(215)** and the ylide **213**, provided that the free acid is converted to its salt by excess alkali.[270,284–286]

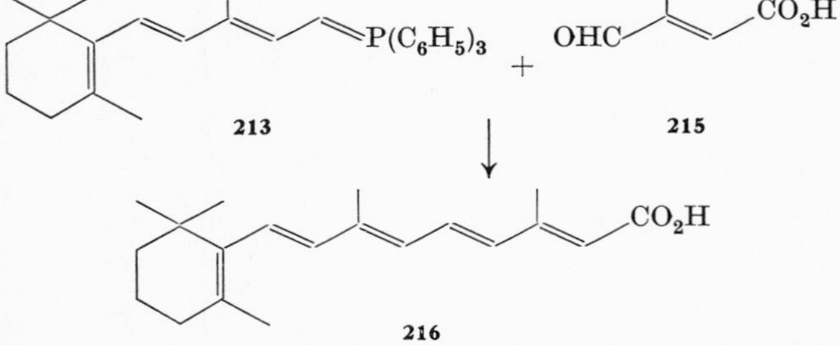

$P(C_6H_5)_3$ OHC $CO_2H$

**213** **215**

$CO_2H$

**216**

[282] Wittig and Pommer, Ger. pat. 950,552 (to BASF) [*C.A.*, **53**, 436g (1959)].
[283] Pommer, Sarnecki, and Wittig, Ger. pat. 1,025,869 (to BASF) [*C.A.*, **54**, 22713b (1960)].
[284] Pommer and Sarnecki, Ger. pat. 1,046,612 (to BASF) [*C.A.*, **55**, 5573a (1961)].
[285] Pommer and Sarnecki, Ger. pat. 1,059,900 (to BASF) [*C.A.*, **55**, 14511a (1961)].
[286] Pommer and Sarnecki, Ger. pat. 1,068,702 (to BASF) [*C.A.*, **55**, 10812c (1961)].

The same scheme was used to synthesize vitamin A acetate **(217)**,[270,284,285] vitamin A methyl ether **(209)**,[286a] and desoxy vitamin A or axerophthene **(218)**.[270,284,285,287]

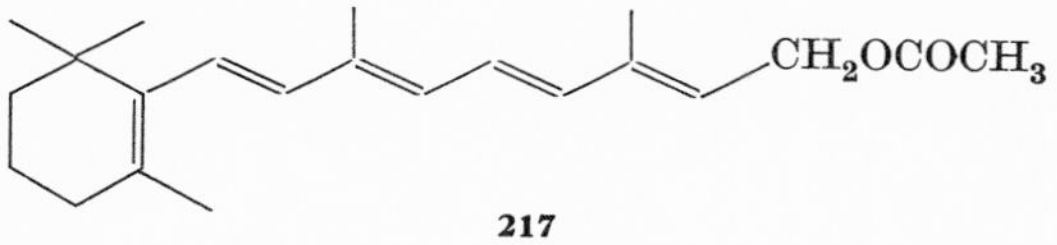

$CH_2OCOCH_3$

**217**

$CH_3$

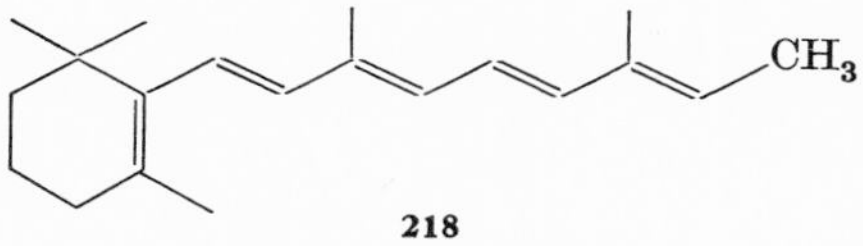

**218**

The synthesis of vitamin $A_2$ and its derivatives according to the scheme $C_{15} + C_5 = C_{20}$ is illustrated by the preparation of vitamin $A_2$ acid methyl ester **(219)**.[244]

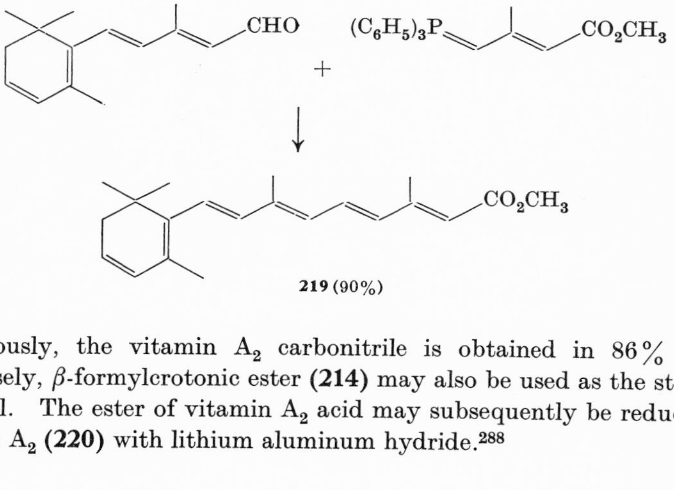

CHO $\quad (C_6H_5)_3P$ $CO_2CH_3$

+

↓

$CO_2CH_3$

**219** (90%)

Analogously, the vitamin $A_2$ carbonitrile is obtained in 86% yield. Conversely, $\beta$-formylcrotonic ester **(214)** may also be used as the starting material. The ester of vitamin $A_2$ acid may subsequently be reduced to vitamin $A_2$ **(220)** with lithium aluminum hydride.[288]

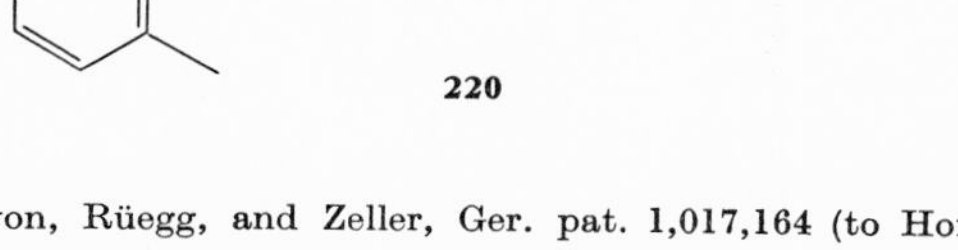

$CH_2OH$

**220**

[286a] Isler, Montavon, Rüegg, and Zeller, Ger. pat. 1,017,164 (to Hoffmann-La Roche) [*C.A.*, **53**, 18982d (1959)].

[287] Pommer and Wittig, Ger. pat. 1,029,366 (to BASF) [*C.A.*, **54**, 22713b (1960)].

[288] Schwieter, Planta, Rüegg, and Isler, *Helv. Chim. Acta*, **45**, 541 (1962).

The synthesis of the vitamin A skeleton according to the scheme $C_{13} + C_7 = C_{20}$ is illustrated by two syntheses of axerophthene **(218)**.[209,240,287,289]

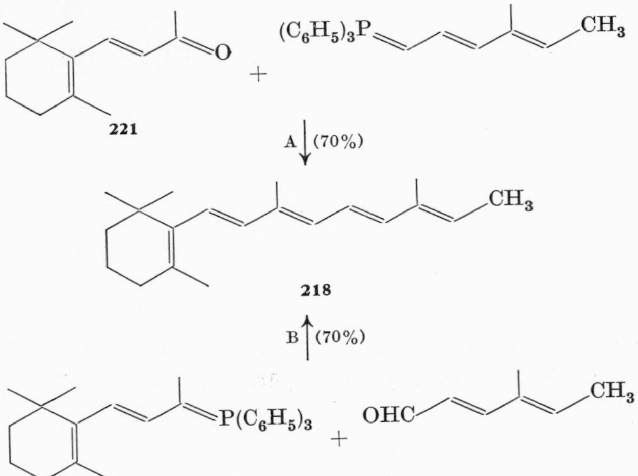

$(C_6H_5)_3P$ $CH_3$

**221**

A (70%)

$CH_3$

**218**

B (70%)

$P(C_6H_5)_3$ OHC $CH_3$

**222**

Path A, which starts with β-ionone **(221)**, was also used for the preparation of vitamin A methyl ether **(209)** and of vitamin A acetate **(217)**.[290] Vitamin A acid **(216)**[291,292] and its ethyl ester **(212*b*)**[289,292] were obtained in excellent yields via path B starting from β-ionylidenetriphenylphosphorane **(222)**.

Finally, it may be mentioned that vitamin A acid ethyl ester **(212*b*)** was prepared from β-cyclogeranylidenetriphenylphosphorane **(223)** according to scheme $C_{10} + C_{10} = C_{20}$.[293] Analogously, axerophthene **(218)** was prepared in 60% yield.[293]

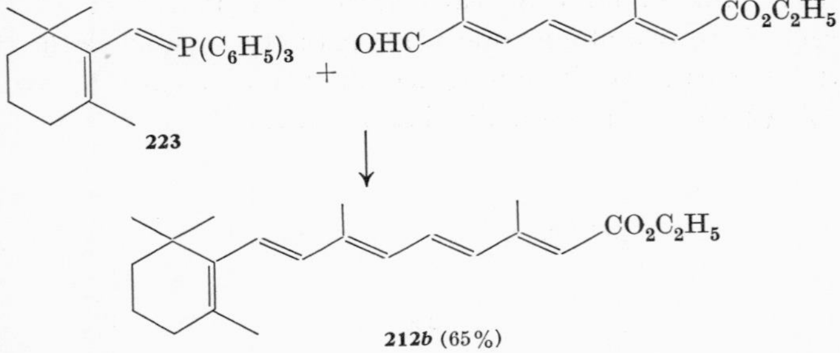

$P(C_6H_5)_3$ OHC $CO_2C_2H_5$

**223**

$CO_2C_2H_5$

**212*b*** (65%)

289 Pommer and Wittig, Ger. pat. 1,001,256 (to BASF) [*C.A.*, **53,** 13081e (1959)].
290 Wittig, Pommer, and Hartwig, Ger. pat. 957,942 (to BASF) [*C.A.*, **53,** 7232g (1959)].
291 Pommer and Sarnecki, Ger. pat. 1,035,647 (to BASF) [*C.A.*, **54,** 12187h (1960)].
292 Pommer and Sarnecki, Ger. pat. 1,050,763 (to BASF) [*C.A.*, **55,** 5572g (1961)].
293 Pommer and Wittig, Ger. pat. 951,212 (to BASF) [*C.A.*, **53,** 437d (1959)].

Higher homologs (**224**; $n = 2, 3$) of vitamin A acid and carotene acids of the form **225** and **226** ($n = 0, 1, 2, 3$) have also been synthesized by means of the Wittig reaction.

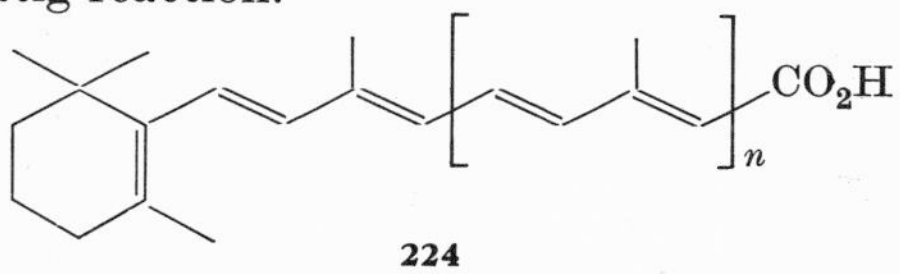

$CO_2H$

**224**

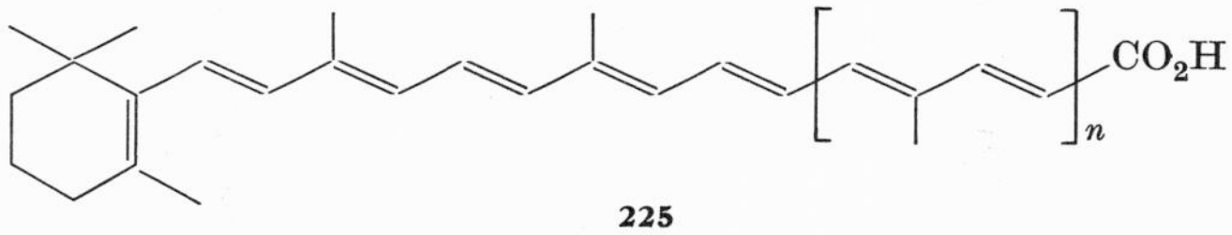

$CO_2H$

**225**

$CO_2H$

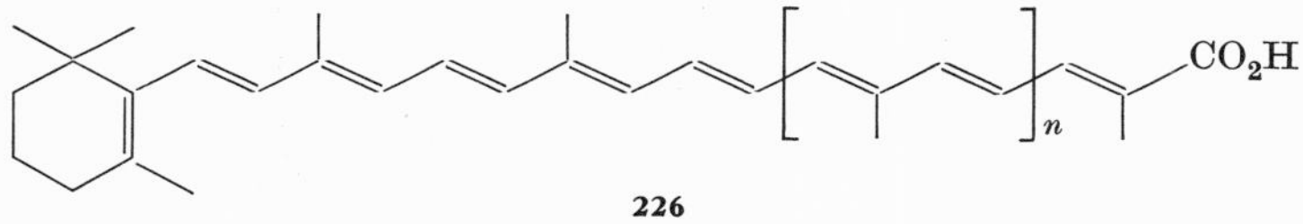

**226**

Homoisoprenovitamin A acid (**224**; $n = 2$), for instance, was prepared by four different methods.[240,266,269,270,294]

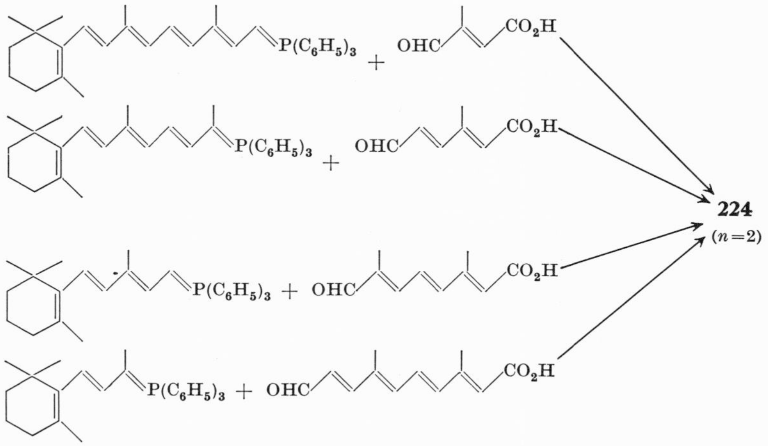

$P(C_6H_5)_3$ + OHC $CO_2H$

$P(C_6H_5)_3$ + OHC $CO_2H$

$P(C_6H_5)_3$ + OHC $CO_2H$

$P(C_6H_5)_3$ + OHC $CO_2H$

→ **224** ($n=2$)

Of the many syntheses of polyenecarboxylic acids (**225** and **226**),[295–298] mention is made only of the preparation of torularhodin (**226**; $n = 3$),

[294] Pommer and Sarnecki, Ger. pat. 1,070,173 (to BASF) [*C.A.*, **55,** 11332h (1961)].

[295] Isler, Guex, Rüegg, Ryser, Saucy, Schwieter, Walter, and Winterstein, *Helv. Chim. Acta*, **42,** 864 (1959).

[296] Rüegg, Guex, Montavon, Schwieter, Saucy, and Isler, *Angew. Chem.*, **71,** 80 (1959).

[297] Guex, Rüegg, Isler, and Ryser, Ger. pat. 1,088,951 (to Hoffmann-La Roche), 1958; corresponds to Brit. pat. 850,137 [*C.A.*, **55,** 17541e (1961)].

[298] Guex, Isler, Rüegg, and Ryser, Ger. pat. 1,096,349 (to Hoffmann-La Roche), 1959; corresponds to Brit. pat. 875,713 [*C.A.*, **56,** 7372b (1962)].

CHO + $(C_6H_5)_3P{=}C(CH_3){-}CO_2CH_3$ → $CO_2CH_3$ → **226** $(n=3)$

$C{\equiv}C$

CHO + $(C_6H_5)_3P{=}$ … $CO_2CH_3$ ←

which definitely established the structure of the acid pigment of near yeast (*Torula rubra*).[295–297] (See p. 366.)

Diesters of norbixin such as methylbixin **(228)** have been prepared from crocetin dialdehyde **(227)** and carbalkoxymethylene triphenylphosphoranes according to the scheme $C_2 + C_{20} + C_2 = C_{24}$.[39,231] Methylbixin can

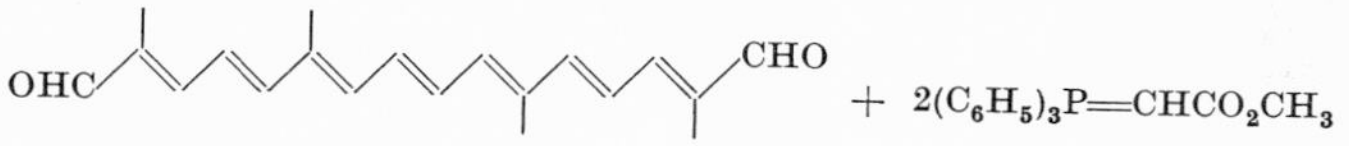

OHC … CHO + $2(C_6H_5)_3P{=}CHCO_2CH_3$

**227**

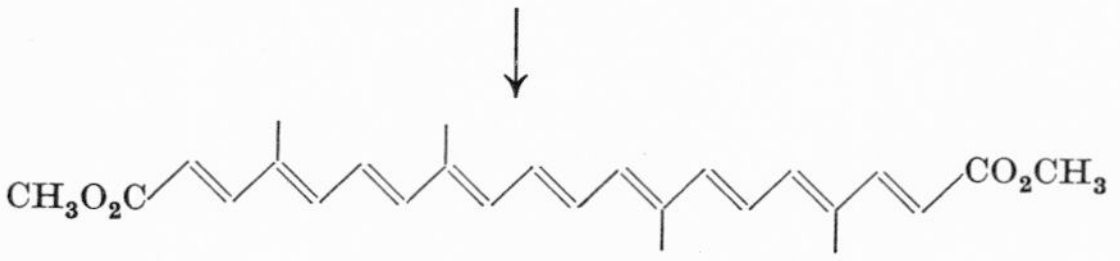

↓

$CH_3O_2C$ … $CO_2CH_3$

**228** (80%)

also be prepared in 76% yield by the scheme $C_5 + C_{14} + C_5 = C_{24}$.[233,234]

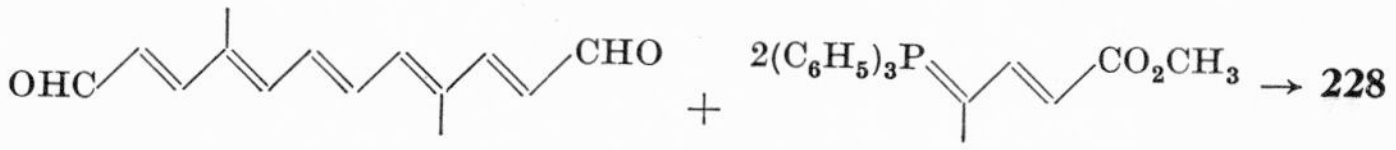

OHC … CHO + $2(C_6H_5)_3P$ … $CO_2CH_3$ → **228**

Crocin, the yellow pigment of saffron, is the digentiobiose ester of the $C_{20}$-dicarboxylic acid crocetin. Diesters of crocetin, such as crocetin dimethyl ester **(229)**, can be synthesized by the same methods as those used for the preparation of norbixin, either according to scheme $C_3 + C_{14} + C_3 = C_{20}$[39,231] or $C_5 + C_{10} + C_5 = C_{20}$.[232,235]

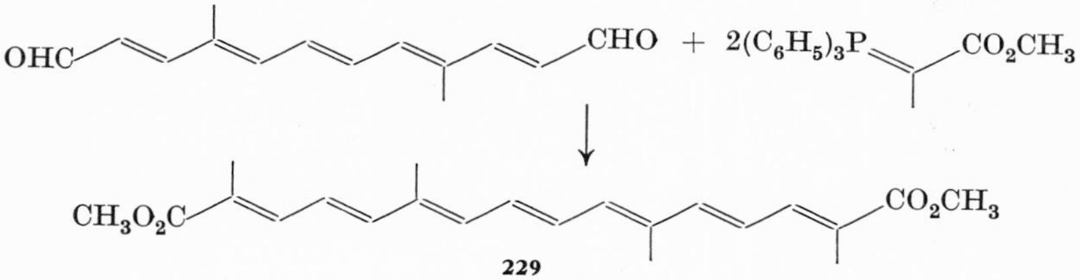

OHC … CHO + $2(C_6H_5)_3P$ … $CO_2CH_3$

↓

$CH_3O_2C$ … $CO_2CH_3$

**229**

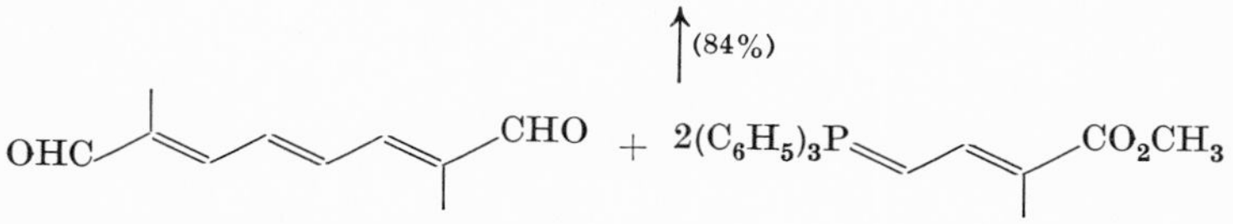

↑ (84%)

OHC … CHO + $2(C_6H_5)_3P$ … $CO_2CH_3$

Finally, mention should be made of the synthesis of squalene, a triterpene related to the carotenoids, which is important as an intermediate in the biosynthesis of steroids. Squalene **(232)** was synthesized by the reaction of geranylacetone **(230)** with the bis-ylide **231** according to scheme $C_{13} + C_4 + C_{13} = C_{30}$.[74,299–301]

299 Trippett, *Chem. Ind.* (*London*), **1956,** 80.

300 Dicker and Whiting, *Chem. Ind.* (*London*), **1956,** 351.

301 Dicker and Whiting, *J. Chem. Soc.*, **1958,** 1994.

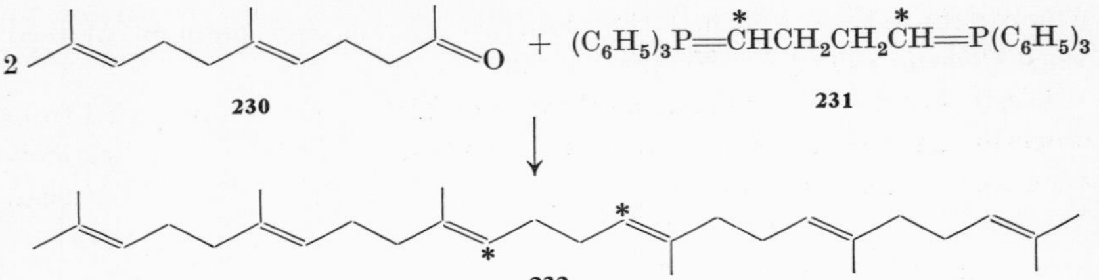

$$2\ \mathbf{230} + (C_6H_5)_3P{=}\overset{*}{C}HCH_2CH_2\overset{*}{C}H{=}P(C_6H_5)_3\ \mathbf{231} \longrightarrow \mathbf{232}$$

The all-*trans*-squalene **(232)** was isolated in 12% yield from the mixture of *cis* and *trans* isomers by way of the thiourea clathrate.[300,301] Analogously, radioactively labeled squalene ($^{14}C$-labeled at $C_{11}$ and $C_{14}$) was prepared in 50% yield from the $^{14}C$-labeled bis-ylide **231**.[302]

## Steroids and Vitamin D

The method of introducing exocylcic double bonds by means of the Wittig reaction has been used in steroid chemistry. The preparation of 3-methylenecholestane **(234)** from cholestan-3-one **(233)** and methylenetriphenylphosphorane was the first instance of a conversion of a steroid ketone to a methylenesteroid by this method.[303] This synthesis was

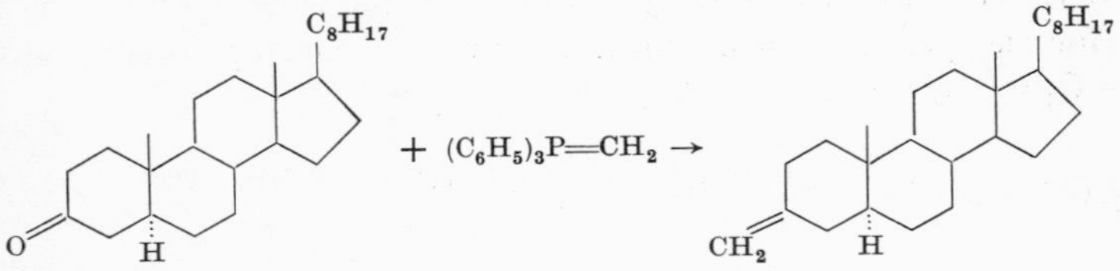

$$\mathbf{233} + (C_6H_5)_3P{=}CH_2 \rightarrow \mathbf{234}$$

followed by the preparation of a whole series of methylenesteroids in similar fashion, in the course of which it was found that excess methylenetriphenylphosphorane increased the yields.[254] 17-Methyleneandrostan-3$\beta$-ol **(236)** was obtained in 58% yield from androstan-3$\beta$-ol-17-one **(235)** by using a five-fold excess of methylenephosphorane, but in only 32% yield by employing 3 moles of ylide.[254] The yield in this reaction was

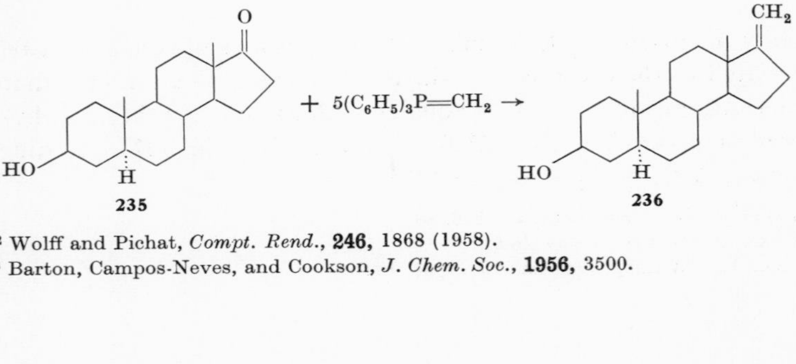

$$\mathbf{235} + 5(C_6H_5)_3P{=}CH_2 \rightarrow \mathbf{236}$$

[302] Wolff and Pichat, *Compt. Rend.*, **246,** 1868 (1958).

[303] Barton, Campos-Neves, and Cookson, *J. Chem. Soc.*, **1956,** 3500.

also increased when the 3-hydroxyl group was blocked by conversion to the tetrahydropyranyl ether. The increased yield, however, was offset by losses in the formation and cleavage of the ether.

$\alpha,\beta$-Unsaturated steroid ketones have also been used as starting materials. $\Delta^4$-Cholesten-3-one **(237)**, for instance, gives 3-methylene-$\Delta^4$-cholestene **(238)** in 80% yield.[254]

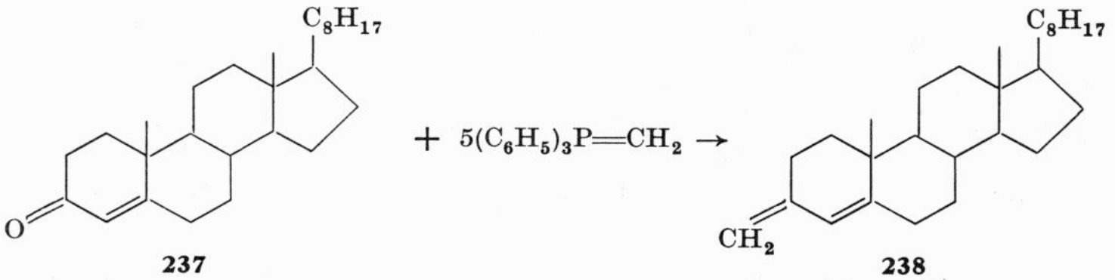

$C_8H_{17}$ O **237** + $5(C_6H_5)_3P{=}CH_2$ → $C_8H_{17}$ $CH_2$ **238**

Acetate esters are cleaved by ylides, as illustrated by the preparation of 7-methylenecholesterol **(240)** from 7-ketocholesteryl acetate **(239)**.[254]

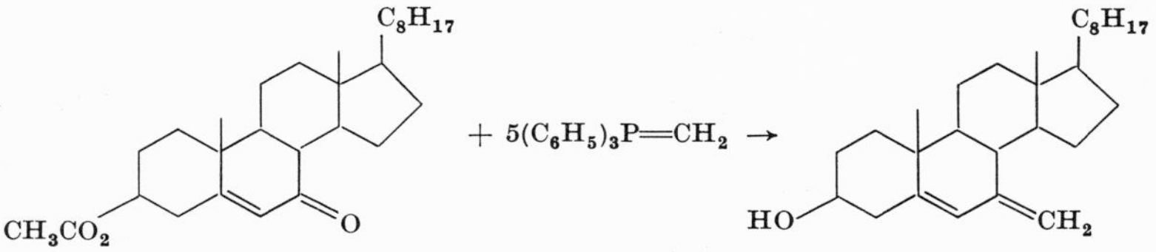

$C_8H_{17}$ $CH_3CO_2$ O **239** + $5(C_6H_5)_3P{=}CH_2$ → $C_8H_{17}$ HO $CH_2$ **240**

The use of methoxymethylenetriphenylphosphorane in place of methylenetriphenylphosphorane in these syntheses leads to exocyclic enol ethers that can be saponified to the corresponding formyl compounds. An example is the synthesis of 3-formyl-5$\alpha$,22$\beta$,25D-spirostane **(242)** from tigogenone **(241)**[246]

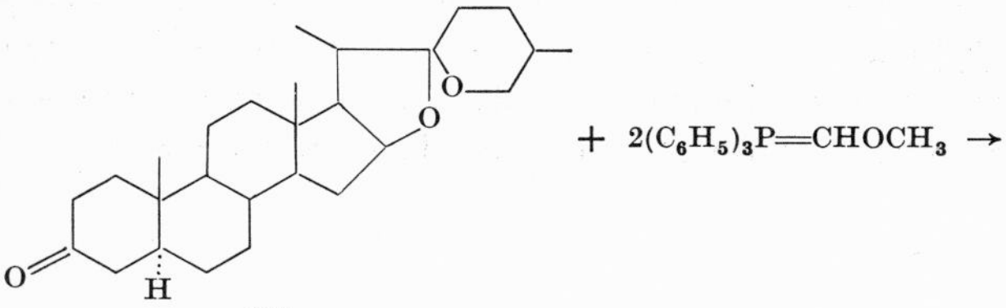

O O O H **241** + $2(C_6H_5)_3P{=}CHOCH_3$ →

O O $CH_3OCH$ H (85%) $\xrightarrow[(HClO_4)]{H^{\oplus}}$ O O OHC H **242** (Almost quant.)

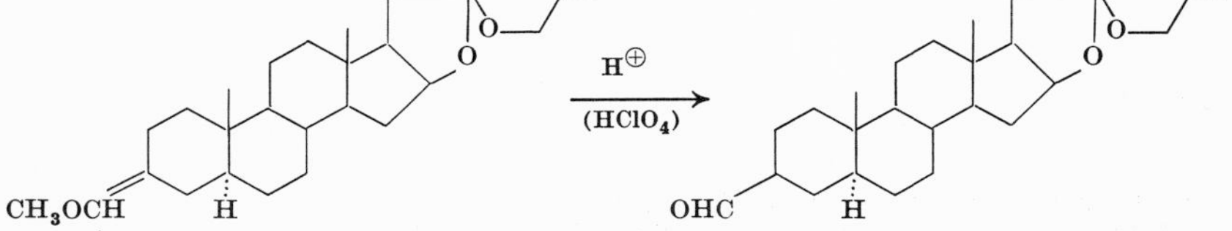

Monomethylene compounds can be isolated from some steroid diketones. The 3-keto group in androstane-3,7-dione **(243)** is the more reactive, and 3-methyleneandrostan-17-one **(244)** can be obtained. Further action of methylenetriphenylphosphorane forms 3,17-dimethyleneandrostane **(245)**.[252]

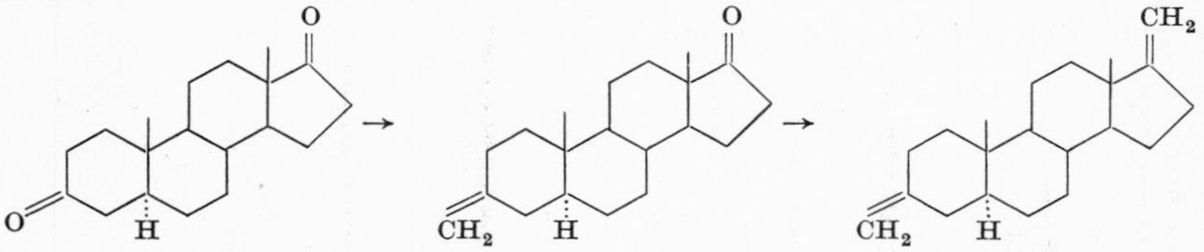

**243** → **244** (55%) → **245** (76%)

Steroid diketones containing keto groups of comparable reactivities give the dimethylene compounds directly, as illustrated by the formation of 3,20-dimethyleneallopregnane **(247)** in 75% yield from allopregnane-3,20-dione **(246)**.[252]

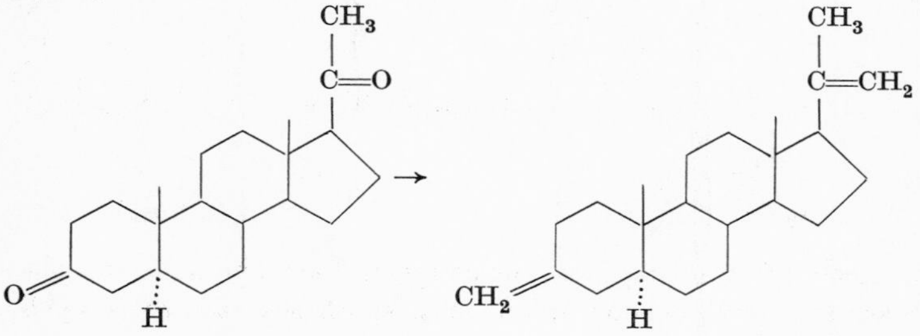

**246** → **247**

The same method has been used frequently to introduce a methylene group into the side chain of steroids. 24-Ketocholesteryl acetate **(248)** was converted in 70% yield to 24-methylenecholesterol **(249)**[304,304a] which proved to be identical with chalinasterol isolated from sponges and sea anemones, *Zoanthus proteus*, whose structure had been in dispute.[304]

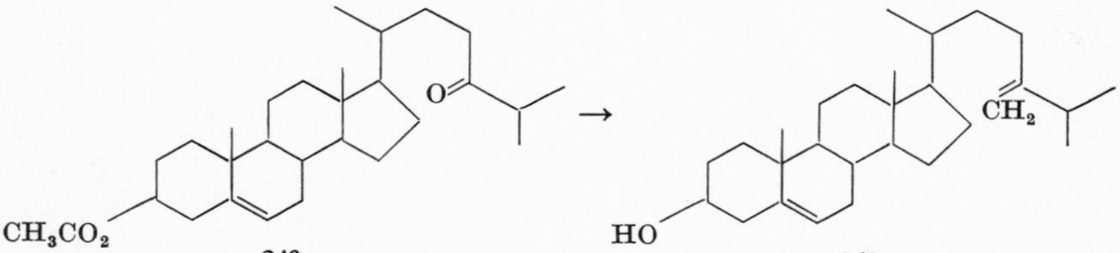

**248** → **249**

[304] Bergmann and Dusza, *Ann.*, **603**, 36 (1957).

[304a] Idler and Fagerlund, *J. Am. Chem. Soc.*, **79**, 1988 (1957).

24-Dehydrocholesterol **(251)**, isolated from acorn barnacle (*Balanus glandula*), was synthesized from 3$\beta$-acetoxy-5-cholenaldehyde **(250)**.[305]

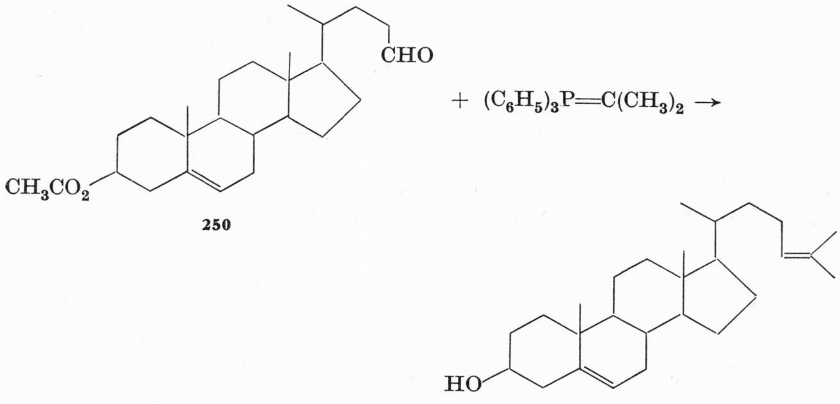

CHO

+ $(C_6H_5)_3P{=}C(CH_3)_2$ →

$CH_3CO_2$

**250**

HO

**251**

Similarly, 29-isofucosterol **(252)** was obtained from 24-ketocholesteryl acetate or 24-ketocholesteryl tetrahydropyranyl ether **(248)**.[306,307]

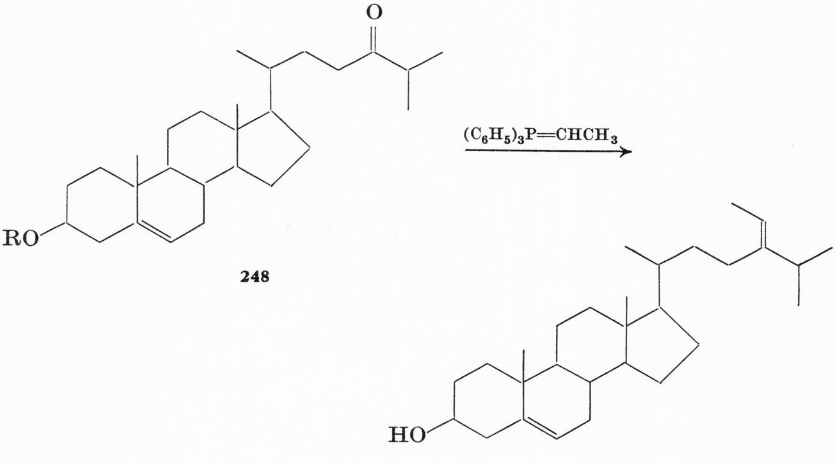

O

$(C_6H_5)_3P{=}CHCH_3$ →

RO

**248**

HO

**252**

The usefulness of the Wittig reaction for the synthesis of vitamin D was tested first in the preparation of model trienes. The ketone **253**

[305] Fagerlund and Idler, *J. Am. Chem. Soc.*, **79**, 6473 (1957).

[306] Dusza, *J. Org. Chem.*, **25**, 93 (1960).

[307] Fagerlund and Idler, *J. Fisheries Res. Board Can.*, **17**, 597 (1960) [*C.A.*, **55**, 2730h (1961)].

could be converted to the triene **254** which contains a double bond system typical of calciferol.[308,309]

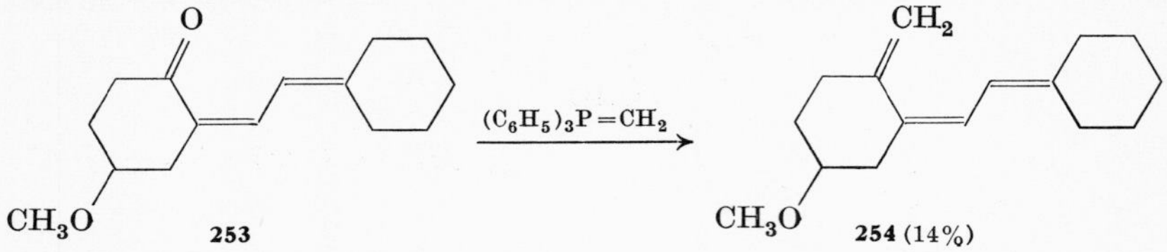

$CH_3O$ **253** $\xrightarrow{(C_6H_5)_3P=CH_2}$ $CH_3O$ **254** (14%)

The unsubstituted model substance **256** was prepared analogously from the cyclohexanone derivative **255**.[69,183,184,310,311]

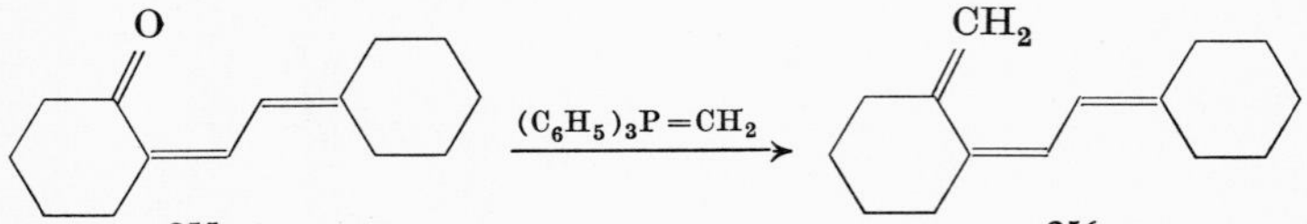

**255** $\xrightarrow{(C_6H_5)_3P=CH_2}$ **256**

A biologically active homolog of vitamin $D_2$, **257**, whose activity almost reached that of calciferol, was synthesized by the following route.[312]

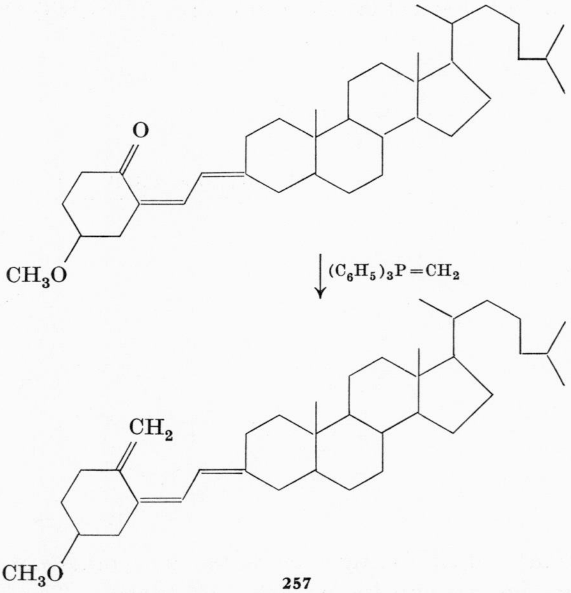

$CH_3O$ $\xrightarrow{(C_6H_5)_3P=CH_2}$ $CH_3O$ **257**

Systematic investigations led eventually to the total synthesis of

[308] Milas, Li-Chin Chiang, Priesing, Hyatt, and Peters, *J. Am. Chem. Soc.*, **77**, 4180 (1955).

[309] Milas and Priesing, *J. Am. Chem. Soc.*, **79**, 6295 (1957).

[310] Harrison and Lythgoe, *Proc. Chem. Soc.*, **1957**, 261.

[311] Harrison and Lythgoe, *J. Chem. Soc.*, **1958**, 837.

[312] Milas and Priesing, *J. Am. Chem. Soc.*, **79**, 3610 (1957).

vitamins $D_2$ (calciferol) and $D_3$, which involved the use of the Wittig reaction in three of the steps.[70,158,313–321] The final steps of the synthesis are shown in the accompanying formulation. It is interesting to note that

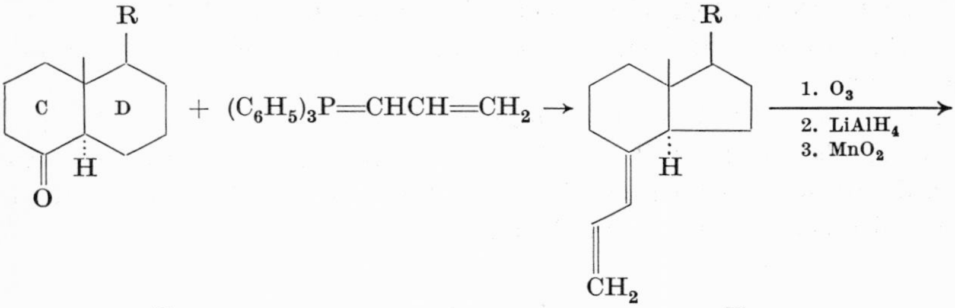

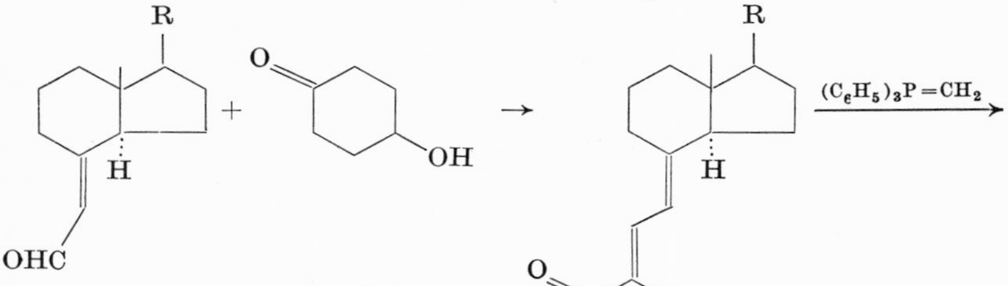

$+ (C_6H_5)_3P{=}CHCH{=}CH_2 \rightarrow$

1. $O_3$
2. $LiAlH_4$
3. $MnO_2$

258

$(C_6H_5)_3P{=}CH_2$

259

$h\nu$

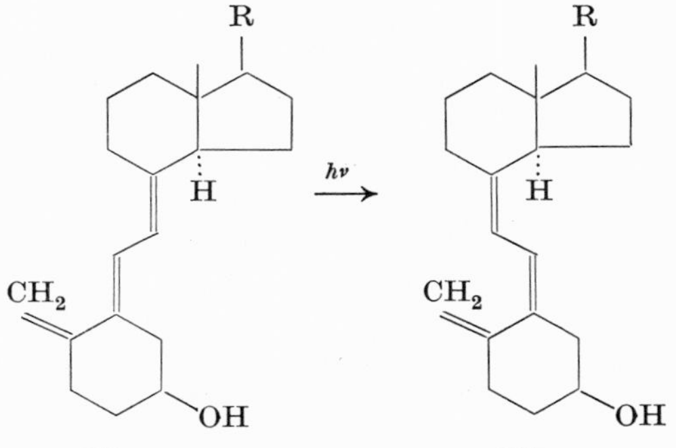

260 261

(Vitamin $D_2$: $R = C_9H_{17} = -CH(CH_3)CH{=}CHCH(CH_3)CH(CH_3)_2$;
Vitamin $D_3$: $R = C_8H_{17} = -CH(CH_3)CH_2CH_2CH_2CH(CH_3)_2$ )

313 Inhoffen, Kath, and Brückner, *Angew. Chem.*, **67**, 276 (1955).
314 Inhoffen and Irmscher, *Chem. Ber.*, **89**, 1833 (1956).
315 Inhoffen, Quinkert, and Hess, *Naturwiss.*, **44**, 11 (1957).
316 Inhoffen, Kath, Sticherling, and Brückner, *Ann.*, **603**, 25 (1957).
317 Inhoffen, Quinkert, Schütz, Kampe, and Domagk, *Chem. Ber.*, **90**, 664 (1957).
318 Inhoffen, Quinkert, and Schütz, *Chem. Ber.*, **90**, 1283 (1957).
319 Inhoffen, Irmscher, Hirschfeld, Stache, and Kreutzer, *Chem. Ber.*, **91**, 2309 (1958).
320 Inhoffen, Burkhardt, and Quinkert, *Chem. Ber.*, **92**, 1564 (1959).
321 Inhoffen, Irmscher, Friedrich, Kampe, and Berges, *Chem. Ber.*, **92**, 1772 (1959).

in the first step the *trans* configuration of the CD ring system is preserved. The condensation of the aldehyde **258** with *p*-hydroxycyclohexanone gives a mixture of epimers **259** from which the 3$\beta$-isomer could be isolated by chromatography. The 3$\beta$-5,6-*trans*-vitamin D **(260)** is then converted to vitamins $D_2$ and $D_3$ **(261)**, respectively, by photoisomerization.

A mixture of calciferol **(261)** and epi-calciferol **(263)** was also obtained by a similar method.[310,311] The photoisomerization, however, was carried out with the ketone **259**, and the mixture of epimers **262** so obtained was subjected to the Wittig reaction.

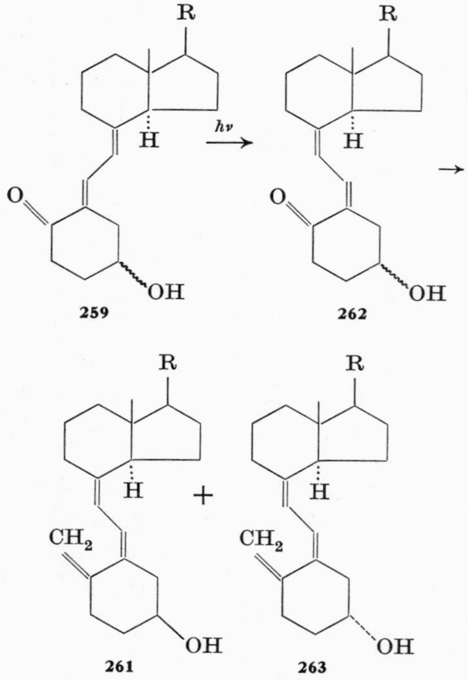

R R H $\xrightarrow{h\nu}$ H → O O OH OH

**259** **262**

R R H + H $CH_2$ $CH_2$ OH OH

**261** **263**

($R = C_9H_{17}$)

## Other Natural Products

The Wittig reaction has also been instrumental in the synthesis of a number of other natural products that do not belong to any of the three groups discussed above. Again, the possibility of introducing exocyclic double bonds by means of methylenetriphenylphosphorane is of decisive importance in these syntheses. The alkaloid 1-methylenepyrrolizidine

**(265)**, for example, can be synthesized in 63% yield from 1-pyrrolizidone **(264)**.[322]

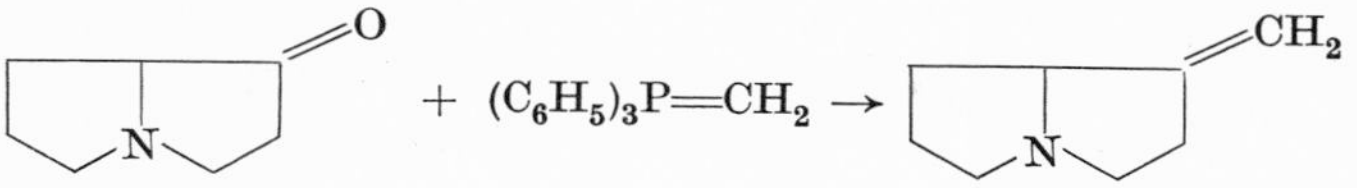

$$\mathbf{264} + (C_6H_5)_3P{=}CH_2 \rightarrow \mathbf{265}$$

In many cases the exocyclic double bond is subsequently hydrogenated catalytically, so that the over-all effect is the transformation

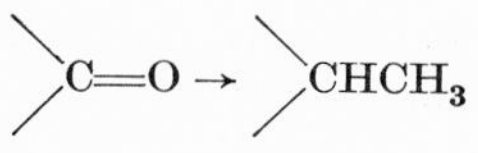

$$>C{=}O \rightarrow >CHCH_3$$

Thus 6-methyldihydrodesoxycodeine **(268)** was prepared from dihydrocodeinone **(266)** by way of 6-methylenedihydrodesoxycodeine **(267)**.[175]

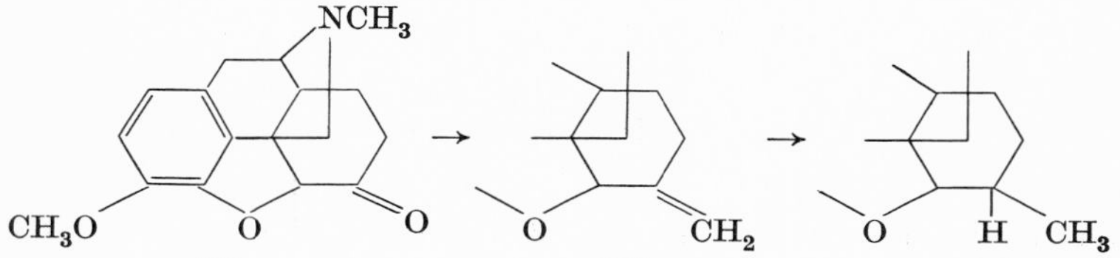

$NCH_3$, $CH_3O$, O, O — **266** → O, $CH_2$ — **267** → O, H, $CH_3$ — **268**

Similarly β-patchoulene **(270)** was synthesized in 38% over-all yield from the ketone **269**.[176]

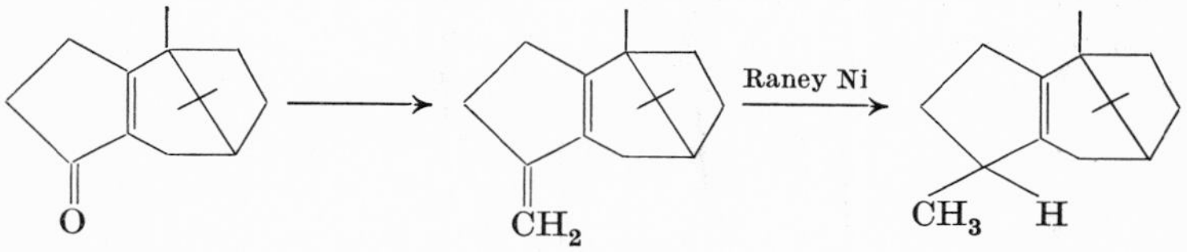

O — **269** → $CH_2$ —Raney Ni→ $CH_3$, H — **270**

The selective reduction of an exocyclic double bond can also be done with lithium in diethylamine, as illustrated by the synthesis of α-amyrin from glycyrrhetic acid, a β-amyrin derivative of known configuration, which elucidated the structure and configuration of that triterpene.[323]

Dihydrotremetone **(273)** was synthesized from 2-acetylbenzofuran **(271)** via 2-isopropenylbenzofuran **(272)**.[324]

[322] Kochetkov, Likhosherstov, and Kritsyn, *Tetrahedron Letters*, **1961,** 92.
[323] Corey and Cantrall, *J. Am. Chem. Soc.*, **80,** 499 (1958).
[324] DeGraw and Bonner, *Tetrahedron*, **18,** 1311 (1962).

$$\underset{\textbf{271}}{\text{(benzofuran-2-yl)}-COCH_3} \longrightarrow \underset{\textbf{272}}{\text{(2,3-dihydrobenzofuran-2-yl)}-C(=CH_2)CH_3} \xrightarrow{\text{3 steps}}$$

$$\underset{\textbf{273}}{CH_3O\text{-(2,3-dihydrobenzofuran-2-yl)}-CH(CH_3)_2}$$

The royal jelly of the honey bee, an α,β-unsaturated acid of structure **276,** was synthesized from glutaraldehyde **(274).**[228,229] The intermediate 6-(methoxycarbonyl)-5-hexenal **(275)** was converted to 9-oxo-2-decenoic acid **(276)** in four steps. The *trans* isomer (royal jelly) could be isolated in a total yield of 12–15%.

$$\underset{\textbf{274}}{OHC(CH_2)_3CHO} + (C_6H_5)_3P{=}CHCO_2CH_3 \longrightarrow$$

$$\underset{\textbf{275}\ (83\%)}{OHC(CH_2)_3CH{=}CHCO_2CH_3} \xrightarrow{\text{4 steps}}$$

$$\underset{\textbf{276}}{CH_3CO(CH_2)_5CH{=}CHCO_2H}$$

Finally, the important syntheses of insect sex attractants should be mentioned.[161,325–327] Bombycol, the sex attractant of the female silk moth (*Bombyx mori L.*), was found to be hexadeca-*trans*-10,*cis*-12-dien-1-ol **(280).** All possible *cis-trans* isomers of the hexadecadienol were prepared by different routes using the Wittig reaction. Bombycol itself was synthesized from 9-(ethoxycarbonyl)nonanal **(277)** and the ylide **278** according to the scheme $C_{10} + C_6 = C_{16}$.[325] The *trans* ester was isolated

$$\underset{\textbf{277}}{C_2H_5O_2C(CH_2)_8CHO} + \underset{\textbf{278}}{(C_6H_5)_3P{=}CHC{\equiv}C(CH_2)_2CH_3} \xrightarrow{46\%}$$

$$\underset{\textbf{279}}{C_2H_5O_2C(CH_2)_8CH{=}CHC{\equiv}C(CH_2)_2CH_3} \rightarrow$$

$$\underset{\textbf{280}}{HOCH_2(CH_2)_8CH{=}CH{-}CH{=}CH(CH_2)_2CH_3}\ (\text{10-}trans,\ \text{12-}cis)$$

[325] Butenandt, Hecker, Hopp, and Koch, *Ann.*, **658,** 39 (1962).

[326] Butenandt, Truscheit, Eiter, and Hecker, Ger. pat. 1,096,345 (to Farbenf. Bayer) [*C.A.*, **57,** 4546f (1962)].

[327] Truscheit, Eiter, Butenandt, and Hecker, Ger. pat. 1,138,037 (to Farbenf. Bayer) [*C.A.*, **58,** 6694f (1963)].

from the mixture of *cis* and *trans* isomers **279** by means of its urea clathrate and converted to bombycol **(280)** by partial hydrogenation followed by lithium aluminum hydride reduction.

An alternative route to bombycol **(280)** from 9-(methoxycarbonyl)-nonanal **(277*a*)** that involves two Wittig reactions is shown in the accompanying formulations.[161,326,327]

$$\underset{\mathbf{277a}}{CH_3O_2C(CH_2)_8CHO} + (C_6H_5)_3P{=}CHCO_2CH_3 \xrightarrow{67\%}$$

$$CH_3O_2C(CH_2)_8CH{=}CHCO_2CH_3 \text{ (trans)} \longrightarrow HOCH_2(CH_2)_8CH{=}CHCHO \text{ (trans)} \longrightarrow$$

$$CH_3CO_2CH_2(CH_2)_8CH{=}CHCHO \text{ (trans)} + (C_6H_5)_3P{=}CH(CH_2)_2CH_3 \xrightarrow{71\%}$$

$$CH_3CO_2CH_2(CH_2)_8CH{=}CH{-}CH{=}CH(CH_2)_2CH_3 \text{ (10-trans, 12-cis)} \xrightarrow{OH^{\ominus}}$$

$$\underset{\mathbf{280}}{HOCH_2(CH_2)_8CH{=}CH{-}CH{=}CH(CH_2)_2CH_3} \text{ (10-trans, 12-cis)}$$

By similar methods, the remaining hexadeca-10,12-dien-1-ols,[161,325–327] as well as other unsaturated aliphatic alcohols, have been synthesized.[327,328]

The elegant syntheses of oleic acid and other naturally occurring unsaturated fatty acids containing *cis* double bonds have been mentioned previously.[114]

## MODIFIED PROCEDURES AND RELATED REACTIONS

The pyrolysis of $\beta$-ketoalkylidene triphenylphosphoranes under certain conditions gives acetylenes by an intramolecular Wittig reaction.[41,329–331] The reaction is successful only if neither $R_1$ nor $R_2$ is

[328] Truscheit and Eiter, Austrian pat. 220,133 (to Farbenf. Bayer), 1960.

[329] Gough and Trippett, *Proc. Chem. Soc.*, **1961,** 302.

[330] Gough and Trippett, *J. Chem. Soc.*, **1962,** 2333.

[331] Märkl, *Chem. Ber.*, **94,** 3005 (1961).

hydrogen. In addition, $R_1$ or $R_2$ must be a substituent capable of resonance, such as a phenyl, ester, or nitrile group. If these conditions are

$$(C_6H_5)_3P{=}C(R_1){-}C({=}O)R_2 \longrightarrow (C_6H_5)_3\overset{\oplus}{P}{-}C(R_1){=}C(O^{\ominus})R_2 \xrightarrow{280^\circ} (C_6H_5)_3PO + R_1C{\equiv}CR_2$$

met, acetylenic compounds are obtained in excellent yields. Thus acetylcarbethoxymethylenetriphenylphosphorane (**281**) could be converted to ethyl 2-butynoate (**282**) in 91% yield.[329]

$$\underset{\mathbf{281}}{(C_6H_5)_3\overset{\oplus}{P}{-}C(CO_2C_2H_5){=}C(O^{\ominus})CH_3} \xrightarrow{280^\circ/10\ \text{mm.}} \underset{\mathbf{282}}{CH_3C{\equiv}CCO_2C_2H_5}$$

A large number of $\alpha,\beta$-acetylenecarboxylic esters has been synthesized in this manner.[331] In analogous fashion phenylpropiolonitrile is formed by pyrolysis of the phosphorane **283**.[329]

$$\underset{\mathbf{283}}{(C_6H_5)_3\overset{\oplus}{P}{-}C(CN){=}C(O^{\ominus})C_6H_5} \rightarrow \underset{(85\%)}{C_6H_5C{\equiv}CCN}$$

In recent years, practical and theoretical considerations have frequently led to modifications of the Wittig reagents, either by substituting the phenyl groups on phosphorus by other substituents or by replacing the phosphorus by other central atoms.

The use of phosphoranes containing basic substituents such as the acetylmethylene-*p*-dimethylaminophenyldiphenylphosphorane (**284**) is recommended because the corresponding phosphine oxide can be readily separated from the other reaction products by extraction of the ether solution with dilute acid.[192]

$$p\text{-}(CH_3)_2NC_6H_4P(C_6H_5)_2{=}CHCOCH_3 + C_6H_5CHO \xrightarrow{78\%}$$

**284**

$$p\text{-}(CH_3)_2NC_6H_4\overset{\overset{\displaystyle O}{\|}}{P}(C_6H_5)_2 + C_6H_5CH{=}CHCOCH_3$$

Replacement of the phenyl groups on phosphorus by electron-releasing substituents will result in decreased yields in reactions of normal ylides which are not resonance-stabilized. As discussed previously, the second step of the Wittig reaction, decomposition of the betaine, is rate-determining in these cases, and increased electron density on the phosphorus will have an unfavorable effect. For example, the high yields which can be obtained with methylenetriphenylphosphorane could, in no case, be duplicated with the methylene phosphoranes **66*a*–*f*** (see p. 311).[155]

Aliphatic substituents on phosphorus, such as methyl[102] or cyclohexyl groups,[28] also decrease the yields. The situation is different for the resonance-stabilized phosphoranes of the second group, for which the first step of the Wittig reaction is rate-determining. Electron-releasing groups on phosphorus have a favorable effect because they increase the nucleophilicity of the ylides. Thus fluorenylidenetri-*n*-butylphosphorane always gives better yields with aldehydes and ketones than does fluorenylidenetriphenylphosphorane.[29] Increased reactivity has been observed also in other resonance-stabilized phosphoranes in which the phenyl groups on phosphorus had been replaced by alkyl groups.[28,58,212,222] In addition, alkyl groups on phosphorus influence the stereochemistry of the Wittig reaction, the *trans* isomers being formed almost exclusively.[28]

Negatively charged oxygen on phosphorus is much more effective than alkyl groups. Metallated phosphine oxides such as **285** are considerably more nucleophilic than the normal alkylidene triphenylphosphoranes, since replacement of a phenyl group by the electron-releasing oxygen results in a strongly increased contribution of the limiting structure **285*b*** to the resonance hybrid.[223–225,332,333]

$$\underset{\textbf{285}\boldsymbol{a}}{(C_6H_5)_2\overset{\overset{\displaystyle O^{\ominus}}{|}}{P}{=}CHC_6H_5} \leftrightarrow \underset{\textbf{285}\boldsymbol{b}}{(C_6H_5)_2\overset{\overset{\displaystyle O}{\|}}{P}{-}\overset{\ominus}{C}HC_6H_5}$$

Reaction of benzophenone with **285**, for instance, gives triphenylethylene in 70% yield.[224] The other product, diphenylphosphinate ion,

[332] Horner, Hoffmann, and Wippel, Ger. pat. 1,079,030 (to Farbw. Hoechst), 1958.

[333] Horner, Hoffmann, and Klink, Ger. pat. 1,138,757 (to Farbw. Hoechst) [*C.A.*, **58**, 9143 (1963)].

is water-soluble. Unlike the normal diphenylalkylphosphine oxides,

$$(C_6H_5)_2\overset{O^{\ominus}}{\overset{|}{P}}{=}CHC_6H_5 + (C_6H_5)_2C{=}O \rightarrow (C_6H_5)_2C{=}CHC_6H_5 + (C_6H_5)_2\overset{O^{\ominus}}{\overset{|}{P}}{=}O$$

**285a**

which are readily obtained by the alkaline decomposition of the corresponding triphenylalkylphosphonium salts, phosphine oxides containing groups capable of resonance cannot be prepared by this route. Since these phosphine oxides are not easily obtainable, their reactions have so far attained only theoretical importance. Although normal diphenylalkylphosphine oxides can be metallated and used for the preparation of olefins, they are inferior to the ylides.

Alkylphosphonic acid esters, which can be obtained from triethyl phosphite by an Arbuzow reaction, can in certain cases be metallated and made to react with carbonyl compounds to give olefins in a reaction similar to that of phosphine oxides.[223–226, 332–334] Thus ethyl bromoacetate reacts with triethyl phosphite to give diethyl carbethoxymethyl phosphonate **(286)** in good yield. The phosphonate **286** can be converted to the anion **287** with sodium hydride in glycol dimethyl ether. Unlike the corresponding phosphorane, which will react with ketones only under forcing conditions, the anion **287** is strongly nucleophilic and reacts with cyclohexanone at room temperature within a few minutes to give ethyl cyclohexylidene acetate **(132)** in 70% yield.[226] The diethyl phosphate

$$(C_2H_5O)_3P + BrCH_2CO_2C_2H_5 \longrightarrow C_2H_5Br + (C_2H_5O)_2\underset{\underset{O}{\|}}{P}{-}CH_2CO_2C_2H_5$$

**286**

$\xrightarrow{NaH}$

$$(C_2H_5O)_2\underset{\underset{^{\ominus}O}{|}}{P}{=}CHCO_2C_2H_5 \longleftrightarrow (C_2H_5O)_2\underset{\underset{O}{\|}}{P}{-}\overset{\ominus}{C}HCO_2C_2H_5 \xrightarrow{\text{cyclohexanone}}$$

**287a** **287b**

$$\text{(cyclohexylidene)}{=}CHCO_2C_2H_5 + (C_2H_5O)_2\underset{\underset{O^{\ominus}}{|}}{P}{=}O$$

**132** **288**

anion **(288)** can be easily removed because it is water-soluble.

[334] Huisman, *Chem. Weekblad*, **59**, 133 (1963).

The first instance of an asymmetric induction in a Wittig reaction was observed when the reaction was used with the optically active ethyl (*p*-menth-3-yloxycarbonylmethyl)phosphonate **(289)**.[335] The levorotatory isomer of the substituted cyclohexylideneacetic acid **290** was formed predominantly.

$$(C_2H_5O)_2\overset{O}{\overset{\|}{P}}CH_2CO_2\text{-menthyl} \xrightarrow{NaH} (C_2H_5O)_2\overset{O^{\ominus}}{\overset{|}{P}}{=}CHCO_2\text{-menthyl}$$

**289**

$$\downarrow \text{R-}C_6H_9{=}O$$

$$\text{R-}C_6H_9{=}CHCO_2H \xleftarrow{H^{\oplus}} \text{R-}C_6H_9{=}CHCO_2\text{-menthyl}$$

**290** $[R = CH_3,\ C(CH_3)_3]$

Sodium amide[223,336] and potassium *t*-butoxide[224,225,332,333] have also been used in place of sodium hydride. Potassium *t*-butoxide has the disadvantage that it leads to resinification with aliphatic ketones such as acetone and cyclohexanone.[225]

The metallated diethyl carbethoxymethylphosphonate **(287)** proved to be superior to carbethoxymethylenetriphenylphosphorane in its reactivity toward other ketones also[37,103] and has been used for the synthesis of a series of $\alpha,\beta$-unsaturated esters.[226,336,337] $\alpha,\beta$-Unsaturated nitriles can be obtained in an analogous manner;[336,338,339] such nitriles are accessible from cyanomethylenetriphenylphosphorane only if aldehydes are used as the carbonyl component. In general, the phosphonate ester reaction will be the method of choice for the interaction of resonance-stabilized reagents with ketones. In all other cases the normal alkylidene triphenylphosphoranes are to be preferred, in accord with the theory of the mechanism of the Wittig reaction. Alkylphosphonic acid esters, for instance, are completely unsuitable for the preparation of olefins.[225]

The phosphonate ester method has found application, particularly in industry,[262,263] in the synthesis of $\alpha,\beta$-unsaturated carboxylic esters,

[335] Tömösközi and Janzsó, *Chem. Ind.* (*London*), **1962,** 2085.

[336] Takahashi, Fujiwara, and Ohta, *Bull. Chem. Soc. Japan,* **35,** 1498 (1962).

[337] Stilz and Pommer, Ger. pat. 1,109,671 (to BASF) [*C.A.*, **56,** 8571a (1962)].

[338] Stilz and Pommer, Ger. pat. 1,108,208 (to BASF) [*C.A.*, **56,** 11422e (1962)].

[339] Pommer and Stilz, Ger. pat. 1,116,652 (to BASF) [*C.A.*, **57,** 2267d (1962)].

nitriles, and stilbenes.[340] Distyrylbenzenes[333,341–345] and polyenes[346] have also been prepared from the corresponding bifunctional phosphonate esters.

In an analogous manner phenyldiethoxyphosphine **(291)** can be converted to phosphinic acid ethyl esters such as ethyl cyanomethylphenylphosphinate **(292)**, which is related to both the phosphine oxides and the phosphinate esters and thus undergoes the same reactions. For instance, it reacts with *β*-ionone to give *β*-ionylideneacetonitrile **(293)** in 87% yield.[347] The phosphinic acid esters have therefore been used for

$$\underset{\textbf{291}}{(C_2H_5O)_2PC_6H_5} + ClCH_2CN \rightarrow \underset{\textbf{292}}{(C_2H_5O)(C_6H_5)P(=O)CH_2CN} \rightarrow$$

$$(C_2H_5O)(C_6H_5)P(O^{\ominus})\!=\!CHCN + \beta\text{-ionone} \rightarrow \underset{\textbf{293}}{\beta\text{-ionylideneacetonitrile}}$$

the same olefin-forming reactions as the phosphonate esters.[225,348–352]

The central atom of the Wittig reagents itself has also been varied. The stability of fluorenylidenetriphenylarsorane **(294)**, which can be obtained by the action of aqueous sodium hydroxide on the arsonium salt, is similar to that of the corresponding phosphorane; decomposition into triphenylarsine oxide and fluorene occurs only on prolonged heating with aqueous alcoholic sodium hydroxide. Its reactivity toward carbonyl compounds is about the same as that of fluorenylidenetri-*n*-butylphosphorane, since it reacts even with such a poor electrophile as *p*-dimethylaminobenzaldehyde to give *p*-dimethylaminobenzalfluorene **(295)** in 97% yield.[353]

[340] Seus and Wilson, *J. Org. Chem.*, **26**, 5243 (1961).
[341] Stilz and Pommer, Ger. pat. 1,108,219 (to BASF) [*C.A.*, **57**, 732e (1962)].
[342] Pommer, Stilz, and Stolp, Ger. pat. 1,108,220 (to BASF) [*C.A.*, **56**, 14165d (1962)].
[343] Stilz, Pommer, and König, Ger. pat. 1,112,072 (to BASF) [*C.A.*, **56**, 2378d (1962)].
[344] Stilz and Pommer, Ger. pat. 1,124,949 (to BASF) [*C.A.* **57**, 3358c (1962)].
[345] Stilz, Stolp, and Pommer, Ger. pat. 1,129,947 (to BASF) [*C.A.*, **57**, 16502f (1962)].
[346] Stilz and Pommer, Ger. pat. 1,092,472 (to BASF) [*C.A.*, **56**, 413b (1962)].
[347] Pommer and Stilz, Ger. pat. 1,116,653 (to BASF) [*C.A.*, **57**, 2267d (1962)].
[348] Stilz and Pommer, Ger. pat. 1,116,654 (to BASF) [*C.A.*, **57**, 737a (1962)].
[349] Stilz and Pommer, Ger. pat. 1,117,122 (to BASF) [*C.A.*, **57**, 2143e (1962)].
[350] Pommer and Stilz, Ger. pat. 1,117,580 (to BASF) [*C.A.*, **57**, 4591d (1962)].
[351] Stilz and Pommer, Ger. pat. 1,122,065 (to BASF) [*C.A.*, **57**, 3358a (1962)].
[352] Pommer and Stilz, Ger. pat. 1,122,524 (to BASF) [*C.A.*, **57**, 3484h (1962)].
[353] Johnson, *J. Org. Chem.*, **25**, 183 (1960).

$$\text{(fluorenylidene)}=As(C_6H_5)_3\ (\mathbf{294}) + p\text{-}(CH_3)_2NC_6H_4CHO \rightarrow \text{(fluorenylidene)}=CHC_6H_4N(CH_3)_2\text{-}p\ (\mathbf{295}) + (C_6H_5)_3AsO$$

The betaine **296** obtained from methylenetriphenylarsorane and benzophenone can decompose by two different paths.[354] After acid hydrolysis,

$$(C_6H_5)_3As{=}CH_2 + (C_6H_5)_2C{=}O \rightarrow (C_6H_5)_3\overset{\oplus}{As}{-}CH_2{-}C(C_6H_5)_2{-}O^{\ominus}\ (\mathbf{296})$$

$$\mathbf{296} \xrightarrow{A} (C_6H_5)_3AsO + (C_6H_5)_2C{=}CH_2$$

$$\mathbf{296} \xrightarrow{B} (C_6H_5)_3As + (C_6H_5)_2C\overset{O}{\frown}CH_2 \xrightarrow{H^{\oplus}} (C_6H_5)_2CHCHO$$

the products from path A and path B are obtained in about a 1:3 ratio. With antimony as the central atom, the reaction proceeds almost exclusively by path B and 1,1-diphenylethylene and triphenylstibine oxide are formed only in traces.[354]

Such an abnormal decomposition of a betaine will occur even in a resonance-stabilized ylide if sulfur is used as the central atom.[355,356] In

$$\text{(fluorenylidene)}=S(CH_3)_2 + RCHO \rightarrow \text{(fluorenyl)}(\overset{\oplus}{S}(CH_3)_2)\text{—}CH(R)O^{\ominus} \rightarrow \text{spiro[fluorene-9,2'-oxirane] with } C(R)(H)$$

[354] Henry and Wittig, *J. Am. Chem. Soc.*, **82**, 563 (1960).

[355] Johnson and LaCount, *Chem. Ind.* (*London*), **1958**, 1440.

[356] Johnson and LaCount, *J. Am. Chem. Soc.*, **83**, 417 (1961).

addition, a Sommelet rearrangement has been observed.[356] Epoxides have been obtained also from other sulfur ylides and carbonyl compounds.[357–359] Dimethylsulfoxonium methylide **(297)**, which has recently been prepared from trimethylsulfoxonium halides, has proved to be particularly useful; epoxides were obtained in 70–90% yields from carbonyl compounds.[360]

$$(CH_3)_3\overset{\oplus}{S}{=}O\ X^{\ominus} \xrightarrow{NaH} \underset{\mathbf{297}}{(CH_3)_2\overset{O}{\overset{\|}{S}}{=}CH_2} \xrightarrow{R_1R_2C{=}O} R_1R_2\overset{}{C}\overset{O}{—}CH_2 + (CH_3)_2SO$$

With benzalacetophenone the cyclopropane derivative **298** was obtained in 95% yield.[360]

$$\underset{\mathbf{297}}{(CH_3)_2\overset{O}{\overset{\|}{S}}{=}CH_2} + C_6H_5CH{=}CHCOC_6H_5 \rightarrow \underset{\mathbf{298}}{C_6H_5CH\underset{CH_2}{—}CHCOC_6H_5}$$

The methylene group in methylene triphenylphosphoranes has also been replaced by isoelectronic groups. Triphenylphosphine imine **(299)**, for instance, reacts with benzophenone to give benzophenone imine **(300)**.[361] Reaction with an α-keto ester gives an α-imino ester which, on catalytic hydrogenation, is converted to an α-amino ester as illustrated by the synthesis of the alanine ester **301**.[362]

$$(C_6H_5)_3\overset{\oplus}{P}NH_2\ Cl^{\ominus} \xrightarrow{NaNH_2} \underset{\mathbf{299}}{(C_6H_5)_3P{=}NH} \xrightarrow{(C_6H_5)_2CO} \underset{\mathbf{300}}{(C_6H_5)_2C{=}NH}$$

$$\mathbf{299} \xrightarrow{CH_3COCO_2CH_3} \underset{\underset{NH}{\|}}{CH_3CCO_2CH_3} \xrightarrow{H_2} \underset{\mathbf{301}}{\underset{\underset{NH_2}{|}}{CH_3CHCO_2CH_3}}$$

Substituted triphenylphosphine imines such as N-phenyltriphenylphosphineimine **(302)** have previously been made to react with a number

[357] Franzen and Driessen, *Tetrahedron Letters*, **1962**, 661.
[358] Corey and Chaykovsky, *J. Am. Chem. Soc.*, **84**, 3782 (1962).
[359] Corey and Chaykovsky, *Tetrahedron Letters*, **1963**, 169.
[360] Corey and Chaykovsky, *J. Am. Chem. Soc.*, **84**, 867 (1962).
[361] Appel and Hauss, *Chem. Ber.*, **93**, 405 (1960).
[362] Appel and Hauss, *Z. Anorg. Allgem. Chem.*, **311**, 290 (1961).

of carbonyl compounds,[8,363,364] as illustrated by the conversion of diphenylketene to N-phenyldiphenylketene imine **(303)**.

$$\underset{\mathbf{302}}{(C_6H_5)_3P{=}NC_6H_5} + (C_6H_5)_2C{=}C{=}O \rightarrow \underset{\mathbf{303}}{(C_6H_5)_2C{=}C{=}NC_6H_5}$$

Diethyl phosphoric acid amide anions **(304)**[365] and even N-sulfinyl derivatives **(305)**[366] undergo the corresponding reaction. In the latter case sulfur dioxide is formed as a by-product.

$$\underset{\mathbf{304}}{(C_2H_5O)_2\overset{O^{\ominus}}{\overset{|}{P}}{=}NR} \qquad \underset{\mathbf{305}}{R{-}N{=}S{=}O}$$

Wittig reagents have also been made to react with a number of compounds that undergo reactions similar to those of carbonyl compounds. For example, thiobenzophenone reacts with methylenetriphenylphosphorane to give 1,1-diphenylethylene.[13,138]

$$(C_6H_5)_3P{=}CH_2 + (C_6H_5)_2C{=}S \rightarrow \underset{(76\%)}{(C_6H_5)_2C{=}CH_2} + \underset{(82\%)}{(C_6H_5)_3PS}$$

Carbon disulfide is converted to thioketenes; thus fluorenylidenetriphenylphosphorane gives the dimeric thioketene **306** in 68% yield.[367]

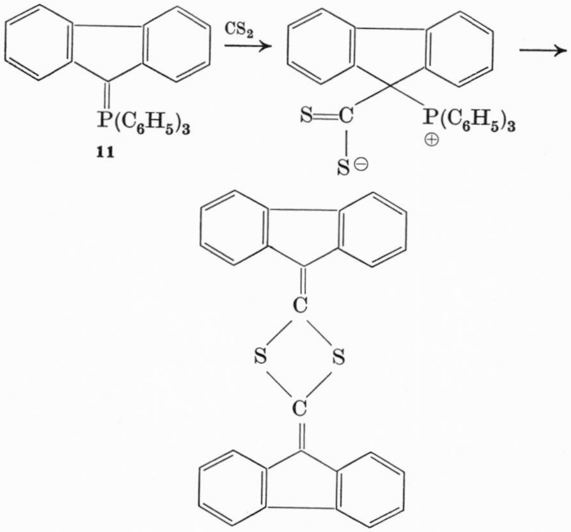

$$\underset{\mathbf{11}}{\text{fluorenylidene}{=}P(C_6H_5)_3} \xrightarrow{CS_2} \text{fluorene-9-}C(S^{\ominus}){=}S,\ 9\text{-}\overset{\oplus}{P}(C_6H_5)_3 \longrightarrow$$

$$\underset{\mathbf{306}}{\text{fluorenylidene}{=}C\!<\!\begin{smallmatrix}S\\S\end{smallmatrix}\!>\!C{=}\text{fluorenylidene}}$$

[363] Staudinger and Hauser, *Helv. Chim. Acta*, **4**, 861 (1921).
[364] Staudinger and Meyer, *Ber.*, **53**, 72 (1920).
[365] Wadsworth, Jr., and Emmons, *J. Am. Chem. Soc.*, **84**, 1316 (1962).
[366] Kresze and Albrecht, *Angew. Chem.*, **74**, 781 (1962).
[367] Schönberg, Frese, and Brosowski, *Chem. Ber.*, **95**, 3077 (1962).

With isocyanates, the expected ketene imines are obtained only in a few cases.[7,8]

$$(C_6H_5)_3P{=}C(C_6H_5)_2 + C_6H_5N{=}C{=}O \rightarrow C_6H_5N{=}C{=}C(C_6H_5)_2$$

Methylenetriphenylphosphorane or alkylidene phosphoranes, which contain a hydrogen atom on the methylene group, undergo a Michael addition to the C=N linkage of phenyl isocyanate.[41]

$$(C_6H_5)_3P{=}CH_2 + 2C_6H_5N{=}C{=}O \longrightarrow C_6H_5NH\overset{O}{\overset{\|}{C}}\underset{\underset{P(C_6H_5)_3}{\|}}{C}\overset{O}{\overset{\|}{C}}NHC_6H_5$$

$$(C_6H_5)_3P{=}CHCO_2C_2H_5 + C_6H_5N{=}C{=}O \xrightarrow{90\%} (C_6H_5)_3P{=}\underset{\underset{O=CNHC_6H_5}{|}}{C}CO_2C_2H_5$$

Nitrosobenzene reacts with alkylidene triphenylphosphoranes to give Schiff bases (azomethines) which can be hydrolyzed to give aldehydes or ketones.[13,185,368] For instance, benzylidenetriphenylphosphorane is converted to benzaldehyde by way of benzalaniline. This is a general

$$(C_6H_5)_3P{=}CHC_6H_5 + C_6H_5N{=}O \rightarrow C_6H_5N{=}CHC_6H_5 \xrightarrow{H^\oplus} C_6H_5CHO + C_6H_5NH_3^\oplus$$

method for the preparation of carbonyl compounds. Geranylidenetriphenylphosphorane, for example, has been converted to citral.

Resonance-stabilized alkylidene phosphoranes react analogously with nitrosobenzene, as illustrated by the conversion of diphenylmethylenetriphenylphosphorane to benzophenone anil **(307)**.[369]

$$(C_6H_5)_3P{=}C(C_6H_5)_2 + C_6H_5N{=}O \rightarrow \underset{\mathbf{307}}{C_6H_5N{=}C(C_6H_5)_2}$$

Nitromethylene triphenylphosphoranes give rise to products containing a carbon-nitrogen triple bond in a reaction which is similar to the formation of acetylenes by an intramolecular Wittig reaction of acylmethylene triphenylphosphoranes. For example, nitrobenzylidenetriphenylphosphorane **(308)**, in the presence of triphenylphosphine, gives benzonitrile in 75% yield.[45]

[368] Wittig, Schöllkopf, and Pommer, Ger. pat. 1,048,568 (to BASF) [*C.A.*, **55,** 4576h (1961)].

[369] Schönberg and Brosowski, *Chem. Ber.*, **92,** 2602 (1959).

$$(C_6H_5)_3P{=}C(C_6H_5)-\overset{\oplus}{N}(=O)-O^{\ominus} \longleftrightarrow (C_6H_5)_3\overset{\oplus}{P}-C(C_6H_5){=}\overset{\oplus}{N}(O^{\ominus})_2$$

**308a** **308b**

$$\downarrow$$

$$C_6H_5C{\equiv}\overset{\oplus}{N}-\overset{\ominus}{O} \xrightarrow{(C_6H_5)_3P} C_6H_5CN$$

Finally, mention should be made of the reaction of alkylidene phosphoranes with epoxides for which an intermediate betaine similar to that occurring in the Wittig reaction has been proposed.[370,371] Styrene oxide and other epoxides react with carbethoxymethylenetriphenylphosphorane to give cyclopropanecarboxylic esters.

$$(C_6H_5)_3P{=}CHCO_2C_2H_5 + RCH\overset{O}{\frown}CH_2 \rightarrow (C_6H_5)_3\overset{\oplus}{P}-CHCO_2C_2H_5-CH_2-CHR-O^{\ominus} \rightarrow$$

$$(C_6H_5)_3P-CHCO_2C_2H_5,\ CH_2,\ O-CH(R) \text{ (cyclic)} \rightarrow RCH-CHCO_2C_2H_5 \text{ (bridged by } CH_2\text{)}$$

Ethyl norcaranecarboxylate **(309)** is formed in 56% yield from cyclohexene oxide.[371] Metallated phosphonate esters[226] and phosphine

$$\text{cyclohexene oxide} + (C_6H_5)_3P{=}CHCO_2C_2H_5 \rightarrow \text{bicyclo}>CHCO_2C_2H_5$$

**309**

oxides[372] undergo analogous reactions.

Methylethylphenylbenzylidenephosphorane **(310)**, on the other hand, reacts with styrene oxide to give mainly benzylacetophenone **(311)** and only a small amount of 1,2-diphenylcyclopropane. A possible mechanism is shown on p. 388.[373,374]

[370] Denney and Boskin, *J. Am. Chem. Soc.*, **81**, 6330 (1959).
[371] Denney, Vill, and Boskin, *J. Am. Chem. Soc.*, **84**, 3944 (1962).
[372] Horner, Hoffmann, and Toscano, *Chem. Ber.*, **95**, 536 (1962).
[373] McEwen and Wolf, *J. Am. Chem. Soc.*, **84**, 676 (1962).
[374] McEwen, Bladé-Font, and Vander Werf, *J. Am. Chem. Soc.*, **84**, 677 (1962).

$$CH_3CH_2\text{—}\overset{CH_3}{\underset{C_6H_5}{P}}\text{=}CHC_6H_5 + \underset{\diagdown O \diagup}{CH_2\text{—}CHC_6H_5} \rightarrow CH_3CH_2\text{—}\overset{\oplus\ CH_3}{\underset{C_6H_5}{P}}\text{—}\underset{C_6H_5}{CH}\text{—}CH_2\overset{O^{\ominus}}{CHC_6H_5}$$

310

$$\rightleftharpoons$$

$$CH_3CH_2\text{—}\overset{CH_3}{\underset{C_6H_5}{P}} + \underset{\diagdown CH_2 \diagup}{C_6H_5CH\text{—}\overset{OH}{C}C_6H_5} \leftarrow CH_3CH_2\text{—}\overset{\oplus\ CH_3}{\underset{C_6H_5}{P}}\text{—}\underset{C_6H_5}{\overset{\ominus}{C}H}\text{—}CH_2\text{—}\overset{OH}{C}C_6H_5$$

$$\downarrow$$

$$C_6H_5CH_2CH_2\overset{O}{\overset{\|}{C}}C_6H_5$$

311

## EXPERIMENTAL CONDITIONS

### Preparation of Phosphonium Salts

Triphenylalkylphosphonium salts are generally prepared from triphenylphosphine and alkyl halides.

$$(C_6H_5)_3P + RX \rightarrow (C_6H_5)_3\overset{\oplus}{P}R \quad X^{\ominus}$$

The rate of reaction decreases in the series RI > RBr > RCl. The alkyl bromides are used most frequently because they are often more readily accessible than the alkyl iodides. In general, three methods may be used for this reaction.

1. Liquid halides may be made to react with triphenylphosphine without the use of a solvent.
2. The reaction of solid halides with triphenylphosphine may be carried out in the melt.
3. Equimolar quantities of triphenylphosphine and an alkyl halide are allowed to react in a suitable solvent.

The first two modes of preparation are customarily used only if the third method fails to give the desired phosphonium salts. For example, 1,4-dibromobutane and triphenylphosphine react in benzene to give the monophosphonium salt **312** in quantitative yield,[74] whereas heating the reactants without a solvent gives rise to the bis-phosphonium salt **313** in

90% yield.[52] The bis-phosphonium salt **313** can also be prepared from

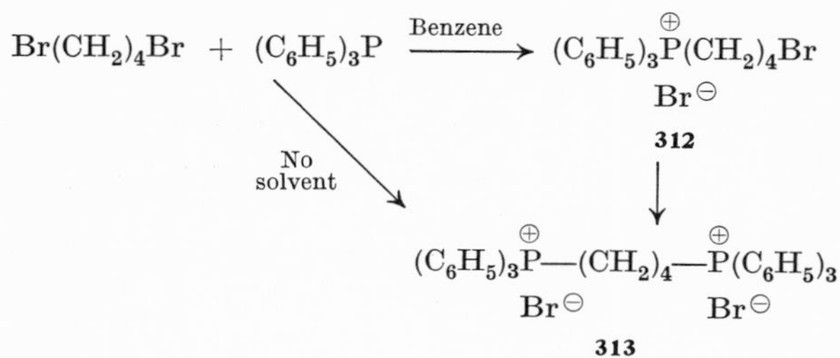


the monophosphonium salt **312** and triphenylphosphine either in the melt[74] or in a different solvent.[301] In solvents of low polarity, such as benzene, xylene, or diethyl ether, bis-halides generally form the pure monophosphonium salts if they are insoluble in the solvent.[375] Strongly polar solvents, such as nitrobenzene, acetonitrile, or dimethylformamide, however, keep the mono salt in solution so that reaction with a second equivalent of triphenylphosphine can take place. For example, reaction of 1,4-dibromobutane with 2 moles of triphenylphosphine in acetonitrile furnishes the bis-phosphonium salt **313** directly in quantitative yield.[225] Suitable choice of solvents will thus make it generally unnecessary to carry out the reaction in the melt. In chloroform, mixtures of mono- and di-phosphonium salts are often formed.

In addition to its polarity, the boiling point of a solvent is also of great importance since it determines the maximum reaction temperature unless the reaction is carried out under pressure. The interaction of $\omega,\omega'$-dibromoxylenes with triphenylphosphine proceeds much less readily in ether than in xylene. Some alkyl halides, however, give good yields only if ether is the solvent.[241] In such reactions it is advantageous to carry out the reaction in a sealed tube at 100–120°.[71,72,241]

Nitromethane,[41,48,155,182] formic acid,[225] acetic acid,[376] ethyl acetate,[236,377] ethanol,[54,79] acetone,[29] butanone or cyclohexanone,[301] benzonitrile,[159] toluene,[378] and tetrahydrofuran[69,209,287] have been used as solvents. The salts precipitate either spontaneously or on addition of diethyl ether.

Alkyl bromides may undergo an allylic rearrangement during the quaternization, as illustrated by the reaction of 1-bromo-2-methylene-cyclohexane.[70,158]

[375] Friedrich and Henning, *Chem. Ber.*, **92,** 2756 (1959).
[376] Brit. pat. 812,268 (to Hoffmann-La Roche) [*C.A.*, **53,** 18983f (1959)].
[377] Isler, Chopard-dit-Jean, Montavon, Rüegg, and Zeller, *Helv. Chim. Acta*, **40,** 1256 (1957).
[378] Bergmann and Dusza, *J. Org. Chem.*, **23,** 1245 (1958).

$$\text{2-bromo-1-methylenecyclohexane} + (C_6H_5)_3P \rightarrow \text{1-cyclohexenyl-}CH_2\overset{\oplus}{P}(C_6H_5)_3\ Br^{\ominus}$$

Complications may also occur in the interaction of triphenylphosphine with α-halocarbonyl compounds. Whereas primary α-bromo- and α-chloro-ketones give the desired products,[31] reaction of secondary and tertiary α-halocarbonyl compounds with triphenylphosphine in the presence of water or alcohol leads to the dehalogenated carbonyl compounds.[379–381]

$$(C_6H_5)_3P + BrCH_2COC_6H_5 \longrightarrow (C_6H_5)_3\overset{\oplus}{P}CH_2COC_6H_5\ Br^{\ominus}$$

$$(C_6H_5)_3P + ClCH_2COCH_3 \longrightarrow (C_6H_5)_3\overset{\oplus}{P}CH_2COCH_3\ Cl^{\ominus}$$

$$\text{2-bromocyclohexanone} + (C_6H_5)_3P \xrightarrow{ROH} \text{cyclohexanone} + (C_6H_5)_3PO + RBr$$

$$C_6H_5CCl_2COC_6H_5 + (C_6H_5)_3P \xrightarrow{H_2O} C_6H_5CHClCOC_6H_5 + (C_6H_5)_3PO + HCl$$

Formation of a triple bond may occur in the absence of water or alcohol.[382] Not only α-halo ketones but also 1-bromo-1-nitroalkanes and N-bromoamides react with triphenylphosphine in this way.[382]

$$C_6H_5CHClCOC_6H_5 + (C_6H_5)_3P \xrightarrow{90\%} C_6H_5C{\equiv}CC_6H_5 + (C_6H_5)_3PO + HCl$$

$$C_6H_5CHBrNO_2 + 2(C_6H_5)_3P \xrightarrow{75\%} C_6H_5C{\equiv}N + (C_6H_5)_3PO + (C_6H_5)_3PO{\cdot}HBr$$

$$C_6H_5CONHBr + (C_6H_5)_3P \longrightarrow C_6H_5C{\equiv}N + (C_6H_5)_3PO{\cdot}HBr$$

Trichloroacetamides react with phosphines and phosphites to give trichlorovinylamines.[383]

$$Cl_3CCONR_2 + R'_3P \rightarrow Cl_2C{=}C(Cl)(NR_2) + R'_3PO$$

[379] Trippett, *J. Chem. Soc.*, **1962**, 2337.

[380] Borowitz and Grossman, *Tetrahedron Letters*, **1962**, 471.

[381] Hoffmann and Diehr, *Tetrahedron Letters*, **1962**, 583.

[382] Trippett and Walker, *J. Chem. Soc.*, **1960**, 2976.

[383] Speziale and Freeman, *J. Am. Chem. Soc.*, **82**, 903 (1960).

The mechanism of these reactions may involve initial removal of a halogen cation by triphenylphosphine;[384] for instance,

$$\mathrm{Cl_3CCONR_2 + R'_3P \rightarrow R'_3\overset{\oplus}{P}Cl + Cl_2C{=}C(O^{\ominus})(NR_2)} \rightarrow$$

$$\mathrm{[R'_3P \cdots Cl,\ O\ \text{cyclic intermediate with}\ Cl_2C{=}C(NR_2)] \rightarrow Cl_2C{=}C(Cl)(NR_2) + R'_3PO}$$

Such a course appears to be followed in the reaction of phenacyl bromide with triphenylphosphine in methanol, whereas phenacyl chloride in methanol undergoes simple displacement of chloride to form the phosphonium salt.[385]

$$\mathrm{(C_6H_5)_3\overset{\oplus}{P}CH_2COC_6H_5\ \ Cl^{\ominus}}$$

Halomethylphosphonium salts prepared from dihalomethanes and triphenylphosphine are often contaminated by the bis-phosphonium salt.[81]

$$\mathrm{3(C_6H_5)_3P + 2CH_2Br_2 \rightarrow \underset{(40\%)}{(C_6H_5)_3\overset{\oplus}{P}CH_2Br\ Br^{\ominus}} + \underset{(20\%)}{(C_6H_5)_3\overset{\oplus}{P}{-}CH_2{-}\overset{\oplus}{P}(C_6H_5)_3\ \ 2Br^{\ominus}}}$$

The pure compounds can be prepared from the corresponding hydroxymethylphosphonium salts with thionyl chloride or a phosphorus pentahalide.[84,92]

$$\mathrm{(C_6H_5)_3\overset{\oplus}{P}CH_2OH\ Cl^{\ominus} \xrightarrow[63\%]{SOCl_2} (C_6H_5)_3\overset{\oplus}{P}CH_2Cl\ Cl^{\ominus}}$$

$$\mathrm{(C_6H_5)_3\overset{\oplus}{P}CH_2OH\ Br^{\ominus} \xrightarrow[74\%]{PBr_5} (C_6H_5)_3\overset{\oplus}{P}CH_2Br\ Br^{\ominus}}$$

Mention is made of an elegant method for the preparation of phosphonium salts from alcohols, which are made to react either with

[384] Speziale and Smith, *J. Am. Chem. Soc.*, **84**, 1868 (1962).

[385] Borowitz and Virkhaus, *J. Am. Chem. Soc.*, **85**, 2184 (1963).

triphenylphosphine hydrohalides or with triphenylphosphine in the presence of proton donors such as hydrogen halides or sulfuric acid.[386]

$$\text{ROH} + (C_6H_5)_3\overset{\oplus}{P}H\ X^{\ominus} \searrow$$
$$(C_6H_5)_3\overset{\oplus}{P}R\ X^{\ominus} + H_2O$$
$$\text{ROH} + (C_6H_5)_3P + HX \nearrow$$

The water is removed either by azeotropic distillation or with silica gel or acetic anhydride. Thus β-ionol and triphenylphosphine hydrochloride react in acetonitrile to give β-ionyltriphenylphosphonium chloride **(314)**.[291]

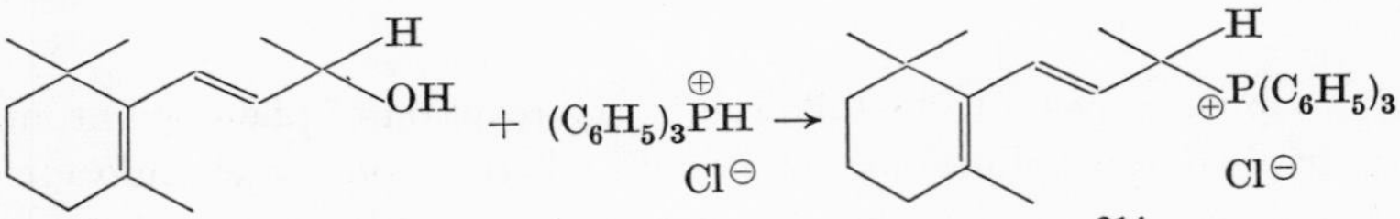

$$+ (C_6H_5)_3\overset{\oplus}{P}H\ Cl^{\ominus} \rightarrow$$

H, OH; H, $\oplus P(C_6H_5)_3$, $Cl^{\ominus}$

**314**

Axerophthyltriphenylphosphonium chloride **(315)** is obtained in quantitative yield by the reaction of vitamin A with triphenylphosphine and hydrogen chloride in ethanol.[128,266]

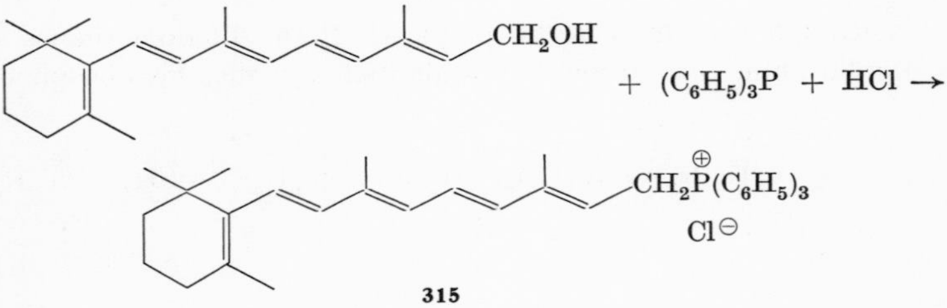

$CH_2OH$ $+ (C_6H_5)_3P + HCl \rightarrow$

$CH_2\overset{\oplus}{P}(C_6H_5)_3$ $Cl^{\ominus}$

**315**

Starting from vinyl-β-ionol **(316)**, (β-ionylideneëthyl)triphenylphosphonium halides **(317)** are obtained by an allylic rearrangement.[264,268,285,286,294,387]

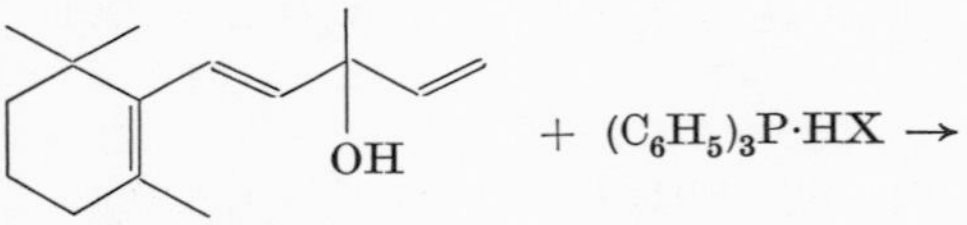

OH $+ (C_6H_5)_3P{\cdot}HX \rightarrow$

**316**

$CH_2\overset{\oplus}{P}(C_6H_5)_3$ $X^{\ominus}$

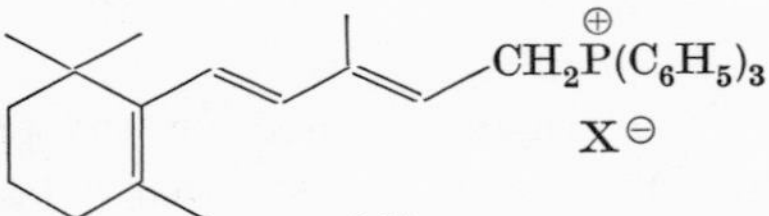

**317**

[386] Sarnecki and Pommer, Ger. pat. 1,046,046 (to BASF) (*Chem. Zentr.*, **1959**, 13003).
[387] Sarnecki and Pommer, Ger. pat. 1,060,386 (to BASF) [*C.A.*, **55**, 4577b (1961)].

The addition of triphenylphosphine hydrohalides to conjugated systems also leads to phosphonium salts.[240,266,269,284,292]

$$+ \ (C_6H_5)_3P{\cdot}HX \rightarrow$$

$$CH_2\overset{\oplus}{P}(C_6H_5)_3 \quad X^{\ominus}$$

**317**

Addition may also occur to a monoölefin provided that it is sufficiently activated.[388]

$$(C_6H_5)_3P \ + \ CH_2{=}CHR \rightleftharpoons (C_6H_5)_3\overset{\oplus}{P}CH_2\overset{\ominus}{C}HR \xrightarrow{HX} (C_6H_5)_3\overset{\oplus}{P}{-}CH_2CH_2R \quad X^{\ominus}$$

$$(R = CO_2H, CO_2CH_3, CONH_2, CN, NO_2)$$

Similarly, the reaction of triphenylphosphine with paraformaldehyde in the presence of hydrogen chloride gives rise to hydroxymethyltriphenylphosphonium chloride **(318)**.[92]

$$(C_6H_5)_3P \ + \ CH_2O \ + \ HCl \rightarrow (C_6H_5)_3\overset{\oplus}{P}CH_2OH \quad Cl^{\ominus}$$

**318**

Thorough washing with benzene usually suffices to purify the phosphonium salts. Purer products are obtained by recrystallization from higher alcohols or tetrahydrofuran or by precipitating the salts from their solutions in chloroform, methanol, or other suitable solvents by addition of ether, acetone, or ethyl acetate. Some phosphonium salts are hygroscopic, or they crystallize with the inclusion of solvent molecules and must be dried well under vacuum before they can be used.

## The Wittig Reaction

As a rule, the Wittig reagents are not isolated; immediately following their generation, they are allowed to react with carbonyl compounds in the same reaction vessel.

**The Organometallic Method.** In this method, phenyllithium or *n*-butyllithium is the usual proton acceptor, and diethyl ether or tetrahydrofuran the solvent. For smaller runs, Schlenk tubes (Fig. 1) have proved to be useful; they are sealed and, if necessary, heated in a water

[388] Hoffmann, *Chem. Ber.*, **94**, 1331 (1961).

bath. Larger runs are made in a three-necked flask fitted with stirrer, reflux condenser, addition funnel, and nitrogen inlet tube. If higher reaction temperatures are necessary, the ether must be replaced by a higher-boiling solvent.

The carefully dried and pulverized phosphonium salt is suspended in, for instance, diethyl ether or tetrahydrofuran and an equivalent of a diethyl ether solution of *n*-butyillthium or, preferably, phenyllithium is added under nitrogen. Ylide formation, which usually occurs instantaneously, is evidenced by disappearance of the phosphonium salt with the formation of a red or orange solution.

The ylide solution or suspension so obtained is then treated with the carbonyl compound. As a rule, the phosphonium betaine precipitates immediately. Decomposition of the betaine often occurs in the cold, especially if the newly formed double bond is resonance-stabilized. Heating to 60–70° may be necessary.

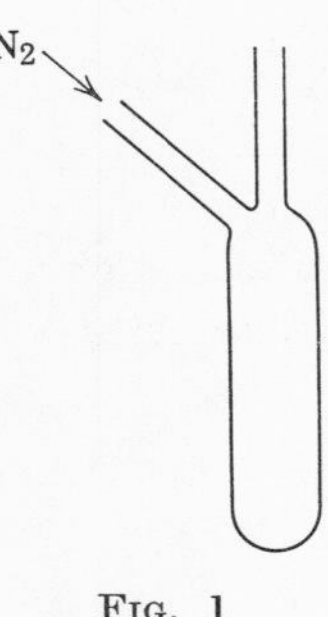


Fig. 1

When diethyl ether is the solvent, the triphenylphosphine oxide precipitates as a crystalline adduct with 1 mole of lithium halide and may be removed by filtration. The filtrate contains the olefin.

Other solvents which keep triphenylphosphine oxide in solution are best removed by distillation after the reaction is complete. The residue is then extracted with petroleum ether in which triphenylphosphine oxide and lithium halides are insoluble.

**The Alkoxide Method.** This method uses the alkali metal alkoxides as proton acceptors and the corresponding alcohols as solvents. Either the phosphonium salt and the carbonyl compound are dissolved in alcohol and then treated with one equivalent of an alcoholic solution of the alkali metal alkoxide, or the components are added successively or simultaneously to the alkoxide solution.

With sparingly soluble carbonyl compounds, it is best to dilute the alcohol with a suitable solvent such as methylene chloride or dimethylformamide. Since base is used in the reaction, its consumption may be followed acidimetrically. Aldehydes and ketones that resinify in the presence of alkoxides should not be subjected to the alkoxide procedure but can readily be used in a modified organometallic method.[389]

The isolation procedure depends on the properties of the resulting olefin. If the olefin is insoluble, it can be isolated by filtration. Triphenylphosphine oxide remains in solution on the addition of up to 40% water and can thus be separated from olefins that are insoluble in

[389] Hauser, Brooks, Miles, Raymond, and Butler, *J. Org. Chem.*, **28**, 372 (1963).

aqueous alcohol. The solvent may be removed by vacuum distillation and the olefin isolated from the residue by distillation, crystallization, chromatography, or extraction with low-boiling petroleum ether.

**Reactions with Resonance-Stabilized Phosphoranes.** Acylmethylene and carbalkoxymethylene triphenylphosphoranes are stable toward water and precipitate as crystalline solids on addition of a base to an aqueous solution of the corresponding phosphonium salt. They may be recrystallized from a mixture of ethyl acetate and petroleum ether. Their reaction with carbonyl compounds may be carried out by heating in a suitable solvent such as benzene or tetrahydrofuran. It is not necessary to isolate the phosphorane; the reaction may be carried out just as well by any of the previously described methods.

## EXPERIMENTAL PROCEDURES

**Methylenecyclohexane.**[2] *Method A (Using n-Butyllithium).* This preparation is described in *Organic Syntheses*, **40,** 66 (1960).

*Method B (Using Sodium Hydride in Dimethyl Sulfoxide).*[63] Sodium hydride (0.10 mole as a 55% dispersion in mineral oil) in a 300-ml. three-necked flask is washed with several portions of *n*-pentane to remove the mineral oil. The flask is then equipped with rubber stoppers, a reflux condenser fitted with a three-way stopcock, and a magnetic stirrer. The system is alternately evacuated and filled with nitrogen; 50 ml. of dimethyl sulfoxide is introduced via syringe, and the mixture is heated at 75–80° for ca. 45 minutes, or until the evolution of hydrogen ceases. The resulting solution of methylsulfinyl carbanion is cooled in an ice-water bath, and 35.7 g. (0.10 mole) of methyltriphenylphosphonium bromide in 100 ml. of warm dimethyl sulfoxide is added. The resulting dark red solution of the ylide is stirred at room temperature for 10 minutes before use.

Freshly distilled cyclohexanone, 10.8 g. (0.11 mole), is added to 0.10 mole of methylenetriphenylphosphorane, and the reaction mixture is stirred at room temperature for 30 minutes and immediately distilled under reduced pressure to give 8.10 g. (86.3%) of methylenecyclohexane, b.p. 42°/105 mm., which is collected in a solid carbon dioxide trap.

**2,2′-Distyrylbiphenyl.**[191] To a suspension of methylenetriphenylphosphorane, prepared from 0.1 mole of triphenylmethylphosphonium bromide and 0.12 mole of phenyllithium in 200 ml. of diethyl ether, is added, under nitrogen, 13.5 g. (37.3 mmoles) of 2,2′-dibenzoylbiphenyl, and the mixture is shaken at room temperature for 3 days. The adduct of triphenylphosphine oxide and lithium bromide is removed by centrifugation and the ethereal solution washed with water. The ether is removed and the residue crystallized twice from ethanol to give 11.4 g. (85.5%) of 2,2′-distyrylbiphenyl, m.p. 100–101°.

***o*-Divinylbenzene.**[52] Triphenylmethylphosphonium bromide (55 mmoles) is added to a solution of 2.34 g. (60 mmoles) of sodium amide in 300 ml. of liquid ammonia. The ammonia is removed, the residue is dissolved in 200 ml. of absolute diethyl ether and heated under reflux for 30 minutes. To the orange-yellow solution of methylenetriphenylphosphorane is added with stirring over a period of 15 minutes a solution of 3.75g. (28 mmoles) of *o*-phthalaldehyde in 100 ml. of absolute diethyl ether. The mixture is then heated under reflux for 2 hours. The filtered solution is concentrated to 50 ml. and again filtered. From the filtrate, 2.7 g. (75%) of *o*-divinylbenzene, b.p. 75–78°/14 mm., $n_D^{20}$ 1.5765, is obtained by distillation.

**4,4′-Divinylbiphenyl.**[187] *A. Biphenyl-4,4′-bis-(methyltriphenylphosphonium chloride).* A mixture of 6.3 g. of 4,4′-bis(chloromethyl)biphenyl, 14.2 g. of triphenylphosphine, and 50 ml. of dimethylformamide is heated with stirring under reflux for 3 hours. The phosphonium salt begins to precipitate after about 20 minutes. After the reaction mixture has cooled, the salt is collected by filtration, washed with absolute benzene, and dried under vacuum at 80°. The yield is 85%.

*B. 4,4′-Divinylbiphenyl.* To a solution of 4 g. of the phosphonium salt in 120 ml. of absolute ethanol are added successively 20 ml. of a 15% solution of formaldehyde in ethanol and 40 ml. of 0.3*M* ethanolic lithium ethoxide. After about 10 minutes the yellow-orange mixture deposits 4,4′-divinylbiphenyl in the form of colorless, shining leaflets. The mixture is diluted with 50 ml. of water and allowed to stand in a refrigerator overnight. The precipitate is collected by filtration and washed with 50% ethanol, giving 4,4′-divinylbiphenyl, m.p. 153°, in 80% yield.

**Benzylidenecyclopentane.**[185] Triphenylcyclopentylphosphonium iodide (18.3 g., 40 mmoles) in 60 ml. of absolute ether is treated under nitrogen with 40 mmoles of *n*-butyllithium. Solution of the salt and formation of the dark red cyclopentylidenetriphenylphosphorane occur within 2 hours. A solution of 4.3 g. (40.5 mmoles) of benzaldehyde in 10 ml. of absolute ether is added dropwise with cooling, and the mixture is then allowed to stand at about 30° for 10 hours. The clear solution is decanted from the triphenylphosphine oxide, shaken three times with 20-ml. portions of 40% aqueous sodium bisulfite solution, and then washed with water until neutral. Removal of the ether from the dried (calcium chloride) solution followed by vacuum distillation gives 4.1 g. (65%) of benzylidenecyclopentane, b.p. 123–124°/17 mm., $n_D^{20}$ 1.5770.

**Lycopene.**[275] To a suspension of 100 g. of geranyltriphenylphosphonium bromide in 1 l. of absolute ether is added with stirring 200 ml. of a 1*N* ethereal phenyllithium solution. After it has been stirred for 1 hour, the dark red solution of geranylidenetriphenylphosphorane is treated within 5 minutes with a solution of 20 g. of crocetin dialdehyde in 500 ml. of anhydrous methylene chloride. The mixture is then stirred for 15 minutes at 30° and heated under reflux for 5 hours. Methanol (600 ml.) is added in one portion to the warm solution, which is then cooled to 10° with stirring. The crystalline mass is collected by filtration under carbon dioxide. The crude lycopene is dissolved in 300 ml. of acid-free methylene chloride at a temperature not higher than 40° and then reprecipitated with 500 ml. of methanol. The yield of lycopene, m.p. 172–173°, is 25 g.

***trans*-4-Nitro-4′-methoxystilbene.**[166] Butyllithium (0.85 g., 0.013 mole) is added under nitrogen to a suspension of 4.3 g. (0.01 mole) of *p*-nitrobenzyltriphenylphosphonium chloride in benzene and the mixture stirred for 2 hours. Anisaldehyde (1.63 g., 0.012 mole) is added, and the mixture is stirred at room temperature for 4 hours and then diluted with petroleum ether. The dark precipitate is collected and crystallized from ethanol, giving 2.23 g. (89%) of *trans*-4-nitro-4′-methoxystilbene in the form of yellow crystals, m.p. 131–132.5°.

**1,4-Diphenyl-1,3-butadiene.**[66] To a solution of sodium ethoxide prepared from 1.5 g. (65.3 mmoles) of sodium and 100 ml. of absolute ethanol is added 18.8 g. (48.3 mmoles) of triphenylbenzylphosphonium chloride. The yellow mixture is stirred vigorously for 5 minutes, and an ethanolic solution of 6.6 g. (50 mmoles) of cinnamaldehyde is added dropwise, resulting in a disappearance of the color. In the course of 30 minutes more at room temperature the mixture assumes a tan coloration. The amount of 1,4-diphenylbutadiene, m.p. 150°, isolated is 8.4 g. (84%).

**1,2-Distyrylbenzene.**[201a] *A. o-Xylylene-bis-(triphenylphosphonium bromide).* A solution of 66.1 g. (0.25 mole) of *o*-xylylene dibromide and 142.5 g. (0.55 mole) of triphenylphosphine in 500 ml. of dimethylformamide is heated under reflux for 3 hours. The colorless salt, which starts precipitating within 10–15 minutes, is collected by filtration after cooling and washed successively with dimethylformamide and ether. The yield of phosphonium salt, m.p. >340°, is 175.9 g. (89%).

*B. 1,2-Distyrylbenzene.* To a solution of 42.5 g. (0.054 mole) of the bis-phosphonium salt and 12.6 g. (0.119 mole) of benzaldehyde in 150 ml. of absolute ethanol is added 500 ml. of an ethanolic 0.4*M* solution of lithium ethoxide. The mixture is allowed to stand at room temperature for 30 minutes and is then heated under reflux for 2 hours. The orange-red solution is concentrated under vacuum to 100 ml., and 300 ml. of water is added; the yellow oil that separates is extracted with ether. The ether is removed, and the residual oil is chromatographed on aluminum oxide. Elution with low-boiling petroleum ether and recrystallization from aqueous ethanol give 12.7 g. (84%) of 1,2-distyrylbenzene in the form of colorless needles, m.p. 117–119°.

***p*-bis-(4-Carbomethoxystyryl)benzene.**[193] To a solution of 54 g. (0.121 mole) of 4-carbomethoxybenzyltriphenylphosphonium chloride and 6 g. (0.0448 mole) of terephthalaldehyde in 600 ml. of ethanol is added 600 ml. of 0.2*M* ethanolic lithium ethoxide. A pale yellow crystalline precipitate forms immediately. Water (600 ml.) is added after 30 minutes, the precipitate is collected by filtration and washed with 200 ml. of 60% ethanol. The yield of *p*-bis-(4-carbomethoxystyryl)benzene, m.p. 290°, is 13 g. (73%).

***p*-bis-(4-Phenylbutadienyl)benzene.** *Method A. From p-*$C_6H_4[CH{=}P(C_6H_5)_3]_2$.[210] A solution of lithium ethoxide prepared from 1.74 g. (0.25 mole) of lithium wire, and 1 *l.* of ethanol is added to a solution of 70 g. (0.10 mole) of *p*-xylylene-bis(triphenylphosphonium chloride) and 35 g. (0.26 mole) of cinnamaldehyde in 250 ml. of ethanol. After the reaction mixture has been left at room temperature for 12 hours, the yellow precipitate is collected by filtration, washed with 300 ml. of 60% ethanol, and dried under vacuum at 70°. The yield is 29–32 g. (87–95%). The crude product may be converted to the all-*trans* isomer with iodine in xylene; after crystallization from dimethylformamide, it melts at 290–293°.

*Method B. From Terephthalaldehyde.*[66] To a solution of sodium ethoxide prepared from 1 g. (43.5 mmoles) of sodium and 120 ml. of absolute ethanol are added successively 19 g. (41.4 mmoles) of triphenylcinnamylphosphonium bromide and a solution of 3 g. (22.4 mmoles) of terephthalaldehyde in 40 ml. of absolute ethanol. The mixture is heated to 80° for 20 minutes, and the solids are collected by filtration and washed with water and a little ethanol. On crystallization from xylene, 6.0 g. (87%) of *p*-bis-(4-phenylbutadienyl)benzene, m.p. 278–279°, is obtained in the form of yellow needles.

**1-Chloro-2,6-dimethyl-1,5-heptadiene.**[390] To a mixture of 346 g. of triphenylchloromethylphosphonium chloride, 86 g. of piperidine (previously dried over sodium), and 400 ml. of diethyl ether a solution of 1 mole of phenyllithium in 900 ml. of diethyl ether is added dropwise at room temperature over a period of 3 hours. After 1 hour, a solution of 126 g. of 2-methylhept-2-en-6-one in 100 ml. of diethyl ether is added over a period of 2½ hours and the mixture is stirred overnight. The crystalline precipitate is collected by filtration and washed with ether. The ether

[390] G. Köbrich and W. E. Breckoff, unpublished research.

and piperidine are removed from the filtrate by distillation, low-boiling petroleum ether is added to the residue, and the triphenylphosphine oxide, which precipitates after a few hours in a refrigerator, is removed. Chromatography on aluminum oxide followed by distillation gives 134 g. (85%) of 1-chloro-2,6-dimethyl-1,5-heptadiene as a colorless liquid, b.p. 69–72°/14 mm., $n_D^{20}$ 1.4687.

**Methoxymethylenecyclohexane.**[246,248] To a suspension of 21.25 g. (62 mmoles) of methoxymethyltriphenylphosphonium chloride in 200 ml. of absolute ether is added under nitrogen at room temperature a solution of 60 mmoles of phenyllithium in ether. The reaction is slightly exothermic, and part of the orange-yellow methoxymethylenetriphenylphosphorane precipitates from the dark red solution. After 10 minutes the mixture is cooled to −30°, and 5.88 g. (60 mmoles) of cyclohexanone in 10 ml. of ether is added dropwise. The triphenylphosphine oxide is removed by filtration after 15 hours at room temperature. Distillation of the filtrate gives 5.4 g. (71%) of methoxymethylenecyclohexane, b.p. 74°/48 mm., $n_D^{20}$ 1.4707.

A small sample is heated in ether with 70% perchloric acid, and the hexahydrobenzaldehyde so obtained is converted to its 2,4-dinitrophenylhydrazone, m.p. 168–169°, in 84% yield.

**2-Benzyl-5-phenyl-2,4-pentadienoic Acid.**[40] *A. Benzylcarbomethoxymethylenetriphenylphosphorane.* Benzyl bromide (1.5 ml.) is added to a solution of 5 g. of carbomethoxymethylenetriphenylphosphorane in ethyl acetate, and the mixture is heated under reflux for 5 hours. The precipitate of triphenylcarbomethoxymethylphosphonium bromide (2.69 g., 87%) is removed by filtration, the filtrate is concentrated under vacuum, and the residue is digested several times with ethyl acetate. The yield of ylide, m.p. 186–187°, is 2.8 g. (75%).

*B. 2-Benzyl-5-phenyl-2,4-pentadienoic Acid.* A mixture of 10 g. of the ylide, 3.23 ml. of cinnamaldehyde, and 400 ml. of ethyl acetate is heated under reflux under nitrogen for 24 hours. The solvent is removed from the extracts, and the residue is heated under reflux with 10% aqueous methanolic potassium hydroxide for 1 hour. After cooling, a ten-fold amount of water is added, and the remaining triphenylphosphine oxide that precipitates is removed by filtration. The filtrate is concentrated to 100 ml., cooled, acidified with sulfuric acid, and extracted with ether. The dienoic acid (5.2 g., 83.3%) is isolated from the dried ether extracts in the form of cream-colored needles, m.p. 180–181°.

**Phenylpropiolic Acid.** *Method A. Via α-Bromocinnamic Acid.*[117] (*a*) Bromocarbomethoxymethylenetriphenylphosphorane. To a vigorously stirred solution of 3.35 g. (10 mmoles) of carbomethoxymethylenetriphenylphosphorane in 50 ml. of benzene is added 0.8 g. (5 mmoles) of bromine. The precipitated phosphonium salt is removed by filtration, and the filtrate is concentrated to dryness. The oily residue solidifies after a short period to give 1.6 g. (77%) of the crystalline bromo ylide.

(*b*) Phenylpropiolic Acid. A mixture of 4.15 g. of the bromo ylide, 1.1 g. of benzaldehyde, and 30 ml. of benzene is heated under reflux for 2 hours. The solvent is removed by distillation, and the residue is heated under reflux with 4 g. of potassium hydroxide in 30 ml. of methanol for 4 hours. The methanol is removed under vacuum, and the residue is extracted with 50 ml. of water. Acidification of the aqueous solution gives a precipitate of 1.2 g. (82% yield) of phenylpropiolic acid in the form of fine colorless needles, m.p. 136–138°.

*Method B. By Intramolecular Wittig Reaction.*[330,331] (*a*) Benzoylcarbomethoxymethylenetriphenylphosphorane. Carbomethoxymethylenetriphenylphosphorane (6.7 g., 20 mmoles) is dissolved in 50 ml. of dry benzene with warming. To the cooled solution is added dropwise with vigorous stirring a solution of 10 mmoles of benzoyl chloride in 10 ml. of benzene. The mixture is left at room temperature for 3 hours, and the precipitated phosphonium salt is removed by filtration. The ylide, m.p. 136–137°, is isolated from the filtrate in 74% yield.

(*b*) Phenylpropiolic Acid. The ylide (1.0 g.) is heated to 220–260° at 0.05 mm. pressure for 15 minutes. The distillate is dissolved in methanol, 3 ml. of concentrated sodium hydroxide solution is added, and the mixture is allowed to stand at room temperature for 24 hours. The methanol is removed under vacuum, and 30 ml. of water is added. The triphenylphosphine oxide is removed by filtration, and the filtrate is acidified with a small amount of concentrated hydrochloric acid, giving phenylpropiolic acid, m.p. 135–136°, in 73% yield.

**Ethyl Cyclohexylideneacetate.**[226] To a suspension of 2.4 g. (0.05 mole) of sodium hydride in 100 ml. of dry glycol dimethyl ether is added dropwise at room temperature 11.2 g. (0.05 mole) of diethyl carbethoxymethylphosphonate. The mixture is stirred for 1 hour until gas evolution stops, and then 4.9 g. (0.05 mole) of cyclohexanone is added dropwise at a rate to keep the temperature below 30°. The mixture is stirred at room temperature for 15 minutes, during which time a viscous oil separates. A large excess of water is added, the ether layer is separated, dried with magnesium sulfate, and distilled under vacuum to give 5.8 g. (70%) of ethyl cyclohexylideneacetate, b.p. 88–90°/10 mm., $n_D^{25}$ 1.4704.

## TABULAR SURVEY

Tables I and II list mainly those phosphonium salts, $[(C_6H_5)_3PR]X$, that have been used as starting materials for the Wittig reaction. The entries are arranged in the order of increasing number of carbon atoms in the alkyl group R. Within each group of salts having the same number of carbon atoms in the alkyl group, the entries follow the Beilstein system.

Tables III, IV, and V also list alkylidene phosphoranes that have not been used in Wittig reactions with carbonyl compounds.

Table VI deals only with those ylides that have been used to prepare olefins. The entries are arranged in the order of increasing number of carbon atoms in the alkylidene portion of the ylide. Under each ylide the olefins prepared from it are listed in the order of increasing number of carbon atoms in the olefin. In this table are also included olefins for whose preparation no details were given in the original reports.

In all the tables, yields are entered only when they were reported in percentages in the original reference.

The literature has been reviewed to January 1, 1963, and some later work has been included. Work reported in patents has been included only when the patent description was detailed. Because of the nature of the topic being reviewed, it is difficult to avoid missing occasional isolated examples of the Wittig reaction. It is hoped that these omissions are few.

TABLE I

PHOSPHONIUM SALTS FROM TRIPHENYLPHOSPHINE AND ALKYL HALIDES

*A. Monophosphonium Salts*

| Carbon Atoms in R | R and X in $[(C_6H_5)_3PR]X$: R | X | Solvent, Time, Temperature, °C. | Yield, % | References |
|---|---|---|---|---|---|
| $C_1$ | $CH_3$ | Br | None, 2 days, 20 | 75 | 2 |
| | | | $C_6H_6$, 4 days, 20 | 89 | 308, 309 |
| | $CHCl_2$ | Br | None, 4 hr., reflux | — | 94 |
| | $CH_2Br$ | Br | None, 18 hr., 60 | 40 | 81 |
| | $CHBr_2$ | Br | None, reflux | 85 | 391 |
| | $CH_2NO_2$ | Br | $C_6H_6$, 0 | — | 41 |
| $C_2$ | $C_2H_5$ | Br | None, 15 hr., 120 | 96 | 178 |
| | | | $C_6H_6$, 20 hr., 135 | 89 | 164 |
| | | | $C_6H_5CH_3$, 1 day, 105 | 90 | 378 |
| | | | $C_6H_6$, 2 days, 20 | — | 307 |
| | $CH_2CH_2Br$ | Br | Xylene, reflux | 49 | 375 |
| | $CH_2CH_2OH$ | Cl | None, 2 hr., reflux | 80 | 392, 393 |
| | $CH_2OCH_3$ | Cl | $C_6H_6$, 60 hr., 50 | 86 | 92, 247 |
| | $CH_2SCH_3$ | Cl | $C_6H_6$, 30 hr., 80 | 77 | 92 |
| | $CH_2CHO$ | Cl | $CHCl_3$, 5 hr., reflux; 12 hr., 20 | — | 33, 58 |
| | $CH_2CONH_2$ | Cl | $C_6H_6$, 8 hr., reflux | — | 243 |
| | | | $CH_3NO_2$, 30 hr., reflux | — | 41 |
| | $CH_2CN$ | Cl | $C_6H_6$, 8 hr., reflux | Quant. | 42, 243 |
| | | | $CH_3NO_2$, 5 hr., reflux | — | 41 |
| | | Br | $C_6H_6$, 2 weeks, 20 | 95 | 43 |
| $C_3$ | $CH(CH_3)_2$ | Br | None, 24 hr., 150 | — | 305 |
| | | I | None, 20 hr., 100 | 94 | 178 |
| | $CH_2CH_2CH_2Br$ | Br | $C_6H_6$, reflux | 63 | 375 |

| | | | | | |
|---|---|---|---|---|---|
| | $CH_2CH{=}CH_2$ | Br | $C_6H_6$, 12 hr., 20; 1 hr., reflux | 92 | 2 |
| | $CH_2CH_2OCH_3$ | Br | $C_6H_6$, 8 hr., reflux | 80 | 393 |
| | $CH_2COCH_3$ | Cl | $CHCl_3$, 45 min., reflux | — | 31 |
| | $CH_2CO_2CH_3$ | Br | $C_6H_6$, 12 hr., 20 | — | 39 |
| $C_4$ | $C_4H_9$-*n* | Br | $CH_3NO_2$, 16 hr., reflux | 79 | 182 |
| | | | $C_6H_6$, 4 hr., reflux | 61 | 130, 375 |
| | $(CH_2)_4Cl$ | Cl | $C_6H_6$, reflux | 6 | 375 |
| | $(CH_2)_4Br$ | Br | $C_6H_6$, 14 hr., 130 | Quant. | 74 |
| | $CH_2CH{=}CHCH_3$ | Br | $C_6H_6$, 48 hr., 20 | 84 | 181 |
| | | | $C_6H_6$, 30 min., reflux | 89 | 328 |
| | $CH_2C(CH_3){=}CH_2$ | Cl | $C_6H_6$, 7 days, reflux | — | 394 |
| | $CH_2CH{=}CHCH_2Br$ | Br | $C_6H_6$, 14 hr., reflux | 84–88 | 230 |
| | $CH_2CH_2OCOCH_3$ | Cl | None, 4 hr., reflux | — | 392 |
| | $CH_2CO_2C_2H_5$ | Cl | $C_6H_6$, 5 hr., reflux | — | 243 |
| | | $Cl \cdot 2H_2O$ | $C_6H_6$, 7 days, 20 | 89 | 395 |
| | | Br | $C_6H_6$, 30 min., reflux | — | 3 |
| | | | $C_6H_6$, 12 hr., 20 | — | 39, 56, 60 |
| | | | None, 5 min., 80 | — | 392 |
| | $CH(CH_3)CO_2CH_3$ | Br | $C_6H_6$, 2 hr., 70 | — | 39, 56, 164, 231 |
| $C_5$ | $CH_2CH_2CH(CH_3)_2$ | Br | Xylene, 15 hr., 160 | 58 | 321 |
| | | I | $C_6H_5CH_3$, 1 day, 115 | — | 378 |
| | $CH_2CH{=}CHC_2H_5$ | Br | $(C_2H_5)_2O$, 2 days, 20 | — | 260 |
| | $CH_2CH{=}C(CH_3)_2$ | Br | $C_6H_6$, 20 hr., 20 | — | 265 |
| | $CH_2OC_4H_9$-*n* | Cl | $C_6H_6$, 15 hr., reflux | 76 | 248 |
| | $(CH_2)_3OC_2H_5$ | I | $C_6H_6$, 3 hr., reflux | — | 138 |
| | $CH_2CH_2CO_2C_2H_5$ | Br | None, 5 min., 100 | — | 392 |
| | $CH(CH_3)CO_2C_2H_5$ | Br | $C_6H_6$, 2 hr., 70 | — | 39 |
| | $CH_2CH{=}CHCO_2CH_3$ | Br | $C_6H_6$, 1 day, 20 | 96 | 232, 234, 235 |

*Note:* References 391 to 404 are on p. 490.

TABLE I—*Continued*

PHOSPHONIUM SALTS FROM TRIPHENYLPHOSPHINE AND ALKYL HALIDES

*A. Monophosphonium Salts—Continued*

| Carbon Atoms in R | R and X in $[(C_6H_5)_3PR]X$ | X | Solvent, Time, Temperature, °C. | Yield, % | References |
|---|---|---|---|---|---|
| $C_5$ (*contd.*) | [cyclopentyl structure] | I | None, 5 hr., 115 | 40 | 185 |
| | [cyclopentadienyl structure] | Br | $CH_2Cl_2$, 6 hr., reflux | — | 46, 47 |
| $C_6$ | $C_6H_{13}$-*n* | Br | — | — | 327 |
| | $CH_2CH{=}CHC_3H_7$-*n* | Br | $C_6H_6$, 20–30 hr., 20 | 96 | 161, 325 |
| | | | $C_6H_6$, 30 min., reflux | 94 | 326 |
| | $CH_2CH_2CH{=}C(CH_3)_2$ | I | $C_6H_6$, 3 hr., reflux | 88 | 182 |
| | $(CH_2)_3C(CH_3){=}CH_2$ | I | $C_6H_6$, 3 hr., reflux | 89 | 182 |
| | $CH_2C{\equiv}CC_3H_7$-*n* | Br | $C_6H_6$, few days, 20 | 95 | 325 |
| | $CH_2(CH{=}CH)_2CH_3$ | Br | $C_6H_6$, 2 days, 20 | 90 | 181 |
| | | | $C_6H_6$, 30 min., reflux | Quant. | 328 |
| | $CH_2CH{=}C(CH_3)C{\equiv}CH$ | Br | $CH_3CO_2C_2H_5$, 1 day, 20 | — | 377 |
| | | | $CH_3CO_2H$ | — | 376 |
| | $CH_2C(CH_3){=}CHCH_2OCH_3$ | Br | — | — | 286*a* |
| | $CH_2C(CH_3){=}CHCO_2CH_3$ | Br | $(C_2H_5)_2O$, 12 hr., 20 | — | 282 |
| | $CH_2CH{=}C(CH_3)CO_2CH_3$ | Br | $C_6H_6$, 1 day, 20 | 77–82 | 232, 235 |
| | $CH(CH_3)CH{=}CHCO_2CH_3$ | Br | $C_6H_6$, 2–3 days, 20 | 71–77 | 233, 234 |
| | [cyclohexyl structure] | I | None, 5 hr., 140; 30 hr., 170 | 72 | 185 |

| | | | | | |
|---|---|---|---|---|---|
| $C_7$ | $(CH_2)_5CH{=}CH_2$ | Br | $(C_2H_5)_2O$, 2 days, 100 | — | 160 |
| | $CH_2CH{=}CHC_4H_9$-*n* | Br | $C_6H_6$ | 70 | 258 |
| | $CH_2CH{=}C(CH_3)CH(OC_2H_5)_2$ | Br | $(C_2H_5)_2O$, 10 hr., 20; 4 hr., reflux | 70 | 216 |
| | $CH_2CH{=}CHC(CH_3){=}CHCH_3$ | Br | THF, 12 hr., 20 | — | 287 |
| | $CH_2C{\equiv}CC(CH_3){=}CHCH_3$ | Br | $C_6H_6$, 2 hr., 40 | 66 | 209 |
| | $CH_2$-(cyclohex-1-enyl) | Br | $C_6H_6$, 1 day, 20 | 59 | 158, 314 |
| | $CH_2C_6H_5$ | Cl | None | Quant. | 375 |
| | $CH_2C_6H_3Cl_2$-2,4 | Cl | DMF, 1 hr., 20; 1 hr., 100 | — | 201 |
| | $CH_2C_6H_4NO_2$-*o* | Cl | $CHCl_3$, 12 hr., 20 | 68 | 220 |
| | | Br | $C_6H_6$, reflux | 79 | 375 |
| | $CH_2C_6H_4NO_2$-*p* | Cl | $C_6H_6$, 2 hr., reflux | 58 | 166 |
| | | Br | $CHCl_3$, few min., 100 | 80 | 220 |
| | | | Xylene, 26 hr., reflux | 95 | 210 |
| | $CH_2OC_6H_5$ | Cl | $(C_2H_5)_2O$ | 80 | 248 |
| $C_8$ | $CH_2CH_2CH_2O$-(tetrahydropyran-2-yl) | Br | $(C_2H_5)_2O$, 36 hr., 100 | — | 71 |
| | $CH_2CH{=}C(CH_3)CH{=}CHCO_2CH_3$ | Br | $CH_3CO_2C_2H_5$, 1 day, 20 | — | 236 |
| | $CH_2CH{=}$(cyclohexylidene) | Br | $C_6H_6$, 12 hr., 20 | — | 183, 184 |
| | | | $C_6H_6$, 2 days, 20 | — | 158 |
| | $CH_2CH{=}$(2-hydroxycyclohexylidene) | Br | THF, 12 hr., 20 | — | 69 |

*Note:* Reference 391 to 404 are on p. 490.

TABLE I—*Continued*

PHOSPHONIUM SALTS FROM TRIPHENYLPHOSPHINE AND ALKYL HALIDES

*A. Monophosphonium Salts—Continued*

| Carbon Atoms in R | R and X in $[(C_6H_5)_3PR]X$ | X | Solvent, Time, Temperature, °C. | Yield, % | References |
|---|---|---|---|---|---|
| $C_8$ (*contd.*) | $CH_2C_6H_4CH_3$-*m* | Br | None | Quant. | 375 |
| | $CH_2C_6H_4CH_3$-*p* | Cl | Xylene, 12 hr., reflux | 85–90 | 195 |
| | | Br | DMF, 3 hr., reflux | 76 | 210 |
| | $CH_2C_6H_4CH_2Br$-*o* | Br | $C_6H_6$, 8 hr., reflux | 92 | 201*a* |
| | $CH_2C_6H_4CH_2Br$-*m* | Br | Xylene, reflux | 38 | 375 |
| | $CH_2C_6H_4CH_2Br$-*p* | Br | Xylene, reflux | 94 | 375 |
| | $CH_2C_6H_4OCH_3$-*p* | Cl | $C_6H_6$, 4 hr., reflux | 62 | 166 |
| | $CH_2OC_6H_4CH_3$-*p* | Cl | $(C_2H_5)_2O$ | 96 | 248 |
| | $CH_2C_6H_3O_2CH_2$-3,4 | Br | DMF, 30 min., 60 | — | 201 |
| | $CH_2C_6H_4CHO$-*p* | Cl | $CH_3NO_2$, 2 days, 80 | 84 | 396 |
| | $CH_2COC_6H_5$ | Br | $CHCl_3$ | — | 31 |
| $C_9$ | $CH_2CH{=}CHC_6H_{13}$-*n* | Br | $C_6H_6$ | 44 | 257 |
| | $CH_2CH{=}CH(CH_2)_4CH{=}CH_2$ | Br | $(C_2H_5)_2O$, several days, 20 | 60 | 160 |
| | $CH_2CH{=}$(cyclohexylidene, 2-$CH_2OH$) | Br | THF/$(C_2H_5)_2O$, 12 hr., 20 | — | 69 |
| | $CH_2CH{=}CHC_6H_5$ | Cl | Xylene, 12 hr., reflux | 91–93 | 210 |
| | | Br | $C_6H_6$, 2 hr., reflux | 82 | 138 |
| | | | None | Quant. | 375 |

| | | | | | |
|---|---|---|---|---|---|
| | 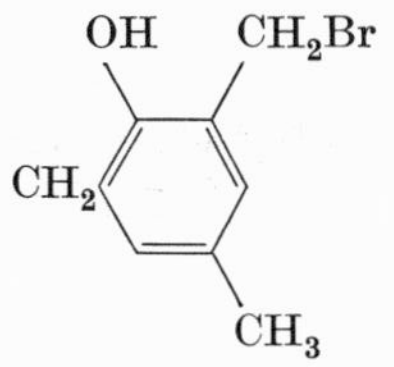 [structure: OH, $CH_2Br$, $CH_2$, $CH_3$] | Br | $C_6H_6$, reflux | 97 | 375 |
| | $CH_2C_6H_3(OCH_3)_2$-2,4 | Br | DMF, 1 hr., 20; 30 min., 60 | — | 201 |
| | $CH_2C_6H_3(OCH_3)(CHO)$-2,4 | Cl | Xylene, 20 hr., reflux | 97 | 208 |
| | $CH(C_6H_5)COCH_3$ | Br | $C_6H_6$, 12 hr., reflux | — | 330 |
| | $CH_2CH_2COC_6H_5$ | Br | $C_6H_6$, 3 hr., reflux | 69 | 203 |
| | $CH_2C_6H_4CO_2CH_3$-*p* | Cl | Xylene, 1 day, reflux | 92 | 193 |
| | | | DMF or $(CH_3)_2SO$, 3 hr., 80 | — | 199, 200 |
| $C_{10}$ | $CH_2CH{=}C(CH_3)CH_2C_5H_{11}$-*i* | Br | $C_6H_6$, 2 days, 20 | — | 276 |
| | $CH_2CH{=}C(CH_3)CH_2CH_2CH{=}C(CH_3)_2$ | Br | $C_6H_6$, 2 days, 20 | — | 275, 276 |
| | | | $C_6H_6$, 6–8 hr., 20 | — | 397 |
| | | | $C_6H_6$, 30 min., reflux | — | 328 |
| | $CH_2CH_2CH_2(C{\equiv}C)_2C_3H_7$-*n* | Br | $(C_2H_5)_2O$, 12 hr., 120 | 90 | 241 |
| | $CH_2CH{=}CH(C{\equiv}C)_2CH{=}CHCH_3$ | Br | $C_6H_6$, 12 hr., 20 | 77 | 259 |
| | 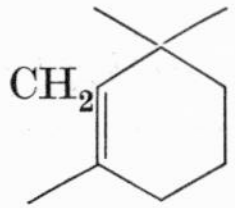 [structure: $CH_2$] | Br | $C_6H_6$, 2 days, 20 | — | 275 |
| | | | $C_6H_6$, 1 hr., 60 | — | 209 |
| | | | $C_6H_6$, 30 min., reflux | Quant. | 328 |
| $C_{11}$ | $CH_2(CH{=}CH)_2C_6H_{13}$-*n* | Br | $(C_2H_5)_2O$, 12 hr., 20 | — | 181 |
| | $CH_2CH{=}CH(C{\equiv}C)_2C_4H_9$-*t* | Br | $(C_2H_5)_2O$, 1 day, 20 | 70 | 180 |
| | $CH_2C_{10}H_7$-$\beta$ | Br | Xylene, reflux | 93 | 197 |
| | $(CH_2)_4COC_6H_5$ | Br | None, 88 hr., 100 | 71 | 202 |

*Note:* References 391 to 404 are on p. 490.

TABLE I—*Continued*

PHOSPHONIUM SALTS FROM TRIPHENYLPHOSPHINE AND ALKYL HALIDES

*A. Monophosphonium Salts—Continued*

| Carbon Atoms in R | R and in $[(C_6H_5)_3PR]X$ | X | Solvent, Time, Temperature, °C. | Yield, % | References |
|---|---|---|---|---|---|
| $C_{12}$ | $C_{12}H_{25}$-*n* | Br | $C_6H_5CN$, 3 hr., 140–150 | — | 159 |
| | 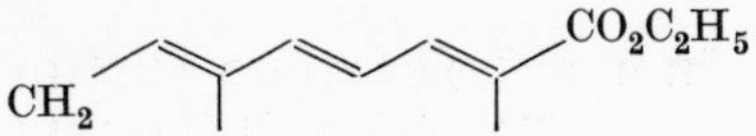 $CH_2$ … $CO_2C_2H_5$ | Br | $CH_3CO_2C_2H_5$, 1 day, 20 | — | 236 |
| | 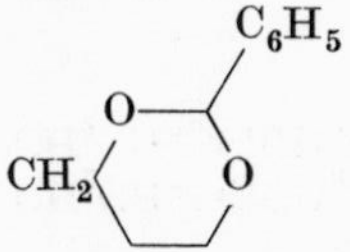 $CH_2$ … O, O, $C_6H_5$ | I | $(C_2H_5)_2O$, 3 days, 100 | 88 | 72 |
| | 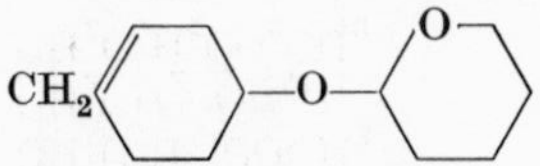 $CH_2$ … O … O | Br | $C_6H_6$, 2 days, 20 | — | 70 |
| $C_{13}$ | 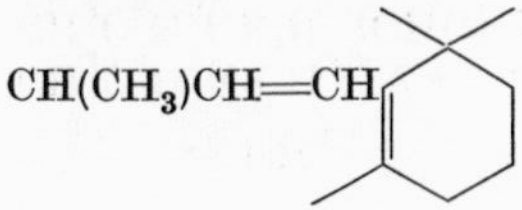 $CH(CH_3)CH{=}CH$ … | Br | THF, 2 hr., reflux | — | 209 |
| | | | $(C_2H_5)_2O$, 5 hr., 20; 30 min., 80–90 | — | 289 |
| | $CH(C_6H_5)_2$ | Br | None, 100 | 50 | 88 |
| | $CH(C_6H_5)C_6H_4NO_2$-*p* | Br | Xylene, 5 hr., reflux | — | 44 |

| | | | | | |
|---|---|---|---|---|---|
| | 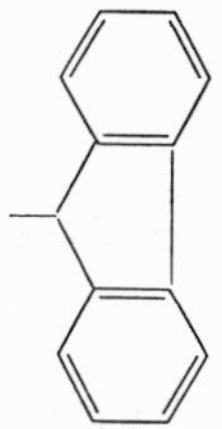 9-fluorenyl (structure) | Br | $C_6H_6$, 1 hr., reflux | 50 | 88 |
| | | | $CH_3NO_2$, 2 hr., 20 | — | 48 |
| | $CH(OC_6H_5)_2$ | Cl | $C_6H_6$, 8 hr., 20 | 82 | 87 |
| $C_{14}$ | $CH{=}C(C_6H_5)_2$ | Br | $(C_2H_5)_2O$* | 7 | 123 |
| $C_{15}$ | $CH_2CH{=}C(CH_3)CH_2CH_2$– 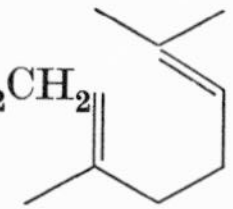 (structure) | Br | $C_6H_6$, 12 hr., reflux | — | 265 |
| | $CH_2CH{=}C(CH_3)CH{=}CH$– 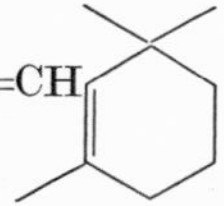 (structure) | Cl | DMF, 12 hr., 20 | — | 283 |
| | | Br | DMF, $C_6H_6$, or $(C_2H_5)_2O$, 1 day, 20 | — | 54, 283, 287 |
| | | | $C_2H_5OH$, 1 day, 20 | — | 54 |
| | $CH_2C_6H_4(CH{=}CHC_6H_5)$-*m* | Br | $C_6H_6$, reflux | 46 | 375 |
| | $CH_2C_6H_4(CH{=}CHC_6H_5)$-*p* | Cl | Xylene, 10 hr., reflux | 80 | 195 |
| | | Br | Xylene, 3–4 hr., reflux | 87–90 | 196 |
| | $CH_2CH{=}C(C_6H_5)_2$ | Br | Xylene, 4 hr., reflux | 93–95 | 211 |

*Note:* References 391 to 401 are on p. 490.

* The halide $(C_6H_5)_2C{=}CHBr$ in ether was converted to the Grignard reagent, THF was added, and the solution was added to triphenylphosphine in ether. Oxygen was bubbled through the reaction mixture which was then decomposed with hydrobromic acid.

TABLE I—*Continued*

PHOSPHONIUM SALTS FROM TRIPHENYLPHOSPHINE AND ALKYL HALIDES

*A. Monophosphonium Salts—Continued*

| Carbon Atoms in R | R and X in $[(C_6H_5)_3PR]X$ | | Solvent, Time, Temperature, °C. | Yield, % | References |
|---|---|---|---|---|---|
| $C_{16}$ | 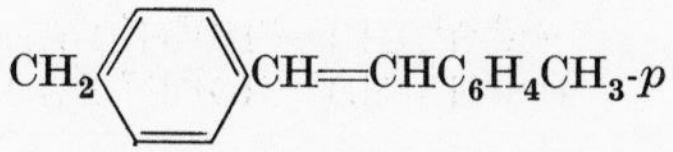 $CH_2C_6H_4CH{=}CHC_6H_4CH_3$-$p$ (structure: $CH_2$–benzene ring–$CH{=}CHC_6H_4CH_3$-$p$) | Br | Xylene, 3–4 hr., reflux | 83–85 | 196 |
| $C_{17}$ | $CH_2C_6H_4(CH{=}CHCH{=}CHC_6H_5)$-$m$ | Br | $C_6H_6$, reflux | 10 | 375 |
| | 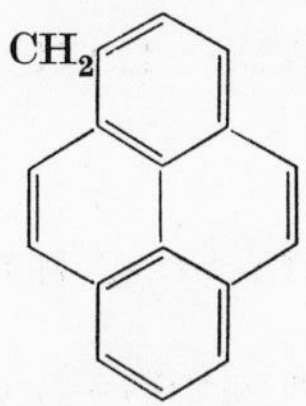 $CH_2$ (pyrenyl structure) | Br | Xylene, 2 hr., reflux | 80 | 197 |
| $C_{18}$ | $CH_2CO_2C_{16}H_{33}$-$n$ | Br | $C_6H_6$, 12 hr., 20 | — | 39 |
| $C_{19}$ | $CH(CH_3)CO_2C_{16}H_{33}$-$n$ | Br | $C_6H_6$, 2 hr., 70 | — | 39 |
| $C_{20}$ | 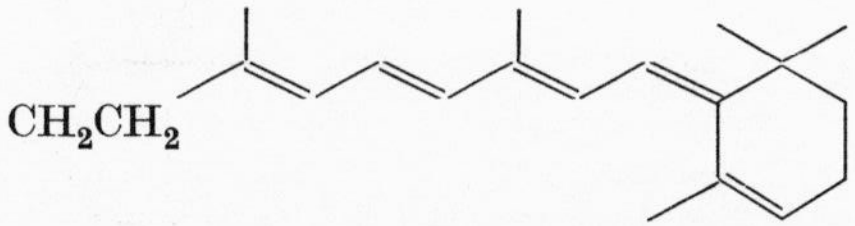 $CH_2CH_2$ (retinyl-type polyene structure) | Br | $C_6H_6$, 4 days, 20 | — | 267 |
| $C_{23}$ | 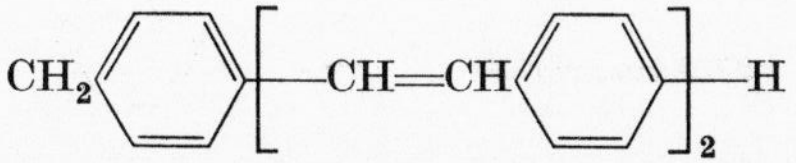 $CH_2C_6H_4[CH{=}CHC_6H_4]_2H$ (structure: $CH_2$–ring–[–$CH{=}CH$–ring–]$_2$–H) | Br | Xylene, 2 hr., reflux | 75–80 | 196 |

*B. Bis-phosphonium Salts*

| Carbon Atoms in R | —R— in $[(C_6H_5)_3P—R—P(C_6H_5)_3]X_2$ | and X | Solvent, Time, Temperature, °C. | Yield, % | References |
|---|---|---|---|---|---|
| $C_1$ | $CH_2$ | Br | None, 150 | 40 | 81 |
| $C_2$ | $CH_2CH_2$ | Br | None, 30 min., 140 | 82 | 52 |
| | | | $C_6H_6$, 4 days, reflux | — | 210 |
| | $CH_2OCH_2$ | Br | — | — | 190 |
| $C_3$ | $CH_2CH_2CH_2$ | Br | None, 30 min., 180 | 77 | 52 |
| | | | DMF, 5 days, 80–90 | 54 | 375 |
| | | | DMF, 3 hr., reflux | 94 | 225 |
| | $CH_2COCH_2$ | Cl | $CHCl_3$, 2 hr., reflux | 8 | 79 |
| $C_4$ | $CH_2CH_2CH_2CH_2$ | Cl | None, 25 hr., 160 | 51 | 185 |
| | | Br | None, 30 min., 250 | 90 | 52, 74, 302 |
| | | | Cyclohexanone, 48 hr., reflux | 77 | 300, 301 |
| | | | $CH_3CN$, 36 hr., 90 | Quant. | 225 |
| | $CH_2CH=CHCH_2$ | Cl | Xylene, 60 hr., reflux | 65 | 210 |
| | | | THF, 16 hr., reflux | 94 | 209 |
| | | Br | None, 1 hr., 250 | 41 | 230 |
| | | | DMF | — | 212 |
| $C_5$ | $CH_2(CH_2)_3CH_2$ | Br | $CH_3CN$, 60 hr., 90 | 54 | 225 |
| $C_6$ | $CH_2(CH_2)_4CH_2$ | Br | $HCO_2H$, 12 hr., 110 | 74 | 225 |
| | $CH_2CO_2CH_2CH_2OCOCH_2$ | Cl | $C_2H_5OH$, 1 day, reflux | 98 | 79 |
| $C_8$ | o-$C_6H_4(CH_2)_2$ (structure: $CH_2$ $CH_2$ on adjacent ring positions of benzene) | Br | DMF, 3 hr., reflux | 89 | 201*a* |

*Note:* References 391 to 404 are on p. 490.

TABLE I—*Continued*

*B. Bis-phosphonium Salts—Continued*

| Carbon Atoms in R | —R— and X in $[(C_6H_5)_3P—R—P(C_6H_5)_3)]X_2$ | X | Solvent, Time, Temperature, °C. | Yield, % | References |
|---|---|---|---|---|---|
| $C_8$ (*contd.*) |  $CH_2$, $CH_2$ | Cl | DMF, 3 hr., reflux | 93–98 | 193 |
| | | Br | DMF, 30 min., 150 | 89 | 375 |
| | | | $C_6H_5NO_2$, 30 min., 150 | 84 | 375 |
| $C_9$ | 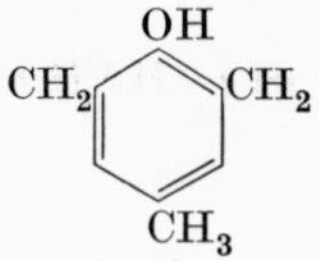 OH, $CH_2$, $CH_2$, $CH_3$ | Br | DMF, 3 hr., 150 | 95 | 375 |
| | | | $C_6H_5NO_2$, 15 min., 150 | 94 | 375 |
| $C_{10}$ | 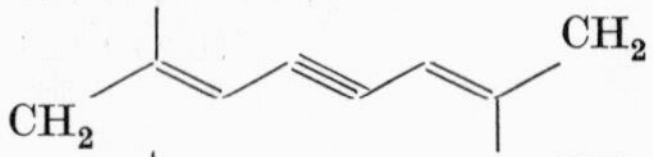 $CH_2$, $CH_2$ | Br | $C_6H_6$, 6 hr., 40–45; 12 hr., 20 | 90 | 80, 271 |
| | 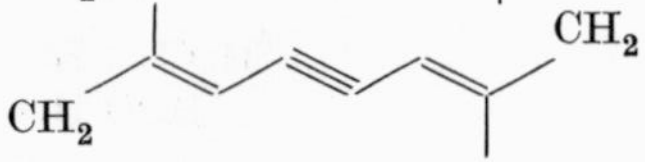 $CH_2$, $CH_2$ | Br | $C_6H_6$, 6 hr., 40–45; 12 hr., 20 | — | 80, 271 |
| | 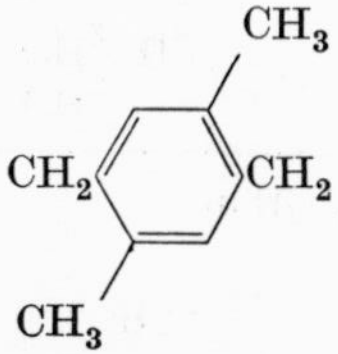 $CH_3$, $CH_2$, $CH_2$, $CH_3$ | Cl | DMF, 2.5 hr., reflux | 94 | 193 |

| | | | | | |
|---|---|---|---|---|---|
| $C_{14}$ | $CH_2C_6H_4$–$C_6H_4CH_2$ | Cl | DMF, 3 hr., reflux | 85 | 187 |
| $C_{16}$ | $CH_2C_6H_4CH{=}CHC_6H_4CH_2$ | Br | DMF, 3 hr., reflux | 75–80 | 187 |
| $C_{24}$ | $CH_2C_6H_4CH{=}CHC_6H_4CH{=}CHC_6H_4CH_2$ | Br | DMF, 16 hr., reflux | 56 | 201*a* |

*Note:* References 391 to 404 are on p. 490.

TABLE II

PHOSPHONIUM SALTS FROM TRIPHENYLPHOSPHONIUM HALIDES, $(C_6H_5)_3PHX$, AND ALCOHOLS OR POLYENES

*A. Monophosphonium Salts*

| Carbon Atoms in R | R and in $[(C_6H_5)_3PR]X$ | X | Solvent, Time, Temperature, °C. | References |
|---|---|---|---|---|
| $C_4$ | $C_4H_9$-*n* | Br | None | 386 |
| $C_{10}$ | $CH_2CH{=}C(CH_3)CH_2CH_2CH{=}C(CH_3)_2$ | Cl | DMF, 1 day, 20 | 274 |
| | | Br | THF, 14 hr., 20 | 368 |
| | | | DMF, 12 hr., 20 | 274 |
| | $CH_2CH{=}C(CH_3)CH{=}CHCH{=}C(CH_3)_2$ | Br | | 279 |
| | $CH_2$ | Br | THF, 1 day, 20 | 368 |
| | | | DMF, 12 hr., 20 | 274 |
| $C_{13}$ | $CH(CH_3)CH{=}CH$ | Cl | $CH_3CN$, 1 hr., 20 | 272, 291 |
| | | | DMF, 15 hr., 20* | 240 |
| | | | DMF, 1 hr., 0*,† | 240 |
| | | Br | Dimethyltetrahydrofuran, 6 hr., 20 | 291 |
| | | | DMF, 12 hr., 20* | 240, 292 |
| | | | $CH_3CN$* | 240 |
| | | | $CH_3CO_2C_2H_5$, 2 days, 20* | 240 |

| | | | | |
|---|---|---|---|---|
| $C_{15}$ | $CH_2CH{=}C(CH_3)CH{=}CH$–(2,6,6-trimethylcyclohex-1-enyl) | Cl | THF, 6 hr., 20 | 285, 387 |
| | | | THF, 30 hr., 5† | 287 |
| | | | $CH_3OH$, 10 hr., 20 | 285 |
| | | | $CH_3OH$, 20 hr., 20† | 264, 286 |
| | | | $CH_3OH$, 18 hr., 20*,† | 284 |
| | | | $C_2H_5OH$, 6 hr., 20 | 268, 285, 294 |
| | | | $CH_3CN$, 12 hr., 20 | 270 |
| | | | $CH_3CN$, 2 days, 20* | 269 |
| | | | DMF, 12 hr., 20 | 270 |
| | | | DMF, 3 hr., 20† | 270 |
| | | | DMF, 12 hr., 10* | 284 |
| | | | DMF, 3 hr., 20*,† | 269 |
| | | Br | None, 15 hr., 20 | 285 |
| | | | CTHF,60 hr., 20 | 387 |
| | | | $DH_3OH$, 36 hr., 20 | 270 |
| | | | MF, 12–24 hr., 20 | 268, 270, 285, 294 |
| | | | DMF, 36 hr., 20* | 269, 284 |
| | $CH_2CH{=}C(CH_3)CH{=}CH$–(2,6,6-trimethylcyclohex-1-enyl) | Br | DMF, 6 hr., 20 | 270 |
| | $CH_2CH{=}C(CH_3)CH{=}CH$–(2,6,6-trimethylcyclohex-2-enyl) | Cl | $CH_3OH$, 3 hr., 40; 50 hr., 20† | 264 |

*Note:* References 391 to 404 are on p. 490.
* The polyene corresponding to the alcohol was used.
† Instead of $(C_6H_5)_3PHX$, triphenylphosphine and the hydrogen halide, HX, were used.

TABLE II—*Continued*

Phosphonium Salts from Triphenylphosphonium Halides, $(C_6H_5)_3PHX$, and Alcohols or Polyenes

*A. Monophosphonium Salts—Continued*

| Carbon Atoms in R | R and in $[(C_6H_5)_3PR]X$ | X | Solvent, Time, Temperature, °C. | References |
|---|---|---|---|---|
| $C_{15}$ (*contd.*) | 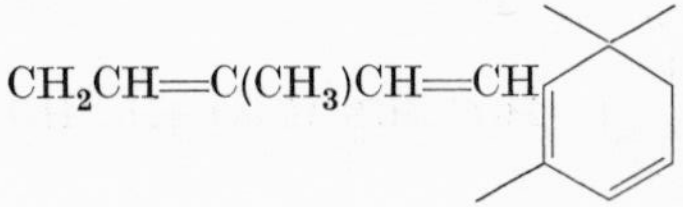 $CH_2CH{=}C(CH_3)CH{=}CH$ | Br | $CH_3OH$, 92 hr., 20 | 288 |
| | 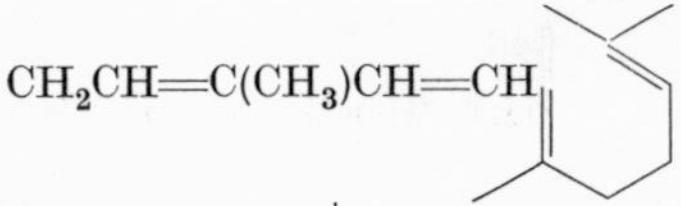 $CH_2CH{=}C(CH_3)CH{=}CH$ | Cl | $CH_3OH$, 60 hr., 20† | 264 |
| $C_{18}$ | 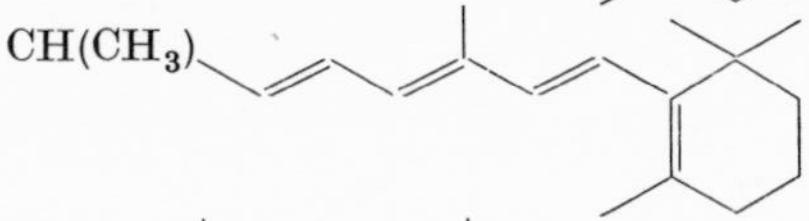 $CH(CH_3)$ | Cl | $CH_3CN$, 2 hr., 20 | 270 |
| $C_{20}$ | 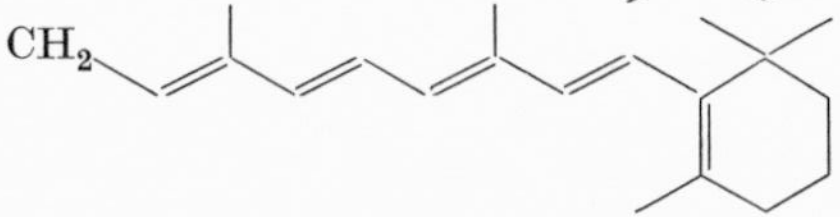 $CH_2$ | Cl | DMF, 8 hr., 0 | 266 |
| | | | DMF, 12 hr., 10* | 266 |
| | | | DMF, 12 hr., 5† | 266 |
| | | | $C_2H_5OH$, 1 day, 20* | 266 |
| | | | $C_2H_5OH$, 1 day, 20† | 266 |

*B. Bis-phosphonium Salt*

| Carbon Atoms in —R— | —R— and in $[(C_6H_5)_3P—R—P(C_6H_5)_3]X_2$ | X | Solvent, Time, Temperature, °C. | References |
|---|---|---|---|---|
| $C_4$ | $CH_2CH{=}CHCH_2$ | Br | $C_6H_6$† | 386 |

*Note:* References 391 to 404 are on p. 490.

* The polyene corresponding to the alcohol was used.

† Instead of $(C_6H_5)_3PHX$, triphenylphosphine and the hydrogen halide, HX, were used.

## TABLE III

### Alkylidene Phosphoranes Prepared from Phosphonium Salts and Bases

### A. *Monophosphoranes*

| Carbon Atoms in Alkylidene Group | Phosphorane (Yield, %) | Base, Solvent | Time, Temperature, °C. | References |
|---|---|---|---|---|
| $C_1$ | $(C_6H_5)_3P{=}CH_2$ | $LiC_6H_5$, $(C_2H_5)_2O$ | 3 hr., 20 | 2, 303, 304*a*, 316 |
| | | | 5 days, 20 | 308, 309 |
| | | | 2 hr., reflux | 182 |
| | | $LiC_4H_9$-*n*, $(C_2H_5)_2O$ | 30 min., 20 | 41, 174 |
| | | | 1 hr., 20 | 304, 398 |
| | | | 2 hr., 20 | 130, 252, 254, 319 |
| | | | 4 hr., 20 | 183, 317 |
| | | $LiC_4H_9$-*n*, THF | 30 min., 20 | 310, 311 |
| | | $NaC(C_6H_5)_3$, $(C_2H_5)_2O$ | | 64 |
| | | $NaNH_2$, liq. $NH_3$ | | 52 |
| | $(C_6H_5)_3P{=}CHCl$ | $LiC_6H_5$, $(C_2H_5)_2O$ | | 96 |
| | $(C_6H_5)_3P{=}CHBr$ | Lithium piperidide, $(C_2H_5)_2O$ | | 84 |
| $C_2$ | $(C_6H_5)_3P{=}CHCH_3$ | $LiC_6H_5$, $(C_2H_5)_2O$ | 1 hr., 20 | 164 |
| | | | 3 hr., 20 | 178, 307 |
| | | $LiC_4H_9$-*n*, $(C_2H_5)_2O$ | 1 hr., 0 | 327 |
| | | | 1 hr., 20 | 41, 306 |
| | | $NaC(C_6H_5)_3$, $(C_2H_5)_2O$ | | 64 |
| | | $Na(CH_2SOCH_3)$, DMSO | | 62 |
| | $(C_6H_5)_3P{=}CHOCH_3$ | $LiC_6H_5$, $(C_2H_5)_2O$ | 10 min., 20 | 92, 246–248 |
| | | $LiC_6H_5$, THF | | 92 |
| | | $LiC_4H_9$-*n*, $(C_2H_5)_2O$ | 2 hr., −50 | 92 |
| | | $NaOCH_3$, $CH_3OH$ | | 92 |
| | | $NaOC_2H_5$, $C_2H_5OH$ | | 92, 247, 248 |
| | $(C_6H_5)_3P{=}CHSCH_3$ | $LiC_6H_5$, $(C_2H_5)_2O$ | | 92 |
| | | $LiC_6H_5$, THF | 15 min., 20 | 92 |

*Note:* References 391 to 404 are on p. 490.

TABLE III—*Continued*

ALKYLIDENE PHOSPHORANES PREPARED FROM PHOSPHONIUM SALTS AND BASES

*A. Monophosphoranes—Continued*

| Carbon Atoms in Alkylidene Group | Phosphorane (Yield, %) | Base, Solvent | Time, Temperature, °C. | References |
|---|---|---|---|---|
| $C_2$ (*contd.*) | $(C_6H_5)_3P{=}CHCHO$ | $NaOH$, $H_2O$ | | 33, 58 |
| | | $(C_2H_5)_3N$, $C_2H_5OH$ | | 33, 58 |
| | $(C_6H_5)_3P{=}CHCONH_2$ | $NaOH$, $H_2O$ | 0 | 41 |
| | | $NaOC_2H_5$, $C_2H_5OH$ | 10 min., 40–50 | 243 |
| | $(C_6H_5)_3P{=}CHCN$ (85) | $NaOH$, $H_2O$ | 1 hr., 20 | 41–43 |
| | | $NaOC_2H_5$, $C_2H_5OH$ | 30 min., 20 | 113, 243 |
| $C_3$ | $(C_6H_5)_3P{=}CHC_2H_5$ | $LiC_4H_9$-*n*, $(C_2H_5)_2O$ | | 113 |
| | $(C_6H_5)_3P{=}C(CH_3)_2$ | $LiC_6H_5$, $(C_2H_5)_2O$ | 2 hr., 20 | 178 |
| | | $LiC_4H_9$-*n*, $(C_2H_5)_2O$ | 12 hr., 20 | 305 |
| | | | 2 days, 20 | 150 |
| | | $NaC(C_6H_5)_3$, $(C_2H_5)_2O$ | | 64 |
| | $(C_6H_5)_3P{=}CHCH{=}CH_2$ | $LiC_6H_5$, $(C_2H_5)_2O$ | 2–3 hr., 20 | 2 |
| | | $LiC_4H_9$-*n*, $(C_2H_5)_2O$ | 2 hr., 20 | 259, 318 |
| | $(C_6H_5)_3P{=}CHCH_2CH_2O^{\ominus}$ | $LiC_4H_9$-*n*, $(C_2H_5)_2O$ | | 16 |
| | $(C_6H_5)_3P{=}CHCH_2OCH_3$ | $LiC_6H_5$, $(C_2H_5)_2O$ | | 281 |
| | $(C_6H_5)_3P{=}C(CH_3)CHO$ | $NaOH$, $H_2O$ | | 33, 58 |
| | $(C_6H_5)_3P{=}CHCOCH_3$ (70) | $NaOH$, $H_2O$ | | 60 |
| | | $Na_2CO_3$, $H_2O$ | 8 hr., 20 | 31 |
| | $(C_6H_5)_3P{=}CClCOCH_3$ (46) | Pyridine, $CH_2Cl_2$ | −70 | 60 |
| | $(C_6H_5)_3P{=}CHCO_2CH_3$ (65) | $NaOH$, $H_2O$ | | 39, 56, 297 |
| | | $NaOCH_3$, $CH_2Cl_2$ | 30 min., 20 | 39, 231, 295, 297 |
| $C_4$ | $(C_6H_5)_3P{=}CHC_3H_7$-*n* | $LiC_6H_5$, $(C_2H_5)_2O$ | 2 hr., reflux | 182 |
| | | $LiC_4H_9$-*n*, $(C_2H_5)_2O$ | 2 hr., 20 | 130 |
| | | $LiC_4H_9$-*n*, pet. ether | 1 hr., 20 | 161, 327 |
| | $(C_6H_5)_3P{=}CH(CH_2)_3Br$ | $LiC_6H_5$, $(C_2H_5)_2O$ | | 74 |
| | | $NaOC_2H_5$, $C_2H_5OH$ | | 74 |
| | $(C_6H_5)_3P{=}CHCH{=}CHCH_3$ | $LiC_4H_9$-*n*, $(C_2H_5)_2$ | 10 min. to 3 hr., 20 | 71, 181, 328, 399 |

| | | | | |
|---|---|---|---|---|
| | $(C_6H_5)_3P{=}CHC(CH_3){=}CH_2$ | $LiC_6H_5$, THF | 1 hr., 20 | 394 |
| | $(C_6H_5)_3P{=}CHCO_2C_2H_5$ (75) | NaOH, $H_2O$ | | 39, 56 |
| | | $NaOCH_3$, $CH_2Cl_2$ | 30 min., 20 | 39 |
| | | $NaOC_2H_5$, $C_2H_5OH$ | 10 min., 20 | 3, 34, 243 |
| | $(C_6H_5)_3P{=}C(CH_3)CO_2CH_3$ (60–70) | NaOH, $H_2O$ | | 39, 56, 297 |
| | (44) | KOH, $H_2O$ | | 164 |
| | | $NaOCH_3$, $CH_2Cl_2$ | 30 min., 20 | 39, 231, 295–297 |
| | $(C_6H_5)_3P{=}CClCO_2C_2H_5$ (80) | $(C_2H_5)_3N$, $CH_2Cl_2$ | −70 | 60 |
| | $(C_6H_5)_3P{=}CBrCO_2C_2H_5$ (69) | NaOH, $H_2O$ | | 60 |
| $C_5$ | $(C_6H_5)_3P{=}CHCH_2C_3H_7$-*i* | $LiC_4H_9$-*n*, $(C_2H_5)_2O$ | | 378 |
| | | | 30 min., 20 | 320 |
| | | | 2 hr., 20 | 321 |
| | $(C_6H_5)_3P{=}CHCH{=}CHC_2H_5$ | $LiC_4H_9$-*n*, $(C_2H_5)_2O$ | 30 min., 20 | 260 |
| | $(C_6H_5)_3P{=}CHCH{=}C(CH_3)_2$ | $NaOC_2H_5$, $CH_2Cl_2$ | 20 | 265 |
| | $(C_6H_5)_3P{=}CHOC_4H_9$-*n* | $LiC_4H_9$-*n*, $C_6H_6$ | 10 min., −10 | 248 |
| | | $LiC_6H_5$, $(C_2H_5)_2O$ | 10 min., −40 | 248 |
| | | $NaOC_2H_5$, $C_2H_5OH$ | | 248 |
| | $(C_6H_5)_3P{=}CHCH_2CH_2OC_2H_5$ | $LiC_6H_5$, $(C_2H_5)_2O$ | 4 hr., 20 | 138 |
| | $(C_6H_5)_3P{=}C(CH_3)CO_2C_2H_5$ (75–80) | NaOH, $H_2O$ | | 39, 56 |
| | | $NaOCH_3$, $CH_2Cl_2$ | 30 min., 20 | 39 |
| | $(C_6H_5)_3P{=}CHCH{=}CHCO_2CH_3$ (70) | NaOH, $H_2O$ | | 232, 234, 235 |
| | | $NaOCH_3$, $CH_3OH$ | 15 min., 20 | 242 |
| | $(C_6H_5)_3P{=}C_5H_8$ | $LiC_4H_9$-*n*, $(C_2H_5)_2O$ | 2 hr., 20 | 185 |
| | $(C_6H_5)_3P{=}C_5H_4$ (41) | NaOH, $H_2O$ | | 46, 47 |
| $C_6$ | $(C_6H_5)_3P{=}CHC_5H_{11}$-*n* | $LiC_4H_9$-*n*, $(C_2H_5)_2O$ | 1 hr., 0 | 327 |
| | $(C_6H_5)_3P{=}CHCH{=}CHC_3H_7$-*n* | $NaOC_2H_5$, $C_2H_5OH$ | | 325 |
| | | $LiC_4H_9$-*n*, $(C_2H_5)_2O$ | 30 min., 0 | 161 |
| | | | 2–3 hr., 20 | 326 |
| | | $LiC_4H_9$-*n*, pet. ether | 1–3 hr., 20 | 326 |
| | $(C_6H_5)_3P{=}CHCH_2CH{=}C(CH_3)_2$ | $LiC_6H_5$, $(C_2H_5)_2O$ | 2 hr., reflux | 182 |
| | $(C_6H_5)_3P{=}CH(CH_2)_2C(CH_3){=}CH_2$ | $LiC_6H_5$, $(C_2H_5)_2O$ | 2 hr., reflux | 182 |

*Note:* References 391 to 404 are on p. 490.

TABLE III—*Continued*

ALKYLIDENE PHOSPHORANES PREPARED FROM PHOSPHONIUM SALTS AND BASES

*A. Monophosphoranes—Continued*

| Carbon Atoms in Alkylidene Group | Phosphorane (Yield, %) | Base, Solvent | Time, Temperature °C. | References |
|---|---|---|---|---|
| $C_6$ (*contd.*) | $(C_6H_5)_3P{=}CHC{\equiv}CC_3H_7$-*n* | $NaOC_2H_5$, $C_2H_5OH$ | | 325 |
| | $(C_6H_5)_3P{=}CH(CH{=}CH)_2CH_3$ | $LiC_4H_9$-*n*, $(C_2H_5)_2O$ | 30 min., 20 | 181 |
| | | | 3 hr., 20 | 328 |
| | $(C_6H_5)_3P{=}CHCH{=}C(CH_3)C{\equiv}CH$ | $LiC_6H_5$, $(C_2H_5)_2O$ | 1 hr., 20 | 377 |
| | | $LiOC_2H_5$, $(C_2H_5)_2O$ | 3 hr., 20 | 376 |
| | $(C_6H_5)_3P{=}CHC(CH_3){=}CHCH_2OCH_3$ | $LiC_6H_5$, $C_6H_6$ | | 286*a* |
| | $(C_6H_5)_3P{=}CHC(CH_3){=}CHCO_2CH_3$ | $NaOC_2H_5$, $C_2H_5OH$ | 4 min., 20 | 282 |
| | | $NaC{\equiv}CH$, DMF | 2 hr., 20 | 55 |
| | | $NaNH_2$, $C_6H_6$ | 1 day, 20 | 55 |
| | $(C_6H_5)_3P{=}CHCH{=}C(CH_3)CO_2CH_3$ (67) | NaOH, $H_2O$ | | 232, 235 |
| | | $NaOCH_3$, $CH_2Cl_2$ | | 297 |
| | $(C_6H_5)_3P{=}C(CH_3)CH{=}CHCO_2CH_3$ (67) | NaOH, $H_2O$ | | 233, 234 |
| | $(C_6H_5)_3P{=}$(cyclohexylidene) | $LiC_4H_9$-*n*, $(C_2H_5)_2O$ | 5–6 hr., 20 | 113, 185 |
| $C_7$ | $(C_6H_5)_3P{=}CHC_6H_{13}$-*n* | $LiC_4H_9$-*n*, $(C_2H_5)_2O$ | | 188, 189 |
| | $(C_6H_5)_3P{=}CH(CH_2)_4CH{=}CH_2$ | $LiC_4H_9$-*n*, $(C_2H_5)_2O$ | 1 hr., 20 | 160 |
| | $(C_6H_5)_3P{=}CHCH{=}CHC_4H_9$-*n* | $LiC_4H_9$-*n*, $(C_2H_5)_2O$ | 2 hr., 20 | 258 |
| | $(C_6H_5)_3P{=}CHCH{=}CHC(CH_3){=}CHCH_3$ | $LiC_4H_9$-*n*, $(C_2H_5)_2O$ | 4 hr., 20 | 287 |
| | | $NaNH_2$, $C_6H_6$ | 50 hr., 20 | 287 |
| | | $NaOCH_3$, DMF | | 287 |
| | $(C_6H_5)_3P{=}CHC{\equiv}CC(CH_3){=}CHCH_3$ | $LiC_6H_5$, $(C_2H_5)_2O$ | | 209 |
| | $(C_6H_5)_3P{=}CHCH{=}C(CH_3)CH(OC_2H_5)_2$ | $NaOCH_3$, $CH_3OH$ | | 216 |
| | $(C_6H_5)_3P{=}CH$(cyclohex-1-enyl) | $LiC_4H_9$-*n*, $(C_2H_5)_2O$ | 2 hr., 20 | 314 |
| | | $LiC_6H_5$, $(C_2H_5)_2O$ | 2 hr., 20 | 158 |

| | Ylide | Base, Solvent | Time, Temp. | Refs. |
|---|---|---|---|---|
| | $(C_6H_5)_3P{=}CHC_6H_5$ | $LiC_4H_9$-*n*, $(C_2H_5)_2O$ | 30 min., 20 | 41, 213 |
| | | $LiC_6H_5$, $(C_2H_5)_2O$ | 20 | 2, 3, 194, 209 |
| | | $LiC_6H_5$, THF | | 209 |
| | | $C_2H_5MgBr$, THF | 3 hr., 20 | 209 |
| | | Na, $(C_2H_5)_2O$ | 2 days, reflux | 61 |
| | | $NaNH_2$, $C_6H_6$ | 16 hr., 20 | 53 |
| | | $NaNH_2$, $NH_3$ liq. | 1 day, −50 | 53 |
| | | $NaC{\equiv}CH$, DMF | | 113 |
| | | $NaOCH_3$, DMF | | 201 |
| | | $LiOC_2H_5$, $C_2H_5OH$ | 20 | 196, 198 |
| | | $NaOC_2H_5$, $C_2H_5OH$ | 10 min., 20 | 3, 66, 243 |
| | $(C_6H_5)_3P{=}CHC_6H_4Cl$-*p* | $NaNH_2$, $C_6H_6$ | 2 days, 20 | 113 |
| | $(C_6H_5)_3P{=}CHC_6H_3Cl_2$-2,4 | $NaOCH_3$, DMF | | 201 |
| | $(C_6H_5)_3P{=}CHC_6H_4NO_2$-*o* | NaOH, $C_2H_5OH/H_2O$ | 0 | 220 |
| | | $Na_2CO_3$, $C_2H_5OH/H_2O$ | | 220 |
| | $(C_6H_5)_3P{=}CHC_6H_4NO_2$-*m* | $LiOC_2H_5$, $C_2H_5OH$ | | 210 |
| | $(C_6H_5)_3P{=}CHC_6H_4NO_2$-*p* | $LiC_4H_9$-*n*, $C_6H_6$ | 2 hr., 20 | 166 |
| | | $NaOC_2H_5$, $C_2H_5OH$ | | 66 |
| | | NaOH, $C_2H_5OH/H_2O$ | 0 | 220 |
| | | $Na_2CO_3$, $C_2H_5OH/H_2O$ | | 220 |
| | $(C_6H_5)_3P{=}CHOC_6H_5$ | $LiC_6H_5$, $(C_2H_5)_2O$ | 20 min., 20 | 248 |
| $C_8$ | $(C_6H_5)_3P{=}CHCH_2CH_2O$-(tetrahydropyran-2-yl) | $LiC_4H_9$-*n*, $(C_2H_5)_2O$ | 10 min., 20 | 71 |
| | $(C_6H_5)_3P{=}C(C_3H_7\text{-}n)COC_3H_7$-*n* | NaOH, $H_2O$ | 0 | 330 |
| | $(C_6H_5)_3P{=}CHCH{=}C(CH_3)CH{=}CHCO_2CH_3$ | $NaOCH_3$, $CH_2Cl_2$ | 1 hr., 20 | 236, 297 |
| | $(C_6H_5)_3P{=}CHC{\equiv}CC(CH_3){=}CHCH_2OCH_3$ | $LiC_6H_5$, $(C_2H_5)_2O$ | 10 min., 20 | 290 |
| | $(C_6H_5)_3P{=}CHCH{=}$(cyclohexylidene) | $LiC_4H_9$-*n*, $(C_2H_5)_2O$ | 12 hr., 20 | 183, 184 |
| | | $LiC_6H_5$., $(C_2H_5)_2O$ | 15 hr., 20 | 158 |
| | $(C_6H_5)_3P{=}CHCH{=}$(2-$O^{\ominus}$-cyclohexylidene) | $LiC_4H_9$-*n*, THF/$(C_2H_5)_2O$ | −65 | 69 |

*Note:* References 391 to 404 are on p. 490.

## TABLE III—*Continued*

### Alkylidene Phosphoranes Prepared from Phosphonium Salts and Bases

#### *A. Monophosphoranes—Continued*

| Carbon Atoms in Alkylidene Group | Phosphorane (Yield, %) | Base, Solvent | Time, Temperature °C. | References |
|---|---|---|---|---|
| $C_8$ (*contd.*) | $(C_6H_5)_3P{=}CHC_6H_4CH_3$-*m* | $NaOC_2H_5$, $C_2H_5OH$ | | 66 |
| | $(C_6H_5)_3P{=}CHC_6H_4CH_3$-*p* | $LiOC_2H_5$, $C_2H_5OH$ | 20 | 195, 196 |
| | $(C_6H_5)_3P{=}CHC_6H_4OCH_3$-*p* | $LiC_4H_9$-*n*, $C_6H_6$ | | 166 |
| | | $C_2H_5MgBr$, THF | 5 hr., 20 | 113 |
| | $(C_6H_5)_3P{=}CHOC_6H_4CH_3$-*p* | $LiC_6H_5$, $(C_2H_5)_2O$ | 20 min., 20 | 248 |
| | | $KOC_4H_9$-*t*, *t*-$C_4H_9OH$ | 2 hr., 65 | 248 |
| | $(C_6H_5)_3P{=}CHC_6H_3O_2CH_2$-3,4 | $NaOCH_3$, DMF | | 201 |
| | $(C_6H_5)_3P{=}CHC_6H_4CHO$-*p* | $NaOC_2H_5$, $C_2H_5OH$ | | 396 |
| | $(C_6H_5)_3P{=}CHCOC_6H_5$ (87) | NaOH, $H_2O$ | | 60 |
| | | $Na_2CO_3$, $H_2O$ | 15 hr., 20 | 31 |
| | $(C_6H_5)_3P{=}CClCOC_6H_5$ (87) | NaOH, $H_2O$ | | 60 |
| | (92) | Pyridine, $CH_2Cl_2$ | 30 min., −70 | 60 |
| | $(C_6H_5)_3P{=}CBrCOC_6H_5$ (84) | KOH, $CH_3OH/H_2O$ | | 60 |
| $C_9$ | $(C_6H_5)_3P{=}CHCH{=}CHC_6H_{13}$-*n* | $LiC_4H_9$, $(C_2H_5)_2O$ | 2 hr., 20 | 257 |
| | $(C_6H_5)_3P{=}CHCH{=}CH(CH_2)_4CH{=}CH_2$ | $LiC_4H_9$-*n*, $(C_2H_5)_2O$ | 10 min., 20 | 160 |
| | $(C_6H_5)_3P{=}CHC{\equiv}CC(CH_3){=}CHCH_2OCOCH_3$ | $NaOC_2H_5$, DMF | 20 | 290 |
| | $(C_6H_5)_3P{=}CHCH{=}C_6H_9(CH_2O^{\ominus})$-2 | $LiC_4H_9$-*n*, $THF/(C_2H_5)_2O$ | −25 | 69 |
| | $(C_6H_5)_3P{=}CHCH{=}CHC_6H_5$ | $LiC_4H_9$, $(C_2H_5)_2O$ | 30 min., 20 | 213 |
| | | $LiC_6H_5$, $(C_2H_5)_2O$ | | 138 |
| | | $LiOC_2H_5$, $C_2H_5OH$ | | 210, 211 |
| | | $NaOC_2H_5$, $C_2H_5OH$ | | 66 |
| | $(C_6H_5)_3P{=}CHC_6H_3(OCH_3)_2$-2,4 | $NaOCH_3$, DMF | | 201 |
| | $(C_6H_5)_3P{=}CHC_6H_3(OCH_3)(CHO)$-2,4 | $LiOC_2H_5$, $C_2H_5OH$ | | 208 |
| | $(C_6H_5)_3P{=}C(C_6H_5)COCH_3$ | NaOH, $H_2O$ | | 58, 330 |
| | $(C_6H_5)_3P{=}CHCH_2COC_6H_5$ | $LiC_6H_5$, $(C_2H_5)_2O/C_6H_6$ | | 203 |

| | | | | |
|---|---|---|---|---|
| | $(C_6H_5)_3P{=}CHC_6H_4CO_2CH_3$-*p* | $LiOC_2H_5$, $C_2H_5OH$ | | 193, 210 |
| | | $NaOCH_3$, DMF or $(CH_3)_2SO$ | | 199, 200 |
| | $(C_6H_5)_3P{=}CHC_6H_4N(CH_3)_2$-*p* | $LiOC_2H_5$, $C_2H_5OH$ | | 210 |
| $C_{10}$ | $(C_6H_5)_3P{=}CH$ | $LiC_6H_5$, $(C_2H_5)_2O$ | | 276, 277 |
| | $(C_6H_5)_3P{=}CH$ | $LiC_6H_5$, $(C_2H_5)_2O$ | 1–2 hr., 20 | 264, 275, 328 |
| | | $NaOCH_3$, DMF | 10 | 274, 368 |
| | $(C_6H_5)_3P{=}CH$ | $NaOCH_3$, $CH_3OH$ | | 279 |
| | OH<br>$(C_6H_5)_3P{=}CH$ | $NaOCH_3$, $CH_3OH$ | | 279 |
| | $(C_6H_5)_3P{=}CHCH_2CH_2(C{\equiv}C)_2C_3H_7$-*n* | $LiC_4H_9$-*n*, $(C_2H_5)_2O$ | 1 hr., 20 | 241 |
| | $(C_6H_5)_3P{=}CHCH{=}CH(C{\equiv}C)_2CH{=}CHCH_3$ | $LiC_4H_9$-*n*, $(C_2H_5)_2O$ | | 259 |
| | $(C_6H_5)_3P{=}CH$ | $LiCH_3$, THF | 1 hr., 20 | 209 |
| | | $LiC_4H_9$-*n* | 3 hr., 20 | 328 |
| | | $LiC_6H_5$, $(C_2H_5)_2O$ | 1 hr., 20 | 209, 275, 293 |
| | | $LiC_6H_5$, THF | | 368 |
| | | $NaC{\equiv}CH$, DMF | 1 hr., 20 | 113, 293 |
| | | $NaOCH_3$, DMF | | 274 |
| | | $NaOC_2H_5$, DMF | 20 | 293 |
| $C_{11}$ | $(C_6H_5)_3P{=}CH(CH{=}CH)_2C_6H_{13}$-*n* | $LiC_4H_9$-*n*, $(C_2H_5)_2O$ | 30 min., 20 | 181 |
| | $(C_6H_5)_3P{=}CHCH{=}CH(C{\equiv}C)_2C_4H_9$-*t* | $LiC_4H_9$-*n*, $(C_2H_5)_2O$ | 15 min., 20 | 180 |
| | $(C_6H_5)_3P{=}CHCH{=}$ | $LiC_6H_5$, $(C_2H_5)_2O$ | | 214 |
| | $(C_6H_5)_3P{=}CHC_{10}H_7$-$\beta$ | $NaOC_2H_5$, $C_2H_5OH$ | 20 | 197 |

*Note:* References 391 to 404 are on p. 490.

## TABLE III—*Continued*

### Alkylidene Phosphoranes Prepared from Phosphonium Salts and Bases

#### *A. Monophosphoranes—Continued*

| Carbon Atoms in Alkylidene Group | Phosphorane (Yield, %) | Base, Solvent | Time, Temperature, °C. | References |
|---|---|---|---|---|
| $C_{11}$ (*contd.*) | 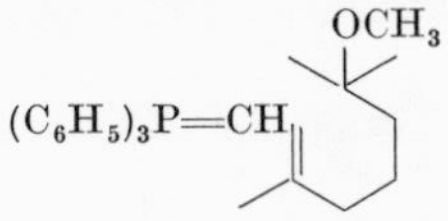 $(C_6H_5)_3P{=}CH$ (structure, $OCH_3$) | $NaOCH_3$, $CH_3OH$ | | 278 |
| | 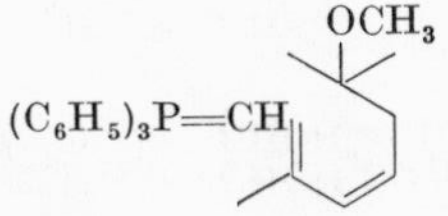 $(C_6H_5)_3P{=}CH$ (structure, $OCH_3$) | $NaOCH_3$, $CH_3OH$ | | 278 |
| | $(C_6H_5)_3P{=}CH(CH_2)_3COC_6H_5$ | $NaOC_2H_5$, $C_2H_5OH$ | | 202 |
| | $(C_6H_5)_3P{=}CH(CH_2)_7CO_2C_2H_5$ | $NaNH_2$, THF | 20 | 170, 171 |
| | | $NaOC_2H_5$, THF or DMF | | 170, 171 |
| | 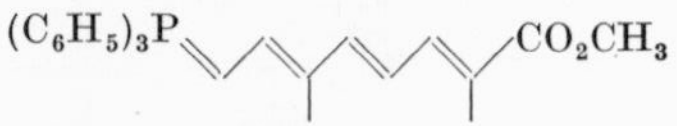 $(C_6H_5)_3P$ (structure) $CO_2CH_3$ | $NaOCH_3$, $CH_2Cl_2$ | | 297, 298 |
| $C_{12}$ | $(C_6H_5)_3P{=}CHC_{11}H_{23}$-*n* | $LiC_4H_9$-*n*, $(C_2H_5)_2O$ | | 159 |
| | | $LiC_6H_5$, THF/$(C_2H_5)_2O$ | 20 | 113 |
| | 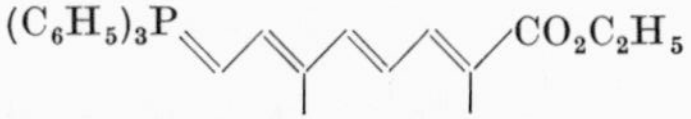 $(C_6H_5)_3P$ (structure) $CO_2C_2H_5$ | $NaOCH_3$, $CH_2Cl_2$ | 1 hr., 20 | 236 |
| | 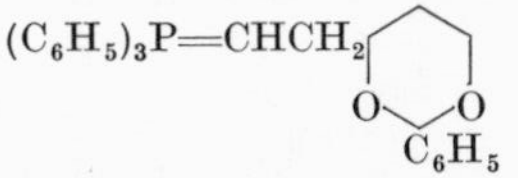 $(C_6H_5)_3P{=}CHCH_2$ (structure, O, O, $C_6H_5$) | $LiC_4H_9$-*n*, $(C_2H_5)_2O$ | 15 min., 20 | 72 |
| | 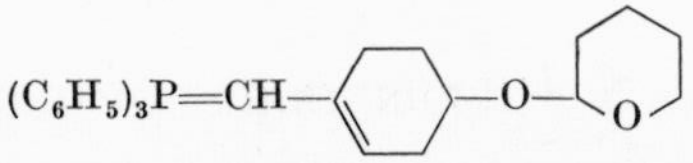 $(C_6H_5)_3P{=}CH$— (structure, O, O) | $LiC_6H_5$, $(C_2H_5)_2O$ | 17 hr., 20 | 70 |

| | | | | |
|---|---|---|---|---|
| $C_{13}$ | $(C_6H_5)_3P{=}CH(CH_2)_9CO_2C_2H_5$ | $NaNH_2$, THF | | 170, 171 |
| | | $NaOC_2H_5$, THF or DMF | | 170, 171 |
| | $(C_6H_5)_3P{=}C(CH_3)CH{=}CH$ 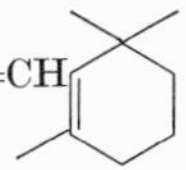 | $LiCH_3$, THF | 2 hr., 20 | 209 |
| | | $LiC_6H_5$, $(C_2H_5)_2O$ | | 240 |
| | | $C_2H_5MgBr$, THF | 15 min., 0 | 240 |
| | | $NaC{\equiv}CH$, DMF | 1 hr., 20 | 289 |
| | | $NaNH_2$, $C_6H_6$ | 12 hr., 20 | 240 |
| | | $NaOCH_3$, $C_6H_6$ | | 240 |
| | | $NaOCH_3$, DMF | | 289 |
| | | $NaOCH_3$, DMF | 30 min. or 3 hr., 20 | 113, 291 |
| | $(C_6H_5)_3P{=}C(C_6H_5)_2$ | $NaC(C_6H_5)_3$, $(C_2H_5)_2O$ | 1 day, 20 | 64 |
| | | $NaOC_2H_5$, $(C_2H_5)_2O$ | | 88 |
| | $(C_6H_5)_3P{=}C(C_6H_5)C_6H_4NO_2$-*p* | $NaOC_2H_5$, $C_2H_5OH$ | | 44 |
| | $(C_6H_5)_3P{=}$ 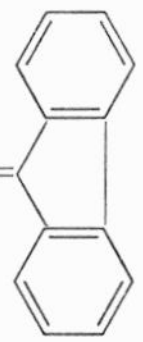 | $NH_4OH$, $C_2H_5OH$ | | 48 |
| | | $NaOH$, $CHCl_3/H_2O$ | | 88 |
| | $(C_6H_5)_3P{=}C(OC_6H_5)_2$ | $LiC_6H_5$, $(C_2H_5)_2O$ | −40 | 87 |
| $C_{14}$ | $(C_6H_5)_3P{=}C{=}C(C_6H_5)_2$ | $LiC_6H_5$, $(C_2H_5)_2O$ | 1 hr., 20 | 123 |
| $C_{15}$ | $(C_6H_5)_3P{=}CHCH{=}C(CH_3)CH_2CH_2$ 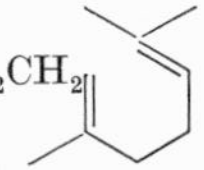 | $NaOCH_3$, DMF | 0 | 265 |
| | $(C_6H_5)_3P{=}CHCH{=}C(CH_3)CH{=}CH$ 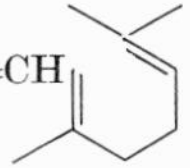 | $NaOC_2H_5$, DMF | | 264 |

*Note:* References 391 to 404 are on p. 490.

## TABLE III—*Continued*

### Alkylidene Phosphoranes Prepared from Phosphonium Salts and Bases

### A. *Monophosphoranes—Continued*

| Carbon Atoms in Alkylidene Group | Phosphorane (Yield, %) | Base, Solvent | Time, Temperature, °C. | References |
|---|---|---|---|---|
| $C_{15}$ (*contd.*) | | | | |
| | $(C_6H_5)_3P{=}CHCH{=}C(CH_3)CH{=}CH$ | $LiC_6H_5$, $(C_2H_5)_2O$ | 20 min., 20 | 54, 283 |
| | | $NaC{\equiv}CH$, DMF | 1 day, 20 | 54, 283 |
| | | | 2 days, −5 | 270 |
| | | $NaNH_2$, $C_6H_6$ | 1 day, 20 | 54, 283 |
| | | $NaOCH_3$, $CH_3OH$ | | 285 |
| | | $NaOCH_3$, DMF | 10 min., 20 | 283 |
| | | $NaOC_2H_5$, $C_2H_5OH$ | 10 min., 20 | 54 |
| | | $NaOC_2H_5$, DMF | | 283 |
| | | $KOC_4H_9$-*t*, DMF | | 270 |
| | | NaOH, $CH_3OH$ or $C_2H_5OH$, or DMF | | 269, 270, 294 |
| | | KOH, $CH_3OH$ | | 284 |
| | $(C_6H_5)_3P{=}CHCH{=}C(CH_3)CH{=}CH$ | $NaOCH_3$, DMF | | 264 |
| | $(C_6H_5)_3P{=}CHCH{=}C(CH_3)CH{=}CH$ | $NaOC_2H_5$, $C_2H_5OH$ | | 288 |
| | $(C_6H_5)_3P{=}CHC_6H_4(CH{=}CHC_6H_5)$-*p* | $LiOC_2H_5$, $C_2H_5OH$ | 20 | 195, 196 |
| | $(C_6H_5)_3P{=}CHCH{=}C(C_6H_5)_2$ | $LiOC_2H_5$, $C_2H_5OH$ | | 211 |
| $C_{16}$ | $(C_6H_5)_3P{=}CH$–$C_6H_4$–$CH{=}CHC_6H_4CH_3$-*p* | $LiOC_2H_5$, $C_2H_5OH$ | 20 | 196 |

| | | | | |
|---|---|---|---|---|
| $C_{17}$ | 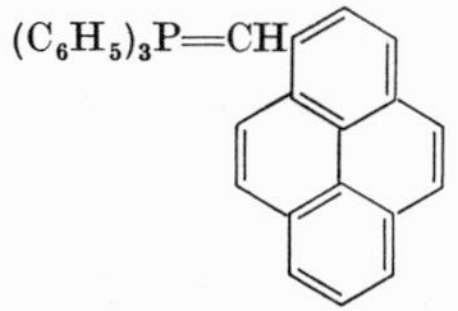 $(C_6H_5)_3P{=}CH$ | $NaOC_2H_5$, $C_2H_5OH$ | 20 | 197 |
| $C_{18}$ | 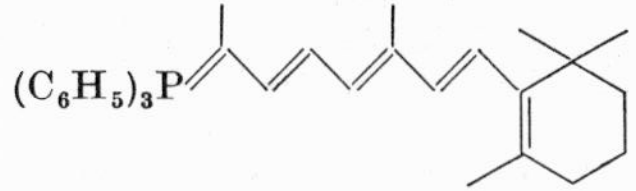 $(C_6H_5)_3P$ | $NaOC_2H_5$, $C_2H_5OH/DMF$<br>$NaOH$, $H_2O$ | | 270<br>39 |
| | $(C_6H_5)_3P{=}CHCO_2C_{16}H_{33}\text{-}n$ | $NaOCH_3$, $CH_2Cl_2$ | 30 min., 20 | 39 |
| $C_{19}$ | $(C_6H_5)_3P{=}C(CH_3)CO_2C_{16}H_{33}\text{-}n$ | $NaOCH_3$, $CH_2Cl_2$ | 30 min., 20 | 39 |
| $C_{20}$ | 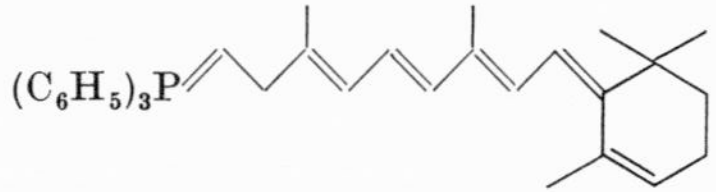 $(C_6H_5)_3P$ | $LiC_4H_9\text{-}n$, $(C_2H_5)_2O$ | 90 min., 20 | 267 |
| | 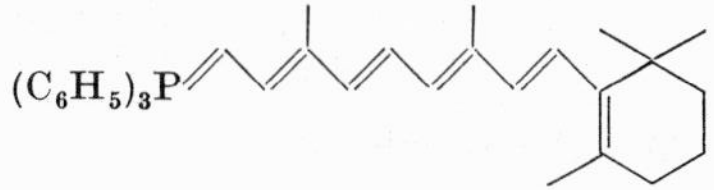 $(C_6H_5)_3P$ | $NaOCH_3$, DMF<br>KOH, $CH_3OH$ or $C_2H_5OH$ | | 266<br>266 |
| $C_{23}$ | 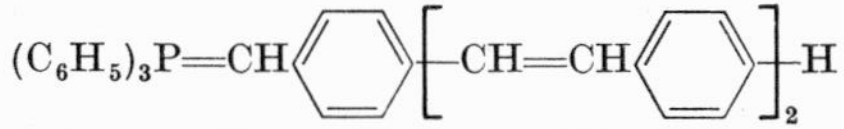 $(C_6H_5)_3P{=}CH$ [ $CH{=}CH$ ]$_2$ H | $LiOC_2H_5$, $C_2H_5OH$ | | 196, 198 |

*Note:* References 391 to 404 are on p. 490.

TABLE III—*Continued*

ALKYLIDENE PHOSPHORANES PREPARED FROM PHOSPHONIUM SALTS AND BASES

*B. Bis-phosphoranes*

| Carbon Atoms in Alkylidene Group | Phosphorane (Yield, %) | Base, Solvent | Time, Temperature, C°. | References |
|---|---|---|---|---|
| $C_1$ | $(C_6H_5)_3P{=}C{=}P(C_6H_5)_3$ (70) | K, diglyme | 20 min., reflux | 81 |
| $C_2$ | $(C_6H_5)_3P{=}CHOCH{=}P(C_6H_5)_3$ | $NaOCH_3$, $CH_3OH$ | | 190 |
| $C_3$ | $(C_6H_5)_3P{=}CHCH_2CH{=}P(C_6H_5)_3$ | $LiC_6H_5$, $(C_2H_5)_2O$ | 2 hr., 20 | 52 |
| | | $NaNH_2$, liq. $NH_3$ | | 52 |
| $C_4$ | $(C_6H_5)_3P{=}CHCH_2CH_2CH{=}P(C_6H_5)_3$ | $LiCH_3$, THF | 4 hr., 20 | 113 |
| | | $LiC_4H_9$-*n*, $(C_2H_5)_2O$ | 1 hr., 20 | 74 |
| | | $LiC_4H_9$-*n*, THF | | 299–301 |
| | | $LiC_6H_5$, $(C_2H_5)_2O$ | 1–2 hr., 20 | 74, 185 |
| | | $NaNH_2$, liq. $NH_3$ | | 52 |
| | | $NaOC_2H_5$, $C_2H_5OH$ | | 74 |
| | | $KOC_4H_9$-*t*, *t*-$C_4H_9OH$ | 1 day, 20 | 302 |
| | $(C_6H_5)_3P{=}CHCH{=}CHCH{=}P(C_6H_5)_3$ | $LiC_6H_5$, THF | 30 min., 20 | 209 |
| | | $LiOC_2H_5$, $C_2H_5OH$ | | 212 |
| $C_5$ | $(C_6H_5)_3P{=}CH(CH_2)_3CH{=}P(C_6H_5)_3$ | $LiC_4H_9$-*n*, $(C_2H_5)_2O$ | | 188, 189 |
| $C_6$ | $(C_6H_5)_3P{=}CHCO_2CH_2CH_2OCOCH{=}P(C_6H_5)_3$ (67) | $Na_2CO_3$, $H_2O$ | 2 hr., 20 | 79 |
| $C_8$ | *o*-$C_6H_4[CH{=}P(C_6H_5)_3]_2$ | $LiOC_2H_5$, $C_2H_5OH$ | | 201*a* |
| | *p*-$C_6H_4[CH{=}P(C_6H_5)_3]_2$ | $LiC_4H_9$-*n*, $(C_2H_5)_2O$ | | 188, 189 |
| | | $LiOC_2H_5$, $C_2H_5OH$ | | 193, 195, 210 |
| | | $NaOC_2H_5$, $C_2H_5OH$ | | 66 |
| $C_{10}$ | $(C_6H_5)_3P{=}CHC(CH_3){=}CHCH{=}CHCH{=}C(CH_3)CH{=}P(C_6H_5)_3$ | $LiC_6H_5$, $(C_2H_5)_2O$ | 30 min., 20 | 80, 271 |
| | $(C_6H_5)_3P{=}CHC(CH_3){=}CH{-}C{\equiv}C{-}CH{=}C(CH_3)CH{=}P(C_6H_5)_3$ | $LiC_6H$, $(C_2H_5)_2O$ | 30 min., 20 | 80, 271 |

| | | | |
|---|---|---|---|
| | $(C_6H_5)_3P{=}CH$–$C_6H_2(CH_3)_2$–$CH{=}P(C_6H_5)_3$ | $LiOC_2H_5$, $C_2H_5OH$ | 193, 210 |
| $C_{14}$ | $(C_6H_5)_3P{=}CH$–$C_6H_4$–$C_6H_4$–$CH{=}P(C_6H_5)_3$ | $LiOC_2H_5$, $C_2H_5OH$ | 187 |
| $C_{16}$ | $(C_6H_5)_3P{=}CH$–$C_6H_4$–$CH{=}CH$–$C_6H_4$–$CH{=}P(C_6H_5)_3$ | $LiOC_2H_5$, $C_2H_5OH$ | 187, 196 |
| $C_{24}$ | $(C_6H_5)_3P{=}CH$–$C_6H_4$–$CH{=}CH$–$C_6H_4$–$CH{=}CH$–$C_6H_4$–$CH{=}P(C_6H_5)_3$ | $LiOC_2H_5$, $C_2H_5OH$ | 201*a* |

*Note:* References 391 to 404 are on p. 490.

TABLE IV

ALKYLIDENE PHOSPHORANES PREPARED FROM YLIDES

| Carbon Atoms in Alkylidene Group | Phosphorane | Reactants (Solvent, Time, Temperature, °C.) | Yield, % | References |
|---|---|---|---|---|
| $C_2$ | $(C_6H_5)_3P{=}CHCHO$ | $(C_6H_5)_3P{=}CH_2 + HCO_2C_2H_5$ | 67 | 33 |
| | $(C_6H_5)_3P{=}CClCHO$ | $(C_6H_5)_3P{=}CHCHO + C_6H_5ICl_2$ | 66 | 59 |
| | $(C_6H_5)_3P{=}CBrCHO$ | $(C_6H_5)_3P{=}CHCHO + Br_2$ | 83 | 59 |
| $C_3$ | $(C_6H_5)_3P{=}C(CH_3)CHO$ | $(C_6H_5)_3P{=}CHCH_3 + HCO_2C_2H_5$ | — | 33 |
| | $(C_6H_5)_3P{=}CHCOCH_3$ | $(C_6H_5)_3P{=}CH_2 + CH_3COSC_2H_5$ | 78 | 32 |
| | $(C_6H_5)_3P{=}CClCOCH_3$ | $(C_6H_5)_3P{=}CHCOCH_3 + Cl_2(CH_2Cl_2$/pyridine, −70) | 46 | 60 |
| | | $(C_6H_5)_3P{=}CHCOCH_3 + C_6H_5ICl_2$ | 88 | 59 |
| | $(C_6H_5)_3P{=}CBrCOCH_3$ | $(C_6H_5)_3P{=}CHCOCH_3 + Br_2$ | 87 | 59 |
| | $(C_6H_5)_3P{=}CHCO_2CH_3$ | $(C_6H_5)_3P{=}CH_2 + ClCO_2CH_3$ | 80 | 111 |
| | $(C_6H_5)_3P{=}CClCO_2CH_3$ | $(C_6H_5)_3P{=}CHCO_2CH_3 + C_6H_5ICl_2$ ($C_6H_6$, 30 min., 10) | 95 | 117 |
| | $(C_6H_5)_3P{=}CBrCO_2CH_3$ | $(C_6H_5)_3P{=}CHCO_2CH_3 + Br_2$ ($C_6H_6$) | 96 | 59 |
| | $(C_6H_5)_3P{=}CICO_2CH_3$ | $(C_6H_5)_3P{=}CHCO_2CH_3 + I_2$ ($CH_3OH$) | 87 | 117 |
| $C_4$ | $(C_6H_5)_3P{=}CClCOC_2H_5$ | $(C_6H_5)_3P{=}CHCOC_2H_5 + C_6H_5ICl_2$ | 94 | 59 |
| | $(C_6H_5)_3P{=}CBrCOC_2H_5$ | $(C_6H_5)_3P{=}CHCOC_2H_5 + Br_2$ | 97 | 59 |
| | $(C_6H_5)_3P{=}C(CH_3)CO_2CH_3$ | $(C_6H_5)_3P{=}CHCH_3 + ClCO_2CH_3$ | 95 | 111 |
| | $(C_6H_5)_3P{=}CClCO_2C_2H_5$ | $(C_6H_5)_3P{=}CHCO_2C_2H_5 + Cl_2$ ($CH_2Cl_2/(C_2H_5)_3N$, −70) | 80 | 60 |
| | $(C_6H_5)_3P{=}CBrCO_2C_2H_5$ | $(C_6H_5)_3P{=}CHCO_2C_2H_5 + Br_2$ ($CH_2Cl_2$/NaOH, −70) | 69 | 60 |
| $C_5$ | $(C_6H_5)_3P{=}C(CO_2CH_3)C_2H_5$ | $(C_6H_5)_3P{=}CHC_2H_5 + ClCO_2CH_3$ | 88 | 111 |
| | $(C_6H_5)_3P{=}C(COCH_3)CO_2CH_3$ | $(C_6H_5)_3P{=}CHCO_2CH_3 + CH_3COCl$ ($C_6H_6$, 3 hr., 20) | 97 | 331 |
| | $(C_6H_5)_3P{=}C(CH_2CN)CO_2CH_3$ | $(C_6H_5)_3P{=}CHCO_2CH_3 + BrCH_2CN$ ($CH_3CO_2C_2H_5$, 4–5 hr., reflux) | 89 | 40 |
| $C_6$ | $(C_6H_5)_3P{=}C(COCH_3)C_3H_7$-$n$ | $(C_6H_5)_3P{=}CHC_3H_7$-$n$ + $CH_3COSC_2H_5$ | 62 | 32 |
| | $(C_6H_5)_3P{=}C(COC_3H_7$-$n)CH_3$ | $(C_6H_5)_3P{=}CHCH_3$ + $n$-$C_3H_7COSC_2H_5$ | 54 | 32 |
| | $(C_6H_5)_3P{=}C(CO_2CH_3)C_3H_7$-$n$ | $(C_6H_5)_3P{=}CHC_3H_7$-$n$ + $ClCO_2CH_3$ | 96 | 111 |
| | $(C_6H_5)_3P{=}C(CH_2CH{=}CH_2)CO_2CH_3$ | $(C_6H_5)_3P{=}CHCO_2CH_3 + CH_2{=}CHCH_2Br$ ($CH_3CO_2C_2H_5$), 4–5 hr., reflux) | 93 | 40 |
| | $(C_6H_5)_3P{=}C(COC_3H_7$-$n)CN$ | $(C_6H_5)_3P{=}CHCN$ + $n$-$C_3H_7COCl$ | 81 | 330 |
| | $(C_6H_5)_3P{=}C(CH_2CO_2CH_3)CO_2CH_3$ | $(C_6H_5)_3P{=}CHCO_2CH_3 + BrCH_2CO_2CH_3$ ($CH_3CO_2C_2H_5$, 4–5 hr., reflux) | 98 | 40 |

| | | | | |
|---|---|---|---|---|
| $C_8$ | $(C_6H_5)_3P{=}C(COC_3H_7\text{-}n)C_3H_7\text{-}n$ | $(C_6H_5)_3P{=}CHC_3H_7\text{-}n + n\text{-}C_3H_7COCl$ | — | 330 |
| | $(C_6H_5)_3P{=}C(COC_5H_{11}\text{-}n)CH_3$ | $(C_6H_5)_3P{=}CHCH_3 + n\text{-}C_5H_{11}COCl$ | 63 | 32 |
| | $(C_6H_5)_3P{=}CHCOC_6H_5$ | $(C_6H_5)_3P{=}CH_2 + C_6H_5COSC_2H_5$ | 80 | 32 |
| | $(C_6H_5)_3P{=}CHCOC_6H_4NO_2\text{-}p$ | $(C_6H_5)_3P{=}CH_2 + p\text{-}O_2NC_6H_4COCl$ | 93 | 32 |
| | $(C_6H_5)_3P{=}CClCOC_6H_5$ | $(C_6H_5)_3P{=}CHCOC_6H_5 + Cl_2$ ($CH_2Cl_2$/pyridine, −70) | 92 | 60 |
| | | $(C_6H_5)_3P{=}CHCOC_6H_5 + t\text{-}C_4H_9OCl$ ($CH_2Cl_2$, 30 min., −70) | 93 | 60 |
| | | $(C_6H_5)_3P{=}CHCOC_6H_5 + C_6H_5ICl_2$ | 87 | 59 |
| | $(C_6H_5)_3P{=}CBrCOC_6H_5$ | $(C_6H_5)_3P{=}CHCOC_6H_5 + Br_2$ | 91 | 59 |
| | | ($CH_2Cl_2$/KOH, −70) | 84 | 60 |
| | $(C_6H_5)_3P{=}C(COC_3H_7\text{-}n)CO_2C_2H_5$ | $(C_6H_5)_3P{=}CHCO_2C_2H_5 + n\text{-}C_3H_7COCl$ | — | 330 |
| | $(C_6H_5)_3P{=}C(COCH{=}CHCH_3)CO_2C_2H_5$ | $(C_6H_5)_3P{=}CHCO_2C_2H_5 + CH_3CH{=}CHCOCl$ | 57 | 329, 330 |
| | $(C_6H_5)_3P{=}C(COC_4H_3O\text{-}2)CO_2CH_3$ | $(C_6H_5)_3P{=}CHCO_2CH_3$ + 2-furoyl chloride (furan ring–COCl, drawn) ($C_6H_6$, 3 hr., 20) | Quant. | 331 |
| $C_9$ | $(C_6H_5)_3P{=}CHCOCH_2C_6H_5$ | $(C_6H_5)_3P{=}CH_2 + C_6H_5CH_2COSC_2H_5$ | 42 | 32 |
| | $(C_6H_5)_3P{=}C(COCH_3)C_6H_5$ | $(C_6H_5)_3P{=}CHC_6H_5 + CH_3COSC_2H_5$ | 68 | 32 |
| | $(C_6H_5)_3P{=}C(COC_6H_5)CH_3$ | $(C_6H_5)_3P{=}CHCH_3 + C_6H_5COSC_2H_5$ | 93 | 32 |
| | $(C_6H_5)_3P{=}C(CO_2CH_3)C_6H_5$ | $(C_6H_5)_3P{=}CHC_6H_5 + ClCO_2CH_3$ | — | 111 |
| | $(C_6H_5)_3P{=}C(COC_6H_5)CN$ | $(C_6H_5)_3P{=}CHCN + C_6H_5COCl$ ($C_6H_6$, 12 hr., 20) | 68 | 329, 330 |
| $C_{10}$ | $(C_6H_5)_3P{=}C(CO_2C_2H_5)(COCH{=}CHCH{=}CHCH_3)$ | $(C_6H_5)_3P{=}CHCO_2C_2H_5 + CH_3CH{=}CHCH{=}CHCOCl$ | 74 | 330 |
| | $(C_6H_5)_3P{=}CHCOCH_2CH_2C_6H_5$ | $(C_6H_5)_3P{=}CH_2 + C_6H_5CH_2CH_2COSC_2H_5$ | 80 | 32 |
| | $(C_6H_5)_3P{=}C(CH_2C_6H_5)CO_2CH_3$ | $(C_6H_5)_3P{=}CHCO_2CH_3 + C_6H_5CH_2Br$ ($CH_3CO_2C_2H_5$, 4–5 hr., reflux) | 75 | 40 |
| | $(C_6H_5)_3P{=}C(COC_6H_5)CO_2CH_3$ | $(C_6H_5)_3P{=}CHCO_2CH_3 + C_6H_5COCl$ ($C_6H_6$, 3 hr., 20) | 83 | 32, 331 |
| | $(C_6H_5)_3P{=}C(COC_6H_4Cl\text{-}o)CO_2CH_3$ | $(C_6H_5)_3P{=}CHCO_2CH_3 + o\text{-}ClC_6H_4COCl$ ($C_6H_6$, 3 hr., 20) | Quant. | 331 |
| | $(C_6H_5)_3P{=}C(COC_6H_4Cl\text{-}p)CO_2CH_3$ | $(C_6H_5)_3P{=}CHCO_2CH_3 + p\text{-}ClC_6H_4COCl$ ($C_6H_6$, 3 hr., 20) | 98 | 331 |
| | $(C_6H_5)_3P{=}C(COC_6H_4NO_2\text{-}m)CO_2CH_3$ | $(C_6H_5)_3P{=}CHCO_2CH_3 + m\text{-}O_2NC_6H_4COCl$ ($C_6H_6$, 3 hr., 20) | 98 | 331 |

*Note:* References 391 to 404 are on p. 490.

TABLE IV—*Continued*

ALKYLIDENE PHOSPHORANES PREPARED FROM YLIDES

| Carbon Atoms in Alkylidene Group | Phosphorane | Reactants (Solvent, Time, Temperature, °C.) | Yield, % | References |
|---|---|---|---|---|
| $C_{11}$ | $(C_6H_5)_3P{=}C(CH_3)COCH_2CH_2C_6H_5$ | $(C_6H_5)_3P{=}CHCH_3 + C_6H_5CH_2CH_2COSC_2H_5$ | 76 | 32 |
| | $(C_6H_5)_3P{=}C(CH_3)COCH{=}CHC_6H_5$ | $(C_6H_5)_3P{=}CHCH_3 + C_6H_5CH{=}CHCOSC_2H_5$ | 70 | 32 |
| | $(C_6H_5)_3P{=}C(COCH_2C_6H_5)CO_2CH_3$ | $(C_6H_5)_3P{=}CHCO_2CH_3 + C_6H_5CH_2COCl$ ($C_6H_6$, 3 hr., 20) | 77 | 331 |
| | $(C_6H_5)_3P{=}C(COC_6H_4CH_3\text{-}p)CO_2CH_3$ | $(C_6H_5)_3P{=}CHCO_2CH_3 + p\text{-}CH_3C_6H_4COCl$ ($C_6H_6$, 3 hr., 20) | 95 | 331 |
| | $(C_6H_5)_3P{=}C(COC_6H_4OCH_3\text{-}o)CO_2CH_3$ | $(C_6H_5)_3P{=}CHCO_2CH_3 + o\text{-}CH_3OC_6H_4COCl$ ($C_6H_6$, 3 hr., 20) | Quant. | 331 |
| | $(C_6H_5)_3P{=}C(COC_6H_4OCH_3\text{-}m)CO_2CH_3$ | $(C_6H_5)_3P{=}CHCO_2CH_3 + m\text{-}CH_3OC_6H_4COCl$ ($C_6H_6$, 3 hr., 20) | Quant. | 331 |
| $C_{12}$ | $(C_6H_5)_3P{=}C(CH_2CH{=}CHC_6H_5)CO_2CH_3$ | $(C_6H_5)_3P{=}CHCO_2CH_3 + C_6H_5CH{=}CHCH_2Br$ ($CH_3CO_2C_2H_5$, 4–5 hr., reflux) | 81 | 40 |
| | $(C_6H_5)_3P{=}C(CO_2CH_3)COCH{=}CHC_6H_5$ | $(C_6H_5)_3P{=}CHCO_2CH_3 + C_6H_5CH{=}CHCOCl$ ($C_6H_6$, 3 hr., 20) | Quant. | 331 |
| $C_{13}$ | $(C_6H_5)_3P{=}C(C_3H_7\text{-}n)COCH_2CH_2C_6H_5$ | $(C_6H_5)_3P{=}CHC_3H_7\text{-}n + C_6H_5CH_2CH_2COSC_2H_5$ | 63 | 32 |
| $C_{14}$ | $(C_6H_5)_3P{=}C(COC_6H_5)C_6H_5$ | $(C_6H_5)_3P{=}CHC_6H_5 + C_6H_5COSC_2H_5$ | 58 | 32 |
| | $(C_6H_5)_3P{=}C(COC_{10}H_7\text{-}\alpha)CO_2CH_3$ | $(C_6H_5)_3P{=}CHCO_2CH_3 + \alpha\text{-}C_{10}H_7COCl$ | 78 | 331 |

*Note:* References 391 to 404 are on p. 490.

# TABLE V

## Alkylidene Phosphoranes Prepared Directly From Triphenylphosphine

| Alkylidenephosphorane | Reagents: $(C_6H_5)_3P$ and | Solvent, Time, Temperature, °C. | References |
|---|---|---|---|
| $(C_6H_5)_3P{=}CH_2$ | $CH_2N_2 + Cu^{1+}$ | THF | 90 |
| $(C_6H_5)_3P{=}CHCl$ | $CH_2Cl_2 + LiC_4H_9\text{-}n$ | −60 | 91, 92 |
| | | $(C_2H_5)_2O$, −30 | 95, 96 |
| $(C_6H_5)_3P{=}CF_2$ | $CF_2Cl_2$ or $CF_2Br_2$ | | 98 |
| $(C_6H_5)_3P{=}CCl_2$ | $CHCl_3 + KOC_4H_9\text{-}t$ | Pentane, 0–5 | 93 |
| | | Heptane, 0 | 94 |
| | $CCl_4$ | 2 hr., 60 | 98 |
| $(C_6H_5)_3P{=}CBr_2$ | $CHBr_3 + KOC_4H_9\text{-}t$ | Heptane, 0 | 94 |
| | $CBr_4$ | $CH_2Cl_2$ | 82 |
| $(C_6H_5)_3P{=}C_6H_3(=O)(OH)$ | $O{=}C_6H_4{=}O$ | $C_6H_6$, 20 hr., 20 | 99 |
| $(C_6H_5)_3P{=}CHC_6H_5$ | $C_6H_5CHN_2 + Cu^{1+}$ | THF | 90 |

*Note:* References 391 to 404 are on p. 490.

# TABLE VI

## Olefins Prepared by the Wittig Reaction

### *A. Mono-ylides as Starting Materials*

| | Ylide | Product | Yield, % | Solvent, Time, Temperature, °C. | References |
|---|---|---|---|---|---|
| $C_1$ | $(C_6H_5)_3P{=}CH_2$ | $n$-$C_4H_9OCH{=}CH_2$ | | $(C_2H_5)_2O$, 12 hr., 20; 3 hr., reflux | 113 |
| | | $n$-$C_4H_9C(CH_3){=}CH_2$ | 12 | $(C_2H_5)_2O$, 4 hr., 60–75 | 182 |
| | | 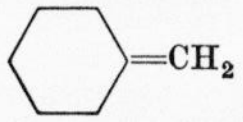 $=CH_2$ | 52 | $(C_2H_5)_2O$, 3 hr., 65 | 2 |
| | | 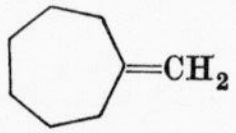 $=CH_2$ | 5–10 | | 173 |
| | | $C_6H_5CH{=}CH_2$ | 67 | $(C_2H_5)_2O$, 1 hr., 20; 12 hr., 65 | 2 |
| | |  N, $=CH_2$ | 63 | $(C_2H_5)_2O$, 6 hr., reflux; 2 days, 20 | 322 |
| | | $(CH_3)_2C{=}CHCH_2CH_2C(CH_3){=}CH_2$ | 10 | $(C_2H_5)_2O$, 4 hr., 60–75 | 182 |
| | | $CH_2{=}C(CH_3)(CH_2)_3C(CH_3){=}CH_2$ | 11 | $(C_2H_5)_2O$, 4 hr., 60–75 | 182 |
| | | 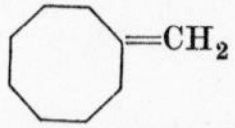 $=CH_2$ | | | 173 |
| | | $C_6H_5C(CH_3){=}CH_2$ | 74 | $(C_2H_5)_2O$, 3 hr., 65 | 2 |
| | | $n$-$C_7H_{15}C(CH_3){=}CH_2$ | | | 401 |
| | | $C_6H_5CH{=}CHCH{=}CH_2$ | 69 | $(C_2H_5)_2O$, 1 hr., 20; 5 hr., 65 | 2 |
| | | $o$-$C_6H_4(CH{=}CH_2)_2$ | 75 | $(C_2H_5)_2O$, 2 hr., reflux | 52 |
| | | $(t$-$C_4H_9)_2CHCH{=}CH_2$ | 76 | $(C_2H_5)_2O$/THF, 12 hr., 20 | 402 |
| | | 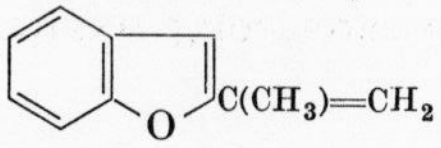 O, $C(CH_3){=}CH_2$ | | THF, 90 min., reflux | 324 |
| | | 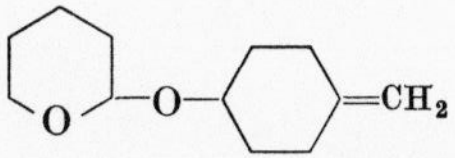 O, —O—, $=CH_2$ | | $(C_2H_5)_2O$, 1 hr., 20; 2 hr., 65 | 70 |

| Compound | Yield | Conditions | Refs. |
|---|---|---|---|
| 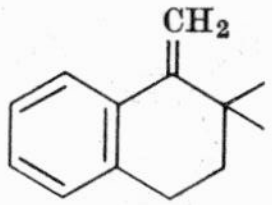 $CH_2$ |  83 | $(C_2H_5)_2O$, 40 hr., 20 | 174 |
| 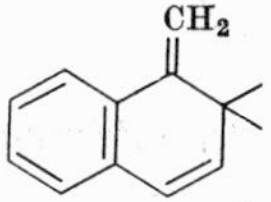 $CH_2$ | 61 | $(C_2H_5)_2O$, 40 hr., 20 | 174 |
| 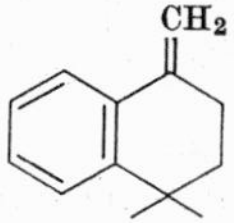 $CH_2$ | 22 | $(C_2H_5)_2O$, 70 hr., 60 | 174 |
| 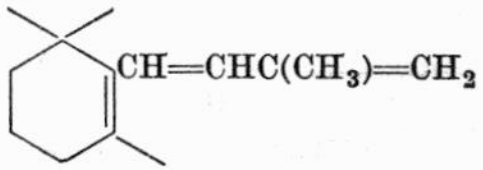 $CH{=}CHC(CH_3){=}CH_2$ | 61 | $(C_2H_5)_2O$, 1 day, 60 | 393 |
| $(C_6H_5)_2C{=}CH_2$ | 84<br>62 | $(C_2H_5)_2O$, 20<br>$CH_3OCH_2CH_2OCH_3$, 4 hr., reflux | 1<br>90 |
| $p$-$O_2NC_6H_4C(C_6H_5){=}CH_2$ | 62 | $(C_2H_5)_2O$, 3 hr., 65 | 2 |
| 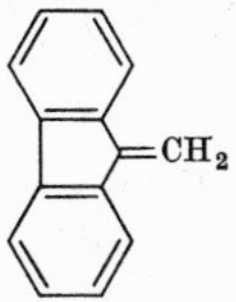 $=CH_2$ | 6 | $(C_2H_5)_2O$, 36 hr., 65 | 130, 185 |
| 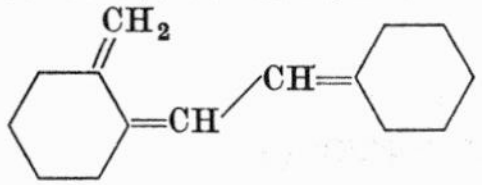 $CH_2$, $=CH$, $CH=$ |  | THF, 7 hr., 20 | 310, 311 |

*Note:* References 391 to 404 are on p. 490.

## TABLE VI—*Continued*

### Olefins Prepared by the Wittig Reaction

*A. Mono-ylides as Starting Materials—Continued*

| Ylide | Product | Yield, % | Solvent, Time, Temperature, °C. | References |
|---|---|---|---|---|
| $C_1$ $(C_6H_5)_3P{=}CH_2$ (*contd.*) | 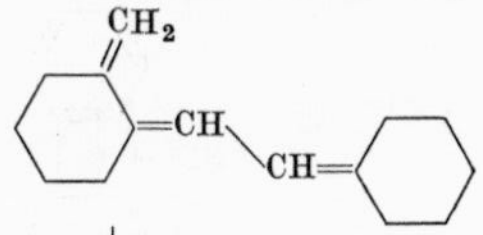 $CH_2$, $=CH$, $CH=$ | | $(C_2H_5)_2O$, 12 hr., 20 | 183 |
| | 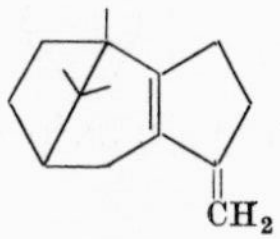 $CH_2$ | | THF | 176 |
| | 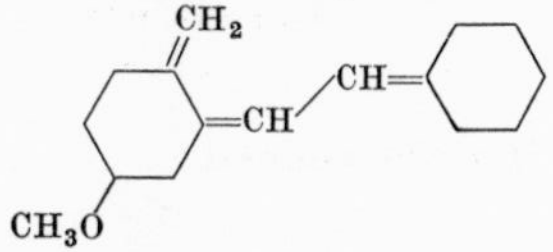 $CH_2$, $=CH$, $CH=$, $CH_3O$ | 14 | $(C_2H_5)_2O$, 3 hr., 65 | 308, 309 |
| | $(C_6H_5)_2C{=}CHCH{=}CH_2$ | 54 | $(C_2H_5)_2O$, 5 hr., 20; 12 hr., 45 | 138 |
| | 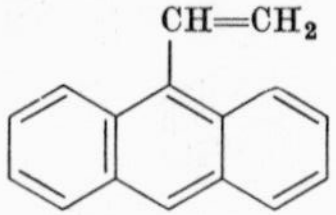 $CH{=}CH_2$ | | $(C_2H_5)_2O$, 12 hr., 65 | 186 |
| | $[p\text{-}(CH_3)_2NC_6H_4]_2C{=}CH_2$ | 66 | $(C_2H_5)_2O$, 1 hr., 20; 3 hr., 65 | 2 |
| | 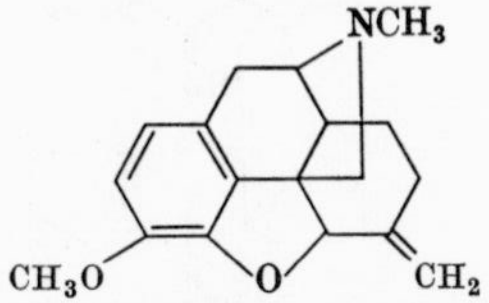 $NCH_3$, $CH_3O$, O, $CH_2$ | 84 | THF, 38 hr., reflux | 175 |

| | | | |
|---|---|---|---|
| $CH_3[CH(CH_3)CH_2CH_2CH_2]_3C(CH_3){=}CH_2$ | | | 403 |
| 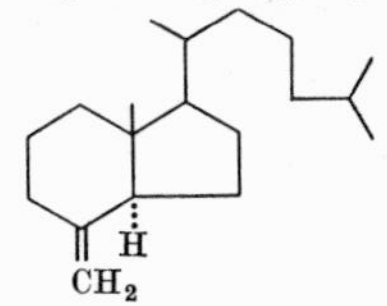 | 67 | $(C_2H_5)_2O$, 1.5 hr., 20; 3.5 hr., 65 | 317 |
| 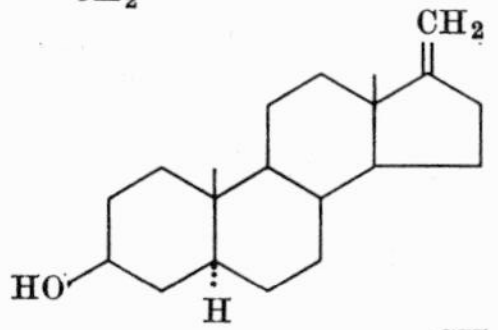 | 58 | THF, 6 hr., reflux | 254 |
| 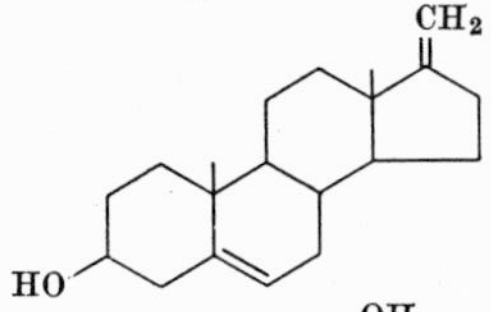 | 36 | THF, 6 hr., reflux | 254 |
| 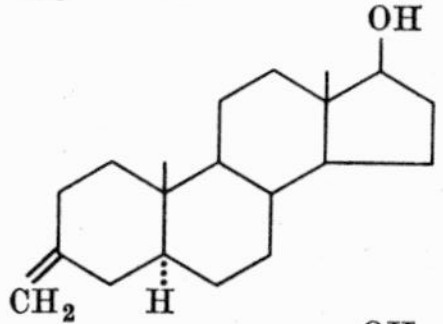 | 38 | THF, 6 hr., reflux | 254 |
| 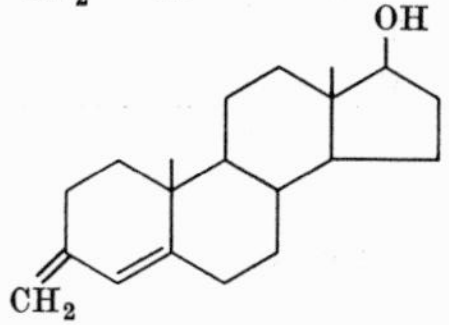 | 57 | THF, 6 hr., reflux | 254 |

*Note:* References 391 to 404 are on p. 490.

## TABLE VI—*Continued*

### Olefins Prepared by the Wittig Reaction

*A. Mono-ylides as Starting Materials—Continued*

| Ylide | Product | Yield, % | Solvent, Time, Temperature, °C. | References |
|---|---|---|---|---|
| $C_1$ $(C_6H_5)_3P{=}CH_2$ (*contd.*) | O; $CH_2$; H | 55 | THF, 6 hr., reflux | 252 |
| | 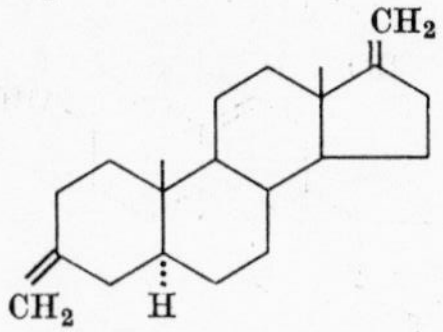 $CH_2$; $CH_2$; H | 76 | THF, 6 hr., reflux | 252 |
| | 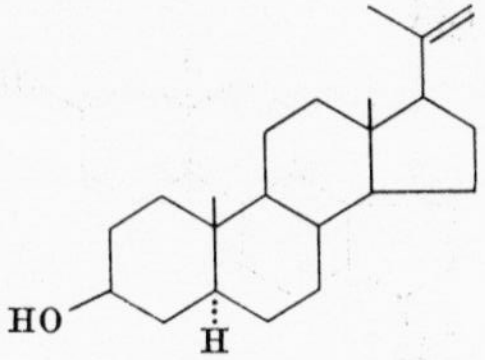 HO; H | 55 | THF, 6 hr., reflux | 252 |
| | 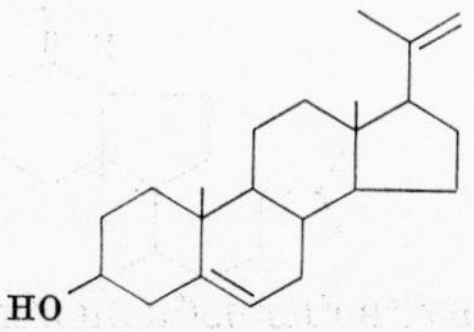 HO | 35 | THF, 6 hr., reflux | 254, 400 |

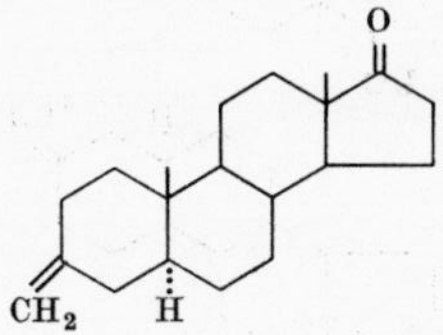

| | | | |
|---|---|---|---|
| 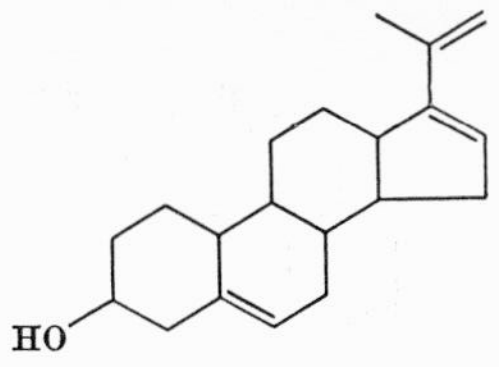 | | | 400 |
| 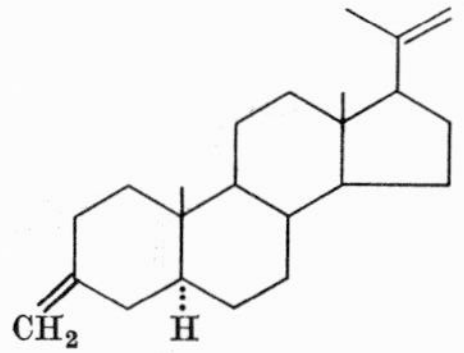 | 75 | THF, 6 hr., reflux | 252 |
| 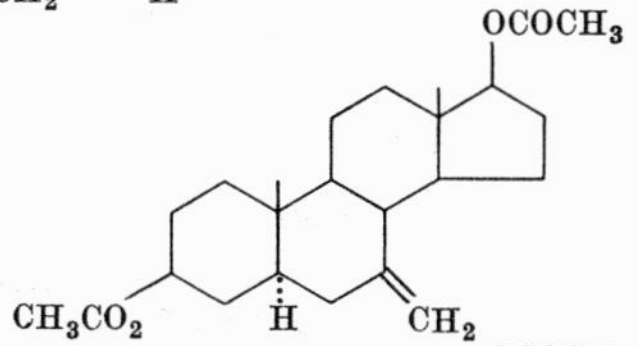 | 17 | THF, 6 hr., reflux | 252 |
| 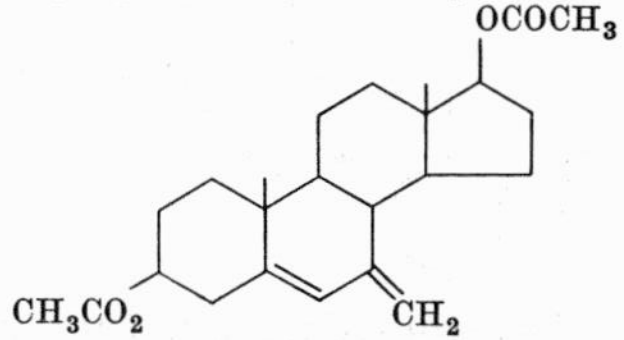 | 26 | THF, 6 hr., reflux | 252 |
| 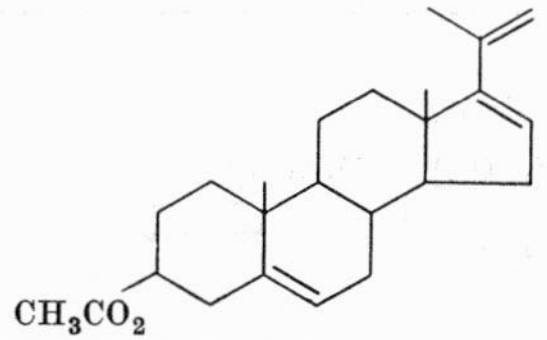 | 48 | THF, 6 hr., reflux | 254 |

*Note:* References 391 to 404 are on p. 490.

## TABLE VI—*Continued*

### Olefins Prepared by the Wittig Reaction

*A. Mono-ylides as Starting Materials—Continued*

| | Ylide | Product | Yield, % | Solvent, Time, Temperature, °C. | References |
|---|---|---|---|---|---|
| $C_1$ | $(C_6H_5)_3P{=}CH_2$ (*contd.*) | $(C_6H_5)_3CCH_2CH(CH_3)C(CH_3){=}CH_2$ | 71 | $(C_2H_5)_2O$, 8 hr., 60 | 2 |
| | | 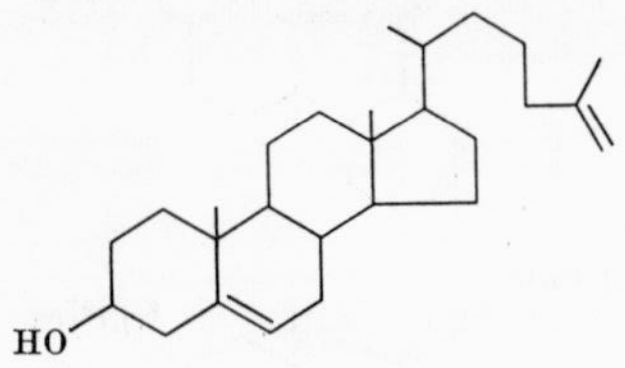 [structure: HO-substituted steroid] | 53 | $(C_2H_5)_2O$, 1 hr., 20; 5 hr., 65 | 304*a*, 398 |
| | | 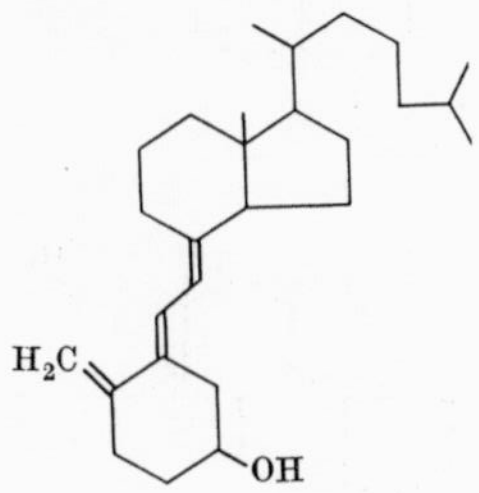 [structure: $H_2C$, OH] | 55 | $(C_2H_5)_2O$, 1 hr., 20; 3 hr., reflux | 319 |
| | | 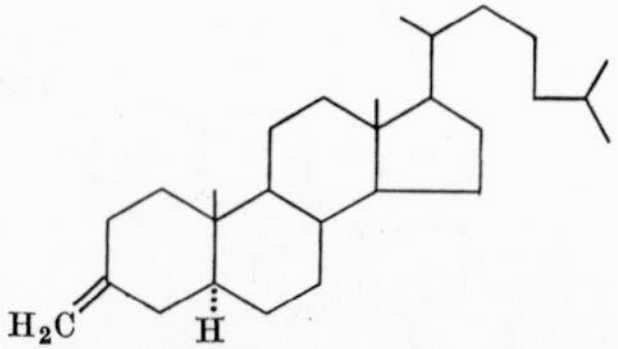 [structure: $H_2C$, H] | 69 | $(C_2H_5)_2O$, 12 hr., reflux<br>THF, 6 hr., reflux | 303<br>254 |

| Product | Yield (%) | Conditions | Refs. |
|---|---|---|---|
| 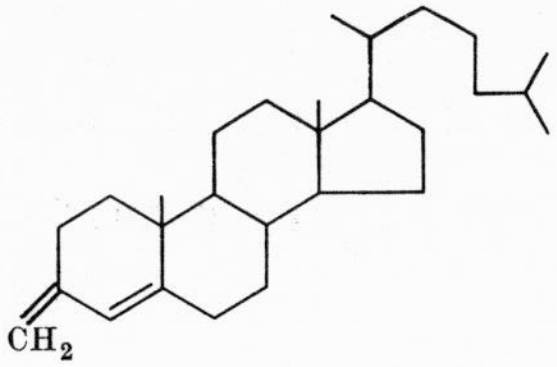 $CH_2$ | 80 | THF, 6 hr., reflux | 254 |
| 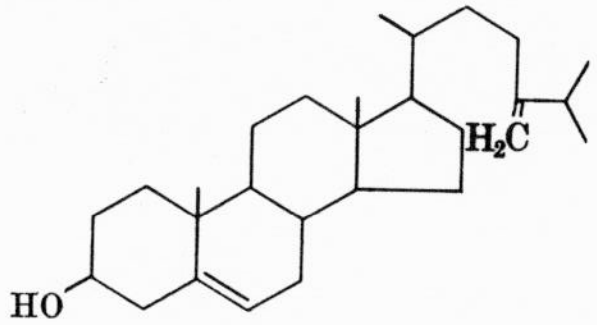 HO, $H_2C$ | | $(C_2H_5)_2O$, 1 hr., 20; 3 hr., 65 | 304 |
| 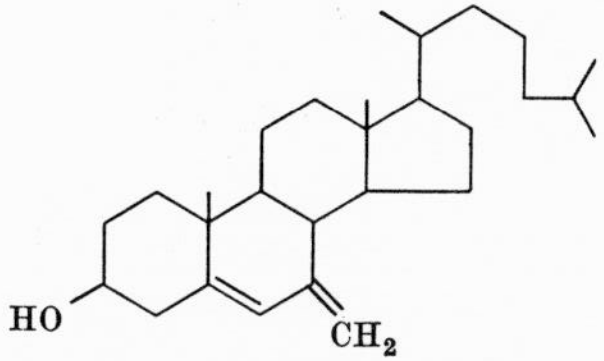 HO, $CH_2$ | 40 | THF, 6 hr., reflux | 254 |
| 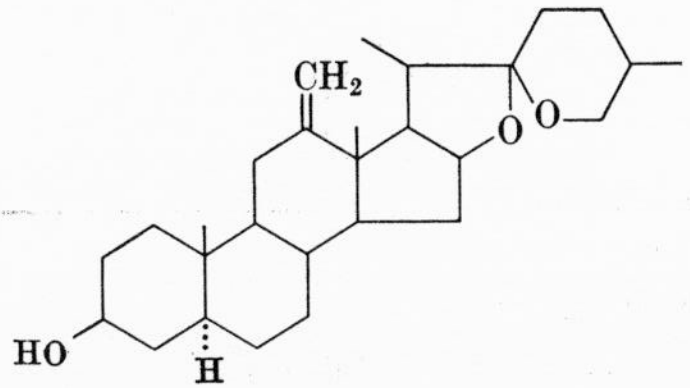 $CH_2$, O, O, HO, H | 51 | THF, 6 hr., reflux | 254 |

*Note:* References 391 to 404 are on p. 490.

## TABLE VI—*Continued*

### Olefins Prepared by the Wittig Reaction

*A. Mono-ylides as Starting Materials—Continued*

| Ylide | Product | Yield, % | Solvent, Time, Temperature, °C. | References |
|---|---|---|---|---|
| $C_1$ $(C_6H_5)_3P{=}CH_2$ (*contd.*) | | | | |
| | 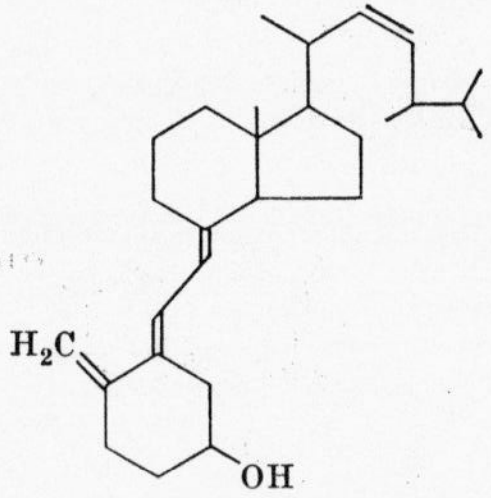 [structure: $H_2C$, OH] | 45 | $(C_2H_5)_2O$, 1 hr., 20; 3 hr., reflux | 319 |
| | 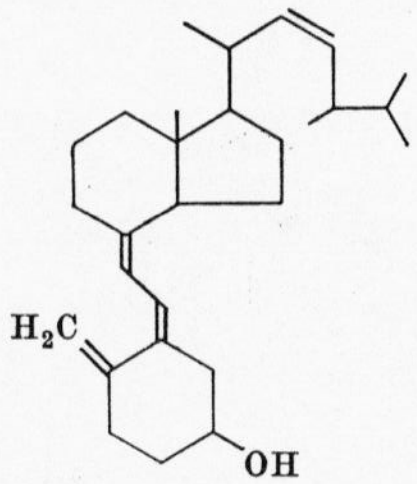 [structure: $H_2C$, OH] | | THF, 7 hr., 20 | 310, 311 |
| | 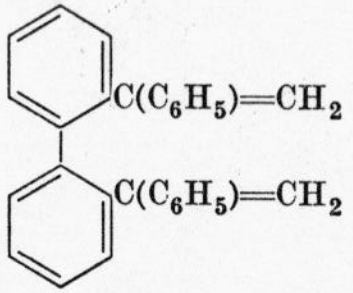 [structure: biphenyl with $C(C_6H_5){=}CH_2$ and $C(C_6H_5){=}CH_2$] | 86 | $(C_2H_5)_2O$, 3 days, 20 | 191 |

| Structure | Yield (%) | Conditions | Refs. |
|---|---|---|---|
| 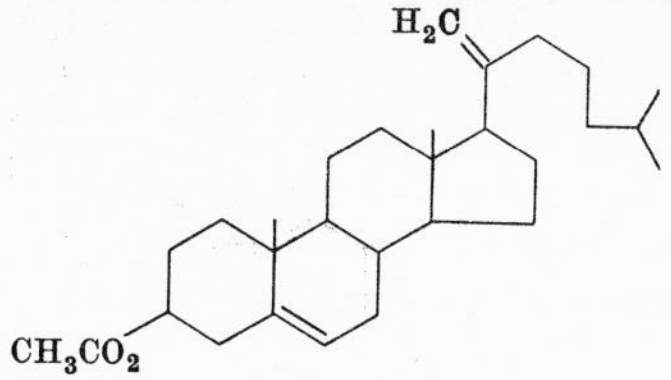 $CH_3CO_2$, $H_2C$ | 54 | THF, 6 hr., reflux | 252 |
| 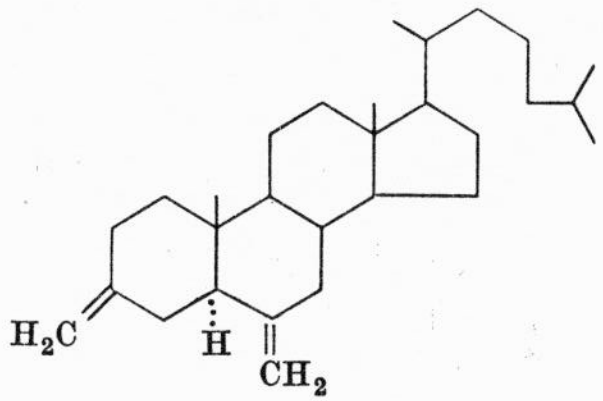 $H_2C$, H, $CH_2$ | 51 | THF, 6 hr., reflux | 252 |
| 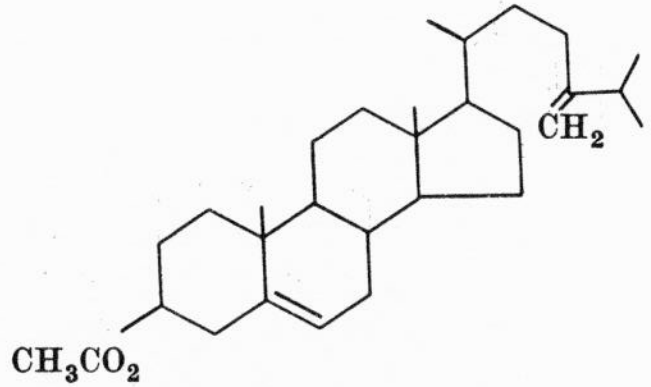 $CH_3CO_2$, $CH_2$ | 71 | $(C_2H_5)_2O$, 1 hr., 20; 5 hr., 65 | 304*a* |
| 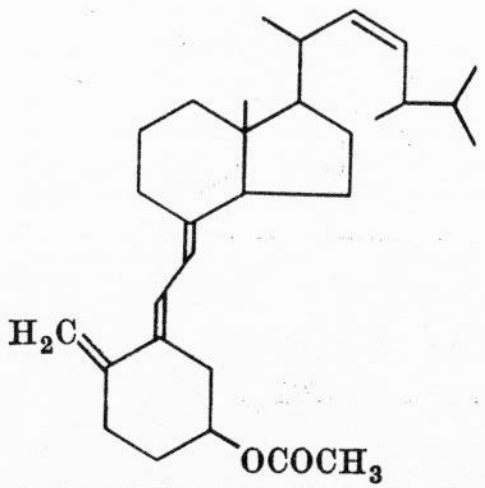 $H_2C$, $OCOCH_3$ | 65 | $(C_2H_5)_2O$, 1 hr., 20; 3 hr., reflux | 316 |

*Note:* References 391 to 404 are on p. 490.

TABLE VI—*Continued*

OLEFINS PREPARED BY THE WITTIG REACTION

*A. Mono-ylides as Starting Materials—Continued*

| Ylide | Product | Yield, % | Solvent, Time, Temperature, °C. | References |
|---|---|---|---|---|
| $C_1$ $(C_6H_5)_3P{=}CH_2$ (*contd.*) | | | | |
| | 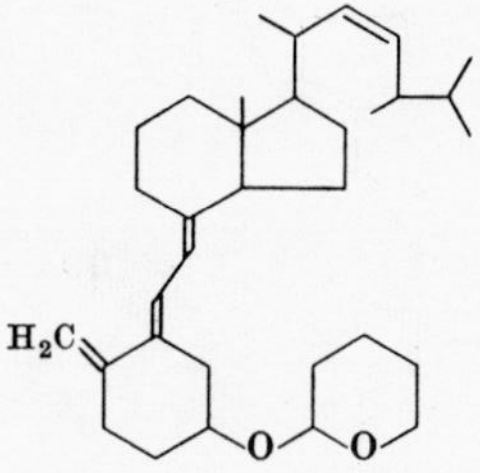 [structure: $H_2C$, O, O] | 28 | $(C_2H_5)_2O$, 90 min., 60; 15 hr., 20 | 313 |
| | 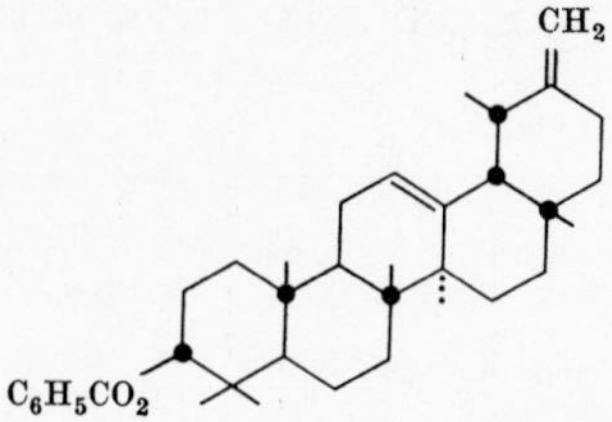 [structure: $CH_2$, $C_6H_5CO_2$] | | | 323 |
| | 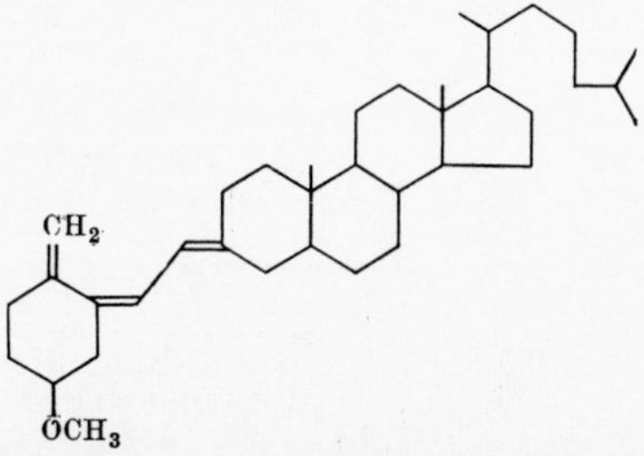 [structure: $CH_2$, $OCH_3$] | 23 | $(C_2H_5)_2O$ | 312 |

| | | | | |
|---|---|---|---|---|
| $(C_6H_5)_3P{=}CF_2$ | $C_6H_5CH{=}CF_2$ | | | 98 |
| $(C_6H_5)_3P{=}CHCl$ | $(C_2H_5)_2C{=}CHCl$ | 20 | THF, 6 hr., reflux | 96 |
| | $i$-$C_4H_9C(CH_3){=}CHCl$ | 27 | THF, 6 hr., reflux | 96 |
| | $C_6H_{10}{=}CHCl$ (cyclohexylidene) | 30 | THF, 6 hr., reflux | 96 |
| | $C_6H_5C(CH_3){=}CHCl$ | 46 | THF, 6 hr., reflux | 96 |
| | $(C_6H_5)_2C{=}CHCl$ | 20 | $(C_2H_5)_2O$, −50; 3 days, 20 | 91,92 |
| $(C_6H_5)_3P{=}CFCl$ | $(C_6H_5)_2C{=}CFCl$ | 40 | Heptane, 30 min., 40–50; 2 days, 20 | 94 |
| $(C_6H_5)_3P{=}CCl_2$ | $C_6H_{10}{=}CCl_2$ (cyclohexylidene) | 33 | Heptane, 30 min., 40–50; 2 days, 20 | 94 |
| | $C_6H_5CH{=}CCl_2$ | 48 | Heptane, 30 min., 40–50; 2 days, 20 | 94 |
| | | 72 | $CCl_4$, 2 hr., 60 | 98 |
| | 2,6-$Cl_2C_6H_3CH{=}CCl_2$ | 46 | Heptane, 30 min., 40–50; 2 days, 20 | 94 |
| | 3,4-$Cl_2C_6H_3CH{=}CCl_2$ | 43 | Heptane, 30 min., 40–50; 2 days, 20 | 94 |
| | $p$-$(CH_3)_2NC_6H_4CH{=}CCl_2$ | 81 | Heptane, 30 min., 40–50; 2 days, 20 | 94 |
| | $p$-$O_2NC_6H_4CH{=}CCl_2$ | 83 | Heptane, 30 min., 40–50; 2 days, 20 | 94 |
| | $C_6H_5CH{=}CHCH{=}CCl_2$ | 77 | Heptane, 30 min., 40–50; 2 days, 20 | 94 |
| | $n$-$C_{11}H_{23}CH{=}CCl_2$ | 29 | Heptane, 30 min., 40–50; 2 days, 20 | 94 |
| | $(C_6H_5)_2C{=}CCl_2$ | 59 | Heptane, 30 min., 40–50; 2 days, 20 | 94 |
| | | 46 | Pentane/$(C_2H_5)_2O$ | 93 |
| | | 78 | $CCl_4$, 4 hr., 60 | 98 |
| $(C_6H_5)_3P{=}CHBr$ | (2,6,6-trimethylcyclohex-1-enyl)$CH{=}CHC(CH_3){=}CHBr$ | 70 | $(C_2H_5)_2O$ | 84 |
| $(C_6H_5)_3P{=}CBr_2$ | $C_6H_5CH{=}CBr_2$ | 42 | Heptane, 30 min., 40–50; 2 days, 20 | 94 |
| | | 84 | $CH_2Cl_2$ | 82 |
| | $(C_6H_5)_2C{=}CBr_2$ | 9 | Heptane, 30 min., 40–50; 2 days, 20 | 94 |

*Note:* References 391 to 404 are on p. 490.

## TABLE VI—*Continued*

### Olefins Prepared by the Wittig Reaction

*A. Mono-ylides as Starting Materials—Continued*

| | Ylide | Product | Yield, % | Solvent, Time, Temperature, °C. | References |
|---|---|---|---|---|---|
| $C_2$ | $(C_6H_5)_3P{=}CHCH_3$ | $CH_3O_2CC(CH_3){=}CHCH_3$ | 21 | $(C_2H_5)_2O$, 12 hr., reflux | 164 |
| | | $i\text{-}C_3H_7C(CH_3){=}CHCH_3$ | | $(C_2H_5)_2O$, 1 hr., 20; 3 hr., 65 | 306 |
| | | 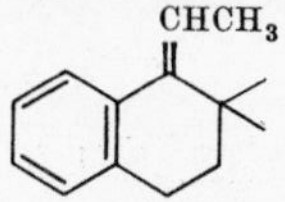 CHCH₃ (structure) | | $(C_2H_5)_2O$, 46 hr., 20 | 174 |
| | | 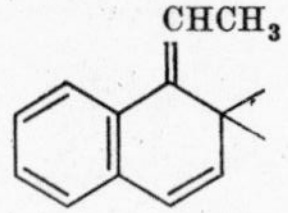 CHCH₃ (structure) | 70 | $(C_2H_5)_2O$, 7 days, 20 | 174 |
| | | 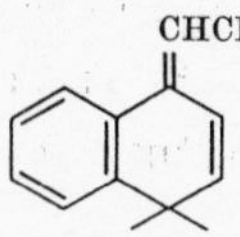 CHCH₃ (structure) | 51 | $(C_2H_5)_2O$, 50 hr., 60 | 174 |
| | | $(C_6H_5)_2C{=}CHCH_3$ | 98 | $(CH_3)_2SO$, 3 hr., 25 | 62 |
| | | $CH_3CO_2CH_2(CH_2)_8CH{=}CHCH{=}CHCH_3$ | 67 | $(C_2H_5)_2O$, 0 | 327 |
| | | $C_6H_5(C{\equiv}C)_3(CH{=}CH)_2CH_3$ | 5 | $(C_2H_5)_2O$, 1 hr., reflux | 261 |
| | | $(C_6H_5)_3CCH_2CH{=}CHCH_3$ | 58 | $(C_2H_5)_2O$, 3 hr., 60 | 178 |
| | | 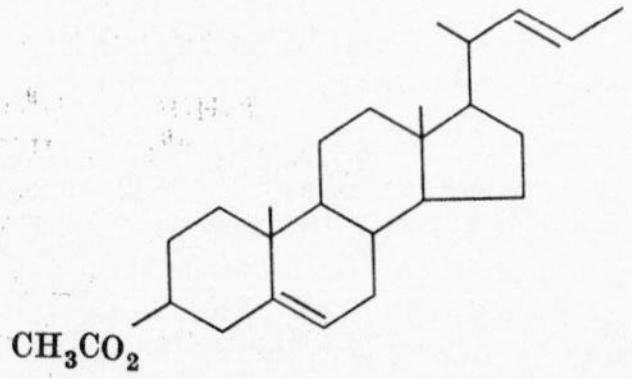 $CH_3CO_2$ (steroid structure) | | $(C_2H_5)_2O$, 1 hr., 20; 15 hr., 65 | 378 |

| | | | | |
|---|---|---|---|---|
| | 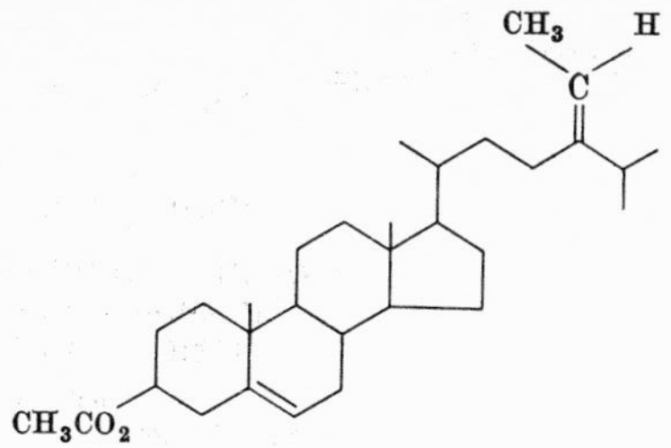 $CH_3CO_2$ (steroid structure; $CH_3$, H, C) | | $(C_2H_5)_2O$, 1 hr., 20; 5 hr., 65 | 307 |
| | 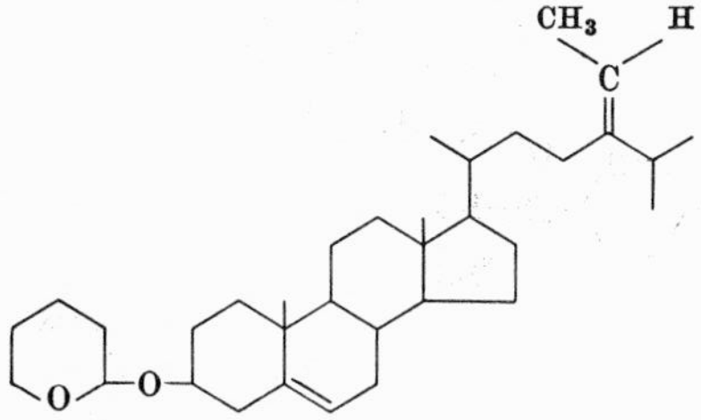 (steroid structure with tetrahydropyranyl ether; O, O; $CH_3$, H, C) | | $(C_2H_5)_2O$, 1 hr., 20; 16 hr., 65 | 306 |
| $(C_6H_5)_3P{=}CHOCH_3$ | $(C_2H_5)_2C{=}CHOCH_3$ | 57 | $(C_2H_5)_2O$, 5 hr., 20; 15 hr., 50 | 248 |
| |  (cyclopentylidene)$=CHOCH_3$ | 15 | $(C_2H_5)_2O$, −40; 30 hr., 20 | 248 |
| | 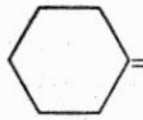 (cyclohexylidene)$=CHOCH_3$ | 71 | $(C_2H_5)_2O$, −30; 15 hr., 20 | 246, 248 |
| | $C_6H_5CH{=}CHOCH_3$ | 40 | $(C_2H_5)_2O$, 2 hr., 20 | 247 |
| | | 61 | THF, 20 | 92 |
| | (cyclooctylidene)$=CHOCH_3$ | 62 | $C_2H_5OH$, 5 hr., 65 | 92 |
| | 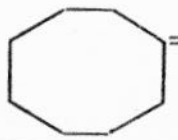 | 62 | $C_2H_5OH$, 15 hr., 50 | 248 |

*Note:* References 391 to 404 are on p. 490.

## TABLE VI—*Continued*

### Olefins Prepared by the Wittig Reaction

*A. Mono-ylides as Starting Materials—Continued*

| | Ylide | Product | Yield, % | Solvent, Time, Temperature, °C. | References |
|---|---|---|---|---|---|
| $C_2$ | $(C_6H_5)_3P{=}CHOCH_3$ (*contd.*) | $C_6H_5C(CH_3){=}CHOCH_3$ | 40 | $(C_2H_5)_2O$ | 246 |
| | | $C_6H_5CH{=}CHCH{=}CHOCH_3$ | 50 | $C_2H_5OH$, 1 day, 50 | 92 |
| | | $o$-$C_6H_4(CH{=}CHOCH_3)_2$ | 55 | $C_2H_5OH$, 60 hr., 50 | 92 |
| | | $(C_6H_5)_2C{=}CHOCH_3$ | 83 | $(C_2H_5)_2O$, 2 hr., 20 | 92, 247 |
| | | 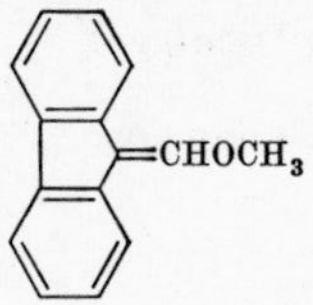 $=CHOCH_3$ | 60 | $CH_3OH$, 53 hr., 50 | 92 |
| | | 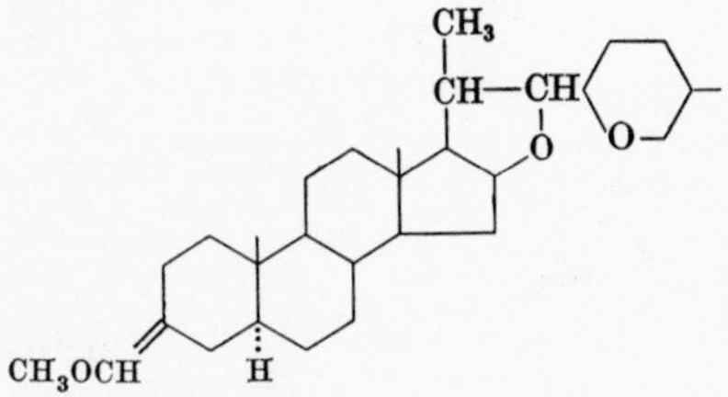 $CH_3$ CH—CH O O $CH_3OCH$ H | 85 | $(C_2H_5)_2O$ | 246 |
| | $(C_6H_5)_3P{=}CHSCH_3$ | $C_6H_5CH{=}CHSCH_3$ | 70 | $THF/(C_2H_5)_2O$, 60 hr., 55 | 92 |
| | | $(C_6H_5)_2C{=}CHSCH_3$ | 84 | $(C_2H_5)_2O$, 40 hr., 20 | 92 |
| | $(C_6H_5)_3P{=}CHCHO$ | $n$-$C_6H_{13}CH{=}CHCHO$ | 81 | $C_6H_6$, 1 day, reflux | 33, 58 |
| | | $C_6H_5CH{=}CHCHO$ | 60 | $C_6H_6$, 1 day, reflux | 33, 58 |
| | | 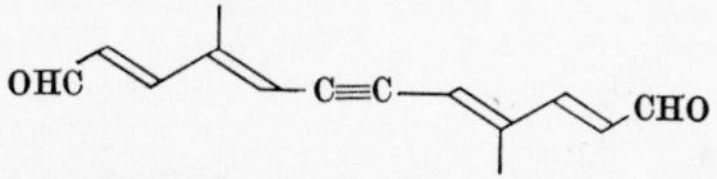 OHC C≡C CHO | 15 | $C_6H_6$, 16 hr., reflux | 33, 58 |

| | | | | | |
|---|---|---|---|---|---|
| | $(C_6H_5)_3P{=}CClCHO$ | $C_6H_5CH{=}CClCHO$ | 52 | $CH_6$, 20 hr., reflux | 59 |
| | $(C_6H_5)_3P{=}CBrCHO$ | $C_6H_5CH{=}CBrCHO$ | 34 | $C_6H_6$, 20 hr., reflux | 59 |
| | $(C_6H_5)_3P{=}CHCONH_2$ | $CH_3CH{=}CHCH{=}CHCONH_2$ | | $C_2H_5OH$ | 243 |
| | | $n$-$C_6H_{13}CH{=}CHCONH_2$ | 14 | $C_6H_6$, 20 hr., reflux | 58 |
| | | $C_6H_5CH{=}CHCONH_2$ | | $C_6H_6$, 20 hr., reflux | 58 |
| | $(C_6H_5)_3P{=}CHCN$ | $C_2H_5OCH{=}CHCN$ | | $C_2H_5OH$, 2 days, 20 | 113 |
| | | $C_6H_5CH{=}CHCN$ | 84 | $C_6H_6$, 4 hr., reflux | 43, 58 |
| | | $o$-$ClC_6H_4CH{=}CHCN$ | 73 | $C_6H_6$, 4 hr., reflux | 43 |
| | | $p$-$ClC_6H_4CH{=}CHCN$ | 78 | $C_6H_6$, 4 hr., reflux | 43 |
| | | $m$-$O_2NC_6H_4CH{=}CHCN$ | 72 | $C_6H_6$, 6–10 hr., 50–60 | 42 |
| | | $p$-$O_2NC_6H_4CH{=}CHCN$ | 74 | $C_6H_6$, 6–10 hr., 50–60 | 42 |
| | | 2,4,6-$(O_2N)_3C_6H_2CH{=}CHCN$ | 71 | $C_6H_6$, 6–10 hr., 50–60 | 42 |
| | | $p$-$CH_3OC_6H_4CH{=}CHCN$ | 71 | $C_6H_6$, 4 hr., reflux | 43 |
| | | 3-$CH_3O$-4-$HOC_6H_3CH{=}CHCN$ | 85 | $C_6H_6$, 4 hr., reflux | 43 |
| | | 3,4-$(CH_2O_2)C_6H_3CH{=}CHCN$ | 82 | $C_6H_6$, 4 hr., reflux | 43 |
| | | 3,4-$(CH_3O)_2C_6H_3CH{=}CHCN$ | 92 | $C_6H_6$, 4 hr., reflux | 43 |
| | | 2,3,4-$(CH_3O)_3C_6H_2CH{=}CHCN$ | 72 | $C_6H_6$, 4 hr., reflux | 43 |
| | | 3,4,5-$(CH_3O)_3C_6H_2CH{=}CHCN$ | 99 | $C_6H_6$, 4 hr., reflux | 43 |
| $C_3$ | $(C_6H_5)_3P{=}CHC_2H_5$ | $C_2H_5CH{=}CHOC_2H_5$ | | $(C_2H_5)_2O$, 12 hr., 20 | 113 |
| | $(C_6H_5)_3P{=}C(CH_3)_2$ | $(CH_3)_2C{=}CHOC_2H_5$ | | $(C_2H_5)_2O$, 12 hr., 20 | 113 |
| | | $(t$-$C_4H_9)_2CHCH{=}C(CH_3)_2$ | 64 | $(C_2H_5)_2O$/THF, 12 hr., 20 | 402 |
| | | $(C_6H_5)_2C{=}C(CH_3)_2$ | 76 | $(C_2H_5)_2O$, 6 hr., 20 | 185 |
| | | $(C_6H_5)_2C{=}C{=}C(CH_3)_2$ | 64 | None, 30 min., 145–160 | 150 |
| | | 2,4,6-$(CH_3)_3C_6H_2(C_6H_5)C{=}C{=}C(CH_3)_2$ | 50 | None, 30 min., 145–160 | 150 |
| | | $(C_6H_5)_3CCH_2CH{=}C(CH_3)_2$ | 68 | $(C_2H_5)_2O$ | 178 |
| | | 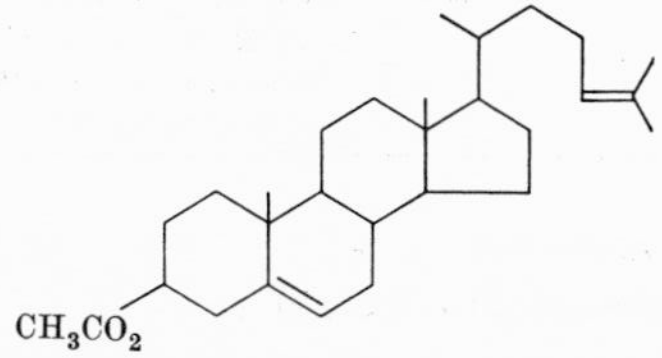 $CH_3CO_2$ | | $(C_2H_5)_2O$, 1 hr., 20; 6 hr., 65 | 305 |

*Note:* References 391 to 404 are on p. 490.

# TABLE VI—*Continued*

## Olefins Prepared by the Wittig Reaction

### *A. Mono-ylides as Starting Materials—Continued*

| Ylide | Product | Yield, % | Solvent, Time, Temperature, °C. | References |
|---|---|---|---|---|
| $C_3$ (*contd.*) | | | | |
| $(C_6H_5)_3P{=}CHCH{=}CH_2$ | $HC{\equiv}CCH{=}CHCH{=}CH_2$ | | $(C_2H_5)_2O$, 2 hr., 20 | 179 |
| | $C_6H_5CH{=}CHCH{=}CH_2$ | 58 | $(C_2H_5)_2O$, 2 hr., 20; 2 hr., 65 | 2 |
| | $C_6H_5N(CH_3)CH{=}CHCH{=}CH_2$ | 20 | $(C_2H_5)_2O$, 2 hr., 20; 2 hr., 65 | 255 |
| | $CH_3CH{=}CH(C{\equiv}C)_2(CH{=}CH)_2CH{=}CH_2$ | 50 | $(C_2H_5)_2O$, 2 hr., 20 | 259 |
| | 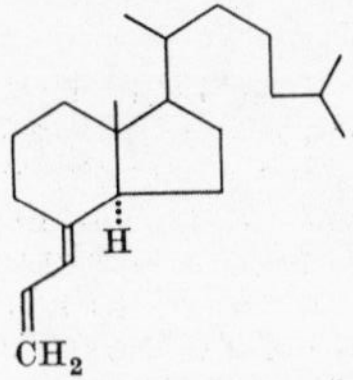 | | $(C_2H_5)_2O$, 2 days, 20 | 318 |
| $(C_6H_5)_3P{=}CHCH_2CH_2O^{\ominus}$ | $C_6H_5CH{=}CHCH_2CH_2OH$ | 65 | $(C_2H_5)_2O$ | 16 |
| $(C_6H_5)_3P{=}CHCH_2OCH_3$ | 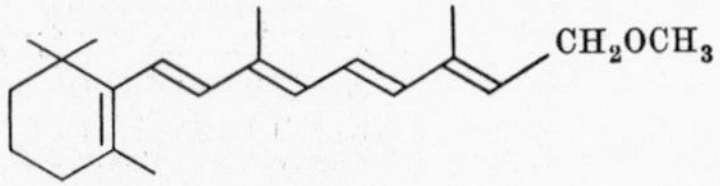 | 10 | $(C_2H_5)_2O$, 5–6 hr., reflux | 281 |
| $(C_6H_5)_3P{=}C(CH_3)CHO$ | $n$-$C_6H_{13}CH{=}C(CH_3)CHO$ | 72 | $C_6H_6$, 1 day, reflux | 33, 58 |
| | $C_6H_5CH{=}C(CH_3)CHO$ | 60 | $C_6H_6$, 1 day, reflux | 33, 58 |
| $(C_6H_5)_3P{=}CHCOCH_3$ | $(CF_3)_2C{=}CHCOCH_3$ | 93 | $(C_2H_5)_2O$ | 219 |
| | $C_6H_5CH{=}CHCOCH_3$ | 76 | $C_6H_6$, 3 days, reflux | 32 |
| | | | THF, 2 days, reflux | 31 |
| | $m$-$O_2NC_6H_4CH{=}CHCOCH_3$ | 80 | $C_6H_6$, 6–10 hr., 50–60 | 42 |
| | $p$-$O_2NC_6H_4CH{=}CHCOCH_3$ | 92 | $C_6H_6$, 6–10 hr., 50–60 | 42 |
| | 2,4,6-$(O_2N)_3C_6H_2CH{=}CHCOCH_3$ | 78 | $C_6H_6$, 6–10 hr., 50–60 | 42 |

| | | | | |
|---|---|---|---|---|
| | 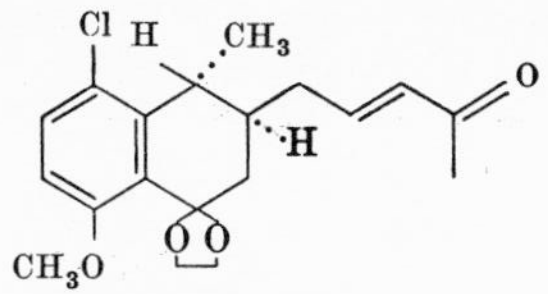 (structure: Cl, H, $CH_3$, H, $CH_3O$, O O) | | THF, 36 hr., reflux | 404 |
| $(C_6H_5)_3P{=}CClCOCH_3$ | $C_6H_5CH{=}CClCOCH_3$ | 98 | None, 30 hr., 90–95 | 60 |
| | $p$-$O_2NC_6H_4CH{=}CClCOCH_3$ | 48 | $C_6H_6$, 35 hr., reflux | 59 |
| $(C_6H_5)_3P{=}CBrCOCH_3$ | $CCl_2{=}CClCH{=}CBrCOCH_3$ | 59 | $C_6H_6$, 8 hr., reflux | 59 |
| $(C_6H_5)_3P{=}CHCO_2CH_3$ | $OHC(CH_2)_3CH{=}CHCO_2CH_3$ | 83 | None, 90 min., 80–90 | 228, 229 |
| | 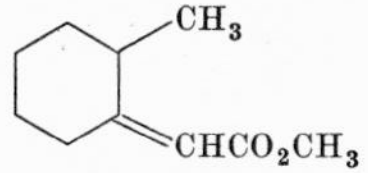 (structure: $CH_3$, $CHCO_2CH_3$) | 30 | None, 150 | 36 |
| | $o$-$O_2NC_6H_4CH{=}CHCO_2CH_3$ | | $C_6H_6$, 4 hr., reflux | 162 |
| | $m$-$O_2NC_6H_4CH{=}CHCO_2CH_3$ | | $C_6H_6$, 10 hr., reflux | 162 |
| | $p$-$O_2NC_6H_4CH{=}CHCO_2CH_3$ | | $C_6H_6$, 10 hr., reflux | 162 |
| | 2,4,6-$(O_2N)_3C_6H_2CH{=}CHCO_2CH_3$ | | $C_6H_6$, 10 hr., reflux | 162 |
| | 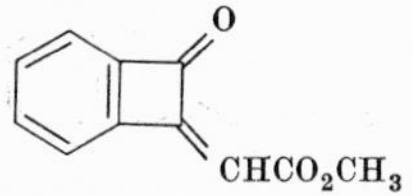 (structure: O, $CHCO_2CH_3$) | 93 | $CH_2Cl_2$, 8 hr., 20 | 227 |
| | $o$-$O_2NC_6H_4CH{=}CHCH{=}CHCO_2CH_3$ | | $C_6H_6$, 10 hr., reflux | 162 |
| | $CH_3O_2C(CH_2)_8CH{=}CHCO_2CH_3$ | 95 | $C_6H_6$, 6 hr., reflux | 161, 327 |
| | 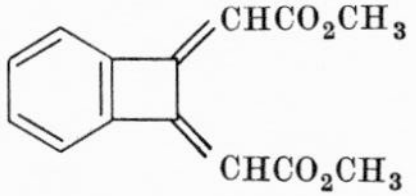 (structure: $CHCO_2CH_3$, $CHCO_2CH_3$) | 85 | $CH_2Cl_2$, 12 hr., 20 | 227 |
| | 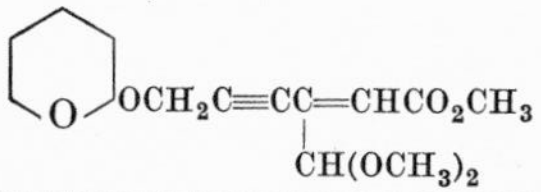 (structure: O) $OCH_2C{\equiv}CC{=}CHCO_2CH_3$, $CH(OCH_3)_2$ | 60 | $C_6H_6$, 1 day, reflux | 237 |

*Note:* References 391 to 404 are on p. 490.

## TABLE VI—*Continued*

### Olefins Prepared by the Wittig Reaction

*A. Mono-ylides as Starting Materials—Continued*

| Ylide | Product | Yield, % | Solvent, Time, Temperature, °C. | References |
|---|---|---|---|---|
| $C_3$ $(C_6H_5)_3P{=}CHCO_2CH_3$ (*contd.*) | $CH_3O$, $CH_3O$, N, $C_2H_5$, $CHCO_2CH_3$ | 58 | Xylene, reflux | 36 |
| | | | None, 150 | 36 |
| | 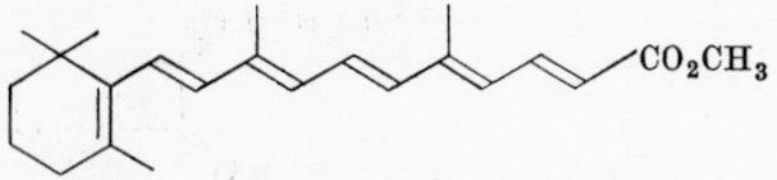 $CO_2CH_3$ | | $CH_2Cl_2$, 5 hr., reflux | 298 |
| | 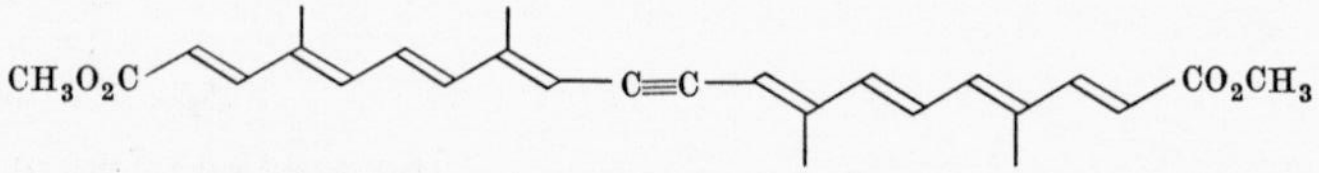 $CH_3O_2C$ … $CO_2CH_3$ | | $CH_2Cl_2$, 5 hr., reflux | 231 |
| | | 80 | $C_6H_6$, 6 hr., reflux | 39 |
| | $CH_3O_2C$ … $C{\equiv}C$ … $CO_2CH_3$ | | $CH_2Cl_2$, 5 hr., reflux | 231 |
| | 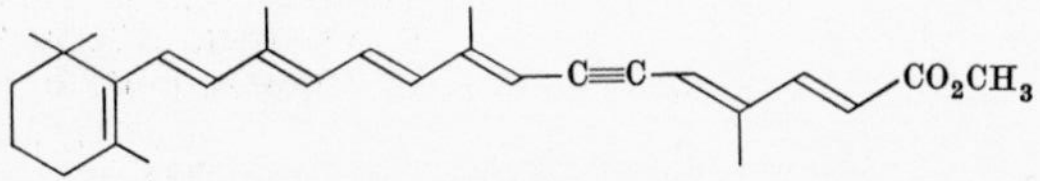 $C{\equiv}C$ … $CO_2CH_3$ | | $CH_2Cl_2$, 5 hr., reflux | 295–297 |
| | 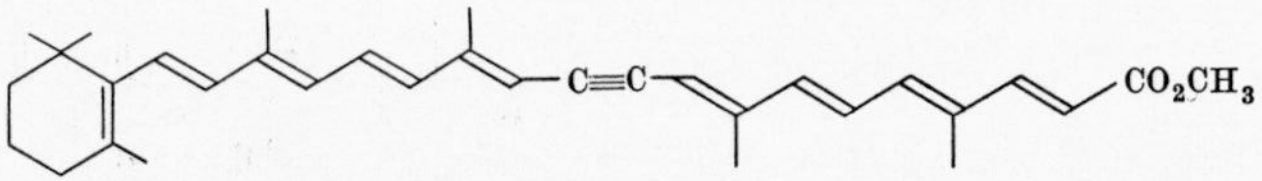 $C{\equiv}C$ … $CO_2CH_3$ | | $CH_2Cl_2$, 5 hr., reflux | 295–297 |

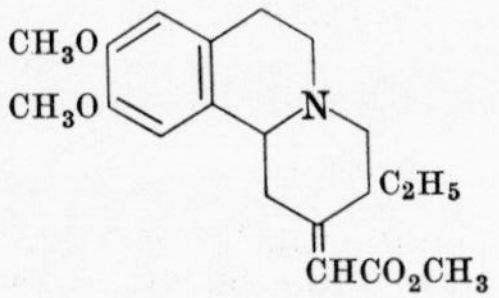

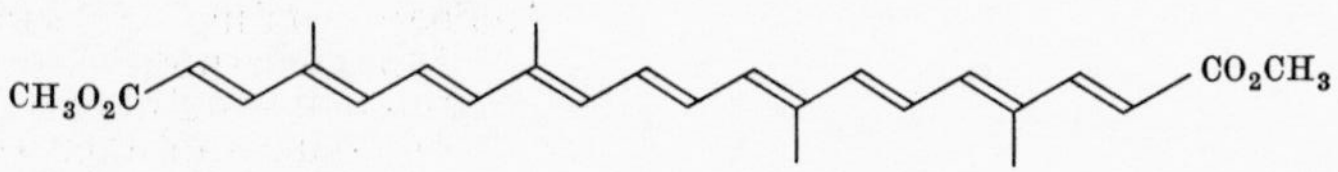

| | | | | | |
|---|---|---|---|---|---|
| | | 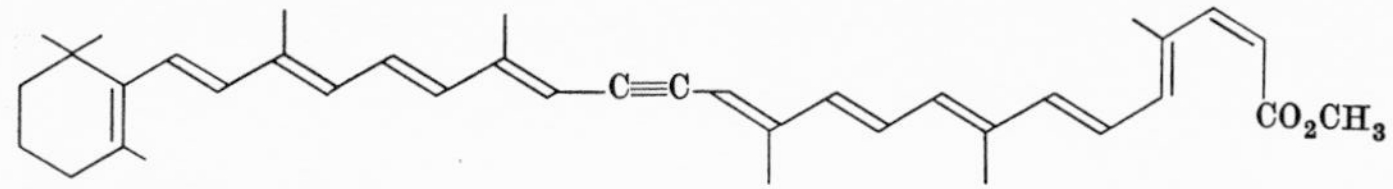 | | $CH_2Cl_2$, 5 hr., reflux | 295–297 |
| | $(C_6H_5)_3P{=}CClCO_2CH_3$ | $CCl_3CH{=}CClCO_2CH_3$ | 72 | $C_6H_6$, 3 hr., reflux | 117 |
| | | $CCl_2{=}CHCH{=}CClCO_2CH_3$ | 79 | $C_6H_6$, 3 hr., reflux | 117 |
| | | $CCl_2{=}CClCH{=}CClCO_2CH_3$ | 95 | $C_6H_6$, 3 hr., reflux | 117 |
| | | $C_6H_5CH{=}CClCO_2CH_3$ | 72 | $C_6H_6$, 3 hr., reflux | 117 |
| | | *o*-$HOC_6H_4CH{=}CClCO_2CH_3$ | 71 | $C_6H_6$, 3 hr., reflux | 117 |
| | | $C_6H_5CH{=}CHCH{=}CClCO_2CH_3$ | 78 | $C_6H_6$, 3 hr., reflux | 117 |
| | $(C_6H_5)_3P{=}CBrCO_2CH_3$ | $CH_3CH{=}CBrCO_2CH_3$ | 61 | $C_6H_6$, 3 hr., reflux | 117 |
| | | $CCl_2{=}CClCH{=}CBrCO_2CH_3$ | 94 | $C_6H_6$, 3 hr., reflux | 117 |
| | | $CH_3CH{=}CHCH{=}CBrCO_2CH_3$ | 84 | $C_6H_6$, 3 hr., reflux | 117 |
| | | 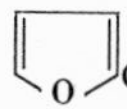 $CH{=}CBrCO_2CH_3$ | 78 | $C_6H_6$, 3 hr., reflux | 117 |
| | | $C_6H_5CH{=}CBrCO_2CH_3$ | 71 | $C_6H_6$, 3 hr., reflux | 117 |
| | | *o*-$HOC_6H_4CH{=}CBrCO_2CH_3$ | 64 | $C_6H_6$, 3 hr., reflux | 117 |
| | | $C_6H_5CH{=}CHCH{=}CBrCO_2CH_3$ | Quant. | $C_6H_6$, 3 hr., reflux | 117 |
| $C_4$ | $(C_6H_5)_3P{=}CHC_3H_7$-*n* | $(CH_3)_2C{=}CHC_3H_7$-*n* | 20 | $(C_2H_5)_2O$, 4 hr., 60–75 | 182 |
| | | 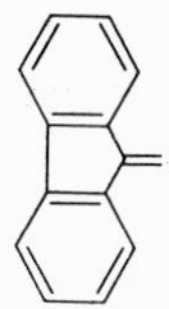 $=CHC_3H_7$-*n* | | $(C_2H_5)_2O$ | 130 |
| | | $CH_3CO_2(CH_2)_9(CH{=}CH)_2C_3H_7$-*n* | 72 | Pet. ether, 1 hr., −20; 3 hr., 0; 3 hr., 20 | 161, 327 |
| | $(C_6H_5)_3P{=}CH(CH_2)_3Br$ | $C_6H_5CH{=}CH(CH_2)_3Br$ | 74 | $C_2H_5OH$, 3 days, 20 | 74 |
| | $(C_6H_5)_3P{=}CHCH{=}CHCH_3$ | $HC{\equiv}CCH{=}CHCH{=}CHCH_3$ | | $(C_2H_5)_2O$, 1 hr., 20 | 399 |
| | | $HC{\equiv}C(CH{=}CH)_3CH_3$ | 50 | $(C_2H_5)_2O$, 1 hr., 20 | 71 |
| | | $CH_3CO_2(CH_2)_9CH{=}CHCH{=}CHCH_3$ | 68 | $(C_2H_5)_2O$, 0 | 328 |
| | | $CH_3O_2C(CH_2)_9CH{=}CHCH{=}CHCH_3$ | 69 | $(C_2H_5)_2O$, 4 hr., 0; 3 hr., 20 | 328 |
| | | $CH_3(CH{=}CH)_3(C{\equiv}C)_2(CH{=}CH)_3CH_3$ | | $(C_2H_5)_2O$/THF, 30 min., 20 | 181 |

*Note:* References 391 to 404 are on p. 490.

# TABLE VI—*Continued*

## Olefins Prepared by the Wittig Reaction

### *A. Mono-ylides as Starting Materials—Continued*

| Ylide | Product | Yield, % | Solvent, Time, Temperature, °C. | References |
|---|---|---|---|---|
| $C_4$ (*contd.*) | | | | |
| $(C_6H_5)_3P{=}CHC(CH_3){=}CH_2$ | $i\text{-}C_3H_7CH{=}CHC(CH_3){=}CH_2$ | | THF, 4 hr., reflux | 394 |
| | cyclohexylidene$=CHC(CH_3){=}CH_2$ | | THF, 4 hr., reflux | 394 |
| $(C_6H_5)_3P{=}CClCOC_2H_5$ | $CCl_2{=}CClCH{=}CClCOC_2H_5$ | 53 | $C_6H_6$, 8 hr., reflux | 59 |
| $(C_6H_5)_3P{=}CBrCOC_2H_5$ | $CCl_2{=}CClCH{=}CBrCOC_2H_5$ | 87 | $C_6H_6$, 8 hr., reflux | 59 |
| $(C_6H_5)_3P{=}CHCO_2C_2H_5$ | $C_2H_5CH{=}CHCO_2C_2H_5$ | 49 | $C_6H_6$, 6 hr., reflux | 220 |
| | $(CH_3)_2C{=}CHCO_2C_2H_5$ | 45 | None, 1 day, 100 | 35 |
| | $CH_2FC(CH_3){=}CHCO_2C_2H_5$ | | | 218 |
| | $CH_3CHBrCH{=}CHCO_2C_2H_5$ | 65 | $C_6H_6$, 6 hr., reflux | 220 |
| | $CH_2{=}CHCH{=}CHCO_2C_2H_5$ | 34 | $C_6H_6$, 6 hr., reflux | 220 |
| | $C_2H_5C(CH_3){=}CHCO_2C_2H_5$ | 42 | None, 1 day, 100 | 35 |
| | $CH_3(CH{=}CH)_2CO_2C_2H_5$ | 80 | $C_6H_6$, 6 hr., reflux | 220 |
| | $C_2H_5OCH_2CH_2CH{=}CHCO_2C_2H_5$ | 50 | $C_6H_6$, 6 hr., reflux | 220 |
| | cyclopentylidene$=CHCO_2C_2H_5$ | 43 | None, 10 hr., 170 | 35 |
| | 2-furyl$CH{=}CHCO_2C_2H_5$ | 51 | $C_6H_6$, 6 hr., reflux | 220 |
| | $CHCO_2C_2H_5{=}CH{-}HCO{-}H_2CO$ (1,2-O-isopropylidene, $C(CH_3)_2$) | 75 | | 238 |
| | $CH_3(CH{=}CH)_3CO_2C_2H_5$ | 87 | $C_6H_6$, 6 hr., reflux | 220 |

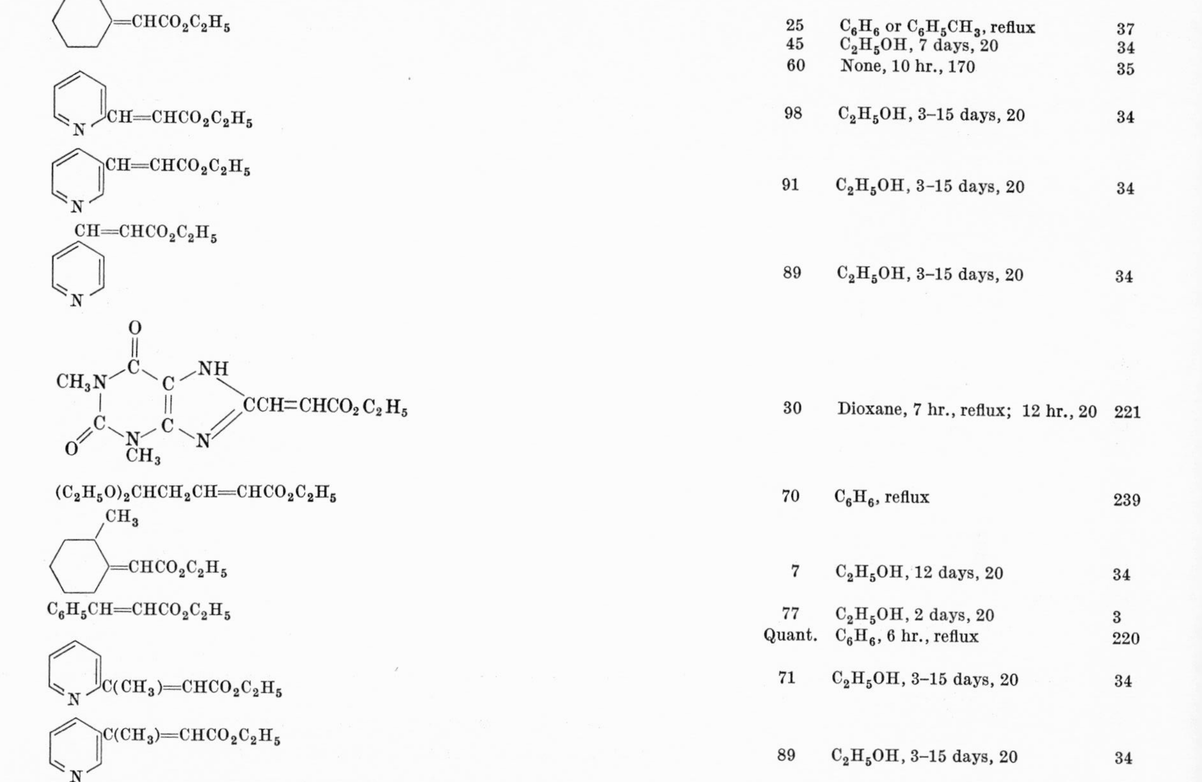

| Compound | Yield (%) | Conditions | Ref. |
|---|---|---|---|
| | 25 | $C_6H_6$ or $C_6H_5CH_3$, reflux | 37 |
| | 45 | $C_2H_5OH$, 7 days, 20 | 34 |
| | 60 | None, 10 hr., 170 | 35 |
| | 98 | $C_2H_5OH$, 3–15 days, 20 | 34 |
| | 91 | $C_2H_5OH$, 3–15 days, 20 | 34 |
| | 89 | $C_2H_5OH$, 3–15 days, 20 | 34 |
| | 30 | Dioxane, 7 hr., reflux; 12 hr., 20 | 221 |
| $(C_2H_5O)_2CHCH_2CH{=}CHCO_2C_2H_5$ | 70 | $C_6H_6$, reflux | 239 |
| | 7 | $C_2H_5OH$, 12 days, 20 | 34 |
| $C_6H_5CH{=}CHCO_2C_2H_5$ | 77 | $C_2H_5OH$, 2 days, 20 | 3 |
| | Quant. | $C_6H_6$, 6 hr., reflux | 220 |
| | 71 | $C_2H_5OH$, 3–15 days, 20 | 34 |
| | 89 | $C_2H_5OH$, 3–15 days, 20 | 34 |

*Note:* References 391 to 404 are on p. 490.

# TABLE VI—*Continued*

## Olefins Prepared by the Wittig Reaction

### *A. Mono-ylides as Starting Materials—Continued*

| | Ylide | Product | Yield, % | Solvent, Time, Temperature, °C. | References |
|---|---|---|---|---|---|
| $C_4$ | $(C_6H_5)_3P{=}CHCO_2C_2H_5$ (*contd.*) | 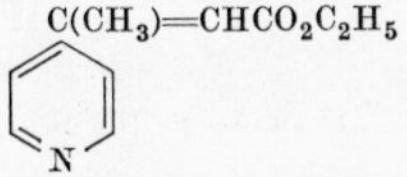 pyridyl–$C(CH_3){=}CHCO_2C_2H_5$ | 90 | $C_2H_5OH$, 3–15 days, 20 | 34 |
| | | $CH_3(CH{=}CH)_4CO_2C_2H_5$ | 82 | $C_6H_6$, 6 hr., reflux | 220 |
| | | 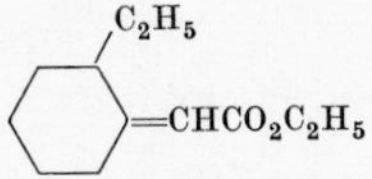 2-$C_2H_5$-cyclohexylidene$={CHCO_2C_2H_5}$ | 6 | $C_2H_5OH$, 12 days, 20 | 34 |
| | | $C_6H_5C(CH_3){=}CHCO_2C_2H_5$ | 58 | None, 10 hr., 170 | 35 |
| | | $CH_2{=}C(CH_3)C{\equiv}CCH{=}C(CH_3)CH{=}CHCO_2C_2H_5$ | 65 | | 239 |
| | | $C_6H_5CH{=}CHCH{=}CHCO_2C_2H_5$ | 71 | $C_6H_6$, 6 hr., reflux | 220 |
| | | $CH_3(CH{=}CH)_5CO_2C_2H_5$ | 88 | $C_6H_6$, 6 hr., reflux | 220 |
| | | $(CH_3)_2C{=}CH(CH_2)_2C(CH_3){=}CHCH{=}CHCO_2C_2H_5$ | 68 | $C_6H_6$, 6 hr., reflux | 220 |
| | | $C_6H_5O_2CCH{=}C(CH_3)CH{=}CHCO_2C_2H_5$ | | | 217 |
| | | $CH_3CO_2(CH_2)_9CH{=}CHCO_2C_2H_5$ | Quant. | $C_6H_6$, 6 hr., reflux | 327 |
| | | $C_2H_5O_2C(CH{=}CH)_5CO_2C_2H_5$ | | $C_6H_6$, 6 hr., reflux | 163 |
| | | 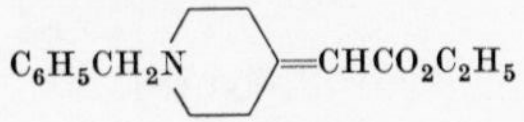 $C_6H_5CH_2N$(piperidylidene)$={CHCO_2C_2H_5}$ | 80 | $C_2H_5OH$, 3–15 days, 20 | 34 |
| | | $(C_6H_5)_2C{=}CHCO_2C_2H_5$ | 54 | None, 10 hr., 170 | 35 |
| | | 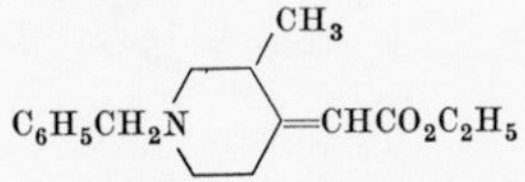 $C_6H_5CH_2N$(3-$CH_3$-piperidylidene)$={CHCO_2C_2H_5}$ | 21 | $C_2H_5OH$, 3–15 days, 20 | 34 |
| | | 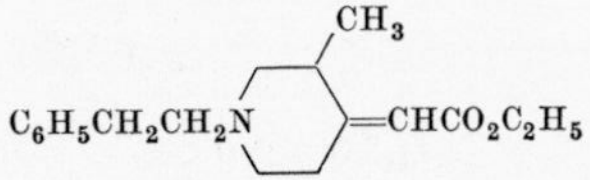 $C_6H_5CH_2CH_2N$(3-$CH_3$-piperidylidene)$={CHCO_2C_2H_5}$ | 19 | $C_2H_5OH$, 3–15 days, 20 | 34 |

| | | | | |
|---|---|---|---|---|
| | 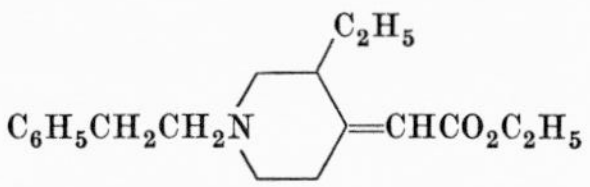 $C_6H_5CH_2CH_2N$ (piperidine ring, 3-$C_2H_5$) $=CHCO_2C_2H_5$ | 20 | $C_2H_5OH$, 3–15 days, 20 | 34 |
| | 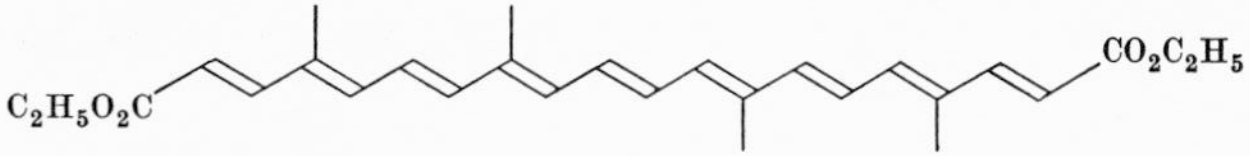 $C_2H_5O_2C$ … $CO_2C_2H_5$ | | $CH_2Cl_2$, 5 hr., reflux | 39 |
| $(C_6H_5)_3P{=}C(CH_3)CO_2CH_3$ | $CH_3CH{=}C(CH_3)CO_2CH_3$ | 90 | $CH_2Cl_2$, 12 hr., 20 | 164 |
| | $CH_2{=}CHCH{=}C(CH_3)CO_2CH_3$ | 60 | $CH_2Cl_2$, 3.5 hr., reflux | 164 |
| | $CH_3O_2C$ … $CO_2CH_3$ | | $C_6H_6$, 6 hr., reflux | 39 |
| | 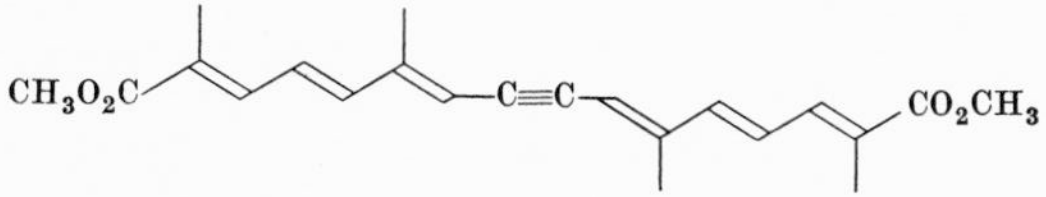 | | $CH_2Cl_2$, 5 hr., reflux | 231 |
| | $CH_3O_2C$ … $C{\equiv}C$ … $CO_2CH_3$ | | $CH_2Cl_2$, 5 hr., reflux | 231 |
| | 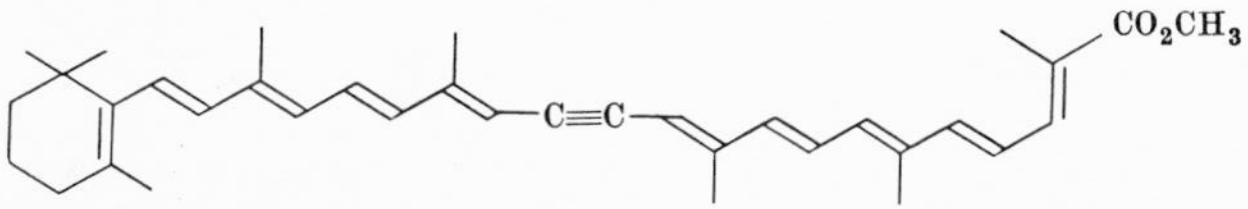 … $C{\equiv}C$ … $CO_2CH_3$ | | $CH_2Cl_2$, 6 hr., reflux | 295–297 |
| | … $C{\equiv}C$ … $CO_2CH_3$ | | $CH_2Cl_2$, 6 hr., reflux | 295–297 |
| | 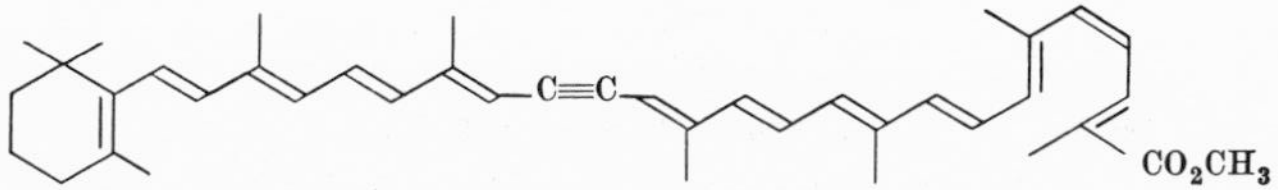 … $C{\equiv}C$ … $CO_2CH_3$ | | $CH_2Cl_2$, 6 hr., reflux | 295–297 |
| $(C_6H_5)_3P{=}CClCO_2C_2H_5$ | $C_6H_5CH{=}CClCO_2C_2H_5$ | 41 | None, 12 hr. | 60 |

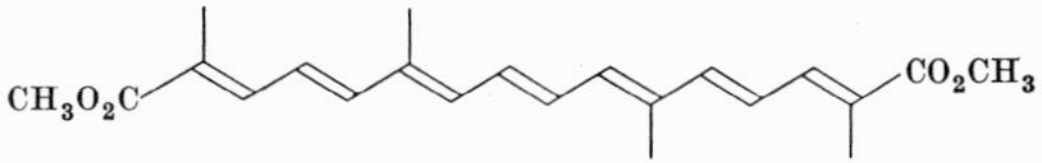

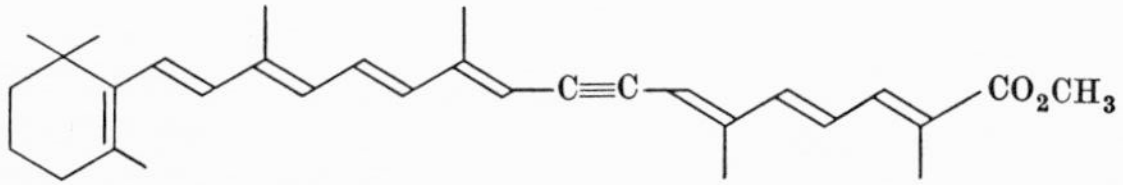

*Note:* References 391 to 404 are on p. 490.

TABLE VI—*Continued*

OLEFINS PREPARED BY THE WITTIG REACTION

*A. Mono-ylides as Starting Materials—Continued*

| | Ylide | Product | Yield, % | Solvent, Time, Temperature, °C. | References |
|---|---|---|---|---|---|
| $C_5$ | $(C_6H_5)_3P{=}CHC_4H_9$-$i$ | 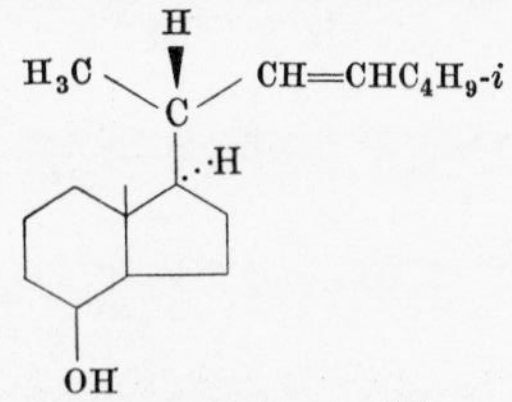 $H_3C$, H, C, $CH{=}CHC_4H_9$-$i$, H, OH (structure) | 60 | Dioxane, 12 hr., 110 | 321 |
| | | 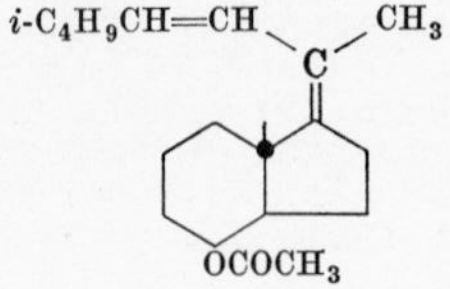 $i$-$C_4H_9CH{=}CH$, C, $CH_3$, $OCOCH_3$ (structure) | | Dioxane, 11 hr., 100 | 320 |
| | | 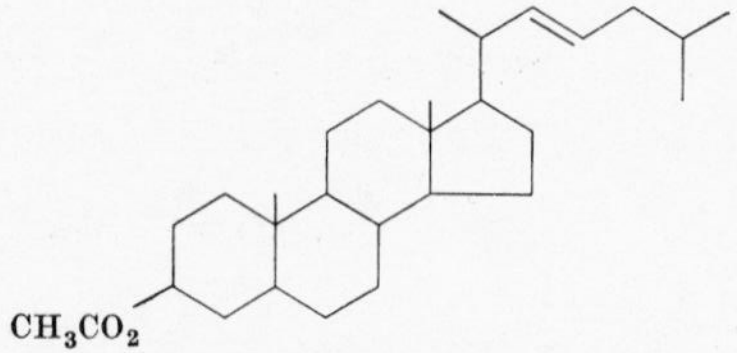 $CH_3CO_2$ (structure) | | $(C_2H_5)_2O$, 1 hr., 20; 15 hr., 65 | 378 |
| | $(C_6H_5)_3P{=}CHCH{=}CHC_2H_5$ | $HC{\equiv}C(CH{=}CH)_2C_2H_5$ | | $(C_2H_5)_2O$, 30 min., 80 | 260 |
| | $(C_6H_5)_3P{=}CHCH{=}C(CH_3)_2$ | 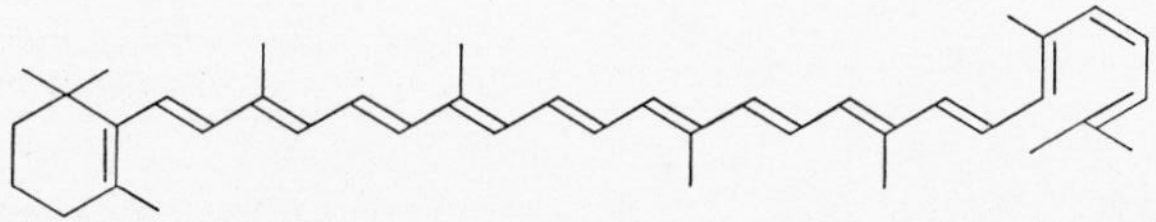 (structure) | | $CH_2Cl_2$, 20 hr., 20 | 265 |

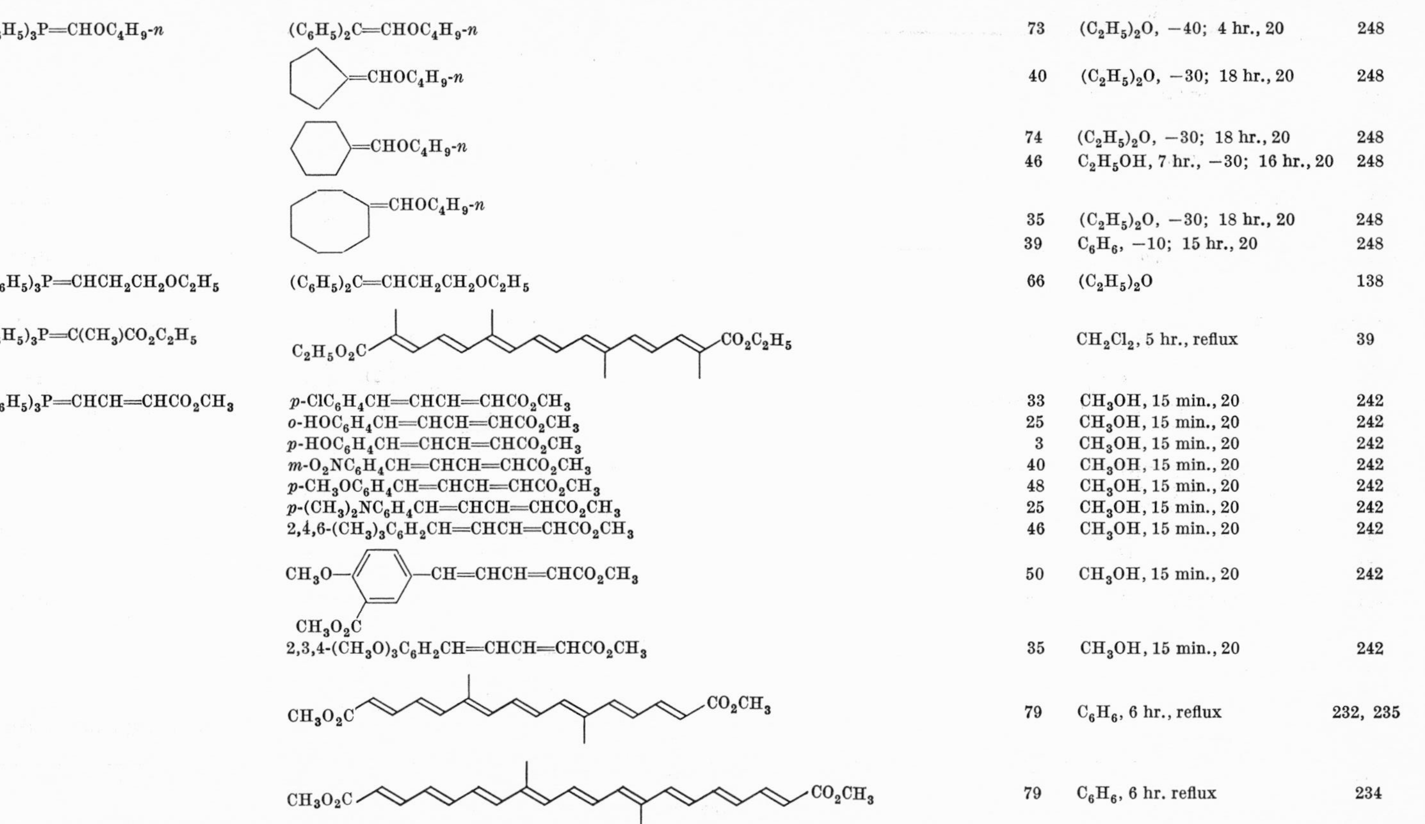

| | | | | |
|---|---|---|---|---|
| $(C_6H_5)_3P{=}CHOC_4H_9\text{-}n$ | $(C_6H_5)_2C{=}CHOC_4H_9\text{-}n$ | 73 | $(C_2H_5)_2O$, −40; 4 hr., 20 | 248 |
| | (cyclopentylidene)$=CHOC_4H_9\text{-}n$ | 40 | $(C_2H_5)_2O$, −30; 18 hr., 20 | 248 |
| | (cyclohexylidene)$=CHOC_4H_9\text{-}n$ | 74 | $(C_2H_5)_2O$, −30; 18 hr., 20 | 248 |
| | | 46 | $C_2H_5OH$, 7 hr., −30; 16 hr., 20 | 248 |
| | (cyclooctylidene)$=CHOC_4H_9\text{-}n$ | 35 | $(C_2H_5)_2O$, −30; 18 hr., 20 | 248 |
| | | 39 | $C_6H_6$, −10; 15 hr., 20 | 248 |
| $(C_6H_5)_3P{=}CHCH_2CH_2OC_2H_5$ | $(C_6H_5)_2C{=}CHCH_2CH_2OC_2H_5$ | 66 | $(C_2H_5)_2O$ | 138 |
| $(C_6H_5)_3P{=}C(CH_3)CO_2C_2H_5$ | $C_2H_5O_2C$–(polyene)–$CO_2C_2H_5$ | | $CH_2Cl_2$, 5 hr., reflux | 39 |
| $(C_6H_5)_3P{=}CHCH{=}CHCO_2CH_3$ | $p\text{-}ClC_6H_4CH{=}CHCH{=}CHCO_2CH_3$ | 33 | $CH_3OH$, 15 min., 20 | 242 |
| | $o\text{-}HOC_6H_4CH{=}CHCH{=}CHCO_2CH_3$ | 25 | $CH_3OH$, 15 min., 20 | 242 |
| | $p\text{-}HOC_6H_4CH{=}CHCH{=}CHCO_2CH_3$ | 3 | $CH_3OH$, 15 min., 20 | 242 |
| | $m\text{-}O_2NC_6H_4CH{=}CHCH{=}CHCO_2CH_3$ | 40 | $CH_3OH$, 15 min., 20 | 242 |
| | $p\text{-}CH_3OC_6H_4CH{=}CHCH{=}CHCO_2CH_3$ | 48 | $CH_3OH$, 15 min., 20 | 242 |
| | $p\text{-}(CH_3)_2NC_6H_4CH{=}CHCH{=}CHCO_2CH_3$ | 25 | $CH_3OH$, 15 min., 20 | 242 |
| | $2,4,6\text{-}(CH_3)_3C_6H_2CH{=}CHCH{=}CHCO_2CH_3$ | 46 | $CH_3OH$, 15 min., 20 | 242 |
| | $CH_3O$–$C_6H_3(CO_2CH_3)$–$CH{=}CHCH{=}CHCO_2CH_3$ | 50 | $CH_3OH$, 15 min., 20 | 242 |
| | $2,3,4\text{-}(CH_3O)_3C_6H_2CH{=}CHCH{=}CHCO_2CH_3$ | 35 | $CH_3OH$, 15 min., 20 | 242 |
| | $CH_3O_2C$–(polyene)–$CO_2CH_3$ | 79 | $C_6H_6$, 6 hr., reflux | 232, 235 |
| | $CH_3O_2C$–(polyene)–$CO_2CH_3$ | 79 | $C_6H_6$, 6 hr. reflux | 234 |

*Note:* References 391 to 404 are on p. 490.

## TABLE VI—*Continued*

## Olefins Prepared by the Wittig Reaction

*A. Mono-ylides as Starting Materials—Continued*

| | Ylide | Product | Yield, % | Solvent, Time, Temperature, °C. | References |
|---|---|---|---|---|---|
| $C_5$ (*contd.*) | $(C_6H_5)_3P{=}CHC(CH_3){=}CHCN$ | CN | 86 | $CH_3CN$, 1 hr., 20; 2.5 hr., 80 | 244 |
| | $(C_6H_5)_3P{=}$(cyclopentylidene) | $C_6H_5CH{=}$(cyclopentylidene) | 65 | $(C_2H_5)_2O$, 10 hr., 30 | 185 |
| $C_6$ | $(C_6H_5)_3P{=}CHC_5H_{11}$-*n* | $CH_3CO_2(CH_2)_9CH{=}CHCH{=}CHC_5H_{11}$-*n* | 65 | $(C_2H_5)_2O$, 0 | 327 |
| | $(C_6H_5)_3P{=}CHCH{=}CHC_3H_7$-*n* | $CH_3O_2C(CH_2)_8CH{=}CHCH{=}CHC_3H_7$-*n* | 68 | $(C_2H_5)_2O$, 2 hr., 0; 1 hr., 20 | 161 |
| | | | 75 | Pet. ether, 3–12 hr., 0 | 326 |
| | | $CH_3CO_2(CH_2)_9CH{=}CHCH{=}CHC_3H_7$-*n* | 70 | Pet. ether, 2 hr., 0; 1 hr., 20 | 161, 326 |
| | | $C_2H_5O_2C(CH_2)_8CH{=}CHCH{=}CHC_3H_7$-*n* | | $C_2H_5OH$, 4 hr., reflux | 325 |
| | $(C_6H_5)_3P{=}CHCH_2CH{=}C(CH_3)_2$ | $(CH_3)_2C{=}CHCH_2CH{=}C(CH_3)_2$ | 15 | $(C_2H_5)_2O$, 4 hr., 60–75 | 182 |
| | $(C_6H_5)_3P{=}CH(CH_2)_2C(CH_3){=}CH_2$ | $(CH_3)_2C{=}CH(CH_2)_2C(CH_3){=}CH_2$ | 12 | $(C_2H_5)_2O$, 4 hr., 60–75 | 182 |
| | $(C_6H_5)_3P{=}CHC{\equiv}CC_3H_7$-*n* | $C_2H_5O_2C(CH_2)_8CH{=}CHC{\equiv}CC_3H_7$-*n* | 47 | $C_2H_5OH$, 1 day, reflux | 325 |
| | $(C_6H_5)_3P{=}CH(CH{=}CH)_2CH_3$ | $CH_3CO_2(CH_2)_9(CH{=}CH)_3CH_3$ | 76 | $(C_2H_5)_2O$, 0 | 328 |
| | | $CH_3(CH{=}CH)_4C{\equiv}C(CH{=}CH)_4CH_3$ | | $(C_2H_5)_2O$/THF, 30 min., 20 | 181 |
| | | $CH_3(CH{=}CH)_4(C{\equiv}C)_2(CH{=}CH)_4CH_3$ | | $(C_2H_5)_2O$/THF, 15 min., 20 | 181 |
| | $(C_6H_5)_3P{=}CHCH{=}C(CH_3)C{\equiv}CH$ | $HC{\equiv}CC(CH_3){=}CHCH{=}CHCH{=}C(CH_3)C{\equiv}CH$ | | $(C_2H_5)_2O$, 2 hr., reflux | 376, 377 |
| | $(C_6H_5)_3P{=}CHC(CH_3){=}CHCH_2OCH_3$ | $CH_2OCH_3$ | 10 | $C_6H_6$, 6 hr., 60 | 286*a* |
| | $(C_6H_5)_3P{=}CHCH{=}CHCO_2C_2H_5$ | $C_2H_5O_2C(CH{=}CH)_7CO_2C_2H_5$ | | $C_6H_6$, 5 hr., reflux; 12 hr., 20 | 163 |
| | $(C_6H_5)_3P{=}CHC(CH_3){=}CHCO_2CH_3$ | $CO_2CH_3$ | | $C_2H_5OH$, 2 hr., 20 | 282 |
| | | | | $C_6H_6$, 2 hr., 20; 30 min., 60 | 55 |
| | | | | DMF, 2 hr., 20 | 55 |
| | | $CO_2CH_3$ | 87 | $CH_3CN$, 1 hr., 20; 3 hr., 85 | 244 |
| | | | 90 | Tetramethylene sulfone, 1 hr., 20; 2 hr., 80 | 244 |

| | Ylide | Product | Yield (%) | Conditions | Refs. |
|---|---|---|---|---|---|
| | $(C_6H_5)_3P{=}CHCH{=}C(CH_3)CO_2CH_3$ | $C_6H_5(CH{=}CH)_2CH{=}C(CH_3)CO_2CH_3$ | 70 | $C_6H_6$, 6 hr., reflux | 232, 235 |
| | | 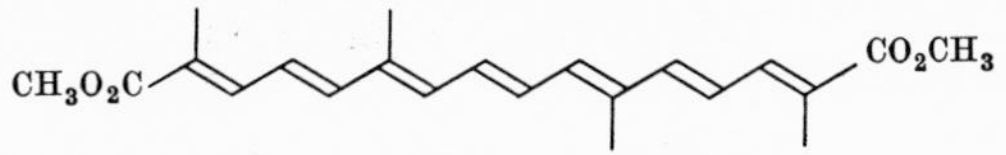 $CH_3O_2C$ … $CO_2CH_3$ | 84 | $C_6H_6$, 6 hr., reflux | 232, 235 |
| | | 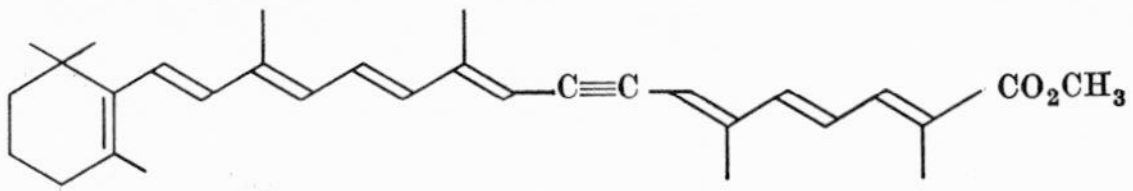 … $C{\equiv}C$ … $CO_2CH_3$ | | $CH_2Cl_2$, 6 hr., reflux | 297 |
| | $(C_6H_5)P{=}C(CH_3)CH{=}CHCO_2CH_3$ | $CH_3O_2C$ … $CO_2CH_3$ | 76 | $C_6H_6$, 6 hr., reflux | 233, 234 |
| | $(C_6H_5)_3P{=}C(CO_2CH_3)CH_2CO_2CH_3$ | $C_6H_5CH{=}C(CO_2CH_3)CH_2CO_2CH_3$ | 40 | $C_6H_6$, 12 hr., reflux | 40 |
| | $(C_6H_5)_3P{=}$cyclohexylidene  | $C_2H_5OCH{=}$cyclohexylidene 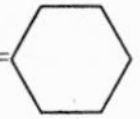 | | $(C_2H_5)_2O$, 1 day, 20; 3 hr., reflux | 113 |
| | | $C_6H_5CH{=}$cyclohexylidene 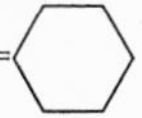 | 70 | $(C_2H_5)_2O$, 12 hr., 20; 8 hr., 50 | 185 |
| $C_7$ | $(C_6H_5)_3P{=}CHC_6H_{13}$-$n$ | $CH_2{=}CHC_6H_{13}$-$n$ | 41 | $(C_2H_5)_2O$, 10–12 hr., 20 | 188, 189 |
| | $(C_6H_5)_3P{=}CH(CH_2)_4CH{=}CH_2$ | $CH_3(C{\equiv}C)_3(CH{=}CH)_2(CH_2)_4CH{=}CH_2$ | | $(C_2H_5)_2O$, 2 hr., reflux | 160 |
| | $(C_6H_5)_3P{=}CHCH{=}CHC_4H_9$-$n$ | $HC{\equiv}C(CH{=}CH)_3C_4H_9$-$n$ | 60 | $(C_2H_5)_2O$, 1 hr., 0 | 258 |
| | $(C_6H_5)_3P{=}CHCH{=}CHC(CH_3){=}CHCH_3$ | … $CH_3$ | | $(C_2H_5)_2O$, 2 hr., 20 | 287 |
| | | | | $C_6H_6$, 20 hr., 20 | 287 |
| | | | | DMF, 16 hr., 20 | 287 |
| | $(C_6H_5)_3P{=}CHC{\equiv}CC(CH_3){=}CHCH_3$ | … $C{\equiv}C$ … $CH_3$ | 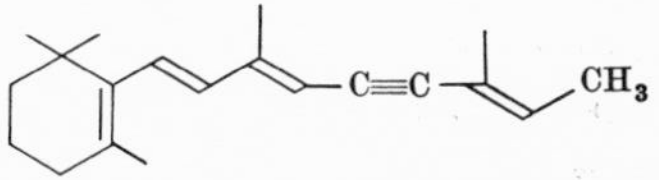 | THF, 16 hr., 20 | 209 |

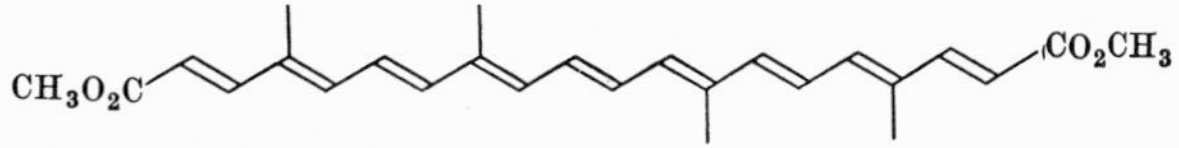

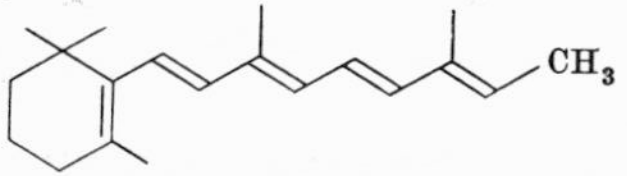

*Note:* References 391 to 404 are on p. 490.

## TABLE VI—*Continued*

### Olefins Prepared by the Wittig Reaction

*A. Mono-ylides as Starting Materials—Continued*

| Ylide | Product | Yield, % | Solvent, Time, Temperature, °C. | References |
|---|---|---|---|---|
| $C_7$ (*contd.*) | | | | |
| $(C_6H_5)_3P{=}CHCH{=}C(CH_3)CH(OC_2H_5)_2$ | 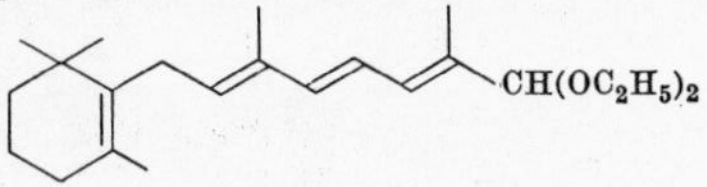 $CH(OC_2H_5)_2$ | 79 | $CH_3OH$, 15 min., 55; 8 hr., 60 | 216 |
| 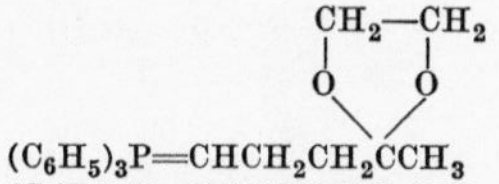 $CH_2—CH_2$ / O O / $(C_6H_5)_3P{=}CHCH_2CH_2CCH_3$ | $CH_2FC(CH_3){=}CHCH_2CH_2COCH_3$ | | | 218 |
| $(C_6H_5)_3P{=}CH(CH_2)_3CO_2C_2H_5$ | $n\text{-}C_{16}H_{33}CH{=}CH(CH_2)_3CO_2C_2H_5$ | 50 | DMF | 114 |
| 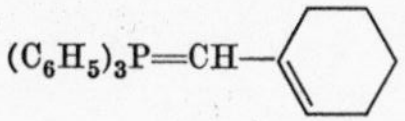 $(C_6H_5)_3P{=}CH—$ | 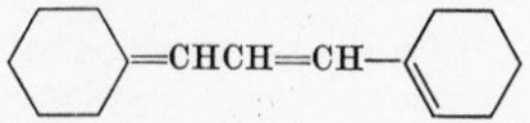 $=CHCH{=}CH—$ | | $(C_2H_5)_2O$, 3 hr., 65 | 158 |
| | 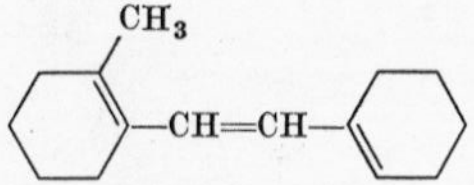 $CH_3$ / $—CH{=}CH—$ | 89 | $(C_2H_5)_2O$, 1 hr., 20; 3 hr., 65 | 314 |
| $(C_6H_5)_3P{=}CHC_6H_5$ | $C_6H_5CH{=}CH_2$ | 64–75 | $(C_2H_5)_2O$ | 188, 189 |
| | $C_6H_5CH{=}CHOC_2H_5$ | | DMF, 5 hr., 60 | 113 |
| | 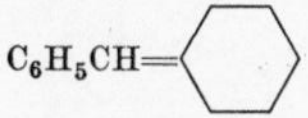 $C_6H_5CH{=}$ | 60 | $(C_2H_5)_2O$, 4 hr., 20; 1 day, 70 | 3 |
| | $C_6H_5CH{=}CHC_6H_5$ | 82 | $(C_2H_5)_2O$, 2 days, 20 | 2 |
| | | 28 | THF, 2 hr., reflux | 90 |
| | | 76 | $C_2H_5OH$, 1 day, 20 | 3 |
| | $p\text{-}ClCH_2C_6H_4CH{=}CHC_6H_5$ | | $C_2H_5OH$, 1 hr., 20 | 196 |
| | $p\text{-}BrCH_2C_6H_4CH{=}CHC_6H_5$ | 70–75 | $C_2H_5OH$, 1 hr., 20 | 196 |

| | | | | |
|---|---|---|---|---|
| | $C_6H_5CH{=}CHCH{=}CHC_6H_5$ | 78 | $(C_2H_5)_2O$, 3 hr., reflux | 209 |
| | | | THF, 12 hr., 20 | 209 |
| | | | THF, 1 hr., 60 | 209 |
| | | 84 | $C_2H_5OH$, 30 min., 20 | 66 |
| | $m$-$HOC_6H_4CH{=}CHCH{=}CHC_6H_5$ | 31 | $C_2H_5OH$, 3 days, 20 | 66 |
| | $p$-$HOC_6H_4CH{=}CHCH{=}CHC_6H_5$ | 43 | $C_2H_5OH$, 3 days, 20 | 66 |
| | 3-$CH_3O$-4-$CH_3C_6H_3CH{=}CHC_6H_5$ | 56 | $(C_2H_5)_2O$, 30 min., reflux | 194 |
| | $p$-$CH_3OC_6H_4CH{=}CHCH{=}CHC_6H_5$ | 67 | $C_2H_5OH$, 8 hr., 50 | 66 |
| | $(C_6H_5)_2C{=}CHC_6H_5$ | 46 | $(C_2H_5)_2O$, 2 days, 75 | 3 |
| | | 23 | THF, 5 hr., reflux | 90 |
| | 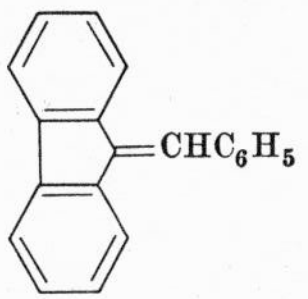 [fluorenylidene structure]$=CHC_6H_5$ | 75 | $(C_2H_5)_2O$, 1 hr., 20 | 3 |
| | $p$-$O_2NC_6H_4C(C_6H_5){=}CHC_6H_5$ | 47 | $(C_2H_5)_2O$, 10 hr., 75 | 3 |
| | 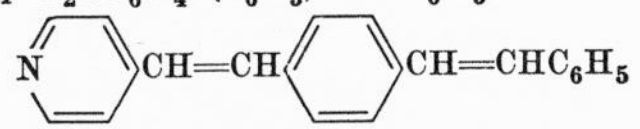 [pyridyl structure, N] $CH{=}CH$ [phenylene structure] $CH{=}CHC_6H_5$ | 52 | $C_2H_5OH$, 2 hr., 20 | 198 |
| | $m$-$C_6H_4(CH{=}CHC_6H_5)_2$ | 27 | $C_2H_5OH$, 1 day, 20 | 66 |
| | 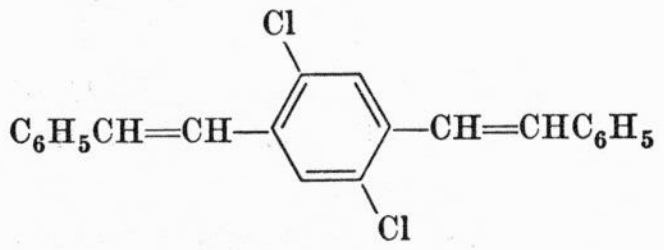 $C_6H_5CH{=}CH$—[dichlorophenylene structure, Cl, Cl]—$CH{=}CHC_6H_5$ | | DMF | 201 |
| | $C_6H_5CH{=}CHC(C_6H_5){=}CHC_6H_5$ | | $(C_2H_5)_2O$/THF, 1 hr., 20 | 213 |
| | $C_6H_5CH{=}C(CH{=}CHC_6H_5)_2$ | | $(C_2H_5)_2O$/THF, 1 hr., 20 | 213 |
| | 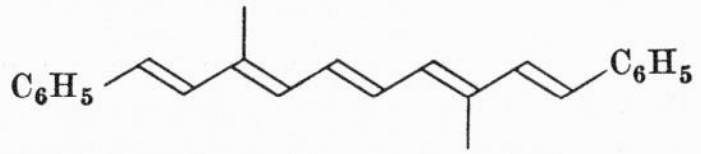 $C_6H_5$ [dimethyl polyene structure] $C_6H_5$ | | THF, 5 hr., 20 | 209 |
| | $C_6H_5CH{=}C(CH{=}CHCH{=}CHC_6H_5)_2$ | | $(C_2H_5)_2O$/THF, 1 hr. 20 | 213 |
| $(C_6H_5)_3P{=}CHC_6H_4Cl$-$p$ | $C_2H_5OCH{=}CHC_6H_4Cl$-$p$ | | $C_6H_6$, 2 days, 20 | 113 |

*Note:* References 391 to 404 are on p. 490.

## TABLE VI—*Continued*

### Olefins Prepared by the Wittig Reaction

*A. Mono-ylides as Starting Materials—Continued*

| | Ylide | Product | Yield, % | Solvent, Time, Temperature, °C. | References |
|---|---|---|---|---|---|
| $C_7$ (*contd.*) | | | | | |
| | $(C_6H_5)_3P{=}CHC_6H_3Cl_2\text{-}2,4$ | $p\text{-}C_6H_4(CH{=}CH\text{-}C_6H_3Cl_2\text{-}2,4)_2$ | | DMF | 201 |
| | $(C_6H_5)_3P{=}CHC_6H_4NO_2\text{-}m$ | $C_6H_5CH{=}CHCH{=}CHC_6H_4NO_2\text{-}m$ | 65 | $C_2H_5OH$, 20 | 210 |
| | $(C_6H_5)_3P{=}CHC_6H_4NO_2\text{-}p$ | $p\text{-}CH_3OC_6H_4CH{=}CHC_6H_4NO_2\text{-}p$ | 89 | $C_6H_6$, 4 hr., 20 | 166 |
| | | $m\text{-}C_6H_4(CH{=}CH\text{-}C_6H_4NO_2\text{-}p)_2$ | 51 | $C_2H_5OH$, 2 hr., 40 | 66 |
| | $(C_6H_5)_3P{=}CHOC_6H_5$ | $C_6H_{10}{=}CHOC_6H_5$ (cyclohexylidene) | 68 | $(C_2H_5)_2O$, 20 | 248 |
| | | $C_6H_5CH{=}CHOC_6H_5$ | 56 | $(C_2H_5)_2O$, 20 | 248 |
| | | $(C_6H_5)_2C{=}CHOC_6H_5$ | 65 | $(C_2H_5)_2O$, 20 | 248 |
| $C_8$ | $(C_6H_5)_3P{=}CHCH_2CH_2O$-(2-tetrahydropyranyl) | $CH_3(C{\equiv}C)_2(CH{=}CH)_3CH_2CH_2OH$ | | $(C_2H_5)_2O$, 3 hr., reflux | 71 |
| | $(C_6H_5)_3P{=}CHC{\equiv}CC(CH_3){=}CHCH_2OCH_3$ | (2,6,6-trimethyl-1-cyclohexenyl)$CH{=}CHC(CH_3){=}CH\text{-}C{\equiv}C\text{-}C(CH_3){=}CHCH_2OCH_3$ | 10 | $(C_2H_5)_2O$, 10 min., 20 | 290 |
| | $(C_6H_5)_3P{=}C(CH_3)COC_5H_{11}\text{-}n$ | $C_6H_5CH{=}C(CH_3)COC_5H_{11}\text{-}n$ | 40 | $C_6H_6$, 1 day, reflux | 32 |
| | $(C_6H_5)_3P{=}CHCH{=}C(CH_3)CH{=}CHCO_2CH_3$ | $CH_3O_2CCH{=}CHC(CH_3){=}CHCH{=}CHC(CH_3){=}CHCH{=}CHC(CH_3){=}CHCH{=}CHCH{=}C(CH_3)CH{=}CHCH{=}C(CH_3)CH{=}CHCH{=}C(CH_3)CH{=}CHCO_2CH_3$ | | $CH_2Cl_2$; 1 hr., 20; 5 hr., reflux | 236 |

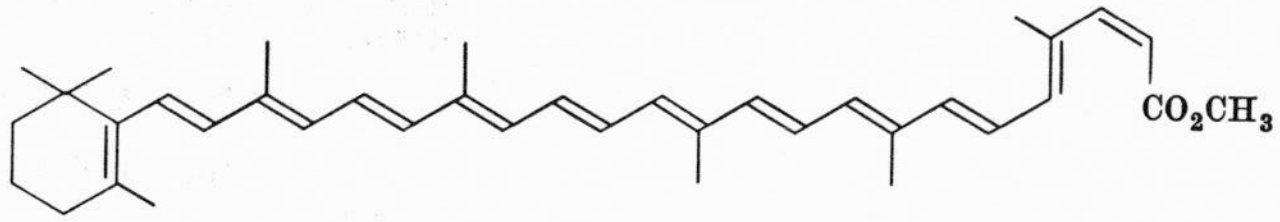

| | | | | |
|---|---|---|---|---|
| | $CO_2CH_3$ | | $CH_2Cl_2$, 6 hr., reflux | 297 |
| | C≡C $CO_2CH_3$ | | $CH_2Cl_2$, 6 hr., reflux | 297 |
| $(C_6H_5)_3P$=CHCH= | =CHCH= | | THF, 2 hr., reflux | 183, 184 |
| | $CH_2N(CH_3)_2$ =CHCH= | | $(C_2H_5)_2O$ 1 day, 20 | 183, 184 |
| | | | $(C_2H_5)_2O$, 1 hr., 20; 3 hr., 65 | 158 |
| $(C_6H_5)_3P$=CHCH= $O^{\ominus}$ | HO =CHCH= | | THF, −65; 18 hr., 20 | 69 |
| $(C_6H_5)_3P$=$CHC_6H_4CH_3$-*m* | $C_6H_5CH$=CHCH=$CHC_6H_4CH_3$-*m* | 23 | $C_2H_5OH$, 3 hr., 20; 15 min., 70 | 66 |
| | *m*-$C_6H_4$(CH=CH– $CH_3$$)_2$ | 11 | $C_2H_5OH$, 2 hr., 80 | 66 |
| $(C_6H_5)_3P$=$CHC_6H_4CH_3$-*p* | $C_6H_5C$≡C– CH=$CHC_6H_4CH_3$-*p* | 75 | $C_2H_5OH$, 2 hr., 20 | 195 |
| $(C_6H_5)_3P$=$CHC_6H_4OCH_3$-*p* | *p*-$BrCH_2C_6H_4CH$=$CHC_6H_4CH_3$-*p* | | $C_2H_5OH$, 30 min., 20 | 196 |
| | $C_2H_5OCH$=$CHC_6H_4OCH_3$-*p* | | THF, 1 day, 20 | 113 |
| | *p*-$O_2NC_6H_4CH$=$CHC_6H_4OCH_3$-*p* | 89 | $C_6H_6$, 4 hr., 20 | 166 |

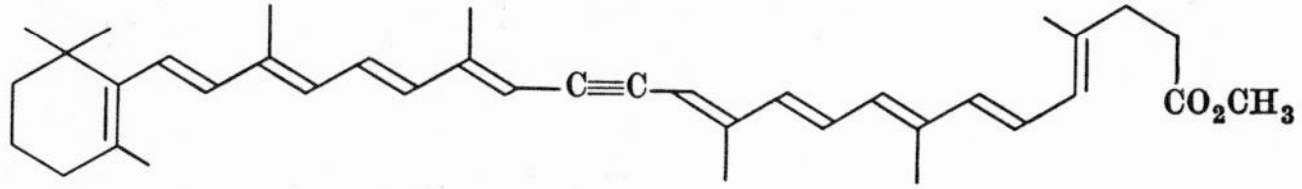

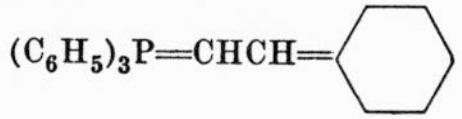

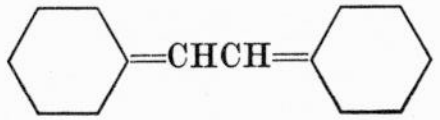

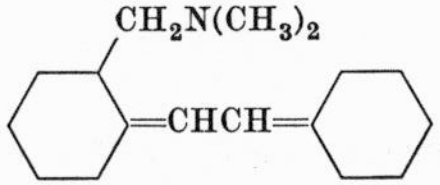

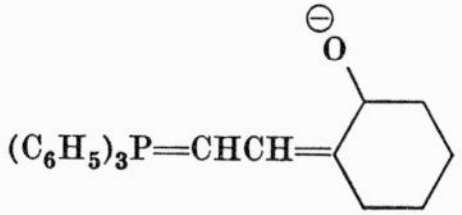

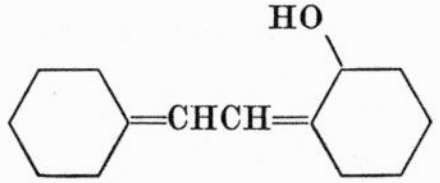

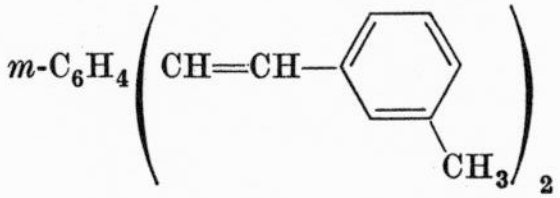

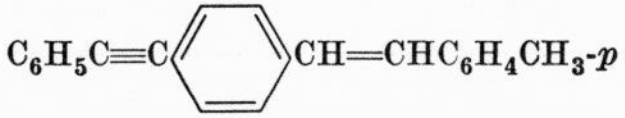

*Note:* References 391 to 404 are on p. 490.

# TABLE VI—*Continued*

## Olefins Prepared by the Wittig Reaction

### *A. Mono-ylides as Starting Materials—Continued*

| Ylide | Product | Yield, % | Solvent, Time, Temperature, °C. | References |
|---|---|---|---|---|
| $C_8$ (*contd.*) | | | | |
| $(C_6H_5)_3P{=}CHC_6H_3(O_2CH_2)$-3,4 | $p$-$C_6H_4(CH{=}CH{-}C_6H_3(O_2CH_2))_2$ | | DMF | 201 |
| $(C_6H_5)_3P{=}CHOC_6H_4CH_3$-$p$ | $(C_2H_5)_2C{=}CHOC_6H_4CH_3$-$p$ | 72 | $(C_2H_5)_2O$, 20 | 248 |
| | $C_6H_{10}{=}CHOC_6H_4CH_3$-$p$ (cyclohexylidene) | 82 | $(C_2H_5)_2O$, 20 | 248 |
| | | 58 | $t$-$C_4H_9OH$, 8 hr., 65 | 248 |
| | $C_6H_5CH{=}CHOC_6H_4CH_3$-$p$ | 75 | $(C_2H_5)_2O$, 20 | 248 |
| | | 64 | $t$-$C_4H_9OH$, 8 hr., 65 | 248 |
| | $(C_6H_5)_2C{=}CHOC_6H_4CH_3$-$p$ | 67 | $(C_2H_5)_2O$, 20 | 248 |
| | | 61 | $t$-$C_4H_9OH$, 8 hr., 65 | 248 |
| | (2,6,6-trimethylcyclohex-1-enyl)$CH{=}CHC(CH_3){=}CHOC_6H_4CH_3$-$p$ | 73 | $(C_2H_5)_2O$, 2 hr., 20 | 248 |
| | (2,6,6-trimethylcyclohex-2-enyl)$CH{=}CHC(CH_3){=}CHOC_6H_4CH_3$-$p$ | 75 | $(C_2H_5)_2O$, 2 hr., 20 | 248 |
| | $o$-$C_6H_4(CH{=}CHOC_6H_4CH_3)_2$ | 62 | $(C_2H_5)_2O$, 20 | 248 |
| $(C_6H_5)_3P{=}CHCOC_6H_5$ | $C_6H_5CH{=}CHCOC_6H_5$ | 70 | $C_6H_6$, 3 days, reflux | 32 |
| | | | THF, 30 hr., reflux | 31 |
| $(C_6H_5)_3P{=}CClCOC_6H_5$ | $C_6H_5CH{=}CClCOC_6H_5$ | Quant. | None, 2 days, 100 | 60 |
| | $p$-$O_2NC_6H_4CH{=}CClCOC_6H_5$ | 64 | $C_6H_6$, 35 hr., reflux | 59 |
| $(C_6H_5)_3P{=}CBrCOC_6H_5$ | $p$-$O_2NC_6H_4CH{=}CBrCOC_6H_5$ | 60 | $C_6H_6$, 35 hr., reflux | 59 |

| | Ylide | Product | Yield (%) | Conditions | Refs. |
|---|---|---|---|---|---|
| $C_9$ | $(C_6H_5)_3P{=}CHCH{=}CHC_6H_{13}\text{-}n$ | $HC{\equiv}C(CH{=}CH)_2C_6H_{13}\text{-}n$ | 43 | $(C_2H_5)_2O$, 2 hr., 20 | 257 |
| | $(C_6H_5)_3P{=}CHCH{=}CH(CH_2)_4CH{=}CH_2$ | $HC{\equiv}C(CH{=}CH)_2(CH_2)_4CH{=}CH_2$ | 53 | $(C_2H_5)_2O$, 1 hr., 20 | 160 |
| | $(C_6H_5)_3P{=}CH(CH_2)_5CO_2C_2H_5$ | $CH_3(CH_2)_9CH{=}CH(CH_2)_5CO_2C_2H_5$ | 59 | DMF | 114 |
| | $(C_6H_5)_3P{=}CHC{\equiv}CC(CH_3){=}CHCH_2OCOCH_3$ | 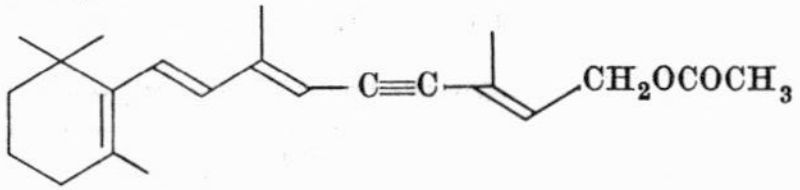 $C{\equiv}C$ ... $CH_2OCOCH_3$ | | DMF, 5 hr., 20 | 290 |
| | 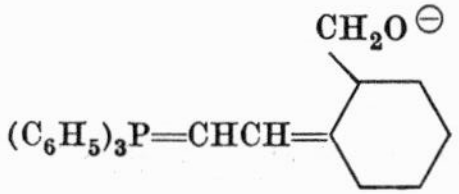 $(C_6H_5)_3P{=}CHCH{=}$ ($CH_2O^{\ominus}$) | 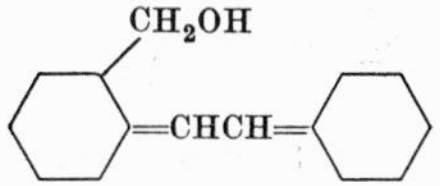 ($CH_2OH$) $=CHCH=$ | | THF, −25; 12 hr., 20 | 69 |
| | $(C_6H_5)_3P{=}CHCH{=}CHC_6H_5$ | $C_6H_5CH{=}CHCH{=}CHC_6H_5$ | 63 | $C_2H_5OH$, 20 | 210 |
| | | $p\text{-}CH_3C_6H_4CH{=}CHCH{=}CHC_6H_5$ | 76 | $C_2H_5OH$, 20 | 210 |
| | | $p\text{-}CH_3OC_6H_4CH{=}CHCH{=}CHC_6H_5$ | 63 | $C_2H_5OH$, 20 | 210 |
| | | $p\text{-}CH_3CONHC_6H_4CH{=}CHCH{=}CHC_6H_5$ | 61 | $C_2H_5OH$, 20 | 210 |
| | | $(C_6H_5)_2C{=}CHCH{=}CHC_6H_5$ | 64 | $(C_2H_5)_2O$, 12 hr., 50 | 138 |
| | | $(C_6H_5)_2C{=}CHCH{=}CHCH{=}CHC_6H_5$ | 61 | $(C_2H_5)_2O$, 12 hr., 50 | 138 |
| | | $C_6H_5CH{=}CHC(C_6H_5){=}CHCH{=}CHC_6H_5$ | | $(C_2H_5)_2O$/THF, 1 hr., 20 | 213 |
| | | 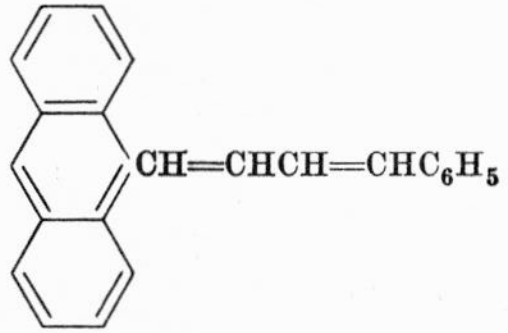 $CH{=}CHCH{=}CHC_6H_5$ | 63 | $C_2H_5OH$, 2 hr., 20 | 211 |
| | | 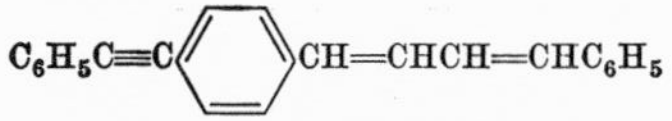 $C_6H_5C{\equiv}C$ ... $CH{=}CHCH{=}CHC_6H_5$ | 38–40 | $C_2H_5OH$, 2 hr., 20 | 211 |
| | | $m\text{-}C_6H_4(CH{=}CHCH{=}CHC_6H_5)_2$ | 29 | $C_2H_5OH$, 4 hr., 70 | 66 |
| | | $p\text{-}C_6H_4(CH{=}CHCH{=}CHC_6H_5)_2$ | 87 | $C_2H_5OH$, 20 min., 80 | 66 |

*Note:* References 391 to 404 are on p. 490.

## TABLE VI—*Continued*

### Olefins Prepared by the Wittig Reaction

*A. Mono-ylides as Starting Materials—Continued*

| Ylide | Product | Yield, % | Solvent, Time, Temperature, °C. | References |
|---|---|---|---|---|
| $C_9$ (*contd.*) | | | | |
| $(C_6H_5)_3P{=}CHCH{=}CHC_6H_5$ | 1-Pyrenyl-$CH{=}CHCH{=}CHC_6H_5$ | 62 | $C_2H_5OH$, 2 hr., 20 | 211 |
| | $C_6H_5(CH{=}CH)_2C(C_6H_5){=}CHCH{=}CHC_6H_5$ | | THF, 15 min., 20; 45 min., reflux | 213 |
| | $(C_6H_5CH{=}CH)_2C{=}CHCH{=}CHC_6H_5$ | | $(C_2H_5)_2O$/THF, 1 hr., 20 | 213 |
| | $C_6H_5(CH{=}CH)_2C(CH{=}CHC_6H_5){=}CHCH{=}CHC_6H_5$ | | $(C_2H_5)_2O$/THF, 1 hr., 20 | 213 |
| | $(C_6H_5CH{=}CHCH{=}CH)_2C{=}CHCH{=}CHC_6H_5$ | | $(C_2H_5)_2O$/THF, 1 hr., 20 | 213 |
| $(C_6H_5)_3P{=}CHC_6H_3(OCH_3)_2$-2,4 | $p$-$C_6H_4(CH{=}CHC_6H_3(OCH_3)_2\text{-}2,4)_2$ | | DMF | 201 |
| $(C_6H_5)_3P{=}CHCH_2COC_6H_5$ | 1,4-Diphenyl-1,3-cyclohexadiene ($C_6H_5$–$C_6H_6$–$C_6H_5$) | 12 | $C_6H_6$, 30 hr., reflux | 203 |
| $(C_6H_5)_3P{=}CHC_6H_4CO_2CH_3$-$p$ | $C_6H_5CH{=}CHCH{=}CHC_6H_4CO_2CH_3$-$p$ | 38 | $C_2H_5OH$, 20 | 210 |
| | $p$-$C_6H_4(CH{=}CHC_6H_4CO_2CH_3)_2$ | 73 | $C_2H_5OH$, 30 min., 20<br>DMF or $(CH_3)_2SO$ | 193<br>199, 200 |
| $(C_6H_5)_3P{=}CHC_6H_4N(CH_3)_2$-$p$ | $C_6H_5CH{=}CHCH{=}CHC_6H_4N(CH_3)_2$-$p$ | 66 | $C_2H_5OH$, 20 | 210 |

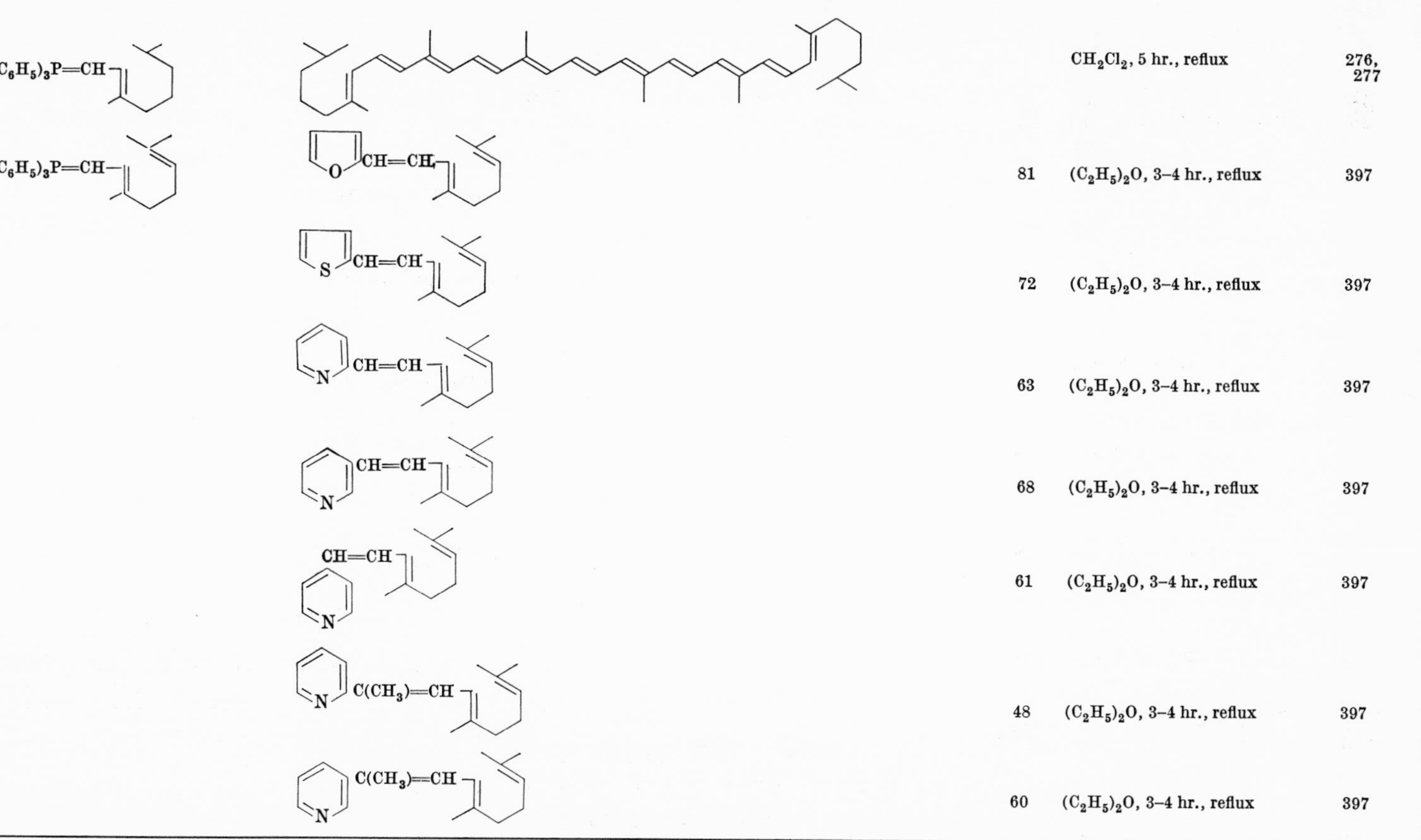

| | Ylide | Yield (%) | Conditions | Refs. |
|---|---|---|---|---|
| $C_{10}$ | $(C_6H_5)_3P{=}CH-$ | | $CH_2Cl_2$, 5 hr., reflux | 276, 277 |
| | $(C_6H_5)_3P{=}CH-$ | 81 | $(C_2H_5)_2O$, 3–4 hr., reflux | 397 |
| | | 72 | $(C_2H_5)_2O$, 3–4 hr., reflux | 397 |
| | | 63 | $(C_2H_5)_2O$, 3–4 hr., reflux | 397 |
| | | 68 | $(C_2H_5)_2O$, 3–4 hr., reflux | 397 |
| | | 61 | $(C_2H_5)_2O$, 3–4 hr., reflux | 397 |
| | | 48 | $(C_2H_5)_2O$, 3–4 hr., reflux | 397 |
| | | 60 | $(C_2H_5)_2O$, 3–4 hr., reflux | 397 |

*Note:* References 391 to 404 are on p. 490.

## TABLE VI—*Continued*

### Olefins Prepared by the Wittig Reaction

*A. Mono-ylides as Starting Materials—Continued*

| Ylide | Product | Yield, % | Solvent, Time, Temperature, °C. | References |
|---|---|---|---|---|
| $C_{10}$ 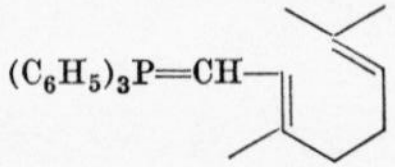 $(C_6H_5)_3P{=}CH-$ (*contd.*) | 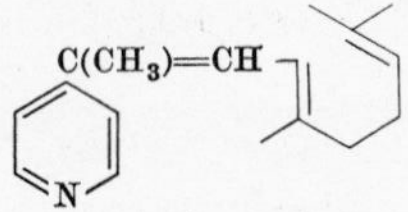 $C(CH_3){=}CH-$ (pyridyl) | 53 | $(C_2H_5)_2O$, 3–4 hr., reflux | 397 |
| | 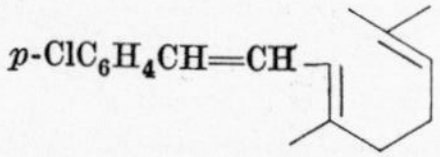 $p\text{-}ClC_6H_4CH{=}CH-$ | 62 | $(C_2H_5)_2O$, 3–4 hr., reflux | 397 |
| | 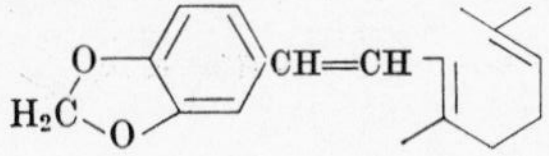 $H_2C$ (O, O) $CH{=}CH-$ | 59 | $(C_2H_5)_2O$, 3–4 hr., reflux | 397 |
| | 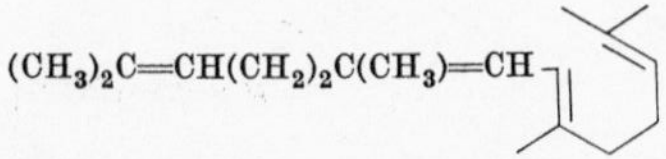 $(CH_3)_2C{=}CH(CH_2)_2C(CH_3){=}CH-$ | 43 | $(C_2H_5)_2O$, 3–4 hr., reflux | 397 |
| | 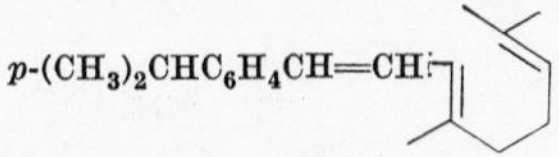 $p\text{-}(CH_3)_2CHC_6H_4CH{=}CH-$ | 75 | $(C_2H_5)_2O$, 3–4 hr., reflux | 397 |
| | 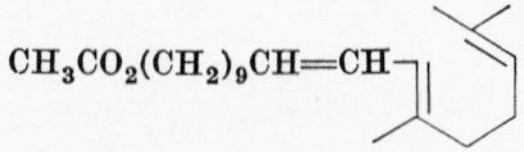 $CH_3CO_2(CH_2)_9CH{=}CH-$ | 53 | $(C_2H_5)_2O$, 5 hr., 0; 6–7 hr., 20 | 328 |
| | 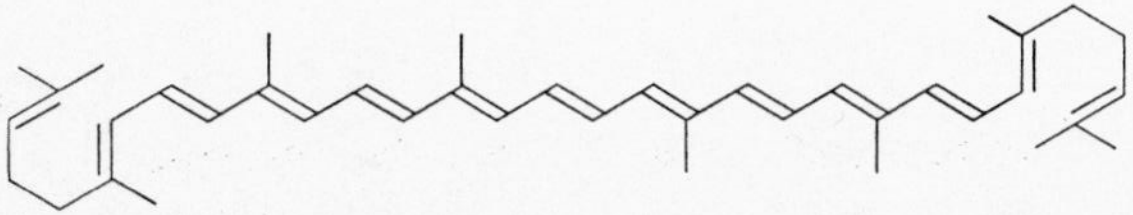 | 70 | $CH_2Cl_2$, 15 min., 30; 5 hr., reflux<br>DMF, 2 hr., 20 | 275, 276<br>274 |

| | | | | |
|---|---|---|---|---|
| | 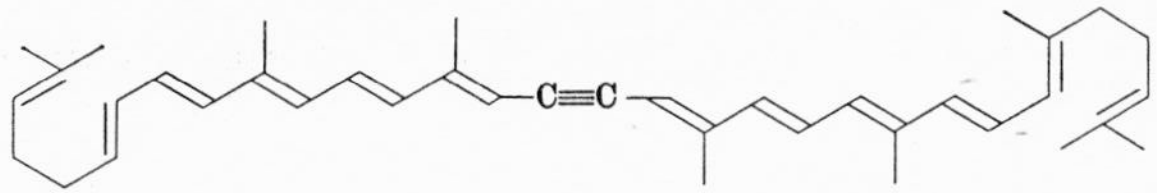 [structure] | | $CH_2Cl_2$, 15 min., 30; 5 hr., reflux<br>DMF, 2 hr., 20 | 275, 276<br>274 |
| | 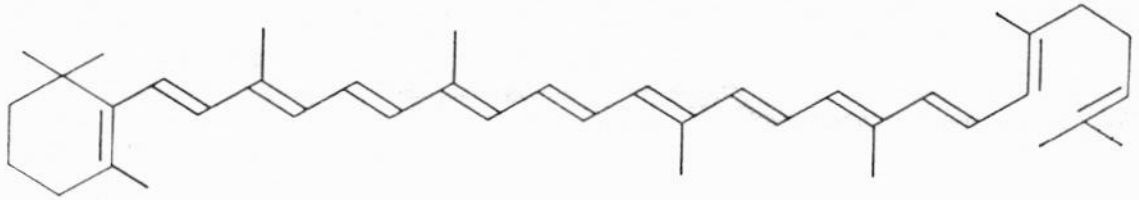 [structure] | | $CH_2Cl_2$, 5 hr., reflux | 264 |
| $(C_6H_5)_3P{=}CH$–[structure] 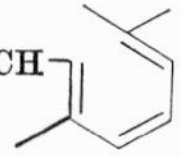 | 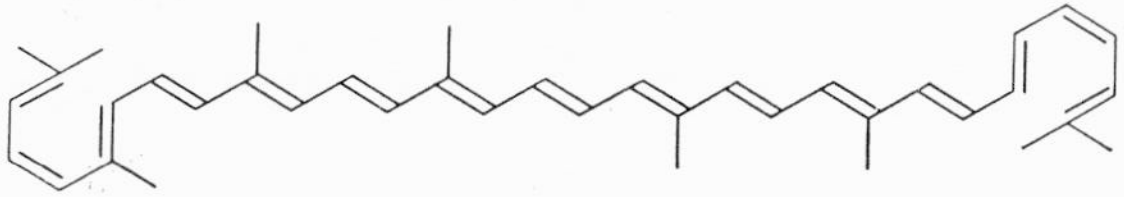 [structure] | 70 | $CH_3OH$ | 279 |
| $(C_6H_5)_3P{=}CH$–[structure] 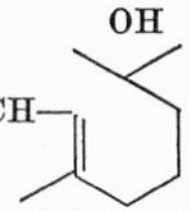 | 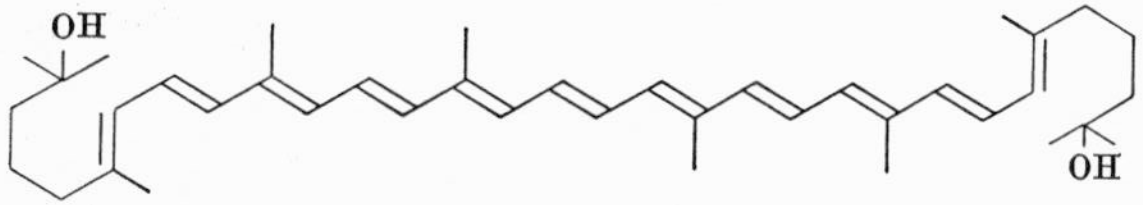 [structure] | 55 | $CH_3OH$ | 279 |
| $(C_6H_5)_3P{=}CHCH_2CH_2(C{\equiv}C)_2C_3H_7$-*n* | $CH_3O_2C(CH{=}CH)_2CH_2CH_2(C{\equiv}C)_2C_3H_7$-*n* | 47 | $(C_2H_5)_2O$, 2 hr., reflux | 241 |
| $(C_6H_5)_3P{=}CHCH{=}CH(C{\equiv}C)_2CH{=}CHCH_3$ | $CH_2{=}CH(CH{=}CH)_2(C{\equiv}C)_2CH{=}CHCH_3$ | 1 | $(C_2H_5)_2O$, 2 days 0 | 259 |
| $(C_6H_5)_3P{=}CH(CH_2)_6CO_2C_2H_5$ | *n*-$C_7H_{15}CH{=}CH(CH_2)_6CO_2C_2H_5$ | 59 | DMF | 114 |
| $(C_6H_5)_3P{=}CH$–[structure] 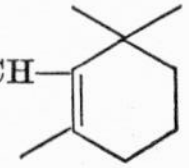 | 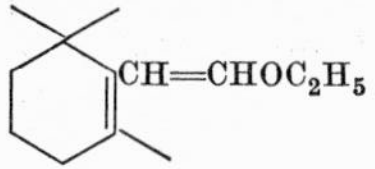 [structure]$CH{=}CHOC_2H_5$ | | DMF, 5 hr., 60 | 113 |
| | 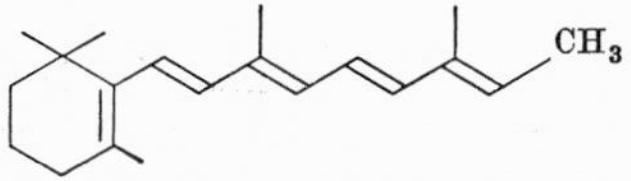 [structure]$CH_3$ | 60 | $(C_2H_5)_2O$, 2 hr., reflux<br>DMF, 5 hr., 20 | 293<br>293 |
| | 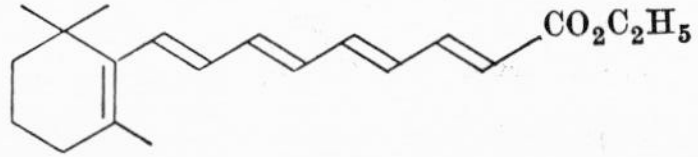 [structure]$CO_2C_2H_5$ | 65 | DMF/$(C_2H_5)_2O$, 2 hr., 20 | 293 |

*Note:* References 391 to 404 are on p. 490.

TABLE VI—*Continued*

OLEFINS PREPARED BY THE WITTIG REACTION

*A. Mono-ylides as Starting Materials—Continued*

| | Ylide | Product | Yield, % | Solvent, Time, Temperature, °C. | References |
|---|---|---|---|---|---|
| $C_{10}$ | $(C_6H_5)_3P{=}CH$— 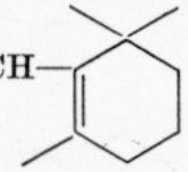 (*contd.*) | 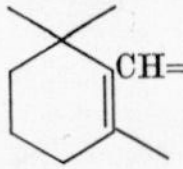 $CH{=}CH(CH_2)_8CO_2CH_3$ | 53 | $(C_2H_5)_2O$, 3 hr., 0; 6 hr., 20; 1 hr., reflux | 328 |
| | | 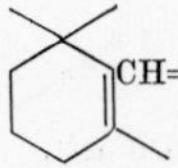 $CH{=}CH(CH_2)_8CH_2OCOCH_3$ | 68 | $(C_2H_5)_2O$, 0 | 328 |
| | | 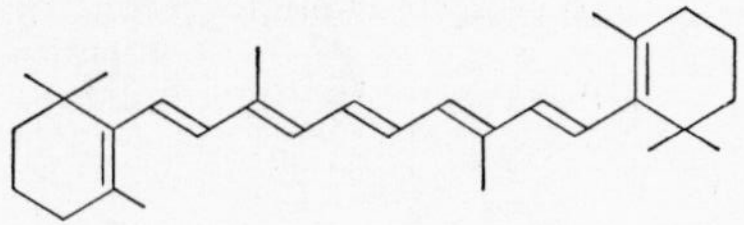 | | THF, 1 hr., 60 | 209 |
| | | 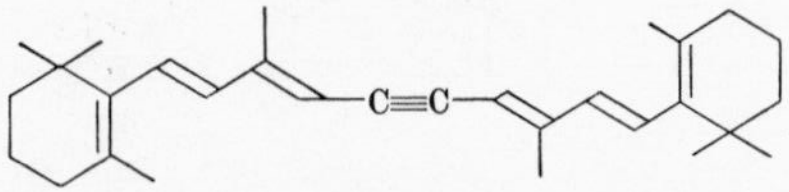 | | THF, 1 hr., 60 | 209 |
| | | 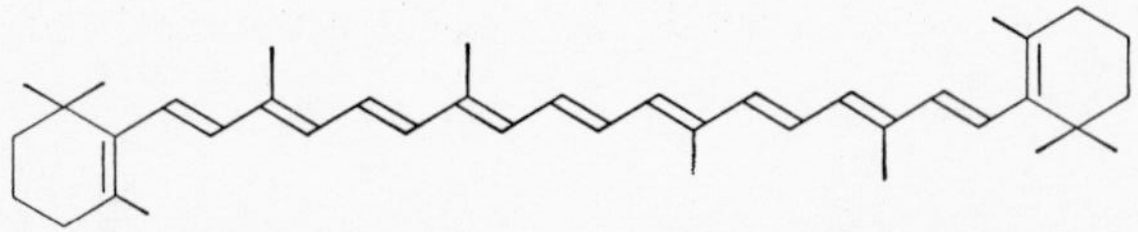 | 3 | $CH_2Cl_2$, 15 min., 30; 5 hr., reflux<br>DMF, 1 hr., 20 | 273<br>274 |
| | | 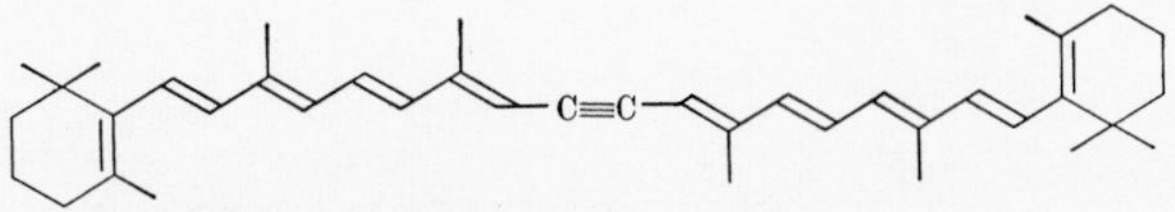 | 3 | $C_6H_6$, 6 hr., 60<br>$(C_2H_5)_2O/C_6H_6$, 1 hr., 40 | 273<br>275 |
| | $(C_6H_5)_3P{=}C(CH_2C_6H_5)CO_2CH_3$ | $C_6H_5CH{=}C(CH_2C_6H_5)CO_2CH_3$ | 42 | $C_6H_6$, 10 hr., reflux | 40 |

| | | | | | |
|---|---|---|---|---|---|
| $C_{11}$ | $(C_6H_5)_3P{=}CH(CH{=}CH)_2C_6H_{13}$-*n* | *n*-$C_6H_{13}(CH{=}CH)_4C{\equiv}CC{\equiv}C(CH{=}CH)_4C_6H_{13}$-*n* | | $(C_2H_5)_2O/THF$, 20 min., 20 | 181 |
| | $(C_6H_5)_3P{=}CHCH{=}CH(C{\equiv}C)_2C_4H_9$-*t* | $HC{\equiv}C(CH{=}CH)_2(C{\equiv}C)_2C_4H_9$-*t* | | $(C_2H_5)_2O$, 30 min., 20 | 180 |
| | | *t*-$C_4H_9(C{\equiv}C)_2(CH{=}CH)_2(C{\equiv}C)_2C_4H_9$-*t* | 19 | $(C_2H_5)_2O$, 30 min., 20 | 180 |
| | 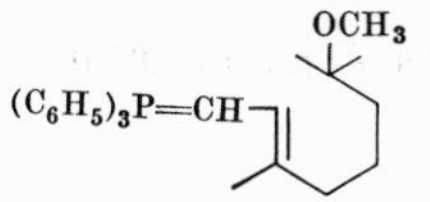 $(C_6H_5)_3P{=}CH$ ($OCH_3$) | 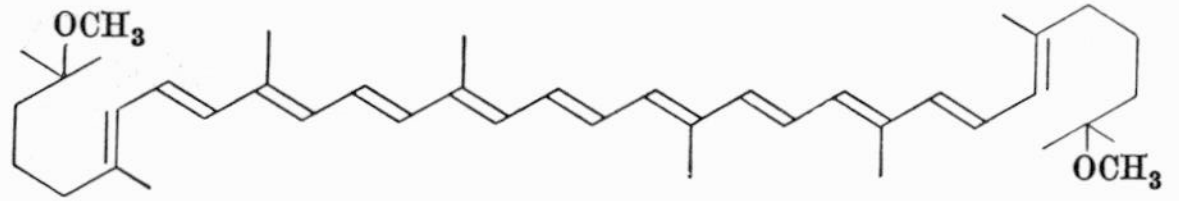 $OCH_3$ … $OCH_3$ | | $CH_3OH$ | 278 |
| | 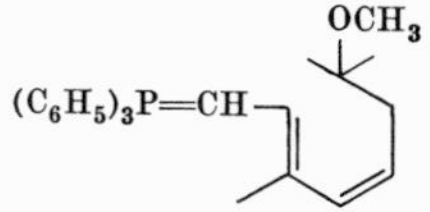 $(C_6H_5)_3P{=}CH$ ($OCH_3$) | 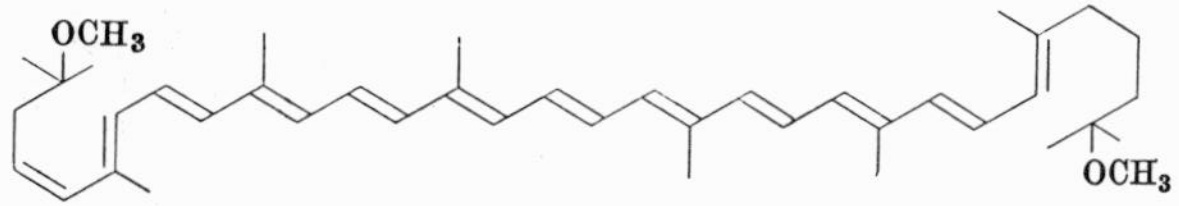 $OCH_3$ … $OCH_3$ | | $CH_3OH$ | 278 |
| | $(C_6H_5)_3P{=}CH(CH_2)_7CO_2C_2H_5$ | *n*-$C_6H_{13}CH{=}CH(CH_2)_7CO_2C_2H_5$ | 64 | DMF, 20 | 114, 170, 171 |
| | | *n*-$C_8H_{17}CH{=}CH(CH_2)_7CO_2C_2H_5$ | 73 | DMF, 20 | 114, 170, 171 |
| | | *n*-$C_7H_{15}C(CH_3){=}CH(CH_2)_7CO_2C_2H_5$ | 26 | DMF | 114 |
| | | *n*-$C_4H_9(CH{=}CH)_3(CH_2)_7CO_2C_2H_5$ | | DMF | 114 |
| | | *n*-$C_8H_{17}C(CH_3){=}CH(CH_2)_7CO_2C_2H_5$ | 32 | DMF | 114 |
| | | *n*-$C_{10}H_{21}CH{=}CH(CH_2)_7CO_2C_2H_5$ | 60 | DMF, 20 | 114, 170, 171 |
| | | *n*-$C_{16}H_{33}CH{=}CH(CH_2)_7CO_2C_2H_5$ | 44 | DMF | 114 |
| | 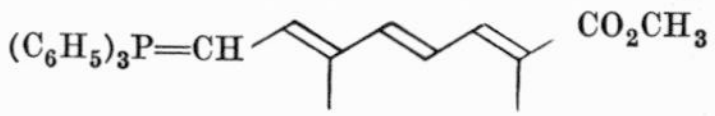 $(C_6H_5)_3P{=}CH$ … $CO_2CH_3$ | 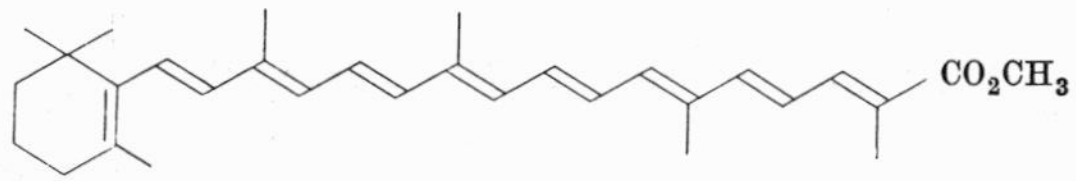 $CO_2CH_3$ | | $CH_2Cl_2$, 5 hr., reflux | 298 |
| | | 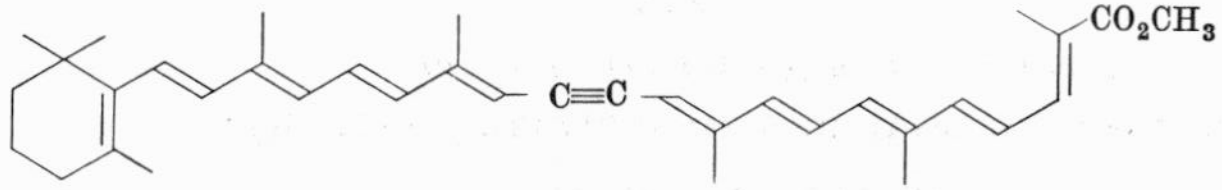 $C{\equiv}C$ … $CO_2CH_3$ | | $CH_2Cl_2$, 6 hr., reflux | 297 |

*Note:* References 391 to 404 are on p. 490.

## TABLE VI—*Continued*

### Olefins Prepared by the Wittig Reaction

*A. Mono-ylides as Starting Materials—Continued*

| | Ylide | Product | Yield, % | Solvent, Time, Temperature, °C. | References |
|---|---|---|---|---|---|
| $C_{11}$ | $(C_6H_5)_3P{=}CH$–(polyene)–$CO_2CH_3$ (*contd.*) 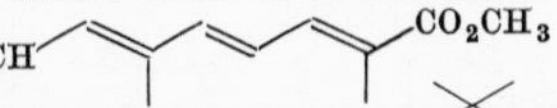 | 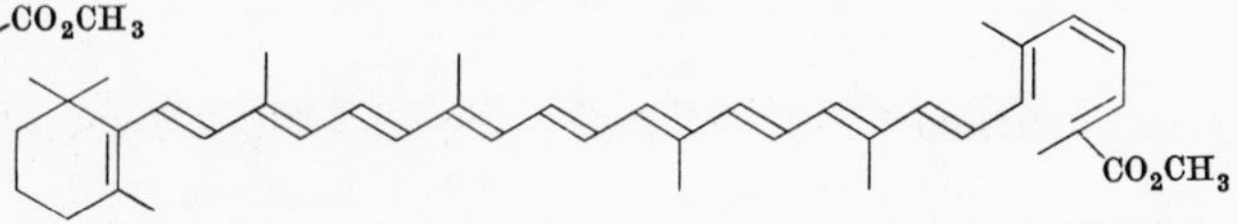 (polyene)–$CO_2CH_3$ | | $CH_2Cl_2$, 6 hr., reflux | 297 |
| | | 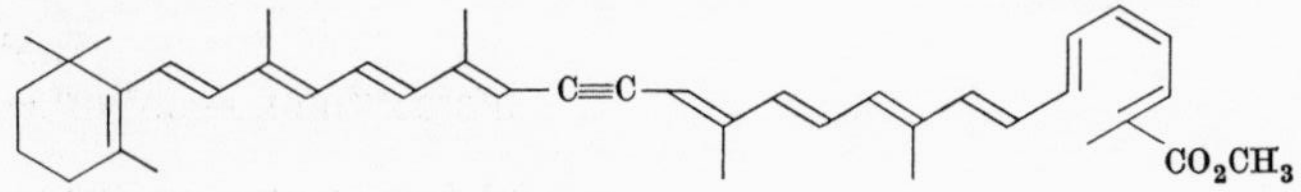 C≡C (polyene)–$CO_2CH_3$ | | $CH_2Cl_2$, 6 hr., reflux | 297 |
| | $(C_6H_5)_3P{=}CHCH{=}$ 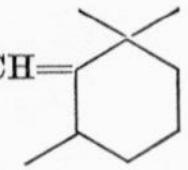 | 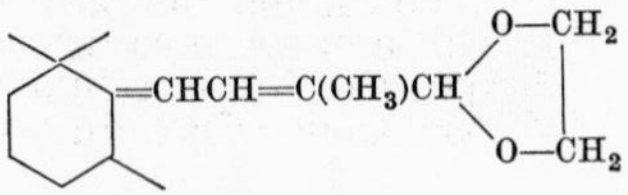 $=CHCH{=}C(CH_3)CH$ (O–$CH_2$, O–$CH_2$) | | $(C_2H_5)_2O/C_6H_6$, 20 min., 20; 6 hr., 60 | 214 |
| | $(C_6H_5)_3P{=}CHC_{10}H_7$-$\beta$ | $C_6H_5CH{=}CHC_{10}H_7$-$\beta$ | 26 | $C_2H_5OH$, 23 hr., 20 | 197 |
| | | $\beta$-$C_{10}H_7CH{=}CHC_{10}H_7$-$\beta$ | 38 | $C_2H_5OH$, 23 hr., 20 | 197 |
| | | 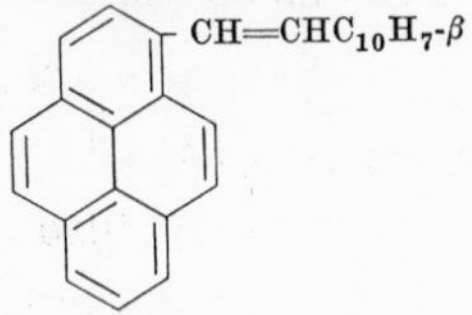 (pyrenyl)–$CH{=}CHC_{10}H_7$-$\beta$ | 61 | $C_2H_5OH$, 23 hr., 20 | 197 |
| | $(C_6H_5)_3P{=}CH(CH_2)_3COC_6H_5$ | 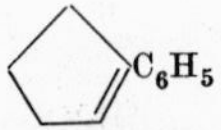 (cyclopentenyl)–$C_6H_5$ | 24 | $C_2H_5OH$, 30 min., reflux | 202 |
| | $(C_6H_5)_3P{=}C(CH_3)COCH_2CH_2C_6H_5$ | $C_6H_5CH{=}C(CH_3)COCH_2CH_2C_6H_5$ | 64 | $C_6H_6$, 3 days, reflux | 32 |
| | $(C_6H_5)_3P{=}C(CH_3)COCH{=}CHC_6H_5$ | $C_6H_5CH{=}C(CH_3)COCH{=}CHC_6H_5$ | 81 | $C_6H_6$, 4 days, reflux | 32 |

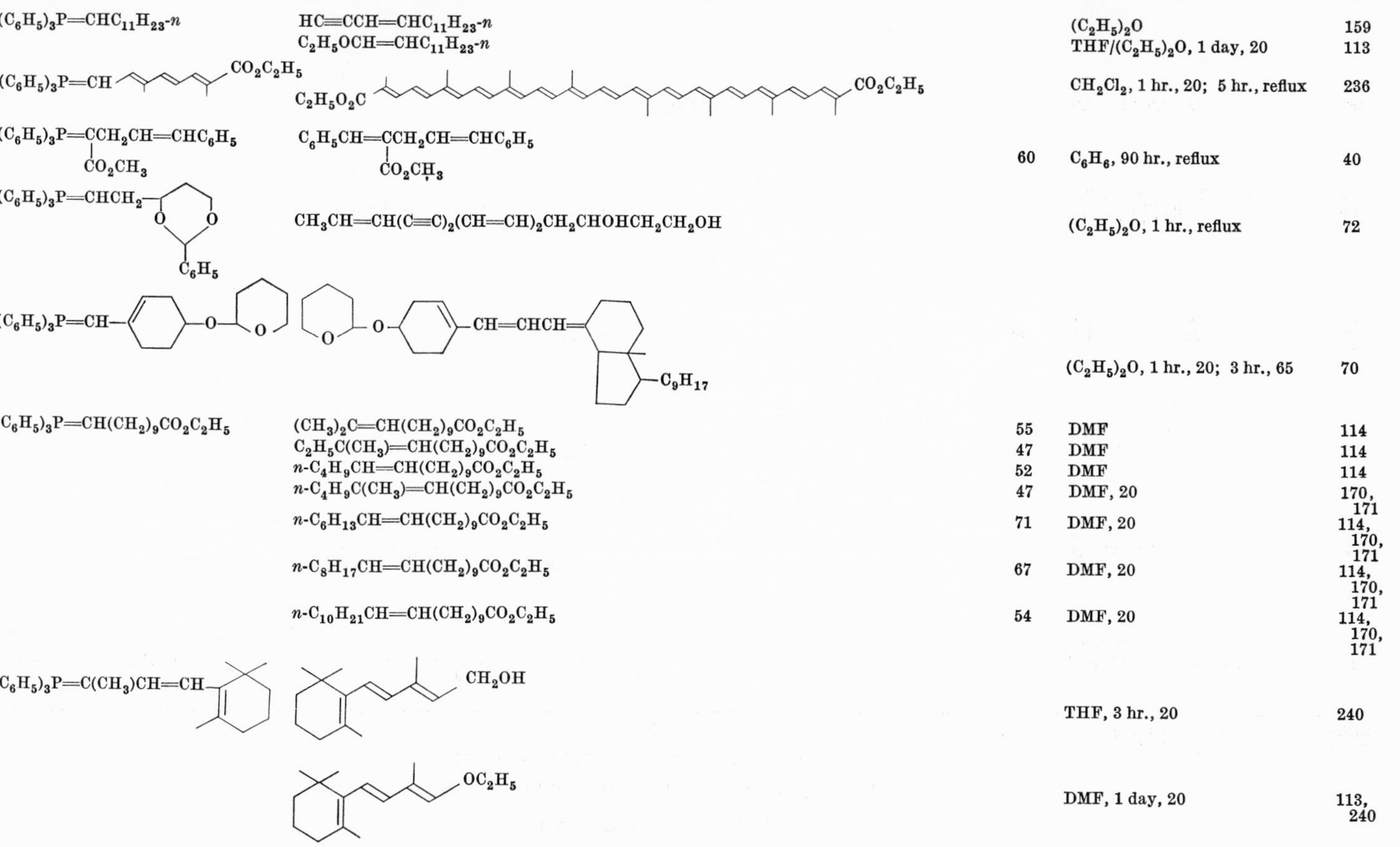

| | Ylide | Product | Yield (%) | Conditions | Refs. |
|---|---|---|---|---|---|
| $C_{12}$ | $(C_6H_5)_3P{=}CHC_{11}H_{23}\text{-}n$ | $HC{\equiv}CCH{=}CHC_{11}H_{23}\text{-}n$ | | $(C_2H_5)_2O$ | 159 |
| | | $C_2H_5OCH{=}CHC_{11}H_{23}\text{-}n$ | | THF/$(C_2H_5)_2O$, 1 day, 20 | 113 |
| | $(C_6H_5)_3P{=}CH$– ... –$CO_2C_2H_5$ | $C_2H_5O_2C$– ... –$CO_2C_2H_5$ | | $CH_2Cl_2$, 1 hr., 20; 5 hr., reflux | 236 |
| | $(C_6H_5)_3P{=}C(CO_2CH_3)CH_2CH{=}CHC_6H_5$ | $C_6H_5CH{=}C(CO_2CH_3)CH_2CH{=}CHC_6H_5$ | 60 | $C_6H_6$, 90 hr., reflux | 40 |
| | $(C_6H_5)_3P{=}CHCH_2$– (2-$C_6H_5$-1,3-dioxan-4-yl) | $CH_3CH{=}CH(C{\equiv}C)_2(CH{=}CH)_2CH_2CHOHCH_2CH_2OH$ | | $(C_2H_5)_2O$, 1 hr., reflux | 72 |
| | $(C_6H_5)_3P{=}CH$– (structure) | (structure) –$CH{=}CHCH{=}$ (structure) –$C_9H_{17}$ | | $(C_2H_5)_2O$, 1 hr., 20; 3 hr., 65 | 70 |
| $C_{13}$ | $(C_6H_5)_3P{=}CH(CH_2)_9CO_2C_2H_5$ | $(CH_3)_2C{=}CH(CH_2)_9CO_2C_2H_5$ | 55 | DMF | 114 |
| | | $C_2H_5C(CH_3){=}CH(CH_2)_9CO_2C_2H_5$ | 47 | DMF | 114 |
| | | $n\text{-}C_4H_9CH{=}CH(CH_2)_9CO_2C_2H_5$ | 52 | DMF | 114 |
| | | $n\text{-}C_4H_9C(CH_3){=}CH(CH_2)_9CO_2C_2H_5$ | 47 | DMF, 20 | 170, 171 |
| | | $n\text{-}C_6H_{13}CH{=}CH(CH_2)_9CO_2C_2H_5$ | 71 | DMF, 20 | 114, 170, 171 |
| | | $n\text{-}C_8H_{17}CH{=}CH(CH_2)_9CO_2C_2H_5$ | 67 | DMF, 20 | 114, 170, 171 |
| | | $n\text{-}C_{10}H_{21}CH{=}CH(CH_2)_9CO_2C_2H_5$ | 54 | DMF, 20 | 114, 170, 171 |
| | $(C_6H_5)_3P{=}C(CH_3)CH{=}CH$– (structure) | (structure) $CH_2OH$ | | THF, 3 hr., 20 | 240 |
| | | (structure) $OC_2H_5$ | | DMF, 1 day, 20 | 113, 240 |

*Note:* References 391 to 404 are on p. 490.

## TABLE VI—*Continued*

### Olefins Prepared by the Wittig Reaction

*A. Mono-ylides as Starting Materials—Continued*

| | Ylide | Product | Yield, % | Solvent, Time, Temperature, °C. | References |
|---|---|---|---|---|---|
| $C_{13}$ | $(C_6H_5)_3P{=}C(CH_3)CH{=}CH-$ 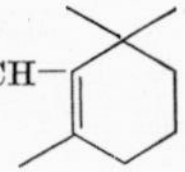 (*contd.*) | 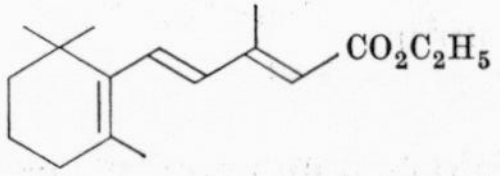 $CO_2C_2H_5$ | | $(C_2H_5)_2O$, 2 hr., 20; 6 hr., reflux | 240 |
| | | 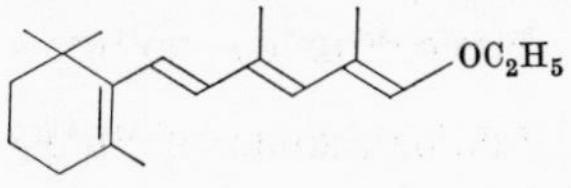 $OC_2H_5$ | | DMF, 2 hr., 20 | 240 |
| | | 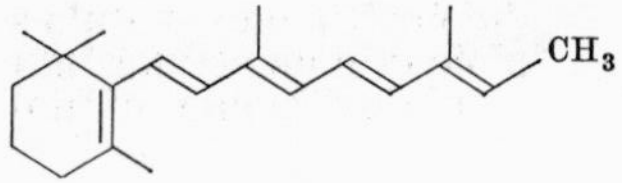 $CH_3$ | 70 | $C_6H_6$, 1 hr., 20<br>DMF, 2 hr., 20 | 240<br>289 |
| | | 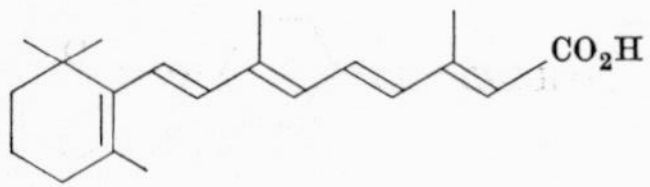 $CO_2H$ | 80 | DMF, 30 min., 20<br>DMF, −40; 5 min., −10 | 291<br>292 |
| | | 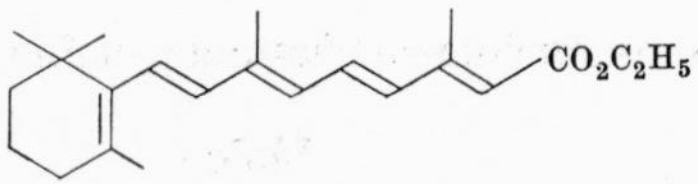 $CO_2C_2H_5$ | 70 | DMF, 2 hr., 20<br>DMF, 1 hr., 70<br>DMF, −40; 5 min., −10 | 289<br>289<br>292 |
| | | 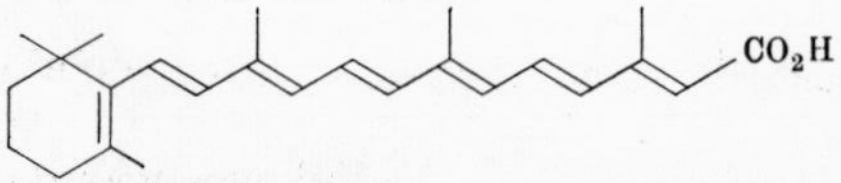 $CO_2H$ | | DMF, 15 min., 20 | 240 |

| | | | | |
|---|---|---|---|---|
| | 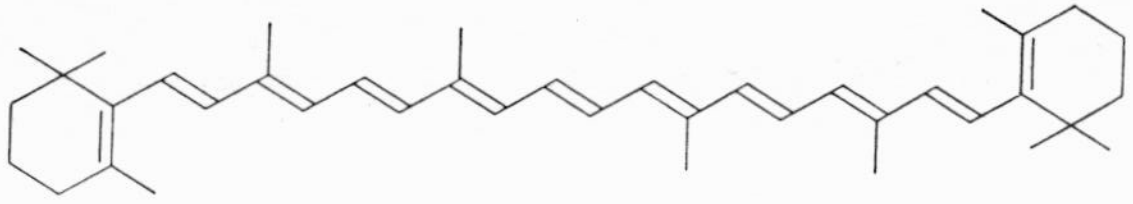 | | THF, 2 hr., 60 | 209 |
| | 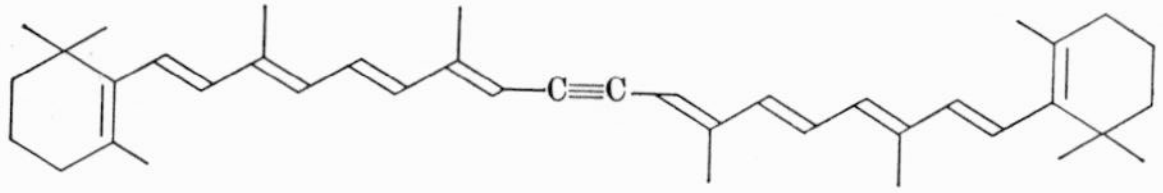 —C≡C— | | DMF, 1 hr., 20 | 240, 272 |
| $(C_6H_5)_3P=$ 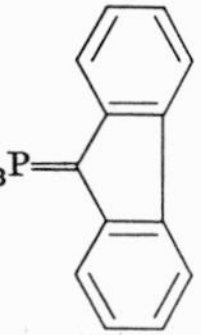 | $C_6H_5CH=$ 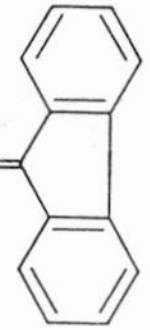 | 84 | $CHCl_3$, 3 hr., reflux | 49 |
| | $p$-$O_2NC_6H_4CH=$ 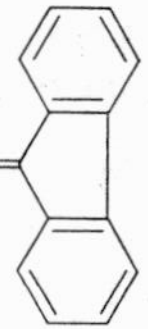 | 96 | $CHCl_3$, 3 hr., reflux | 49 |
| | $p$-$CH_3OC_6H_4CH=$ 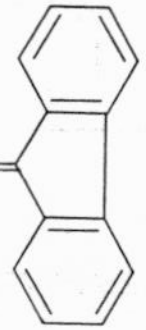 | 37 | $CHCl_3$, 3 hr., reflux | 49 |

*Note:* References 391 to 404 are on p. 490.

## TABLE VI—*Continued*

### Olefins Prepared by the Wittig Reaction

*A. Mono-ylides as Starting Materials—Continued*

| | Ylide | Product | Yield, % | Solvent, Time, Temperature, °C. | References |
|---|---|---|---|---|---|
| $C_{13}$ | $(C_6H_5)_3P{=}$(9-fluorenylidene) (*contd.*) 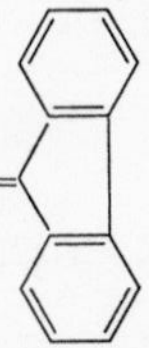 | 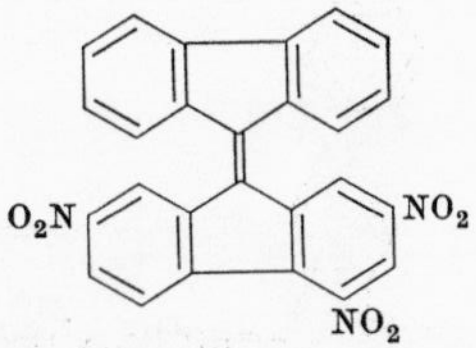 2,4,7-trinitro-9,9'-bifluorenylidene ($O_2N$, $NO_2$, $NO_2$) | Quant. | $CHCl_3$, 3 hr., reflux | 49 |
| | $(C_6H_5)_3P{=}C(C_3H_7\text{-}n)COCH_2CH_2C_6H_5$ | $C_6H_5CH{=}C(C_3H_7\text{-}n)COCH_2CH_2C_6H_5$ | 60 | $C_6H_6$, 3 days, reflux | 32 |
| $C_{14}$ | $(C_6H_5)_3P{=}C{=}C(C_6H_5)_2$ | $(C_6H_5)_2C{=}C{=}C(C_6H_5)_2$ | 46 | THF, 12 hr., reflux | 123 |
| $C_{15}$ | $(C_6H_5)_3P{=}CHCH{=}C(CH_3)CH_2CH_2$–(2,6,6-trimethylcyclohexenyl) 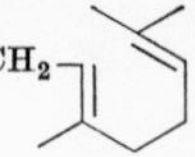 | 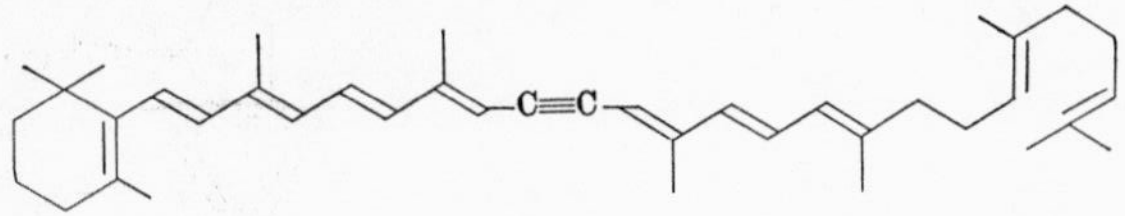 (polyene with central $C{\equiv}C$) | | DMF, 12 hr., 0 | 265 |
| | $(C_6H_5)_3P{=}CHCH{=}C(CH_3)CH{=}CH$–(2,6,6-trimethylcyclohexenyl) 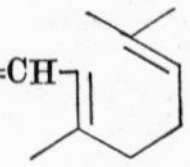 | 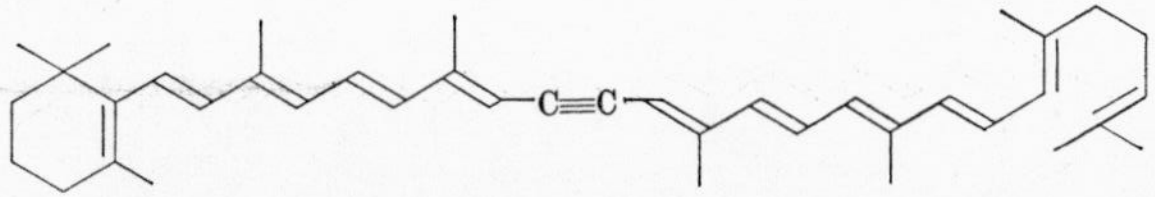 (polyene with central $C{\equiv}C$) | | DMF, 16 hr., 20 | 264 |

| Reactant | Product | Yield (%) | Conditions | Refs. |
| --- | --- | --- | --- | --- |
| $(C_6H_5)_3P{=}CHCH{=}C(CH_3)CH{=}CH$–(2,6,6-trimethylcyclohexenyl) 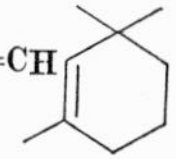 | | | | |
| | 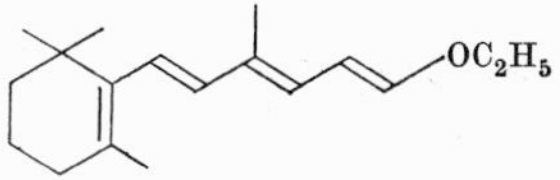 …$OC_2H_5$ | | DMF, 12 hr., 20 | 269, 270 |
| | 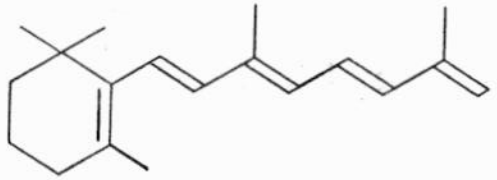 | | DMF, 3 hr., 20 | 269, 270 |
| | 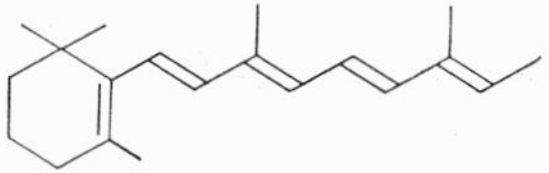 | 20 | DMF, 8 hr., 20 | 287 |
| | | | DMF, 2 hr., 20 | 270, 284 |
| | | | $C_2H_5OH$, 1 hr., −15; 12 hr,. 20 | 285 |
| | 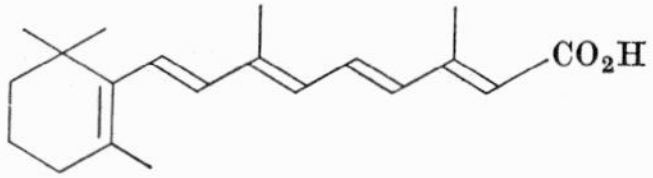 …$CO_2H$ | | DMF, 4 hr., 0 | 284 |
| | | | $CH_3OH$, 1 hr., 20 | 270, 286 |
| | | | $C_2H_5OH$,1 hr., −15; 12 hr., 20 | 285 |
| | 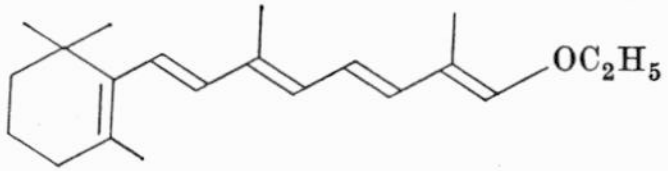 …$OC_2H_5$ | | $CH_3OH/C_6H_6$, 5 hr., 20 | 269 |
| | | | $CH_3OH$, 10 hr., 20 | 270 |
| | 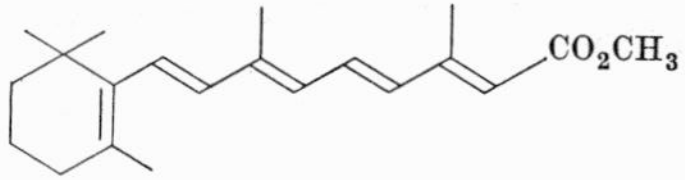 …$CO_2CH_3$ | | None, 3 hr., 5 | 285 |
| | | | $CH_3OH$, −20; 5 hr., 0 | 286 |
| | 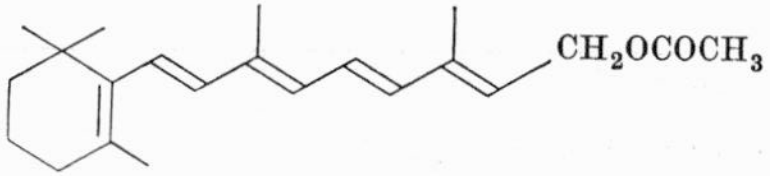 …$CH_2OCOCH_3$ | | DMF, 5 hr., 0 | 270, 284 |
| | | | $CH_3OH$, 2 hr., −10; 12 hr., 20 | 285 |

*Note:* References 391 to 404 are on p. 490.

## TABLE VI—*Continued*

## Olefins Prepared by the Wittig Reaction

### A. *Mono-ylides as Starting Materials—Continued*

| | Ylide | Product | Yield, % | Solvent, Time, Temperature, °C. | References |
|---|---|---|---|---|---|
| $C_{15}$ | $(C_6H_5)_3P{=}CHCH{=}C(CH_3)CH{=}CH$ 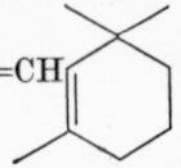 (*contd.*) | | | | |
| | | 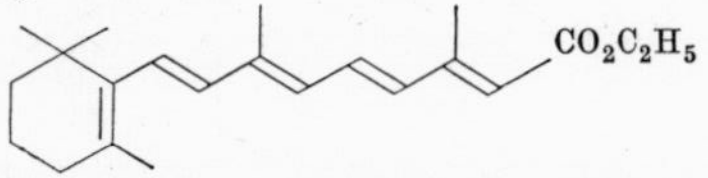 $CO_2C_2H_5$ | 70 | $(C_2H_5)_2O$ | 283 |
| | | | | THF, 30 min., −30; 18 hr., 20 | 285 |
| | | | | $C_6H_6$ | 283 |
| | | | | DMF, 30 min., −10; 2 hr., 20 | 284 |
| | | | | DMF, 6 hr., 20 | 283 |
| | | | | $CH_3OH$, 1 hr., 20 | 284 |
| | | | | $CH_3OH$, 45 min., −30; 4 hr., 20 | 285 |
| | | | | $C_2H_5OH$, 30 min., −10; 2 hr., 20 | 285 |
| | | 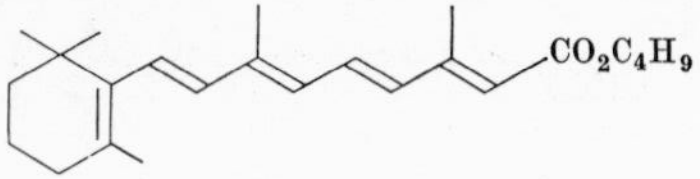 $CO_2C_4H_9$ | | DMF, 10 hr., 20 | 270 |
| | | 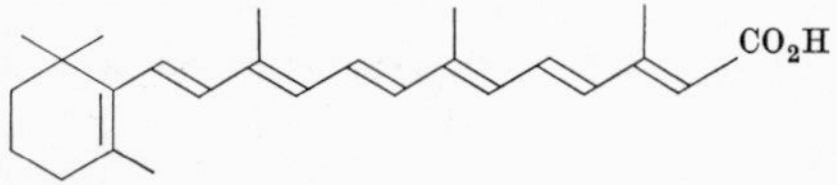 $CO_2H$ | | DMF, 15 min., 20 | 269, 294 |
| | | | | $CH_3OH$, 15 min., 20 | 270 |
| | | | | $C_2H_5OH$, 1 hr., 20 | 294 |
| | | 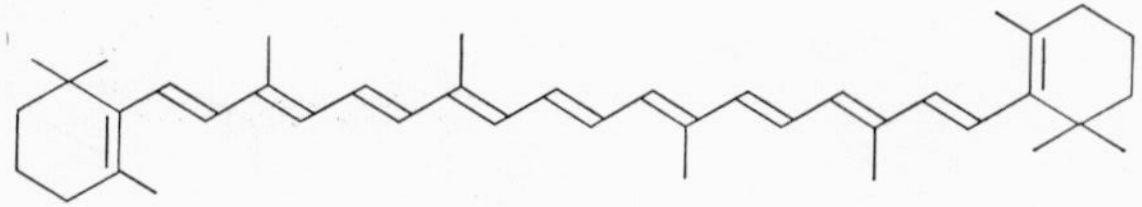 | | $(C_2H_5)_2O$ or $C_6H_6$, 10 hr., 20 | 54 |
| | | | | DMF, 1–3 hr., 20 | 268–270 |
| | | 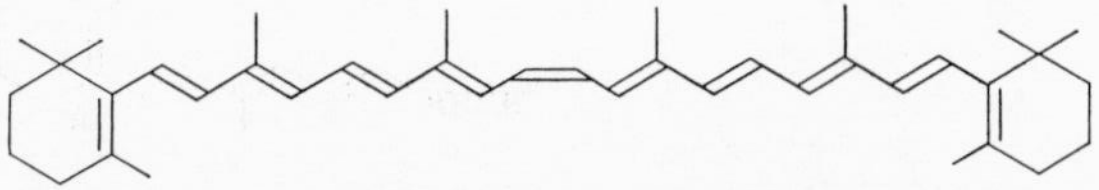 | | $C_2H_5OH$/DMF, 3 hr., 20 | 268 |
| | | 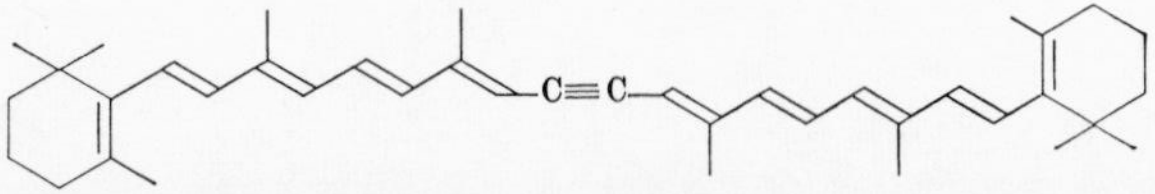 C≡C | | DMF, 1–2 hr., 20 | 268, 270 |
| | | | | DMF, 20 hr., 20 | 264 |
| | | | | $C_2H_5OH$, 2 hr., 20; 10 min., reflux | 54 |

| Ylide | Product | Yield (%) | Conditions | Refs. |
|---|---|---|---|---|
| $(C_6H_5)_3P{=}CHCH{=}C(CH_3)CH{=}CH-$ (2,6,6-trimethylcyclohexenyl) | | | | |
| 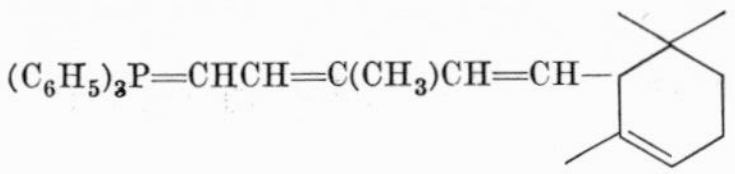 | 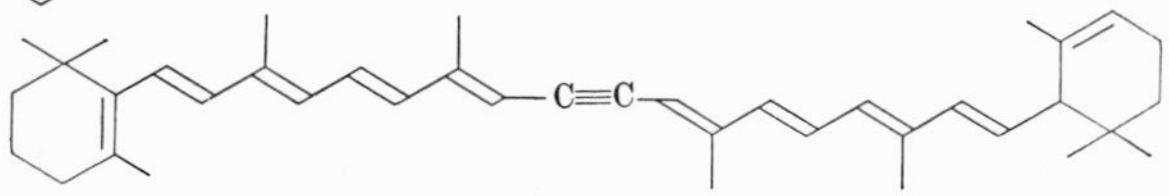 (bis-trimethylcyclohexenyl polyene) $-C{\equiv}C-$ | | DMF, 16 hr., 20 | 264 |
| $(C_6H_5)_3P{=}CHCH{=}C(CH_3)CH{=}CH-$ (trimethylcyclohexadienyl) | | | | |
| 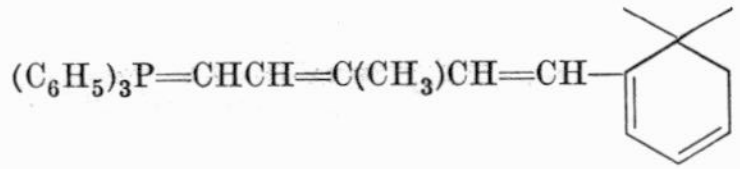 | 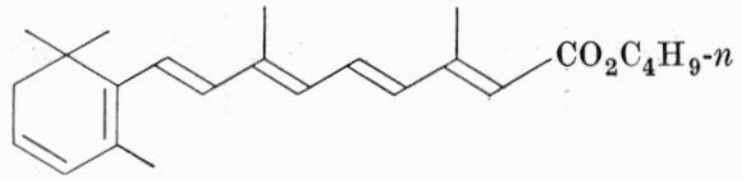 (trimethylcyclohexadienyl polyene) $CO_2C_4H_9$-$n$ | | $C_2H_5OH$, 10 min., −10; 2 hr., 20 | 288 |
| $(C_6H_5)_3P{=}CHCH{=}C(C_6H_5)_2$ | $C_6H_5CH{=}CHCH{=}C(C_6H_5)_2$ | 72 | $C_2H_5OH$, 2 hr., 20 | 211 |
| | 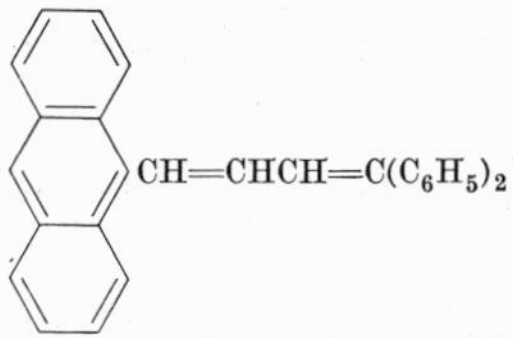 (9-anthryl)$CH{=}CHCH{=}C(C_6H_5)_2$ | 62 | $C_2H_5OH$, 2 hr., 20 | 211 |
| | $p$-$C_6H_4[CH{=}CHCH{=}C(C_6H_5)_2]_2$ | 57 | $C_2H_5OH$, 2 hr., 20 | 211 |
| 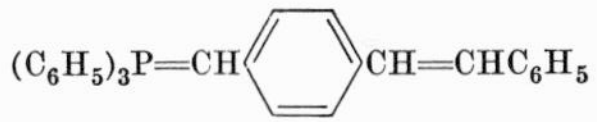 $(C_6H_5)_3P{=}CH$-$C_6H_4$-$CH{=}CHC_6H_5$ |  $CH_2{=}CH$-$C_6H_4$-$CH{=}CHC_6H_5$ | 72–73 | $C_2H_5OH$, 20 | 187 |
| | 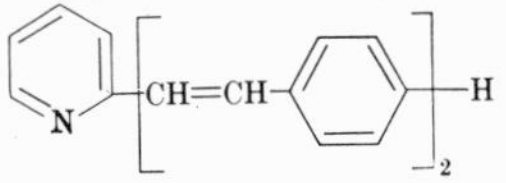 2-pyridyl$[CH{=}CH$-$C_6H_4]_2$-$H$ | 54 | $C_2H_5OH$, 2 hr., 20 | 198 |
| | 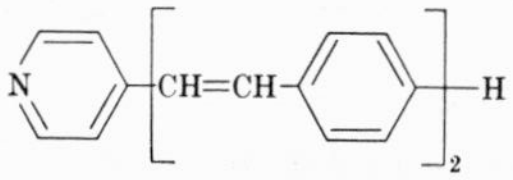 4-pyridyl$[CH{=}CH$-$C_6H_4]_2$-$H$ | | $C_2H_5OH$, 2 hr., 20 | 198 |

*Note:* References 391 to 404 are on p. 490.

## TABLE VI—*Continued*

### OLEFINS PREPARED BY THE WITTIG REACTION

*A. Mono-ylides as Starting Materials—Continued*

| Ylide | Product | Yield, % | Solvent, Time, Temperature, °C. | References |
|---|---|---|---|---|
| $C_{15}$ $(C_6H_5)_3P{=}CHC_6H_4CH{=}CHC_6H_5$ (*contd.*) | | | | |
| | $NC_5H_4[CH{=}CHC_6H_4]_3H$ | 45 | $C_2H_5OH$, 2 hr., 20 | 198 |
| | $p\text{-}BrCH_2C_6H_4CH{=}CHC_6H_4CH{=}CHC_6H_5$ | 63–65 | $C_2H_5OH$, 1 hr., 20 | 196 |
| | $C_6H_5C{\equiv}CC_6H_4CH{=}CHC_6H_4CH{=}CHC_6H_5$ | | $C_2H_5OH$, 2 hr., 20 | 195 |
| | $C_6H_5CH{=}CHC_6H_4CH{=}CHCH{=}CHC_6H_4CH{=}CHC_6H_5$ | 38–40 | $C_2H_5OH$, 2 hr., 20 | 211 |
| $C_{16}$ $(C_6H_5)_3P{=}CHC_6H_4CH{=}CHC_6H_4CH_3\text{-}p$ | | | | |
| | $p\text{-}CH_3C_6H_4[CH{=}CHC_6H_4]_3CH_3$ | 57 | $C_2H_5OH$ | 196 |
| $C_{17}$ $(C_6H_5)_3P{=}CH\text{-}C_{16}H_9$ (1-pyrenyl) | $C_{16}H_9CH{=}CHC_{16}H_9$ (1,2-di-1-pyrenylethylene) | 91 | $C_2H_5OH$ | 197 |
| $C_{18}$ $(C_6H_5)_3P{=}CHCO_2C_{16}H_{33}\text{-}n$ | $n\text{-}C_{16}H_{33}O_2C\text{-}(\text{polyene})\text{-}CO_2C_{16}H_{33}\text{-}n$ | | $H_2Cl_2$, 5 hr., reflux | 39 |

|  | Reagent | Product | Conditions | Refs. |
|---|---|---|---|---|
|  | 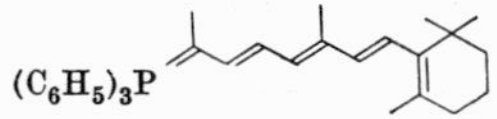 $(C_6H_5)_3P$ | 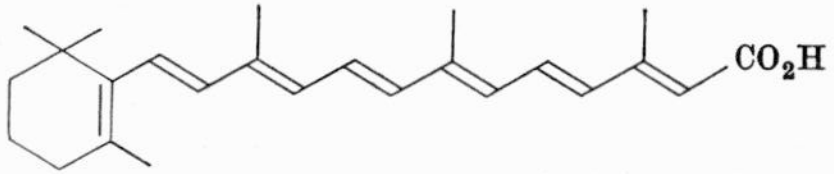 $CO_2H$ | DMF, 1 hr., 20 | 270 |
| $C_{19}$ | $(C_6H_5)_3P{=}C(CH_3)CO_2C_{16}H_{33}$-*n* | 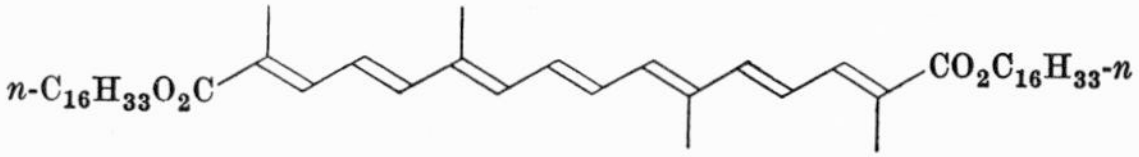 *n*-$C_{16}H_{33}O_2C$ … $CO_2C_{16}H_{33}$-*n* | $CH_2Cl_2$, 5 hr., reflux | 39 |
| $C_{20}$ | 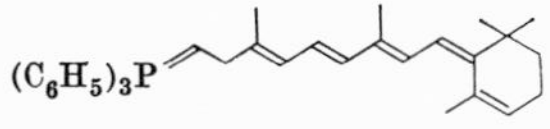 $(C_6H_5)_3P$ | 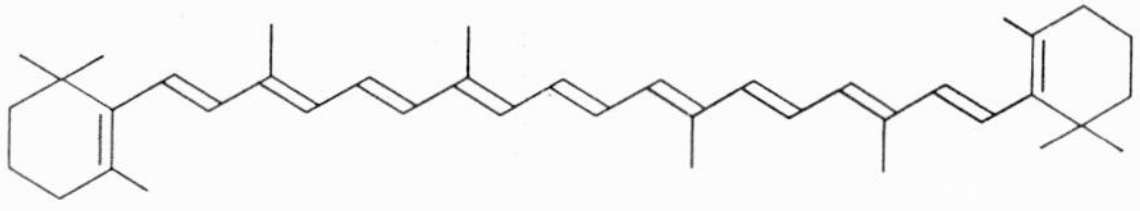  | $(C_2H_5)_2O$, 3 hr., 20 | 267 |
|  | 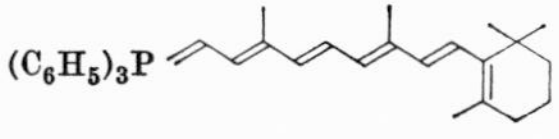 $(C_6H_5)_3P$ | 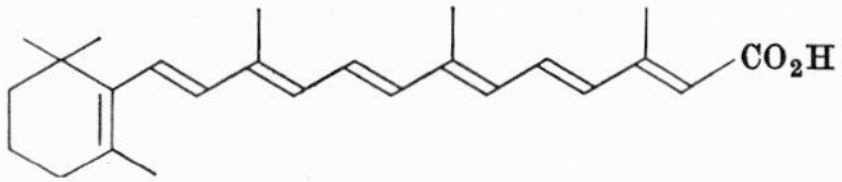 $CO_2H$ | DMF, 15 min., 20 | 266 |
|  |  | 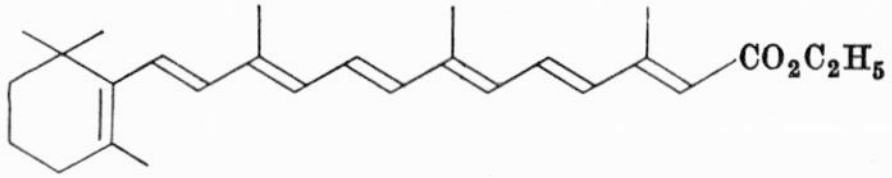 $CO_2C_2H_5$ | DMF, 3 hr., 0; 1 hr., 20 | 266 |
|  |  | 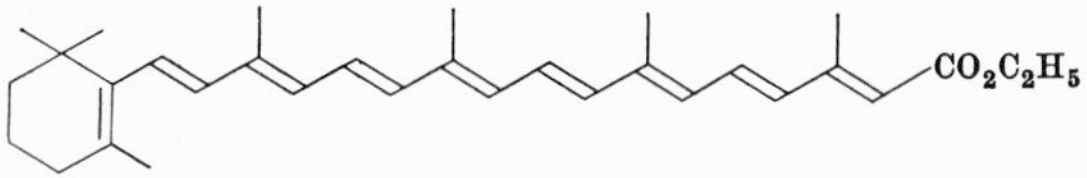 $CO_2C_2H_5$ | $CH_3OH$, 15 hr., 20 | 266 |
|  |  | 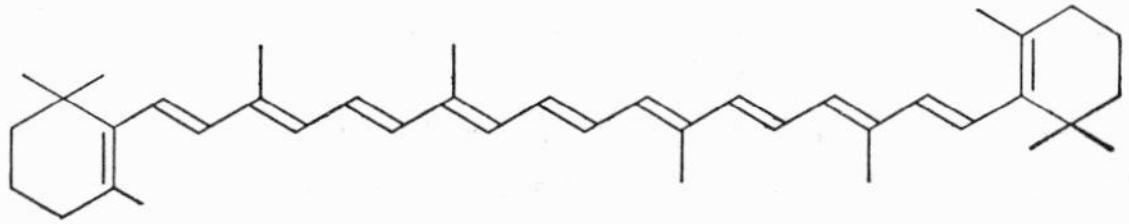  | $C_2H_5OH$, 2 hr., 20 | 266 |

*Note:* References 391 to 404 are on p. 490.

## TABLE VI—*Continued*

### Olefins Prepared by the Wittig Reaction

*A. Mono-ylides as Starting Materials—Continued*

| | Ylide | Product | Yield, % | Solvent, Time, Temperature, °C. | References |
|---|---|---|---|---|---|
| $C_{23}$ | $(C_6H_5)_3P{=}CH{-}[C_6H_4{-}CH{=}CH{-}C_6H_4]_2{-}H$ | | | | |
| | | 2-$C_5H_4N{-}[CH{=}CH{-}C_6H_4]_3{-}H$ | 50 | $C_2H_5OH$, 2 hr., 20 | 198 |
| | | 4-$C_5H_4N{-}[CH{=}CH{-}C_6H_4]_3{-}H$ | | $C_2H_5OH$, 2 hr., 20 | 198 |
| | | $CH_3C_6H_4{-}[CH{=}CH{-}C_6H_4]_4{-}H$ | | $C_2H_5OH$, 2 hr., 40 | 196 |
| | | $C_6H_5{-}[CH{=}CH{-}C_6H_4]_6{-}H$ | 42 | $C_2H_5OH$, 2 hr., 40 | 196 |
| | | $C_6H_5{-}[CH{=}CH{-}C_6H_4]_7{-}H$ | 32 | $C_2H_5OH$, 2 hr., 40 | 196 |

*B. Bis-ylides as Starting Materials*

| | Ylide | Product | Yield, % | Solvent, Time, Temperature, °C. | References |
|---|---|---|---|---|---|
| $C_2$ | $(C_6H_5)_3P{=}CHOCH{=}P(C_6H_5)_3$ | 3-Benzoxepin | | $CH_3OH$ | 190 |

| | | | | | |
|---|---|---|---|---|---|
| $C_3$ | $(C_6H_5)_3P{=}CHCH_2CH{=}P(C_6H_5)_3$ | 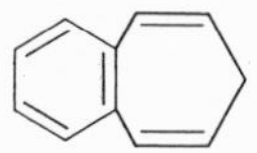 | 28 | $(C_2H_5)_2O$/THF, 6 hr., reflux | 52 |
| | | $C_6H_5CH{=}CHCH_2CH{=}CHC_6H_5$ | 65 | $(C_2H_5)_2O$, 10 hr., 65 | 52 |
| $C_4$ | $(C_6H_5)_3P{=}CHCH_2CH_2CH{=}P(C_6H_5)_3$ | $C_2H_5OCH{=}CH(CH_2)_2CH{=}CHOC_2H_5$ | | THF, 2 days, 20 | 113 |
| | | 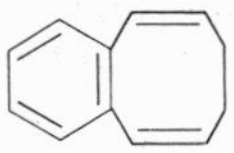 | 18 | $(C_2H_5)_2O$/THF, 20 hr., 70 | 52 |
| | | $C_6H_5CH{=}CHCH_2CH_2CH{=}CHC_6H_5$ | 75 | $C_2H_5OH$, 8 days, 20 | 74 |
| | | 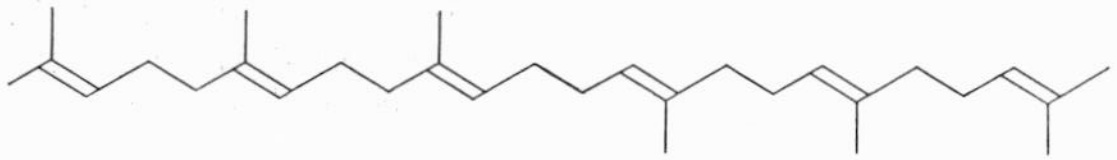 | 6 | $(C_2H_5)_2O$, 2 hr., 20; 3 hr., 60 | 74 |
| | | | 13 | THF, 1 hr., reflux | 300, 301 |
| | | | 35 | THF | 299 |
| | | | 26 | $C_2H_5OH$, 5 days, 20 | 74 |
| | | | 50 | $t$-$C_4H_9OH$, 5 hr., 25 | 302 |
| | | $(C_6H_5)_2C{=}CHCH_2CH_2CH{=}C(C_6H_5)_2$ | 17 | $(C_2H_5)_2O$, 36 hr., 20; 3 hr., 60 | 74 |
| | | | 68 | $(C_2H_5)_2O$, 14 hr., 65 | 185 |
| | | | 15 | $(C_2H_5)_2O$/THF, 1 hr., reflux | 52 |
| | $(C_6H_5)_3P{=}CHCH{=}CHCH{=}P(C_6H_5)_3$ | $C_6H_5(CH{=}CH)_3C_6H_5$ | | $C_2H_5OH$ | 212 |
| | | 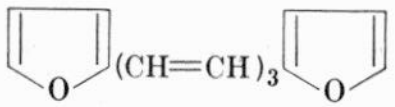 | | $C_2H_5OH$ | 212 |
| | | $p$-$CH_3OC_6H_4(CH{=}CH)_3C_6H_4OCH_3$-$p$ | | $C_2H_5OH$ | 212 |
| | | $C_6H_5(CH{=}CH)_5C_6H_5$ | | THF, 4 hr., reflux | 209 |
| | | | | $C_2H_5OH$ | 212 |
| | | 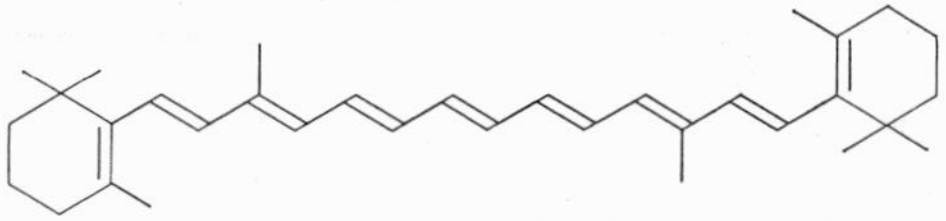 | | $C_2H_5OH$ | 212 |

*Note:* References 391 to 404 are on p. 490.

# TABLE VI—*Continued*

## Olefins Prepared by the Wittig Reaction

### *B. Bis-ylides as Starting Materials—Continued*

| | Ylide | Product | Yield, % | Solvent, Time, Temperature, °C. | References |
|---|---|---|---|---|---|
| $C_5$ | $(C_6H_5)_3P{=}CH(CH_2)_3CH{=}P(C_6H_5)_3$ | $CH_2{=}CH(CH_2)_3CH{=}CH_2$ | 45 | $(C_2H_5)_2O$ | 189, 188 |
| $C_8$ | $o$-$C_6H_4[CH{=}P(C_6H_5)_3]_2$ | $o$-$C_6H_4(CH{=}CHC_6H_5)_2$ | 84 | $C_2H_5OH$, 30 min., 20; 2 hr., reflux | 201*a* |
| | | $o$-$C_6H_4(CH{=}CHC_6H_4CH_3$-$o)_2$ | 53 | $C_2H_5OH$, 30 min., 20; 4 hr., reflux | 201*a* |
| | $p$-$C_6H_4[CH{=}P(C_6H_5)_3]_2$ | $p$-$C_6H_4(CH{=}CH_2)_2$ | 42 | $(C_2H_5)_2O$ | 188, 189 |
| | | $p$-$C_6H_4(CH{=}CH$-2-$C_5H_4N)_2$ | 56 | $C_2H_5OH$, 2 hr., 20 | 198 |
| | | $p$-$C_6H_4(CH{=}CHC_6H_5)_2$ | 32 | $C_2H_5OH$, 1 hr., 80 | 66 |
| | | | 86 | $C_2H_5OH$, 2 hr., 20 | 193 |
| | | $p$-$C_6H_4(CH{=}CHC_6H_4Cl$-$p)_2$ | 13 | $C_2H_5OH$, 1 day, 20 | 193 |
| | | $p$-$C_6H_4(CH{=}CHC_6H_4NO_2$-$m)_2$ | 83 | $C_2H_5OH$, 1 day, 20 | 193 |
| | | $p$-$C_6H_4(CH{=}CHC_6H_4NO_2$-$p)_2$ | 60 | $C_2H_5OH$, 1 day, 20 | 193 |
| | | $p$-$C_6H_4(CH{=}CHC_6H_4CH_3$-$p)_2$ | 81 | $C_2H_5OH$, 12 hr., 20 | 193 |
| | | $p$-$C_6H_4(CH{=}CHC_6H_4OCH_3$-$p)_2$ | 61 | $C_2H_5OH$, 12 hr., 20 | 193 |
| | | $p$-$C_6H_4(CH{=}CHCH{=}CHC_6H_5)_2$ | 88 | $C_2H_5OH$, 12 hr., 20 | 210 |
| | | | 70 | $C_2H_5OH$, 12 hr., 80 | 66 |
| | | $p$-$C_6H_4(CH{=}CHCH{=}CHC_6H_4NO_2$-$m)_2$ | 65 | $C_2H_5OH$, 12 hr., 20 | 210 |
| | | $p$-$C_6H_4(CH{=}CHC_6H_4NHCOCH_3$-$p)_2$ | 72 | $C_2H_5OH$, 12 hr., 20 | 193 |
| | | $p$-$C_6H_4[CH{=}CHC(CH_3){=}CHC_6H_5]_2$ | 83 | $C_2H_5OH$, 12 hr., 20 | 210 |
| | | $p$-$C_6H_4(CH{=}CHCH{=}CHC_6H_4CH_3$-$p)_2$ | Quant. | $C_2H_5OH$, 12 hr., 20 | 210 |
| | | $p$-$C_6H_4(CH{=}CHCH{=}CHC_6H_4OCH_3$-$p)_2$ | 13 | $C_2H_5OH$, 12 hr., 20 | 210 |
| | | $p$-$C_6H_4[CH{=}CHCH{=}CHC_6H_4N(CH_3)_2$-$p]_2$ | Quant. | $C_2H_5OH$, 12 hr., 20 | 210 |
| | | $p$-$C_6H_4(CH{=}CH$-$p$-$C_6H_4$-$CH{=}CH$-4-$C_5H_4N)_2$ | 35 | $C_2H_5OH$, 2 hr., 20 | 198 |

| | Reactant | Product | Yield (%) | Conditions | Refs. |
|---|---|---|---|---|---|
| | | $p\text{-}C_6H_4(CH{=}CH\text{-}C_6H_4\text{-}CH{=}CHC_6H_5)_2$ | | $C_2H_5OH$, 2 hr., 40 | 196 |
| | | $p\text{-}C_6H_4(CH{=}CH\text{-}C_6H_4\text{-}C{\equiv}CC_6H_5)_2$ | 75–78 | $C_2H_5OH$, 2 hr., 40 | 195 |
| | | $p\text{-}C_6H_4(CH{=}CH\text{-}C_6H_4\text{-}CH{=}CHC_6H_4CH_3\text{-}p)_2$ | | $C_2H_5OH$ | 196 |
| | | $OHC\text{-}C_6H_4\text{-}[CH{=}CH\text{-}C_6H_4]_9\text{-}CHO$ | Quant. | $C_2H_5OH$ | 208 |
| $C_{10}$ | 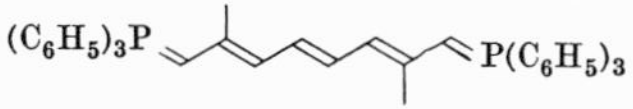 $(C_6H_5)_3P{=}\ldots{=}P(C_6H_5)_3$ | 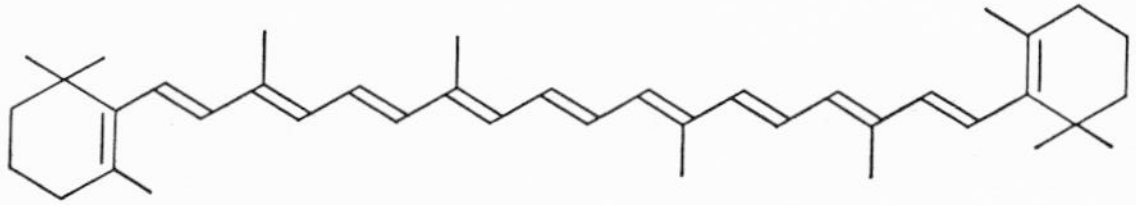 | 51 | $(C_2H_5)_2O$, 8 hr., reflux | 80, 271 |
| | | 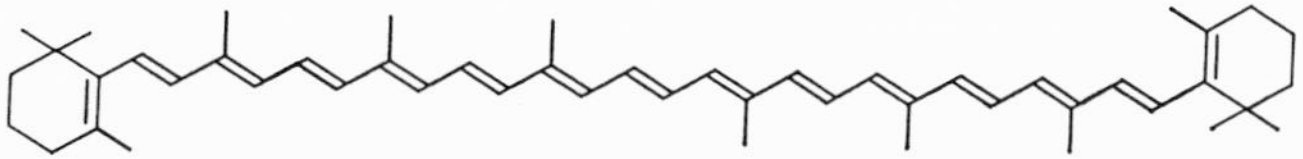 | 38 | $(C_2H_5)_2O$, 8 hr., reflux | 80, 271 |
| | 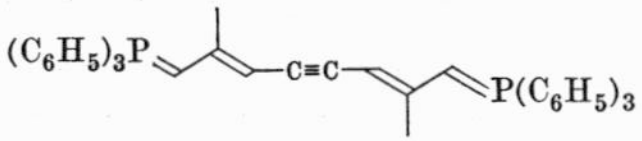 $(C_6H_5)_3P{=}\ldots C{\equiv}C\ldots{=}P(C_6H_5)_3$ | 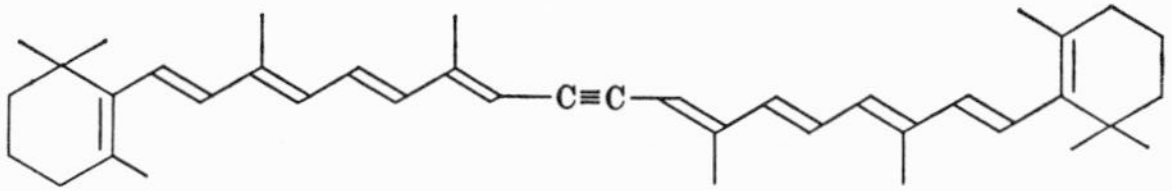 C≡C | 55 | $(C_2H_5)_2O$, 8 hr., reflux | 80, 271 |
| | | 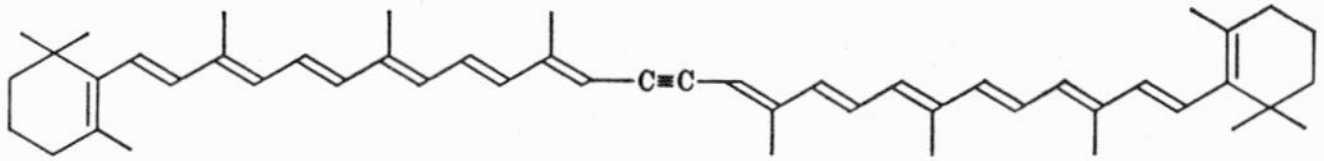 C≡C | 42 | $(C_2H_5)_2O$, 8 hr., reflux | 80, 271 |

*Note:* References 391 to 404 are on p. 490.

## TABLE VI—*Continued*

### Olefins Prepared by the Wittig Reaction

*B. Bis-ylides as Starting Materials—Continued*

| Ylide | Product | Yield, % | Solvent, Time, Temperature, °C. | References |
|---|---|---|---|---|
| $C_{10}$ (*contd.*) $(C_6H_5)_3P{=}CH{-}C_6H_2(CH_3)_2{-}CH{=}P(C_6H_5)_3$ (2,5-dimethyl-p-phenylene) | | | | |
| | $C_6H_5CH{=}CH{-}C_6H_2(CH_3)_2{-}CH{=}CHC_6H_5$ | 36 | $C_2H_5OH$, 2 hr., 20 | 193 |
| | $C_6H_5CH{=}CHCH{=}CH{-}C_6H_2(CH_3)_2{-}CH{=}CHCH{=}CHC_6H_5$ | 75 | $C_2H_5OH$, 12 hr., 20 | 210 |
| $C_{14}$ $(C_6H_5)_3P{=}CH{-}C_6H_4{-}C_6H_4{-}CH{=}P(C_6H_5)_3$ | | | | |
| | $CH_2{=}CH{-}C_6H_4{-}C_6H_4{-}CH{=}CH_2$ | 80 | $C_2H_5OH$, 20 | 187 |
| $C_{16}$ $(C_6H_5)_3P{=}CH{-}C_6H_4{-}CH{=}CH{-}C_6H_4{-}CH{=}P(C_6H_5)_3$ | | | | |
| | $CH_2{=}CH{-}C_6H_4{-}CH{=}CH{-}C_6H_4{-}CH{=}CH_2$ | 65–70 | $C_2H_5OH$, 20 | 187 |
| | $2\text{-}C_5H_4N{-}[CH{=}CH{-}C_6H_4]_2{-}CH{=}CH{-}2\text{-}C_5H_4N$ | 56 | $C_2H_5OH$, 2 hr., 20 | 198 |

| | | | | | |
|---|---|---|---|---|---|
| | | 4-$NC_5H_4$–[CH=CH–$C_6H_4$-p]$_2$–CH=CH–$C_5H_4N$-4 | | $C_2H_5OH$, 2 hr., 20 | 198 |
| | | 4-$NC_5H_4$–[CH=CH–$C_6H_4$-p]$_4$–CH=CH–$C_5H_4N$-4 | 27 | $C_2H_5OH$, 2 hr., 20 | 198 |
| | | $CH_3$–$C_6H_4$-p–[CH=CH–$C_6H_4$-p]$_5$–$CH_3$ | 46–48 | $C_2H_5OH$ | 196 |
| $C_{24}$ | o-[$(C_6H_5)_3P$=CH–$C_6H_4$–CH=CH–$C_6H_4$]–CH=CH–$C_6H_4$–CH=P$(C_6H_5)_3$-o (o-phenylene rings linked by CH=CH; terminal CH=P$(C_6H_5)_3$, CH=P$(C_6H_5)_3$) | Cyclic tetramer: four o-phenylene rings joined by four CH=CH bridges | 34 | $C_2H_5OH$, 14 hr., reflux | 201*a* |

*Note:* References 391 to 404 are on p. 490.

## REFERENCES

[391] Ramirez and McKelvie, *J. Am. Chem. Soc.*, **79**, 5829 (1957).
[392] Aksnes, *Acta Chem. Scand.*, **15**, 438 (1961).
[393] G. Rapp, Doctoral Dissertation, Universität Tübingen, 1955.
[394] Wolinsky, Chollar, and Baird, *J. Am. Chem. Soc.*, **84**, 2775 (1962).
[395] Considine, *J. Org. Chem.*, **27**, 647 (1962).
[396] A. Haag, Doctoral Dissertation, Universität Heidelberg, 1962.
[397] Dallacker, Kornfeld, and Lipp, *Monatsh. Chem.*, **91**, 688 (1960).
[398] Bergmann and Dusza, *J. Org. Chem.*, **23**, 459 (1958).
[399] Bohlmann and Herbst, *Chem. Ber.*, **91**, 1631 (1958).
[400] Dusza and Bergmann, *J. Org. Chem.*, **25**, 79 (1960).
[401] Oliver, Smith, and Fenning, *Chem. Ind.* (*London*), **1959**, 1575.
[402] Bothner-By, Naar-Colin, and Günther, *J. Am. Chem. Soc.*, **84**, 2748 (1962).
[403] Burrell, Jackman, and Weedon, *Proc. Chem. Soc.*, **1959**, 263.
[404] Muxfeldt and Rogalski, Ger. pat. 1,129,480 (to Farbw. Hoechst) [*C.A.*, **57**, 11132e (1962)].

# AUTHOR INDEX, VOLUMES 1–14

# CHAPTER INDEX, VOLUMES 1–14

# SUBJECT INDEX, VOLUME 14

Since the tables of contents of the individual chapters provide a quite complete index, only those items not readily found from the contents pages are indexed here.

Numbers in **boldface** type refer to experimental procedures.